HIGH POLYMERS

Vol. I: **Collected Papers of W. H. Carothers on High Polymeric Substances**
Edited by H. Mark and G. S. Whitby

Vol. II: **Physical Chemistry of High Polymeric Systems**
Second Edition. By H. Mark and A. V. Tobolsky

Vol. III: **Mechanism of Polymer Reactions**
By G. M. Burnett

Vol. V: **Cellulose and Cellulose Derivatives** (in five parts)
Parts I–III: *Second Edition.* Edited by Emil Ott, H. M. Spurlin, and M. W. Grafflin
Parts IV and V: Edited by N. M. Bikales and L. Segal

Vol. VI: **Mechanical Behavior of High Polymers**
By Turner Alfrey, Jr.

Vol. VII: **Phenoplasts: Their Structure, Properties and Chemical Technology**
By T. S. Carswell

Vol. IX: **Emulsion Polymerization**
By F. A. Bovey, I. M. Kolthoff, A. I. Medalia, and E. J. Meehan

Vol. X: **Polymer Processes**
Edited by Calvin E. Schildknecht

Vol. XII: **Analytical Chemistry of Polymers** (in three parts)
Edited by Gordon M. Kline

Vol. XIII: **Polyethers** (in three parts)
Edited by Norman G. Gaylord

Vol. XIV: **Polyesters** (in two parts)
Edited by Norman G. Gaylord

Vol. XV: **Radiation Chemistry of Polymeric Systems**
By Adolphe Chapiro

Vol. XVI: **Polyurethanes: Chemistry and Technology** (in two parts)
By J. H. Saunders and K. C. Frisch
Part III: Analytical Chemistry of the Polyurethanes
By D. J. Daniel and H. B. Staley

Vol. XVII: **Configurational Statistics of Polymeric Chains**
By M. V. Volkenstein. Translated by Serge N. Timasheff and Marina J. Timasheff

Vol. XVIII: **Copolymerization**
Edited by George E. Ham

Vol. XIX: **Chemical Reactions of Polymers**
Edited by E. M. Fettes

Vol. XX: **Crystalline Olefin Polymers** (in two parts)
Edited by R. A. V. Raff and K. W. Doak

Vol. XXI: **Macromolecules in Solution**
By Herbert Morawetz

Vol. XXII: **Conformations of Macromolecules**
By T. M. Birshtein and O. B. Ptitsyn. Translated by Serge N. Timasheff and Marina J. Timasheff

Vol. XXIII: **Polymer Chemistry of Synthetic Elastomers** (in two parts)
Edited by Joseph P. Kennedy and Erik G. M. Törnqvist

Vol. XXIV: **Vinyl Monomers** (in three parts)
Edited by Edward C. Leonard

Vol. XXV: **Fluoropolymers**
Edited by Leo A. Wall

HIGH POLYMERS

A SERIES OF MONOGRAPHS ON THE CHEMISTRY, PHYSICS, AND TECHNOLOGY OF HIGH POLYMERIC SUBSTANCES

VOLUME V

PART V

CELLULOSE AND CELLULOSE DERIVATIVES

EDITED BY
NORBERT M. BIKALES
Consulting Chemist
Livingston, New Jersey

AND
LEON SEGAL
Southern Regional Research Laboratory
U.S. Department of Agriculture
New Orleans, Louisiana

PART
V

WILEY-INTERSCIENCE
A DIVISION OF JOHN WILEY & SONS, INC.
NEW YORK · LONDON · SYDNEY · TORONTO

Library of Congress Catalog Card Number: 79–136710

ISBN O 471 39038-0

Printed in the United States of America

10 9 8 7 6 5 4 3 2 1

CONTRIBUTORS

Jett C. Arthur, Jr., Southern Regional Research Laboratory, U.S. Department of Agriculture, New Orleans, Louisiana. *Reactions Induced by High-Energy Radiation* (XVII-H) and *Electron Spin Resonance Studies* (XVII-I).

Robert H. Barker, Clemson University, Clemson, South Carolina. *Nuclear Magnetic Resonance Spectroscopy* (XIII-D).

O. A. Battista, FMC Corporation, Princeton, New Jersey. *Microcrystalline Cellulose* (XIX-G).

P. J. Baugh, University of Salford, Salford, Lancashire, England. *Photochemical Degradation* (XVIII-D).

Leo J. Bernardin, Kimberly-Clark Corporation, Neenah, Wisconsin. *Paper from Chemically Modified Cellulose* (XIX-D).

Norbert M. Bikales, Consulting Chemist, Livingston, New Jersey. *Ethers Derived from α,β-Unsaturated Compounds* (XVII-D).

J. Blackwell, Case Western Reserve University, Cleveland, Ohio. *Structure Studies* (XIII-A.1) and *Effect of Treatments Involving Microstructure* (XIII-A.2).

G. C. Bockno, American Viscose Division, FMC Corporation, Marcus Hook, Pennsylvania. *High-Wet-Modulus and Other High-Tenacity Rayons* (XIX-B).

R. J. Brewer, Tennessee Eastman Company, Kingsport, Tennessee. *Fractionation* (XIV-E).

Wyn Brown, Institute of Physical Chemistry, University of Uppsala, Uppsala, Sweden. *Molecular Conformation and Dimensions* (XIV-G).

J. Ross Colvin, National Research Council of Canada, Ottawa, Ontario, Canada. *Structure and Formation of the Microfibril* (XVI-B).

G. C. Daul, ITT Rayonier, Inc., Whippany, New Jersey. *Films* (XIX-E).

George L. Drake, Jr., Southern Regional Research Laboratory, U.S. Department of Agriculture, New Orleans, Louisiana. *Flame-Resistant Textiles* (XIX-I).

Ø. Ellefsen, Norwegian Pulp and Paper Research Institute, Oslo (Vinderen), Norway. *Polymorphic Forms* (XIII-C.2) and *Submicroscopical Investigations* (XIII-F).

L. S. Gal'braikh, Moscow Textile Institute, Moscow, U.S.S.R. *Derivatives with Unusual Functional Groups* (XVII-F).

R. E. Glegg, Eastman Kodak Company, Rochester, New York. *Fractionation* (XIV-E).

Herman B. Goldstein, Sun Chemical Corporation, Wood River Junction, Rhode Island. *Durable-Press Treatments* (XIX-A).

James N. Grant, Southern Regional Research Laboratory, U.S. Department of Agriculture, New Orleans, Louisiana. *Mechanical Behavior of Fibers* (XV-A).

John D. Guthrie, Southern Regional Research Laboratory, U.S. Department of Agriculture, New Orleans, Louisiana. *Ion-Exchange Celluloses* (XIX-H).

W. Z. Hassid, University of California, Berkeley, California. *Enzymatic Synthesis* (XVI-A).

Gordon D. Hiatt, Eastman Kodak Company, Rochester, New York. *Esters* (XVII-B).

H. B. Hopfenberg, North Carolina State University, Raleigh, North Carolina. *Graft Copolymers* (XVII-G).

Georg Jayme, Institute of Macromolecular Chemistry, Technical University of Darmstadt, Germany. *New Solvents* (XIV-A).

D. W. Jones, University of Bradford, Bradford, England. *Structure Studies* (XIII-C.1).

Robert E. Kesting, Chemical Systems Incorporated, Santa Ana, California. *Flat Membranes* (XIX-F.1).

Frank J. Kilzer, Cabrillo College, Aptos, California. *Thermal Degradation* (XVIII-C).

B. J. Lipps, The Dow Chemical Company, Walnut Creek, California. *Hollow-Fiber Membranes* (XIX-F.2).

H. I. Mahon, The Dow Chemical Company, Walnut Creek, California. *Hollow-Fiber Membranes* (XIX-F.2).

Mary Mandels, U.S. Army Natick Laboratories, Natick, Massachusetts. *Enzymatic Degradation* (XVIII-E).

Jack Mann, Carrington Plastics Laboratories, Shell Research Limited, Urmston, Manchester, England. *Deuteration and Tritiation* (XIII-B).

R. H. Marchessault, University of Montreal, Montreal, Quebec, Canada. *Structure Studies* (XIII-A.1).

R. L. Mitchell, ITT Rayonier, Inc., New York, New York. *Films* (XIX-E).

W. R. Moore, University of Bradford, Bradford, England. *Concentrated Solutions* (XIV-F).

F. F. Morehead, Consultant, West Chester, Pennsylvania. *Microscopical Investigations* (XIII-E).

Robert T. O'Connor, Southern Regional Research Laboratory, U.S. Department of Agriculture, New Orleans, Louisiana. *Analysis of Chemically Modified Cotton* (XIII-A.3).

Glyn O. Phillips, University of Salford, Salford, Lancashire, England. *Photochemical Degradation* (XVIII-D).

Robert A. Pittman, Southern Regional Research Laboratory, U.S. Department of Agriculture, New Orleans, Louisiana. *Nuclear Magnetic Resonance Spectroscopy* (XIII-D).

William J. Rebel, Eastman Kodak Company, Rochester, New York. *Esters* (XVII-B).

E. T. Reese, U.S. Army Natick Laboratories, Natick, Massachusetts. *Enzymatic Degradation* (XVIII-E).

Wilson A. Reeves, Southern Regional Research Laboratory, U.S. Department of Agriculture, New Orleans, Louisiana. *Flame-Resistant Textiles* (XIX-I).

G. N. Richards, The University College of Townsville, Pimlico, Townsville, Queensland, Australia. *Alkaline Degradation* (XVIII-B).

Z. A. Rogovin, Moscow Textile Institute, Moscow, U.S.S.R. *Derivatives with Unusual Functional Groups* (XVII-F) and *Biologically Active Cellulosic Materials* (XIX-J).

A. B. Savage, The Dow Chemical Company, Midland, Michigan. *Ethers* (XVII-C).

J. W. Schappel, American Viscose Division, FMC Corporation, Marcus Hook, Pennsylvania. *High-Wet-Modulus and Other High-Tenacity* Rayons (XIX-B).

Leon Segal, Southern Regional Research Laboratory, U.S. Department of Agriculture, New Orleans, Louisiana. *Effect of Morphology on Reactivity* (XVII-A).

A. Sharples, Arthur D. Little Research Institute, Inveresk, Midlothian, Scotland. *Acid Hydrolysis and Alcoholysis* (XVIII-A).

C. L. Smart, Celanese Research Company Laboratories, Summit, New Jersey. *Cellulose Triacetate Fibers* (XIX-C).

V. T. Stannett, North Carolina State University, Raleigh, North Carolina. *Graft Copolymers* (XVII-G).

Sanford S. Sternstein, Rensselaer Polytechnic Institute, Troy, New York. *Micromechanics of Fiber Networks* (XV-B.2).

Lars-Olaf Sundelöf, Institute of Physical Chemistry, University of Uppsala, Uppsala, Sweden. *Light Scattering* (XIV-B).

L. J. Tanghe, Eastman Kodak Company, Rochester, New York. *Fractionation* (XIV-E).

Harold Tarkow, Forest Products Laboratory, U.S. Department of Agriculture, Madison, Wisconsin. *Dimensional Stabilization* (XIX-K).

Giuliana C. Tesoro, Burlington Industries, Inc., New York, New York. *Crosslinked Cellulose* (XVII-E).

Norman S. Thompson, The Institute of Paper Chemistry, Appleton, Wisconsin. *Pulping Processes* (XIX-L).

B. Andvord Tønnesen, Royal Norwegian Council for Scientific and Industrial Research, Oslo (Blinderen), Norway. *Polymorphic Forms* (XIII-C.2) and *Submicroscopical Investigations* (XIII-F).

V. W. Tripp, Southern Regional Research Laboratory, U.S. Department of Agriculture, New Orleans, Louisiana. *Measurement of Crystallinity* (XIII-G).

J. A. Van den Akker, The Institute of Paper Chemistry, Appleton, Wisconsin. *Tensile Straining* (XV-B.1).

Hans Vink, Institute of Physical Chemistry, University of Uppsala, Uppsala, Sweden. *Osmometry* (XIV-C) and *Viscometry* (XIV-D).

A. D. Virnik, Moscow Textile Institute, Moscow, U.S.S.R. *Biologically Active Cellulosic Materials* (XIX-J).

Kyle Ward, Jr., The Institute of Paper Chemistry, Appleton, Wisconsin. *Paper from Chemically Modified Cellulose* (XIX-D).

J. O. Warwicker, The Cotton Silk and Man-Made Fibres Research Association, Shirley Institute, Manchester, England. *Swelling* (XIII-H).

John J. Willard, J. P. Stevens & Company, Inc., Garfield, New Jersey. *Crosslinked Cellulose* (XVII-E).

Carl N. Zellner, Consultant, New Hope, Pennsylvania. *Cellulose Triacetate Fibers* (XIX-C).

EMIL OTT
May 19, 1902–September 29, 1963

EMIL OTT AND CELLULOSE CHEMISTRY

Without Emil Ott, we would not have *Cellulose and Cellulose Derivatives.* He did no experimental work in the cellulose or any other polymer field. As the first director of research for Hercules, he had to cope with reorganization and extension of the company research effort, under very adverse wartime conditions. Yet he acquired a well-deserved reputation as an expert in the cellulose field in the brief period from 1939 to 1943. This, in retrospect, sounds very implausible. However, it seemed to be the most natural thing in the world to those of us who collaborated with Emil. I shall attempt to describe this collaboration, but it must be remembered that it was thirty years ago, and my memory of our thoughts and motivations may be wide of the mark.

Emil had received his degree from the Eidgenossische Technische Hochschule (E.T.H.) in Zürich in 1927, worked briefly with Stauffer, and was at Johns Hopkins University until 1933. He then came to Hercules and became head of the research division of the Hercules Research Center in 1937. I soon learned to value his opinions very highly. In the spring of 1938 I was preparing a paper on the distribution of substituents in cellulose derivatives and asked Emil to help me with the discussion. He invited me to his home, so that we would have plenty of time.

First, we toured the garden. I can still remember a marvelous clump of Russell lupines. I can never get them to do so well. Emil had a great fund of information about flowers, which I turned to good account in my own plantings some three years later. I cite this episode to illustrate the fact that Emil had a drive to understand, as completely as possible, everything he came in contact with.

Then, we went into the house. Dorothy Ott made sure that refreshment was at hand, sat down and talked a few moments, then discreetly withdrew. Emil read my manuscript, caught a few careless errors, then asked, "Why do you feel that it is necessary to go to such length and to assemble so much corroborating data? The whole thing is so obvious." I was able to assert with conviction my feeling that my best effort would be only partially successful. So it was. A text published in 1953 contained many erroneous arguments based on misconceptions of the nature of partially substituted cellulose derivatives. Even now, papers on heterogeneous reactions of cellulose tend to neglect the proposition that statistical considerations will govern the fate of that fraction of the cellulose which is momentarily available, while the crystalline regions are being chewed up.

Emil Ott became director of research of Hercules in 1939. He believed in directing research. He would check with all of us in the cellulose field, individually and in meetings, about the nature of problems and the chances for their solution. He immediately recognized the important questions and could get to the heart of the matter more rapidly than the rest of us. Then, a program would be adopted that was satisfactory to all concerned.

In these meetings we often had to point out that a certain proposition might not be supported by much of the literature, yet we were sure of our ground. The great heresies of cellulose science had not all died down. Biosan anhydride was dead, but cementing substance lived on. Meyer and Mark had never intended the concept of micellar solutions to be anything but a hypothesis for discussion, but others were to defend the concept for many years. The concept of weak linkages at regular intervals was very much alive. We also had the particle theory.

It is possible that Emil thought, "If these people know so much, why don't they publish"? We did, of course, publish the results of our own work. We did not intend to combat what we thought to be erroneous views. Perhaps this is the background for the fact that Emil was soon giving papers at meetings, ACS local sections, and on tours to foreign countries. He, of course, got much assistance from the rest of us. So far as I know he never originated a concept, yet he immediately recognized the broad implications of a suggestion. I must emphasize, also, that he presented material that he thoroughly understood and believed. I think that I speak for my colleagues when I say that we were glad to help Emil in this somewhat thankless job, which we did not wish to do ourselves.

The next step was "the cellulose book" (*Cellulose and Cellulose Derivatives*, first edition). Emil justified this effort by saying, "Nothing will help the preeminence of Hercules cellulose research so much as a concerted effort to summarize all available knowledge." It was to be the product of experts in the various fields. Emil himself made the final decision as to invited authors, but he had the rest of us take the major load of examining the contributions. If we felt changes were in order, he would check our position and help us in contacts with the author. In other words, the whole book was edited. I would welcome the same attitude to the myriad of collaborative treatises appearing at the present time.

The publication of the first edition of *Cellulose and Cellulose Derivatives* in 1943 marked the high point of Emil's interest in cellulose. He gave papers for a few more years. However, when the second edition (Parts I–III) became necessary, he turned responsibility over to Mildred Grafflin and myself, for all but the most important decisions of policy.

Emil left Hercules shortly thereafter and had little to do with cellulose. I talked with him in 1963, a few months before his early death. A major part of the discussion was devoted to our collaboration on "the cellulose book." I believe that this episode in his busy and varied career was remembered with as much pleasure as anything else. To us who worked with him he is best remembered for his keen questions and his stimulating mind.

Harold M. Spurlin

January 12, 1970
Hercules Research Center
Wilmington, Delaware

PREFACE

With the publication of the first edition in 1943, *Cellulose and Cellulose Derivatives* established itself as the most complete and authoritative description of the field in the English language. The well-deserved reputation of this book was further enhanced when the second completely revised and augmented edition was issued in three parts in 1954–1955 under the very able editorship of Emil Ott, Harold M. Spurlin, and Mildred W. Grafflin. A whole generation of cellulose chemists, including ourselves, became acquainted with the field by reading and rereading this book.

By the mid-1960s it became apparent that *Cellulose and Cellulose Derivatives* was gradually becoming outdated. This was the case not so much because the book presented information that was obsolete, but rather because it was becoming incomplete. It lacked many of the important developments that had occurred since the early 1950s.

Among the subjects that increasingly demanded the attention of cellulose chemists were such topics as structural studies by new instrumental techniques (e.g., by nuclear magnetic resonance), studies of solution properties in new metallic solvents, crosslinking reactions, reactions induced by high-energy radiation, thermal and photolytic degradation, as well as the preparation of graft copolymers and of derivatives with unusual functional groups, microcrystalline cellulose, membranes, paper from chemically modified cellulose, and flame- and wrinkle-resistant textiles. Even in the traditional fields (esters, ethers, regenerated cellulose, etc.) some significant developments had taken place; examples are the intensive study of cellulose triacetate fibers, cyanoethylcellulose, and high-wet-modulus rayons. The need arose, therefore, for a supplement to the second edition to cover the important advances in cellulose science and technology that had taken place since its publication.

In 1963 Emil Ott unfortunately passed away. Since the other editors of the second edition were occupied with new duties, we felt that, unless someone undertook the task of bringing the work up to date, it would soon lose its usefulness. We therefore conceived the plan to issue supplementary volumes, closely tied in with Parts I, II, and III of the second edition, to cover only new developments. In this way we concentrated on those aspects that challenge contemporary cellulose chemists. We retained the general format of Parts I, II, and III, and especially of having chapters contributed by outstanding specialists. We also felt it convenient to number chapters consecutively from Part III, so as to avoid confusion with the earlier parts.

A work that is so closely tied to recent developments depends to a great extent on critical advice from colleagues. We are especially pleased to acknowledge the contributions of Robert H. Marchessault and Edmund H. Immergut, as well as those of Herman F. Mark, Kyle Ward, Jr., C. M. Conrad, W. A. Sisson, V. W. Tripp, and E. I. Valko. We note with regret the passing of W. R. Moore, which occurred shortly after he prepared Chapter XIV-F; the excellence of this chapter is in keeping with the high quality of his life's work.

Finally, but most important, we express our sincerest thanks to the contributors whose names appear herein. They shared our enthusiasm, and their wholehearted cooperation made it possible to produce this work. Any praise for the contents should be addressed to them.

Norbert M. Bikales
Leon Segal

September 1970

CONTENTS

Part V

Cellulose and Cellulose Derivatives

PART V

Chapter XVII

DERIVATIVES OF CELLULOSE

A. EFFECT OF MORPHOLOGY ON REACTIVITY

Leon Segal

The interest in the effect of the morphology of cellulose on its reactivity arises from numerous observations that samples which appear to be chemically identical differ, in fact, greatly in ease of reaction. This difference in reactivity must therefore be related to differences in *physical structure*. Cellulose is commonly found in the form of films and fibers, but these forms are only two facets of its morphology. These are actually the ultimate in the *gross structure*, which is that which can be seen with the naked eye and in the optical microscope. Below this lies the *fine structure* extending down to the molecular level, and in this region one begins to find the effects that are similar to the steric hindrance effect of the simpler molecules of organic chemistry (see also Chapter XIII). The term "morphology," then, means more than merely the type of structure recognized by botanists. In addition to the form and structure of the fiber and film, the morphology of cellulose includes factors of fine structure, such as fibrils and micelles, accessible and nonaccessible regions, amorphous and crystalline areas, and other supramolecular aggregates. Obviously these factors can exist only when cellulose is in the solid state, and hence considerations of the effect of morphology on the reactivity of cellulose are restricted to *heterogeneous systems*. By heterogeneous systems is meant those systems of reactions in which the gross structure of the starting material is retained, whether it be fiber or film; there are at least two phases always present, a factor that complicates kinetic studies. The various basic chemical reactions of cellulose – the organic chemistry of the basic molecule – are dealt with elsewhere (see Part II, Chapter IX and the subsequent sections of this Chapter XVII).

The terms "reactivity" and "accessibility" have been used to describe the ease with which cellulose can react. More properly, however, reactivity is the ability of the chain molecule to react with other molecules, whereas accessibility defines the ease by which the functional groups of the chain molecule can be reached by the reactant molecules. Strictly speaking, then, this discussion would seem to deal mostly with accessibility, the steric hindrance effect of the gross and fine structures when cellulose is exposed to reactive molecules. The functional groups of the cellulose molecule retain their own individual reactivities; for example, the primary hydroxyl groups of cellulose are more reactive than secondary hydroxyl groups in most cases, but secondary groups may be reacted before the primary ones if the latter are rendered nonaccessible because of their steric relationships in the fine and gross structures of the cellulosic material. Cellulose enters into the usual reactions of organic chemistry (esterification, etherification, etc.), but in all cases in which cellulose remains as a solid phase in the reaction system one must keep in mind that the *extent* of these reactions may be dictated by the morphology of the form in which the cellulose molecules are combined in the solid state. Several other aspects of this effect have already been discussed in Section A.2 of Chapter IX in Part II.

The products of chemical reaction generally enhance the usefulness of cellulose because more desirable properties are obtained. In the instance of one reaction, however, desirable properties do not usually result. That reaction is solvolysis, which breaks the linkages between the monomeric units, thereby causing decomposition of the cellulose chain molecules (oxidation also can do this). With most celluloses only a short period of hydrolysis is needed to break enough linkages for the material to become useless, and hydrolysis can be carried to the extent that the monomer is the principal product. Solvolysis is useful, though because by it much information is obtained regarding the fine structure of the polymer. Indeed, solvolysis, particularly hydrolysis, is highly sensitive to morphology, so that various aspects of solvolysis—for example, reaction rates—are taken as indices of accessibility, reactivity, crystallinity, and the like (see also Part I, Chapter III-C; pp. 260–270 of Chapter IV-B; Chapter XVIII-A; and Chapter XIX-G).

1. General Considerations

The Haworth formula (see p. 66 in Chapter III) is an incomplete diagram of the molecular structure because it fails to display the spatial arrangements of the atoms of the glucose units, the true conformation of the pyranose ring, and the spatial arrangement of the hydroxyl groups. These factors, which play highly important roles in the morphology of cellulose and its effects on cellulose reactions, are more properly considered when the molecule is diagrammed as in structure 1, where the chair conformation of the glucose units is displayed as well as the equatorial positions of the hydroxyl groups.

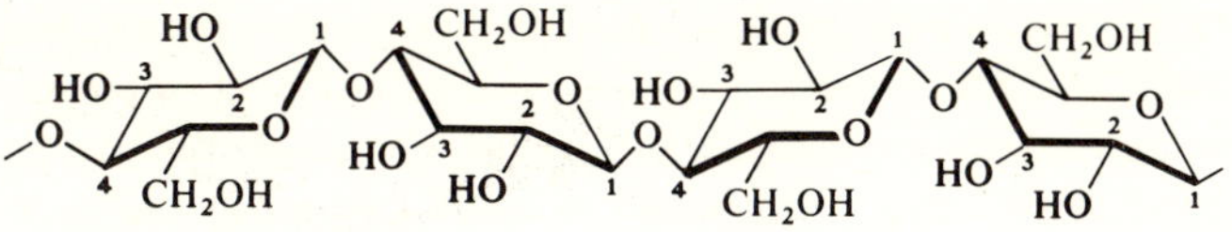 (1)

1

For more detailed information on this and other considerations of the molecular structure of cellulose the reader is referred to the extensive descriptions given in Sections A and B of Chapter III; in Chapter IV; in Sections A.1, B, C, and F of Chapter XIII; and in references 1 and 2.

With its three hydroxyl groups per anhydroglucose unit and the chair conformation of the glucose rings, cellulose has the capability of forming many hydrogen bonds along the length of the polymer chain. These bonds, combined with other secondary valence forces (principally the van der Waals attraction), bind together portions of the molecular chains into various degrees of lateral order, ranging from perfect geometrical packing of the crystal lattice to random conditions. These regions, above the molecular level now and constituting the smallest units of the so-called fine structure, are not sharply defined but gradually merge into one another. One chain molecule, because of its length, can pass through varying degrees of lateral order several times along its length. Studies of the crystalline structure are reviewed in great detail in Chapters IV-B, XIII-C.1, and XIII-F; such studies, however, are far from closed. Frey-Wyssling (3) questions the directions of hydrogen bonding assumed in the presently accepted crystal lattice of cellulose, claiming that it does not fit in with conclusions based on plant physiology. Hess (4) presents data that he interprets as showing that the crystalline and amorphous areas are arranged in neat, orderly rows. These elements of structure are among the factors that have a bearing on the reactivity of cellulosic materials.

Chemical substitution on the hydroxyl groups and attack on the glucosidic linkages that connect one monomer unit to the other is restricted according to their location in the overall morphology. Thus the groups located in the amorphous regions react readily in all chemical reactions, being in a highly accessible environment. But in the crystalline regions, where there is close packing and strong interchain bonding, these groups are not readily accessible to reactant molecules and, indeed, are completely inaccessible to some (5).

Certain factors of the fine structure—such as the degree of perfection of the lattice, the number and size of the crystalline regions, and the position or orientation of the crystalline regions with respect to the gross structure—depend on the origin and previous history of the cellulosic material. Other elements of the fine structure, for example, the crystal lattice type, are influenced by chemical treatments. There are at present at least four accepted crystalline modifications of cellulose, designated as celluloses I, II, III, and IV (6) (see also Chapters IV-B and XIII-C.2).

As with any chemical compound that exhibits polymorphism, these materials are the same insofar as the constituent molecule is concerned: there is only unmodified cellulose present. They differ from one another in that the planes of the anhydroglucose units of one chain molecule in the regions of close parallel alignment lie at different positions. These differences in position are evidenced by x-ray diffraction. Infrared-absorption spectroscopy also differentiates the polymorphic forms because group vibrations are affected by changes in the positions of the anhydroglucose units and of the chain molecules. Although these crystalline modifications differ in the parameters of their crystalline unit cells, as shown by x-ray diffraction, there is no direct experimental evidence showing clearly that the crystal lattice type per se has significant influence on the chemical reactivity of cellulose or on the accessibility of the reactive groups concerned. In all cases in which lattice transformations are involved and in which enhancement of reactivity or accessibility is observed, particularly in the case of cellulose II, such enhancement is more closely connected with a reduction in the size of the crystallites, an increase in the amorphous areas, and a decrease in the density of the gross structure, which always accompanies such transformations. There is, however, some evidence that chain length, or degree of polymerization (DP), influences reactivity The next elements of fine structure may be considered to be the elementary fibril and the microfibril, which are described in Section B of Chapter XVI. In addition to these, there are the micelles, or fibrils (7). These structures, which are visible under the electron microscope, have been studied extensively by Rånby (8) and by Rollins and Tripp (9). The published electron micrographs at this level of structure display to a marked extent visible evidence of chemical reaction, which is not usually apparent in the gross structure.

It is the gross structure that is most familiar—cotton, linen, and ramie fibers; rayon filaments; cellophane film; wood pulp. These are all varied among themselves. The cotton fiber is a single cell, but it is complex, being built up of many layers that spiral around one another in opposite directions. The long ramie fiber, like linen and wood pulp, is not a single cell but is built up of many small cells, with much intercellular matter present. In all of these fibers, single-cell cotton or multicell ramie, many micellar strands are associated to form layers (10). In rayons, on the other hand, where cellulose has been dissolved and regenerated, the structure is not as complex. The crystalline and amorphous regions are still present, but with more of the latter and with smaller crystalline regions. In viscose rayon a skin and a core develop, but "all skin" and "all core" rayon can be produced by proper control in the manufacturing process. Richardson (11) has reviewed this subject very thoroughly. Reviews of the fiber structure of some of the more important sources of cellulose are given in Chapter IV-C, and Rollins (12) has dealt in detail with the cotton fiber. The reader is referred to these reviews because the subject matter is pertinent to this discourse, but the extensiveness of the subject precludes going into the details of it here.

When native cellulose—cotton and ramie fibers, bacterial and animal membranes, purified wood pulp, and the like—is mercerized, not only are there changes in fine structure (i.e., in the crystal lattice, crystallite size, and degree of crystallinity) but there are also marked changes in gross structure. Such changes have been illustrated for cotton by Rollins (12). A mercerized cellulosic fiber is therefore of different total morphology than the fiber prior to mercerization (13), and this affects reactivity. Cellulosic fibers and film are swollen by water and other reagents, and solutions of other reagents [reviewed in Part I, Chapter IV-B.2, and by Heuser (14)]; changes in chemical reactivity of the polymer are nevertheless observed, although only the morphology has been changed.

2. Etherification and Esterification

Cumberbirch (15) has brought forward many cases in which the morphology of the cellulosic material has marked effects on the extent of chemical reactions, as, for example, in the work of Sitch (16), who reported that when cotton, mercerized cotton, cuprammonium rayon, hydrocellulose, and a cotton ground in a ball mill were methylated with ethereal diazomethane, methoxyl contents of 7.0, 14.9, 18.6, 5.9, and 20.6%, respectively, were obtained. Sitch concluded that the degree of methylation produced is greatly affected by the crystallinity of the cellulose. Sitch's results confirmed earlier work of Reeves and Thompson (17) with native and mercerized cottons. Ten years before Sitch, however, Reeves and Sisson (18) applied diazomethane to air-dried cottons and related the methoxyl contents with the fine structure as revealed by x-rays. Whereas Sitch used celluloses of mixed prehistory (i.e., given various chemical treatments and subjected to a physical process), Reeves and Sisson utilized only one cellulosic material that was given several chemical treatments. Changes in fine structure were induced by treatment with solutions of sodium hydroxide at various temperatures, giving partially and completely mercerized cottons (crystal lattices of celluloses I and II and of cellulose II only, respectively). With native cotton 5.6–7.9% methoxyl was obtained, with partially mercerized cotton 8.7–9.8% methoxyl, and with mercerized cotton 10.7–12.6% methoxyl. The application of boiling water or hot glycerol to never-dried, caustic-treated cotton resulted in what was described as a reversion to the native state, but these conditions are recognized today as productive of crystalline cellulose IV. Nevertheless, Reeves and Sisson found that the methoxyl contents of cottons treated with hot glycerol (nearly all converted) were in the range 6.3–7.4%; with boiling water (partially converted) the methoxyl contents were 9.2–9.9%.

There is no simple correlation of the extent of reaction with lattice type.

Reeves and Sisson did not report to what extent the crystalline lattice was influenced by the methylation, but there is little reason to expect ethereal diazomethane to penetrate the crystalline regions of cellulose. What is of significance here, but apparently not considered, is the reduction of crystallinity (or increase of accessibility), the reduction of crystallite size, the increase in fiber cross-sectional area as related to the treatments. Valko (14) has commented on this at great length. The loss of reactivity as a result of the drying and the heat treatment is a consequence of the treatment, not the lattice conversion. This is the "hornification" effect, discussed in Chapter IX (pp. 693, 695, 696), which arises from the formation of stronger bonds between the mobile cellulose chains of the swollen structure as it collapses in drying. Heuser (19) discusses the deleterious effect of drying on the reactivity of cotton that had been mercerized in order to enhance reactivity Loeb and Segal (20) have compared the reactivity of cotton decrystallized with anhydrous ethylamine with that of untreated and mercerized cottons, using the degree of acetylation as the criterion of reactivity. They too, found that full reactivity could be sustained only if drying was avoided after the initial treatments, which in their systems employed non-aqueous solvents.

Because of the commercial importance of cellulose acetate, a staggering amount of work has been done to establish optimum conditions for the acetylation reaction. All results point to the fact that the reactivity of the cellulose being esterified is a function of the accessibility of the hydroxyl groups under the conditions being used (5, 21) (see also Section B of this Chapter XVII) . In the acetylation of cellulose a factor of prime importance to reactivity is the *moisture content* of the material. Water does not play any part in the acetylation reaction. Indeed, in that reaction water is not wanted, and the conditions are such as to remove it from the system as fast as it is formed. But water affects the morphology of the cellulosic material in both the fine and gross structures in such a way that the hydroxyl groups of the cellulose chain molecules are rendered more accessible to the reactants. Water, however, does not affect the crystalline material, whereas other reagents (e.g., ethylamine) affect both amorphous and crystalline cellulose.

As already pointed out, intermolecular hydrogen bonding through the hydroxyl groups effectively crosslinks the chains to form the fine structure. Water, ethylamine, the aliphatic diamines, and other compounds (the so-called swelling agents) supply the necessary energy to break the hydrogen bonds, but not enough to bring the cellulose molecule into solution. These bonds reform with the swelling agent. What happens afterward depends on the strength of the hydrogen bond in the particular system involved and what is done to the system.

Urquhart (22), who has studied the cellulose-water system very extensively, explains clearly the manner in which the swelling agent affects the morphology of the cellulose. In industrial practice it has long been known that, in order to preserve high reactivity, swollen cellulose must not be allowed to dry; where wet, water-free cellulose is needed, water is removed by means of other solvents. In acetylation, for example, acetic acid is commonly used. The work of Staudinger and co-workers (23, 24) is classic in this respect, well illustrating the effect of morphology on the reactivity of the polymer. In these studies, the following changes in morphology were examined:

1. Native fibers of different gross morphology (cotton, ramie, linen, and hemp).
2. Native fibers, mercerized and dried.
3. Both dried native and dried mercerized fibers made to "include" various solvents. This latter was accomplished by thoroughly wetting the fiber with water and then completely extracting the water, ending with the desired solvent from which the fiber is dried. Solvent exchange is resorted to when a water-immiscible solvent is to be the final solvent. The dry "inclusion" cellulose normally contains several percent of the final solvent, which can be removed only by rewetting with water or some other swelling solvent.

Table 1 presents acetylation data for "acetic acid-inclusion" cellulose taken from Staudinger's work. The variation of acetyl contents in column 2 shows an effect of the original gross structure. The reduced acetyl contents of column 3 illustrate the deactivation or "hornification" effect of drying a swollen cellulose structure. The data in columns 4 and 5 show that inclusion markedly reactivates the polymer. The expected increased reactivity of mercerized cellulose is found in the data of column 5. Data of this sort are even more dramatic when milder acetylating conditions are used, as shown in Table 2 for "pyridine-inclusion" cellulose. A simplified explanation of these effects can be gathered from the pictorial presentation of Figure 1. The closely-bonded chains of native cellulose

TABLE 1

Acetylation of Cellulose Fibers with Benzene–Acetic Anhydride–Sulfuric Acid Reagent for 24 hr at 60°C (23, 24)

	Acetyl content (%)			
Fiber	Dried, native	Dried, mercerized	Dry, "acetic acid inclusion"	
			Native	Mercerized
Cotton	2.0	0.9	41.8	44.9
Ramie	2.1	1.5	40.3	43.6
Linen	1.2	2.2	38.9	41.4
Hemp	3.3	2.3	40.3	41.3

TABLE 2

Acetylation of Cellulose Fibers with Pyridine–Acetic Anhydride Reagent for 24 hr at 60°C (23, 24)

Fiber	Acetyl content, (%)			
	Dried, native	Dried, mercerized	Dry, "pyridine inclusion"	
			Native	Mercerized
Cotton	4.3	0.7	7.9	14.2
Ramie	2.1	0.9	7.2	15.6
Linen	1.3	0.4	7.6	15.7
Hemp	1.4	0.6	9.2	16.3

are loosened by the mercerization process to give the reactive, never-dried, water-wet mercerized cellulose (Fig. 1*a*). Reactivity is lost when the structure is dried because strong bonding is restored (Fig. 1*b*). Wetting with water breaks the bonds again, but only in the noncrystalline regions. The space between the chains occupied by the water molecules is taken up by the solvent molecules on "inclusion," and some of these solvent molecules are trapped in the fine and gross structures, separating the polymer chain molecules when the solvent is driven off (Fig. 1*c*). The "inclusion" cellulose is reactive by virtue of the chain separation, which hinders the reformation of hydrogen bonding, thereby keeping the hydroxyl groups of the cellulose molecule in a more accessible condition for the entering reactant.

Sebille-Anthoine (25) prepared a "benzene-inclusion" cellulose from mercerized, bleached cotton linters in which the distention of the structure by the benzene was such that a translucent mass was obtained. With this "inclusion" cellulose and the anhydrides of acetic, propionic, and butyric acids, the triesters were obtained in about 30 min. With the dry, mercerized linters there was practically no reaction. Sharkov, Dmitrieva, and Potapova (26) noted the same thing regarding "benzene-inclusion" cellulose and acetylation, but they found no reactivity difference between this cellulose and the original material during solvolysis.

One of the most striking demonstrations of the effect of morphology on cellulose reactivity has been given by Demint and Hoffpauir (27), who also used acetylation (acetic anhydride–acetic acid–perchloric acid reagent). In their study the effect of gross and fine structures is separated; that is, they acetylated sliver lap (a web of fibers before spinning), ground sliver lap, yarn, and fabric, all from the same lot of cotton in one series of experiments, and in another series the yarn alone was subjected to various treatments before acetylation. Thus in the

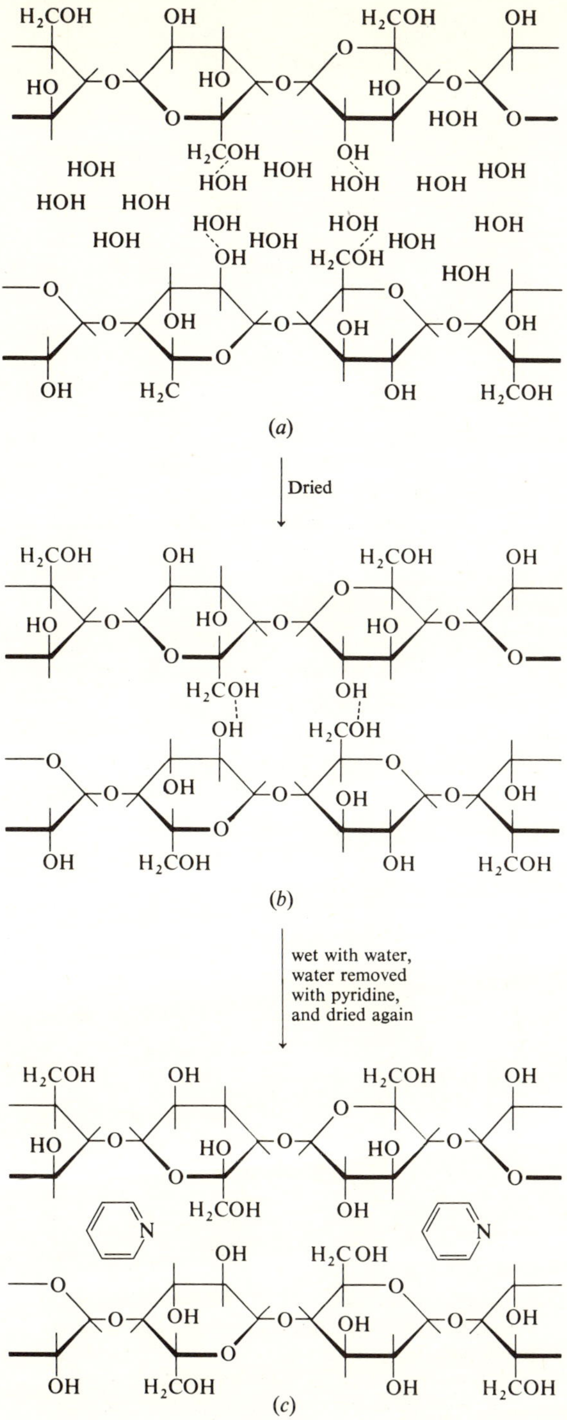

Fig. 1. Structural considerations of "inclusion" cellulose: (*a*) mercerized cotton, never dried and water-wet (reactive); (*b*) mercerized cellulose, dry (unreactive); (*c*) "pyridine-inclusion" cellulose, dry (reactive).

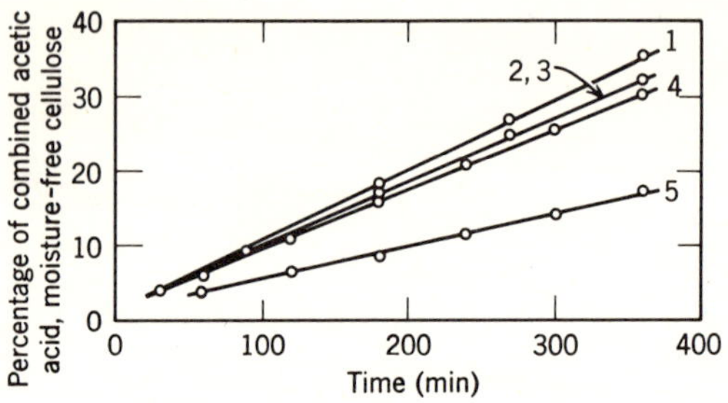

Fig. 2. Effect of textile form on the rate of acetylation (27): (1) 7s/3 yarn; (2) sliver lap (1-in. cut); (3) 34s/2 yarn; (4) fabric; (5) ground sliver lap. (Reproduced by permission of *Textile Research Journal*.)

first series only gross structure is changed, whereas in the second series the fine structure is influenced. The gross-structure effects are observed in the plotted data of Figure 2. The faster rate shown by the heavier 7s/3 yarn and the slower rate of the ground sliver lap (relative to the other three forms) has been postulated as possibly arising from greater retention of heat of reaction by the bulkier structures so that the acetylation proceeds at a higher temperature (27), but there has been no experimental proof of this.

Much more impressive changes in rate are observed when fine structure is affected (Fig. 3). The "hornification" effect already mentioned in connection

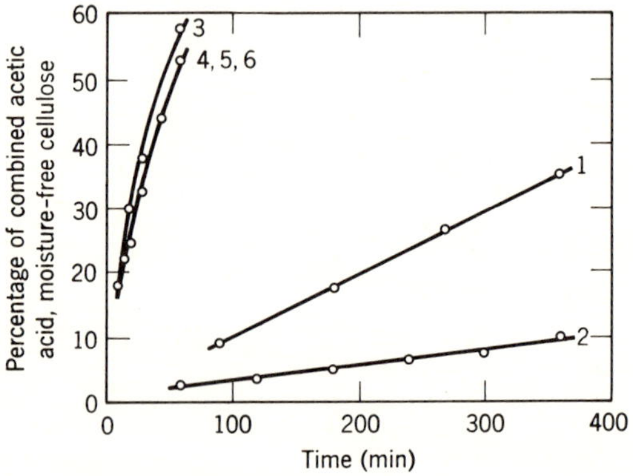

Fig. 3. Effect of wetting of 7s/3 yarn on rate of acetylation after drying (27): (1) raw; (2) excessively dried; (3) wet by Soxhlet extraction, first cycle; (4) rewetted, sixth cycle; (5) rewetted, 12th cycle; (6) wet by soaking. (Reproduced by permission of *Textile Research Journal*.)

with mercerized cellulose is markedly evident even when only exposure to water is involved (curve 2). The drastic drying conditions, forcing out more water from between the chains, cause extensive bonding, which reduces the accessibility of the hydroxyl groups. On the other hand, water-wetting followed by air-drying enhances reactivity (curves 3, 4, 5, and 6) because chain separation is maintained by water molecules. The cellulose–HOH–cellulose bonding is more easily broken by the entering reactant than the cellulose–cellulose bond. The effect on cellulose reactivity of repeated wetting by water and drying in the nitration of dewaxed cotton linters has been studied by Brown and co-workers (28), whose work confirms this marked behavior where only the noncrystalline regions are first involved.

The effects of swelling agents other than water are apparent in Figure 4. The data plotted here show without question what a tremendous effect distention of the fine structure has on the reactivity of the polymer molecule by making the hydroxyl groups more accessible. Curve 1 in Figure 4 is the same as that in Figures 2 and 3. Of the five pretreated yarns in Figure 4, four were air-dried before acetylation (curves 2, 3, 4, and 6) and one was never dried (curve 5). Where the same swelling pretreatment was involved but drying conditions were different (curves 2 and 5), Demint and Hoffpauir have confirmed the findings of earlier workers (20) regarding the retention of reactivity on keeping the fine structure distended with a solvent.

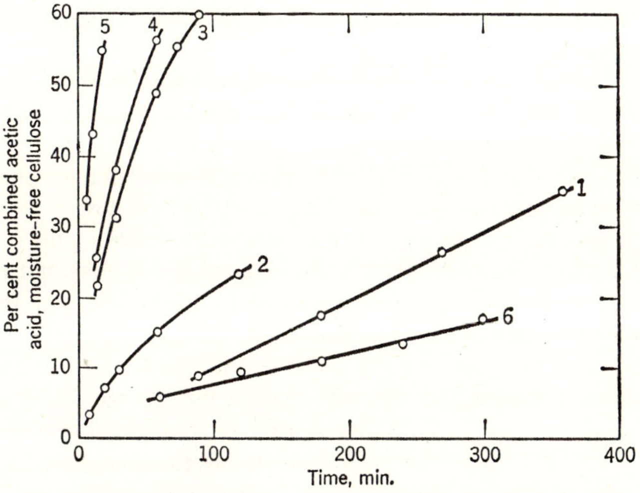

Fig. 4. Effect of applying various swelling agents to 7s/3 yarn prior to acetylation (27): (1) raw yarn; (2) decrystallized yarn; (3) alcohol- and water-extracted yarn; (4) scoured yarn; (5) never-dried, decrystallized yarn, washed with ice water and exchanged to acetic acid; (6) mercerized yarn. (Reproduced by permission of *Textile Research Journal*.)

The findings of Demint and Hoffpauir, however, also pose the question as to why the dry, mercerized yarn should be so much more unreactive than the raw yarn (curve 6 versus curve 1). The structures of both yarns have been exposed to high degrees of swelling, so that the degree of crystallinity, as well as the crystallite size, of each pretreated yarn has been reduced from that of the raw yarn. The explanation probably lies in differences that exist between the two swelling systems, differences that cause the fine structures to be affected differently. Thus in mercerization, which is carried out in an aqueous system with sodium hydroxide, the presently accepted concept holds that the sodium ion with its sheath of water molecules penetrates the fine structure, enters the crystalline regions, and forms new crystalline compounds. It is the energy content of this highly hydrated ion that causes the extensive changes. Removing the sodium hydroxide is accomplished with water, which remains in the crystalline regions and thus permeates the entire polymer system. When the water molecules are removed on drying, the collapse of the swollen structure is far-reaching. With the cellulose chains rearranged because of the extent of swelling, thus relieving strains, more hydrogen bonds form between the chains to "lock up" the fine structure more tightly than it was prior to mercerization.

Decrystallization, on the other hand, is accomplished with anhydrous ethylamine. There is no hydrated ion or molecule penetrating the fine structure, and, without water being present, an entirely different crystalline complex results. Breaking this complex by removing the ethylamine with cold water, as was done with the cellulose in Demint and Hoffpauir's work, does not follow the same scheme as in mercerization. The amine diffuses from the crystalline regions to the water in the amorphous regions and that external to the gross structure, whereas the swollen structure simultaneously shrinks. Water does not occupy space in the crystalline regions, and when the swollen structure collapses on drying, the extent and the type of hydrogen bonding set up between the cellulose molecules are not like those in the mercerized cellulose.

The changes on morphology brought about by water in the amorphous regions alone have marked influence on reactivity; this holds not only with respect to the amount of water present but also for the prehistory of the sample. Here "prehistory" refers to prior treatments, such as purification, that the sample may have been subjected to and the direction from which the present moisture content was reached. This effect can be seen in Figure 5 (29). Whether water is leaving or entering the cellulosic structure has great importance on the extent of reaction. Thus from Figure 5 one readily concludes that, for the same moisture content, the cellulose is more accessible for reaction if the water molecules are leaving the cellulosic structure. This follows from what has already been discussed, because until the water molecules have broken the cellulose–cellulose hydrogen bonds (in the noncrystalline regions) the entering reactant finds strong interchain bonds present. There is some evidence in Figure 5*A* and *B*

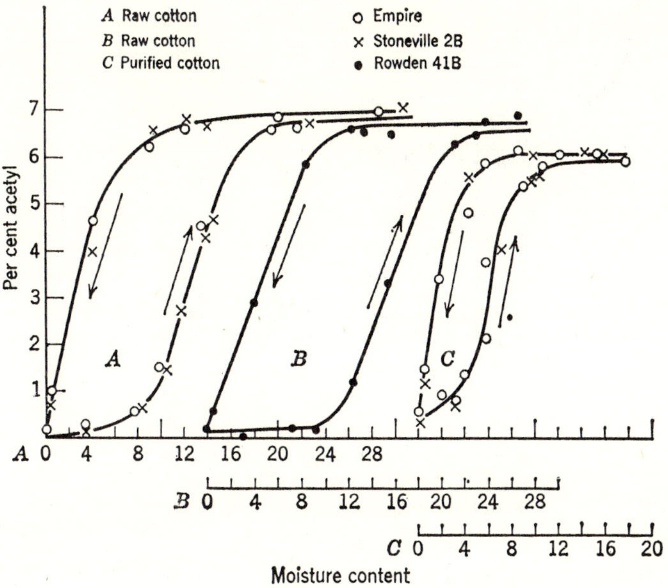

Fig. 5. Influence of moisture and previous sorption history on extent of acetylation (29). (Reproduced by permission of *Textile Research Journal*.)

that botanical morphology affects the chemical reactivity of the polymer chain. Thus the data in Figure 5*B* were obtained from a much coarser fiber than that in Figure 5*A*. The difference has been postulated as possibly due to the smaller area per gram and the greater diameter of the coarser cotton fiber, which requires more extensive swelling to reach a given degree of accessibility. Removal of waxes and noncellulosic constituents from cotton fibers has a marked effect: less moisture in the fiber is required in order to attain the same level of substitution as in the raw fiber (Fig. 5*C*).

Freezing of alkali-swollen celluloses in dilute sodium hydroxide solutions has been found by Nikitin (30) to have a profound effect on the fine structure of the cellulose. Hygroscopicity of the treated cellulose and its solubility in alkali solutions are markedly enhanced. Nikitin assumes that the formation of ice crystals in the already swollen submicroscopic structure opens it still further, and this assists subsequent disruption of interchain forces by the alkali solution on thawing. In one freezing and one thawing operation with 8–12% sodium hydroxide solution at −8 to −10°C up to 11% of cotton linters and up to 27% of purified sulfite wood pulp could be dissolved in caustic solutions. This could be enhanced by repeated freeze–thaw operations (30, pp. 200 and 201). Tsvetaeva, Petropavlovskii, and Ionanova (31) utilized freezing as a pretreatment

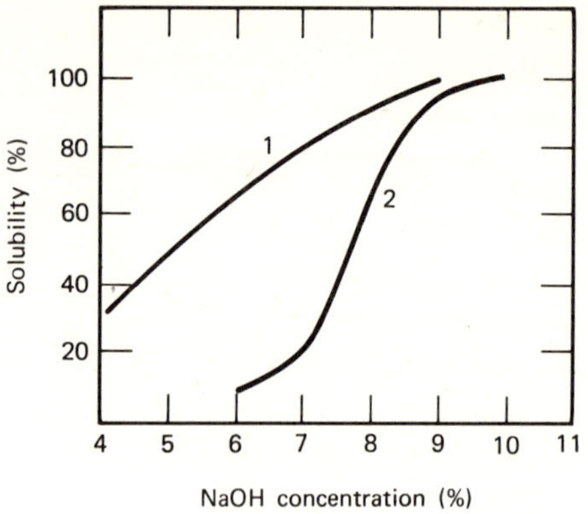

Fig. 6. Effect of freezing sulfite wood pulp in sodium hydroxide solutions on the solubility of the partially xanthated esters (31): (1) solubility before freezing; (2) solubility after freezing.

in their studies of the supramolecular structure of cellulose on its solubility during emulsion xanthation. The drastic effect of this pretreatment is very obvious from the difference in the solubility curves of a sulfite wood pulp handled in this manner (Fig. 6).

Blouin and co-workers (29) have indicated that botanical morphology has an effect when only cottons are involved. Other workers have found differences in behavior when celluloses from different sources were investigated.

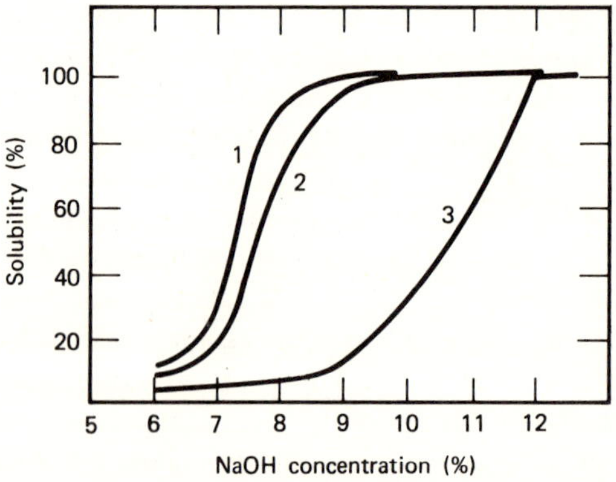

Fig. 7. Solubility of partially xanthated celluloses in sodium hydroxide solutions (31): (1) sulfite wood pulp; (2) sulfite wood pulp of another grade; (3) cotton linters.

Petitpas (32) reported that different concentrations of sodium hydroxide are required to obtain x-ray diagrams characteristic of mercerized cellulose when cotton, ramie, and wood pulp are treated with sodium hydroxide solutions. Rånby (33) found similar results with animal cellulose (tunicin), cotton cellulose, and wood cellulose–animal cellulose requiring a more concentrated caustic solution and wood cellulose the least. Rånby also found bacterial cellulose (from *Acetobacter xylinum*) to have properties very similar to those of cotton cellulose. Tunicin enters into all cellulose reactions, but these are governed by differences in accessibility caused by the morphological structure of the tunicin (34). Tsvetaeva and co-workers (31) similarly report that the solubility of partially xanthated celluloses in sodium hydroxide solutions of various concentrations is markedly influenced by the nature of the initial cellulose. This is clearly seen in the curves of Figure 7, where it is to be noted that caustic solutions of higher concentrations are required to affect the cotton linters to the same extent as wood pulps. This parallels the findings of Rånby in his x-ray studies of the sodium hydroxide-cellulose systems.

Additional information on reactivity in the formation of cellulose derivatives may be found in the succeeding sections of this Chapter XVII.

3. Hydrolysis

When cellulose is heated with aqueous mineral acids, the polymer chain is attacked at the glucosidic oxygen atoms connecting the glucose units, and the reaction ultimately ends with the monomer glucose. However, cellulose morphology influences the rate of reaction very highly in easily distinguishable and characteristic ways in the early stages of this reaction (35–37) (see also Figure 2 in Chapter XVIII-A).

Nickerson and Habrle (36) concluded that their data on the amount of cellulose hydrolyzed by boiling 2.5 *N* sulfuric acid indicate that all the substrates may represent molecular order ranging from highly disordered and readily accessible to well-ordered and less readily hydrolyzed. Philipp, Nelson, and Ziifle (38) pursued this concept to develop a simplified method for estimating the relative proportions of crystalline and amorphous cellulose present in any one cellulosic material. The heart of that method are the hydrolysis-rate curves obtained for cellulosic materials, two of which are shown in Figure 8. The morphological parameters involved in this application of hydrolysis concern the arrangement of the cellulose chains, as already indicated by Nickerson and Habrle. In the areas of rapid attack by the acid (the curved portions in Figure 8) the cellulose chains are visualized as lying in disorder with respect to one another (x-ray amorphous material). Where the cellulose is more resistant to acid attack (the linear portions of the rate curves in Figure 8), the polymer chain molecules are pictured as being in close-packed order (characteristic x-ray diffraction). The

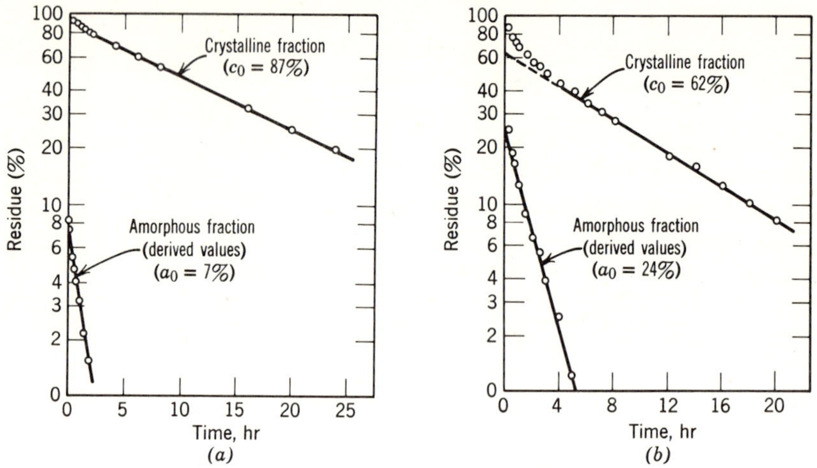

Fig. 8. Weight-loss curves of cellulosic material hydrolyzed with hydrochloric acid (38): (*a*) cotton linters, 100°C, 6 *N* HCl; (*b*) viscose rayon, 100°C, 4 *N* HCl.

chemically resistant hydrocellulose is only slowly attacked by continued, prolonged exposure to strong acid. This is shown in the work of Nelson and Tripp (39), who carried out hydrolysis for up to 600 hr, using several acid concentrations and two temperatures to show how little the resistant hydrocellulose is affected (Fig. 9). Their conclusions verified those of earlier workers that each cellulosic material has a characteristic leveling-off degree of polymerization that is related to the length of the crystallite. A practical consequence of this effect is discussed in Section G of Chapter XIX.

It may be thought that once hydrocelluloses have been produced, the influence of the morphological differences of the original cellulosic materials should disappear. This does not seem to happen. For example, hydrocelluloses from cotton linters and viscose rayon retain the large difference in moisture sorption displayed by the initial materials (39). Millett and co-workers (35) also found that moisture-sorption differences are retained by the hydrocelluloses (Fig. 10). However, their experiments were carried to great length so that the fact could be seen that moisture sorption leveled off at limiting values for each cellulose used.

Sharples (40) prepared hydrocelluloses from various cellulosic materials, including regenerated, ramie, cotton, bacterial, and animal celluloses. Rate measurements made on the hydrocelluloses further exposed to hot hydrochloric acid showed that weight losses still followed the order expected on the basis of crystallinity, crystallite size, and accessibility of the original, unhydrolyzed material. One may conclude from Sharples' data that continued chemical attack on the hydrocelluloses occurs at the ends of the crystalline particles.

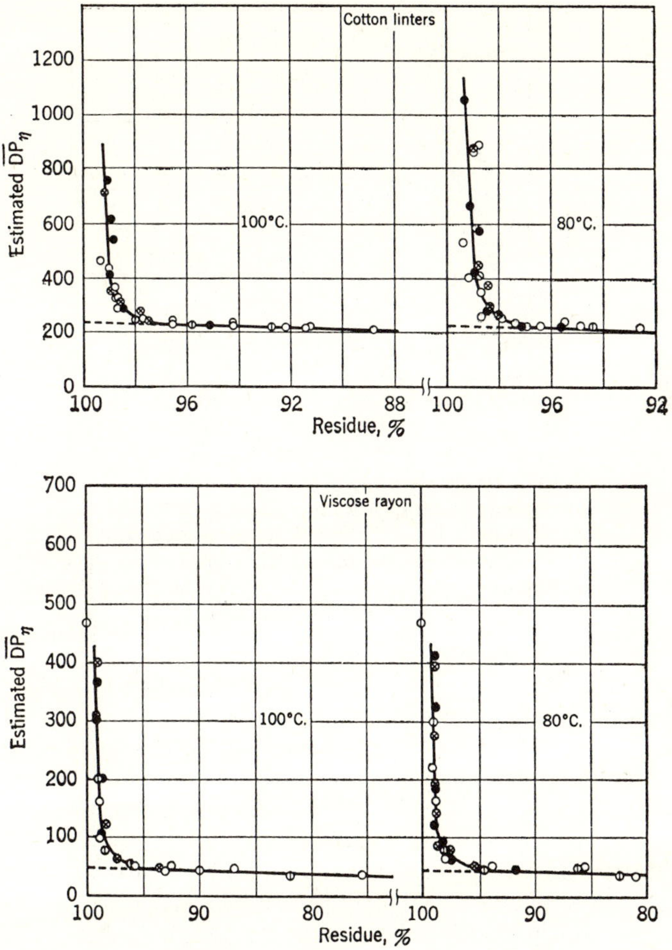

Fig. 9. Relation between degree of polymerization and residual hydrocellulose after prolonged hydrolysis of cotton linters and viscose rayon (39). Concentration of hydrochloric acid: ○ 2.5 *N*; ⦶ 1.0 *N*; ● 0.1 *N*; ⊗ 0.01 *N*.

The attack of acid on crystalline cellulose is not without some anomalies. Segal and co-workers (41) found that immersion of cotton cellulose III in boiling water for 2 hr caused a reversion of its lattice to that of cellulose I. But the same cellulose III exposed similarly to 6 *N* hydrochloric acid was found by Segal and Nelson (42) to be still almost pure cellulose III. Somehow the acid molecules almost completely blocked the lattice conversion. Acid at somewhat lower

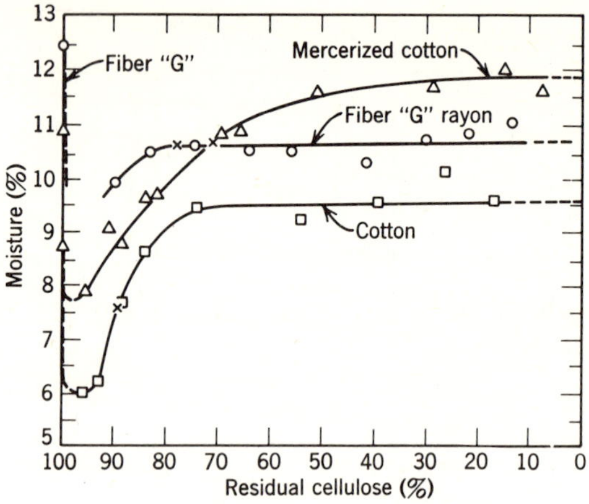

Fig. 10. Moisture sorption (at 68.5% RH) of three cellulosic materials as a function of the extent of hydrolysis (35). The x marks indicate the point at which all easily hydrolyzable material has been removed. (Reproduced by permission of the American Chemical Society, the copyright owner.)

concentrations was slightly less effective in preventing this reversion. Microscopic examination of the cellulose III hydrocelluloses after treatment with 2 *N* hydrochloric acid showed fairly large, easily recognizable fiber fragments. Much smaller fragments, still identifiable as segments of fibers, remained after treatment with 6 *N* hydrochloric acid. It is quite possible that these fiber sections, retaining morphological structure, are merely the framework of the original fiber, consisting of the more resistant portions of the structure, but to settle this point is beyond the ability of present microscopic techniques.

Hydrolysis is discussed further in Section A of Chapter XVIII.

4. Thermal Treatments

It is well known that heating cellulose in any form to elevated temperatures results in browning, charring, and loss of structure. This type of thermal degradation is discussed in Chapter XVIII-C. At lower temperatures, *below* those producing thermal decomposition, the effects of heat become manifest in more subtle ways, and there is some experimental evidence to show that the extent of such changes is influenced by morphology. These changes for the most part are hardly detected by the usual chemical means, but rather show up as changes in tensile strength and in the fluidity of cellulose solutions or, relatedly, in the degree of polymerization. Some of this is touched on in Part I, pages 176–180.

Conrad, Tripp, and Mares (43) made a very detailed study of the effect of heat alone on cotton and rayon tire cords at various stages of processing and in tire sections. Temperatures ranged as high as 150°C for periods up to 1024 hr with the oven atmosphere maintained at 40% relative humidity. For the cotton tire cord 100% strength loss was found to occur when 0.3% of the glucose links were broken, whereas the corresponding breakage for rayon was 0.6%. Exactly why there is this difference in behavior toward heat alone is not clear unless crystallite size, degree of crystallinity, accessibility, or lateral-order distribution are the factors contributing to this effect. Similar evidence for differences in behavior between cotton and rayon tire cords was obtained by Waller, Bass, and Roseveare (44). Data can be seen in Part I, page 178. Thermal degradation is attributed principally to oxidation, and it seems that the degree of accessibility conferred by morphology is an important factor contributing to the difference in behavior.

The above conclusion relating to morphology is further supported by the findings of Philipp, Baudisch, and Gaudig (45) in their study of the effect of supermolecular fiber structure on the thermal degradation of cellulose in dry nitrogen. Differences were found in weight loss and carboxyl contents for cotton linters, normal viscose rayon, and high-tenacity rayon when subjected to temperatures of up to 300°C over periods of up to 100 hr.

The work of Wiegerink (46) shows that mercerized cotton has a lower rate of degradation with temperature than purified cotton or viscose rayon. This finding confirms the already mentioned "hornification" of mercerized cotton, which reduces its chemical reactivity. This behavior of mercerized cotton is also verified by the work of Lewis (47), who subjected rayon, cotton, and mercerized-cotton tire cords to alternate heating (130°C for 16 hr) and cooling for a total of 48 cycles. Regardless of the test conditions, standard or oven-dry, the mercerized-cotton cord showed the least loss in specific strength (tenacity). No significant conclusion can be drawn from the data obtained on the other cords because of variations in the test conditions.

Sharkov, Dmitrieva, and Potapova (26) found that "inclusion cellulose" as well as swollen celluloses are affected by heat. In their work heat was applied by placing the material in kerosene heated to 200°C for 3 hr after which the sample was hydrolyzed. Hydrolysis of wood cellulose before heating gave a glucose yield of 7%, which liquid-ammonia treatment of the cellulose raised to 18% and sodium hydroxide treatment raised to 12%. After heating, the glucose yield was 12% for the original wood cellulose, 31% for the ammonia-treated cellulose, and 28% for the sample treated with sodium hydroxide. "Inclusion cellulose" prepared directly from cellulose wet with liquid ammonia, gave glucose yields of 16–22% before heating, 30–36% after heating, except for three samples that gave values of 39, 49, and 57%. These three results were obtained with carbon tetrachloride, chloroform, and dichloroethane, respectively, and are attributed to the ability of these solvents to penetrate the ammonia-wet structure more deeply than the other solvents and thus to retard subsequent close packing of the cellulose when it is dried. "Inclusion cellulose" prepared from samples treated with sodium hydroxide solutions gave glucose yields of 11% before heating and 24–39% after heating. All of this indicates that the stability of cellulose to heat has been lowered, which results from loosening of the macromolecular packing of the chain molecules.

A discussion of the thermal degradation of cellulose may be found in Section C of Chapter XVIII.

References

1. J. Honeyman, Ed., *Recent Advances in the Chemistry of Cellulose and Starch*, Interscience, New York, 1959.
2. A. H. Nissan, G. K. Gunther, and S. S. Sternstein, in N.M. Bikales, Ed., *Encyclopedia of Polymer Science and Technology*, Vol. 3, Interscience, New York, 1965, pp 131–226.
3. A. Frey-Wyssling, *Biochem. Biophys. Acta,* **18**, 166 (1955).
4. K. Hess, *Chim. Ind. (Paris)*, **80**, 129 (1958).
5. F. Happey, *J. Textile Inst.*, **41**, T381 (1950).
6. H. J. Marrinan, in J. Honeyman, Ed., *Recent Advances in the Chemistry of Cellulose and Starch,* Interscience, New York, 1959, pp. 147–186.
7. C. W. Hock, *J. Polymer Sci.*, **8**, 425 (1952).
8. B. G. Rånby, *Tappi,* **35**, 53 (1952); inaugural dissertation, University of Uppsala, Sweden, 1952; *Svensk Papperstidn.*, **57**, 9 (1954).
9. M. L. Rollins and V. W. Tripp, *Forest Prod. J.*, **11**, 493 (1961).
10. E. Heuser, *Chemistry of Cellulose,* Wiley, New York, 1944, pp. 7–33; P. H. Hermans, *Physics and Chemistry of Cellulose Fibres,* Elsevier, New York, 1949, pp. 163–172; J. C. Arthur, Jr., in N.M. Bikales, Ed., *Encyclopedia of Polymer Science and Technology,* Vol. 4, Interscience, New York, 1966 pp. 244–272.
11. W. A. Richardson, *Rept. Progr. Appl. Chem.*, **41**, 366 (1956).
12. M. L. Rollins, *Anal. Chem.*, **26**, 718 (1954).
13. E. I. Valko, in E. Ott, Ed., *Cellulose and Cellulose Derivatives* Vol. V, *High Polymers*, 1st ed., Interscience, New York, 1946, pp. 324–340; P. H. Hermans, *Physics and Chemistry of Cellulose Fibres*, Elsevier, New York, 1949, pp. 143–155.
14. E. Heuser, *Chemistry of Cellulose*, Wiley, New York, 1944, pp. 42–85.
15. R. J. E. Cumberbirch, *Rept. Progr. Appl. Chem.*, **40**, 451 (1955).
16. D. A. Sitch, *J. Textile Inst.*, **44**, T407 (1953); *Shirley Institute Mem.*, **26**, 89 (1952–1953).
17. R. E. Reeves and H. J. Thompson, *Contrib. Boyce Thompson Inst.*, **11**, 55 (1939).
18. R. E. Reeves and W. A. Sisson, *Contrib. Boyce Thompson Inst.*, **13**, 11 (1943).
19. E. Heuser, *Chemistry of Cellulose,* Wiley, New York, 1944, pp. 101, 113, and 246.
20. L. Loeb and L. Segal, *Textile Res. J.*, **24**, 654 (1954).
21. E. Heuser, *Chemistry of Cellulose,* Wiley, New York, 1944, pp. 240–243; 247–255.
22. A. R. Urquhart, *Textile Res. J.*, **28**, 159 (1958).

23. H. Staudinger, K. H. In den Birken, and M. Staudinger, *Makromol. Chem.*, **9**, 148 (1953).
24. H. Staudinger and W. Döhle, *Makromol. Chem.*, **9**, 188, 190 (1953).
25. J. Sebille-Anthoine, *Compt. Rend.*, **252**, 3797 (1961).
26. V. I. Sharkov, O. A. Dmitrieva, and N. P. Potapova, *J. Appl. Chem.* USSR (English translation) **34**, 1079 (1961).
27. R. J. Demint and C. L. Hoffpauir, *Textile Res. J.*, **27**, 290 (1957).
28. H. Rosenthal and R. K. Brown, *Pulp Paper Mag. Canada,* **51**, 99 (1950); W. J. Brickman, H. B. Dunford, E. M. Tory, J. L. Morrison, and R. K. Brown, *Can. J. Chem.*, **31**, 550 (1953); T. P. Yin and R. K. Brown, *Can. J. Chem.*, **37**, 444 (1959).
29. F. A. Blouin, R. E. Reeves, and C. L. Hoffpauir, *Textile Res. J.*, **26**, 272 (1956).
30. N. I. Nikitin, *The Chemistry of Cellulose and Wood,* Academy of Sciences, USSR, 1962; English translation, Israel Program for Scientific Translations, Jerusalem, 1966, pp. 133, 200–201.
31. I. P. Tsvetaeva, G. A. Petropavlovskii, and T. N. Ionanova, *J. Appl. Chem. USSR* (English translation), **39**, 1446 (1966).
32. T. Petitpas, *Mém. Serv. Chim. Etat (Paris),* **37**, 271 (1952); through *Chem. Abstr.*, **48**, 9682*a* (1954).
33. B. G. Rånby, *Arkiv Kemi,* **4**, 241, 249 (1952).
34. H. Krässig, *Makromol. Chem.*, **26**, 17 (1958).
35. M. A. Millett, W. E. Moore, and J. F. Saeman, *Ind. Eng. Chem.*, **46**, 1493 (1954).
36. R. F. Nickerson and J. A. Habrle, *Ind. Eng. Chem.*, **39**, 1507 (1947); W. E. Roseveare, *Ind. Eng. Chem.*, **44**, 168 (1952).
37. H. J. Philipp, M. L. Nelson, and H. M. Ziifle, *Textile Res. J.*, **17**, 585 (1947).
38. M. L. Nelson, *J. Polymer Sci.*, **43**, 351 (1960).
39. M. L. Nelson and V. W. Tripp, *J. Polymer Sci.*, **10**, 577 (1953).
40. A. Sharples, *Trans. Faraday Soc.*, **54**, 913 (1958).
41. L. Segal, L. Loeb, and J. J. Creely, *J. Polymer Sci.*, **13**, 193 (1954).
42. L. Segal and M. L. Nelson, *J. Am. Chem. Soc.*, **76**, 4626 (1954).
43. C. M. Conrad, V. M. Tripp, and T. Mares, *Textile Res. J.*, **21**, 726 (1951).
44. R. C. Waller, K. Bass, and W. E. Roseveare, *Ind. Eng. Chem.*, **40**, 138 (1948).
45. B. Philipp, J. Baudisch, and A. Gaudig, *Faserforsch. u. Textiltech.*, **18**, 9 (1967).
46. J. G. Wiegerink, *J. Res. Natl. Bur. St.*, **25**, 435 (1940).
47. W. S. Lewis, *Textile Res. J.*, **17**, 431 (1947).

B. ESTERS

Gordon D. Hiatt and William J. Rebel

Because it is a polyhydroxyl alcohol, cellulose can be esterified by a variety of acids and anhydrides, and the resulting esters can be partially or totally hydrolvzed. This section is concerned with the esterification and deesterification of cellulose. The esters discussed include both inorganic and organic esters. Some contain both inorganic and organic ester groups, and these are treated as inorganic esters. The reader is also referred to the extensive discussion of this subject in Parts I, II, and III of this work, and to Chapters XIX- C and XIX- F.

1. Inorganic Esters

(a) CELLULOSE NITRATE

Cellulose nitrate is by far the most important inorganic ester of cellulose. Most of the recent efforts have been directed toward developing a continuous nitrating process, and the search continues for a new low-cost nitration-grade cellulose.

(1) Preparation of Cellulose Nitrate

Cellulose for Nitration. During World War I attempts were made to prepare cellulose nitrate from wood pulp rather than cotton (1). By World War II the use of wood pulp for cellulose nitrate was well established, but pulps give lower yields and an inferior product. There have been few recent developments in the nitration of wood pulps.

Brooks (2) has reported that cotton and cellulosic material from wood and straw can be nitrated by a two-stage process to give a cellulose nitrate of good uniformity, stability, and viscosity, but the yield was generally lower for pulp than for linters. The yield was found to be proportional to the α-cellulose content (3). In addition the degree of polymerization (DP) was generally lower for pulp cellulose nitrates. These defects were attributed to pentosans present in the pulps (4).

Nitration Media. Although considerable effort has been made to develop new nitration media, the nitric acid–sulfuric acid–water system is still the only one that has achieved commercial importance. Other nitrating agents have been employed to prepare cellulose nitrate with special properties. Most modified procedures have been previously reported and can be summarized in the following categories:

1. Use of pure nitric acid.
2. Replacement of sulfuric acid by phosphoric acid, with or without phosphorus pentoxide, with and without the addition of oxides of nitrogen.
3. Conventional nitric–sulfuric mixed acids containing organic diluents (usually chlorinated hydrocarbons).
4. Nitric acid with nitrate salts present.

A number of preparations of cellulose nitrate without degradation have been reported. Bennett and Timell (5) reported the preparation of a completely substituted cellulose nitrate with the same DP as the original cellulose. A mixture containing nitric acid, acetic acid, and acetic anhydride in the proportions 43 : 32 : 25 nitrated cotton linters at 0°C to the maximum nitration level (14.14% nitrogen) with no apparent degradation. However, prolonged exposure to the nitration medium caused a decrease in DP. Cellulose nitrate can be prepared without degradation (6) by esterification with mixtures of anhydrous nitric acid and chlorinated hydrocarbons at low temperatures with very short reaction times. Concentrated nitric acid dissociates in inert solvents as follows:

$$HNO_3 \rightleftharpoons NO_2^+ + OH^- \tag{1}$$

Cellulose nitrate (13.8–14.0% nitrogen) prepared in this manner has an extraordinary stability and is readily soluble in appropriate solvents.

Mechanism of Nitration. Cellulose nitration may be compared with a typical equilibrium esterification; an acid and an alcohol react, and the products are an ester and water. Since this is an equilibrium reaction, it is necessary to remove the water in order to force the reaction to completion.

Numerous investigators have studied the nitric acid–sulfuric acid–water system. Evidence has been presented in favor of the NO_2^+ ion as the nitrating agent (7). It is generally believed that the only role of the sulfuric acid is to remove the water formed during the reaction. Two mechanisms have been advanced to explain the manner of nitration (7). In the first the nitration reagent must penetrate the cellulose micelles, and nitration proceeds progressively from the outside to the inside of the micelles (heterogeneous nitration). The second mechanism involves homogeneous nitration, in which the nitrating agent penetrates slowly but uniformly into the entire fiber structure and all molecules react rapidly at about the same time; this latter mechanism has been favored.

Sakata and Komatsu (8) suggested that during the initial period of nitration with a mixture of nitric and sulfuric acids the reaction is nonuniform but becomes uniform with progressing nitration. These authors used infrared spectra and x-ray diffraction, in conjunction with deuteration of partly reacted cellulose, to show that the accessibility of hydroxyl groups increases as nitration by a mixture of nitric and phosphoric acids proceeds. The mechanism of nitration by a mixture of nitric and phosphoric acids may be considered as follows (8):

> The reagent penetrates into the crystallite, and most of the OH groups on the cellulose chains are uniformly nitrated. For nitration with HNO_3–H_2SO_4 the initial increase of accessibility of the OH groups is not greater than that of the HNO_3–H_3PO_4 system, and the spectrum of the unaccessible OH group is similar to that of unreacted cellulose.

This indicates the presence of a region that does not participate in the reaction. Toward the end of the nitration the spectrum becomes similar to that of the nitric acid–phosphoric acid mixture. These results support the mechanism in which the reaction is nonuniform during the early stages of nitration with a mixture of nitric and sulfuric acids but becomes uniform with progressing nitration. The x-ray results showed good correlation with the infrared observations.

Stability and Stabilization. The instability of cellulose nitrate is usually attributed to retained traces of nitrating acid or a gradual cleavage of the unstable sulfate groups (9). A series of patents assigned to the Hercules Powder Company claims that stable, acetone-soluble, sulfur-free esters are obtained by reacting cellulose with a mixture containing 45–94% (by weight) nitric acid, 3.3–34% magnesium nitrate, and 2.7–21% water. The magnesium nitrate : water ratio is 1.2 : 1 to 2.2 : 1, and the reaction temperature is 15–70°C (10, 11). Cellulose nitrate prepared by this method can be further heat stabilized by blending with 0.2–2.0% diphenylamine and 0.02–1.0% of an organic acid having a dissociation constant of less than 2×10^{-2}, such as citric, stearic, or tartaric acid (12, 13). Arthen, Coleman, and Hardy (14) have stabilized cellulose nitrate against degradation by ultraviolet radiation by adding a carboxybenzophenone, such as 2′-carboxy-2-hydroxy-4-methoxybenzophenone.

(2) Manufacture of Cellulose Nitrate

Although a number of methods have been used to prepare cellulose nitrate, all these procedures employ a batch process. A great deal of effort has been directed recently toward developing a continuous process.

Plunkett (15, 16) has patented a process that involves continuously forming a slurry of cellulose in the nitrating mixture, advancing the slurry in an alternately ascending and descending bath, agitating the slurry, and continuously withdrawing a slurry of cellulose nitrate in the spent nitrating mixture from the final reaction zone. McMillan and Plunkett (17–19) recovered most of the spent nitrating mixture by centrifuging; the cellulose nitrate was centrifugally washed in a series of aqueous nitric acid washes of decreasing nitric acid concentration, and each discharge was recycled to the preceding wash. The spent nitrating mixture was fortified with nitric acid and magnesium nitrate and recycled. Cellulose nitrate has also been prepared in sheet (20, 21) or particle (22, 23) form.

Ramsey (24) has described a continuous process in which long strips of cellulose are drawn through a supporting tube in a bath of nitric acid, phosphoric acid, and water. Diels and Orth (25) have employed a continuous process that subjects the cellulose to a preliminary treatment with spent nitric acid. The prenitrated cellulose is then reacted with nitrating acid, washed, and stabilized. Supersonic waves were used to obtain a highly dispersed state during nitration and stabilization. Frejacques (26) has regenerated the spent water-containing nitrating mixture from the reaction of cellulose with a mixture of nitric acid and a low-molecular-weight chlorinated hydrocarbon.

(b) CELLULOSE ESTERS OF ACIDS CONTAINING SULFUR

Since the introduction of the use of a mixture of sulfuric acid and an alcohol (27, 28) for the preparation of water-soluble cellulose sulfate, a great deal of commercial interest has ensued. Klug (29, 30) has prepared a uniform water-soluble cellulose sulfate without excessive cellulose degradation. Activated cellulose (1 mole) containing 20–30% water was immersed in a bath containing 3.5–15 moles of sulfuric acid, 0.3–1 mole of a C_3–C_5 primary or secondary alcohol, and an inert, volatile organic diluent, such as carbon tetrachloride or toluene. The resulting slurry contained less than 8% cellulose by weight and less than 10% water based on the combined weight of sulfuric acid and water. The slurry was agitated at −5 to 10°C until cellulose sulfate formed.

Petropavlovskii and Krunchak (31) investigated the esterification of cellulose with mixtures of concentrated sulfuric acid and aliphatic alcohols. The effects of such variables as the chain length of the alcohol, the sulfuric acid : alcohol molar ratio, the structure of the alcohol, and temperature and time on the course of esterification were studied. They found that for a sulfuric acid–propanol system (1 : 1 to 3.5 : 1) the sulfur content was directly proportional to the amount of acid in the mixture. Further increase in the acid-to-alcohol ratio did not increase the degree of esterification. The degree of substitution (DS) never exceeded that of a monoester (DS = 1). The concentration of water greatly influenced the reaction rate, and the ability of alkyl sulfates to transesterify with cellulose decreased with increasing chain length of the alcohol. With higher alcohols the esterifying agent is sulfuric acid itself. Cellulose sulfates (32) of low DS were obtained from a mixture of *n*-propanol and sulfuric acid. At DS values of up to 0.1 the sulfates were heterogeneous and contained up to 20% of a nonesterified fraction.

Read and Bilton (33) have prepared an undegraded water-soluble cellulose sulfate from wood pulp and oleum containing 5–20% free sulfur trioxide dissolved in liquid sulfur dioxide at –25 to –10°C for 15–20 min. Similarly, Asami and Tokura (34) have prepared cellulose sulfates from cotton and sulfur trioxide, chlorosulfonic acid, or 98% sulfuric acid in liquid sulfur dioxide; they obtained high yields with high DS by sulfating cotton in a heterogeneous medium, but the DS depended on the esterifying agent. The maximum DS for sulfuric anhydride, chlorosulfonic acid, and sulfuric acid was 2.2, 1.8, and 0.9, respectively. With chlorosulfonic acid the DS increases with increasing concentration of the reagent. Schweiger (35) claimed that the usual difficulties of severe degradation and low DS, characteristic of this method, can be overcome if sulfur trioxide dissolved in dimethylformamide is used as the sulfating agent.

Cellulose sulfate has also been prepared by using a mixture of sulfuric acid, a carboxylic acid (containing four to six carbon atoms), and water until undissolved cellulose hydrogen sulfate with a combined sulfur content of at least 3% was obtained (36). The sodium salts of these hydrogen sulfates were water soluble. Water-soluble mixed salts of cellulose sulfates, cellulose acylate sulfates, and alkyl sulfates have been prepared by reacting the cellulosic starting material with acetic anhydride, acetic acid, sodium sulfate, sulfuric acid, and calcium acetate (37).

Touey and Kiefer (38) have employed a cellulose activated with sodium hydroxide to prepare sodium cellulose sulfate under anhydrous conditions. The cellulose was treated with a mixture of sulfuric acid, propionic acid, and acetic anhydride. The sodium salt of a weak organic acid was added to this mass with thorough mixing, and the products were separated from the esterification mixture and dried. These authors have also prepared a water-soluble salt of cellulose acetate propionate sulfate (39). Cellulose derivatives have been sulfated by Hiatt and Rowley (40, 41); lower-fatty-acid cellulose esters or ethylcellulose were sulfated in a bath that was not acid to crystal-violet indicator and included at least 50% excess of the theoretical amount of acetic anhydride and an alkali-metal salt of acetylsulfuric acid.

In an effort to determine the most effective pretreatment methods Klein and Snowden (42) first prepared cellulose sulfonate intermediates by the esterification of cotton with methanesulfonyl or *p*-toluenesulfonyl chloride; the cellulose sulfonates undergo exchange reactions with amines, halogens, phenols, substituted phosphoric acids, thiocyanates, and thiophenols (see Section F of this Chapter XVII.)

(c) CELLULOSE ESTERS OF ACIDS CONTAINING PHOSPHORUS

Because of their potential use as flame retardants for textiles (see Chapter XIX-I), there has been considerable interest in the preparation of cellulose phosphates. When molten urea and phosphoric acid were heated with cellulose (up to 170°C), the chief product was cellulose monoammonium monohydrogen phosphate (43). Factors influencing the phosphorylation of cellulose with urea-phosphoric acid were investigated by Katsuura and Nonaka (44). An optimum molar ratio of urea to phosphoric acid was found to give a high DS. Higher temperatures and shorter treatments with a constant molar ratio gave cellulose phosphates of higher DS with less degradation. The reaction rate was slow, and even with repeated treatments (three times) a cellulose phosphate with only 9.03% phosphorus was obtained. Katsuura and Inagaki (45) have found that phosphoric acid undergoes polycondensation in the presence of isocyanic acid, which is formed from urea under the reaction conditions. This may explain, at least in part, the low DS of cellulose phosphate prepared by this method. Touey (46) prepared stable, essentially undegraded, water-soluble cellulose acid phosphates of high molecular weight by esterifying cellulose in an anhydrous solution of phosphoric acid (such as orthophosphoric acid), phosphorus pentoxide, and an alcohol diluent.

Organic phosphites have been prepared as intermediates in the synthesis of organic phosphonates by the Arbuzov rearrangement. Yuldashev and Muratova (47) have employed this reaction to prepare cellulose phosphonates from trialkyl phosphites and chlorodeoxycellulose. Chlorodeoxycellulose (16% Cl), which was obtained by the reaction of thionyl chloride with cotton linters, was heated with triethyl phosphite for 5-20 hr to give cellulose phosphonate. (There was only a negligible amount of trivalent phosphorus remaining.) Phosphorylated celluloses were prepared from dialkyl and trialkyl phosphites (48) and di-β-chloroethyl, di-β-fluoroethyl, and diphenyl phosphites (49). Cellulose phosphonates were also prepared with anhydrides of methylphosphonic and phenylphosphonic acids (50). When the reactions were carried out in dimethylformamide at 130-140°C, the products contained a maximum of 10% phosphorus and there was little degradation of cellulose.

Cellulose phosphites have been prepared by the transesterification of dimethyl phosphite with cellulose (51). The product is sensitive to hydrolysis. Monosubstituted and disubstituted esters of phosphorous acid and cellulose were prepared in this manner (52, 53). Cellulose can be phosphorylated with orthophosphorous acid by three different methods (54); the resulting phosphites contain up to 17% phosphorus, and the reactive site is apparently the primary cellulosic hydroxyl. Predvoditelev, Nifant'ev, and Rogovin (55) reported a different method for the synthesis of cellulose phosphites, using a mixed anhydride of methyl phosphorous and acetic acids (obtained in high yield by reacting monomethyl phosphite with acetyl chloride in the presence of triethylamine). However, even under the best of conditions the esters contained a maximum of only 7.5% phosphorus.

A number of other phosphorus-containing esters, such as cellulose phosphinates and phosphonites, have been prepared. Kiselev and Danilov (56) obtained mainly cellulose dialkylphosphinates from the reaction of cellulose with dialkylphosphinic acid chlorides in the presence of a tertiary amine and an organic solvent medium. The synthesis of cellulose phosphonites has also been reported (57).

(d) CELLULOSE ESTERS OF BORIC ACID

Using a considerable excess of the trivalent boron compound, Gertsev (58) has reacted alkyl borates with cellulose in a heterogeneous medium. The resulting cellulose polyborates were rapidly hydrolyzed and easily underwent alcoholysis.

2. Cellulose Acetate

Cellulose acetate is prepared commercially in a variety of DS, ranging from the primary acetate (triacetate (see also Chapter XIX-C)) to the acetone-soluble acetate (secondary acetate). Although cellulose acetate has been produced for over 50 years, the search continues for new sources and refined grades of acetylation cellulose, improved methods of acetylation (59), as well as reduced cost of manufacture.

(a) METHOD OF ACETYLATION

With the exception of cellulose triacetate, which may be manufactured by fibrous acetylation, cellulose acetate is made by a solution process.

(1) Fibrous Process

Fibrous acetylation differs from the solution process in that a sufficient quantity of a liquid that is a nonsolvent for cellulose acetate is added to the acetylation medium, the reaction product is prevented from dissolving, and the fibrous structure is retained. Geurden (60) compared fibrous and solution processes for the production of cellulose triacetate, taking into account the mechanical form of the cellulose and the effect of pretreatment. The principal advantages of the heterogeneous (fibrous) process are as follows:

1. Since the acetylation mixture maintains its fluidity through the esterification, the temperature can be more readily controlled.
2. The large amount of liquid can act as a heat sump in case of a sudden exothermic reaction.
3. Simple and cheap equipment can be used because stirring can be eliminated.
4. The ester does not have to be precipitated; it is easily stabilized and dried.

5. Since the anhydride can be recovered by distillation, the savings in concentration and reconversion can be substantial.

However, the heterogeneous process does have the following disadvantages:

1. Longer esterification time is needed.
2. Larger amounts of esterifying mixture are required.
3. No satisfactory heterogeneous hydrolysis process exists. Homogeneous hydrolysis must be employed to prepare a hydrolyzed acetate with acceptable acetone solubility.

The cellulose is usually given a treatment with acetic acid, with or without catalyst, to improve sorption. The sorption rates (61) of acetylation catalysts on various types of cellulose in glacial acetic acid were measured at different concentrations and temperatures. The apparent heats of sorption decreased in the order perchloric acid, sulfuric acid, zinc chloride. The authors concluded that perchloric acid combines with cellulose by both chemical and hydrogen bonding, sulfuric acid combines by hydrogen bonding, and zinc chloride combines with cellulose by van der Waals forces.

Solvents mixed with acetic acid and acetic anhydride (62) influence the rate of acetylation of cellulose. Solvents containing ether oxygen, carbonyl, or nitrile combine with the catalyst (perchloric acid) by hydrogen bonding; this decreases sorption, and the rate of acetylation also decreases. On the other hand, hydrocarbons and chlorinated hydrocarbons are inert toward perchloric acid, and their lower dielectric constant increases the dissociation of the perchloric-acetic acid complex, causing an increase in sorption and acetylation rate.

Fibrous acetylation also takes place when cellulose fibers are treated with acetic anhydride vapors. Kido and Suzuki (63) have investigated the vapor-phase acetylation of cotton linters and linters tissue paper at various temperatures with a variety of catalysts, such as zinc chloride, sulfuric acid, or perchloric acid. Cellulose triacetate was obtained. However, x-ray diffraction of the resulting acetate indicated that, regardless of catalyst used, cellulose triacetate I was obtained at temperatures below 75–80°C, and another cellulose triacetate, II, was produced at temperatures above 80–85°C.

Kido et al. (64) studied the effect of kind of catalyst, type of cellulose, and concentration of acetic anhydride on the acetylation rate. Best results were obtained with cellulose that was swollen with acetic acid solution, and zinc chloride was the preferred catalyst. The rate constants and activation energies for various zinc chloride concentrations were determined; the vapor pressure of acetic anhydride had almost no direct bearing on the reaction rate. These results indicate that, in the vapor-phase reaction, acetic anhydride vapor condenses in the fiber phase, and the cellulose reacts with the resulting liquid. Fibrous cellulose acetate prepared in this manner (65) has a DS of 3, and both films and fibers of it are brittle due to crystallinity and molecular symmetry. Hydrolysis of the primary acetate has been suggested (65) to rectify this defect. Another approach (66) to the alleviation of brittleness is to break up the molecular symmetry, and apparently this can be achieved by vapor-phase acetylation with a mixture of acetic and propionic or butyric anhydrides.

Saito and Sugeta (67) have studied the temperature rise of fibers during vapor-phase acetylation. They found the internal temperature of fibers to be nearly uniform and equal to the temperature of the surrounding reaction gas. The temperature rise was attributed to the heat transfer from the gas passing through the fiber plus the heat of adsorption. Isozaki and Ishikawa (68) pretreated viscose rayon with an aqueous solution containing sulfamic acid plus urea, acetylated with acetic anhydride vapors at 50 and 100°C, and obtained high acetylation rates and fibers with good tensile strengths.

Patents were issued to Rayonier (69, 70) for the vapor-phase acetylation of swollen cellulose fibers. Dry cellulose is treated with acetic acid without a catalyst, and the excess acetic acid is removed. The swollen fiber is saturated with liquid acetic anhydride at temperatures up to 140°C and is then exposed to superheated acetic anhydride vapor at temperatures up to 220°C in the absence of catalyst. A maximum of 30% acetyl has been obtained.

The tendency of cellulose to absorb water is decreased when acetylation of the noncrystalline cellulose approaches completion, leaving the crystalline cellulose unchanged (71). Gibson and Shoemaker (72) described a method for acetylating paper. Strong paper with excellent moisture resistance is made by partially acetylating cellulose (paper). Partially acetylated paper (73) is manufactured (Sandoz Ltd.) by esterifying fibrous cellulose, grinding the material before esterification has reached the desired degree, and completing the esterification (to the desired degree) before fabricating into strips, sheets, and layers. Paper of good strength and dimensional stability is prepared from partially acetylated wood-pulp fibers (74). In this process kraft pulp is activated, and the acid-wet pulp from the activation is acetylated to an apparent acetyl content of 15-25% in a bath containing acetic anhydride, acetic acid, and a catalyst (see also Section D in Chapter XIX).

(2) Solution Process

Almost all cellulose acetate, with the exception of fibrous triacetate, is manufactured by a solution process, and sulfuric acid is the most commonly used catalyst (75, 76). However, the producers of cellulose acetate use different processes (77), varying the size of equipment, liquid-to-cellulose ratio, solvent, temperature, and catalyst concentration. Even the same manufacturer employs modifications in order to obtain cellulose acetates with different properties or in order to use different grades of cellulose.

Acetylation of cellulose is a heterogeneous reaction, and the course of the reaction was found to be controlled by the rates of diffusion of the reagents into the cellulose fibers (75). Therefore the chemical and physical properties of the cellulose are of paramount importance to both the acetylation rate and the quality of the product.

Cellulose for Acetylation. In evaluating cellulose for acetylation two basic considerations must be taken into account: (a) the ability of the cellulose to undergo smooth, uniform acetylation and (b) the quality of the product. Unfortunately the suitability of the cellulose cannot be predicted entirely from its chemical composition (78). The processing history, which often determines physical properties not revealed by analytical techniques, can be most important.

The choice of cellulose depends on the particular esterification procedure, but there are certain minimum requirements of acetylation-grade cellulose:

1. The cellulose must be adequately swollen in the activation treatment. Although proper acetylation cannot be obtained without sufficient swelling, it does not follow that highly swollen cellulose ensures good acetylation.
2. There must be a good catalyst penetration and sufficient, but not excessive, chain-length reduction during the pretreatment.
3. The rate of acetylation must be commercially feasible with a minimum of unacetylated residue so that the product can be easily filtered.

Acetylation-grade cotton linters are preferred to wood pulps because lint esters have better ester color and solution clarity.

Because of price advantage and the fact that the quality of pulps has improved, most of the cellulose acetate produced today is made from wood pulp. Yarns made from wood-pulp acetates have desirable elongation properties (78). Typical specifications are shown in Table 1. Sodium, calcium, and

TABLE 1
Typical Specifications for Acetylation-Grade Pulp (76)

α-Cellulose(%)	95.6 (minimum)
Moisture (%)	5–8 (off supplier's dryer)
Pentosans (%)	2.1 (maximum)
2½% cuprammonium viscosity (centipoises)	1100–4000 (intrinsic viscosity 5.5–7.5 dl/g)
Ether extractable (%)	0.15 (maximum)
Ash (%)	0.08 (maximum)
Iron (ppm)	10 (maximum)
Trial acetylation:	
Haze (ppm)	100 (maximum)
Color (ppm)	600 (maximum)

magnesium carbonates and sulfates are commonly found in commercial acetylation-grade pulps (79). Since these salts increase the activation time of the cellulose, a low ash content is desirable.

A pulp that does not meet these specifications is not necessarily unsuitable for the manufacture of a particular cellulose acetate. The pulp might be used for esterification by a different procedure. Quality and price determine which pulps are used, with pulps that give satisfactory quality at the lowest price being selected. Malm (76) has tried to correlate the different end uses with critical properties (Table 2).

TABLE 2
Uses and Critical Properties of Cellulose Acetate (76)

	Use[a]			
Property	Yarn	Photographic film	Plastics	Lacquers
Color (absence of)	I	–	C	I
Haze	–	–	I	I
False viscosity	I	I	–	I
Filterability	C	I	I	I
Adhesion	–	C	–	–
Radioactive contamination	–	C	–	–
Price	C	I	I	I

[a]Abbreviations: I–important; C–critical.

Until recently acetylation-grade pulps were made exculsively by the sulfite (acid) process. Previously dissolving pulps derived from a prehydrolyzed kraft process had been unable to break the stringent purity and processing barriers established for cellulose esterification reactions (80). Because of price advantage and the ever increasing pollution problems from sulfite pulping, prehydrolyzed kraft pulps are now being examined. In addition it is claimed (80) that acetate-grade yarn made from pine-based kraft pulp has superior extrusion stability, yielding more uniform filaments. A recent patent (81) claims a breakthrough in processing and describes a continuous process for commercial refinement of kraft prehydrolysis pulp in a digester. A new 500-ton/day plant based on this process is running at rated capacity (80). Kosaya and Chechurina (82) have also demonstrated that prehydrolysis kraft pulps, when manufactured under certain conditions, are fully suitable for refining and acetylation. A method for preparing acylation pulp from kraft wood pulps has been patented (83). After being treated with strong aqueous alkali containing a boron compound at moderate temperature, the pulp is treated with dilute aqueous alkali at elevated temperature for the purpose of extracting undesired materials.

The crystallinity of cellulose is also an important factor in acetylation. Glegg and co-workers (84) investigated the relative reactivities of celluloses I and II to heterogeneous acetylation. They demonstrated the presence of unreacted cellulose I even after 40% acetyl content had been reached. This may explain why the manufacture of cellulose esters from cellulose I requires complete esterification before the esters are hydrolyzed to the desired acetyl level. These authors also found distinct differences between the acetylation reactivity of celluloses I and II in favor of cellulose I, indicating the importance of avoiding the use of cellulose II during the refining of cellulose that is to be used for acetylation involving activation with acetic acid.

Activation. It is usually necessary to activate cellulose before acetylation. Activation may involve two steps, swelling and sorption. In the laboratory excellent esterification conditions result if cellulose is swollen in water and filtered, and the water is replaced by acetic acid (85). The cellulose is then treated with acetic acid containing a catalyst (sulfuric acid). The primary reasons for activation are to swell the fibers and to promote a controlled breakdown of the chain length by the catalyst. Water is an effective swelling agent because of its ability to rupture hydrogen bonds within the interchain network (86), increase surface area (87), and accelerate the diffusion of reagents into cellulose (88). These effects are not unrelated. In commercial practice cellulose is generally treated with acetic acid with part of the total amount of catalyst under a multitude of conditions. The amount of activation that is necessary depends on the source and the history of the cellulose. The efficiency of activation is increased by increase in temperature, increase in water and catalyst concentration, and decrease in the ratio of acetic acid to cellulose (89). Since cellulose contains residual water, a decrease in the ratio of acetic acid to cellulose presumably promotes the efficiency of activation because of an increase in the water concentration, which increases swelling and surface area. Kuznetsova and Klenkova (87) have shown that a decrease in the ratio of acetic acid to cellulose causes an increase in surface area. By using nitrogen vapor sorption they found that the internal surface areas of cellulose for large and small amounts of acetic acid were about 10 and 15 m^2/g, respectively.

Sulfuric acid is an effective catalyst in degrading cellulose because it is sorbed by the cellulose (90). By means of autoradiography, Ishikawa (91) was able to show that sulfuric acid (^{35}S) is absorbed in the accessible regions within the wood-pulp fiber in an acetic acid medium. However, the combined effects of water and sulfuric acid in acetic acid are complex because water reduces the acidity of sulfuric acid and reacts with it to form hydrates (90). Hydrate formation and low ionization of sulfuric acid were shown by the freezing points and conductivities of the solutions. As a consequence, when uniformly swollen cellulose is used, increased water content decreases the rate of cellulose breakdown. On the other hand, with partly swollen cellulose, increased water concentration increases the uniformity of sulfuric acid sorption and the average rate of cellulose breakdown.

Acetylation. Acetylation of cellulose with a sulfuric acid catalyst and a modest excess of acetic anhydride in acetic acid has been used with modifications over the last 20 years, and no evidence has been presented for an unequivocally better method of acetylation.

In a typical commercial acetylation (77) the mixer is charged with cellulose and acetic acid. After the fibers become swollen, part of the catalyst is added; it is distributed throughout the fibers and lowers the degree of polymerization (pretreatment). Because acetylation is an exothermic reaction, the mixture must be cooled before the addition of anhydride. Cold acetic anhydride is added. Any water present in the system immediately reacts with the anhydride, liberating heat, and some acetylation takes place. After recooling, the bulk of the catalyst is added, and another exothermic reaction results. Although sulfation is faster, both sulfation and acetylation take place, producing a clear solution. After the solution has attained the desired viscosity, an aqueous acetic acid solution is added to initiate hydrolysis.

It has been reported that high reactivity of cellulose to acetylation is obtained when there is rapid and uniform sorption of sulfuric acid (89). Although the quantity of sorbed acid is important, the uniformity of distribution of the sorbed acid was found to be even more important (88, 89). In an attempt to evaluate acetylation reactivity, Zaitseva and co-workers (92, 93) found that, after pretreatment, cotton linters were more homogeneous than pulps with respect to fiber-width distribution. Rosenthal and White (94) found that wood pulps reacted faster than cotton linters when blends of pulp and linters were acetylated by means of a high-dilution procedure in the presence of sulfuric acid catalyst. Tests have been developed to measure the uniformity of cellulose, its rate of activation and its acetylation activity (95). These tests are based on the measurement of the moisture content of cellulose, sulfuric acid sorption, viscosity of pretreated cellulose, adiabatic acetylation time-temperature relationship, and light scattering and transmission of acetylation mixtures.

In earlier work it was reported (96) that sulfuric acid combines rapidly and quantitatively with cellulose on the addition of acetic anhydride, whereas acetylation is a much slower reaction. Since a preliminary investigation (97) revealed that primary hydroxyl groups react faster than secondary hydroxyl groups, Malm and Tanghe (98) performed a detailed investigation of the chemical reactions in making cellulose acetate. Because of the experimental difficulties involved in working with fibrous cellulose (98), these authors used hydrolyzed cellulose acetate in solution to study the introduction of sulfate during acetylation. Sulfation was carried out on two samples of cellulose acetate. Both samples contained approximately 40% acetyl; in one sample 50% of the hydroxyl was primary, and in the second sample only 8% was primary. Both samples were treated under identical conditions, as shown in Table 3, and were precipitated from solution by pouring into methanol.

At room temperature all the sulfuric acid combined with cellulose acetate in both cases. In the low-primary-hydroxyl sample (8%) more sulfuric acid combined than could be accommodated by the primary hydroxyl present. At lower temperatures the reaction was slower, and less sulfate was found in the sample with the lower amount of primary hydroxyl. This supports preferential sulfation of the primary hydroxyl groups.

TABLE 3
Sulfation of Cellulose Acetate by Sulfuric Acid in the Presence of Acetic Anhydride (98)[a]

T (°C)	Solvent	Sulfur % — Primary OH in starting material: High	Low
25	AcOH	1.10	1.13
0	AcOH + CH_2Cl_2	0.99	0.85
−10	AcOH + CH_2Cl_2	0.73	0.46
−20	AcOH + CH_2Cl_2	0.68	0.40
Maximum for reaction of all sulfuric acid		1.14	1.14
Maximum for reaction of all primary hydroxyls		2.7	0.43

[a] This material has been used by courtesy of the American Chemical Society.

Since the process of acetylation was reviewed (59) a number of modifications have appeared in the literature. Malm, Tanghe, and Schmitt (99) have examined a number of catalysts for acetylation, including methanedisulfonic, methanesulfonic, perchloric, sulfoacetic, and sulfuric acids. Sulfuric acid gave more uniform acetates than the other catalysts. Tanghe and Brewer (100, 101) have investigated the reaction of sulfuric acid and acetic anhydride to determine the extent of conversion of sulfuric acid to acetylsulfuric acid in various mixtures of acetic anhydride and acid. They were able to show that both sulfuric acid and acetylsulfuric acid could be the active species combining with cellulose.

A typical laboratory cellulose acetylation mixture contains 1.5 kg of cellulose, 4.5 kg of acetic anhydride, 9 kg of acetic acid, and 60 ml of concentrated sulfuric acid. When the cellulose is added to this mixture, it would seem that acetylsulfuric acid, *not* free sulfuric acid, reacts initially with the cellulose (100, 101). However, in commercial acetylations the major part of the catalyst is added last and is found to be almost completely combined with the cellulose within 3 min. In view of this, Tanghe and Brewer have proposed the following scheme for commercial acetylation:

$$R_{cell}(OH)_3 + H_2SO_4 + 3(CH_3CO)_2O \longrightarrow R_{cell}\begin{cases} OSO_2OH \\ (OCOCH_3)_2 \end{cases} + 4\,CH_3COOH \qquad (2)$$

These authors propose that the sulfuric acid does not remain in solution but is mostly combined with the cellulose, and the product goes into solution.

The use of a titanium compound, such as potassium titanium oxalate, as a catalyst has been reported (102). Trifluoroacetic anhydride was found to be an effective esterification promoter for the preparation of partially acetylated cellulose acetate under mild reaction conditions (103). Trifluoroacetic anhydride functions as an impelling agent, and no additional catalyst is required. Trifluoroacetic acid has also been used in esterification (104). Presumably, this acid swells the cellulose, ruptures hydrogen bonding in the cellulose, and reacts to form an intermediate addition compound. None of these is practical as a catalyst for the preparation of cellulose acetate. Because of its high efficiency and low cost, sulfuric acid remains the overwhelming choice as the catalyst for acetylation.

Only a nominal amount of work on acetylation media has appeared in the recent literature. Although acetylation in acetic anhydride–pyridine or acetic anhydride–potassium acetate takes place, the reaction rate is too slow for commercial use. However, Blume and Swezey (105) have acetylated cellulose without degradation by means of acetic anhydride–potassium acetate, with dialkylamides or γ-lactones being used as the esterification medium. A dramatic increase, by a factor of 20–30, in reaction rate was attributed to the dielectric constant of the medium. These liquids are also effective in the acid-catalyzed acetylation of cellulose with acetic anhydride. The use of methylene or ethylene chloride as the acetylation medium has been mentioned in the patent literature (106). Malm, Tanghe, and Glegg (107) have described a variety of catalysts in conjunction with a large number of solvents, such as dioxane, ethylene chloride, ethyl acetate, isopropyl acetate, or acetic acid, as esterification media.

In an excellent small-scale procedure (108) cellulose is acetylated in a glass bottle. This allows observation of the solution and measurement of the haze and color by visual comparison with standards. A laboratory method permits the preparation of cellulose triacetate in a heterogeneous medium (109).

Hydrolysis. The acetylation reaction is stopped by the addition of water, which destroys the remaining acetic anhydride and promotes rapid desulfation. In order not to precipitate the cellulose acetate, the water is added in the form of aqueous acetic acid (59).

Providing sufficient water has been added, the resulting solution undergoes hydrolysis, which removes some of the acetyl groups. For a given catalyst concentration, increasing the temperature will increase the rate of hydrolysis. Although increasing the percentage of water has only a slight accelerating effect on hydrolysis, there is less degradation when a high concentration of water is present in the hydrolysis solution. The concentration of water in the hydrolysis bath influences the ratio of primary to secondary hydroxyl groups in the acetate (59). In the presence of more than 2% water the hydrolysis products contain increasing amounts of primary hydroxyl groups as the concentration of water increases (110). On the other hand, an increase in the acetic acid concentration enhances reacetylation of the more reactive hydroxyls and results in a product of lower primary hydroxyl content.

Malm and Tanghe (98) have determined the optimum conditions for the desulfation and deacetylation of cellulose acetate. These authors found that the rate of combined sulfate removal is strongly influenced by the concentration of water in the acetic acid solution. They quickly added 2 volumes of acetic acid containing sufficient water to provide 0.3% water in the solvent after mixing with a solution of cellulose acetate containing 0.85% combined sulfur. Another portion was treated in a similar manner to contain 1% water. Samples of cellulose acetate were isolated at intervals of up to 24 hr and analyzed for sulfur (see Fig. 1). With 0.3% water there was a sudden drop in combined sulfur during the first few minutes. With 1% water the decrease was not so rapid at first, but lower sulfur content was achieved later. After 2 hr the minimum amount of sulfur was found for a water concentration of about 0.5% in the final solution.

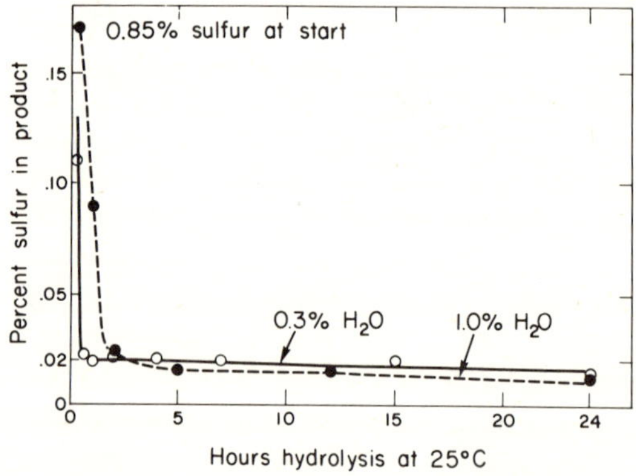

Fig. 1. Loss of combined sulfate from cellulose acetate in acetic acid solution in the presence of small amounts of water. (Reproduced from *Industrial and Engineering Chemistry*, 47, 997 (1955) by permission of the American Chemical Society, the copyright owner.)

Delsemme (111) performed a kinetic study on the desulfation of cellulose acetate after the addition of water at the completion of the esterification reaction. He found that desulfation is an equilibrium reaction and that the equilibrium can be displaced. By setting the water content of the solvent at a minimum value in order to obtain rapid desulfation and then neutralizing most of the free sulfuric acid and raising the temperature to 50–55°C, he was able to prepare cellulose esters of improved stability in the laboratory and pilot plant without the addition of a stabilizer.

The small amounts of water that are so effective for desulfation are not adequate for deacetylation. Normally 5–20% water in the final solution is used for hydrolysis of commercial acetates. With less water degradation is a serious problem, and with more water at a particular temperature the cellulose acetate may begin to precipitate. Secondary cellulose acetates can also be made by the hydrolysis of primary acetates that were acetylated in chlorinated-hydrocarbon media (106). An excess of water, 75–160%, based on residual acetic anhydride is added, and the sulfuric acid that is liberated is neutralized by the addition of magnesium carbonate. Most of the solvent is then removed by distillation. The use of other bases, such as magnesium acetate in conjunction with sodium acetate, has been reported (112).

When acetylation is performed in a bottle, it is also customary to add aqueous acetic acid to destroy the excess acetic anhydride and to remove combined sulfate. Sufficient 1 *N* hydrochloric acid is added to increase the rate of hydrolysis. The resulting dope contains about 10% water (85).

Crane (113) has prepared cellulose triesters of excellent stability by the esterification of cellulose with a bath containing a fatty acid anhydride, diluent, and sulfuric acid catalyst. After neutralization of the catalyst with an excess of magnesium carbonate under anhydrous conditions, the bath was heated to 95–175°C, whereby the combined sulfate was replaced by the acyl group to give the triacyl ester. A refinement of this process involves removal of the liquid under reduced pressure (114). In a similar manner Malm and Crane (115) neutralized 50–90% of the sulfuric acid and maintained the anhydrous mass at 20–60°C, which imparts low hydroxyl and sulfur in the the cellulose ester. The resulting stable triester can be hydrolyzed in solution, or it can be separated from the mass by precipitation after the addition of an aqueous lower fatty acid that converts the anhydride present to lower fatty acid.

The hydrolysis of simple aliphatic acid esters has been studied (116). Factors that influence the breakdown of cellulose acetate, propionate, and butyrate during hydrolysis have been reported. An increase in water concentration caused a decrease in the rates of hydrolysis of the cellulose butyrate and propionate but not the acetate. It also caused a decrease in the rate of viscosity reduction. With different esters in acetic acid–water solutions the rate of hydrolysis and of viscosity reduction was found to decrease in the order acetyl, propionyl, *n*-butyryl. When the same ester was hydrolyzed with different acyl groups in the solvent, the order was reversed, even when the results were corrected for reesterification. "The relative rates of hydrolysis and viscosity reduction of cellulose acetate in acetic acid–water, cellulose propionate in propionic acid–water, and cellulose butyrate in butyric acid–water depend on the concentration of the water in the solution" (116).

Far-hydrolyzed cellulose acetates have been prepared by the hydrolysis of cellulose acetates in acetic acid–water solutions at 40–60°C with sulfuric acid catalyst (117). The resulting esters were stable, colorless, water-soluble or nearly water-soluble polymers of relatively high molecular weight.

It is possible to depolymerize cellulose esters without affecting the DS. Low-molecular-weight cellulose triacetate has been prepared through the depolymerization of higher-molecular-weight cellulose triacetate in solution containing boron trifluoride etherate complex catalyst under anhydrous conditions (118, 119). The recovered cellulose triacetate is used in the manufacture of packaging films, which require good clarity, flexibility, and strength. This process is applicable to other cellulose triesters. A patent (120) was issued for a process to manufacture low-molecular-weight esters or ethers from high-molecular-weight derivatives by anhydrous depolymerization of the powdered triester in the presence of hydrogen halide catalyst.

Purification. Control tests are made by turbidimetric titration during the course of the hydrolysis to measure the DS. When the proper degree of hydrolysis has been obtained, the cellulose acetate is precipitated from solution. It is most important that the acetate have the proper texture for thorough washing to remove the free acid; however, it is equally important that the acetic acid in the precipitation medium be recovered at a reasonable cost. The expense encountered in the recovery of acetic acid increases with dilution. In order to remove the last traces of acetic acid and catalyst, the cellulose acetate must be washed with water that is free of iron and dissolved salts to ensure good color and clarity in the product (59). Malm and Crane (121) precipitated cellulose acetates from the solutions in which they were hydrolyzed by adding water containing magnesium sulfate (10%, based on the acetic acid present). The cellulose acetate obtained was in good physical condition, and the acetic acid, which contained less water, could be recovered economically.

Instability of cellulose acetate was traced to sulfate in the unstabilized ester. Keirstead and Myers (122) performed ion-exchange studies on precipitated and fibrous cellulose acetates and showed that the equilibrium reached during ion exchange is affected by the pH and the concentration of the salt solution used. Furthermore, the thermal stability of the ester varies with the sulfate concentration. Therefore the stabilizers used rely on removal of combined sulfate, conversion of the sulfate to a salt, or both effects. The use of a base in the hydrolysis medium to neutralize free and combined sulfate has been covered in the preceding discussion of hydrolysis. Other stabilizers have been used. Cellulose esters can be stabilized against discoloration and degradation by the incorporation of certain compounds, such as chlorothiophenol (123) or a neopentyl phosphite (124). The use of potassium acid oxalate in combination with a phenol and magnesium stearate (125), certain *o*- and *p*-alkoxyphenylglycidyl ethers (126) and dialkyl esters of 3,3′-thiodipropionic acid (127) have been cited as heat stabilizers. Carbodiimides are reported to be effective stabilizers against both heat and light for cellulose acetates (128).

Viscosity. Because cellulose esters are produced to fall within certain viscosity specifications, it is imperative to know what factors during esterification and hydrolysis affect viscosity. In the preparation of cellulose acetate it was found that for a given catalyst lower catalyst concentration favors less viscosity reduction; lower temperatures and higher anhydride concentrations also favor higher viscosity (107, 129). If these were the only factors involved, the manufacturer would employ lower temperatures and higher anhydride concentrations, and these would be balanced against the proper amount of catalyst. Since low temperatures retard the rate of esterification, and because acetic anhydride is a comparatively expensive reagent, moderate temperatures and relatively low levels of acetic anhydride are used.

Cellulose acetates with low solution viscosities are now commercially available (130). These low-molecular-weight materials are soluble in a large number of solvents and solvent blends, and because lacquers with high solids content can be prepared, the net result is that fewer lacquer coats are required to produce the desired film thickness. These acetates are compatible with numerous plasticizers, and the resulting coatings are tough and resistant to chemicals and marring. Blends with melamine resin were found to give highly desirable coatings for wood (130).

The viscosities of concentrated acetone solutions of cellulose acetate prepared from high-quality cotton linters can be predicted from the intrinsic viscosities of their dilute solutions. This relationship (131) does not apply strictly for acetates prepared from wood pulps, since in concentrated solutions they exhibit higher viscosities than the predicted values, which vary with the processing history of the wood cellulose. This difference between experimental values and those predicted from dilute solutions of linter acetates is referred to as "false viscosity." A cellulose ester is said to have a false viscosity when it has a higher viscosity in concentrated solution than a similar ester of the same intrinsic viscosity (132). False viscosity has been attributed to certain impurities, such as hemicellulose (133). It has been suggested that orientation and association of the cellulose in the fiber before and after acetylation may also be involved (131).

Malm and Tanghe (132) have observed false viscosity for cellulose acetates made from different kinds of cellulose by the same acetylation process or from the same cellulose by different acetylation processes. They found that the latter type could be eliminated or minimized by employing one or more of the following steps:

1. Carrying out as much as possible of the chain-length degradation before, rather than after, esterification.
2. Increasing the temperature during the chain-length breakdown after acetylation.
3. Increasing the solvent concentration in the acetylation bath.
4. Replacing part of the acetic acid with methylene chloride during acetylation.
5. Increasing the amount of sulfuric acid catalyst.

In comparing acetates from several sources of cellulose these authors noted that false viscosities correlated with the carboxyl content and also with the mannose content of the original celluloses. When cotton linters were oxidized to the carboxyl content of a standard wood pulp, the false viscosity matched that of the standard-pulp ester in 97.5% acetone but not in other solvents.

Malm and Tanghe (132) also tested the effect of mannan. When 0.5% of ivory nut mannan was added to cotton linters and the mixture was acetylated, false viscosity was detected and matched that of a standard-pulp ester in all solvents tested, However, when 1% mannan was added, the false viscosity increased still further but was below, equal to, or above that of the ester from a high-false-viscosity pulp, depending on the solvent used. Other workers have cited mannan as a cause of false viscosity (133, 134).

In order to study the effect of hemicellulose (mannan and xylan) on the properties of cellulose acetate solution acetylation-grade cotton linters were cooked in kraft or sulfite liquor in the presence of hardwood or softwood chips (135). Absorbed hemicellulose was determined by paper chromatography, and although the mannan content was found to correlate with false viscosity, xylan also had an effect, particularly the portion that was resistant to alkaline extraction. False viscosity caused by hemicellulose, both mannan and xylan, varies with the history of the cellulose and the method of digestion.

(b) MECHANISM OF ANHYDROUS ACETYLATION AND HYDROLYSIS

The manufacture of cellulose acetate, based on the acetylation of cellulose with acetic anhydride and a sulfuric acid catalyst in acetic acid, can be divided into two parts, acetylation and hydrolysis.

Michie (136) studied the rate of acetylation of activated cellulose and cellulose that had not been previously activated. By using a special technique for measuring the temperature rise under adiabatic conditions, it was found that cellulose not previously treated was acetylated in a reaction of approximately first order (observed order 0.8). Within limits the rate was not influenced by the concentration of sulfuric acid and was virtually independent of the level of the excess acetic anhydride. Cellulose that had been activated by swelling in water also was acetylated by an approximately first-order reaction (with respect to cellulose), but there was a threefold rate increase. Furthermore, the rate was proportional to the concentrations of acetic anhydride and sulfuric acid. Michie (136) claims that swelling with water causes the cellulose fiber structure to open and thus renders the crystalline material accessible to the reagents.

Barabash, Rosenthal, and White (137) have shown that the accessibility of cellulose as determined by a chromium trioxide oxidation method correlates well with the measurements of acetylation rate. In some earlier work Hiller (138) studied the reaction curves for cellulose with acetic acid and water. Although the reaction was slow and limited, an analysis of the reaction products led him to postulate that most of the hydroxyl groups (about 88%) could not be acetylated because of the energy barrier due to hydrogen bonding. The equilibrium constant for the acetylation of these groups is highly unfavorable. Part of the remaining hydroxyl groups appear to react according to the kinetics of homogeneous acetic acid acetylation, and they must be in freely accessible locations, such as on the fiber surfaces. The remaining hydroxyl groups react very slowly, at a rate that probably depends on the diffusion of acetic acid (or water) into the cellulose fibers. Increasing the temperature (86, 139) increases the efficiency of activation through better swelling of fibers and enhances their accessibility to the acetylating agent. The energy of activation of diffusion with perchloric acid catalyst was calculated (139) and found to exceed the usually observed value of 2.5 kcal/mole. Raising the temperature helps to overcome this energy barrier.

During a pretreatment with acetic acid containing sulfuric acid controlled breakdown of chain length takes place. It has been shown by use of ^{35}S that sulfuric acid is selectively absorbed in the accessible regions (91). Heath (140) has determined the maximum sorption at room temperature, approximately 3 g of sulfuric acid per 100 g of cotton linters, in a medium of glacial acetic acid containing sulfuric acid. The highly swelled and partly degraded cellulose is very receptive to the acetylating acid.

Herdle and Wahtera (141) also studied the accessibility of cellulose to sulfuric acid. They obtained a nearly linear relationship by plotting sulfuric acid sorption versus DS in partially acetylated celluloses of low acetyl contents. The intercepts of these lines on the DS axis are a measure of the accessibility of the parent cellulose to sulfuric acid. They found that parallel lines are obtained for different types of cellulose (cotton linters, sulfite pulp, mercerized pulp), showing that sorption is reduced by the same amount for a given DS in all cellulose, regardless of original accessibility. The maximum sorption found for highly accessible celluloses was 1 mole of sulfuric acid per anhydroglucose unit.

Sulfuric acid combines rapidly and quantitatively with cellulose on the addition of acetic anhydride (59, 96, 98); acetylation is a much slower reaction, and the primary hydroxyl groups react faster than the secondary groups (97, 98). The higher reactivity of the primary hydroxyl groups has been established by a number of workers (97, 142, 143). Staudinger and Döhle (142) acetylated mercerized cellulose with pyridine and acetic anhydride in a cyclohexane medium and interpreted the kinetics in favor of rapid acetylation of primary hydroxyl groups, slow acetylation of one of the secondary groups, and very gradual acetylation of the last group. Malm and co-workers (97) found that catalysts had a more pronounced effect on the reactivity of the secondary hydroxyl groups than on that of the primary hydroxyl groups. In the uncatalyzed reaction with acetic acid at room temperature the primary hydroxyl groups reacted 10 times as fast as the secondary ones, whereas they reacted only 2.5 times as fast in the presence of sulfuric acid catalyst. Fedorova and Rogovin (143), by means of partial acetylation under somewhat different reaction conditions, also found that primary hydroxyl groups were more reactive than the secondary hydroxyl groups. However, they found that esterification was not selective with respect to the primary hydroxyl groups, and even in the early stage of the reaction some secondary hydroxyl groups were esterified, although at a much lower rate.

Some recent work suggests the formation of an intermediate as the first step in acetylation (144–148). Laamanen and Sihtola (144) proposed that the electrical conductivity of acetylation baths depends on the concentration of a strongly dissociated intermediate (monoacetylsulfuric or acetylperchloric acid). The formation of this intermediate is through the reaction of acetic anhydride with the catalyst. They demonstrated the existence of monoacetylsulfuric acid (145), which can react further with excess acetic anhydride to form diacetylsulfuric acid. The monoacetylsulfuric acid, which dissociates into CH_3CO^+ and HSO_4^-, is the active agent in the acetylation medium. They detected the presence of the acetylium ion by nuclear-magnetic-resonance (NMR) spectroscopy. On the basis of sorption of radioactive sulfuric acid by cellulose, Sihtola and co-workers (147) support the theory of formation of monoacetylsulfuric acid as the first step in the acetylation of cellulose. Howard and Parikh (148) prepared cellulose triacetate under anhydrous conditions using zinc chloride and perchloric acid as the catalysts. They report that the acetylium ion, CH_3CO^+, was shown to be the effective acetylating agent. Akim (146) studied the conductivity of acetylating mixtures and showed that the mechanism of catalysis with sulfuric acid depends on the composition of the acetylating bath and the reactivity of the pulp.

Wlodarski (149) has performed a series of acetylations under various experimental conditions to determine the correctness of the assumption that the reaction between cellulose and sulfuric acid is an intermediary reaction in acetylation. Unless acetic anhydride was present there was no significant reaction of sulfuric acid with cellulose; sulfoacetic acid gave practically no reaction even in the presence of acetic anhydride. Similarly, there was no significant reaction when the cellulose was not activated.

Tanghe and Brewer (100, 101) were able to show that both sulfuric and acetylsulfuric acids could be the active species, depending on the sulfuric acid concentration and the order of addition. The formation of acetylsulfuric acid is favored by a high concentration of anhydride and a high ratio of acetic anhydride to sulfuric acid. Since in commercial acetylations the major part of the catalyst is added last, and it (sulfuric acid) is found to be almost completely combined with the cellulose within 3 min, Tanghe and Brewer propose that sulfuric acid is not in solution but sorbed or combined with the cellulose and aids the solubility of the product. The role of acid catalysts in the manufacture of cellulose acetate has also been discussed by Rosenthal (150). The heat of acetylation has been determined (106 cal/g of cotton per DS) (151). Although the heat is released more slowly with raw cotton than with purified cotton, the experimental procedure gives reproducible results. The structural properties of highly acetylated cotton cellulose prepared by perchloric acid-catalyzed fibrous acetylation have been investigated (152, 153). Esterification appears to proceed by a diffusion-controlled reaction, and above a DS of 2 the x-ray diffraction pattern disappears (152). The gross morphology, as viewed by electron microscopy, does not change appreciably during progressive acetylation, but the crystalline structure decreases progressively with substitution and is independent of the method of preparation.

After the esterification is completed, water is added to destroy excess anhydride and to start hydrolysis. An equilibrium exists between cellulose acetate and its hydrolysis bath. The addition of more water favors hydrolysis. and the addition of acid promotes reesterification (59). Hiller (154) has determined the kinetics and energy changes that occur during the homogenous hydrolysis of cellulose acetate in aqueous acetic acid. The hydrolysis involves a fast second-order reaction of the primary groups and a slower second-order reaction of the secondary groups. Petropavlovskii and Rakhmanberdiev (155) have studied the kinetics of extensive hydrolysis of cellulose acetate. In the early stage of hydrolysis in the presence of a small amount of water the reaction is regarded by these authors as second order. Subsequently, with excess water, the reaction becomes irreversible. The rate constants for the overall hydrolysis were calculated, and the energy of activation was found to be approximately 16.6 kcal/mole. The specific factors that affect the rate and degree of hydrolysis have been discussed in subsection 2(a)(2), "Solution Process," and hydrolysis, including saponification, is discussed further in the following section.

(c) REACTIONS OF THE ESTER GROUPS

In theory cellulose esters can undergo any of the reactions of simple esters. However, in practice they undergo only hydrolysis and transesterification. In addition, partially hydrolyzed cellulose esters can be reesterified,

The hydrolysis of cellulose esters can be acid or base catalyzed. Just as with simple carboxylic acid esters, acid-catalyzed hydrolysis is a reversible reaction, and saponification is irreversible. In addition to the sulfuric acid–catalyzed hydrolysis of cellulose acetate in aqueous acetic acid, acid-catalyzed hydrolysis has been carried out in the presence of a number of organic solvents, including acetone, benzene, dioxane, ethanol, and trichloroethane (156). It was suggested that organic solvents increase the permeability of cellulose esters to the hydrolyzing acid solution. Turner (157) has partially deacetylated cellulose esters by heating the ester in an alcohol (such as methanol, ethanol, or ethylene glycol) under pressure to at least 180°C to dissolve the ester. This temperature is maintained until the desired degree of deacetylation is attained. A kinetic study of the hydrolysis of cellulose acetate 23–95°C was made in the pH range of 2–10 (158). In what appeared to be homogeneous conditions, the data indicated a pseudo-first-order reaction.

In an attempt to evaluate the effects of steric hindrance of substituted cellulose acetates (phenylacetate, trimethylacetate, and triphenylacetate), esters were prepared by a heterogeneous procedure and saponified (159). These esters showed outstanding resistance to saponification, attributed to steric hindrance and a reduced tendency to ionize. Zilahi and Oswald (160) have prepared fibrous cellulose acetate (36.4% acetyl), and this ester was partially saponified with boiling 0.1 *M* sodium carbonate. Although the initial rate of saponification was low, it increased thereafter. They concluded that saponification begins at the surface of the individual fibers and progresses into the interior, regardless of the heterogeneous microstructure.

Bailey and co-workers (161) also studied the topochemical mechanisms involved in the deacetylation of partially acetylated fibrous cellulose. With 0.5 *N* sodium hydroxide solution saponification progressed annularly from the surface, with little or no selectivity with respect to amorphous and crystalline regions or with respect to acetyl or primary or secondary hydroxyl groups. However, when 2 *N* hydrochloric acid was used (161), the amorphous phase was essentially deacetylated before the acetylated crystalline region was attacked. The use of x-ray and infrared techniques on partially hydrolyzed cellulose acetates of the same acetyl content showed that the properties depended on the reagent used for deacetylation and on the acetyl content before deacetylation.

The relative rates of deacetylation in glacial acetic acid with pyridine, 80% acetic acid plus 2% sulfuric acid, and 95% acetic acid plus 2% sulfuric acid have been examined by Klimova and Petushkova (162). The highest deacetylation rate was found for 80% acetic acid with 2% sulfuric acid. In glacial acetic acid with pyridine and 95% acetic acid with 2% sulfuric acid hydrolysis is accompanied by reesterificication of primary hydroxyl groups; thus the overall increase in total free hydroxyl groups is retarded.

Reesterification and transesterification are integral parts of the esterification-hydrolysis procedure. During the acetylation of cellulose sulfuric acid combines rapidly to form cellulose sulfate (98). Sulfation is followed by acetylation and transesterification of the sulfate to give the acetate.

Reesterification during the hydrolysis of cellulose acetate has been investigated (110). When acetone-soluble cellulose acetate was treated with 99.8% acetic acid containing a catalyst, the resulting ester had a low primary-hydroxyl content. If more than 2% water was employed in the hydrolysis medium, the products contained increasing percentages of primary hydroxyl groups as the concentration of water was increased. It was demonstrated that reesterification is responsible for low primary hydroxyl content. When hydrolysis was carried out in solvents (acetone, dioxane, and methyl Cellosolve) where there was no reesterification, products contained a high percentage of primary hydroxyl groups regardless of the amount of water present.

Transesterification occurs when cellulose triacetate is treated with methanol or ethanol in chloroform in the presence of benzenesulfonic acid (163) to give methyl and ethyl glycosides, respectively. Another example of transesterification occurs during acetylation.

(d) TRENDS IN THE MANUFACTURE OF CELLULOSE ACETATE

Most cellulose acetate is still produced by batch process, solution or fibrous. Although there have been few recent developments in the solution process, a number of improvements in the fibrous process have been claimed (165–169).

In order to reduce production costs a continuous process is most desirable. Patents that describe methods of pretreating cellulose for continuous acetylation have been issued (170, 171), and both heterogeneous and homogeneous continuous acetylations have been reported. A number of authors have published on continuous homogeneous acetylation (171–178). One process utilizes continuous passage of the reaction mixture (cellulose, acetic anhydride, catalyst, and solvent) through a number or successive reaction zones (172, 173). The reaction mixture, which undergoes agitation in each zone, is continuously withdrawn from each zone in partially reacted condition and passed to succeeding zones for further reaction. In a similar manner cellulose acetate was prepared by passing the esterification mixture through a tubular esterification zone, in which the mixture is forced through a number of screens of successively smaller aperture size (174). Other homogeneous methods that resemble these processes have been reported (171, 175–177), and these methods may be used to prepare other esters of cellulose as well as cellulose acetate.

Genevray and Robin (178) described the continuous-acetylation process (solution) of Société Rhodiacéta (in France). As compared with the batch process, the main advantages are as follows:

1. Saving of labor.
2. Reduction in the amount of acetic anhydride.
3. Possibility of complete automation.
4. Larger capacity for the same size plant.
5. Increased use of cellulose blends.

The principal disadvantage, aside from the need for different equipment, is the increased power consumption for this highly mechanized process.

Continuous heterogeneous acetylation has been carried out by incorporating a nonsolvent for cellulose acetate in the acetylation mixture (179). A British patent (180) has been granted for a continuous method of fibrous acetylation in the gaseous phase.

The recent rate of growth of cellulose acetate production is not impressive, but because of huge volume the small percentage gain should amount to about 4% or 40 million lb annually through 1971. Total United States production of cellulose esters amounted to 1.01 billion lb in 1968 (181, 182).

3. Other Cellulose Esters of Aliphatic Acids

Because of the instability (59) of cellulose formate, there has been little interest in this ester. In addition to cellulose acetate most of the simple aliphatic acid esters of cellulose have been prepared. Cellulose esters of lower, normal aliphatic acids (C_2 through C_5) were synthesized by Malm and co-workers (183).

(a) CELLULOSE PROPIONATE

Cellulose propionate may be prepared by esterification of cellulose with propionic anhydride in the presence of acid catalysts (59, 77) in a manner similar to that used for the manufacture of cellulose acetate. Diluents, such as methylene chloride, were used by Terasaki (184) to prepare cellulose tripropionates in solution. Cellulose sulfate is formed first, and transesterification to propionate follows. Terasaki (185) performed kinetic measurements on the reaction. Data on solubility in acetone indicate that, when sulfuric acid is used as the catalyst, the esterification proceeds heterogeneously from the surface of the cellulose fibers. Cellulose propionate has been prepared by Malm and co-workers (183) through the reaction of propionyl chloride with a suspension of cellulose in a mixture of pyridine and dioxane.

Low-viscosity cellulose propionate has also been prepared (186). In many respects this ester (3% acetyl, 45% propionyl) resembles other cellulose esters, but its films have the added features of low odor, high melting point, and excellent surface hardness. In addition low-viscosity propionate has high tolerance for alcohols and aromatics as diluents.

Cellulose propionate has also been prepared (187) by the reaction of cellulose in a base-free esterification bath containing propionyl chloride and a reaction medium at least half of which is dimethylformamide, the remainder being an inert solvent. This method may be used for the preparation of other cellulose esters. The rates of esterification of cellulose by lower aliphatic anhydrides have been measured (188). Although an exact comparison of these reaction rates is not possible, the rates were found to depend on the catalyst and on whether the reaction was carried out in solution or in suspension. Degradation accompanying esterification became more severe in the following order: acetic, propionic, *n*-butyric, isobutyric anhydrides. The hydrolysis of cellulose propionate is similar to that of cellulose acetate.

(b) CELLULOSE BUTYRATES

Cellulose *n*-butyrate may be prepared by esterification with anhydride and a catalyst, such as sulfuric acid, in a manner similar to the preparation of the propionate (59). Unless the cellulose is given an efficient activation and the esterification is well controlled, degradation may become a serious problem (59, 188).

Since esterification of cellulose with isobutyric anhydride is sluggish, the cellulose must be highly activated. A number of methods have been reported for the activation of cellulose that is to be used in the preparation of cellulose isobutyrates. In one of these (189) the catalyst is uniformly distributed throughout the cellulose by activating the cellulose with a swelling agent containing the catalyst. After removal of the swelling agent by acid displacement, the activated cellulose, containing isbutyric acid, is ready for esterification. This cellulose is thoroughly mixed with a solution of isobutyrylsulfuric acid in isobutyric acid and esterified in a bath containing isobutyric anhydride. Malm and Blanchard (190) have activated cellulose by soaking it in water. The water is displaced with acetic acid and the acetic acid with isobutyric acid. A modification of this method (191) is used to produce cellulose isobutyrate that is not degraded excessively. After the water has been displaced with acetic acid, the cellulose is treated with a solution of sulfuric acid catalyst in acetic acid, and then substantially all of the acetic acid is displaced with isobutyric acid. The activated cellulose, which contains sulfuric acid uniformly distributed throughout, is ready for acylation, and no additional catalyst is necessary.

In addition to the conventional method (use of anhydride with sulfuric acid catalyst) for preparing cellulose butyrate and cellulose isobutyrate (59, 183, 188), the reaction of cellulose in an esterification bath containing acid chloride, pyridine, and a reaction medium of dioxane (183) or in a base-free bath consisting of acid chloride, dimethylformamide, and an inert solvent has been reported (187).

In order to obtain cellulose isobutyrate with desirable properties, it is often necessary to employ special techniques. Malm and Blanchard (192) treated cellulose in an esterification bath containing isobutyric anhydride and sulfuric acid catalyst. They were able to prepare cellulose isobutyrate with a minimum amount of degradation by keeping the temperature from rising above 40°C. This was accomplished by adding the cellulose to the esterification mixture over a period of 10–45 min. Cellulose isobutyrate-propionate of low viscosity, high uniformity, and low combined sulfur was prepared by the esterification of cellulose in a mixture containing isobutyric anhydride and sulfuric acid (193). When the esterification was 80–90% complete, 0.5–2 parts by weight of propionic anhydride per part of cellulose were added, and the esterification was continued. Cellulose wet with isobutyric acid from the activation process is esterified in a bath containing isobutyric anhydride, methylene chloride, and perchloric acid catalyst (190). Similarly cellulose can be esterified (194) in a bath comprised of isobutyric anhydride and zinc chloride catalyst at a temperature of 40–75°C.

Several methods for the hydrolysis of cellulose butyrates in solution have been reported (195–197). In an unusual method of hydrolysis (196) a solution of cellulose isobutyrate prepared by esterification with isobutyric anhydride and sulfuric acid catalyst is diluted with propionic acid. Propionic, rather than acetic, acid is used to keep the ester in solution. Aqueous propionic acid, sufficient to destroy 95% of the anhydride, is added to the solution; after this sufficient aqueous acetic acid to give a solution containing 5–10% of water is added with careful temperature control. Normal butyric and isobutyric acid esters, containing at least 20% butyryl, were hydrolyzed by allowing the esters to stand in a solution of acetone, methanol, and water in the presence of sodium hydroxide until a hydroxyl content of 0.5–7% was obtained (197).

Alcohol-soluble cellulose butyrate is available (198, 199). This ester contains about 47% butyryl, 1.6% acetyl, and 4.5% hydroxyl. With such a high hydroxyl content, alcohol-soluble butyrate might be crosslinked with urea-formaldehyde or polyisocyanates.

The low odor of lacquers containing alcohol solvent makes formulations of alcohol-soluble butyrate attractive in the household-commodity market. It also shows promise for use in printing inks because conventional solvents often attack rubber applicator rolls.

(c) CELLULOSE ESTERS OF HIGHER ALIPHATIC ACIDS

With the exception of cellulose pelargonate (C_9), all the cellulose aliphatic acid esters from valeric through capric (C_5–C_{10}) have been prepared by the addition of the appropriate acid chloride to a suspension of cellulose in a mixture of pyridine and dioxane (183). Cellulose valerate (200) has also been synthesized by the conventional procedure employing the acid anhydride and sulfuric acid as catalyst. This method is generally not suitable for the preparation of higher aliphatic acid esters of cellulose because of severe degradation. Although higher esters (59) may be prepared by the use of acid chloride, pyridine, and diluent or by use of an impelling agent, such as chloroacetic anhydride, recently published work on these cellulose esters has been scanty. Malm and co-workers (183) have prepared the fully esterified even-numbered cellulose esters of saturated carboxylic acids (C_{12}-C_{18}) by use of the acid chloride-pyridine method.

(d) MIXED ESTERS OF ALIPHATIC ACIDS

The search for mixed esters of cellulose containing acetyl and other groups was stimulated by the fact that many of these mixed esters have potentially desirable properties not possessed by cellulose acetates. Most of the methods for the preparation of mixed esters of cellulose have already been reported (59), but a limited number of commercially useful processes have recently been published.

A type of cellulose acetate-butyrate that has a low solution viscosity, and whose dried films provide tough transparent coatings highly resistant to grease and water, has been found suitable for use as a paper lacquer (201). Malm and Blanchard (202) have reported a method of preparing hydrolyzed cellulose acetate-butyrate that is of good uniformity and whose solution is made free from haze or grain by the use of a diluting technique that provides sufficient water for hydrolysis while keeping the ester in solution.

Cellulose acetate-isobutyrate of desirable viscosity and uniformity with high isobutyryl content was prepared by the reaction of activated cellulose in an esterification bath containing essentially isobutyric anhydride and zinc chloride (203). After substantial uptake of isobutyryl, acetic anhydride is added to complete the esterification. This method has been adapted for the preparation of other cellulose esters containing the isobutyryl group. Cellulose propionate-isobutyrate with good solution properties and high moisture resistance has been prepared by a modification of this procedure (204). Activated cellulose, wet with isobutyric acid, was treated with a mixture of isobutyric anhydride, propionic anhydride, and zinc chloride until a soluble product was obtained.

Cellulose acetate-valerate, propionate-valerate, and butyrate-valerate have been prepared economically by the conventional procedure employing the acid anhydride and sulfuric acid as catalyst (200). They were made by the dehydration of water-activated linters with the corresponding lower fatty acid, followed by esterification with valeric anhydride. The amount of lower acyl group introduced was varied by changing the ratio of lower acid to valeric anhydride in the esterification bath.

4. Esters of Other Organic Acids

(a) CELLULOSE ESTERS OF UNSATURATED CARBOXYLIC ACIDS

Unsaturated esters of cellulose, some of which may be used as intermediates, have been synthesized. Cellulose methacrylate (DS = 2) was prepared by treating cellulose (fibers or fabric) with methacryloyl chloride in dimethylformamide at elevated temperature in the presence of pyridine (205). Cellulose acetate-methacrylate has been prepared by the acetylation of cellulose methacrylate and by reacting cellulose with a conventional acetylation mixture containing methacrylic acid (206, 207) or with an impelling agent (208). Engelmann and Exner (209) synthesized cellulose crotonates and acetate-crotonates that could be crosslinked. Mench and Fulkerson (210) prepared cellulose propionate-crotonate (25% crotonyl).

Because cellulose acetate-sorbates have a number of potential commercial applications (in paper impregnating and laminating, in adhesives, in printing inks, etc.), a number of methods have been tried to prepare these esters (211). Hewson (211) has acylated cellulose with a mixture of acetic anhydride and sorbic acid in methylene chloride in the presence of sulfuric acid catalyst. The resulting cellulose acetate-sorbates were partially hydrolyzed to a DS of 2.2-2.9 with a sorboyl DS of 0.01-0.35. Berenshtein and co-workers (212-214) have prepared cellulose acetate-sorbates by a variety of methods: in acetic acid medium (212), in dioxane (213), and in methylene chloride (214, 215). Cellulose sorbates were synthesized from tosylated cellulose (cotton linters) and sorbic acid chloride in various media, such as benzene with pyridine, dimethylformamide, or dioxane, under a multitude of conditions (216). In no case did the DS exceed 1.

Ioannidis and co-workers (217, 218) have reported the preparation of cellulose acetate-oleates from cellulose or cellulose oleates. The cellulose oleates were prepared by the reaction of cotton linters with oleic acid chloride, and the resulting cellulose oleates were then acetylated by the standard method using sulfuric acid, perchloric acid, or zinc chloride as catalyst. Activated or nonactivated cellulose was acylated by a mixture of acetic anhydride and oleic acid. However, even under favorable conditions, the DS of oleate was less than 0.2. After investigating the effects of the reaction variables (temperature, time, and medium) on DS (219), these authors were able to prepare cellulose oleates with much higher degrees of substitution (DS = 2.2 from regenerated cellulose and DS = 1.7 from cotton linters).

Masaidova and co-workers (220) have prepared cellulose esters containing triple bonds (carbon to carbon). They employed three different methods, the best of which did not give a DS exceeding 1.3.

(b) CELLULOSE ESTERS OF AROMATIC CARBOXYLIC ACIDS

Since the aromatic acid esters of cellulose generally do not have properties with distinct advantages over the more readily prepared aliphatic esters (59), only a limited number of new aromatic esters have been synthesized. Cellulose cinnamate has been prepared by two methods (221). Cellulose was treated with cinnamoyl chloride in the presence of an organic base and organic diluent. When the base was pyridine and the diluent was nitrobenzene, cellulose cinnamates with DS of 2.5-3 could be prepared under optimum conditions. The second method employed nucleophilic substitution of tosylated cellulose with sodium cinnamate. Various cinnamic acid esters of hydroxyalkylcellulose, such as hydroxyethylcellulose cinnamate or *m*-nitrocinnamate, were prepared by Minsk and VanDeusen (222). The appropriate hydroxyalkylcellulose was esterified with the proper cinnamic acid halide.

Purified cotton linters, which were activated by a preliminary treatment with pyridine or dimethylformamide, have been esterified with salicylic acid chloride in the presence of sodium hydroxide and various organic solvents (223).

(c) CELLULOSE ESTERS OF POLYCARBOXYLIC ACIDS

Phthalic acid esters of cellulose have been prepared by esterifying cotton linters in an anhydrous bath consisting of phthalic anhydride, acetic acid, and sodium acetate (224, 225). This method has also been used to prepare cellulose acetate-phthalates and ethylcellulose phthalates by starting with cellulose acetate and ethylcellulose, respectively (226).

Cellulose acetate-phthalates are generally manufactured from partially hydrolyzed cellulose acetate. Klimova and Bershtein (227) esterified cellulose acetate containing 38% acetyl with phthalic anhydride in pyridine for various times at different temperatures. This procedure had been reported earlier (228, 229). Modifications have also been described in which part or all of the pyridine is replaced with solvents (230, 231). In one of these Malm and Crane (232) prepared cellulose acetate phthalate from partially esterified cellulose acetate that was made from the reaction of acetic anhydride and cellulose with sulfuric acid catalyst in acetic acid. After hydrolysis sufficient acetic anhydride to react with the water was added, and the cellulose acetate was esterified with phthalic anhydride and sodium acetate. Sufficient sodium acetate was used to neutralize the sulfuric acid present and to catalyze the phthalation. Similar procedures have been used for the production of other dicarboxylic acid esters of cellulose acetate (226, 233, 234). Another extension of this method employs a two-stage process for the esterification of partially acetylated cellulose with phthalic anhydride (235). It is known that the phthaloyl content of the derivatives is inversely dependent on the reaction temperature, but the rate of introduction is faster at high temperatures. With this in mind, economical batch production is achieved by carrying out the main portion of phthalation at about 80°C while the remainder of phthalation takes place at a lower temperature in the second step.

Malm and Crane (236) have prepared low-viscosity cellulose acetate-phthalates. Cellulose acetate (39% acetyl) was partially hydrolyzed with aqueous acetic acid and sulfuric acid catalyst until the acetyl content was reduced to 26-34% and the intrinsic viscosity did not exceed 0.5dl/g. The sulfuric acid was neutralized in two steps with magnesium acetate, and sodium acetate catalyst was added to the mass. Sufficient acetic anhydride was added to dehydrate the mass before phthalic anhydride was added to effect the desired degree of esterification Gluzman, Zaslavskaya, and Rubtsova (237) examined mixed esters containing 16-20% acetyl and 34-40% phthaloyl groups and soluble in organic solvents and sodium bicarbonate solution. Since these esters were resistant to the action of 0.25% hydrochloric acid, they were suitable for coating medicinal tablets. This work is almost identical with the earlier work of Malm and co-workers (238). After the desired phthaloyl uptake has been reached, the cellulose acetate-phthalates may be precipitated (239) by the addition of water (2-3.5 parts of water per part of acetic acid in the bath). Salts of cellulose acetate-phthalate or other dicarboxylic acid esters of cellulose can be made by dispersing the ester in a mixture of lower alcohol and water and adding the stoichiometric amount of alkali metal or ammonium salt, or aliphatic amine (240).

The method of preparing cellulose acetate-phthalates that starts with partially hydrolyzed cellulose esters has also been used to prepare ethylcellulose phthalates (224-226). Malm and Crane (241) have used this method to prepare low-viscosity ethylcellulose phthalates. The viscosity of the starting ethylcellulose is reduced to the desired degree by heating in acetic acid solution containing sulfuric acid catalyst. Phthalic anhydride and sodium acetate are then added for esterification. Ethylcellulose phthalates of excellent quality for photographic use were prepared by esterification of ethylcellulose in a bath containing such additives as hydroxylamine or hydrazine or their salts in addition to the usual reagents (242).

Methods have been described for the preparation of tricarboxylic acid esters of cellulose (77, 243, 244). In one of these procedures (243) cellulose or a partially hydrolyzed cellulose ester was reacted with trimellitic anhydride in acetic acid containing sodium acetate catalyst to give products containing up to 19% trimellityl.

Cellulose esters of other saturated and unsaturated dicarboxylic acids have been prepared by esterifying cellulose, cellulose ethers, or partially hydrolyzed esters, such as cellulose acetate, in an anhydrous medium containing the appropriate dicarboxylic acid anhydride, acetic acid, and sodium acetate catalyst (224). Cellulose derivatives containing hexahydrophthalate, glutarate, maleate, succinate, and tetrahydrophthalate were made by this method.

(d) CELLULOSE ESTERS CONTAINING OTHER FUNCTIONAL GROUPS

(1) Oxygen Groups

Only a limited number of cellulose esters containing ether linkages have been prepared. Mench and Rowley (245) have hydroxyalkylated cellulose acetate and other cellulose esters by the use of such agents as epichlorohydrin in the presence of boron trifluoride, sulfuric acid, or stannic chloride catalyst. Cellulose furoates were prepared by the reaction of cellulose with furoic acid chloride in an appropriate solvent in the presence of pyridine (246). Rowley and Hiatt (247, 248) have synthesized unsaturated mixed esters of cellulose, such as cellulose crotonate, cellulose butyrate-10-undecenoate, or cellulose acetate-tetrahydrophthalate, and these were treated with peracids to yield epoxidized cellulose esters that retained the solubility characteristics of the original esters.

Cellulose esters of acids containing free hydroxyl (249) or keto groups (250) have been reported. Touey and Kiefer (249) have synthesized cellulose acid citrates that have a free-carboxyl content of 1-15.5% and are insoluble in water, organic solvents, and aqueous alkalies. A number of methods were used to prepare cellulose derivatives containing levulinic acid (250), but even under the best conditions the DS of the levulinic acid portion did not exceed 1.

(2) Nitrogen Groups

The fact that some amine derivatives of cellulose esters are soluble in dilute acid has stimulated commercial interest in their preparation. Cellulose esters that are insoluble in water but soluble in organic solvents and in dilute aqueous acids were prepared by chloroacylating a cellulose ester to give a halogen-containing derivative, such as cellulose acetate-chloroacetate (251, 252). This intermediate is then reacted with a secondary amine (e.g., diethylamine) to form an aminated halogen-free derivative that has desirable solubility properties. Other secondary amines also react with cellulose acetate-chloroacetate by this method to yield the dialkylaminoacetate esters, but the reaction with primary amines results predominantly in the removal of the entire chloroacetyl grouping by amniolysis (253). Cellulose acetate-*N,N*-dimethylaminoacetate (254) made by this method may be used for removable antihalation and antistatic layers on cellulosic photographic film, and cellulose acetate-*N,N*-diethylaminoacetate (255) has been used to coat pharmaceutical tablets.

An amino derivative of cellulose propionate that possesses all the desirable solubility characteristics of cellulose acetate dialkylaminoacetates has been prepared by a much more economical process. Mench and Fulkerson (210) prepared cellulose propionate–3-morpholinobutyrate by the reaction of morpholine with cellulose propionate–crotonate (25% crotonyl). The product is soluble in a variety of organic solvents and in dilute, aqueous acids below a pH of 5.25 but insoluble in water. Possible uses for cellulose propionate–3-morpholinobutyrate include acid-removable film backings, moisture-resistant coatings for medicaments that would be released in the stomach, and as coatings for certain medicaments used in veterinary practice to prevent reaction of the drugs in the stomach of ruminants.

The synthesis of amino acid esters of cellulose has been described (256, 257). The preparation of esters of cellulose and *N*-substituted amino acids to give products with a high DS has also been reported (258). Cellulose acetates containing *N*-substituted succinamates (259) or *N,N*-dimethylsuccinamates (260) were prepared. Cellulose diacetate has been reesterified with ethyl *p*-aminobenzoate (261) and ethyl carbamate (262) to cellulose acetate–*p*-aminobenzoate and cellulose acetate–carbamate, respectively.

(3) Miscellaneous Groups

Several sulfur-containing cellulose esters have been reported, but none of these has any commercial importance. Allewelt (263) prepared thiourethanes of cellulose acetate by the acetylation of cellulose thiourethanes. Sulfobenzoate derivatives of methylcellulose and cellulose acetate have been prepared by the reaction of the appropriate chlorosulfonylbenzoic acid in pyridine with methylcellulose and cellulose acetate, respectively (264).

Chlorine-containing mixed esters of cellulose are known. They may be used as intermediates and subjected to amination. Cellulose acetate–chloroacetate (265), acetate–chlorovalerate (266), and acetate–chloroenanthate (266) have been prepared.

The synthesis of other cellulose derivatives containing amino and other unusual functional groups is discussed in Section F of this Chapter XVII.

5. Graft Copolymers of Cellulose Esters

Graft polymerization on cellulose esters is a relatively new technique. Graft or block copolymers of cellulose esters can be synthesized by means of anionic, oxidation–reduction, or macroradical initiators. Although the DS of the species being grafted is low, the molecular weight of the grafted polymer may be appreciable. Monosaccharide derivatives have been grafted on cellulose diacetate in methylene chloride in the presence of boron trifluoride catalyst (267).

Copolymers of cellulose diacetate and triacetate with poly(2-methyl-5-vinylpyridine) were synthesized in an oxidation–reduction medium (H_2O_2–Fe^{2+}) that contained a 15% aqueous solution of methylvinylpyridine phosphate (268, 269). Cellulose that had been previously oxidized with hypochlorite was graft polymerized with methyl-5-vinylpyridine and the resulting product was acetylated (269, 270).

Acrylates can be graft polymerized on cellulose acetate in a solution of acetic acid or dimethylformamide containing ceric ammonium nitrate initiator (271). The reactivity of the monomer decreases as follows: *n*-butyl acrylate > ethyl acrylate > methyl acrylate. Graft copolymers of methyl methacrylate on cellulose were synthesized in the presence of Ce^{4+} ions, and the resulting cellulose derivatives, which contained residual hydroxyl, were acetylated (272). Kavalyunas and co-workers (273) have prepared graft copolymers of cellulose diacetate in heterogeneous media and cellulose triacetate in solution with poly(methacrylic acid).

Ermolenko and Kaputski (274) have employed a selective oxidant, such as nitrogen dioxide, to convert preferentially the residual primary hydroxyl of cellulose esters to radicals. These radicals are present in low concentration but are uniformly distributed within the cellulose chain at the $C_{(6)}$ carbon of the glucose units. Treatment with phosphoric acid yielded cellulose ester grafts containing up to 13% phosphorus.

Radiation has been used to generate macroradical initiators. Methacrylamide has been grafted to the surface of a cellulose triacetate film through free radicals on the surface of the film (275, 276). These radical initiators were formed by cleavage of the acetate ester groups under the influence of ultraviolet radiation.

Styrene in methanol was graft polymerized on cellulose acetate films that had previously been irradiated by γ-rays in air (277). The use of γ-rays, β-particles, or neutron irradiation to graft polymerize styrene, acrylonitrile, or phenylacetylene with cellulose acetate or triacetate has been reported (278). Reaction conditions affecting graft polymerization of styrene with cellulose acetate were investigated and factors that increase the rate (279) and affect the molecular weight of the grafts (280) were reported.

A detailed discussion of the preparation of graft copolymers of cellulose is given in Section G of this Chapter XVII (see also Section H of this Chapter XVII).

General discussions of cellulose esters—including preparation, properties, industrial processing, and uses—may be found in references 281 and 282.

References*

1. R. G. Woodbridge, Jr., *Ind. Eng. Chem.*, **12**, 380 (1920).
2. R. M. Brooks, Canadian Patent 557,671 (1958) (to Hercules Powder Co.).
3. G. A. Petropavlovskii, M. M. Krunchak, and G. G. Vasil'eva, *Zh. Prikl. Khim.*, **36**, 1799 (1963).

* Many of the references cited here were obtained through the *Abstract Bulletin of the Institute of Paper Chemistry.*

4. F. Krejci, *Sb. Ved. Praci, Vysoka' Skola Chem.-Technol., Pardubice,* **2**, 81 (1963).
5. C. F. Bennett and T. E. Timell, *Svensk Papperstidn.,* **58**, 281 (1955).
6. K. Thinius and W. Thümmler, *Makromol. Chem.,* **99**, 117 (1966).
7. J. Barsha, in E. Ott, H. M. Spurlin, and M. W. Grafflin, Eds., *Cellulose and Cellulose Derivatives*, Part II, 2nd ed., Interscience, New York, 1954, p. 733.
8. H. Sakata and N. Komatsu, *J. Soc. Textile Cellulose Ind. Japan,* **19**, 337 (1963).
9. J. Barsha, op. cit., pp. 739–743.
10. J. L. Bennett, R. M. Brooks, J. G. McMillan, Jr., and W. L. Plunkett, U.S. Patent 2,776,965 (1957) (to Hercules).
11. J. L. Bennett, R. M. Brooks, J. G. McMillan, Jr., and W. L. Plunkett, Canadian Patent 563,545 (1958) (to Hercules).
12. D. S. Bruce and H. M. Spurlin, Canadian Patent 563,546 (1958) (to Hercules).
13. D. S. Bruce and H. M. Spurlin, U.S. Patent 2,776,880 (1957) (to Hercules).
14. F. J. Arthen, R. A. Coleman, and W. B. Hardy, U.S. Patent 3,208,865 (1965) (to American Cyanamid Co.).
15. W. L. Plunkett, Canadian Patent 581,319 (1959) (to Hercules).
16. W. L. Plunkett, U.S. Patent 2,950, 278 (1960) (to Hercules).
17. J. G. McMillan Jr., and W. L. Plunkett, U.S. Patent 2,776,964 (1957) (to Hercules).
18. J. G. McMillan Jr., and W. L. Plunkett, U.S. Patent 2,776,944 (1957) (to Hercules).
19. J. G. McMillan. Jr., and W. L. Plunkett, Canadian Patent 563,548 (1958) (to Hercules).
20. J. G. McMillan, Jr. and W. L. Plunkett, U.S. Patent 2,776,966 (1957) (to Hercules).
21. J. G. McMillan, Jr., and W. L. Plunkett, Canadian Patent 563,549 (1958) (to Hercules).
22. C. M. Reinhardt, U.S. Patent 2,982,643 (1961) (to Olin Mathieson Chemical Co.).
23. R. C. Bergman, Canadian Patent 732,186 (1966) (to E. I. du Pont de Nemours & Co.).
24. W. C. Ramsey, Canadian Patent 536,191, (1957) (to Olin Mathieson).
25. A. Diels and H. Orth, Canadian Patent 586,662 (1959) (to Poudreries Réunies de Belgique).
26. C. Frejacques, U.S. Patent 3,085,086 (1963) (to Ministry of National Defence and Armed Forces).
27. G. Frank, U.S. Patent 2,559,914 (1951); through *Chem. Abstr.,* **45**, 8770 (1951).
28. C. J. Malm and C. L. Crane, U.S. Patent 2, 539, 451 (1951) (to Eastman Kodak Co.); through *Chem. Abstr.,* **45**, 4453 (1951).
29. E. D. Klug, U.S. Patent 2,753,337 (1953) (to Hercules).
30. E. D. Klug, Canadian Patent 542,733 (1957) (to Hercules).
31. G. A. Petropavlovskii and M. M. Krunchak, *Kh. Prikl. Khim.,* **36**, 2506 (1963).
32. G. A. Petropavlovskii, G. G. Vasil'eva, and O. A. Vasil'eva, *Zh. Prikl. Khim.,* **39**, 2053 (1966).
33. D. E. Read and J. A. Bilton, Canadian Patent 535,342 (1957) (to Canadian Industries, Ltd., and Du Pont of Canada, Ltd.).
34. R. Asami and N. Tokura, *J. Chem. Soc. Japan, Ind. Chem. Sect.,* **62**, 1593 (1959).
35. R. G. Schweiger, *Chem. Ind. (London)*, No. 22, 900 (1966).
36. C. J. Malm, M. E. Rowley, and G. D. Hiatt, U.S. Patent 2,969,355 (1961) (to Eastman Kodak).
37. G. D. Hiatt and M. E. Rowley, U.S. Patent 3,335,128 (1967) (to Eastman Kodak).

38. G. P. Touey and J. E. Kiefer, U.S. Patent 2,969,356 (1961) (to Eastman Kodak).
39. G. P. Touey and J. E. Kiefer, U.S. Patent 3,086,007 (1963) (to Eastman Kodak).
40. G. D. Hiatt and M. E. Rowley, U.S. Patent 3,075,962 (1963) (to Eastman Kodak).
41. G. D. Hiatt and M. E. Rowley, U.S. Patent 3,075,963 (1963) (to Eastman Kodak).
42. E. Klein and J. E. Snowden, *Ind. Eng. Chem.*, **50**, 80 (1958).
43. A. C. Nuessle, F. M. Ford, W. P. Hall, and A. L. Lippert, *Textile Res. J.*, **26**, 32 (1956).
44. K. Katsuura and S. Nonaka, *Sen-i Gakkaishi,* **13**, 24 (1957).
45. K. Katsuura and N. Inagaki, *J. Chem. Soc. Japan, Ind. Chem. Sect.*, **69**, 681 (1966).
46. G. P. Touey, U.S. Patent 2,759,924 (1956) (to Eastman Kodak).
47. A Yuldashev and U. M. Muratova, *Dokl. Akad. Nauk Uzl. SSR,* **23**, 42 (1966).
48. A. Yuldashev, U. M. Muratova, and M. A. Askarov, *Vysokomol. Soedin.,* **7**, 1923 (1965).
49. K. A. Petrov, E. E. Nifant'ev, I. I. Sopikova, and M. I. Merkulova, *Tsellyuloza i ee Proizvodnye, Sb. Statei,* 86 (1963).
50. K. A. Petrov, I. I. Sopikova, and E. E. Nifant'ev, *Vysokomol. Soedin.,* **7**, 1667 (1965).
51. D. A. Predvoditelev, M. A. Tyuganova, E. E. Nifant'ev, and Z. A. Rogovin, *Zh. Vses. Khim. Obshchestva im. D.I. Mendeleeva,* **10**, 459 (1965).
52. D. A. Predvoditelev, E. E. Nifant'ev, and Z. A. Rogovin, *Vysokomol. Soedin.,* **7**, 1005 (1965).
53. E. E. Nifant'ev, M. A. Belaventsev, and L. P. Levitan, *Vysokomol. Soedin.,* **7**, 513 (1965).
54. K. A. Petrov, E. E. Nifant'ev, I. I. Sopikova, and M. A. Belaventsev, *Tsellyuloza i ee Proizvodnye, Sb. Statei,* **90** (1963).
55. D. A. Predvoditelev, E. E. Nifant'ev, and Z. A. Rogovin, *Vysokomol. Soedin.,* **8**, 76 (1966).
56. A. D. Kiselev and S. N. Danilov, USSR Patent 159,524 (1962).
57. D. A. Predvoditelev, E. E. Nifant'ev, and Z. A. Rogovin, *Vysokomol. Soedin.,* **7**, 791 (1965).
58. V. V. Gertsev, *Vysokomol. Soedin.,* **9B**, 470 (1967).
59. C. J. Malm and G. D. Hiatt, in E. Ott, H. M. Spurlin, and M. W. Grafflin, Eds., *Cellulose and Cellulose Derivatives*, Part II, 2nd ed. Interscience, New York, 1954, pp. 763–824.
60. J. Geurden, *Pure Appl. Chem.*, **14**, 507 (1967).
61. I. Kido, K. Suzuki, and T. Yoshikawa, *J. Soc. Textile Cellulose Ind. Japan,* **15**, 857 (1959).
62. I. Kido, K. Suzuki, and S. Nishimori, *J. Soc. Textile Cellulose Ind. Japan,* **17**, 132 (1961).
63. I. Kido and K. Suzuki, *J. Soc. Textile Cellulose Ind. Japan,* **15**, 447 (1959).
64. I. Kido, K. Suzuki, and A. Hayashi, *J. Soc. Textile Cellulose Ind. Japan,* **18**, 557 (1962).
65. I. Kido, K. Suzuki, and K. Isojima, *J. Soc. Textile Cellulose Ind. Japan,* **19**, 95 (1963).
66. I. Kido and K. Suzuki, *J. Soc. Textile Cellulose Ind. Japan,* **18**, 913 (1962).
67. S. Saito and S. Sugeta, *J. Soc. Textile Cellulose Ind. Japan,* **18**, 749 (1962).
68. M. Isozaki and N. Ishikawa, *Repts. Govt. Chem. Ind. Research Inst., Tokyo,* **53**, 155 (1958).

69. British Patent 964,197 (1964) (to Rayonier Inc.).
70. D. Edge, Jr., M. D. Fahey, and R. G. Fickey, Canadian Patent 733,013 (1966) (to Rayonier).
71. L. E. Herdle and W. H. Griggs, *Tappi,* **48**, 103A (1965)
72. D. L. Gibson and W. M. Shoemaker, U.S. Patent 2,348,001 (1944) (to National Vulcanized Fibre Co.); through *Chem. Abstr.,* **39**, 191 (1945).
73. British Patent 535,932 (1941) (to Sandoz, Ltd.); through *Chem. Abstr.,* **36**, 1492 (1942).
74. L. E. Herdle and G. C. Gandy, Canadian Patent 649, 270 (1962) (to Canadian Kodak Co., Ltd.).
75. C. J. Malm and G. D. Hiatt, op. cit., p. 775, (Ref. 59).
76. C. J. Malm, *Svensk Papperstidn.,* **64**, 740 (1961).
77. C. J. Malm, *Svensk Kemisk Tidsk.,* **73**, 523 (1961).
78. G. A. Richter, *Tappi,* **40**, 429 (1957).
79. C. J. Malm, K. T. Barkey, D. C. May, and P. I. Abell, *Ind. Eng. Chem.,* **46**, 557 (1954).
80. *Chemical Week*, February 10, 1968, pp. 62–63.
81. A. W. Brinkley, Jr., G. S. Mabrey, and J. W. Gilbert, U.S. Patent 3,294,623 (1966) (to International Paper Co.).
82. G. S. Kosaya and Z. N. Chechurina, *Bumazh. Prom.,* **7**, 9 (1968).
83. R. H. MacClaren and F. L. Wells, U.S. Patent 3,068,141 (1962) (to Eastman Kodak).
84. R. E. Glegg, D. Ingerick, R. R. Parmerter, J. S. T. Salzer, and R. S. Warburton, *J. Polymer Sci.,* A-1, **6**, 745 (1968).
85. L. J. Tanghe, L. B. Genung, and J. W. Mench, in R. L. Whistler, Ed., *Methods in Carbohydrate Chemistry*, Vol. III, Academic Press, New York, 1963, pp. 194–200.
86. A. W. McKenzie and H. G. Higgins, *Holzforschung,* **10**, 150 (1956).
87. E. P. Kuznetsova and N. I. Klenkova, *Zh. Prikl. Khim.,* **37**, 399 (1964).
88. C. J. Malm, K. T. Barkey, D. C. May, and E. B. Lefferts, *Ind. Eng. Chem.,* **44**, 2904 (1952).
89. C. J. Malm and G. D. Hiatt, op. cit., p. 776, (Ref. 59).
90. C. J. Malm, K. T. Barkey, E. B. Lefferts, and R. T. Gielow, *Ind. Eng. Chem.,* **50**, 103 (1958).
91. S. Ishikawa, *Bull. Textile Res. Inst. Japan,* No 58, 1 (1961).
92. A. F. Zaitseva, G. A. Petropavlovskii, and N. G. Bocharnikova, *Zh. Prikl. Khim.,* **38**, 2762 (1965).
93. A. F. Zaitseva, G. A. Petropavlovskii, N. G. Bocharnikova, and G. I. Yurchenko, *Zh. Prikl. Khim.,* **38**, 2768 (1965).
94. A. J. Rosenthal and B. B. White, *Tappi,* **43**, 69 (1960).
95. C. J. Malm, K. T. Barkey, J. T. Schmitt, and D. C. May, *Ind. Eng. Chem.,* **49**, 763 (1957).
96. C. J. Malm, L. J. Tanghe, and B. C. Laird, *Ind. Eng. Chem.,* **38**, 77 (1946).
97. C. J. Malm, L. J. Tanghe, B. C. Laird, and G. D. Smith, *J. Am. Chem. Soc.,* **75**, 80 (1953).
98. C. J. Malm and L. J. Tanghe, *Ind. Eng. Chem.,* **47**, 995 (1955).
99. C. J. Malm, L. J. Tanghe, and J. T. Schmitt, *Ind. Eng. Chem.,* **53**, 363 (1961).

100. L. J. Tanghe and R. J. Brewer, paper presented at the 154th National Meeting of the American Chemical Society, Chicago, September 10–16, 1967.
101. L. J. Tanghe and R. J. Brewer, *Anal. Chem.*, **40**, 350 (1968).
102. G. P. Touey and J. W. Tamblyn, U.S. Patent 2,976,277 (1961) (to Eastman Kodak).
103. C. Hamalainen, R. H. Wade, and M. D. Cruz, *Textile Res. J.*, **29**, 821 (1959).
104. A. L. Geddes, *J. Polymer Sci.*, **22**, 31 (1956).
105. R. C. Blume and F. H. Swezey, *Tappi*, **37**, 481 (1954).
106. H. Bates, J. W. Fisher, and J. R. Smith, U.S. Patent 2,775,585 (1956) (to British Celanese, Ltd.).
107. C. J. Malm, L. J. Tanghe, and R. E. Glegg, *Ind. Eng. Chem.*, **51**, 1483 (1959).
108. J. A. Mitchell, *Tappi*, **40**, 713 (1957).
109. K. Dimov and D. Laleva, *Khim. i Ind. (Sofia)*, **38**, 407 (1966).
110. C. J. Malm, L. J. Tanghe, B. C. Laird, and G. D. Smith, *J. Am. Chem. Soc.*, **74**, 4105 (1952).
111. A Delsemme, *Pure Appl. Chem.*, **14**, 481 (1967).
112. T. M. Andrews, L. G. Reed, and J. L. Thomas, Canadian Patent 537,268 (1957) (to Celanese Corp.).
113. C. L. Crane, U.S. Patent 3,047,561 (1962) (to Eastman Kodak).
114. C. L. Crane, U.S. Patent 3,057,854 (1962) (to Eastman Kodak).
115. C. J. Malm and C. L. Crane, U.S. Patent 3,089,871 (1963) (to Eastman Kodak).
116. C. J. Malm, R. E. Glegg, J. T. Salzer, D. F. Ingerick, and L. J. Tanghe, *Ind. Eng. Chem., Process Des. Develop.*, **5**, 81 (1966).
117. C. J. Malm, K. T. Barkey, M. Salo, and D. C. May, *Ind. Eng. Chem.*, **49**, 79 (1957).
118. British Patent 1,108,046 (1968) (to Celanese Corp.).
119. H. W. Steinmann, U.S. Patent 3,386,932 (1968) (to Celanese Corp.).
120. S. Ouno, M. Tonedachi, S. Koyanagi, and K. Ito, U.S. Patent 3,391,135 (to Shin-Etsu Chemical Industry Co., Ltd.).
121. C. J. Malm and C. L. Crane, U.S. Patent 2,772,267 (1956) (to Eastman Kodak).
122. K. Keirstead and J. Myers, *Can. J. Chem.*, **32**, 566 (1954).
123. H. W. Coover, Jr., and W. C. Wooten, Jr., U.S. Patent 2,917,398 (1959) (to Eastman Kodak).
124. K. Ritchie and J. W. Addleburg, U.S. Patent 3,305,378 (1967) (to Eastman Kodak).
125. R. F. Williams, Jr., and C. S. Lowe, U.S. Patent 2,899,315 (1959) (to Eastman Kodak).
126. A. Bell, U.S. Patent 2,805,170 (1957) (to Eastman Kodak).
127. G. M. Moulds, U.S. Patent 3,314,808 (1967) (to Du Pont).
128. W. Neumann, H. Holtschmidt, O. Bayer, H. Glaesser, and W. Roehm, Belgian Patent 626,176 (1963) (to Farbenfabriken Bayer, A.-G.).
129. C. J. Malm, R. E. Glegg, L. J. Tanghe, and J. Thompson, *Tappi*, **44**, 669 (1961).
130. *Half-Second Acetate*, Technical Data Sheet No. X-182, Eastman Chemical Products, Inc., Kingsport, Tenn.
131. G. A. Richter and L. E. Herdle, *Ind. Eng. Chem.*, **49**, 1451 (1957).
132. C. J. Malm and L. J. Tanghe, *Tappi*, **46**, 629 (1963).
133. H. W. Steinmann and B. B. White, *Tappi*, **37**, 225 (1954).
134. J. K. Watson and D. R. Henderson, *Tappi*, **40**, 686 (1957).

135. F. L. Wells, W. C. Schattner, and A. Walker, Jr., *Tappi,* **46**, 581 (1963).

136. R. I. C. Michie, *Pulp Paper Mag. Can.,* **67**, T461 (1966).

137. E. Barabash, A. J. Rosenthal, and B. B. White, *Tappi,* **38**, 745 (1955).

138. L. A. Hiller, Jr., *J. Polymer Sci.,* **14**, 555 (1954).

139. Y. V. Bytenskii, E. P. Kuznetsova, and N. I. Klenkova, *Zh. Prikl. Khim.,* **39**, 185 (1966).

140. M. A. Heath, *Tappi,* **40**, 156 (1957).

141. L. E. Herdle and W. E. Wahtera, *Tappi,* **51**, 540 (1968).

142. H. Staudinger and W. Döhle, *Makromol. Chem.,* 9, 188 (1953).

143. A. F. Fedorova and Z. A. Rogovin, *Vysokomol. Soedin.,* **5**, 519 (1963).

144. L. Laamanen and H. Sihtola, *Paperi Puu,* **46**, 159 (1964).

145. H. Sihtola and L. Laamanen, *Papier,* **18**, 600 (1964).

146. E. L. Akim, *Pure Appl. Chem.,* **14**, 475 (1967).

147. H. Sihtola, L. Blomberg, M. Johanson, and L. Laamanen, *International Symp. Macromol. Chem.,* Tokyo/Kyoto, Session 10, 26 (1966).

148. P. Howard and R. S. Parikh, *Indian J. Chem.,* **3**, 191 (1965).

149. G. Wlodarski, *Zeszty Nauk. Politech. Lodz., Wlokiennictwo,* No. 9, 5 (1962).

150. A. J. Rosenthal, *Pure Appl. Chem.,* **14**, 535 (1967).

151. L. H. Greathouse, H. J. Janssen, and C. H. Haydel, *Ind. Eng. Chem.,* **50**, 97 (1958).

152. C. M. Conrad, P. Harbrink, and A. L. Murphy, *Textile Res. J.,* **33**, 784 (1963).

153. C. M. Conrad and J. J. Creely, *J. Polymer Sci.,* **58**, 781 (1962).

154. L. A. Hiller, Jr., *J. Polymer Sci.,* **10**, 385 (1953).

155. G. A. Petropavlovskii and G Rakhmanberdiev, *Zh. Prikl. Khim.,* **39**, 237 (1966).

156. V. I. Sharkov, E. N. Garmanova, and N. N. Sintsova, *Sb. Tr. Gos. Nauch.-Issled. Inst. Gidroliz. i Sulf.-Spirt. Prom.,* 9, 66 (1961).

157. H. W. Turner, U.S. Patent 2,836,590 (1958) (to Hercules).

158. K. D. Vos, F. O. Burris, Jr., and R. L. Riley, *J. Appl. Polymer Sci.,* **10**, 825 (1966).

159. D. J. Stanonis and W. D. King, *J. Appl. Polymer Sci.,* 8, 943 (1964).

160. M. Zilahi and L. Oswald, *Faserforsch. u. Textiltech.,* 7, 262 (1956).

161. A. V. Bailey, E. Honold, and E. L. Skau, *Textile Res. J.,* **28**, 861 (1958).

162. O. M. Klimova and L. F. Petushkova, *Zh. Obshch. Khim.,* **27**, 2096 (1957).

163. K. Ward, Jr., Chen-Chuan Tu, and M. Lakstigala, *J. Am. Chem. Soc.,* **76**, 6102 (1954).

164. C. J. Malm and L. W. Blanchard, Jr., U.S. Patent 2,772,266 (1956) (to Eastman Kodak).

165. B. B. White and C. L. Smart, U.S. Patent 2,860,132 (1958) (to Celanese Corp.).

166. E. P. Kuznetsova, *Khim. i Tekhnol. Proizvodn. Tsellyulozy, Vladimir,* 76 (1964).

167. I. Sakurada and Y. Sakaguchi, *Sen-i Gakkaishi,* **13**, 364 (1957).

168. I. Sakurada, Y. Sakaguchi, and K. Nagai, *Sen-i Gakkaishi,* **13**, 292 (1957).

169. I. Sakurada, Y. Sakaguchi, and H. Yamaguchi, *Sen-i Gakkaishi,* **13**, 431 (1957).

170. H. Engelmann, German Patent 1,150,374 (1964) (to Lonza-Werke Elektrochemische Fabriken G. m. b. H.).

171. J. Robin and R. Clévy, U.S. Patent 2,854,446 (to Société Rhodiaceta).

172. K. C. Laughlin, R. J. Osborne, and J. G. Santangelo, U.S. Patent 2,966,485 (1960) (to Celanese Corp.).

173. K. C. Laughlin, R. J. Osborne, and J. G. Santangelo, Canadian Patent 609,900 (1960) (to Celanese Corp.).
174. H. Bates, F. Hindley, and W. Popiolek, U.S. Patent 3,040,027 (1962) (to British Celanese, Ltd.).
175. R. Clévy and J. Robin, U.S. Patent 2,801,237 (1957) (to Société Rhodiaceta).
176. R. Clévy and J. Robin, U.S. Patent 2,854,445 (1958) (to Société Rhodiaceta).
177. R. Clévy and J. Robin, Canadian Patent 605,169 (1960) (to Société Rhodiaceta).
178. H. Genevray and J. Robin, *Pure Appl. Chem.,* **14**, 489 (1967).
179. W. H. Groombridge, J. W. Fisher, and H. Bates, Canadian Patent 530,164 (1956) (to British Celanese).
180. British Patent 930,208 (1963) (to Toho Rayon Co., Ltd.).
181. U.S. Tariff Commission Reports, 1957-1970.
182. *Chemical and Engineering News*, September 4, 1967, p. 71A.
183. C. J. Malm, J. W. Mench, D. L. Kendall, and G. D. Hiatt, *Ind. Eng. Chem.,* **43**, 684 (1951).
184. I. Terasaki, *J. Soc. Textile Cellulose Ind. Japan,* **18**, 331 (1962).
185. I. Terasaki, *J. Soc. Textile Cellulose Ind. Japan,* **18** 336 (1962).
186. *Half-Second Propionate*, Technical Data Sheet No. E-136, Eastman Chemical Products, Inc., Kingsport, Tenn.
187. J. W. Mench, B. Fulkerson, and U. K. Schutt, U.S. Patent 2,992,214 (1961) (to Eastman Kodak).
188. C. J. Malm, L. J. Tanghe, H. M. Herzog, and M. H. Stewart, *Ind. Eng. Chem.,* **50**, 1061 (1958).
189. C. J. Malm, L. J. Tanghe, and L. W. Blanchard, Jr., U.S. Patent 2,790,794 (1957) (to Eastman Kodak).
190. C. J. Malm and L. W. Blanchard, Jr., U.S. Patent 2,801,238 (1957) (to Eastman Kodak).
191. C. J. Malm. L. J. Tanghe, and L. W. Blanchard, Jr., U.S. Patent 2,835,665 (1958) (to Eastman Kodak).
192. C. J. Malm and L. W. Blanchard, Jr., U.S. Patent 2,816,104 (1957) (to Eastman Kodak).
193. C. J. Malm, L. J. Tanghe, and L. W. Blanchard, Jr., U.S. Patent 2,816,105 (1957) (to Eastman Kodak).
194. C. J. Malm and L. W. Blanchard, Jr., U.S. Patent 2,801,999 (1957) (to Eastman Kodak).
195. C. J. Malm and L. J. Tanghe, U.S. Patent 2,801,240 (1957) (to Eastman Kodak).
196. C. J. Malm and L. W. Blanchard, Jr., U.S. Patent 2,816,107 (1957) (to Eastman Kodak).
197. C. J. Malm, L. J. Tanghe, and H. M. Herzog, U.S. Patent 2,816,106 (1957) (to Eastman Kodak).
198. *Alcohol-Soluble Butyrate*, Technical Data Sheet No. X–171, Eastman Chemical Products, Inc., Kingsport, Tenn.
199. M. Salo and L. W. Blanchard, Jr., U.S. Patent 3,370,025 (1968) (to Eastman Kodak).
200. J. W. Mench, B. Fulkerson, and G. D. Hiatt, *Ind. Eng. Chem., Prod. Res. Develop.,* **5**, 110 (1966).
201. Anon., *Chem. Eng.,* **60**, 228 (1953).

202. C. J. Malm and L. W. Blanchard, Jr., U.S. Patent 2,834,776 (1958) (to Eastman Kodak).
203. C. J. Malm and L. W. Blanchard, Jr., U.S. Patent 2,828,303 (1958) (to Eastman Kodak).
204. C. J. Malm and L. W. Blanchard, Jr., U.S. Patent 2,828,304 (1958) (to Eastman Kodak).
205. R. P. Svistunova, B. I. Aikhodzhaev, Y. L. Pogosov, and I. V. Pakhimova, USSR Patent 173,739 (filed June 9, 1964).
206. R. P. Svistunova, B. I. Aikhodzhaev, and Y. L. Pogosov, *Plas. Massy*, No. 6, 57 (1965).
207. K. Matsuzaki and T. Miyata, *J. Chem. Soc. Japan, Ind. Chem.*, **70**, 770 (1967).
208. T. Miyata and K. Matsuzaki, *J. Chem. Soc. Japan, Ind. Chem.*, **70**, 2192 (1967).
209. H. Engelmann and F. Exner, *Makromol. Chem.*, **23**, 233 (1957).
210. J. W. Mench and B. Fulkerson, *Ind. Eng. Chem., Prod. Res. Develop.*, **7**, 2 (1968).
211. W. B. Hewson, Canadian Patent 536,131 (1957) (to Hercules).
212. E. I. Berenshtein, L. V. Gurkovskaya. V. N. Kryazhev, B. I. Aikhodzhaev, Y. L. Pogosov, and G. Sh. Talipov, USSR Patent 188,949 (filed May 10, 1965).
213. E. I. Berenshtein, L. V. Gurkovskaya, Y. L. Pogosov, V. N. Kryazhev, and B. I. Aikhodzhaev, USSR Patent 187,757 (filed October 6, 1965).
214. E. I. Berenshtein, B. I. Aikhodzhaev, Y. L. Pogosov, and G. Sh. Talipov, *Khim. Volokna*, No. 5, 43 (1966).
215. B. I. Aikhodzhaev, G. Sh. Talipov, and E. I. Berenshtein, *Vysokomol. Soedin.*, **9**, 1459 (1967).
216. E. I. Berenshtein, Y. L. Pogosov, and B. I. Aikhodzhaev, *Khim. i Fiz.-Khim. Prirodn. i Sintetich. Polimerov, Akad. Nauk Uzb. SSR, Inst. Khim. Polimerov*, No. 2, 72 (1964).
217. O. K. Ioannidis, Y. L. Pogosov, and B. I. Aikhodzhaev, *Uzbek. Khim. Zh.*, **8**, 76 (1964).
218. O. K. Ioannidis, Y. L. Pogosov, B. I. Aikhodzhaev, R. Rozyakhunov, V. N. Kryazhev, and L. V. Gurkovskaya, *Khim. Volokna*, No. 3, 58 (1966).
219. O. K. Ioannidis, Y. L. Pogosov, and B. I. Aikhodzhaev, *Khim. i Fiz.-Khim. Prirodn. i Sintetich. Polimerov, Akad. Nauk Uzb. SSR, Inst. Khim. Polimerov*, No. 2, 60 (1964).
220. G. S. Masaidova, L. S. Gal'braikh, V. I. Nikitin, and Z. A. Rogovin, *Vysokomol. Soedin*, **9**, 166 (1967).
221. S. T. Shaposhnikova, B. I. Aikhodzhaev, and Y. L. Pogosov, *Khim i Fiz.-Khim. Prirodn. i Sintetich. Polimerov, Akad. Nauk Uzb. SSR, Inst. Khim. Polimerov*, No. 2, 43 (1964).
222. L. M. Minsk and W. P. VanDeusen, Canadian Patent 540,502 (1957) (to Canadian Kodak).
223. E. I. Berenshtein, Y. L. Pogosov, and B. I. Aikhodzhaev, *Khim. i Fiz.-Khim. Prirodn. i Sintetich. Polimerov, Akad. Nauk Uzb. SSR, Inst. Khim. Polimerov*, No. 2, 67 (1964).
224. G. D. Hiatt, J. W. Mench, and J. Emerson, U.S. Patent 2,759,925 (1956) (to Eastman Kodak).
225. G. D. Hiatt, J. W. Mench, and J. Emerson, Canadian Patent 555,945 (1958) (to Eastman Kodak).
226. C. J. Malm, J. W. Mench, B. Fulkerson, and G. D. Hiatt, *Ind. Eng. Chem.*, **49**, 84 (1957).
227. O. M. Klimova and V. A. Bershtein, *Zh. Prikl. Khim.*, **29**, 1849 (1956).

228. C. J. Malm and C. R. Fordyce, *Ind. Eng. Chem.*, **32**, 405 (1940).
229. C. J. Malm and C. E. Waring, U.S. Patent 2,093,462 (1937) (to Eastman Kodak).
230. C. L. Crane and L. W. Blanchard, Jr., U.S. Patent 2,183,982 (1939) (to Eastman Kodak).
231. L. B. Genung, U.S. Patent 2,126,460 (1938) (to Eastman Kodak).
232. C. J. Malm and C. L. Crane, U.S. Patent 2,768,161 (1956) (to Eastman Kodak).
233. G. D. Hiatt and J. Emerson, U.S. Patent 2,794,799 (1957) (to Eastman Kodak).
234. C. L. Crane and J. Emerson, U.S. Patent 2,852,507 (1958) (to Eastman Kodak).
235. J. W. Mench, B. Fulkerson, and G. B. Lapham, U.S. Patent 2,856,399 (1958) (to Eastman Kodak).
236. C. J. Malm and C. L. Crane, U.S. Patent 2,856,400 (1958) (to Eastman Kodak).
237. M. K. Gluzman, R. G. Zaslavskaya, and V. P. Rubtsova, *Zh. Prikl. Khim.*, **39**, 1372 (1966).
238. C. J. Malm, J. Emerson, and G. D. Hiatt, *J. Am. Pharm. Assoc., Sci. Ed.*, **40**, 520 (1951).
239. G. D. Hiatt and C. L. Crane, U.S. Patent 2,852,508 (1958) (to Eastman Kodak).
240. G. D. Hiatt and J. Emerson U.S. Patent 2,865,898 (1958) (to Eastman Kodak).
241. C. J. Malm and C. L. Crane, U.S. Patent 2,753,339 (1956) (to Eastman Kodak).
242. C. L. Crane, U.S. Patent 3,008,593 (1961) (to Eastman Kodak).
243. J. W. Mench, J. Emerson, and M. E. Rowley, British Patent 969,335 (1964) (to Eastman Kodak).
244. J. W. Mench, J. Emerson, and M. E. Rowley, Canadian Patent 677,981 (1964) (to Eastman Kodak).
245. J. W. Mench and M. E. Rowley, U.S. Patent 2,828,305 (1958) (to Eastman Kodak).
246. S. T. Shaposhnikova, Y. L. Pogosov, and B. I. Aikhodzhaev, *Vysokomol. Soedin.*, **7**, 1314 (1965).
247. M. E. Rowley and G. D. Hiatt, *Ind. Eng. Chem., Prod. Res. Develop.*, **4**, 256 (1965).
248. G. D. Hiatt and M. E. Rowley, U.S. Patent 3,347,847 (1967) (to Eastman Kodak).
249. G. P. Touey and J. E. Kiefer, U.S. Patent 2,759,787 (1956) (to Eastman Kodak).
250. T. V. Vladimirova, L. S. Gal'braikh, K. S. Peker, and Z. A. Rogovin, *Vysokomol. Soedin.*, **7**, 786 (1965).
251. J. W. Mench and B. Fulkerson, U.S. Patent 2,861,068 (1958) (to Eastman Kodak).
252. J. W. Mench and B. Fulkerson, U.S. Patent 3,089,870 (1963) (to Eastman Kodak).
253. G. D. Hiatt, J. W. Mench, and B. Fulkerson, *Ind. Eng. Chem., Prod. Res. Develop.*, **3**, 295 (1964).
254. G. F. Nadeau and C. B. Thompson, U.S. Patent 2,801,191 (1957) (to Eastman Kodak).
255. G. D. Hiatt and J. Emerson, U.S. Patent 2,940,901 (1960) (to Eastman Kodak).
256. A. I. Polyakov, V. A. Derevitskaya, and Z. A. Rogovin, *Vysokomol. Soedin.*, **3**, 1027 (1961).
257. Z. A. Rogovin, V. A. Derevitskaya, and Sun' Tun, *J. Polymer Sci.*, **53**, 117 (1961).
258. Lin-Yan', V. A. Derevitskaya, and Z. A. Rogovin, *Vysokomol. Soedin.* **1**, 157 (1959).
259. J. E. Kiefer, G. P. Touey, and J. R. Caldwell, *Ind. Eng. Chem.*, **51**, 1481 (1959).
260. G. P. Touey and J. R. Caldwell, U.S. Patent 2,996,497 (1961) (to Eastman Kodak).
261. I. Matei and M. Grigoras, *Bul. Inst. Politekh. Iasi*, **14**, 145 (1964).

262. I. Matei and M. Grigoras, *Bul. Inst. Politekh. Iasi,* **14**, 123 (1964).
263. A. L. Allewelt, U.S. Patent 2,748,109 (1956) (to American Viscose Corp.).
264. J. Lemmerling, *Makromol. Chem.,* **81**, 137 (1965).
265. C. J. Malm, J. W. Mench, R. F. Williams, Jr., and G. D. Hiatt, *Ind. Eng. Chem.,* **42**, 1547 (1950).
266. W. K. Chang and Z. A. Rogovin, *Vysokomol. Soedin,* **2**, 456 (1960).
267. R. L. Whistler, G. Ruffini, and R. E. Pyler, *J. Polymer Sci.,* [A-1], **6**, 2501 (1968).
268. R. I. Kavalyunas, G. D. Shershneva, R. M. Livshits, and Z. A. Rogovin, *Vysokomol. Soedin.,* **8**, 240 (1966).
269. Z. A. Rogovin, *Pure Appl. Chem.,* **14**, 523 (1967).
270. R. I. Kavalyunas, R. M. Livshits, and Z. A. Rogovin, *Khim. Volokna*, No. 3, 54 (1966).
271. F. Ide, R. Handa, and K. Nakatsuka, *Chem. High Polymers (Tokyo),* **21**, 57 (1964).
272. A. A. Movsum-Zade, E. S. Goryainova, R. M. Livshits, Z. A. Rogovin, and A. A. Konkin, *Vysokomol. Soedin.,* **6**, 1340 (1964).
273. R. I. Kavalyunas, R. M. Livshits, and V. E. Lozhkin, *Vysokomol. Soedin.,* **9**, 672 (1967).
274. I. N. Ermolenko and F. N. Kaputski, *Stärke,* **12**, 311 (1960).
275. S. K. Kudrna, *Vysokomol. Soedin.,* **7**, 557 (1965).
276. S. K. Kudrna, *Vysokomol. Soedin.,* **8**, 1823, 1828 (1966).
277. H. Sobue, K. Matsuzaki, H. Komagata, and A. Ishida, *J. Polymer Sci.,* [C], No. 2, 415 (1963).
278. D. Gagnaire, O. Bernard, and M. Vincendron, *Univ. Grenoble Symp., Chem. Biochem. Lignin*, 1969, pp. 359–367, 388–389.
279. H. Sumitomo, S. Takamuku, and Y. Hachihama, *J. Chem. Soc. Japan, Ind. Chem. Sect.,* **66**, 269 (1963).
280. J. D. Wellons, J. L. Williams, and V. Stannett, *J. Polymer Sci.,* [A-1], **5**, 1341 (1967).
281. G. N. Bruxelles and V. R. Grassie, in N. M. Bikales, Ed., *Encyclopedia of Polymer Science and Technology*, Vol. 3, Interscience, New York, 1965, pp. 307–325.
282. Contributions by R. O. Hill, Jr., B. P. Rouse, Jr., B. S. Sprague, L. I. Horner, and D. J. Stanonis under "Cellulose Esters, Organic," in N. M. Bikales, Ed., *Encyclopedia of Polymer Science and Technology,* Vol. 3, Interscience, New York, 1965, pp. 325–458.

C. ETHERS

A. B. Savage

The purpose of this chapter is not to repeat information already given in Chapter IX-E of Part II, but to supplement it with information developed since the mid-1950s.

During the last decade the cellulose ethers have progressed from materials of largely experimental and developmental importance to products of considerable industrial importance. The most important products are shown in Table 1.

The production of sodium carboxymethylcellulose has steadily increased both in the United States and abroad. Its utility has changed from essentially 90% for drilling muds in the oil industry to a variety of uses, including food additives, paper sizes, and antiredeposition agents in detergents. The production of methylcellulose has also increased. The introduction of (hydroxypropyl)-(methyl)cellulose has greatly broadened the use field of the methylcelluloses. The production of water-soluble hydroxyethylcellulose has approached that of the older water-soluble ethers. Low-degree-of-substitution hydroxyethylcellulose films are being produced (see Section E of Chapter XIX).

The various water-soluble ethers have generally similar properties, but there is only a partial direct competition among them. For example, sodium carboxymethylcellulose and hydroxyethylcellulose do not generally gel from their solutions on being heated. On the other hand, the materials of the methylcellulose group have distinct gelation temperatures. In general the nongelling derivatives will be more useful where high temperatures of use are encountered, but the methylcellulose group will be more useful in applications in which gelation is a distinct advantage.

The only organosoluble ethers produced commercially are ethylcellulose, which has been produced at a constant rate over the last two decades, and cyanoethylcellulose, which is used in small quantities in the manufacture of electroluminescent articles because of its high dielectric constant (see section D of this Chapter XVII).

The primary improvements that have been made during the last decade are in the physical form (e.g., granular) of the ethers and the growing use of surface crosslinking to facilitate dispersion. With the aid of dialdehydes, such as glyoxal,

TABLE 1
Types of Commercial Cellulose Ethers (1)

Cellulose ether	Reagent	Range of utility: Solubility	Range of utility: DS
Sodium carboxymethylcellulose	Chloroacetic acid or sodium salt	Water	0.5–1.2
Methylcellulose	Methyl chloride	Water	1.5–2.4
(Hydroxypropyl)(methyl)cellulose	Methyl chloride and propylene oxide	Water	1.5–2.0
(Hydroxyethyl)(methyl)cellulose	Methyl chloride and ethylene oxide	Water	1.5–2.0
Hydroxyethylcellulose	Ethylene oxide	Water	1.3–3.0[a]
Ethylcellulose	Ethyl chloride	Organic solvents	2.3–2.6
(Ethyl)(hydroxyethyl)cellulose	Ethyl chloride and ethylene oxide	Water	1.4–1.6
(Ethyl)(methyl)cellulose	Ethyl chloride and methyl chloride	Water	1.0–1.3
Sodium (carboxymethyl)-(hydroxyethyl)cellulose	Chloroacetic acid and ethylene oxide	Water	0.3[b], 0.7[c] 0.4[b], 0.3[c]
Cyanoethylcellulose	Acrylonitrile	Organic solvents	2.0

[a] Given in terms of molar substitution.
[b] Substitution value for carboxymethylcellulose.
[c] Substitution value for hydroxyethylcellulose.

products that can be slurried to wet them out without development of viscosity can be prepared, and viscosity will develop rapidly as soon as the pH of the solution is changed.

The trend in alkali cellulose preparation is toward continuous or toward *in situ* preparation by spraying and mixing, or slurrying and pressing (e.g., in a screw press), or by slurrying in an organic liquid. Attention is being directed to the use of wood pulp instead of cotton linters. Although sodium hydroxide is used almost universally in the production of soluble cellulose ethers, it has been shown that in treating small amounts of textiles other bases (such as cesium or rubidium hydroxides) produce a more rapid initial reaction; however, after 30 min of reaction, the rate obtained by using sodium hydroxide appeared to be equally fast (2).

Additional highlights of new information on cellulose ethers, including some corrections to the previously published volumes, are discussed below.

Sodium chloromethanesulfonate has been found to react with alkali cellulose, provided water is removed to drive the reaction to completion; this is not necessary in the case of sodium β-chlorethanesulfonate (3).

Although only the alkyl chlorides are used to make soluble alkyl cellulose ethers commercially, cloth has been treated with relatively nonreactive reagents and reasonable substitutions have been obtained by introducing iodide salts to serve as intermediates (4). The iodide salts react with relatively inactive organic chlorides or bromides to give the corresponding organic iodides. The iodides react more readily with cellulose, and the iodide salts are then re-formed:

$$RCl + R_{cell}(OH)_3 \rightarrow \text{little reaction} \quad (1)$$

$$RCl + NaI \rightarrow RI + NaCl \quad (2)$$

$$RI + R_{cell}(OH)_3 + NaOH \rightarrow R_{cell}(OH)_2OR + NaI + H_2O \quad (3)$$

Advances that have been made in the study of the rates of reaction of organic compounds have been applied to cellulose; the general discussion of reaction mechanisms by Ingold (5) is excellent in this respect. In general alkali cellulose attacks alkyl halides by S_N2 nucleophilic bimolecular mechanisms. If the substituent is large, solvent effects enter in, and the organic liquid used must be suited to the particular reaction. Thus methyl or ethyl chloride reacts by the S_N2 mechanism, but chloroacetic acid approaches an S_N1 mechanism. One may ethylate in ethyl chloride or benzene, but one would prefer isopropyl alcohol or *t*-butyl alcohol as a liquid in carboxymethylation. Postulated mechanisms for ethylation and its side reactions have been verified with the aid of an analog computer (1).

In a note of correction to Part II, page 898, the preparation of 2-methylcellulose was claimed by Sugihara and Wolfrom. They obtained indications of this from a 12% overall yield (6). In another correction, although ethylenimine was said to serve as its own catalyst in reacting with cellulose (in Part II, p. 958), in reality acid catalysis is required. Early German ethylenimine preparations contained sulfuric acid residues that served to catalyze the reaction.

Ethers derived from α, β-unsaturated compounds, especially acrylonitrile, have been the object of much research during the last decade; they are described in Section D of this Chapter XVII. Crosslinked ethers are covered in Section E of this Chapter XVII. Information on some recent developments in the uses of cellulose ethers may be found in Chapter XIX, especially in Sections A, B, D, E, H, and I.

1. Sodium Carboxymethylcellulose

The development of organic-liquid-slurry techniques, in a process and plant that might be used interchangeably to make several ethers, was an important advance. It made possible *in situ* preparation of alkali cellulose, attainment of high degree of substitution (DS) without gelation, and ready purification of highly substituted (DS 1.2) products. Purified products of DS not greater than 0.95 have found use as food additives and where high processing temperatures are encountered.

Nonuniform sodium carboxymethylcellulose is actually desirable in paper processing, for example (see Chapter XIX-D). Here the product may contain fibers, some of which are bound together by gels, but this is actually advantageous in the end use. Such products can be made by the conventional techniques.

Because the reaction mechanism of cellulose with chloroacetic acid or with sodium chloroacetate is intermediate between S_N1 and S_N2, the solvent of choice is likely to consist of branched-chain alcohols. This poses problems of liquid recovery and handling that appear to be trade secrets.

In German cellulose ether production it has been conventional to use the steeping press that is common to viscose technology to prepare the alkali cellulose and then to shred and mill the alkali cellulose with the reactants. The E.I. du Pont de Nemours Company proposed removal of the alkali cellulose from its slurry by continuous drainage through rolls. Kalle A.G. has described a process wherein a slurry is drained and passed through a double-screw mechanism, wherein high pressure and tearing action are exerted by rectangular profiles of the Mapre screws (7). Earlier the Du Pont Company used a screw with a toothed housing that reciprocated along the axis of the screw (8). If a small amount of nonsolvent organic liquid is introduced into such an apparatus, working is facilitated and the chloroacetic acid or salt is readily distributed. Excellent quality is therefore obtained. Fine division of the cellulose permits more uniform mixing. The water will be partitioned between the water phase and the organic phase, so a somewhat polar organic phase is desirable.

(a) SOLUTION PROPERTIES

Sodium carboxymethylcellulose is the sodium salt of a weak carboxylic acid. At a DS of approximately 0.5 the p*K* is 4, at a DS of 0.8 the p*K* is 4.4. At pH 8.25 a dilute solution is at its equivalence point. At pH 7 sodium carboxymethylcellulose is roughly 90% in the salt form, whereas at pH 5 it is roughly 10% in the salt form. The viscosity tends to a maximum at pH 6–8. This is probably due to the uncoiling of the chain because of electrostatic repulsion. The ionization of the free acid is repressed at low pH values, whereas repulsion forces at high pH values are scattered by the large number of sodium ions present. Thus the chains are coiled at the extremes of the pH scale, but uncoiled near pH 6–8 (see also Chapter XIV).

Table 2 shows the molecular weights of sodium carboxymethylcellulose and their relation to solution viscosity.

TABLE 2
Molecular Weights of Commercially Available
Sodium Carboxymethylcellulose (1)[a]

	Technical viscosity				DP value	
DS	Conc. (%)	Grade viscosity (cp)	[η] (dl/g)	Molecular weight	Calculated	Rounded off
0.68	2	17	1.9	24,600	188	200
0.78	2	30	3.0	69,200	296	300
0.77	2	520	4.8	101,000	454	500
0.72	1	1800	8.6	204,000	960	1000

[a]Unpublished data of Hercules, Inc., on its viscosity types. Molecular weights are calculated from the following equation:

$$[\eta] = 1.23 \times 10^{-4} M^{0.91}$$

Solutions of sodium carboxymethylcellulose are pseudoplastic; that is, their apparent viscosities decrease as the shear rate is increased. This is due to deformation of the solvated chains. Many solutions of sodium carboxymethylcellulose are thixotropic; that is, their viscosity is time dependent after the shearing force is applied. At a low shear rate a solution may be a solid gel, yet it may become fluid as the shear is increased.

Ease of solution of sodium carboxymethylcellulose depends on fairly uniform distribution of carboxymethoxyl substitution. Probably the regions that were originally crystalline are of lower substitution than those that were amorphous. The regions of lower substitution may also be due to nonuniformity of substitution or to other causes, such as different reactivities of the hydroxyl group. The crystalline regions may aggregate to form gel centers. When agitated in solution, these centers break down, and the viscosity is decreased, but they may recombine, due to entanglement, and if the shear is decreased, they will be of different size than before. Thus the solution may be of lower or higher viscosity than before, thus accounting for the non-Newtonian behavior.

Solubility is improved by using a stronger solvent, such as dilute sodium hydroxide solution, or by increasing the DS to about 1.2. Nonthixotropic solutions can also be obtained if precautions are taken to ensure reaction uniformity. The method of measuring viscosity must be standardized in order to obtain reproducible results.

Heavy-metal salts, such as those of silver, copper, lead, or zirconium, precipitate sodium carboxymethylcellulose. Aluminum and ferric ions cause dense precipitates or, in some cases, gels. Calcium ions will not precipitate sodium carboxymethylcellulose, but calcium chloride causes haze and lowers the viscosity. Eventually if the concentration of the calcium chloride is increased, the sodium carboxymethylcellulose will be salted out of solution.

Sodium chloride lowers viscosity and increases thixotropy. The viscosity is higher when salt is added to the sodium carboxymethylcellulose solution than when the ether is dissolved in a brine. This is a salting-out effect.

Sodium carboxymethylcellulose is physiologically inert, both internally and topically, but it should not be ingested without adequate amounts of water so that intestinal blockage does not occur.

Enzymatic degradation of a series of sodium carboxymethylcellulose samples of increasing DS showed that stability increased as the substitution was increased. Scission occurred only where three or more unsubstituted anhydroglucose rings were connected in series. Thus protection, by the use of such agents as phenylmercuric acetate, is necessary in most cases.

(b) USES

The uses of sodium carboxymethylcellulose are as follows at the present time: in detergents, 38%; foods, 14%; drilling muds, 13%; textiles, 11%; paper size, 8%; pharmaceuticals, 8%; paints, 3%; and miscellaneous uses, 5% (1).

Sodium carboxymethylcellulose is used as a soil-suspending "antiredeposition" agent in detergents. It is selectively absorbed by the fibers and then ionizes to give to the cloth a negative charge that repels dirt particles. It is used as a laundry and warp size for textiles, in textile printing pastes, in oil-well-drilling fluids, in some emulsion paints, as a surface size for paper and boxboard, and in ceramics and glazes. Purified sodium carboxymethylcellulose is used as a thickener and a protective colloid in foods and as a suspending agent and thickener in lotions and cosmetics.

United States production was about 69 million lb per year in 1970. Most of that produced is the technical product containing salt. Estimated world production was about 152 million lb per year in 1970. The market price is from $0.43 to $0.60 per pound. It is produced in most of the industrial countries of the world. Manufacturers have been listed in reference 1.

2. Methyl Ethers of Cellulose

(a) METHYLCELLULOSE

Methylcellulose is cold-water soluble. Solutions of it generally gel when they are heated. It is manufactured primarily in a DS range of 1.6–2.0. Mixed ethers containing both hydroxyalkoxyl and methoxyl are discussed below in the subsection on (hydroxyalkyl)(methyl)celluloses.

In the German Federal Republic alkali cellulose for methylcellulose is traditionally prepared by the steeping and pressing method that is characteristic of the viscose process. A steeping press containing separation plates and having a movable head to which hydraulic pressure may be applied is used. Sheets of cellulose are placed between the leaves, and 42% aqueous sodium hydroxide is slowly added. After steeping, the press is closed and the sheets are pressed to a 2.6 : 1 to 2.7 : 1 press ratio (i.e., 2.6–2.7 times the weight of the original sheets); the sodium hydroxide : cellulose ratio is therefore approximately 0.7 : 1, and the water : cellulose ratio approximately 0.9 : 1. The sodium hydroxide content tends to be a little higher in the pulp than in the displaced solution. The alkali cellulose is then shredded and aged for several hours in order to reduce the viscosity of the final product. More recently, alkali cellulose has been prepared by slurrying the cellulose and then passing the pressed material through a double-screw Mapre press. Here the high pressure distributes the sodium hydroxide solution more uniformly, and square shoulders tear the cellulose to finer size (7). The resulting alkali cellulose may be methylated in batch equipment or fed by screws into a continuous reactor (9).

In general the quality of methylcellulose is proportional to the uniformity of the alkali cellulose from which it is made.

Methylation and the side reactions that accompany it are apparently S_N2 substitutions. The primary reaction is shown by equation 4. The equilibrium constant is on the order of 5×10^{-10}.

$$R_{cell}(OH)_3 + CH_3Cl \rightleftharpoons R_{cell}(OH)_2OCH_3 + HCl \xrightarrow{NaOH} NaCl + H_2O \qquad (4)$$

Methanol is formed as a by-product according to equation 5.

$$HOH + CH_3Cl \rightleftharpoons CH_3OH + HCl \xrightarrow{NaOH} NaCl + H_2O \qquad (5)$$

The equilibrium constant is on the order of 4×10^{-4}. When the concentration of methanol in the reactor becomes appreciable, formation of methyl ether, a by-product, takes place according to equations 6 and 7. The equilibrium constant is on the order of 2×10^{-3}.

$$CH_3OH \rightleftharpoons CH_3O^- + H^+ \qquad (6)$$

$$CH_3O^- + CH_3Cl \xrightarrow{\text{fast}} CH_3OCH_3 + Cl^- \qquad (7)$$

$$\downarrow NaOH$$

$$NaCl + H_2O$$

The three reactions in order (methylation, methanol formation, dimethyl ether formation) are decreasingly endothermic, but they are made exothermic by the dilution and neutralization of sodium hydroxide. The process as a whole is exothermic.

In general the reactions are carried on in jacketed, agitated, nickel-lined autoclaves. I. G. Farbenindustrie A.G. developed an interesting process wherein the etherification takes place in a nickel pressure vessel and the methyl chloride is circulated by a blower. The vapor from the reactors is cooled to remove methanol and water, while the methyl chloride is returned through a heater to the reactor. The removal of methanol and water reduces the amount of by-product formation, but the overall efficiency is about the same as in processes in which the methanol remains in the reactor and is converted to dimethyl ether. In a typical reaction alkali cellulose, prepared by steeping and pressing, was loaded into the reactor, and a mixture of methyl chloride and cellulose (1.0 : 1) was circulated at 90–105°C and 4.92-kg/cm^2 (70-psig) pressure. When the reaction was completed, a product of DS 1.5 (methoxyl content 25.2%) resulted. The efficiency of methyl chloride utilization was 47% (1).

In the continuous process the alkali cellulose passes into the reactor and out through a rotary backpressure valve that feeds the product to a wash tank. The volatiles that are flashed off are condensed and returned to the feed end of the reactor. The temperature of the reactor is controlled through circulating hot water and by a cooling coil mounted in a dome. The material leaving the wash tank is then fed by a screw to a conventional washing and finishing system. The circulating methyl chloride has a constant dimethyl ether content, and alkylene oxides may also be added as reactants.

(b) (HYDROXYALKYL)(METHYL)CELLULOSES

(Hydroxyethyl)(methyl)cellulose has been manufactured both in England and in the German Federal Republic by the reaction of alkali cellulose first with ethylene oxide and then with methyl chloride. The effect of hydroxyethoxyl substitution is to make the molecule more disordered. Although the hydroxyethoxyl content, as well as DS is only about 0.12–0.17, solubility at a lower methoxyl content, improved solution appearance, and improved salt compatibility result. The gelation temperature may be raised from near 50 to near 70°C.

$$OH^- + \underbrace{CH_2CH_2O}_{} \xrightarrow{\text{slow}} {}^-OCH_2CH_2OH \qquad (8)$$

$$R_{cell}(OH)_3 + {}^-OCH_2CH_2OH \xrightarrow{\text{fast}} R_{cell}(OH)_2OCH_2CH_2OH + OH^- \qquad (9)$$

The rate is proportional to [epoxide] [OH$^-$]. The sodium hydroxide is not consumed. Ethylene glycol and poly(ethylene oxide) are formed as by-products, and the epoxide may react further with the hydroxyl group.

The use of 1,2-propylene oxide in place of ethylene oxide has led to a family of products that are still farther from true methylcellulose in properties. (Hydroxypropyl)(methyl)cellulose was first manufactured in the United States and is now manufactured in Great Britain, in Japan, and in the Soviet Union. Propylene oxide reacts in a more controllable manner than does ethylene oxide, and hydroxypropylation leads to a secondary hydroxyl group on the chain; runaway polymerizations do not occur, and excess propylene oxide may be used as a solvent if desired.

$$OH^- + \underbrace{OCH_2CH}CH_3 \xrightarrow{\text{slow}} {}^-OCH_2CH(OH)CH_3 \qquad (10)$$

$$R_{cell}(OH)_3 + {}^-OCH_2CH(OH)CH_3 \xrightarrow{\text{fast}} R_{cell}(OH)_2OCH_2CH(OH)CH_3 + OH^- \qquad (11)$$

In a typical preparation alkali cellulose containing 1.2–2.8 moles of sodium hydroxide per mole of cellulose is subjected to an alkyl halide of one to two carbon atoms and to an alkylene oxide of two to four carbon atoms at 20–140°C until all of the alkyl halide and from 0.1 to 0.5 parts of the alkylene oxide have reacted. The mixed ether that results yields solutions that have a higher gelation temperature than methylcellulose, ranging from 50 to 100°C, or higher, greatly improved salt compatibility and useful thermal properties. The latter may be due in part to methoxyl capping of some hydroxypropoxyl groups.

(Hydroxypropyl)(methyl)celluloses are more readily dissolved in water than methylcelluloses, when cooling is inadequate. (Hydroxypropyl)(methyl)celluloses are soluble in glacial acetic acid, in methylene chloride–methanol, and in methanol–water mixtures. Thermoplastic (hydroxypropyl)(methyl)cellulose (Methocel®60 HG) is soluble in such hot solvents as glycols, glycol ether, glycerol acetates, and ethanolamines. These hot solvents are also suitable plasticizers for thermoplastic film forming, molding, and extrusion shaping.

Recently a (hydroxypropyl)(methyl)cellulose product that contains close to DS 1.0 methoxyl and close to MS (molar substitution) 1.0 of hydroxypropoxyl has been marketed and has found wide acceptance in the paint industry (10).

(c) DISTRIBUTION OF SUBSTITUENTS

Smith and Smith studied the substituent distribution for methylcellulose and found the ratios of the relative reactivities of the $C_{(2)}$, $C_{(3)}$, and $C_{(6)}$ hydroxyls of the anhydroglucose unit to be on the order of 1.45 : 1 : 1.5, respectively (1, 11).

(d) SOLUTION PROPERTIES

The recommended procedure for dissolving methylcellulose has been to disperse it in hot water and then to cool the dispersion while it is being mixed. Methylcellulose powder cakes and lumps when added to cold water. Granular products or aggregated products disperse more readily but are slow to dissolve. Materials may be slurried in alcohols or other solvents. Surface treatments with surface-active agents, with plasticizers, with water-repelling agents, or with crosslinking agents, such as glyoxal (12, 13), have also been used to make them more readily dispersible. The addition of alkali (14) or alkaline earth (15) metal salts or other bases, such as ammonia, to glyoxal-treated products leads to rapid rates of solution.

When the water sheath is lost from methylcellulose molecules, association and aggregation occur, resulting in gelation. Heating or the addition of salts produces the same effect. The effect of gelation temperature as a function of the extent of substitution is shown in Table 3.

TABLE 3
Relation of Gelation Temperature to DS in Methylcelluloses and to MS in (Hydroxypropyl)(methyl)celluloses (1)

Methoxyl DS	Hydroxypropoxyl MS	Gelation Temperature (°C)
1.8	–	56
1.8	0.15	60
1.7	0.1	70
1.4	0.2	80–90

A methylcellulose with an intrinsic viscosity of 3.8 dl/g and a grade viscosity of 400 cp is fully dissolved at temperature below 10°C and shows little tendency toward dissociation at temperatures below 30°C. Conversely, it does not develop viscosity, if a slurry is cooled, until below 30°. Microscopically, aggregation begins near 40°C on heating or in the presence of appreciable amounts of salts, but precise optical measurements may detect such phenomena at lower temperatures.

While dissolved, the molecules have a water sheath that is hydrogen bonded to them. The dipole moment lies along the axis. The molecules are slightly compressed, in random coil configuration. If the temperature is increased, the bonds are broken, and the molecules are free to rotate. The energy input is moderately positive, and the entropy change strongly so. The molecules, freed from water, associate or bond to one another, evolving a large amount of energy, but the entropy change is weakly negative, so the overall effect is an evolution of heat with but slight entropy change. Curiously, heat is evolved both during solution and during association. Cooling aids solution stability. When the molecules associate, they fold in an accordionlike manner. There is neutralization of the dipole components that lie along the axis, and the transverse components add. Such associated groups contain up to 96% water, so that the composition of a gel centrifuged from a 2% aqueous solution will be similar to that of the solution itself. An associated group can contain as many as 4000 macromolecules, but the effect on the viscosity is an increase of only about 15%. The coils can contain fixed water. Water molecules also associate with external points to yield a slightly cationic mass with a positive exterior dipole. This dipole will collapse in the presence of anionic solutes. The zeta potential, which is already low, will decrease still more and aggregation will occur. Alcohols, on the contrary, stabilize the zeta potential.

Either cations or anions may decrease the hydration of methylcellulose molecules in aqueous solution and may lower the gelation temperature. The cation tolerance seems to be in the following order: $Pb^{2+} > Zn^{2+} > Cu^{2+} > Fe^{3+} > NH_4^+ > Ca^{2+} > Ba^{2+} > K^+ > Mg^{2+} > Na^+ > Al^{3+}$. The order of tolerance for anions is as follows: $I^- > CNS^- >$ borate $> NO_3^- > CO_3^{2-} > Cl^- >$ acetate $>$ tartrate $> SO_4^{2-} > PO_4^{3-}$.

A shift to the left will raise the gel point, a shift to the right lower it. Some data for salt and sucrose tolerance are shown in Table 4.

TABLE 4.
Salt and Sucrose Tolerances of Methylcellulose and (Hydroxypropyl)(methyl)celluloses (1)

Parameter	Methylcellulose		(Hydroxypropyl)(methyl)cellulose			
Methoxyl DS	1.8		1.7		1.5	
Hydroxypropoxyl MS			0.2		0.25	
Viscosity grade (cp)	15	400	50	4000	100	4000
Intrinsic viscosity						
[η] (dl/g)	1.6	3.8	2.2	3.8	2.6	3.8
Additive:[a]	11	7	17	11	19	12
NaCl						
Na_3PO_4	2.9	2.6	3.9	3.5	4.7	4.3
$Al_2(SO_4)_3$	3.1	2.5	4.1	3.6	4.1	3.6
Na_2SO_4	6	4	6	4	6	4
Na_2CO_3	4	3	5	4	4	4
Sucrose	100	65	120	80	160	115

[a] Tolerance measured in grams of salt per 100 grams of 2% solution

Table 5 shows viscosity-concentration data for powdered and granular methylcellulose; the curves that were given in Part II, page 932, are for fibrous methylcellulose. The viscosity is usually due in most part to the longest chains or molecules. The intrinsic viscosity and the related grade viscosity are measures of chain length or degree of polymerization (DP) and hence of molecular weight.

TABLE 5
Viscosity–Concentration Data for Methylcelluloses and (Hydroxyalkyl)(methyl)celluloses (1)

Grade viscosity of 2% solution at 20°C (cp)	Intrinsic viscosity [η] (dl/g)	Concentration (%)						
		1	2	3	4	6	10	12
15	1.6	6	15	60	80	320	2500	7500
100	2.6	14	100	480	1600	13000		
400	3.8	32	400	2400	11000			
1500	5.5	80	1500	12000	30000			
4000	7.3	170	4000	42000				
8000	8.9	300	8000	85000				
15000	10.4	470	15000					

Molecular weights for methylcellulose have been calculated from light-scattering, ultracentrifugation, osmotic-pressure, and viscometric data. The methods have unfortunately only been applied to unfractionated or superficially fractionated materials, so they only partly describe the materials tested. Because factors for calculation differ among different workers and methods, molecular weight values may differ by several hundred percent. For example, weight-average molecular weight can be twice the number-average value. Intrinsic viscosity is readily determined with a dilution viscometer, and so-called molecular weights are of little practical use unless the grade and the intrinsic

TABLE 6
Viscosities of Methylcellulose of Various Molecular Weights, DS 1.8 (1)

Grade viscosity of 2% solution at 20°C (cp)	Intrinsic viscosity [η] (dl/g)	Osmotic number-average degree of polymerization $\overline{DP}_n$	Osmotic molecular weight $\overline{M}_n \times 10^{-3}$
10	1.4	70	13
40	2.05	110	20
100	2.65	140	26
400	3.90	220	41
1500	5.7	340	63
4000	7.5	460	86
8000	9.3	580	110
15000	11.0	650	130
19000	12.0	750	140

Note: weight-average $\overline{DP}_w$ and $\overline{M}_w$ values would be about twice the number-average values.

viscosity are given also. Table 6 shows the relationship between viscosity and number-average molecular weight for methylcellulose.

The intrinsic viscosity of methylcellulose is said to be related to the number-average molecular weight by the following equation:

$$[\eta]_{20°C} = 0.0022(\overline{M}_n)^{0.09} \text{ dl/g.} \tag{12}$$

Table 7 shows the general properties of methylcellulose and (hydroxypropyl)(methyl)cellulose solutions.

TABLE 7
General Properties of Methylcellulose and (Hydroxypropyl)(methyl)cellulose Solutions (1)

Property	Methyl-cellulose	(Hydroxypropyl)(methyl)cellulose	
Methoxyl DS	1.8	1.7	1.5
Hydroxypropoxyl MS		0.2	0.25
Specific gravity (20/4°C):			
1%	1.0012		
5%	1.0017		
10%	1.0245		
Refractive index (2% solution) n_D^{20}	1.336	1.337	1.336
Partial specific volume (cc/g), 4000 cp, $[\eta]$ 3.8 dl/g	0.725	0.734	0.717
Freezing point (°C)	0.0		
Surface tension $\gamma_{25°C}$ (dynes/cm)	47–53	44–50	50–56
Interfacial tension, (paraffin oil)$_{25°C}$ (dynes/cm)	19–23	18–19	26–28
Specific heat (10–25% solution), 20–90°C (cal/g°C)	0.93±0.05		
Gelation temperature, 2% solution (°C)	56	65	80–90

Solutions of methylcellulose of 600-cp viscosity or higher are pseudoplastic: if shear is increased, the apparent viscosity decreases; that is, there is a greater than proportional increase in flow with an increase in shear. Logarithmic plots of shear against force are linear. Methylcellulose solutions are more pseudoplastic than (hydroxyalkyl)(methyl)cellulose solutions. Solutions of low-viscosity grades tend to be thixotropic at high concentrations; that is, at a given shear rate the viscosity decreases as the time of shear increases.

Viscosity and concentration may be related by Philippoff's equation,

$$\eta_{\text{rel}} = \left(1 + [\eta]\frac{c}{8}\right)^8 \tag{13}$$

Viscosity and temperature are related by the expression

$$\log \eta = A + \frac{B}{T} \tag{14}$$

Where A and B are constants, η is the relative viscosity, and T is the absolute temperature in degrees Kelvin.

If methylcellulose is degraded, chain breakage and reduction in intrinsic viscosity result. The number of reducing end groups increases during acid hydrolysis. Methylcellulose is stable in alkaline solution if oxygen is absent, but viscosity loss occurs if oxygen is present. Methylcellulose is nonionic and fairly stable at pH 2–12.

Methylcellulose of DS 2 does not readily support the growth of bacteria that elaborate enzymes capable of degrading cellulose. But the solutions are not bacteriostatic, and, if attack occurs, viscosity loss results. Attack may occur between a substituted and an unsubstituted anhydroglucose ring; thus residual components of the original cellulose provide foci for attack. Good uniformity of substitution is most desirable.

(e) USES

The (hydroxyalkyl)(methyl)celluloses and methylcelluloses have a great many uses, and although the amount used for any particular purpose may be very small, the function of this small quantity may be all-important. Information on the proper selection of materials for various properties desired has been published in reference 1. Methylcellulose is used as a thickener, as a protective colloid, and, under approval, for food uses (described below), but the utility of (hydroxypropyl)(methyl)celluloses is greater, particularly if a dissolving temperature higher than 15°C is required.

In adhesives high-intrinsic-viscosity material provides binding, but (hydroxypropyl)(methyl)cellulose is preferred if higher temperature and ease of dissolving are important factors in selection. Adhesives and binders are used in laminating, leather pasting, and in manufacturing pressure-sensitive tapes. In agriculture low intrinsic-viscosity methylcellulose is used for thickening and dispersing seeds and powders, whereas high-intrinsic-viscosity (hydroxypropyl)(methyl)cellulose is used as a sticker for sprays and dusts. The latter product is also used in asphalt emulsions and calking compounds.

In the manufacture of ceramics low-intrinsic-viscosity methylcellulose reduces flocculation and improves viscosity and shrink resistance; its gelation characteristics permit heavy coats to be applied in each application. The water requirement of refractories is reduced and uniformity is improved, yet waste is decreased. (Hydroxypropyl)(methyl)cellulose is preferred in tile mortars, plastic mixes, and cements. Low-intrinsic-viscosity methylcellulose may be used to control the setting time of construction cements and to provide added initial strength.

(Hydroxypropyl)(methyl)cellulose is preferred in latex and alkyd paints. Advantages include high wet-rub resistance, good flow and leveling, storage and pH stability, color stability, and tolerance of heavy-metal salts. Methylcellulose is used in multicolor lacquers as a thickener. (Hydroxypropyl)(methyl)cellulose, (hydroxybutyl)(methyl)cellulose, and methylcellulose of DS 1.8 or higher are used both in flush-off and in scrape-off paint removers; solubility in water–alcohol–dichloromethane mixtures is required.

(Hydroxypropyl)(methyl)cellulose of low intrinsic viscosity is used in the production of suspension or emulsion polymers, where wetting and dispersing properties are needed. High-intrinsic-viscosity products are used in gluing plywood, thickening and binding water-based printing pastes, and in the textile industry for adhesion, whereas low-intrinsic-viscosity materials are used for sizing and coating. Various types are used in paper treatment, binding pencils and crayons, pharmaceutical preparations, tableting, as well as for weight-reducing pills, bulk laxatives, suspensions, and emulsions. Methylcellulose swells in water to 40 times its normal volume, is nonnutritive, and is not metabolized. It is surface active in ointments. It may also be used as a binder and film former in producing reconstituted tobacco sheets (16).

(Hydroxypropyl)(methyl)cellulose thickens dehydrated fruits and vegetables and makes them more easily reconstituted; it thickens dietetic preparations and stabilizes frozen foods to prevent water separation. It holds together meat patties and provides moisture, which prevents burning.

High-strength, cold-water-soluble film or sheeting can be cast, coated, rolled, or extruded from solutions of methylcellulose or (hydroxypropyl)(methyl)cellulose. Such films are clear and impervious to oils, greases, and other organic liquids; specific formulations may be heat sealed on standard equipment (see also Section E of Chapter XIX).

Methylcellulose may be identified by its infrared-absorption spectrum. The principal bands are 2.92 μ hydroxyl; 3.4 μ weak C–H deformation; 6.86 μ symmetrical methyl deformation; 7.3 μ C–H deformation; 8.8–9.4 μ, CO, strong; 10.62 μ characteristic methyl deformation; 11.7–12 μ, miscellaneous bands (7).

Methoxyl content is determined by the modified Zeisel method (17) hydroxyethoxyl content by the method of Morgan (18), and hydroxypropoxyl content by a modification of the chromic acid oxidation method of Lemieux and Purves (19, 20). Cellulose ethers can be recognized in mixtures by the use of the anthrone method of Dreywood, as modified by Black. In this method a solution of anthrone in concentrated sulfuric acid gives a permanent green color with carbohydrates. The intensity is measured spectrophotometrically (21, 22).

It is estimated that world production of methylcellulose and its derivatives was about 44 million lb per year in 1970. The price range is from $0.69 to $0.84 per pound (see reference 1 for producers).

(f) REGULATIONS FOR FOOD USES

In the United States methylcellulose USP, with a methoxyl content within the limits of 27.5–31.5% on a dry-weight basis, is included in the list of the food additives that are generally recognized as safe and therefore exempt from the Food Additives Amendment of 1958 (23). In 1959 USP-grade methylcellulose was permitted as an optional emulsifying agent under the Definitions and Standards for Identity for French and Salad Dressings (24). In 1960 a food additive regulation in the *Federal Register* (25) permitted the use of (hydroxypropyl)(methyl)cellulose NF (propylene glycol ether of methylcellulose, NF) as an emulsifier, film former, protective colloid, stabilizer, suspending agent, or thickener. In the same year it was recognized as an optional emulsifier in French dressings (26).

In Great Britain "special" grades of methylcellulose and (hydroxypropyl)(methyl)cellulose that have been manufactured in accordance with the Food and Drug Act of 1955 and comply with the *British Pharmacopoeia* and *British Pharmaceutical Codex* may be used in foods. Specifications include the following: arsenic, < 1 ppm; copper, < 15 ppm; lead, < 5 ppm; zinc, < 50 ppm; and the absence of dimethyl sulfate. The products were added to the *Canadian Food and Drug Directorate* and to the *Food Chemicals Codex* (27), (see reference 28 for further details).

In the German Federal Republic cellulose ethers are not permitted in foods (29), although sodium carboxymethylcellulose was formerly permitted in ice creams.

The use of methylcellulose or other colloids in parenteral injections is not recommended. Such colloids may mimic the effects of bacterial polysaccharides and cause hemolytic anemia (30).

3. Hydroxyethylcellulose

When Part II of *Cellulose and Cellulose Derivatives* was written, the development of hydroxyethylcellulose was in the process of moving from a low-substituted material, soluble in aqueous sodium hydroxide solution, to a water-soluble product of higher substitution. Since that time the utility and favorable economics of products of much higher substitution (MS 2.5 or more) have been realized, and hydroxyethylcellulose now enjoys considerable industrial utility.

Hydroxyethylcellulose results from the reaction of ethylene oxide with cellulose in the presence of relatively small amounts of sodium hydroxide, which serves as a catalyst and is not consumed. The use of organic liquids as reaction media and stepwise operation have permitted simple and economical operation.

Hydroxyethylcellulose does not gel in water solutions, although in the presence of high concentrations of salts it may. It is used in the water-soluble range of substitution, from MS 1.8 to 3.0.

(a) PREPARATION

In the preparation of hydroxyethylcellulose the first portion of the reaction may be similar to the first stage of production of (hydroxyethyl)(methyl)cellulose (equations 8 and 9). Handling ethylene oxide involves hazards of explosion and of runaway polymerization. One must avoid contamination of stored epoxide by acids, alkalies, or salts. Excess catalyst and excess epoxide must be avoided. Excellent cooling must be provided.

The terminal hydroxyl of the substituent is primary, and it can easily react with more epoxide to form pendant oxyethylene chains:

$$R_{cell}(OH)_2OCH_2CH_2OH + n\underbrace{CH_2CH_2O} \xrightarrow[\text{fast}]{OH^-} R_{cell}(OH)_2(OCH_2CH_2)_{n+1}(OH) \quad (15)$$

Water competes with cellulose for the reagent:

$$HOH + {}^-OCH_2CH_2OH \xrightarrow{\text{fast}} HOCH_2CH_2OH + OH^- \quad (16)$$

The rate is again proportional to [epoxide] [OH^-]. The by-product is ethylene glycol. Poly(ethylene oxide) may also result. In neutral or acid media different mechanisms are given. The epoxide ring is opened by salts. The reaction mechanism may also depend on the degree of branching of a solvent alcohol, if used.

In a typical preparation 60 parts of 20% aqueous sodium hydroxide is added to 30 parts by weight of shredded cotton linters stirred in 850 parts by volume of *t*-butyl alcohol or acetone, and 17.8 parts of ethylene oxide dissolved in 50 parts of *t*-butyl alcohol or isopropyl alcohol is added. The reaction is carried out at reflux temperature, the liquor is drained, and the product is neutralized with acetic acid and then washed with methanol or acetone to remove sodium acetate. The cellulose remains in particulate form. A stepwise process may be used wherein, after partial reaction, most of the sodium hydroxide is removed by neutralization and washing, and the reaction is completed at high efficiency by having only a small portion of the initial sodium hydroxide present (31).

The best combination of ethylene oxide efficiency and product solubility is obtained by maintaining the sodium hydroxide : cellulose weight ratio in the range from 0.3 : 1 to 1.0 : 1, and the water : cellulose ratio in the range from 1.2 : 1 to 2.5 : 1. If a small amount of ethylene oxide is used at first to yield a low-DS product, the remainder of the reaction may be carried out by using a much lower sodium hydroxide : cellulose ratio than would otherwise be required to give a satisfactory product.

(b) DISTRIBUTION OF SUBSTITUENTS

Because pendant chains are formed, the DS differs from, and is less than the MS. Water-soluble commercial hydroxyethylcellulose products have MS values ranging from 1.8 to 3.0. An MS value of 1.0 is the lower limiting value for water solubility. At the minimum MS the DS is about 0.64 and the unsubstituted anhydroglucose residues may vary from one to six units in length; thus a single molecule can contain both water-soluble hydroxyethylcellulose and unsubstituted cellulose segments. Wirick (32) has proposed a *substitution index*, the percentage of substituted anhydroglucose units, as a measure of enzyme resistance. Scission occurs rapidly at unsubstituted sequences, but more slowly at links of substituted and unsubstituted units; the latter type of breakage is the case for methylcellulose (32).

Substituent distribution has been studied both by Stratta (33) and by Wirick (32). Stratta, studying material of lower substitution, reported slightly lower DS and longer chains than Wirick. Wirick found that the relative reactivities of the three cellulosic hydroxyls and the terminal end group are in the order 3:1:10:10 for the $C_{(2)}$, $C_{(3)}$, and $C_{(6)}$ and terminal hydroxyls, respectively. Stratta found for MS 1.0 the following relative reactivities: $C_{(2)}$, 3.4; $C_{(3)}$, 1; $C_{(6)}$, 13.7 (1, 33). Table 8 presents a combination of the data of Stratta and of Wirick with the older data of Morgan; it is thus a modification of Table 18, page 900, of Part II.

The ratio of MS to DS is the average length of the pendant chains. The DS increases slowly with increased ethylene oxide consumption, whereas the proportion of epoxide that contributes to DS steadily decreases. About one-half the ethylene oxide reacts with the cellulose, and the rest is converted to by-product glycols. The length of the poly(ethylene oxide) chains increases parabolically with MS. Since most preparations of hydroxyethylcellulose have insufficient DS to avoid degradation by cellulolytic enzymes, bacteriostats, such as phenylmercuric acetate, a toxic compound, are generally used.

(c) SOLUTION PROPERTIES

Hydroxyethylcellulose is nonionic, soluble in both hot and cold water, and has good tolerance for electrolytes. Specially dispersible grades are in nonswollen form because of the use of organic liquids during reaction and are readily dispersed, particularly if they have been prewet with a nonsolvent, dry-blended with other powders, or surface-treated with glyoxal or other crosslinking agent. In the latter case viscosity is developed rapidly, if the solution, dispersed at an

TABLE 8
Analysis of Products Obtained at Various Mole Ratios of Ethylene Oxide to Anhydroglucose Units (1, 32, 33)

Moles of ethylene oxide per mole of anhydroglucose	MS	DS		MS/DS (chain length)		Efficiency (%)[a]		Soluble in solvent
		Stratta (33)	Wirick (32)	Stratta (33)	Wirick (32)	MS	DS	
0.30	0.18	0.13		1.39		60	43	7% NaOH
0.50	0.25	0.20		1.25		51	40	7% NaOH
0.75	0.42	0.31		1.36		57	41	Swollen in cold water
1.0	0.50	0.38		1.31		50–69	38	
1.50	0.73	0.48		1.52		48	52	Swollen by some solvents
2.1	1.0	0.66	0.70	1.52	1.35	47	31	Water
2.6	1.35	0.85		1.60		51	32	
2.9	1.4	0.85		1.66		46	29	
3.05	1.44	0.83		1.74		47	27	Nearly soluble in acetic acid, pyridine
3.2	1.5	0.88	0.95	1.7	1.5	47	27	
4.05	1.84	0.96		1.91		46	24	
5.8	2.5	1.17		2.14		43	20	
7.0	3.0	1.33	1.4	2.27	2.1	43	19	
8.4	3.5	1.49		2.36		41	17	
9.7	4.0	1.65	1.6	2.42	2.4	41	17	
10.0	4.1	1.66		2.46		41	16	

[a] Data from Stratta (33).

acid pH, is made alkaline by the addition of ammonium hydroxide or other alkali.

Hydroxyethylcellulose manufacturers present salt-compatibility data according to the classical method of Freundlich: a saturated salt solution is mixed with a colloid, and a "yes–no" type of result is obtained. Thus hydroxyethylcellulose is not precipitated from solution when chloride, dichromate, sulfite, nitrate, or perborate salt is added. It is precipitated by ferrocyanide, phosphate, most sulfates, sulfites, and thiosulfates. Sulfates are more compatible at 10% concentration than at saturation. The method consists of adding 1 ml of a 2% or stronger solution of the polymer to 15 ml of saturated salt solution and observing the presence or the absence of a precipitate.

In most practical cases however, the amount of a cellulose ether in a formulation exceeds the amount of salt, so that the above test is not representative of use conditions, unless brine thickening is involved. Hydroxyethylcellulose is more compatible with salt solutions of moderate concentration than is methylcellulose. Hydroxyethylcellulose is more soluble in solutions that contain multivalent positive ions than in solutions that contain multivalent negative ions (1).

Hydroxyethylcellulose solutions are stable over a wide pH range and a broad range of temperature. Low concentrations of water-soluble esters and ketones have a precipitating effect, whereas alcohols, carboxylic acids, aldehydes, amines, and polyhydric alcohols are tolerated in larger amounts. Dimethylformamide, diethylacetamide, and *N*-acetylethanolamine are solvents of lesser utility. Hydroxyethylcellulose is not thermoplastic.

Hydroxyethylcellulose was fractionated by Brown (34), who used acetone to obtain fractional precipitation from solution. The skew distribution peaked at $[\eta] = 10$ dl/g.

TABLE 9
Viscosity–Concentration Data for
Hydroxyethylcellulose, MS 1.3–2.5 (1)

Concentration %	Viscosity at 20°C (cp) for various grades							
1	2	5	15	30	280	500	2100	4000
2	10	20	100	250	2200	5800	32000	97000
3	20	60	470	1200	3300	38000		
4	50	110	1300	4500				
6	130	500	10000	12000				
8	500	1700						
10	1100	4000						

The viscosity of hydroxyethylcellulose at low shear depends on the molecular size (molecular weight) and on chain interference. Replication of viscosities depends on exact adherence to prescribed test methods. Table 9 shows viscosity-concentration data. Whether the viscosity of a so-called high-viscosity hydroxyethylcellulose will be lower, higher, or the same as that of a "low-viscosity" hydroxyethylcellulose depends on the shear applied. At low molecular weights the effect of shear is less. Viscosities are highest near pH 8 and fall off above pH 10 and below pH 6. A solution that is put through a heating and cooling cycle may have a permanent viscosity loss. The viscosity also depends on the lowest temperature to which a solution has been subjected. Hydroxyethylcellulose has a negative dependence of intrinsic viscosity on temperature, which may be due to the formation of addition compounds with the solvent or to increasing chain flexibility at higher temperatures. Data on apparent solution viscosity as a function of shear rate may be found in reference 1.

Molecular weights, and DP values calculated from them, vary according to the method of determination. Table 10 shows some typical values.

TABLE 10
Molecular Weights and DP Values for Hydroxyethylcellulose (1)

Intrinsic viscosity	Osmotic		Sedimentation		Light scattering	
$[\eta]$ (dl/g)	$M_n \times 10^{-5}$	DP	$M_s \times 10^{-5}$	DP	$M_w \times 10^{-5}$	DP
11.0	3.8	1610	5.45	2310	6.25	2650
9.5	3.1	1310	4.20	1780	5.15	2180
7.75	1.85	550	3.30	1400	3.35	1420
3.95	1.25	370	1.25	530	1.90	805

Hydroxyethylcellulose picks up at 2% moisture in 140 hr at 25°C and reaches a maximum moisture content of 30%. It is sold with a moisture content of 5%. The ash content is 4-6% when it is calculated as sodium carbonate. From 98 to 99.5% of hydroxyethylcellulose should be water soluble. The biological oxygen demand is 11,000 ppm. The surface tension of a 0.1% solution is 63.6 dynes/cm.

The time required to hydrate hydroxyethylcellulose that has been surface crosslinked depends on the pH, being lowest at pH 7-8 (25-35°C).

Because surface-crosslinked material is pH sensitive, it may be slurried at one pH and dissolved when the pH is adjusted.

(d) USES

Hydroxyethylcellulose has been used as a thickener for latex paints. Although it is subject to severe bacterial degradation, it is readily dissolved, contributes to scrub resistance, and gives low-foaming solutions. It also contributes to freeze-thaw stability, flowout, and brushout.

Hydroxyethylcellulose is also used in the emulsion polymerization of vinyl acetate, as a wet-strength aid (with glyoxal) in paper, and in sizing, as a pigment carrier in printing pastes, and in lithography. It is used as a protective colloid in electroplating, as a binder in ceramics, and in adhesives. It is used for temporary protection of rubber and plastics. Alkali-soluble hydroxyethylcellulose has been used for extruded film of good dimensional stability (see Chapter XIX-E). It is also used as a textile size.

Hydroxyethylcellulose is apparently of low oral toxicity, but may inflame the eye or sensitize the skin; hence, if used in cosmetics, it should be carefully tested. It should not be used for parenteral injections. It may be used in food packaging, but not as a food additive.

It is estimated that world production of hydroxyethylcellulose was about 20 million lb in 1970 (see reference 1 for producers).

4. Hydroxypropylcellulose

Hydroxypropylcellulose with an MS value of approximately 4 has recently been introduced on a pilot-plant scale. This material is produced by a modification of the hydroxyethylcellulose process, either in an organic diluent, such as hexane (35, 36), or in excess propylene oxide (37). Mixed (hydroxypropyl)(hydroxyethyl)cellulose can also be made (38). Hydroxypropylcellulose of this type is soluble in cold, but not in hot water; it gels from solution, essentially as a precipitate, at about 40°C. Although hydroxyethoxyl substitution improves water solubility, hydroxypropoxyl substitution improves thermoplasticity (39) and organosolubility. Thus hydroxypropylcellulose may be used as a thickener for organic solutions as well as for water systems, as a stabilizer in emulsion polymerization systems, and in food applications, in which, like (hydroxypropyl)(methyl)cellulose it is approved as an intentional food additive.

The substituent groups of hydroxypropylcellulose contain secondary hydroxyl groups that are available for further reaction. Almost all of the primary hydroxyls of the cellulose are substituted. Hydroxypropylation of cellulose to a low DS does not lead to alkali-soluble products, nor to water-soluble products until the MS is close to 4. The poly(propylene oxide) chains have a negative temperature coefficient of solubility, which causes precipitation from aqueous solution near 40°C. Similar effects are seen in organic solvents. Hydroxypropylcellulose may be soluble in water or in both water and ethanol. Other solvents for the ethanol-soluble types include lower alcohols, glycols, chlorinated aliphatics, and cyclic ethers. It is less soluble in higher alcohols, cycloalkanes, and esters.

Water-cast hydroxypropylcellulose films (MS 4) show morphological structure at all levels from molecular to supermolecular. The molecule itself has a backbone twisted into an irregular 3_1 helix and is stiff and rodlike. Microfibrils form crystallites 470 Å long and 34 Å in diameter and associate into supermolecular rodlike structures (40).

5. Aralkyl and Aryl Cellulose Ethers (1)

Benzylcellulose (phenylmethylcellulose) enjoyed brief commerical interest in Europe (see Part II, p. 949). It is soft, self-plasticized, and has a DS of 2 or less. It is substituted largely in the 6-position and, additionally, in the 2- or the 3-position. The principal problem in preparations is the removal of by-product benzyl alcohol and dibenzyl ether.

Benzhydrylcellulose (diphenylmethylcellulose) requires a special preparative technique; Stanonis and King (41) used 2-lutidine as the base.

$$R_{cell}(OH)_3 + n(C_6H_5)_2CHBr \xrightarrow{\text{2-lutidine}} R_{cell}(OH)_{3-n}[OCH(C_6H_5)_2]_n + nHBr \qquad (17)$$

The linkage is easily cleaved by sulfuric acid.

Tritylcellulose (triphenylmethylcellulose) was first prepared by Helferich and Koester (42), who heated trityl chloride with regenerated cellulose in pyridine and obtained a product of DS 1. The trityl-group substitution can be calculated from elementary analysis. The infrared spectrum was given by O'Connor and others (43); at 6.27 and 6.73 μ, C=C stretch; at 13.20 and 14.35 μ aromatic-ring bonds. The degradation rate was reported by Koz'mina and Koslov (44). The trityl group can be used to block the 6-position in cellulose, or it can be reactively removed to permit substitutions that otherwise would be hindered.

Phenylcellulose can be prepared (45) by replacing the trityl group with sodium phenoxide in phenol solvent. Material with a DS of 0.75–0.8 is partially soluble in trimethylbenzylammonium hydroxide. The product may be converted to a stable nitrate that might find utility in explosives.

References

1. Contributions by A. B. Savage, E. D. Klug, N. M. Bikales, and D. J. Stanonis, under "Cellulose Ethers," in N. M. Bikales, Ed., *Encyclopedia of Polymer Science and Technology*, Vol. 3, Interscience, New York, 1965, p. 459–549.
2. R. H. Wade and C. M. Welch, *Textile Res. J.*, **35** (10), 930 (1965).
3. J. Porath, *Arkiv Kemi*, **11**, 97 (1959).

4. R. Wade, C. M. Welch, and H. P. Bennett, U.S. Patent 3,338,662 (to U.S. Department of Agriculture) (1967).
5. C. K. Ingold, *Structure and Mechanism in Organic Chemistry,* Cornell University Press, Ithaca, N.Y. 1953.
6. J. M. Sugihara and M. L. Wolfrom, *J. Am. Chem. Soc.,* **71,** 3509 (1949).
7. Netherlands Pat. Appl. 6,606,211 (1966) (to Kalle A.G.): French Patent 1,484,825 (to Kalle A.G.) (1967)
8. J. M. Cordrey, S. T. Druckenmiller, and J. E. Lufkin, U.S. Patent 3,088,943 (to E. I. du Pont de Nemours & Co., Inc.) (1963).
9. Belgian Patent 691,814 (1966) (to Kalle A.G.); French Patent 1,516,072 (to Kalle A.G.) (1968).
10. S. M. Rodgers, B. F. Wakeman, and A. B. Savage, U.S. Patent 3,338,082 (1968) (to the Dow Chemical Co.).
11. B. F. Smith and F. Smith, unpublished data, The University of Minnesota, St. Paul.
12. I. Jullander, U.S. Patent 2,879,268 (1959) (to Mo och Dömsjo, AB).
13. J. H. Menkart and R. S. Allen, U.S. Patent 3,072,675 (1963) (to Chemical Development Corporation of Canada).
14. W. C. Chambers and M. Lee, U.S. Patent 3,356,517 (1967) (to Dow).
15. W. Dierichs and W. Sammer, U.S. Patent 3,297,583 (1967) (to Henkel & Cie., G.m.b.H.).
16. R. J. Moshy and H. M. Halter, in N. M. Bikales, Ed., *Encyclopedia of Polymer Science and Technology,* Vol. 7, Interscience, New York, 1967, pp. 223–230.
17. ASTM D1347–54 T.
18. P. W. Morgan, *Ind. Eng. Chem., Anal. Ed.,* **18,** 500 (1946).
19. R. W. Lemieux and C. B. Purves, *Can. J. Res.,* **25B,** 485 (1947).
20. ASTM D2363–65 T.
21. R. Dreywood, *Ind. Eng. Chem., Anal. Ed.,* **18,** 499 (1946).
22. C. H. Black, Jr., *Anal. Chem.,* **23,** 1792 (1951).
23. *Federal Register,* **26,** 21 CFR (Code of Federal Regulations), 938 (January 31, 1961).
24. *Federal Register,* **24,** 21 CFR, 6711 (August 19, 1959).
25. *Federal Register,* **25,** 2 CFR, 8949 (September 17, 1960).
26. *Federal Register,* **25,** 21 CFR, 8498 (September 17, 1960).
27. *Food Chemicals Codex,* 1st ed., National Academy of Sciences–National Research Council, Washington, D.C., 1966.
28. *A Bibliography of Pharmaceutical and Medical Uses of Methocel*®, Dow Chemical Company, Midland, Mich.
29. *Das Lebensmittelrecht,* C. H. Beck'sche Verlagbuchhandlung, Berlin, 1958.
30. F. S. Philips, Sloan Kettering Institute for Cancer Research, New York, personal communication.
31. E. Klug, *Stärke,* **13,** 429 (1961).
32. M. G. Wirick, *J. Polymer Sci.,* [A-1], **6,** 1705 (1968).
33. J. J. Stratta, *Tappi,* **46,** 717 (1963).
34. W. Brown, *Arkiv Kemi,* **18,** 227 (1961).
35. E. D. Klug, U.S. Patent 3,357,971 (1967) (to Hercules).

36. R. G. Bishop, U.S. Patent 3,351,583 (1967) (to Hercules).
37. E. D. Klug, U.S. Patent 3,278,520 (1966) (to Hercules).
38. E. D. Klug, U.S. Patent 3,278,521 (1966) (to Hercules).
39. E. D. Klug, U.S. Patent 3,314,809 (1967) (to Hercules).
40. R. J. Samuels, *J. Polymer Sci.*, [A-2], 7, 1197 (1969).
41. D. J. Stanonis and W. D. King, *Textile Res. J.*, **30**, 802 (1960).
42. B. Helferich and H. Koester, *Ber.*, **57**, 587 (1924).
43. R. T. O'Connor, E. F. DuPre, and E. R. McCall, *Textile Res. J.*, **28**, 542 (1958).
44. O. P. Koz'mina and M. P. Koslov, *Vysokomol. Soedin.*, **5** (7), 1054 (1963); through *Chem. Abstracts.*, **59**, 14187 (1963).
45. Z. A. Rogovin and T. V. Vladimirova, *Khim. Nauk. i Prom.*, **2**, 527 (1957).

D. ETHERS FROM α,β-UNSATURATED COMPOUNDS

Norbert M. Bikales

This chapter deals with the cellulose ethers derived by addition to cellulose of olefins activated by polar substituent groups (see also Part II, pp. 954–958). For a discussion of other ethers see Section C in this Chapter XVII.

Cellulose reacts with various α,β-unsaturated compounds to form substituted ethyl ethers.

$$R_{cell}OH + CH_2 = CRY \rightarrow R_{cell}OCH_2CHRY \qquad (1)$$

The reaction proceeds most readily when R is H and Y is a strongly electron-withdrawing group, such as $-CN$, $-CONH_2$, or $-SO_2R$. The most widely studied reaction has been that of cellulose with acrylonitrile to form β-cyanoethylcellulose.

$$R_{cell}OH + CH_2 = CHCN \overset{OH^-}{\rightleftharpoons} R_{cell}OCH_2CH_2CN \qquad (2)$$

Cyanoethylation requires the presence of a strongly basic catalyst (usually sodium hydroxide), and it is known to be reversible (1, 2). Cyanoethylcelluloses of various degrees of substitution (DS) and differing considerably in properties from each other have been prepared.

Other α,β-unsaturated nitriles, as well as amides, esters, sulfones, and sulfonic acids, have also been reacted. The sulfones in particular have been of interest in connection with crosslinking investigations (see Section E of this Chapter XVII).

1. Ethers from Acrylonitrile

The cyanoethylation of cellulose, in particular cotton, has received much attention during the last two decades. Although already mentioned in a German patent in 1938, its investigation was pioneered in the 1940s by MacGregor in Great Britain and by Houtz and Stallings in the United States (see Chapter IX–E). Beginning in about 1953, a major effort was carried out to bring about the commercialization of partially cyanoethylated cotton. Intensive laboratory investigations were combined with pilot-plant production and commercial development. Cyanoethylated cotton was found to be particularly suited for use in fishing nets, ironing-board covers, sandbags, filter fabrics, enameling duck, and tobacco shade cloth (3). Alkali-soluble derivatives of cyanoethylated cotton may be employed as backing fabric in the Schiffli process for making lace (4). Cyanoethylated cotton and manila are reportedly useful in marine ropes (3, 5). Although these development efforts were highly successful from the technical point of view and aroused much interest in the textile industry, there were no known producers of important quantities of cyanoethylated cotton in 1970. Some research activity is continuing, particularly in the Soviet Union and Japan, but recent published reports do not indicate any significant advance over prior developments.

Although the efforts spent in cyanoethylating cellulose in forms other than cotton were much less intensive, they have nevertheless produced some small-volume, but significant, new products. Cyanoethylated paper is being used as insulator in electrical transformers, where its better heat resistance is a decisive advantage (6, 7). Its electrical properties (5–8) and dimensional stability (9) are also improved.

Highly cyanoethylated cellulose is the preferred material for embedding phosphors in electroluminescent devices (10, 11). It may be plasticized with other cyanoethylated materials (10), such as poly(vinyl alcohol) (12) or hydroxyethylcellulose (13). Films may be used in capacitors (11–14) and as insulation (14). High-DS cyanoethylcellulose may also be used as a binder for photochromic materials (15).

(a) GENERAL FEATURES OF CYANOETHYLATION

(1) Catalysts

The useful catalysts for the cyanoethylation of cellulose are the strong bases, and especially the alkali-metal hydroxides dissolved in water. Although other hydroxides can be used (5, 16), sodium hydroxide has been employed in virtually all studies.

(2) Reversibility

Cyanoethylation is usually carried out at moderate temperatures (e.g., room temperature). This is possible because the reaction is vigorous even under these conditions. Of equal importance, however, is the fact that elevated temperatures tend to reverse the cyanoethylation process, and lower yields will therefore be obtained if excessive heat is applied (1, 2).

The reversibility of the reaction can be demonstrated in many ways; for example, treatment of cellulose with 3,3′-oxydipropionitrile in the presence of alkali produces cyanoethylcellulose (17). Attempts to hydrolyze poly(cyanoethyl vinyl ether) or cyanoethylcellulose with aqueous sodium hydroxide to form the sodium carboxyethyl derivatives cause instead primarily cleavage of the cyanoethyl groups (18–21):

$$\underset{\text{minor product}}{ROCH_2CH_2COONa} \xleftarrow{\text{NaOH}} ROCH_2CH_2CN \xrightarrow{\text{NaOH}} \underset{\text{major products}}{ROH + CH_2{=}CHCN} \qquad (3)$$

Because of the reversibility of the reaction, it is often advantageous to use an excess of reagent to increase the yield, but high concentrations of alkaline catalyst usually have the opposite effect.

(3) Reaction Mechanism

The nitrile group is is a powerful electronegative substituent; that is, it strongly attracts electrons from the double bond of acrylonitrile. The β-carbon atom is thereby rendered relatively positive and the nitrogen atom relatively negative. This is indicated in structure (**1**).

$$\overset{\delta+}{CH_2}{=}CH{-}C{\equiv}\overset{\delta-}{N}$$

1

The molecule is therefore susceptible to attack by nucleophilic reagents at the β-position. These nucleophiles may be anions or neutral molecules with an unshared pair of electrons (1, 2). The probable reaction mechanism for cyanoethylation may be summarized as follows: The strongly basic catalyst abstracts the hydroxyl (active) hydrogen atom of cellulose thereby producing an alkoxide ion (equation 4). The alkoxide ion attacks the relatively positive carbon atom of acrylonitrile, forming the intermediate anion (**2**), which is stabilized by resonance (equation 5). The anion then reacts with a substance capable of donating a proton, in this case water, to produce the final product and to regenerate the hydroxide ion (equation 6).

$$R_{cell}OH + OH^- \rightleftharpoons R_{cell}O^- + H_2O \qquad (4)$$

$$R_{cell}O^- + CH_2{=}CHCN \rightleftharpoons [R_{cell}OCH_2\bar{C}HCN \underset{\mathbf{2}}{\longleftrightarrow} R_{cell}OCH_2CH{=}C{=}\bar{N}] \qquad (5)$$

$$\mathbf{2} + H_2O \rightleftharpoons R_{cell}OCH_2CH_2CN + OH^- \qquad (6)$$

The hydroxide ion can start the process of equation 4 again, thus confirming the need for only a small quantity of catalyst. All steps are assumed to be reversible (1, 2). Kinetic investigations with alcohols (22, 23) indicate that the slow, rate-determining step is equation 5, whereas equation 6 must be very rapid. This is in agreement with the known behavior of anions.

A particularly interesting phenomenon related to cyanoethylation is the anionic polymerization of acrylonitrile. Anion (**2**), should, a priori, be capable of adding another molecule of acrylonitrile (equation 7). The product, anion (3), could then repeat the process to produce eventually a polyacrylonitrile chain (**4**).

$$\mathbf{2} + CH_2{=}CHCN \longrightarrow R_{cell}OCH_2CH(CN)CH_2\bar{C}H(CN) \qquad (7)$$

3

$$R_{cell}OCH_2CH(CN){\left(CH_2CH(CN)\right)_x}H$$

4

The fact that essentially no graft copolymer is formed under most ordinary cyanoethylation conditions provides additional evidence that reaction 6 must be very rapid.

(b) PREPARATION

There are three major processes and numerous variations thereof for producing cyanoethylated cellulose. All are heterogeneous systems in which the cellulose is immersed in the liquid reagents (in the case of highly cyanoethylated cellulose the reaction is heterogeneous for the first part and homogeneous after a sufficiently high DS has been reached). The choice among the alternatives depends largely on the characteristics desired of the product and on the equipment available.

An important economic consideration in selecting a given process is its efficiency in utilizing acrylonitrile. Since all practical processes utilize aqueous sodium hydroxide as the catalyst, significant amounts of acrylonitrile are consumed in side reactions to form primarily 3,3′-oxydipropionitrile

$$2CH_2{=}CHCN + H_2O \xrightarrow{OH^-} O(CH_2CH_2CN)_2 \qquad (8)$$

as well as small amounts of ethylene cyanohydrin ($HOCH_2CH_2CN$), sodium acrylate, etc. The efficiency has been expressed in various ways, for example, as percentage of yield based on acrylonitrile (i.e., the percentage of the acrylonitrile consumed that went into cyanoethylcellulose formation) or by an acrylonitrile efficiency ratio, also referred to in the literature simply as an acrylonitrile ratio, which is the ratio of the weight of acrylonitrile lost in side reactions to the weight of acrylonitrile reacted with cellulose. A low acrylonitrile ratio corresponds, therefore, to a low amount of by-product formation (high yield).

The degree of reaction in cyanoethylated cellulose is usually expressed as the percentage of nitrogen in the cellulose. Fully cyanoethylated cellulose has 13.2% nitrogen, but the content of cyanoethyl groups in the cellulose is not directly

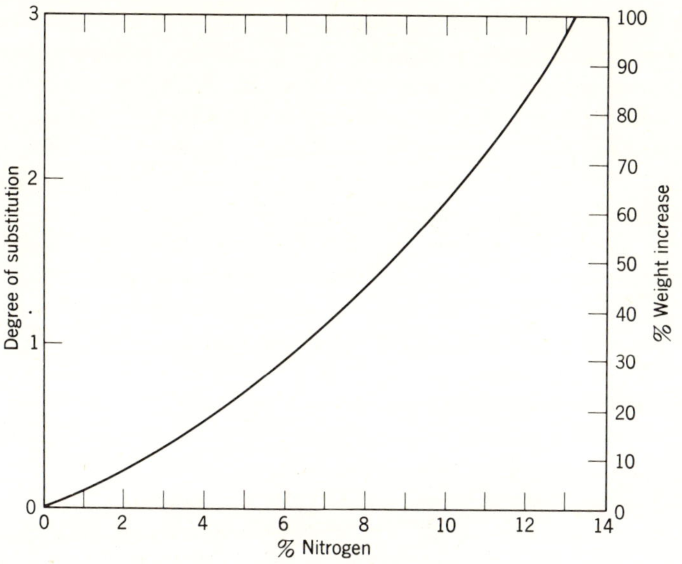

Fig. 1. Relationship between DS, percentage of weight increase, and percentage of nitrogen in cyanoethylcellulose.

proportional to the percentage of nitrogen. An alternative way of expressing extent of reaction is by the DS. In general cyanoethylated cotton has been reacted to about 3.5% nitrogen (DS 0.5) and highly cyanoethylated cellulose to about 12% nitrogen (DS 2.5). The relationship between DS, percentage of weight increase, and percentage of nitrogen in cyanoethylated cellulose is shown in Figure 1.

(1) Two-Step Batch Process

The two-step batch process was the original procedure described by Houtz (24) for the production of highly substituted cyanoethylcellulose that was soluble in organic solvents (including acrylonitrile) and by Stallings (25) for the production of partially cyanoethylated cellulose that still retained the morphology of cellulose. The process was subsequently utilized by workers at the Institute of Textile Technology to produce partially cyanoethylated cotton (26), and it received extensive study in a number of other laboratories.

In essence, the two-step process consists of first impregnating cellulose in a large volume of dilute sodium hydroxide solution, removing the excess alkaline solution until no more than about 1 part of solution is retained per part of dry cellulose, and then immersing the alkali-pretreated cellulose in a large volume of acrylonitrile until the desired extent of reaction is reached. Circulation of liquids through the cellulose is maintained throughout the reaction. The reaction is stopped by neutralization of the alkali. Unused acrylonitrile is then recovered by distillation. It is of interest to note that the cellulose selectively sorbs sodium hydroxide during the pretreatment, so that the alkali concentration is higher during reaction than in the original pretreating solution (27). Typical conditions of reaction are as follows: alkali concentration, 2%; reaction for 60 min at 50°C in about 5–10 parts of acrylonitrile per part of dry cellulose. The cellulose thus obtained has a nitrogen content of 3.5%. The procedure is reproducible and relatively simple to operate and has therefore been used extensively in laboratory investigations, including studies for the treatment of fabrics (28). The method suffers from several disadvantages, most notable of which are slow reaction, high waste of acrylonitrile to by-products (acrylonitrile efficiency ratio of about 5), and the need for a pretreatment step (27).

A number of variations have been made to overcome these deficiencies; for example, the pretreatment may be eliminated (see below under "One-Step Batch Process") or the rate of reaction may be increased and the amount of by-product formation decreased by the use of certain salts in conjunction with alkali (29, 30) (see below under "Use of Salts"). The use of inert diluents (such as hydrocarbons) has been investigated (5, 19), but these tend to complicate the process without large benefits. A two-step process utilizing simple block-and-tackle equipment has been described for use by nontechnical personnel, such as fishermen wanting to treat nets (31).

Many attempts have been made to eliminate entirely the use of water in the cyanoethylation of cellulose so as to avoid the formation of oxydipropionitrile. These efforts have all met with failure. Polyacrylonitrile has been observed to form in some of these cases. Graft copolymerization may also be taking place (32).

(2) Use of Salts

In an effort to increase the reactivity of cellulose in cyanoethylation Bikales discovered that the addition of certain salts to dilute alkali greatly increases the rate of reaction with cellulose at the expense of reagent-consuming side reactions (29). Typical among these are sodium thiocyanate, potassium iodide, or sodium perchlorate when used as nearly saturated aqueous solutions. Under the conditions cited in the preceding subsection, these salts resulted in reaction to about 11% nitrogen (instead of 3.5%), with typical acrylonitrile efficiency ratios of 0.6–0.9 (instead of 5) (see Table 1 and Fig. 2). Other salts, such as sodium xylenesulfonate, sodium benzoate, and sodium iodide, behaved similarly but were not quite as effective. The use of dilute solutions of these salts (Fig. 2) or of solutions (whether dilute or concentrated) of other salts, such as potassium bromide, sodium chloride, or sodium sulfate, did not result in any increase in reactivity.

The effectiveness of the salts has been attributed to swelling of the cellulose

TABLE 1
Increase in the Rate of Cyanoethylation by Salts[a] (29)

Salt	Concentration (molality)	Percentage of nitrogen in product
None	0	3.5
Sodium xylenesulfonate	2.4	8.2
Sodium benzoate	2.8	9.1
Sodium iodide	11.3	9.3
Sodium thiocyanate	15.8	10.9
Sodium perchlorate	8.2	11.0
Potassium iodide	8.3	11.5

[a]Reactions run with 100 parts of 2% sodium hydroxide solution per 100 parts of cotton; alkaline solution nearly saturated with salt; reaction for 45 min at 55°C.

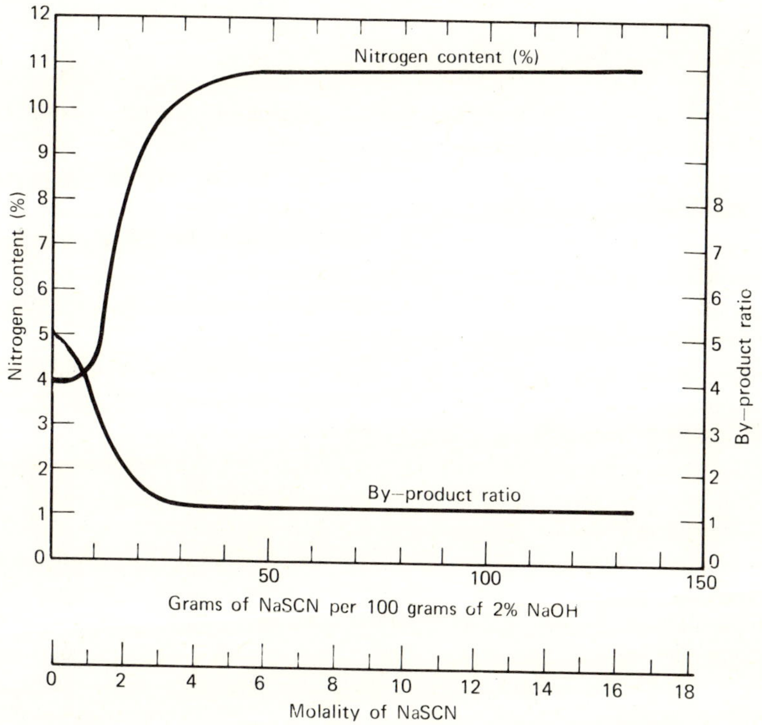

Fig. 2. Effect of sodium thiocyanate concentration on the extent of reaction (expressed as percentage of nitrogen) and on by-product formation (expressed as parts of acrylonitrile going to by-products per part of acrylonitrile reacting with cellulose) (29). Reaction for 45 min at 55°C in the presence of 2% sodium hydroxide.

and has been correlated with the physicochemical properties of the solutions (29). These salts also increase the solubility of acrylonitrile in water (29, 30). Although the use of swelling agents to increase penetration of reagents is well known (see, for example, pp. 335–337, 710, 767, 884, and 888 in Parts I and II), their use in conventional based-catalyzed reactions is not necessary because sodium hydroxide solutions are inexpensive and efficient (see Section C in this Chapter XVII). However, the high concentrations of alkali necessary for optimum swelling cannot be used in reactions with α,β-unsaturated compounds because of the reversibility of the reactions; furthermore, acrylonitrile and cyanoethylcellulose can be hydrolyzed and polymerization can take place under these conditions. Neutral or slightly basic alkali-stable swelling agents are therefore desirable for increasing the reactivity of the cellulose and for decreasing the relative amounts of by-products formed, when dilute solutions of alkali must be maintained (29) (see also subsection 3, "Ethers from Acrylamides").

(3) One-Step Batch Process

The one-step process is a development specifically designed to overcome some of the major defects of the two-step procedure. In an effort to avoid the need for pretreatment it was found (27, 33) that cellulose may be treated with only small amounts of dilute aqueous alkali (about 0.6–1.0 part per part of cellulose) together with the normal amount of acrylonitrile. However, in order to achieve uniformity of cyanoethylation, which is critical to the properties of the product, a dual-temperature procedure is required. The acrylonitrile–aqueous alkali mixture is first allowed to impregnate the cellulose uniformly (e.g., by pumping the liquids through cotton-yarn packages) under conditions in which no significant amount of reaction takes place (about 20 min at room temperature); the temperature is then raised to effect reaction as in the two-step process. Because this process permits exact control over the quantities of water and sodium hydroxide, by-product formation is reduced (27). Further reductions may be effected by the use of relatively high concentrations of sodium hydroxide (8–11%) but at lower temperatures (about 30°C). This variation raises the yield to well over 50% (an acrylonitrile ratio < 1) and permits economical batch production of cyanoethylated cotton (34, 35). A pilot plant was successfully operated over a period of several months, producing over 10 tons of partially cyanoethylated cotton yarn in package dyeing machines (kiers) (34, 36). A flow sheet for the process used, including washing of the treated cotton and recovery of unused acrylonitrile, has been published (2, 34). The recommended treating solution for 12.8 parts of bleached cotton yarn consists of 10.7 parts of 8% sodium hydroxide solution, 42.2 parts of fresh acrylonitrile, 47.0 parts of recycled acrylonitrile (containing 4.8 parts of by-product), and 0.1 part of alkali-stable wetting agent.

The one-step process is particularly well suited for the production of highly

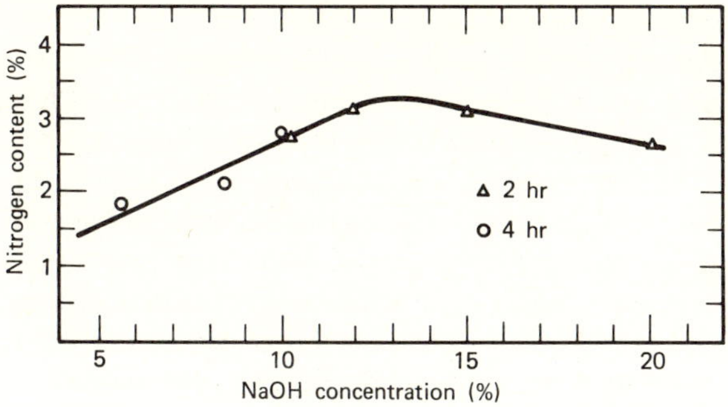

Fig. 3. Effect of the sodium hydroxide concentration on the percentage of nitrogen (38). Reaction conditions: 10 g of cotton; 100 g of sodium hydroxide solution; 3.2 g of acrylonitrile; 15–20°C.

cyanoethylated cellulose; that is, products with sufficiently high substitution to render them soluble in organic solvents, including acrylonitrile itself. In a typical procedure (12) regenerated cellulose is suspended in a large volume of acrylonitrile and a dilute solution of sodium hydroxide is added. The mixture is heated at 50°C for 4 hr, in the course of which the cellulose is gradually dissolved. The viscous solution is neutralized, filtered, and then added dropwise to boiling water to precipitate the product (containing about 12.5% N) and recover unreacted acrylonitrile. Purification may be effected, if desired, by

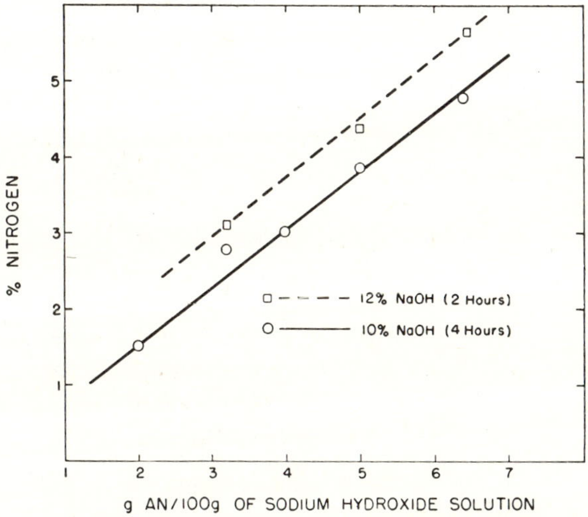

Fig. 4. Effect of the amount of acrylonitrile used on the percentage of nitrogen (38). Reaction conditions: 10 g of cotton; 100 g of sodium hydroxide solution; 15–20°C.

electrodialysis or by extensive washing. When more highly crystalline cellulose is the starting material, the addition of a swelling salt (e.g., sodium thiocyanate) to the alkali is advantageous because it increases the rate of reaction (29).

A variation of the one-step procedure has been used to cyanoethylate paper pulp (37) and cotton (38, 39). It consists of cyanoethylating the cellulosic material in essentially aqueous medium; that is, small amounts of acrylonitrile are used with large amounts of relatively concentrated aqueous alkali, instead of large amounts of the organic with small amounts of the aqueous solution, as described above. The variables have been thoroughly explored, and the process permits cyanoethylation with good efficiency (see, for example, Figs. 3 and 4). However, the product was found to lack the high degree of resistance to microbiological degradation characteristic of cyanoethylated cellulose made by other processes (38).

(4) High-Solids Process

A further development of the latter procedure consists of carrying out the reaction at very high solids contents (9, 40). The process permits reaction to take place with only minimal quantities of both liquid reagents (i.e., sodium hydroxide solution and acrylonitrile). The process is also highly efficient and very simple to operate in standard mixing equipment. No recovery of acrylonitrile is necessary since essentially all of it is consumed. It may be operated simply at room temperature. Typically, 1 part of cellulose is mixed with 1 part of 4% sodium hydroxide solution, followed by 0.24–1 part of acrylonitrile. Mixing is continued for about 4 hr at room temperature. The product obtained by this procedure, unlike that by the aqueous process cited above, was found to have a high degree of resistance to microbiological degradation (5). The high-solids process is particularly well suited to the cyanoethylation of finely divided cellulose, such as paper pulp, to degrees of reaction where the morphology of the cellulose is retained (nitrogen content below 6%) (9, 40, 41). Because of the absence of a visible liquid phase, the method has been referred to as a "dry" process.

A similar procedure may be used for treating cotton fabric. Low levels of about 10% sodium hydroxide solution and of acrylonitrile are padded onto the fabric, which is then tightly wound into a roll and covered with a plastic sheet. Reaction is allowed to proceed for several hours without external application of heat. No unused acrylonitrile needs to be recovered. Although cyanoethylation is reported to take place only on the surface of the cotton, the product is claimed to have a high degree of rot resistance (42).

(5) Continuous Process

A major advance over the batch processes described above is the continuous process for cyanoethylating cotton fabrics (34, 43). It consists of passing alkali-treated cotton into a reaction chamber where hot vapors of acrylonitrile are allowed to contact the fabric for short periods. A cold "leg" of acrylonitrile is used as a vapor seal where the cloth enters the reactor. Vapor losses at the exit are prevented by an acidified dip. After acidification, unused acrylonitrile is removed with a jet of steam. The fabric is then washed in the conventional way to remove salts and nonvolatile reaction by-products. A schematic diagram of this process may be found in references 2 and 34.

The reaction is not a true vapor-phase reaction in the chemical sense; it occurs between the cotton and acrylonitrile condensed on the cotton. An important feature of the process is the maintenance of low levels of water in the reaction chamber by azeotropic removal of excess water with acrylonitrile. This technique permits cyanoethylation to proceed with relatively low losses of acrylonitrile to form 3,3′-oxydipropionitrile. Some water, however, must be present on the cotton to permit cyanoethylation to proceed at an acceptable rate.

Suitable reaction conditions for 4-oz desized fabric are as follows: The fabric is padded with 6% sodium hydroxide solution, squeezed to a "pickup" of solution of 0.65–0.75 part per part of dry cotton, and then exposed to acrylonitrile in the reaction chamber for 3.2–3.6 min at about 72°C (50 m/min). The cotton thus treated contains about 3.5% nitrogen (34).

(6) Forms of Cellulose Treated

Most of the reports on the cyanoethylation of cellulose are concerned with cotton in its various forms, that is, staple, yarn, and fabric. Some work has also been carried out with cotton linters and various wood pulps high in α-cellulose content, primarily for paper applications and the preparation of highly cyanoethylated cellulose. Regenerated cellulose is highly reactive in cyanoethylation and offers some advantages in the preparation of high-DS products. The vegetable fibers (e.g., flax, jute, manila, and sisal) react with greater difficulty.

The derivatives of cellulose may be reacted, provided that not all of the hydroxyl groups have been substituted. The cyanoethylation of ethylcellulose (1, 2, 24), methylcellulose (1, 2), hydroxyethylcellulose (44), sodium carboxymethylcellulose (45), and others has thus been reported. Cellulose xanthate solutions are highly reactive and give, on coagulation, products that differ markedly from ordinary cyanoethylated cellulose in that they are soluble in alkali or even in water, depending on the DS (see Chapter IX–E–9). Simultaneous xanthation and cyanoethylation of alkali cellulose has been claimed to be more economical and to improve the physical properties of regenerated-cellulose fibers (46, 47).

(c) PROPERTIES

Of all the characteristics affecting the properties of cyanoethylcellulose, the degree of substitution is usually considered to be most important. In general cyanoethylated products may be classified into two useful categories: (a) those with only partial cyanoethylation, which still retain the morphology of cellulose but have increased in size (up to about 6% N); (b) those with high levels of substitution (10–13% N), which are soluble in polar organic solvents and exhibit much swelling and agglomeration of fibers (48). The intermediate region has received comparatively little attention, largely because properties have not been sufficiently superior to those of products with about 3–4% nitrogen to warrant the increased cost.

(1) Cyanoethylated Cotton

Partially cyanoethylated cotton (fiber, yarn, and fabric) with a nitrogen content of 3.8 ± 0.5% is similar to ordinary cotton in appearance, "hand," and processing characteristics but differs significantly in that it exhibits the characteristics discussed below.

The treated cotton is permanently *rot- and mildew-proof*; it may be buried in soil inoculated with cellulose-degrading microorganisms for indefinite periods of time and retain its full strength (in some cases even slight increases in strength have been observed). Cyanoethylated cotton and cyanoethylated manila may also be immersed in water for long periods without rotting (3, 34). The degree of resistance to microbial attack increases with increasing nitrogen content and becomes essentially complete at about 2.8–3.5% nitrogen. However, the presence of relatively small amounts of carboxyethyl groups (resulting from the hydrolysis of cyanoethyl groups) has been found to be detrimental. It is therefore important to use cyanoethylation conditions that are as mild as possible. Treatment of cyanoethylated cotton with alkali must also be minimized or avoided in scouring, bleaching, and dyeing (20). Another important aspect is the need for uniformity of cyanoethylation, since the less reacted portions of the cotton may be more susceptible to attack by microorganisms Dyeing has been used as a test for uniformity.

Cyanoethylated cotton has increased resistance to *thermal degradation* under either wet or dry conditions. For example, cyanoethylated fabric (3.6% N) lost only 3.5% of its tensile strength in 15 hr at 160°C, whereas untreated fabric lost 38%. The improvement appears to increase with nitrogen content and to reach a maximum at about 5% nitrogen (34). Thermal degradation of cyanoethylcellulose in oxygen at 150°C results in a rapid decrease in intrinsic viscosity and slow evolution of hydrogen cyanide, acrylonitrile, water, carbon dioxide, and traces of ammonia and formaldehyde. The decomposition appears to be free radical in character (49). Heating to 800°C under nitrogen causes carbonization (50).

Cyanoethylated cotton has greater resistance to *degradation by acids* than untreated cotton. For example, cyanoethylated cloth (3.5% N) lost 44% of its tensile strength in 20 hr at 60°C in 20% sulfuric acid, whereas untreated cloth

TABLE 2
Mechanical Properties of Cyanoethylated Cotton Yarns (48)

DS	Elongation at break (%)	Tensile strength (g/tex[a])	Tenacity (g/tex)	Energy of rupture (g/tex)	Tensile stiffness (g/tex)	Elastic recovery (%) Immediate	Elastic recovery (%) Delayed	Elastic recovery (%) Permanent set	Energy recovered (%)
0[b]	8.54	14.4	14.4	0.64	168	8.7	41.8	49.5	44.5
0[c]	7.63	16.6	16.6	0.58	218	8.0	37.8	54.2	47.4
0.58	7.09	19.7	16.9	0.56	278	10.3	39.2	50.5	48.4
0.75	7.70	24.6	17.8	0.64	319	9.7	40.5	49.8	46.2
1.17	7.49	24.6	17.8	0.66	328	11.0	35.5	53.5	51.0
1.28	8.13	26.3	18.6	0.72	324	10.3	36.3	53.4	50.2
1.51	14.44	23.2	15.5	0.88	161	8.8	30.5	60.7	46.3
2.12	13.56	19.1	11.2	0.70	141	14.0	32.3	53.7	48.3
2.18	15.17	17.4	10.1	0.75	114	13.7	31.0	55.3	52.5
2.38	18.21	16.7	9.4	0.72	92	12.8	32.8	54.4	47.9

[a] Breaking load expressed per tex of untreated control. The tex is a unit of fineness.
[b] Untreated control.
[c] Control immersed overnight in 6% NaOH, then heated in toluene for 60 min.

lost 86%. On the other hand, cyanoethylated cotton is more readily degraded by the action of alkalies (34).

Because of its greater hydrophobicity, cyanoethylated cotton has a lower moisture regain than cotton. A 5.0–5.5% regain is typical, as compared with 7.0–7.5% for cotton (34).

The *mechanical properties* of cyanoethylated cellulose depend not only on the extent of cyanoethylation but also, to some degree, on the method of preparation and the construction of the cotton. It may be stated that, in general, the tensile strength of fibers and yarns is increased to about the same extent as the increase in weight due to reaction (at the lower DS range, which is of most interest in textile applications). In some cases higher tenacities and elongations have been observed (see, for example, Table 2, which is based on the extensive investigations of Conrad and co-workers). For fabrics, on the other hand, decreases in strength, particularly in tear strength, have been found. The reader is referred to the original literature for more details (20, 26, 27, 36, 38, 48, 51–53).

The *crystallinity* of native cotton is retained essentially intact in partially cyanoethylated cotton (DS 0.5) (26, 38, 48, 53), confirming that reaction is confined to the amorphous regions of the cellulose. At a DS of 1.1 the crystalline structure gradually gives way to an amorphous state, which appears to be complete at a DS of 2 (Fig. 5). This DS corresponds to the point at which the cyanoethylated cellulose becomes soluble in organic solvents (see below under "Highly Cyanoethylated Cellulose"). At a DS higher than 2, the material may be annealed at approximately 175°C to produce a new characteristic x-ray pattern (48, 53, 54).

The nitrile group can undergo the usual chemical reactions (e.g., hydrolysis, reduction). As already described above, alkaline treatment results not only in hydrolysis but also in decyanoethylation. The most effective way of converting cyanoethylcellulose to carboxyethylcellulose is by way of the carbamoylethylcellulose (**5**). The latter is conveniently obtained by a mild treatment with hydrogen peroxide in the presence of dilute alkali. Acid treatment then produces the carboxylic derivative (**6**) (55, 56):

$$R_{cell}OCH_2CH_2CN \xrightarrow[pH \sim 9.5]{H_2O_2} \underset{\mathbf{5}}{R_{cell}OCH_2CH_2CONH_2} \xrightarrow{H^+} \underset{\mathbf{6}}{R_{cell}OCH_2CH_2COOH} \qquad (9)$$

The carbamoylethylated cellulose can be crosslinked by treatment with formaldehyde (56). The nitrile group may also be reduced to the amine to give 3-aminopropylcellulose (**7**) (57, 58), by using, for example, catalytic hydrogenation:

$$R_{cell}OCH_2CH_2CN \xrightarrow{(H)} \underset{\mathbf{7}}{R_{cell}OCH_2CH_2CH_2NH_2} \qquad (10)$$

Graft copolymerization can be carried out on cyanoethylated cotton (see also Sections G and H in this Chapter XVII). Treatment with strong oxidizing agents (e.g., chromic acid) causes the formation of water-soluble fibers (59, 60). A most interesting product is obtained by treatment of the nitrile with hydroxylamine:

$$R_{cell}OCH_2CH_2CN \xrightarrow{NH_2OH} R_{cell}OCH_2CH_2C(NH_2){=}NOH \qquad (11)$$

The product is an amidoxime fiber that has strong chelating properties and may be used for the extraction of copper, iron, gold, uranium, etc. (61, 62). The

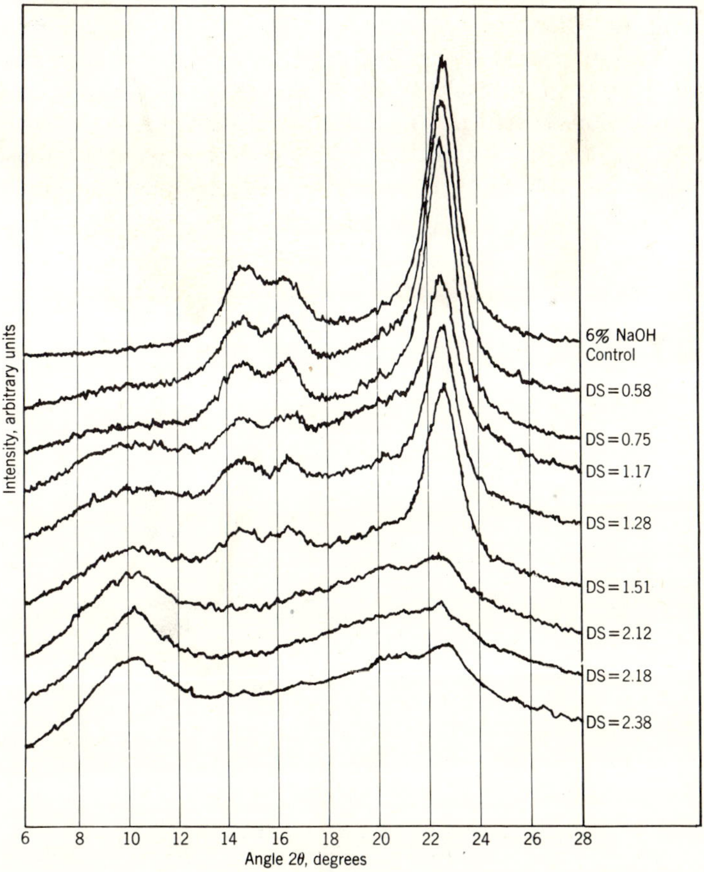

Fig. 5. Radial tracings of x-ray diffraction intensity from cotton yarns cyanoethylated without shrinkage (48). (Courtesy of the Southern Regional Research Laboratory, U.S. Department of Agriculture.)

amidoxime can be treated with formaldehyde to effect crosslinking, with resulting improvements in wrinkle resistance (63).

The unsubstituted hydroxyl groups of cyanoethylcellulose can also be reacted conventionally, especially under acidic or neutral conditions, e.g., to form acetates.

(2) Cyanoethylated Paper

The first report of the cyanoethylation of paper pulp appears to have been by Miller and Flowers (37). They discovered that cyanoethylation to nitrogen contents of about 2.3% increases the dielectric constant of the paper; this enhances the value of the paper when used as spacers in capacitors. Subsequently it was found that very low degrees of cyanoethylation (0.5–1.8% N) also improve the fold (64) and tear (65) strengths of handsheets made from cyanoethylated linters.

In a systematic examination of the cyanoethylation of cellulose for paper applications Morton and Bikales (9) found that satisfactory treatment of paper sheets presented many difficulties that were obviated when cyanoethylation was carried out on the pulp before the papermaking process. The properties of cyanoethylated handsheets were found to depend greatly on the DS within the

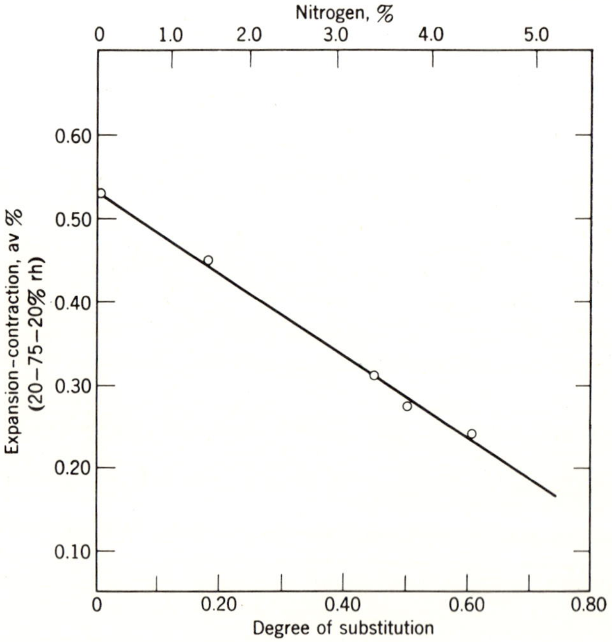

Fig. 6. Dimensional changes in paper as a function of the extent of cyanoethylation (9).

0–5% nitrogen range investigated. In particular it was found that the dimensional stability (expansion–contraction as a function of relative humidity) is substantially improved by cyanoethylation and that the decrease in dimensional changes is directly proportional to the DS (Fig. 6). Blends of cyanoethylated and untreated pulp give paper whose dimensional changes are inversely proportional to the cyanoethylcellulose content. The mechanical properties (e.g., tensile and bursting strengths and folding endurance), however, were adversely affected by the treatment, and the porosity was increased (9). As is the case with cyanoethylated cotton, the treated paper has other advantages, such as improved resistance to heat (2, 6, 7) and rot (5) (see also Section D in Chapter XIX).

(3) Highly Cyanoethylated Cellulose

Highly cyanoethylated cellulose (12–13% N) is available commercially as a fluffy, amorphous, thermoplastic solid. It is readily soluble in acrylonitrile, acetone, dimethylformamide, pyridine, dimethyl sulfoxide, nitromethane, γ-butyrolactone, acetonitrile, and concentrated aqueous solutions of zinc chloride and sodium thiocyanate (5, 24, 66, 67).

The outstanding characteristic of highly cyanoethylated cellulose is its unusually high dielectric constant coupled with a low dissipation factor. Typical electrical properties are shown in Figures 7 and 8. The dielectric constant is considerably higher than that of the common plastic materials. The material may also be molded, although neither its flow properties nor its mechanical

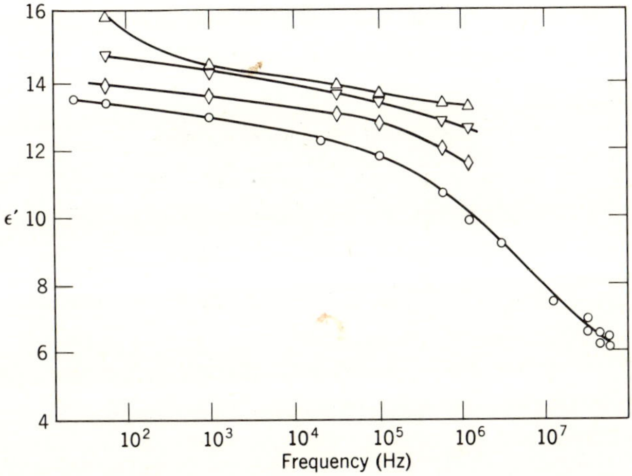

Fig. 7. Dielectric constant of cyanoethylcellulose (66). Key: △, 100°C; ▽, 75°C; ◇, 50°C; ○, 25°C. (Courtesy *Journal of Polymer Science*.)

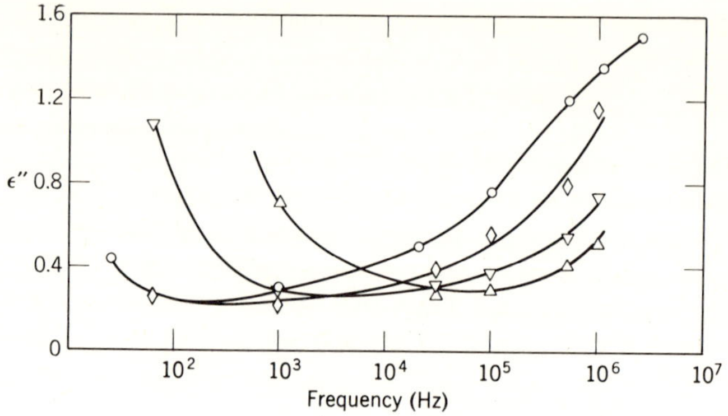

Fig. 8. Loss factor of cyanoethylcellulose (66). Key: △, 100°C; ▽, 75°C; ◇, 50°C; ○, 25°C. (Courtesy *Journal of Polymer Science*.)

properties are particularly outstanding. Highly cyanoethylated cellulose may be plasticized, for example, with dicyanoethyl phthalate (10) or other cyanoethyl compounds (12, 13).

2. Ethers from Related Nitriles

A study of the cyanoalkylation of cellulose with various α,β-unsaturated nitriles has shown the following order of reactivity: acrylonitrile ≫ α-methyleneglutaronitrile ≈ *trans*-crotononitrile > methacrylonitrile (68). Fumaronitrile was also found to react with cellulose, but the extent of reaction was small (69).

3. Ethers from Acrylamides

The base-catalyzed reaction of cellulose with acylamide produces carbamoylethylcellulose (8) provided that the alkali concentration is kept relatively low (< ca. 4%).

$$R_{cell}OH + CH_2{=}CHCONH_2 \xrightleftharpoons{OH^-} \underset{\textbf{8}}{R_{cell}OCH_2CH_2CONH_2} \qquad (12)$$

At higher alkali concentrations the reaction tends to be reversed (regenerating cellulose and acrylamide and/or its hydrolysis products) and hydrolysis of the amide group takes place. In fact the reaction of cellulose with acrylamide in the presence of 20–40% sodium hydroxide solution is an excellent method of preparing alkali salts of carboxyethylcellulose (70, 71).

The reaction of acrylamide with cellulose proceeds much less readily than the similar reaction of acrylonitrile with cellulose (discussed in subsection 1, "Ethers from Acrylonitrile"). The efficiency is relatively low and only low degrees of carbamoylethylation have been obtained, accompanied by considerable carboxyethyl content (72). In a typical treatment of cotton, fabric is first padded to an 89% pickup (i.e., 89 parts of solution per 100 parts of cotton) with an aqueous solution consisting of 50% acrylamide and 4% sodium hydroxide. The pretreated cotton is then heated in an oven for 4 min at 135°C to effect the reaction. The properties of the resulting fabric, containing about 2.4% nitrogen, are not greatly different from untreated cotton, and the process has therefore not aroused any commercial interest (69, 73, 74).

When cellulose is treated with acrylamide and dilute (2%) sodium hydroxide solution but in the presence of neutral salts with ability to swell cellulose (e.g., sodium thiocyanate), the reaction proceeds to such an extent that the product dissolves in the reaction mixture. The carbamoylethylcellulose thus obtained (DS 0.5–0.7) is soluble in water under acidic, neutral, and alkaline conditions, unlike carboxyethylcellulose, which is soluble only in alkali (75).

A similar reaction may be carried out with the *N*-substituted acrylamides (69, 75). The reaction of difunctional acrylamides, such as *N,N'*-methylenebisacrylamide and *N*-methylolacrylamide, has been of interest especially in connection with crosslinking (see, for example, reference 76 and Section E of this Chapter XVII).

An alternative method of preparing carbamoylethylcellulose is the treatment of cyanoethylcellulose with dilute hydrogen peroxide under slightly alkaline conditions (equation 9) (55, 56).

The relative distribution of substituents in a material prepared from acrylamide and cotton linters and containing a DS for carbamoylethyl of 0.48 and a DS for carboxyethyl of 0.30 was as follows: 0.47 : 0.053 : 1.0 for positions 2, 3, and 6, respectively (77).

4. Ethers from Sulfones and Sulfonic Acids

Alkylsulfonylethylcelluloses were first prepared by Schoene and Chambers (78) by the reaction of an alkyl vinyl sulfone in the presence of aqueous sodium hydroxide:

$$R_{cell}OH + CH_2{=}CHSO_2R \rightleftharpoons R_{cell}OCH_2CH_2SO_2R \qquad (13)$$

The products were claimed to swell or dissolve in water or to be water-repellent, depending on the DS and the length of the alkyl chain.

Subsequent investigation showed that the relative reactivity of the three hydroxyl groups in the anhydroglucose unit toward methyl vinyl sulfone under typical conditions is as follows: 0.20 : 0.03 : 1.0 for positions 2, 3, and 6, respectively (79); these values are similar to those found for the reaction of cellulose with acrylamide (77). However, the distribution of substituents varied to some extent with the reaction conditions (80, 81). The reaction may be catalyzed not only by aqueous sodium hvdroxide but also by quaternary ammonium hydroxide groups substituted in the cellulose molecule itself (82). When divinyl sulfone is allowed to react with cellulose in the presence of a secondary amine, a product having structure (**9**) is obtained. Alcohols react similarly (88).

$$R_{cell}OCH_2CH_2SO_2CH_2CH_2NR_2$$

9

Because of the highly reactive nature of vinyl sulfones, saturated derivatives that are convertible *in situ* to the vinyl compounds are generally preferred (83–85). The major interest in vinyl sulfone derivatives of cellulose has been in connection with crosslinking by difunctional compounds; this aspect is discussed in detail in Section E of this Chapter XVII.

Ethylenesulfonic acid (in the form of its alkali salts) reacts with cellulose to give β-sulfoethylcellulose salts, which are water-soluble materials (see p. 958 in Part II). Sulfoalkylcelluloses are usually more conveniently obtained from the appropriate haloalkanesulfonates or sultones (see, for example, reference 86).

5. Ethers from Other α,β-Unsaturated Compounds

The reaction of acrylic acid salts does not proceed readily with cellulose (5, 69); carboxyethylcellulose is preferably made via the cyanoethylcellulose or carbamoylethylcellulose (equation 9). Some evidence has also been obtained that reaction can take place with salts of maleic and itaconic acids, with crotonaldehyde, and methyl acrylate (69). By analogy with related cases (87), it is likely that both polar and steric factors of the reactants play a role in reactivity.

These ethers, with the exception of carboxyethylcellulose, have received only the most cursory examination.

References

1. *The Chemistry of Acrylonitrile*, 2nd ed., American Cyanamid Co., New York, 1959.
2. N. M. Bikales, in A. Standen, Ed., *Encyclopedia of Chemical Technology*, 2nd ed., Vol. 6, Interscience, New York, 1965, p. 634.
3. J. Compton, *Textile Res. J.*, **27**, 222 (1957).
4. R. W. Reinhardt, T. W. Fenner, and J. D. Reid, *Am. Dyestuff Reptr.*, **50**, (19), 67 (September 18, 1961).
5. N. M. Bikales, unpublished results.
6. *Paper Trade J.*, **143** (42), 44 (1959); *Chem. Week*, **85**, 61 (August 15, 1959).
7. N. L. Greenman, H. W. Knudson, and J. C. Leslie, *Proc. Elec. Insulation Conf. (New York)*, December 1959.

8. U. Usmanov, S. N. Kolesov, V. I. Sodovnikova, and B. E. Baklitskii, *Dokl. Akad. Nauk Uzb. SSR,* **20**, (3), 25 (1963); through *Chem. Abstr.*, **60**, 84 (1964).
9. J. L. Morton and N. M. Bikales, *Tappi,* **42** (10), 855 (1959).
10. E. G. Fridrich, U.S. Patents 2,901,652 and 2,918,594 (1959) (to General Electric Co.).
11. *Chem. Week*, **86**, 103 (May 28, 1960).
12. N. M. Bikales and W. O. Fugate, U.S. Patent 3,067,141(1962) (to American Cyanamid Co.).
13. M. J. D'Errico and M. M. Timney, U.S. Patent 3,096,289 (1963) (to American Cyanamid).
14. Westinghouse Electric Corporation, British Patent 825,166 (1959).
15. J. A. Chapoorian, U.S. Patent 3,271,176 (1966) (to American Cyanamid).
16. T. Lukanoff, *Faserforsch. u. Textiltech.,* **16** (2), 77 (1965).
17. G. E. Journeay, U.S. Patent 2,842,541 (1958) (to Monsanto Chemical Co.).
18. J. F. Wright and L. M. Minsk, *J. Am. Chem. Soc.,* **75**, 98 (1953).
19. G. C. Daul, R. M. Reinhardt, and J. D. Reid, *Textile Res. J.,* **25**, 246 (1955).
20. L. W. Mazzeno, R. M. Reinhardt, J. D. Reid, and J. B. Dickson, *Textile Res. J.,* **26**, 597 (1956).
21. T. Lukanoff and B. Philipp, *Zh. Prikl. Khim. (Leningrad),* **42**, (1), 170 (1969); through *Chem. Abstr.,* **70**, 107634 (1969).
22. Y. Ogata and M. Okano, *J. Am. Chem. Soc.,* 78, 5426 (1956).
23. B.-A. Feit and A. Zilkha, *J. Org. Chem.,* **28**, 406 (1963).
24. R. C. Houtz, U.S. Patent 2,375,847 (1945) (to E. I. du Pont de Nemours & Co., Inc.).
25. J. W. Stallings, U.S. Patent 2,473,308 (1949) (to Rohm & Haas Co.).
26. J. Compton, W. H. Martin, B. H. Word, and D. D. Thompson, *Textile Ind.,* **117**, 138A (1953).
27. A. H. Gruber and N. M. Bikales, *Textile Res. J.,* **26**, 67 (1956).
28. L. H. Greathouse, H. J. Janssen, W. N. Berard, and C. H. Haydel, *Ind. Eng. Chem.,* **48**, 1263 (1956).
29. N. M. Bikales, A. H. Gruber, and L. Rapoport, *Ind. Eng. Chem.,* **50**, 87 (1958); N. M. Bikales, Canadian Patent 538,628 (1957) (to American Cyanamid).
30. E. Klein, J. W. Weaver, and B. G. Webre, *Chem. Eng. Data Series,* **2**, (1), 72 (1957).
31. V. E. Wellman and N. M. Bikales, U.S. Patents 3,026,168 and 3,046,775 (1962) (to American Cyanamid).
32. B.-A. Feit, A. Bar-Nun, M. Lahav, and Z. Zilkha, *J. Appl. Polymer Sci.,* **8**, 1869 (1964)
33. N. M. Bikales, A. H. Gruber, and E. L. Carpenter, Canadian Patent 586,257 (1959) (to American Cyanamid).
34. *Cyanoethylation of Cotton*, a joint report by the Institute of Textile Technology, American Cyanamid Co., and Monsanto Chemical Co., September 1956; quoted extensively in reference 36 and in *America's Textile Reporter,* **71**, (1) 8 (1957).
35. N. M. Bikales, A. H. Gruber, and L. Rapoport, U.S. Patent 2,907,625 (1959) (to American Cyanamid).
36. J. Compton, W. H. Martin, B. H. Word, and R. P. Barber, *Textile Res. J.,* **25**, 58 (1955).
37. H. F. Miller and R. G. Flowers, U.S. Patent 2,535,690 (1950) (to General Electric).
38. N. M. Bikales and L. Rapoport, *Textile Res. J.,* **28**, 737 (1958); *Textile Recorder,* **78**, (936), 71 (1961).

39. N. M. Bikales, U.S. Patent 2,857,239 (1958) (to American Cyanamid).
40. E. D. McKay (to Southern Chemical Cotton Co.) and N. M. Bikales (to American Cyanamid), Canadian Patent 622,868 (1961).
41. F. B. Cramer and E. G. Parrish, U.S. Patent 3,266,972 (1966) (to Du Pont).
42. D. M. Gagarine, U.S. Patent 2,904,386 (1959) (to Deering Milliken Research Corporation).
43. S. F. Belt, U.S. Patent 2,860,946 (1958) (to Monsanto).
44. W. O. Fugate and E. C. McClenachan, Belgian Patents 619,830 and 619,831 (1962) (to American Cyanamid).
45. R. M. Reinhardt, J. D. Reid, T. W. Fenner, and R. Y. Mayne, *Textile Res. J.*, **29**, 802 (1959).
46. A. I. Bates, U.S. Patent 3,146,116 (1964) (to FMC Corp.).
47. E. H. Man, U.S. Patent 3,328,892 (1967) (to Du Pont).
48. C. M. Conrad, D. J. Stanonis, J. J. Creely, and P. Harbrink, *J. Appl. Polymer Sci.*, **5**, 163 (1961).
49. O. P. Koz'mina, V. N. Syutkin, P. A. Slavetskaya, and N. S. Danilov, *Vysokomol. Soedin.* **8** (7), 1196 (1966); through *Chem. Abstr.*, **65**, 15589 (1966)
50. G. Machell, U.S. Patent 3,395,970 (1968) (to Deering Milliken).
51. J. N. Grant, L. H. Greathouse, J. D. Reid, and J. Weaver, *Textile Res. J.*, **25**, 76 (1955).
52. C. M. Conrad, *Textile Res. J.*, **29**, 287 (1959)
53. C. M. Conrad, D. J. Stanonis, P. Harbrink, and J. J. Creely, *Textile Res. J.*, **30**, 339 (1960).
54. J. H. MacGregor and C. Pugh, *Proc. XIth Intern. Congr. Pure Appl. Chem. (London)*, **5**, 123–131 (1947) (published in 1953).
55. J. R. Stephens and L. Rapoport, U.S. Patent 2,820,691 (1958) (to American Cyanamid).
56. M. Negishi and N. Aida, *Textile Res. J.*, **29**, 982 (1959)
57. W. H. Schuller and L. C. Beegle, U.S. Patent 3,122,526 (1964) (to American Cyanamid).
58. L. Alexandru, M. Opris, and A. Ciocanel, *J. Polymer Sci.*, **59**, 129 (1962).
59. H. Weisberg, U.S. Patent 2,724,632 (1955) (to Londat Aetz Fabric Company).
60. R. M. Reinhardt, J. D. Reid, and T. W. Fenner, *Ind. Eng. Chem.*. **50**, 83 (1958).
61. *Chem. Eng. News*, **40**, 85 (May 7, 1962).
62. T. V. Vladimirova, L. S. Gal'braikh, Z. A. Rogovin, and G. A. Gabrielyan, *Izv. Vyssh. Ucheb. Zaved., Khim. i Khim. Tekhnol.*, **11** (5), 594 (1968); through *Chem. Abstr.*, **69**, 60120 (1968).
63. M. Negishi, H. Ito, and N. Aida, *Textile Res. J.*, **29**, 990 (1959).
64. J. J. Spadaro, J. J. Janssen, J. L. E. Vix, and F. L. Simons, *Tappi*, **41**, 674 (1958).
65. J. A. Harpham, A. R. Reid, and H. W. Turner, *Tappi*, **41**, 629 (1958).
66. C. W. Lewis and D. H. Hogle, *J. Polymer Sci.*, **21**, 411 (1956).
67. K. W. Saunders and N. M. Bikales, U.S. Patent 3.097,956 (1963) (to American Cyanamid).
68. T. Lukanoff, L. Zielske, and B. Philipp, *Cellul. Chem. Technol. (Jassy)*, **1** (3), 287 (1967).
69. J. W. Frick, W. A. Reeves, and J. D. Guthrie, *Textile Res. J.*, **27**, 92 (1957).
70. L . H. Bock and A. L. Houk, U.S. Patent 2,338,681 (1944) (to Rohm & Haas).

71. C. L. P. Vaughan, U.S. Patent 2,618,633 (1952) (to Hercules Power Company).
72. R. M. Reinhardt and J. S. Bruno, *J. Appl. Polymer Sci.,* **10**, 387 (1966).
73. J. W. Frick, W. A. Reeves, and J. D. Guthrie, *Textile Res. J.,* **27**, 294 (1957).
74. K. M. Decossas, O. J. McMillan, W. A. Reeves, J. D. Guthrie, and E. F. Pollard, *Am. Dyestuff Reptr.,* **48**, 30 (July 13, 1959).
75. N. M. Bikales and J. R. Stephens, U.S. Patent 3,029,232 (1962) (to American Cyanamid).
76. J. L. Gardon, *J. Appl. Polymer Sci.,* **5**, 734 (1961).
77. G. F. Touzinsky, *J. Org. Chem.,* **30**, 426 (1965).
78. D. L. Schoene and V. S. Chambers, U.S. Patent 2,539,704 (1951) (to U.S. Rubber Company).
79. S. P. Rowland, V. O. Cirino, and A. L. Bullock, *Can. J. Chem.,* **44**, 1051 (1966).
80. S. P. Rowland, A. L. Bullock, V. O. Cirino, and C. P. Wade, *Can. J. Chem.,* **46**, 451 (1968).
81. V. O. Cirino, A. L. Bullock, and S. P. Rowland, *J. Polymer Sci.,* **A-1, 7**, 1225 (1969).
82. S. P. Rowland and M. A. F. Brennan, *Abstracts of Papers, 158th National Meeting of the American Chemical Society, New York, Sept. 1969,* paper Cell-38; *Textile Res. J.,* **40**, 101 (1970).
83. G. C. Tesoro, *J. Appl. Polymer Sci.,* **5**, 721 (1961).
84. F. Burkitt, *Chem. Ind. (London),* 587 (1964).
85. A. Süszer, *Israel J. Chem.,* **4**, 123 (1966).
86. G. Natus and E. J. Goethals, *J. Macromol. Sci.-Chem.,* **A2** (3), 489 (1968).
87. M. Friedman and J. S. Wall, *J. Org. Chem.,* **31**, 2888 (1966).
88. C. M. Welch and J. D. Guthrie, *Textile Res. J.,* **31**, 84 (1961).

E. CROSSLINKED CELLULOSE

Giuliana C. Tesoro and John J. Willard

The year 1955, in which Part III of this treatise was published, marks the threshold of major advances in the investigation of cellulose crosslinking. It is significant that at that time only one page was devoted to the subject (pp. 1348–1349) and formaldehyde was the only crosslinking agent recognized. Since then the science and the technology of cellulose crosslinking have made impressive strides, and the section that follows attempts to review the important contributions made by many investigators during the last decade. Most of the work dealing with crosslinked cellulose has been carried out in the context of research on fibers and textile products. In fact the vast industrial technology that has both motivated and utilized these investigations is a textile technology.

For this reason crosslinking reactions of cellulose have been investigated almost exclusively in *heterogeneous* systems with the objective of maintaining the integrity of the cellulose fibrous structure. Some important limitations are therefore apparent in the scope of the knowledge that has been accumulated. Crosslinking reactions expected to proceed slowly (e.g., ether formation with dihalide) have been excluded from consideration or examined only superficially. The direct chemical analysis of reaction products has seldom been attempted, and the isolation of crosslinked-cellulose products has been deemed to be impossible. The reader is also referred to Section A in Chapter XIX for a discussion of the practical applications of crosslinking in the textile industry.

1. Crosslinking Reactions

(a) ETHER FORMATION

(1) Aldehydes

Formaldehyde. Since the observation by Eschalier (1) in 1906 that regenerated cellulose reacts with formaldehyde in the presence of acid to provide improved wet strength, this reaction has been the subject of innumerable investigations. Research efforts have been concerned with changes in the properties of cellulosic substrates resulting from treatments with formaldehyde under various conditions. The study of this reaction has been documented in a large number of patents and publications, and processes employing formaldehyde as a crosslinking agent for cellulose have also been used industrially. The theoretical aspects of the reaction of formaldehyde have received considerable attention, and the results of these fundamental investigations have contributed significantly to a broader understanding of the effect of covalent crosslinks on the properties of cellulosic substrates generally.

The literature on the reaction of cellulose with formaldehyde has been admirably covered in two reviews (2, 3) that appeared about ten years ago, and therefore the present review emphasizes work reported since that time.

Strong evidence has been brought forth in support of the hypothesis that formaldehyde condenses with cellulose with the loss of water to form *intermolecular crosslinks* between macromolecular chains. This reaction of formaldehyde was postulated by Meunier and Guyot (4) in 1929, even before a definition of the polymer structure of cellulose was generally accepted.

$$2R_{cell}OH + HCHO \xrightarrow{H^+} R_{cell}{-}O{-}CH_2{-}O{-}R_{cell} + H_2O \qquad (1)$$

Although the reaction shown in the above equation is certainly important, it has not been possible to establish to what extent adjacent, secondary hydroxyl groups on a single anhydroglucose unit may concurrently condense with formaldehyde to form *intramolecular crosslinks.*

Formaldehyde has been used commercially to provide dimensional stability to rayon fabrics in the 1950s (5), but it has largely been replaced by alternative crosslinking reactants (6). Low-level crosslinking of rayon in the spinning bath (7) is believed to involve at least in part the formation of unstable crosslinks through the xanthate groups (see also Section B in Chapter XIX).

$$R_{cell}O\overset{\overset{\displaystyle S}{\|}}{C}S{-}CH_2{-}S\overset{\overset{\displaystyle S}{\|}}{C}OR_{cell}$$

Conditions of Application. Formaldehyde can be reacted with celluose (a) in substantially anhydrous solvent media, (b) in the wet, or water-swollen, state, (c) by a baking, or "dry-cure," process, and (d) by using formaldehyde vapor. These treatments are sometimes referred to as the Form D, Form W, Form C, and Form V processes, respectively (8). The degree of fiber swelling at the time of reaction affects the rate and extent of reaction obtained, as well as the physical (9) and chemical (10) properties of the reacted substrate. The reaction proceeds very sluggishly in anhydrous solvent, and the reactive species is probably the hydrated form, methylene glycol, $CH_2(OH)_2$ (11). The concentration of formaldehyde affects the structure of the crosslinks formed as well as their number. It has been reported (12) that the use of more concentrated solutions yields longer, polymeric crosslinks and smaller changes in physical properties for a given amount of total combined formaldehyde. Preferential crosslinking in the peripheral regions of cellulosic fibers has been suggested when sulfuric acid was used as catalyst in a wet formaldehyde treatment (13).

Relatively drastic conditions of catalysis are required to bring about the condensation reaction. Formaldehyde is adsorbed reversibly by cellulose from neutral aqueous solutions (14), and the rate of adsorption is doubled by the addition of 5% of a neutral salt, such as potassium chloride (15), but acidic conditions are required to bring about the condensation reaction. It has been reported (16) that condensation products stable to boiling water (pH 7) are produced only with the use of a catalyst whose dissociation constant is 10^{-2} or greater. However, effective catalysis by acids whose dissociation constants are as low as 10^{-3} - 10^{-4} (e.g., lactic and tartaric acids) was later observed (17).

Kinetics of the Reaction. By using several types of cellulose substrate, Roff (18) showed that at all temperatures between 25 and 98°C the rate of the reaction in the presence of water increases rapidly as the pH falls below a critical value of 3 and that the reaction is extremely slow at higher pH. Rate curves reflected two first-order components, which were attributed to different reactivities of different cellulose hydroxyl groups.

An extensive study of the reaction between formaldehyde and cotton, viscose rayon, and Fortisan at temperatures ranging from 110 to 130°C has been described (19). The relationship between formaldehyde concentration and bound formaldehyde was found to be linear for curing times from 5 to 30 min at pH 2.0, 2.2, and 2.4. A more recent study (20) has shown that the high-temperature reaction is a pseudo-first-order one with respect to formaldehyde concentration. A complicated rate function with respect to catalyst concentration was revealed by the data and was interpreted as indicative of heterogeneous catalysis. Typical catalysts for the reaction were identified as Lewis acids, as distinguished from Brønsted acids.

Renewed interest has been shown in a room-temperature reaction between cellulose and formaldehyde, first described by Whinfield in 1936 (21). The reaction employs hydrochloric acid in dehydrating salt solutions, such as calcium chloride. In this system the reaction rate has been shown (22) to be dependent on the concentration of both formaldehyde and calcium chloride, and to go through a maximum at 35% calcium chloride. The rate appeared to be independent of hydrochloric acid concentration at 30°C (23), but it did show dependence at 16°C (24). In the latter study the reaction rate was resolved into two components, reflecting a fast reaction obeying first-order kinetics, followed by a slower, zero-order reaction.

Other Aldehydes. Glyoxal, the simplest dialdehyde, and higher aliphatic dialdehydes can be reacted with cellulose under conditions similar to those used for the reaction of formaldehyde (25). Hydrated dialdehydes are tetrafunctional compounds, and a study (26) of the reaction of glyoxal with cellulose has shown that all possible linkages are formed, namely, the hemiacetal, monoacetal, and diacetal.

Serious strength losses associated with formaldehyde crosslinking of cellulosic fabrics have often been attributed to the short crosslinks formed (one carbon in length). A crosslinking study carried out with dialdehydes of varying chain length [glyoxal (C_2), glutaraldehyde (C_5), and α -hydroxyadipaldehyde (C_6)] revealed relationships of crease recovery to strength similar to those obtained with formaldehyde (27). The chemistry of the reaction suggests that the acetal linkage governs the length of the crosslink and remains unchanged with the use of higher dialdehydes. Thus

$$2R_{cell}OH + OHC{-}R{-}CHO \xrightarrow{H^+} R_{cell}{-}O{-}\underset{}{\overset{\displaystyle RCHO}{\overset{|}{C}}}H{-}O{-}R_{cell} \qquad (2)$$

A systematic study (28) of the reaction of glyoxal with several types of rayon showed that the amount of free aldehyde and hemiacetal groups formed in the reaction increases initially and then decreases with increasing amounts of combined glyoxal.

Condensation products of formaldehyde and unsaturated aldehydes, particularly acrolein, can be reacted with cellulose under acid-cure conditions and provide dimensional stability with modest increases in wrinkle recovery (29).

Acetals. Simple monomeric acetals and polyformals (30) that can release formaldehyde at elevated temperature are relatively inefficient crosslinking agents and have not gained commercial acceptance.

Simple methylene ether bridges can be introduced into cellulose via transacetalization in dry refluxing diethylformal in the presence of a trace of acid catalyst (31).

Crosslinking with methanol hemiformal (55% formaldehyde in methanol) applied to cellulose from the vapor phase, together with formic acid as catalyst, has been reported (32) to give high crease resistance at fixed weight gains of less than 1%. This reagent-catalyst combination has also been applied to cotton fabric pretreated with cyclic amide precursors of known crosslinking agents.

(2) N-*Hydroxymethyl Compounds*

Pioneering work at Tootal, Broadhurst and Lee in England in the late 1920s established that some practical disadvantages noted for formaldehyde crosslinking of cellulose could be overcome by the use of urea-formaldehyde resin precondensates (33). This discovery laid the foundation for an extensive, worldwide technology based on the use of polyfunctional, *N*-hydroxymethyl amides for the treatment or "finishing" of fabrics containing cellulosic fibers to improve their dimensional stability, crease resistance, wash-and-wear performance, and durable-press properties.

Reactants. Under neutral or slightly alkaline conditions formaldehyde condenses reversibly with urea and its derivatives, with triazines, and with amides or carbamates to form the *N*-hydroxymethyl, or methylolated, adducts. Examples are shown in equations 3, 4, and 5 for ureas, triazines, and carbamates, respectively.

$$2CH_2O + HN(R)-\overset{\displaystyle O}{\overset{\|}{C}}-NH(R) \longrightarrow HOCH_2N(R)-\overset{\displaystyle O}{\overset{\|}{C}}-N(R)-CH_2OH \qquad (3)$$

(4)

$$2CH_2O + RO\overset{\displaystyle O}{\overset{\|}{C}}NH_2 \longrightarrow RO\overset{\displaystyle O}{\overset{\|}{C}}N(CH_2OH)_2 \qquad (5)$$

The condensation reaction can also be catalyzed by acid, but the products of alkaline catalysis are generally superior in terms of freedom from by-products and contaminants. The kinetics of the reaction of formaldehyde with amide compounds under conditions of acid and base catalysis have recently been compared (34). Reactivity in methylolation is a function of the basicity of the nitrogen and is related to the tendency toward mesomerism with the adjacent carbonyl group.

Specific reactants that have become important in commercial finishing of cellulosic materials include the following:

Ureas. Dimethylolurea (**1**, formaldehyde-to-urea ratio of 2 : 1) slowly polymerizes in aqueous solution in the presence of acid.

$$HOCH_2NH\overset{\displaystyle O}{\overset{\|}{C}}NHCH_2OH$$

1

In fabric treatment with this compound some polymer is deposited on the fabric surface, causing stiffness. To obtain good stability of the treating solution, minimum release of formaldehyde during and after curing, and improved fabric properties, the use of a formaldehyde-to-urea ratio of about 1.6 is generally preferred (35).

Cyclic ureas. Methylolated cyclic urea derivatives have been the subject of much study and have great practical significance (see Section A Chapter XIX). Of particular importance are the compounds commonly referred to as dimethylolethyleneurea (DMEU, **2**), dimethylolpropyleneurea (DMPU, **3**), and dimethyloldihydroxyethyleneurea (DMDHEU, **4**) (36).

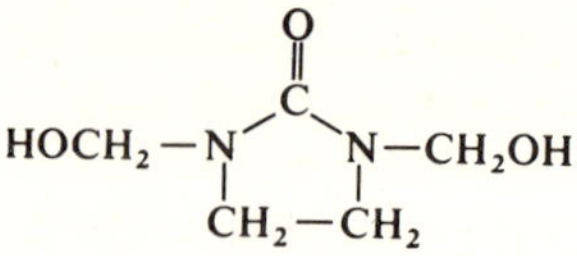

2

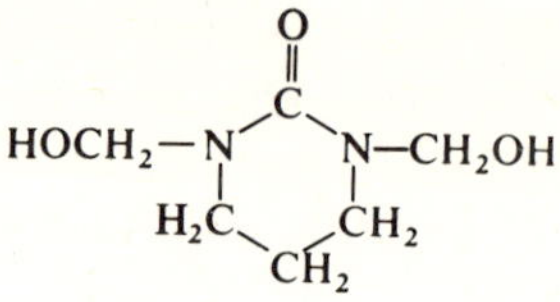

3

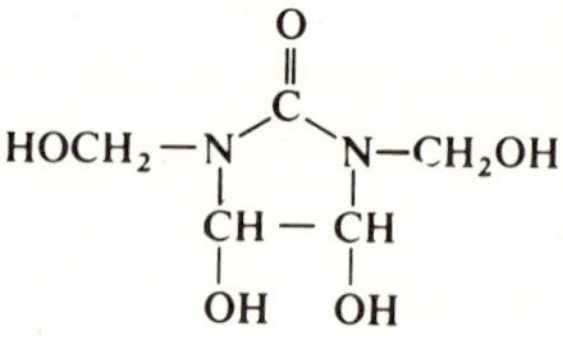

4

DMDHEU is the product obtained by the reaction of formaldehyde, urea, and glyoxal at an acid pH (37). The hydroxyl groups attached to the imidazolidone ring can participate in a crosslinking reaction with cellulose at elevated temperatures, and their presence probably contributes to the interesting combination of properties obtained with this reagent, which is, in effect, *tetrafunctional.*

Other cyclic urea derivatives, such as 1,3-bis(hydroxymethyl)-5-alkylperhydro-2-triazone (**5**) (38) and tetrahydro-3,5-bis(methoxymethyl)-4*H*-1,3,5-oxadiazin-4-one (**6**) (39), have also been utilized as effective crosslinking agents for cellulose.

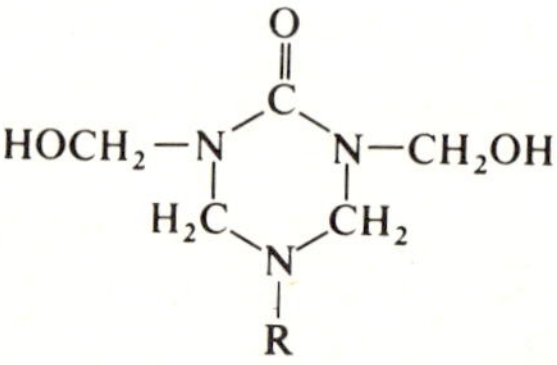

R = alkyl

5

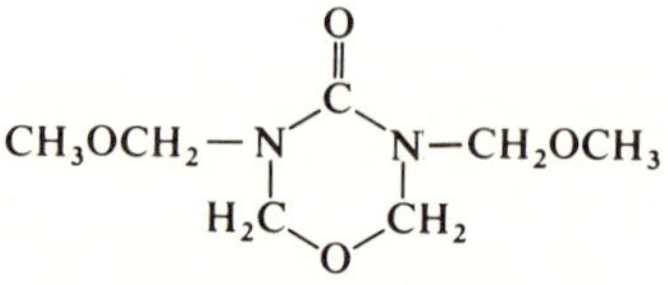

6

Triazines. Methylolated aminotriazines, particularly melamine, were among the earliest reactants used extensively on a commercial scale (40).

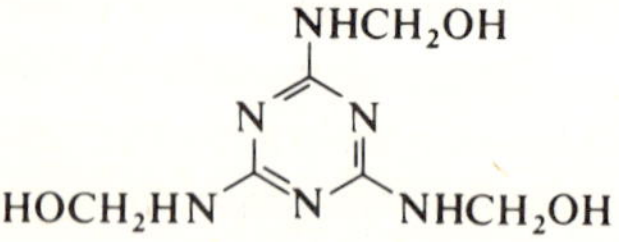

7

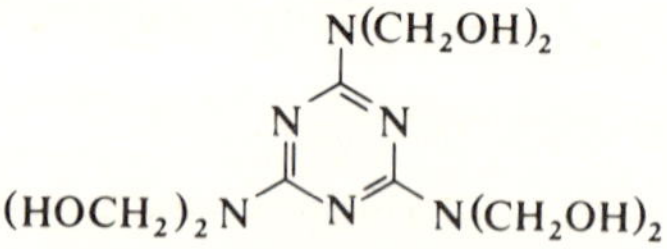

8

Trimethylolmelamine (**7**) and hexakismethylolmelamine (**8**) are prepared by reacting formaldehyde with melamine. These compounds (and a number of related derivatives, including the methoxymethyl analogs) have been the object of considerable study, particularly in the 1950s (40). The polyfunctional methylolated melamines are prone to resin formation by self-condensation. The comparatively high basicity of their nitrogen provides finishes of high durability.

Amides. The crude products of the reaction of simple amides, such as formamide and acetamide, with 2-3 moles of formaldehyde, when applied to fabric in the presence of acid catalysts, provide substantial increases in fabric crease recovery (41). The behavior of the product in acid and alkaline hydrolysis has been interpreted as evidence that the dimethylol adduct of the amide is the reactive species, forming a crosslinked cellulose (**9**).

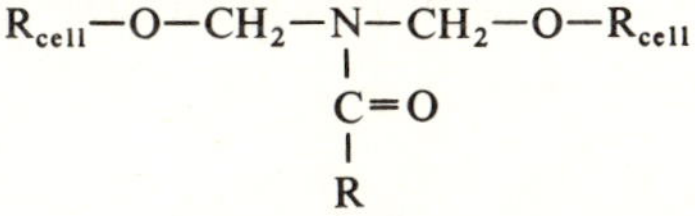

($R = H, CH_3$)

9

By way of contrast, formamide, or acetamide, applied to cellulose in equimolar amounts with formaldehyde, has been claimed (42) to form the singly substituted derivative **10**, but crease recovery is not developed.

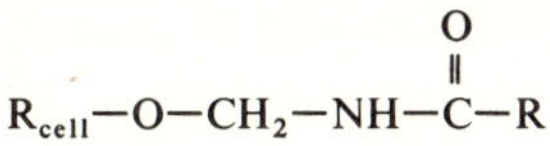

($R = H, CH_3$)

10

The very low resistance to chlorine of cellulose crosslinked with methylolated amide derivatives indicates that an appreciable amount of the singly substituted species is formed, and this has prevented commercial interest in the methylolated amides as crosslinking agents for cellulose.

The acid-catalyzed reaction of dimethylol derivatives of a variety of amides of dicarboxylic acids, including compound **11**,

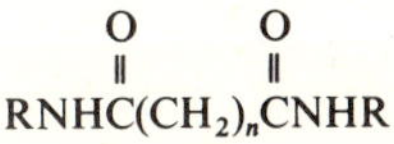

($n = 1$–3, $R = H, CH_3$)

11

has been studied but, in this case also, resistance to chlorine was poor (43).

Carbamates. Investigations at the Southern Regional Research Laboratory of the U.S. Department of Agriculture have resulted in the commercial development of crosslinking agents for cellulose based on methylolated alkyl carbamates. When ethylenebis(*N*-methylol carbamate) (**12**) was used, the finished fabric showed excellent durability to acid and alkaline hydrolysis but was susceptible to chlorine damage.

$$HOCH_2NHC(O)OCH_2CH_2OC(O)NHCH_2OH$$

12

Substitution on the amido hydrogens [e.g., ethylenebis(*N*-methylol-*N*-methyl) carbamate] resulted in reduced reaction efficiency. It was found (44) that effective crosslinking agents can be prepared by formation of bismethylol derivatives of a monocarbamate [$ROCON(CH_2OH)_2$], and *N,N*-dimethylol alkyl carbamates were recently introduced as a new class of crosslinking agents for cellulose.

The reaction of formaldehyde with alkyl carbamates under alkaline conditions has been studied systematically as a function of pH, reaction time, and molar ratio (45). Methyl, ethyl, isopropyl, and hydroxyethyl carbamates are the most useful (46) among the carbamates examined as starting materials for methylolation. Alkoxyethyl carbamates (47) and methylol derivatives of polycarbamates (48) have also been suggested.

As in the case of other *N*-hydroxymethyl reactants, resistance to chlorine damage in the crosslinked fabric is critically dependent on the choice of catalyst and reaction conditions. In the case of *N,N*-dimethylol alkyl carbamates the size and structure of the alkyl group (49) affect reactivity and thus, indirectly, the resistance to chlorine damage.

Application. Methylolated reactants are generally applied to cellulosic substrates by impregnation from aqueous solution together with an acid, or acid-forming, catalyst. The substrate is dried, and the crosslinking reaction is induced by heating at 130–160°C for a few minutes, the preferred conditions of this curing step depending on the substrate, reactant, and catalyst used. For example, in the case of ureas,

$$2R_{cell}OH + HOCH_2{-}N(R)C(O)N(R){-}CH_2OH \longrightarrow R_{cell}{-}OCH_2{-}N(R)C(O)N(R){-}CH_2O{-}R_{cell} \qquad (6)$$

The crosslinking reaction can be carried out in the presence of alkaline catalysts, such as sodium bicarbonate (50), but acid catalysis is employed exclusively in practice. The crosslinking reaction is never as simple as the one depicted above. Polymerization of the reactant can occur; formaldehyde can be split off under the conditions of treatment and react with cellulose, with the formation of methylene ether crosslinks. Among the catalysts commonly employed are the chlorides and nitrates of zinc, magnesium, ammonium, and amines.

The kinetics of the reaction of DMEU, DMDHEU (51), and of DMPU (52) with cellulose have been studied at 45°C in the presence of the nitrate and chloride salts of magnesium and zinc. No simple relationship of reactivity to the basicity of the nitrogen, to the carbonyl wavenumber, or to the rate of methylolation could be established (34).

Polyfunctional methylolated amides that are prone to self-condensation can be insolubilized within cellulosic fibers in the presence of oxidative catalysts, such as hydrogen peroxide, or of acid catalysts under mild conditions. Such reactions, often referred to as wet fixation, are carried out on swollen fibers in the presence of water (53). They result in minimal formation of crosslinking, but they have been combined with crosslinking processes to obtain special effects (54).

Finish Durability. Treatments based on the reaction of urea-formaldehyde condensates with cellulose lack durability to repeated laundering. The development of cyclic urea derivatives and other improved reactants was initially designed to overcome this problem. Furthermore, when active hydrogen is attached to the nitrogen in the *N*-hydroxymethyl reactant, a chloramide can be formed in the presence of hypochlorite bleaching solution. The labile chloramide can release hydrochloric acid during ironing, causing damage to the fabric.

$$\sim\!\!\underset{}{\overset{\displaystyle H}{\overset{|}{N}}}\!-CH_2-O\!\sim + Cl_2 \longrightarrow \sim\!\!\overset{\displaystyle Cl}{\overset{|}{N}}\!-CH_2-O\!\sim + HCl \qquad (7)$$

$$\sim\!\!\overset{\displaystyle H}{\overset{|}{N}}\!-CH_2-O\!\sim + HOCl \longrightarrow \sim\!\!\overset{\displaystyle Cl}{\overset{|}{N}}\!-CH_2-O\!\sim + H_2O \qquad (8)$$

Generally, the choice of catalyst and curing conditions can have a pronounced effect on the chemical stability of a given product in laundering (55). Hydrolysis of crosslinks can lead to loss of formaldehyde and to the formation of chlorine-sensitive secondary amide groups. Thus a product initially resistant to chlorine can become sensitive after laundering. Reactants containing basic nitrogen—for example, triazones and triazines—offer some built-in buffering capacity, with increased resistance to damage resulting from chloramide formation (56) and potential liberation of hydrochloric acid.

(3) *Activated Vinyl Compounds*

The reaction of divinyl sulfone (57) provides the first documented example of a crosslinking reaction of cellulose with a compound containing two activated vinyl groups. The relationship of this reaction to the extensively investigated *cyanoethylation* reaction (see Section D of this Chapter XVII) has suggested that the alkali-catalyzed reaction of activated olefinic compounds containing more than one reactive group could provide new crosslinking systems. Thus the bifunctional compound *N,N'*-methylenebisacrylamide produced some evidence of crosslinking (58).

A number of subsequent investigations have centered on the crosslinking of cellulose with saturated derivatives (or precursors) of divinyl sulfone and of bis(vinylsulfonyl) compounds, in which an elimination reaction generates activated vinyl groups under conditions of alkaline catalysis. The initial motivation was to utilize the high reactivity and apparent efficiency of these crosslinking agents, while avoiding the difficulties associated with the toxic, vesicant, and lachrymatory properties of the vinyl compounds.

Reactions of ionic derivatives of divinyl sulfone (59) and of bis(vinylsulfonyl) compounds (60) were found to proceed readily at room temperature in the presence of alkali. The elimination and addition reactions can be represented by the following equations where Y represents an ionic leaving group:

Elimination:

$$-SO_2CH_2CH_2Y + NaOH \longrightarrow -SO_2CH{=}CH_2 + NaY + H_2O \qquad (9)$$

Addition:

$$-SO_2CH{=}CH_2 + R_{cell}OH \xrightarrow{OH^-} -SO_2CH_2CH_2O{-}R_{cell} \qquad (10)$$

Since both reactions require the presence of hydroxide ion, it is evident that when elimination takes place in the presence of cellulose (*in situ*), the addition reaction can proceed concurrently, provided base is present in excess of the amount required to neutralize the leaving group. Bis(vinylsulfonyl) compounds containing different leaving groups [e.g., $(-\overset{+}{N}C_5H_5)Cl^-$, pyridinium; $-SSO_3Na$, thiosulfate; $-OSO_3Na$, sulfate; $-OOCCH_3$, acetate] have been reported (59, 60). The rate of the elimination reaction observed under given reaction conditions for ionic derivatives of divinyl sulfone ($YCH_2CH_2SO_2CH_2CH_2Y$) was found to depend on the structure of the leaving group Y (61). However, in the crosslinking of cellulose with these compounds the addition reaction to the vinyl groups formed remains the rate-determining step for reagents where the elimination reaction is sufficiently rapid.

The alkali-catalyzed elimination of water or alcohol from bis(hydroxyethyl) sulfone, bis(alkoxyethyl) sulfone, and other bis(oxyethylsulfonyl) compounds takes place at elevated temperatures

$$-SO_2CH_2CH_2OR \xrightarrow{OH^-} -SO_2CH{=}CH_2 + ROH \qquad (11)$$

and the reaction has been used effectively for the crosslinking of cellulose with these reagents under anhydrous conditions (62, 63).

The use of ionic and oxyethyl derivatives of bis(vinylsulfonyl) compounds has resulted in a large number of patents, some of which reflect the commercial utilization of cellulosic fabrics crosslinked with derivatives of bis(vinylsulfonyl) compounds (64–66).

The widely different conditions under which the reaction of the $YCH_2CH_2SO_2-$ and $ROCH_2CH_2SO_2-$ groups discussed above takes place have led to the development of unsymmetrical crosslinking reagents containing these reactive groups. Furthermore, differences in the properties of crosslinked fabrics obtained in the presence and in the absence of water with derivatives of divinyl sulfones have been utilized for a permanent-creasing process based on heat-curing of fabrics crosslinked at room temperature in the swollen state (67).

The concepts utilized for sulfone-activated vinyl compounds have been applied, in part, to the corresponding *sulfoxides* (68, 69), but the activating effect of the sulfoxide group is not sufficient to confer the level of reactivity required for the crosslinking of cellulose under relatively mild reaction conditions. The strong activating effect of the *sulfonium* group, on the other hand, has permitted the use of derivatives of vinylsulfonium compounds (70) in exceedingly rapid crosslinking reactions at room temperature. The disodium salt **(13)** of tris(β-sulfatoethyl)sulfonium inner salt reportedly has been used commercially in Europe for the crosslinking of cellulosic fabrics.

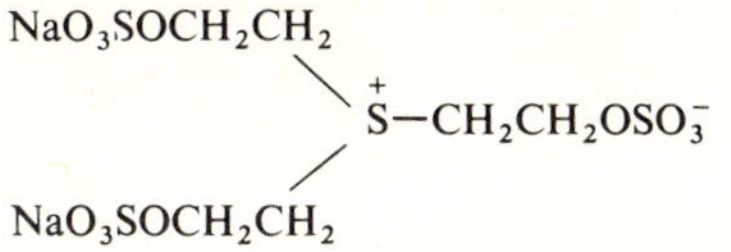

13

Other activated vinyl compounds and their derivatives (such as acrylamides and vinyl ketones) have also been studied, particularly with respect to the relationship of their reactivity to the hydrolytic stability to alkali of the crosslinked cellulose ethers formed from them. The crosslinking of cellulose with reagents containing hydroxyl or alkoxyl groups in a β-position to activating groups other than sulfones has also been reported (71) (see also section D of Chapter XVII).

The products of the base-catalyzed condensation of acetone (72) or other ketones (73) with formaldehyde are related to bis(β-hydroxyethyl) ketones and could thus be considered to be precursors of vinyl ketones. These compounds have been used to provide dimensional stability in cellulose (74); the finished fabric can be made stable to discoloration by a mild oxidation treatment (75).

(4) Ring-Opening Reactions

The reaction of the hydroxyl groups in cellulose with polyfunctional compounds containing epoxy **(14)** or aziridinyl **(15)**

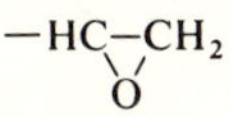

14

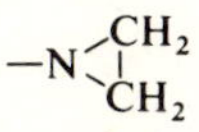

15

groups leads to the formation of crosslinked cellulose through opening of the strained three-membered rings. Generally speaking, the crosslinking reaction is accompanied by the polymerization of the reagent. The polymerized reagent can be removed by washing or extraction, but the crosslinked cellulose is inevitably a complex product containing crosslinked homopolymers, crosslinks of varying length and structure, and grafted polymeric side chains. The complexity of the products obtained is such that the reactions can be illustrated only by schematic representations. The characterization of the products by chemical means has not been attempted, and product compositions and structures have been postulated on the basis of changes in the properties of the cellulose substrate commonly associated with crosslinking.

Aziridinyl Compounds. The reaction of cellulose with tris(aziridinyl)-phosphine oxide (APO, **16**), the reaction product of phosphorus oxychloride and ethylenimine, was first investigated (76) in conjunction with the use of APO as a flame-retardant compound for cotton (see Section I of Chapter XIX).

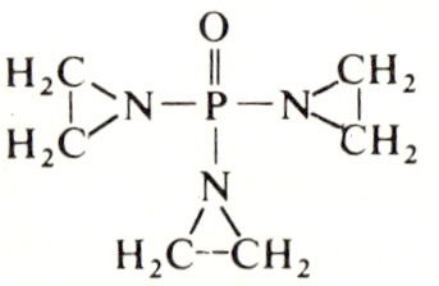

16

The results of its application to cotton as a crosslinking agent and crease-resistant finish (77) suggested that extensive crosslinking takes place in the presence of strongly acidic catalysts, such as zinc fluoroborate. Reactions of cellulose with the reaction products of phosgene with ethylenimine (78), of substituted ureas (**17**) obtained by the reaction of alkylene diisocyanates with ethylenimine, of the urethanes (**18**) obtained by the reaction of bischloro-fomates with ethylenimine, and of amides (**19**) obtained by the reaction of ethylenimine with acid chlorides of dicarboxylic acids have also been investigated (79).

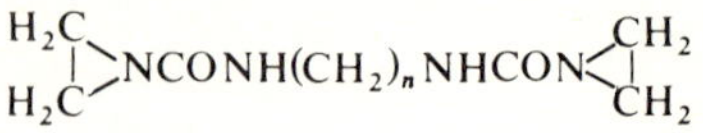

17

18

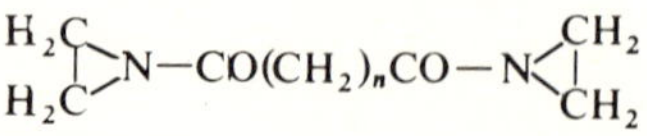

19

Crosslinking efficiency was found to vary with structure, but it was generally lower than that for many other crosslinking agents, undoubtedly due to side reactions arising from polymerization of the aziridinyl compounds. The hexafunctional reagent derived by the reaction of ethylenimine with phosphonitrilic chloride (80) was found to crosslink cellulose in the presence of acid catalysts, although the extent of crosslinking, as indicated by changes in properties, was low. The reaction products (**20**) of ethylenimine with dialdehydes (81) are even less efficient, the complexity of polymerization products being perhaps increased by the presence of hydroxyl groups in the crosslinking reagent.

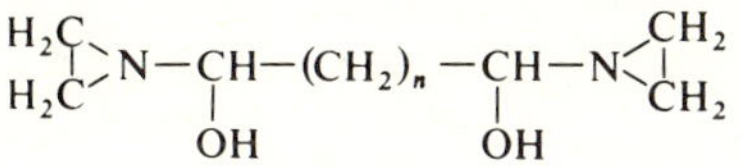

20

It has also been indicated (82) that aziridinyl compounds in which the nitrogen of the aziridinyl ring is an amino nitrogen, as in the case above, polymerize more readily than those of the urea, urethane, or amide type. Reactive aziridinyl groups are consumed in the polymerization, and the reaction with cellulose hydroxyl groups does not take place.

Epoxides. A review of the mechanism of epoxide reactions (83) shows that opening of the epoxide ring (followed by addition of active hydrogen) generally takes place by an ionic mechanism (equation 12) and can be catalyzed by base or by acid.

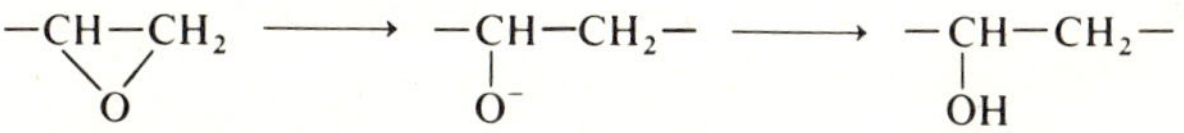

(12)

Furthermore, the facile dehydrohalogenation of α-halohydrins by base (84) (equation 13) suggests that α-halohydrins can be considered as precursors of epoxides in the presence of base.

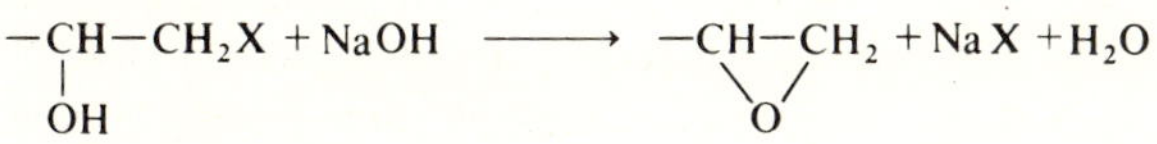

(13)

Polyfunctional epoxides, and epoxide precursors, have been investigated extensively as crosslinking agents for cellulose, by varying the structure of the reagent and the conditions of reaction over a broad range. Since the reaction of an epoxide group with hydroxyl results in the formation of a new hydroxyl group, it is immediately apparent that the newly formed hydroxyl group can also react with epoxide present, and reactions of epoxides with cellulose, particularly crosslinking reactions involving polyfunctional epoxides, are thus complicated by side reactions, including the formation of graft copolymers and polymeric crosslinks.

Epichlorohydrin. Chemical reaction between alkali cellulose and epichlorohydrin was first observed in 1921 (85), but the first indication that crosslinks could be formed in the reaction is to be drawn from the observation (86) that insolubility in cuprammonium hydroxide results from the treatment of cellulosic fabrics with epichlorohydrin or its glycerol halohydrin precursors in the presence of strong alkali. Further evidence that epichlorohydrin and dichloropropanols can crosslink cellulose is to be found in the crease-recovery improvements obtained with these reagents (87); these are used as the basis of a patented process of commercial significance in the early 1960s. The action of epichlorohydrin in the presence of alkali and various salts on the crease recovery of cotton was subsequently investigated extensively by using the reagent in the liquid (88) and in the vapor phases (89). It was generally concluded that crosslinked cellulose is indeed obtained in the base-catalyzed reaction of epihalohydrins with cellulose, although many complicating side reactions lead to the formation of polymeric crosslinks and side chains in addition to the simplest crosslinked product (**21**) postulated.

$$R_{cell}{-}OCH_2CH(OH)CH_2O{-}R_{cell}$$

21

The related reaction of dihalopropanols with cellulose in the presence of base undoubtedly proceeds through an epihalohydrin intermediate (84) and leads to the formation of closely similar, complex products.

The acid-catalyzed reaction of epichlorohydrin vapor with cotton cellulose has been studied (90), and some evidence was obtained that reaction proceeds through the formation of a cellulose chlorohydrin ether, capable of further reaction to form crosslinked products.

Bishalohydrins. The rapid reaction of bischlorohydrins with cellulose in the presence of alkali has been attributed to the formation of intermediate epoxides (91). Evidence was obtained that esterification or etherification of the hydroxyl groups in the chlorohydrins prevents epoxide formation and thus decreases reactivity by several orders of magnitude. Reactive crosslinking agents investigated included 1,3-dichloro-2-propanol and 1,4-dichloro-2,3-butanediol. The dehydrohalogenation of *dl*-2,3-dibromo-1,4-butanediol and of the *dl* and *meso* isomers of 2,3-dichloro-1,4-butanediol on cotton in the presence of several alkaline catalysts has been reported (92) to yield products in which the extent of crosslinking, as indicated by crease-recovery improvement, varies with the catalyst and the reaction medium.

Polyfunctional Epoxides. The first extensive study of crosslinking with polyfunctional expoxides (93) included consideration of polyglycidyl ethers of polyols, of diepoxybutane, and of vinylcyclohexene diepoxide, reacted with cotton under strongly acidic conditions, in the presence of fluoroborates, perchlorates, and sulfates. On the basis of the results of a subsequent investigation (94) the conclusion was reached that in order to be effective crosslinking agents for cellulose under conditions of acid catalysis, bifunctional epoxides must be somewhat water soluble, nonvolatile, and free from interfering steric and polar effects.

Further studies of the reactivity of polyfunctional epoxides of varying structure in the presence of zinc fluoroborate have been reported (95–97). Among the polyfunctional epoxides studied as crosslinking agents for cellulose, 1,2-3,4-diepoxybutane, or butadiene diepoxide, has been most thoroughly investigated. The effect of alkaline and acidic catalysts and of other reaction conditions on the reaction of the *meso* isomer with cotton have been reported (98), and the products obtained from the base-catalyzed reaction of the *dl*-isomer have also been examined (99). The base-catalyzed reaction of the *dl*-isomer in the vapor phase has been reported (89).

The optimum conditions for the crosslinking of cellulose with butadiene diepoxide have been defined on the basis of kinetic studies of the hydrogen-ion-catalyzed hydrolysis of the diepoxides (*dl* and *meso* isomers) and found to include zinc fluoroborate catalysis, and specific ratios of diepoxide to catalyst and water to diepoxide (100, 101). The reaction of cellulose with 1,3-diglycidyl glycerol (102) in the presence of a large number of "curing agents," including acidic catalysts and coreactants, gave seemingly complex crosslinked products. Reaction with *N,N,N*-tris(2,3-epoxypropyl)amine and its precursor, *N,N,N*-tris-(3-chloro-2-hydroxylpropyl)amine (103) under conditions of alkaline catalysis also gave complex, poorly characterized products. The trisepoxides could also be reacted in the presence of acidic catalysts, and in this case also the crosslinking reaction was greatly complicated by polymer formation.

(5) Miscellaneous Alkylation Reactions

The reactivity of saturated, unsubstituted aliphatic halides is not generally sufficient to permit the use of simple dihalides as crosslinking agents for cellulose under reaction conditions that would avoid degradation of the polymer substrate and excessive loss of crosslinking reagent through hydrolysis. The results of a detailed study of the reaction of methylene dibromide with sodium cellulosate and with alkali cellulose (104) at elevated temperature support this generalization. A comparative investigation of bifunctional aliphatic chlorides as crosslinking agents for cellulose (91) has included consideration of unsaturated structures (e.g., 1,4-dichloro-2-butene and 1,4-dichloro-2-butyne) and of dichlorohydrins (e.g., 1,3-dichloro-2-propanol). The expected order of reactivity has been observed, with unsaturation alpha-beta to the halogen increasing the reactivity significantly. The results obtained with the methyl ether of 1,3-dichloro-2-propanol are consistent with those reported for the esters of this compound (105, 106) and indicate that the high reactivity of the chlorohydrins is due to the formation of intermediate epoxides. When hydroxyl groups are replaced by alkoxyl or acyl groups, the reactivity of the dichloride is of the same low order of magnitude as that observed for unsubstituted dihalides.

Systematic investigations of dihalides as crosslinking agents for cellulose, particularly for textile uses, have often been difficult due to the low solubility of the reagents in water. Rath and his co-workers have examined the behavior of cleavable "onium compounds" as water-soluble precursors of dihalides in considerable depth (107) and have compared the effectiveness of a large number of structures as crosslinking reagents for cellulose (108). Bispyridinium compounds obtained from the bischloromethyl ethers of diols were among the most reactive "onium" compounds investigated. Work with pyridinium compounds prepared from bischloromethyl ethers (109) has defined the reaction conditions required to obtain optimum utilization of the reagents in the crosslinking reaction, avoiding side reactions originating from the decomposition of the reagent into formaldehyde, pyridine hydrochloride, and diol.

Disulfonate esters of polymethylene diols $[RSO_2O(CH_2)_nOSO_2R]$ have been reported as potential crosslinking agents for cellulose (110). Evidence of crosslinking was obtained for the bismethyl sulfonyl esters ($R = CH_3-$), whereas the bistolylsulfonyl esters ($R = CH_3C_6H_4-$) were not sufficiently reactive, possibly due to the lack of solubility under the conditions studied.

α,ω-Bis(diazo)alkanes $[N_2CH(CH_2)_nCHN_2]$ have been shown to react with cotton fabric (111) and to improve crease recovery. The reagents are, however, difficult to prepare and unstable.

Interesting results have been obtained in the reaction of cotton with the vinyl ethers of polyols (112). The reaction of a vinyl ether with cellulose is postulated to occur forming an acetal as follows:

$$-CH_2OCH{=}CH_2 + R_{cell}OH \longrightarrow R_{cell}{-}O\underset{\displaystyle CH_3}{\underset{|}{C}}HO{-}CH_2{-} \tag{14}$$

Good experimental evidence was obtained for the formation of acetal crosslinks. The efficiency of the divinyl ethers of diols as crosslinking agents varied over a considerable range. The reaction generally required a strongly acidic catalyst (e.g., hydrochloric acid), elevated temperature, and strictly anhydrous conditions, making it of doubtful practical value. The acetal crosslinks formed have excellent stability to alkali but liberate acetaldehyde quantitatively (2 moles per crosslink) in the presence of acid.

A number of other polyfunctional alkylating agents have been proposed as crosslinking agents for cellulose in the patent literature, but documented evidence of true crosslinking is often lacking. Dihydroxyacetone (113) might be mentioned in this context.

(b) ESTER FORMATION

The modification of cellulose with reagents capable of forming simple diester crosslinks (**22**) has been described in studies dealing with the fundamental aspects of cellulose crosslinking.

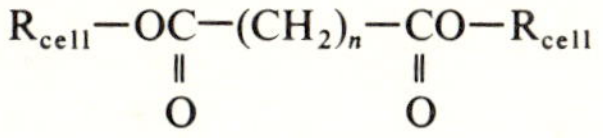

22

In view of the specific conditions required for the formation of the ester linkage and of the limited stability of the products to alkaline hydrolysis, these derivatives have not been considered for commercial use.

On the other hand, the reactions involved in the formation of diester crosslinks are well suited for fundamental studies where the determination of the number of crosslinks present by chemical analysis is desirable. The acid chloride, acid anhydride, or the dicarboxylic acid can be reacted under such conditions that dimerization of the reagent is avoided. Furthermore, the amount of carboxyl-substituted monoester that is formed can be determined readily by simple analytical techniques (114). An important limitation in the study of the diester crosslinks of cellulose must be recognized, however; since methods in which interfering side reactions, particularly acid degradation, are avoided depend on the use of swollen cellulose in nonaqueous solvents, direct comparison of the products with other crosslinked cellulose derivatives prepared in aqueous media or in the dry state is not possible.

Diester crosslinks can be introduced into cotton in several ways with minimal degrees of monosubstitution.

(1) Acid Anhydrides

Phthalic and maleic anhydrides have been reacted with cotton cellulose to a degree of substitution (DS) of about 1 by using benzenesulfonic acid as catalyst. A high proportion (>95%) of diester crosslinks is formed, but appreciable degradation of the cellulose was indicated (115). Succinic anhydride, or the acid in the presence of acetic anhydride in refluxing pyridine, also yielded primarily crosslinked product (116).

(2) Acid Chlorides

Limited use has been made of the dicarboxylic acid chlorides to study the formation of diester crosslinks. It has been shown that the acid chlorides of dicarboxylic acids of varying chain length (from succinyl to sebacoyl) react readily with cotton at room temperature in dimethylformamide (117).

(3) Dicarboxylic Acids

Carboxylic acids can be reacted with cellulose by using trifluoroacetic anhydride as condensing agent, and this elegant procedure was used by Campbell and Francis (118) in a comprehensive study of cellulose containing diester crosslinks of increasing chain length (C_3-C_{22}). Esterification did not take place with oxalic acid, and succinic and glutaric acids showed only slight reactivity. Other dibasic acids reacted readily to give the crosslinked products with cotton fabric that had been water swollen and solvent exchanged to benzene. Ester crosslinking of cellulose with polycarboxylic acids, reacted by curing at elevated temperatures in the presence of acid catalysts, has been examined (119). The products were not well characterized, and appreciable acid damage to the cotton fabric was indicated by irreversible losses in strength.

(4) Diisocyanates

As early as 1942, alkylene diisocyanates were studied (120) as crosslinking agents for viscose rayon in nonaqueous media to provide improved wet and dry strength and to reduce swelling. Similar studies with benzene-included cotton cellulose were described (121, 122) subsequently. More recently it has been shown (123) that water (50-60% optimal moisture content) can exert a catalytic effect on the reaction between hexamethylene diisocyanate and cellulose when the reagent is applied from an inert solvent, such as toluene.

(c) CROSSLINKING WITH UNSYMMETRICAL REAGENTS

Crosslinking reactions in which the functional groups of the polyfunctional reagents differ widely in reactivity and can be linked to cellulose in separate reaction steps offer the possibility of separating the effect of substituents introduced into the cellulose molecule by crosslinking. A generic schematic representation of such stepwise crosslinking reactions is shown in the following equations, where X and Y represent the functional groups of an unsymmetrical, bifunctional crosslinking reagent:

$$R_{cell}OH + X{-}R{-}Y \longrightarrow R_{cell}{-}O{-}R{-}Y \qquad (15)$$

$$R_{cell}{-}O{-}R{-}Y + R_{cell}OH \longrightarrow R_{cell}{-}O{-}R{-}O{-}R_{cell} \qquad (16)$$

Gardon (124) reported the results of a study on the stepwise-crosslinking reaction of cotton with *N*-methylolacrylamide ($HOCH_2NHCOCH{=}CH_2$), proposed earlier by Mantell (125), the hydroxymethyl group being reacted in the presence of acid catalyst and the vinyl group being subsequently reacted in the presence of an alkaline catalyst or a free-radical initiator.

In a later publication (126) Gardon reported additional results concerning the free-radical-catalyzed reaction of the acrylamidomethyl ether of cotton ($R_{cell}-OCH_2NHCOCH{=}CH_2$) formed in the first step (see also Section D of this Chapter XVII).

The use of unsymmetrical reactive sulfones for the stepwise crosslinking of cellulose was investigated by Tesoro et al. (127), who examined the stepwise-crosslinking reaction of reagents **23** and **24** on cotton and rayon fabric, yarn, and fiber and on polymer solutions [viscose, starch, and poly(vinyl alcohol)].

$ROCH_2CH_2SO_2CH{=}CH_2$	$ROCH_2CH_2SO_2CH_2CH_2Y$
($R = H, CH_3$)	(Y = an ionic leaving group)
23	**24**

These authors obtained evidence that the formation of side chains in the first step of the alkali-catalyzed, two-step crosslinking process could be carried out at room temperature without significant complicating side reactions. The formation of crosslinks required elevated temperature in a second step.

Other crosslinking agents containing functional groups of different reactivity have been proposed in the patent literature. However, the results reported do not demonstrate true stepwise formation of crosslinks with these reagents.

The reactions of epichlorohydrin (90), acrolein (128), glycidaldehyde (129), and alkylenimine derivatives of activated sulfones (130) offer possible routes to further investigation of unsymmetrical crosslinking agents. The availability of suitable compounds is necessarily limited by the requirements that reactivity of the functional groups differ sufficiently to permit stepwise reaction and that the second functional group be unaffected under the reaction conditions used to link the reagent molecules to cellulose through the first functional group. The polymerization of acrolein within rayon or cotton fibers gave products characteristic of crosslinked materials, but of poor alkaline stability. Efforts to improve the latter property resulted in the loss of crosslinks, which were probably of the hemiacetal type (131).

(d) CROSSLINKING OF CELLULOSE DERIVATIVES

Studies of crosslinked cellulose derivatives can be either (a) modification of cellulose to introduce new reactive groups, which are then used to form new types of covalent crosslinks, or (b) treatments in which the cellulose is partially modified by simple derivative formation and the new substrate is further modified by the reaction of the remaining hydroxyl groups with known crosslinking systems. In the latter category several investigations have been concerned with the overall effect exerted on physical properties by the chemical modification of the hydroxyl groups of a cellulosic substrate either before or after a crosslinking treatment.

(1) Reactions with Unsaturated Compounds

Narita and co-workers (132) have studied the effect of crosslinking cellulose with *N,N'*-dimethylolitaconamide (**25**), a bifunctional reagent that also contains a vinyl group.

$$CH_2{=}\underset{\displaystyle CONHCH_2OH}{\underset{|}{C}}{-}CH_2CONHCH_2OH$$

25

The proper choice of catalysis conditions permits crosslinking to be accomplished before or after polymerization through the vinyl group. One catalyst—namely, zinc fluoroborate,—brought about both reactions simultaneously.

The gelation of cellulose crotonate esters in dioxane solution has been studied (133) with benzoyl peroxide being used to initiate crosslinking. Several side reactions that accompanied crosslink formation were identified by detailed chemical analysis. Solutions of cellulose acetate-crotonate mixed esters have been crosslinked by the addition of ethylenediamine across the double bond, giving compound **26**.

$$(R_{cell}{-}OCOCH_2{-}\underset{\displaystyle CH_3}{\underset{|}{CH}}{-}NHCH_2{-})_2$$

26

Ammonia reacted similarly, and the inability of *n*-hexylamine to produce a gel was interpreted as evidence for the role of covalent crosslinking in gel formation (134).

(2) Reactions with Mercapto Groups

Reactions of the toluenesulfonyl ester (tosyl) derivative of cellulose with α,ω-*n*-alkanedithiols under alkaline conditions have been used (135) in a study designed to introduce crosslinks of varying length ($n = 2$–10).

$$R_{cell}{-}OTs + NaS{-}(CH_2)_n{-}SNa \longrightarrow R_{cell}{-}S{-}(CH_2)_n{-}S{-}R_{cell} + NaOTs \qquad (17)$$

Displacement of the tosyl or brosyl ester with potassium disulfide, leading to the formation of polymeric sulfur crosslinks, was also recorded.

(3) Reactions with Dialdehyde Cellulose

Cotton cellulose partially oxidized by periodate (136, 137) exhibits high dry crease recovery (138), and this behavior has been attributed to covalent-crosslink formation of the hemialdal type, involving the condensation of the aldehyde groups with hydroxyls in neighboring chains.

The aldehyde groups generated by periodate react with hydrazides under mild conditions. This reaction was used (139) to introduce the disulfide group in crosslinks formed by treatment with acethydrazide disulfide,

$$H_2NNHCOCH_2SSCH_2CONHNH_2$$

(see also Section F in this Chapter XVII).

(e) REVERSIBLE CROSSLINKS

In recent years considerable interest has been shown in reversible crosslinks, or the preparation of crosslinked cellulose derivatives in which the crosslink's structure permits its rupture and reformation as a function of "environment." Reversible crosslinks would, in principle, afford the possibility of inserting permanent folds into fabric that has been crosslinked in a flat configuration and, on the other hand, of moving or removing creases that have been permanently set in a previous crosslinking process (see also Section A of Chapter XIX).

A review of the chemistry of wool shows that disulfide bonds provide effective reversible crosslinks, easily ruptured by mild reducing agents and re-formed by oxidation. Several methods of introducing crosslinks containing disulfide groups into cellulose have been reported in recent years. Schwenker and co-workers (140, 141) reported the preparation of 6-mercaptodeoxycellulose **(29)** and its oxidation to 6,6′-deoxycellulose disulfide **(30)** by the series of reactions shown schematically below.

$$R_{cell}OH + CH_3C_6H_4SO_2Cl \xrightarrow{C_5H_5N} \underset{\mathbf{27}}{R_{cell}{-}OSO_2C_6H_4CH_3} + C_5H_5N{\cdot}HCl \qquad (18)$$

$$\mathbf{27} + KSCOCH_3 \longrightarrow \underset{\mathbf{28}}{R_{cell}{-}SCOCH_3} + KOSO_2C_6H_4CH_3 \qquad (19)$$

$$\mathbf{28} \xrightarrow{OH^-} \underset{\mathbf{29}}{R_{cell}{-}SH} + {}^-OCOCH_3 \qquad (20)$$

$$\underset{\mathbf{29}}{2R_{cell}{-}SH} \underset{[H]}{\overset{[O]}{\rightleftharpoons}} \underset{\mathbf{30}}{R_{cell}{-}SS{-}R_{cell}} \qquad (21)$$

Deoxycellulose derivatives are discussed in detail in Section F of this Chapter XVII.

The preparation of other disulfide-containing crosslinked celluloses has been reported by Rath (142) and Tesoro (143). These authors prepared the cellulose derivatives directly from appropriate *N*-hydroxymethyl amide compounds. The reaction of cellulose with bis(sulfatoethylsulfonylphenyl) disulfide **(31)** was reported by Saur and co-workers (144), whereas Hobart et al. (139) studied the reaction of dialdehyde cotton with acethydrazide disulfide (see the preceding subsection on "Reactions with Dialdehyde Cellulose").

$$\underset{\mathbf{31}}{NaO_3SOCH_2CH_2SO_2C_6H_4{-}S{-}S{-}C_6H_4SO_2CH_2CH_2OSO_3Na}$$

In all instances it was qualitatively shown that the disulfide groups in the crosslinked cellulose could be reduced and reoxidized. The changes in crease recovery resulting from rupture (reduction) and re-formation (oxidation) of the disulfide group in the crosslink were consistent with the documented role of crosslinks in enhancing crease recovery.

However, it has been shown (143) that the reduction and oxidation reactions are far from quantitative and that many crosslinks remain inaccessible to reaction and to analytical determinations. The evidence for reversibility obtained by the reduction and oxidation of crosslinks containing diazo groups and introduced into cellulose by reaction with 4,4′-bis(sulfatoethylsulfonyl)azobenzene (**32**) is even slimmer (145, 146).

$$NaO_3SOCH_2CH_2SO_2C_6H_4N{=}NC_6H_4SO_2CH_2CH_2OSO_3Na$$

32

Only some preliminary data have been published (147) on the reversible formation of ionic crosslinks in carboxymethylcotton neutralized with an epoxy amine.

One would conclude that much further investigation of reversible crosslinks is needed before valid conclusions are reached regarding the value of this concept.

2. Effect of Crosslinks on the Properties of Cellulosic Materials

The far-reaching effects obtained when cellulosic materials, particularly cellulosic fibers, are crosslinked by reaction with polyfunctional reagents have been examined by many investigators. The principal interest has been to produce fabrics that have the ability to recover from creasing deformation, as measured, for example, by the Monsanto wrinkle-recovery test (148). For reasons admirably reviewed by Steele (149) it can no longer be refuted that covalent crosslinking is *an* effective way, even if not the only one, of improving the resiliency of cellulosic substrates. The various physicochemical and mechanical properties of cellulose that are effected by the crosslinking reaction are discussed below.

(a) PROPERTIES OF CROSSLINKED CELLULOSIC MATERIALS

(1) Physicochemical Properties

Solubility and Swelling. A characteristic feature of crosslinked cellulose is that even a small number of intermolecular crosslinks prevents dissolution. A substrate crosslinked in the wet, or swollen state generally exhibits a high degree of swelling but remains insoluble. The presence of bifunctional crosslinking reagents can be distinguished from monosubstituted derivatives by insolubility in cuprammonium hydroxide (150) or other cellulose solvents.

Although crosslinks prevent the dissolution of cellulose, a modified substrate can respond to swelling agents. Aftermercerization has been suggested as a means of improving the strength properties of crosslinked-cotton fabric (151) and particularly of linen (152), which suffers severe loss of strength after resin treatment. As might be expected, however, a high density of crosslinks restricts fiber penetration by alkali. Similarly, crosslinking in the dry state can reduce water imbibition by as much as 50% or more compared with natural cellulose. The state of swelling at the time of crosslinking controls the level of imbibition in the product.

Equilibrium moisture content, or moisture regain, of a substrate under standard conditions (65% relative humidity and 21°C) decreases drastically after crosslinking in the dry state. This reduction is greater than can be explained on the basis of a reduced number of hydroxyl groups. Regain levels of cellulose crosslinked in the wet state are generally higher than those of untreated cellulose, with the crosslinks presumably retaining the swollen, distended structure of the cellulose and enabling it to accommodate more moisture. The effect of crosslinking in the wet state on increasing moisture regain in native and regenerated celluloses has been clearly distinguished from the influence of analogous monofunctional reagents (60).

Chemical Reactivity. As in all topochemical reactions, the chemical reactivity of cellulose is dependent on the diffusion of the reactant into the substrate, and this is strongly influenced by the presence of covalent crosslinks. The factors that reduce moisture regain and water imbibition in crosslinked cellulose also influence reactivity with chemical reagents in a related manner, even in a cellulose-swelling medium.

Kärrholm (153) observed the diffusion of the periodate ion into rayon fibers at 23° C and found that the diffusion constant was reduced from 12.7×10^{-10} to 1.7×10^{-10} by the presence of 2.7% bound formaldehyde reacted in the absence of water. More importantly, the crosslinked cellulose reacted to an extent that was considerably below the theoretical values. The course of periodate oxidation of oxymethylenecellulose prepared by aqueous, solvent, and vapor-phase reactions has been interpreted in terms of the uniformity of crosslink distribution (154). Limited reactivity toward periodate has also been observed for cotton fabric acid-cured with dimethylolurea (155) and trimethylolmelamine (156). The latter reagent, introduced by wet fixation, caused essentially no change in subsequent periodate-oxidation rate. Crosslinked cellulose shows only limited reactivity toward formylation treatments (157).

The extent and the rate at which crosslinked cotton cellulose is oxidized with sodium hypochlorite were found to decrease substantially in the range of 0.3–0.9% bound formaldehyde, larger amounts causing little further change (158). The major changes in the mechanical properties of the treated fabric also occur over this range of substitution. In the case of viscose rayon, however, a study of mechanical properties suggested that the number of crosslinks reaches a maximum at 5% bound formaldehyde, larger amounts giving longer crosslinks (159). The internal surface area of water-swollen cellulose decreased from 116 ro 17.5 m^2/g at combined formaldehyde levels of 1%, as measured by nitrogen adsorption (160). Equilibrium accessibility to a direct dye decreased by 50% when 1% formaldehyde was introduced in the dry state (acid bake), and no reduction in dye accessibility was observed with a comparable product crosslinked in the presence of water.

Whereas crosslinked cellulose maintains some reactivity in aqueous systems, crosslinks strongly curtail reactions in media of lesser swelling power. A crosslinked substrate becomes essentially unreactive toward benzoyl chloride (161) or phenyl isocyanate (162) in pyridine, even under relatively severe conditions of treatment.

Successful methylation treatments have been carried out on cotton containing oxymethylene crosslinks, but reaction rates are considerably reduced, particularly in the latter stages of methylation (163). Certain modified substrates can be nitrated without apparent loss of crosslinks (164).

Microbiological Reactivity. Crosslinking of cellulose provides an effective method for imparting durable resistance to microbial degradation, the level of resistance being far greater than would be anticipated from the degree of substitution. Restricted diffusion of essential enzymes into the fiber substrate probably plays a role in this stabilization. The effectiveness of melamine derivatives in imparting rot resistance to cotton fabric (165) is probably related to the formation of crosslinks.

(2) Mechanical Properties

The changes that crosslinking effects in the mechanical properties of cellulosic substrates are considered in two major categories: (a) resilience, or the ability to recover from bending deformation, and (b) strength properties, or resistance to tensile, tearing, and abrading forces.

Resilience. The extent to which a cellulosic material recovers from creasing and resists wrinkling depends on many factors, including (a) the morphology of the substrate, particularly the degree of crystallinity; (b) fiber length and diameter; (c) fabric construction; (d) the state of swelling at the time of testing; (e) the presence of foreign substances, either internal or as surface deposits; and (f) the presence of covalent crosslinks. Although factors other than the effect of crosslinks are beyond the scope of this treatment, it must be borne in mind that an interplay of many factors governs the performance of a cellulosic material under stress (166, 167).

The restraining action of covalent crosslinks causes a decrease in the stress-induced slippage between the structural elements of the fiber. A crosslinked fabric thus shows reduced extension at break, and increased fiber elasticity occurs at relatively low extensions. The ability of a fabric to recover from deformation is more important than its ability to resist creasing, as these properties are related to practical performance (40).

A fabric crosslinked in the dry or collapsed state shows improved wrinkle recovery (168) in both the wet and the dry states. The common methods of dry crosslinking impart varying degrees of wash-and-wear properties, and the dry recovery obtained in this manner has been shown to be a linear function of wet recovery (169). "Dry" crease recovery has also been shown to be influenced by relative humidity (170).

Crosslinking in the wet, or swollen, state leads to a different arrangement of crosslinks and produces primarily high wet recovery. In this case the crosslinks stabilize the fiber and restrain slippage only when the fiber is in the distended state, the entrance of water causing a buildup of internal pressure. The nature of wet crease recovery in cotton has been reviewed by McKelvey (171). It has been speculated (172) that crosslinking in the wet state leads to more intrachain substitution in cellulose, and the results of microscopical examination are consistent with this view (173). A technique of crosslinking that gives high dry, but low wet, crease recovery has been reported (174).

Efforts to establish a relationship between recovery performance and the number of crosslinks have been impeded by a lack of methods to determine the latter. Increasing extents of crosslinking are reflected in decreasing efficiency in terms of recovery, and, in the range of interest, wrinkle recovery is a linear function of the logarithm of reagent concentration (175). Differences frequently noted in the relative molar efficiencies of crosslinking reagents can probably be attributed to differences in the tendency of reagents toward self-condensation and in the extents of fiber penetration, depending on the molecular size of the reagent.

Strength. Although crosslinked cellulosic material with increased elastic recovery resists deformation and retains the configuration prevailing when the crosslinks were introduced, the stress it can withstand prior to rupture is reduced. Measurements of fiber extensibility, or fiber extension at break, reflect this change. Since the structural elements are not free to realign themselves to better withstand the applied stress, failure occurs at lower stress levels, irrespective of the crosslinking agent. In conventional processes a direct relationship exists between wrinkle recovery and strength loss.

Resistance to breaking, tearing, and abrading forces reflects the response of cellulosic materials to different kinds of stress. These responses are affected by crosslinking to different extents and are also dependent on the substrate. For example, rayon often shows improved tensile strength with certain levels of crosslinking, whereas resistance to flex abrasion can simultaneously be affected adversely (176).

Fiber elasticity and extension at break can be correlated with yarn and fabric properties (40). Torsional forces arising from the twisted configuration of a fiber in fabric form do not appear to be a major factor in tensile failure (167), but the influence of twist on the response to tearing or abrading stresses has not been defined.

Tensile Strength. In the case of fabrics, tensile-strength losses associated with the crosslinking of cellulose (177) have less direct bearing on practical performance than changes in the resistance to abrasion or tearing. Losses in tensile strength can be minimized by crosslinking under tension (178). By this means the crosslinks form with the substrate in the most suitable disposition to accommodate stress.

Although the high elastic modulus of cotton fibers is not generally changed by crosslinking (124), crosslinking under tension causes small but significant increases in the modulus of both rayon (179) and cotton (167). Losses in tensile and tear strength can be minimized by premercerization (180, 181). Abrasion strength also is partially recovered by mercerization subsequent to crosslinking (182). It is possible, however, that the effects observed with postmercerization are a result of alkaline cleavage of crosslinks near fiber surfaces since localization of crosslinks at fiber cores has been shown to confer improved relationships between crease recovery and strength.

Tear Strength. Tear-strength losses (183) associated with crosslinking reflect reductions in fiber elongation on the order of 50% or more. Fabric construction and fiber surface properties are also important factors in determining tear strength. Although crosslinking under tension can reduce losses in tensile strength, the fiber extensibility is further decreased, and tear strength can be affected adversely (184).

Abrasion. The effect of crosslinking on resistance to abrading forces appears to be even more complicated than the effect on tensile and tear strength. The reduction of fiber extensibility is certainly a factor and is proportional to abrasion losses under accelerated test conditions (185). On the other hand, increased elastic recovery can counteract the effect of reduced extensibility under less strenuous conditions of loading. In certain cases loss of abrasion resistance may not be encountered at low loading rates, whereas accelerated testing shows up substantial strength losses. The construction of the fabric (186) is a major factor, as is stiffness or bending moment (187). Polyethylene or silicone fabric softeners often cause moderate increases in abrasion resistance, and this effect can be enhanced by the use of such polymer additives as polyurethanes (188).

Numerous application techniques are presently under study in efforts to provide wash-and-wear cellulosic fabrics with resistance to abrasion. These include attempts to control crosslink location along the fiber radius (162) together with studies of curing techniques milder than the conventional "dry cure" to minimize embrittlement of the cellulose fiber.

(b) NATURE OF THE SUBSTRATE

As mentioned in the preceding discussion, celluloses of different origins exhibit profound differences in response to crosslinking, particularly with regard to the relation between wrinkle-recovery improvement and reduction in strength properties. Although differences observed with a particular polyfunctional reagent are a matter of degree, rather than kind, important differences are observed, especially between native and regenerated celluloses. In addition, the condition of a substrate at the time of crosslinking has an important effect, and variations in reaction conditions do not have the same effect on different fibers.

(1) Origin

The properties of cotton are changed more than those of rayon by a small extent of crosslinking. It is generally believed that this difference is related to the higher crystallinity of cotton, which is on the order of 80%, compared with 60% for regular viscose rayon. Highly crystalline linen or ramie fibers respond similarly to cotton, although strength losses at a given crease-recovery level are generally even more pronounced.

At high levels of crosslinking, rayon exhibits levels of dry recovery comparable to those obtained in cotton, but wet-recovery properties remain poor, a fact that is probably related to the higher molecular disorder in the rayon fiber (149). The tensile and tear strengths of rayon are not severely affected by extensive crosslinking, but abrasion resistance may be severely impaired. The extensibility of untreated rayon is much greater (15–30%) than that of cotton (4–6%), and both kinds of fiber lose approximately 50% of their initial extensibility after crosslinking (176).

The differences in the behavior of rayon and cotton have been interpreted in terms of relative lateral-order distribution (189). Cellulosic fibers contain regions of intermediate disorder that are accessible to water but not to the usual crosslinking agents. The proportion of rayon, relative to cotton, that is accessible to only water and therefore remains extensible when wet is appreciably greater. It is proposed that these areas govern the differences between cotton and rayon under dry and wet conditions.

In a series of publications Rebenfeld described the response to crosslinking of cotton of six different varieties chosen on the basis of divergent fiber characteristics. Single fiber (167) and yarn (190) studies showed that differing changes in the mechanical properties could be related to fibrillar orientation as measured by x-ray angle.

(2) Effect of Physical Construction

The response of cotton fabrics of varying construction (184, 191) to the same crosslinking treatment showed substantial differences in resistance to abrasion, whereas tear-strength losses have been reported to be less dependent on construction (186, 192). Breaking stress and elongation and modulus of fibers crosslinked under different conditions have been related to corresponding properties of fabrics in two constructions (193).

(3) Effect of Chemical Pretreatment

Mercerization prior to crosslinking of cotton or linen was suggested to obtain improved strength at a given level of wrinkle recovery. Effects on breaking tenacity are particularly significant, and resistance to tear and abrasion is also less impaired. Yarn mercerizing is probably more effective than fiber or fabric mercerizing, and tension mercerizing is probably essential for achievement of an improved overall property balance (181). Increased elastic recovery after mercerization (194, 195) (relative to unmercerized cotton) is maintained even after crosslinking (196).

(c) TYPE AND EXTENT OF CROSSLINK MODIFICATION

In addition to the nature of the substrate (discussed above) three factors are known to have an effect on the properties of crosslinked cellulose:

1. Extent of crosslinking, or crosslink density.
2. Chemical structure and length of crosslinks.
3. Location and distribution of crosslinks (in heterogeneous systems).

Some considerations that apply in particular to cotton crosslinking have been reviewed by Tovey (184).

(1) Extent of Crosslinking

The crosslinking theory first postulated by Cameron and Morton (164) has been supported by later investigations of an empirical nature. However, the polyfunctionality of the known reagents employed for reactions with cellulose and the heterogeneous nature of the fibrous substrates studied inherently prevented the definition of *quantitative* relationships between the extent of crosslinking and the properties of the products. In 1957 Cooke and coworkers (197) compared the effect of varying amounts of several polyfunctional reagents on the wrinkle recovery, tensile strength, and tearing strength of cotton fabrics and established a common relationship of molar concentration applied to wrinkle-recovery improvement obtained for the reagents studied. Recognizing that a number of side reactions could and probably did occur when the then known crosslinking agents were employed, several investigators searched for reagents and reactions that would yield crosslinked cellulose either free of by-products or at least susceptible to accurate chemical analysis and definition of the extent of crosslinking obtained from a given amount of reagent. Pyridinium salts of bis(chloromethyl) ethers (109); dimethylolethyleneurea;

dimethylolurea; formaldehyde (198); and *N*-methylolacrylamide were used in these studies on cotton. In all instances wrinkle recovery increased to a maximum or leveling-off value with increasing extent of crosslinking, and concomitant losses in tensile strength and fiber extensibility were found to be of the same order of magnitude for the crosslinking reactions studied.

In cotton maximum wrinkle recovery was reached at degrees of crosslinking calculated from analytical data to be from 0.04–0.05 (109, 198) to 0.08 (197) per anhydroglucose unit, or from 1 crosslink per four anhydroglucose units to 1 crosslink per two anhydroglucose units in the accessible region of the cotton fiber (estimated at about 20%). The theoretical significance of these findings has been discussed by Gardon (124). Additional considerations regarding these relationships are given by Gardon and Steele (199) and O'Brien and Van Loo (200). A somewhat more academic approach by Schwenker and co-workers (140, 141) has been based on the oxidation reaction of 6-mercaptodeoxycellulose, and the results have been consistent with those of prior work.

(2) Crosslink Structure and Length

It has been postulated that short crosslinkages are likely to be more effective than long ones in preventing chain slippage in fibers and thus in improving fiber resilience and/or fabric wrinkle recovery (184). It has also been proposed that the effectiveness of short crosslinkages is inevitably accompanied by greater losses in tensile strength and fiber elongation and that a more favorable ratio between tensile loss and crease-recovery improvement might be obtained in cellulosic fibers by crosslinking with long, flexible molecules rather than with short or rigid ones (201). In fact severe losses in tensile strength and large wrinkle-recovery improvements associated with the use of formaldehyde have often been attributed to the formation of the rigid, one-atom-long methylene crosslink. Although several systematic studies have been carried out on the crosslinking of cellulose, particularly cotton, with a series of bifunctional reagents of varying length and rigidity, the experimental evidence obtained is somewhat contradictory and no definite conclusion regarding the effect of crosslink length appears to be warranted. As Tovey pointed out (184), longer reagent molecules do not necessarily result in the formation of longer crosslinkages, but several among the studies reported nevertheless permit a good definition of crosslink length in the crosslinked products prepared (27, 62, 108, 117, 201–204).

Several studies have been carried out in an attempt to examine the relationship between the functionality of reagents and their effectiveness in the crosslinking reactions. Work with ethylenimine derivatives (205) and polyfunctional, substituted acrylamides (202) led to the conclusion that reagents of functionality greater than 2 are not necessarily more efficient than bifunctional crosslinking agents in changing the properties of crosslinked cellulose.

(3) Crosslink Location and Distribution

As pointed out above, monomeric crosslinking agents can penetrate cellulosic fibers, but they generally do not enter the crystalline structure under the reaction conditions employed. Their reaction is therefore inherently restricted to less ordered regions within the fiber structure. The location of crosslinks within the fiber has been identified as an important factor in the effect of crosslinking on fiber and fabric properties, particularly in the case of crosslinked cotton. Several excellent investigations have been carried out for the purpose of defining the variables that affect crosslink location and of controlling the distribution of crosslinks within the fiber. Relationships between the site of crosslinking and the effect on physical properties have been postulated, even though direct evidence for the occurrence of crosslinking in selected parts of the fiber is meager at best.

The approaches used in studying this problem are based on the concept of fiber swelling: variations in fiber distention, or degree of swelling during the crosslinking reaction, and selective radial distribution of crosslinks through the fiber cross section have been obtained by appropriate use of swelling and nonswelling solvents for the crosslinking agents.

Differences in wrinkle recovery between cotton fabric crosslinked on the one hand in the absence of moisture and, on the other hand, while swollen in alkali solution of mercerizing strength, have been reported and claimed in the patent literature. It has been suggested (3) that the location of reaction sites depends on the extent of fiber distention or swelling at the time of treatment. The first systematic investigation of this concept was reported by Reeves and co-workers (11), who examined the crosslinking of cotton cellulose with formaldehyde at varying degrees of fiber swelling and proposed a theoretical interpretation of the results obtained. This interpretation has remained consistent with the results of later investigations with other reagents and reaction conditions (206) and has been further discussed (199, 207) by several authors. An extensive study of variations in crosslink organization in cotton modified with formaldehyde under widely different conditions of swelling (154) has provided good evidence of the relationships among the degree of swelling at the time of crosslinking, location of crosslinks, and effect of crosslinks on the chemical and physical properties of the crosslinked cotton.

Recently the concept of fiber swelling during crosslinking, as it affects the properties of crosslinked cotton, has been extended to include consideration of "permanent" swelling effects (208) obtained by the simultaneous introduction of bulky substituent side chains and crosslinks, the bulky substituents providing an irreversible distention of the fiber in the crosslinked product.

Crosslink location within the microfibrillar structure can also be controlled by chemical blocking of some hydroxyl groups prior to crosslinking. The reaction of dimethylolethyleneurea with partially acetylated cotton showed that hydroxyl groups most essential in providing wrinkle recovery were acetylated first and were also liberated first during saponification (206).

As a first approximation, crosslinking tends to stabilize fibers in the state in which they are treated. Reactions carried out on dry or collapsed fibers (e.g., in the vapor phase or by heat curing) yield fibers that tend to maintain their collapsed state even on wetting, exhibit high dry resilience (or fabric wrinkle recovery), low moisture regain, and low water imbibition. Fibers reacted in the swollen state tend to maintain their distended structure and exhibit high wet resilience, low dry resilience, high moisture regain, and high water imbibition. If one assumes that the less ordered regions in which crosslinking occurs in fibers are located uniformly along the fiber axis, it is evident that the presence or absence of swelling agents at the time of crosslinking affects the fiber segments uniformly throughout the cross section. Experimental techniques that allow controlled localization of crosslinking modifications in the peripheries or the cores of fibers have also been described.

The concept of radial control of crosslinking as a means of improving the properties of crosslinked fibers was suggested many years ago (209), but experimental investigations designed to yield "skin" and "core" crosslinked fibers have been reported only recently (162, 210, 211). *Skin crosslinking* has been obtained by applying reagent and catalyst from a nonswelling solvent, such as ethanol (42, 210). *Core crosslinking* has been realized by applying reagent and catalyst from water (uniformly), drying, and subsequently applying a catalyst poison from a nonswelling solvent (210), or poisoning the catalyst on the fiber surface with ammonia gas (211), or rinsing the reagent and catalyst from fiber surfaces with a nonswelling solvent (162). Although the study of skin crosslinking and core crosslinking of cellulosic fibers is still in an early stage, the results reported indicate that the losses in abrasion resistance strength and extensibility normally associated with improvements in fiber resilience and fabric wrinkle recovery as a result of crosslinking can, in fact, be reduced by controlling the location of crosslinks within the fiber by appropriate experimental techniques. A similar conclusion was indicated earlier by the observation (212) that an improved relationship of fabric crease recovery to strength can be obtained by coating a cotton fabric with a composition containing reagent and catalyst to attain a segmental distribution of crosslinks in individual fibers.

Variations in reagent size through self-condensation of polymerization have also been considered to be important since, in principle at least, polymerized polyfunctional reagents are formed prior to, or during the occurrence of, the crosslinking reaction. Polymerization of the reagent can yield a product of such proportions that it will not penetrate the fiber and will remain on fiber surfaces. A dependence of the rate and extent of penetration of a reagent on its average molecular weight and molecular-weight distribution can be postulated (11) for those polyfunctional reagents for which chain growth prior to reaction with cellulose hydroxyl groups is a significant factor.

3. Characterization of Crosslinked Cellulose

The introduction of covalent chemical crosslinks within a cellulosic substrate —especially in nonswollen fibers, as with acid-cure techniques—gives reaction products that pose formidable problems in the investigation of structure by direct chemical methods. Crosslinked substrates are penetrated by chemical reagents with difficulty, and it has been common practice to resort to characterization techniques based on indirect criteria.

Reaction products have displayed sufficient chemical stability only in specific instances. Limited information has been brought forth from several laboratories describing structural features of formaldehyde-crosslinked cellulose and some other products containing stable ether crosslinks.

Characterization methods have involved multistep, complex analytical schemes that furnish answers to specific features of the structural relationship of reactant and substrate. For example, attention has been given to the extent of interaction of the crosslinking reagent with cellulose (DS) and the location of substituents in the anhydroglucose unit. No work appears to have been done, however, on the even more difficult problem of defining the distribution of substituent groups along the cellulose macromolecular chain.

(a) CHANGES IN PHYSICOCHEMICAL PROPERTIES

The following diagnostic changes in the physicochemical properties of cellulose that occur as a result of covalent crosslinking have been documented to varying degrees. The interpretation of these changes must be cautious, with adequate consideration for each of the numerous factors affecting the properties of a crosslinked cellulosic material.

(1) Accessibility and Reactivity

It is well known that crosslinking treatments carried out by an acid-cure process limit the reactivity of the cellulosic substrate and its accessibility to dyes and reagents. Limited response to sodium metaperiodate has often been observed (153, 155, 156), and one attempt has been made to interpret accessibility data by comparing oxidation-rate curves on samples containing formaldehyde cross-links introduced under different conditions (213). Additional important information regarding changes in reactivity due to formaldehyde crosslinking has been obtained by studying the oxidation rates of formaldehyde-crosslinked cotton with sodium hypochlorite and hypobromite (158). Formylation in 90% formic acid can also give a measure of accessibility, but only in the case of acid-stable crosslinks, thus excluding crosslinks formed with formaldehyde and *N*-hydroxymethyl reagents (157). A general lack of reactivity has been observed toward esterifying agents in pyridine (161), and a technique involving reactivity toward phenyl isocyanate has been used recently as a measure of the proportion of cellulose "effectively free" of crosslinks in cotton fibers containing radially localized crosslinks (162).

Certain stable crosslinked derivatives can apparently be nitrated with nitric acid and phosphorus pentoxide; acetone insolubility of the nitrated product can be used as a criterion of the presence of crosslinks (164). Inability of sulfuric acid to produce a parchmentizing effect can also be used as a qualitative test for crosslinking (164).

(2) Hydrolysis

Acid-sensitive *N*-hydroxymethyl reagents can be removed hydrolytically from crosslinked products by using 1.5% phosphoric acid buffered with 5% urea (214) without inducing appreciable damage to the cellulose. By selective alkaline hydrolysis of cellulose treated with methylolureas (215, 216), oxymethylene crosslinks can be distinguished from methylene crosslinks. However, the hydrolysis of melamine–formaldehyde resins reacted with cellulose has not provided insight into structural features (216).

In some cases known differences in stability to acid hydrolysis can be used to rule out the presence of certain crosslink types. For example, controlled hydrolysis, coupled with solubility measurements of the hydrolysis products, can rule out the presence of methylene crosslinks in a fabric treated with dimethylolethyleneurea. A molar ratio of near unity was found for formaldehyde and nitrogen, providing strong evidence that under a broad range of catalysis conditions dimethylolethyleneurea reacts to form single-unit crosslinks (203). Earlier work (217) had indicated that dimethylolethyleneurea forms crosslinks averaging 1.4–1.5 residues in length based on analytical values for nitrogen and formaldehyde.

(3) Swelling and Solubility

The swelling effects of *N*-hydroxymethyl reagents reacted in cotton and rayon have been studied, and density changes resulting from resin treatment have been examined (218).

The limited swelling and solubility of crosslinked cellulose in such cellulose solvents as cuprammonium hydroxide is often taken as *prima facie* evidence of covalent crosslinking; swelling behavior is best observed microscopically. A relatively small number of crosslinks can inhibit the penetration of solvent and render the substrate insoluble. It has been reported (26) and confirmed (219) that even one crosslink for each 180 anhydroglucose units can confer insolubility in cupriethylenediamine, demonstrating the extreme sensitivity of the solubility test in indicating the presence of crosslinks.

A method for the quantitative determination of the fraction of a crosslinked sample insoluble in cupriethylenediamine has been described (220). The crosslinks must, of course, be stable to the strongly alkaline cellulose solvent.

Extensive work at the Southern Regional Research Laboratory of the U.S. Department of Agriculture on the effect of solvents on thin cross sections of fibers has provided some insight into the location and distribution of reactants within or between the structural elements of cotton fibers (173). Changes in the swellability of the fiber and its solubility in cellulose solvents caused by crosslinking have been correlated with the simultaneous changes in the mechanical properties of the fibers (221). The nature of the fragmentation patterns obtained by mechanical and ultrasonic disintegration has been described for cellulose treated with several conventional crosslinking agents (222) and has been stated to provide insight into the distribution of crosslinks within the various structural elements of the cotton fiber.

Fiber swelling that occurs when the substrate is embedded in methacrylate polymers has been utilized in recent years as an additional diagnostic aid (223). This technique can reveal the presence and relative amounts of crosslinking that span the lamallae or growth layers of a natural cellulosic fiber. Theoretical explanations for the swelling phenomenon have been proposed (224).

(4) Instrumental Techniques

In 1954 the earliest application of infrared-absorption studies to the problem of cellulose-resin interactions provided strong evidence that melamine-formaldehyde and urea-formaldehyde reactants enter into chemical combination with cellophane films under high-temperature curing conditions (40). Refinements in the techniques of sample preparation and assignments of many specific absorption bands have aided in characterization studies. The spectra of a ground sample can be measured by using the potassium bromide-disk technique (225), but the study of cellulose-reactant interactions is better carried out by using cellophane films that can be scanned directly (226, 227). Infrared techniques have been used to define the structure of cellulose crosslinked with *N*-methylol carbamates (45), dimethylolpropyleneurea (228), and butadiene diepoxide (229) (see also Section A in Chapter XIII). Because of their complexity and insolubility, crosslinked-cellulose products have not been successfully examined by newer and even more powerful analytical tools, such as nuclear magnetic resonance spectroscopy.

(b) ANALYSIS OF CHEMICAL STRUCTURE

Experimental methods for the analysis of the chemical structure of crosslinked cellulose are extremely difficult and time consuming. Furthermore, the significance of results obtained in such analytical investigations is necessarily limited to the specific system studied. The mode of bonding of a specific reagent depends on its reactivity with specific hydroxyl groups in the cellulose molecule, and this in turn is influenced by the concentration of reagent and the reaction conditions. The results of a study of model compounds containing oxymethylene bridges have been interpreted by Jones (230) as evidence for the relative reactivities of specific cellulose hydroxyl groups with formaldehyde. In recent years the chemical structure of crosslinked cellulose has been partly elucidated for cotton treated with trimethylolmelamine (156, 231), for cotton treated with divinyl sulfone (232) and for formaldehyde-crosslinked cotton (163). Continuing work of this nature may ultimately define some direct relationships between the chemical structure of crosslinked cellulose products [e.g., the extent of reaction at $C_{(6)}$, $C_{(2)}$, $C_{(3)}$, length of crosslinks] and specific changes in the properties of fibers and fabrics.

4. Utilization of Crosslinked Cellulose

The practical applications of crosslinked cellulose are discussed in Chapters XIX-A and XIX-D, dealing with textiles and paper, respectively.

References

1. X. Eschalier, *J. Soc. Chem. Ind. (London),* **26**, 821 (1907); British Patent 25,647 (1906).
2. W. J. Roff, *J. Textile Inst.,* **49**, T646 (1958).
3. J. T. Marsh, *J. Soc. Dyers Colourists,* **75**, 244 (1959).
4. L. Meunier and R. Guyot, *Rev. Gen. Colloides,* **7**, 53 (1929); through *Chem. Abstr.,* **23**, 3103 (1929).
5. A. J. Hall, *Text. Merc.,* **130**, 103 (1954).
6. J. J. Press, Ed., *Man-Made Textile Encyclopedia,* Interscience, New York, 1959, p. 650.
7. G. C. Daul and T. E. Muller, *J. Appl. Polymer Sci.,* **12**, 487 (1968).
8. L. H. Chance, R. M. Perkins, and W. A. Reeves, *Am. Dyestuff Reptr.,* **51**, 583 (1962).
9. R. L. Arceneaux, R. A. Fujimoto, J. D. Reid, and R. M. Reinhardt, *Am. Dyestuff Reptr.,* **51**, 559 (1962).
10. S. M. Stark, Jr., and S. P. Rowland, *J. Appl. Polymer Sci.,* **10**, 1777 (1966).
11. W. A. Reeves, R. M. Perkins, and L. H. Chance, *Textile Res. J.,* **30**, 179 (1960).
12. J. D. Guthrie, *Textile Res. J.,* **33**, 955 (1963).
13. J. G. Frick, Jr., A. G. Pierce, and V. W. Tripp, *Textile Res. J.,* **32**, 425 (1962).
14. W. J. Roff, *J. Textile Inst.,* **47**, T309 (1956).
15. W. J. Roff, *J. Textile Inst.,* **52**, T301 (1961).
16. I. J. Gruntfest and D. D. Gagliardi, *Textile Res. J ,* **18**, 643 (1948).
17. S. Matsuazaki and T. Yanasc, *J. Soc. Textile Cellulose Ind. Japan,* **10**, 177, 225 (1954); *ibid.,* **11**, 16, 556, 627 (1955).
18. W. J. Roff, *J. Textile Inst.,* **54**, T281 (1963).
19. H. K. Woo, J. H. Dusenbury, and J. H. Dillon, *Textile Res. J.,* **26**, 745, 761 (1956).

20. E. Klein and B. E. M. Bingham, *Textile Res. J.,* **34**, 585 (1964).
21. J. R. Whinfield, British Patent 462,005 (1936) (to Calico Printers' Association).
22. L. H. Chance, E. K. Leonard, and W. A. Reeves, *Textile Res. J.,* **32**, 481 (1962).
23. K. V. Datye, *Textile Res. J.,* **32**, 883 (1962).
24. K. V. Datye, *Textile Res. J.,* **32**, 825 (1962).
25. C. W. Schroeder, U.S. Patent 2,826,514 (1958) (to Shell Development Co.).
26. F. S. H. Head, *J. Textile Inst.,* **49**, T345 (1958)).
27. M. D. Hurwitz and L. E. Conlon, *Textile Res. J.,* **28**, 257 (1958).
28. T. Fujimura and S. Okamoto, *J. Polymer Sci.,* **51**, 173 (1961).
29. R. F. Fisher and C. W. Smith, U.S. Patent 3,183,054 (1965) (to Shell Oil Co.).
30. J. B. Irvine and B. H. Kress, *Textile Res. J.,* **28**, 148 (1958); B. H. Kress, *Am. Dyestuff Reptr.,* **48**, 33 (1959).
31. R. F. Schwenker and E. Pacsu, *Textile Res. J.,* **32**, 618 (1962).
32. D. D. Gagliardi, W. J. Jutras, and F. B. Shippee, *Textile Res. J.,* **36**, 168 (1966).
33. R. P. Foulds, J. T. Marsh, and F. C. Wood, U.S. Patent 1,734,516 (1929) (to Tootal Broadhurst and Lee Co., Ltd.).
34. H. Petersen, *Textile Res. J.,* **38**, 156 (1968).
35. A. J. Hall, *Textile Finishing,* American Elsevier, New York, 1966.
36. P. K. Shenay and J. W. Pearce, *Am. Dyestuff Reptr.,* **57**, 352 (1968).
37. H. B. Goldstein and J. M. May, *Textile Res J.,* **34**, 325 (1964).
38. A. C. Nuessle, *Am. Dyestuff Reptr.,* **47**, 465 (1958).
39. D. M. Musser and J. H. Velleramp, British Patent 927,166 (1963).
40. T. F. Cooke, J. H. Dusenbury, R. H. Kienle, and E. E. Lineken, *Textile Res. J.,* **24**, 1015 (1954).
41. S. L. Vail, J. G. Frick, Jr., P. J. Murphy, Jr., and J. D. Reid, *Am. Dyestuff Reptr.,* **50**, 437, 466 (1961).
42. H. C. Walter, J. K. Buxbaum, and L. Q. Green, *Textile Res. J.,* **27**, 146 (1957).
43. S. L. Vail and C. M. Moran, *Am. Dyestuff Reptr.,* **53**, 282 (1964).
44. R. L. Arceneaux, J. G. Frick, Jr., J. D. Reid, and G. A. Gautreaux, *Am. Dyestuff Reptr.,* **50**, 849 (1961).
45. J. D. Reid, R. M. Reinhardt, and J. S. Bruno, *Am. Dyestuff Reptr.,* **54**, 485 (1965).
46. S. L. Vail, *Chem. Ind. (London),* No. 8, 305 (1967).
47. H. G. Goodman, Jr., G. H. Lourigan, and C. A. Dupraz, *Textile Res. J.* **37**, 103 (1967).
48. G. C. Tesoro, S. B. Sello, and D. R. Moore, *Textile Res. J.,* **39**, 471 (1969).
49. A. H. Reine, J. D. Reid, and R. M. Reinhardt, *Am. Dyestuff Reptr.,* **55**, 711 (1966).
50. J. T. Marsh, *The Textile Manufacturer,* **91**, 111 (1965); R. M. Reinhardt, J. S. Bruno, and J. G. Frick, Jr., *Textile Res. J.,* **36**, 602 (1966).
51. E. J. Gonzales and R. R. Benerito, *Textile Res. J.,* **35**, 168 (1965).
52. E. J. Gonzales, H. M. Ziifle, R. J. Berni, and R. R. Benerito, *Textile Res. J.,* **37**, 726 (1967).
53. A. Ruperti, *Am. Dyestuff Reptr.,* **50**, 762 (1961).
54. N. R. S. Hollies and N. F. Getchell, *Textile Res. J.,* **37**, 70 (1967).
55. O. C. Bacon, J. E. Smith, and L. E. Hughes, *Am. Dyestuff Reptr.,* **47**, 259 (1958); K. Risch, *Textilveredlung,* **2**, 113 (1967).

56. H. Enders and G. Pusch, *Am. Dyestuff Reptr.*, **49**, 1, 107 (1960).
57. D. L. Schoene and V. S. Chambers, U.S. Patent 2,524,399 (1950) (to U.S. Rubber Co.).
58. J. G. Frick, Jr., W. A. Reeves, and J. D. Guthrie, *Textile Res. J.*, **27**, 92 (1957).
59. G. C. Tesoro, P. Linden, and S. B. Sello, *Textile Res. J.*, **31**, 283 (1961).
60. G. C. Tesoro, *J. Appl. Polymer Sci.*, **5**, 721 (1961).
61. G. C. Tesoro, unpublished work.
62. G. C. Tesoro, *Textile Res. J.*, **32**, 189 (1962).
63. S. H. Foster and I. Tashlick, *Textile Res. J.*, **32**, 574 (1962).
64. G. C. Tesoro, U.S. Patent 3,000,762 (1961) (to J. P. Stevens & Co., Inc.).
65. R. O. Steele and C. Huggett, U.S. Patent 3,218,118 (1965) (to Rohm and Haas Co.).
66. Tootal Broadhurst and Lee Co., Ltd., British Patent 915,572 (1963).
67. G. C. Tesoro and I. Pensa, *Textile Res. J.*, **34**, 960 (1964).
68. D. J. Gale, U.S. Patent 3,240,797 (1966) (to Deering Milliken Research Corp.).
69. Tootal Broadhurst and Lee Co., Ltd., British Patent 1,000,031 (1965).
70. F. H. Burkitt, *Textile Res. J.*, **34**, 336 (1964).
71. J. P. Stevens & Co., Inc., British Patent 993,596 (1965).
72. D. M. Gagarine, U.S. Patent 2,641,591 (1953) (to Dan River Mills, Inc.).
73. Naamlooze Vennotschap (N.V. deBataafsche Petroleum Maatschappij), British Patent 779,092 (1957); through *Chem. Abstr.*, **51**, 18637*c* (1957).
74. W. S. Miller and J. S. Schofield, U.S. Patent 2,711,971 (1955) (to Calico Printers).
75. W. S. Miller and H. M. Duckworth, U.S. Patent 3,198,652 (1965) (to Calico Printers).
76. W. A. Reeves, G. L. Drake, L. H. Chance, and J. D. Guthrie, *Textile Res. J.*, **27**, 260 (1957).
77. G. L. Drake and J. D. Guthrie, *Textile Res. J.*, **29**, 155 (1959).
78. L. H. Chance, R. M. Perkins, and W. A. Reeves, *Textile Res. J.*, **30**, 918 (1960).
79. F. B. Jones, H. G. Hannon, R. K. Leiniger, and R. G. Helligmann, *Textile Res. J.*, **31**, 57 (1961).
80. S. R. Hobart, G. L. Drake, Jr., and J. D. Guthrie, *Am. Dyestuff Reptr.*, **51**, 657 (1962).
81. L. H. Chance, R. M. Perkins, and W. A. Reeves, *Textile Res. J.*, **30**, 306 (1960).
82. G. C. Tesoro, S. B. Sello, and K. Domovs, *Textile Res. J.*, **34**, 523 (1964).
83. R. E. Parker and N. S. Isaacs, *Chem. Rev.*, **59**, 737 (1959).
84. A. Fairbourne, G. P. Gibson, and D. W. Stephens, *J. Chem. Soc.*, 1965 (1932); through *Chem. Abstr.*, **26**, 5068 (1932).
85. H. Dreyfus, British Patent 166,767 (1921).
86. B. J. Brown and L. Steger, British Patent 696,282 (1953).
87. D. M. Gagarine, U.S. Patent 2,985,501 (1961) (to Deering Milliken).
88. J. B. McKelvey, R. R. Benerito, R. J. Berni, and B. G. Burgis, *J. Appl. Polymer Sci.*, **7**, 1371 (1963).
89. T. Francis and M. L. Staples, *Textile Res. J.*, **33**, 583 (1963).
90. G. R. Ferrante, *Textile Res. J.*, **35**, 446 (1965).
91. D. Fiebig, *Melliand Textilber.*, **45**, 1249 (1964).
92. J. B. McKelvey, R. R. Benerito, and R. J. Berni, *J. Appl. Polymer Sci.*, **9**, 181 (1965).
93. C. W. Schroeder and F. E. Condo, *Textile Res. J.*, **27**, 135 (1957).

94. R. Steele, *Textile Res. J.,* **31**, 257 (1961).
95. J. Galligan, A. M. Sookne, J. T. Adams, H. Guest, and J. Lourigan, *Textile Res. J.,* **30**, 208 (1960).
96. R. R. Benerito, B. G. Webre, and J. B. McKelvey, *Textile Res. J.,* **31**, 757 (1961).
97. R. J. Berni, R. R. Benerito, J. B. McKelvey, and L. D. Calamari, *Am. Dyestuff Reptr.,* **54**, 426 (1965).
98. J. B. McKelvey, R. R. Benerito, B. G. Webre, and R. J. Berni, *J. Polymer Sci.,* **51**, 209 (1961).
99. R. R. Benerito, R. J. Berni, J. B. McKelvey, and B. G. Burgis, *J. Polymer Sci.,* A, **1**, 3047 (1963).
100. H. M. Ziifle, R. J. Berni, and R. R. Benerito, *J. Appl. Polymer Sci.,* **9**, 169 (1965).
101. J. B. McKelvey, R. R. Benerito, and R. J. Berni, *J. Appl. Polymer Sci.,* **9**, 191 (1965).
102. R. M. Reinhardt, R. M. Kullman, H. B. Moore, and J. D. Reid, *Textile Res. J.,* **31**, 941 (1961).
103. J. B. McKelvey, R. R. Benerito, R. J. Berni, and B. G. Burgis, *Textile Res. J.,* **33**, 273 (1963).
104. W. Lobunez and J. H. Dusenbury, *J. Appl. Polymer Sci.,* **5**, 686 (1961).
105. V. B. Chipalkatti, R. M. Desai, N. B. Sattur, J. Varghese, and J. C. Patel, *Textile Res. J.,* **33**, 282 (1963).
106. V. B. Chipalkatti, *Textile Res. J.,* **35**, 1049 (1965).
107. H. Rath, *Melliand Textilber.,* **38**, 78, 181, 321, 656, 799, 927, 1166, 1290 (1957).
108. H. Rath and U. Einsele, *Melliand Textilber.,* **40**, 526 (1959).
109. G. C. Tesoro, *Textile Res. J.,* **30**, 192 (1960).
110. E. Jones and J. E. McKay, *J. Soc. Dyers Colourist,* **81**, 359 (1965).
111. E. Jones and J. E. McKay, *J. Soc. Dyers Colourists,* **81**, 319 (1965).
112. D. M. Jones, R. N. Robinson, and N. F. Wood, *J. Soc. Dyers Colourists,* **81**, 108 (1965).
113. J. L. Gardon, U.S. Patent 3,109,695 (1963) (to Rohm and Haas).
114. G. F. Davidson and T. P. Nevel, *J. Textile Inst.,* **39**, T93 (1948).
115. R. Rigamonti and V. Riccio, *Ann. Chim. (Rome),* **42**, 283 (1952) through *Chem. Abstr.,* **46**, 11672*e* (1952).
116. T. C. Allen, *Textile Res. J.,* **34**, 331 (1964).
117. K. Matschat, *Textil Rundschau,* **16**, 580 (1961).
118. H. J. Campbell and T. Francis, *Textile Res. J.,* **35**, 260 (1965).
119. D. D. Gagliardi and F. B. Shippee, *Am. Dyestuff Reptr.,* **52**, 300 (1963).
120. W. E. Hanford and D. F. Holmes, U.S. Patent 2,339,913 (1944) (to E.I. du Pont de Nemours & Co., Inc.).
121. I. P. Eckert and E. Herr, *Kunst u. Zellwolle,* **5–6**, 204 (1943–1947).
122. H. Krässig, *Makromol. Chem.,* **10**, 1 (1953).
123. Kh. U. Usmanov and M. Ya. Pilosov, *Uzbek. Khim. Zh.,* No. 1, 26 (1962).
124. J. L. Gardon, *J. Appl. Polymer Sci.,* **5**, 734 (1961).
125. G. S. Mantell, U.S. Patent 2,837,511 (1958) (to Du Pont); *ibid.*, U.S. Patent 2,837,512 (1958).
126. J. L. Gardon, *J. Polymer Sci.,* [A], **2**, 2657 (1964).
127. G. C. Tesoro and A. Oroslan, *Textile Res. J.,* **33**, 93 (1963).

128. G. V. Lund, U. S. Patent 3,039,167 (1962) (to Courtaulds North America, Inc.).
129. R. N. Suiter, U. S. Patent 3,215,488 (1965) (to Dan River Mills, Inc.).
130. J. P. Stevens & Co., Inc., British Patent 1,043,341 (1966).
131. S. Haworth and J. R. Holker, *J. Soc. Dyers Colourists,* **81**, 212 (1965).
132. S. Narita, M. Date, and K. Ishii, *Textile Res. J.,* **36**, 124 (1966).
133. W. M. Corbett and J. E. McKay, *J. Soc. Dyers Colourists,* **77**, 543 (1961).
134. J. E. McKay and W. Taylor, *J. Soc. Dyers Colourists,* **77**, 547 (1961).
135. A. L. Clingman and R. F. Schwenker, Jr., *J. Polymer Sci.,* C. No. 11, 107 (1965).
136. J. M. Bobbitt, in *Advances in Carbohydrate Chemistry,* Vol. 11, Academic Press, New York, 1956, p. 1.
137. T. P. Nevell, in *Methods in Carbohydrate Chemistry,* Vol. 3, Academic Press, New York, 1963, p. 165.
138. C. H. Mack and W. A. Reeves, *Textile Res. J.,* **31**, 800 (1961).
139. S. R. Hobart, C. H. Mack, and C. P. Wade, *Textile Res. J.,* **36**, 30 (1966).
140. R. F. Schwenker, Jr., L. Lifland, and E. Pacsu, *Textile Res. J.,* **32**, 797 (1962).
141. R. F. Schwenker, Jr., and L. Lifland, *Textile Res. J.,* **33**, 107 (1963).
142. H. Rath, K. Brodereck, W. Möck, and M. Diamantoglou, *Melliand Textilber.,* **1** (4), 97 (1965).
143. G. C. Tesoro, S. B. Sello, and J. J. Willard, *J. Appl. Polymer Sci.,* **12**, 683 (1968).
144. W. Saur, O. A. Stamm, and H. Zollinger, *Textilveredlung,* **1**, 4 (1966).
145. C. H. Mack, H. H. McGregor, and S. R. Hobart, *Textile Res. J.,* **35**, 353 (1965).
146. O. A. Stamm, *Faserforsch. u. Textiltechn.,* **17** (2), 41 (1966).
147. W. A. Mueller and R. Swidler, *Textile Res. J.,* **34**, 656 (1964).
148. ASTM-D1295-67, *Wrinkle Recovery of Woven Textile Fabrics Using the Vertical Strip Apparatus.*
149. R. Steele, *Am. Dyestuff Reptr.,* **54**, 6 (1965).
150. W. A. Reeves, G. L. Drake, Jr., O. J. McMillan, Jr., and J. D. Guthrie, *Textile Res. J.,* **25**, 41 (1955); W. T. Schreiber, *Textile Res. J.,* **27**, 324 (1957).
151. L. W. Mazzeno, Jr., R. M. Reinhardt, A. R. Markezich, and J. D. Reid, *Am. Dyestuff Reptr.,* **46**, 719 (1957).
152. Tootal Broadhurst and Lee Co., Ltd., British Patent 437,361 (1935).
153. M. Kärrholm, *J. Textile Inst.,* **47**, T453 (1956).
154. S. P. Rowland, E. R. Cousins, and D. Mitcham, *J. Appl. Polymer Sci.,* **9**, 3869 (1965).
155. P. C. Mehta and R. D. Mehta, *Textile Res. J.,* **33**, 870 (1963).
156. J. J. Willard, R. Turner, and R. F. Schwenker, Jr., *Textile Res. J.,* **35**, 564 (1965).
157. S. P. Rowland and P. F. Pittman, *Textile Res. J.,* **35**, 421 (1965).
158. M. Lewin and S. Weinstein, *Textile Res. J.,* **37**, 751 (1967).
159. E. M. Kärrholm, *Textile Res. J.,* **25**, 756 (1955).
160. S. R. S. Iyer and L. S. Rao, *Textile Res. J.,* **38**, 668 (1968).
161. J. J. Willard and R. F. Schwenker, Jr., *Textile Res. J.,* **37**, 316 (1967).
162. J. J. Willard, G. C. Tesoro, and E. I. Valko, *Textile Res. J.,* **39**, 413 (1969).
163. S. Patel, J. Rivlin, T. Samuelson, O. A. Stamm, and H. Zollinger, *Textile Res. J.,* **38**, 226 (1968); R. E. Wagner and E. Pacsu, *Textile Res. J.,* **22**, 12 (1952).
164. W. G. Cameron and T. J. Morton, *J. Soc. Dyers Colourists,* **64**, 329 (1948).

165. A. G. Kempton, M. R. Rogers, and A. M. Kaplan, *Am. Dyestuff Reptr.*, **52**, 745 (1963).
166. J. H. Dillon, *Textile Res. J.*, **17**, 207 (1947).
167. L. Rebenfeld, *Textile Res. J.*, **32**, 154 (1962).
168. N. J. Abbott, *Textile Mercury International*, **150**, 18 (Dec. 13, 1963).
169. C. R. Williams, *Am. Dyestuff Reptr.*, **45**, 472 (1956).
170. R. A. Gill and R. Steele, *Textile Res. J.*, **32**, 338 (1962).
171. J. B. McKelvey, *Textile Res. J.*, **34**, 486 (1964).
172. R. Steele, *Textile Res. J.*, **30**, 37 (1960).
173. M. L. Rollins, J. H. Carra, E. J. Gonzales, and R. J. Berni, *Textile Res. J.*, **36**, 185 (1966).
174. R. J. Harper, Jr., J. S. Bruno, and W. A. Reeves, *Textile Res. J.*, **38**, 292 (1968).
175. S. H. Foster, *Textile Res. J.*, **27**, 129 (1957).
176. A. C. Nuessle, M. N. Fineman, and E. O. J. Heiges, *Textile Res. J.*, **25**, 24 (1955).
177. J. T. Marsh, *Textile Recorder*, 60–61 (1965).
178. V. W. Tripp, R. S. Orr, H. M. Zürfle, and C. M. Conrad, *Textile Res. J.*, **28**, 404 (1958).
179. L. Rebenfeld, *Textile Res. J.*, **34**, 168 (1964).
180. W. G. Parks, *Am. Dyestuff Reptr.*, **44**, 779 (1955).
181. A. L. Murphy, M. F. Margavio, A. S. Cooper, Jr., and C. M. Welch, *Textile Res. J*, **38**, 544 (1968).
182. R. M. Reinhardt, R. M. H. Kullman, H. B. Moore, and J. D. Reid, *Am. Dyestuff Reptr.*, **47**, 758 (1958).
183. O. B. Hager, D. D. Gagliardi, and H. B. Walker, *Textile Res. J.*, **17**, 376 (1947).
184. H. Tovey, *Textile Res. J.*, **31**, 185 (1961).
185. D. D. Gagliardi and A. C. Nuessle, *Am. Dyestuff Reptr.*, **40**, 409 (1951).
186. J. N. Grant, F. R. Andrews, L. C. Weiss, and C. B. Hassenboechler, Jr., *Textile Res. J.*, **38**, 217 (1968).
187. R. J. Harper, E. J. Blanchard, J. T. Lofton, and G. A. Gautreaux, *Textile Res. J.*, **37**, 233 (1967).
188. E. J. Blanchard, R. J. Harper, G. A. Gautreaux, and J. D. Reil, *Am. Dyestuff Reptr.*, **57**, 610 (1968).
189. R. H. Marchessault and J. A. Howsmon, *Textile Res. J.*, **27**, 30 (1957).
190. L. Rebenfeld, *Textile Res. J.*, **31**, 311 (1961).
191. W. J. Hamburger, M. M. Platt, and H. M. Morgan, *Textile Res. J.*, **22**, 695 (1952); S. Ikeda and O. Kajima, *J. Soc. Textile Cellulose Ind. Japan*, **11**, 397 (1955).
192. D. D. Gagliardi and I. J. Gruntfest, *Textile Res. J.*, **20**, 180 (1950).
193. R. S. Orr, J. J. Herbert, L. C. Weiss, R. H. Tsoi, and J. N. Grant, *Am. Dyestuff Reptr.*, **54**, 896 (1965).
194. L. Rebenfeld, *Textile Res. J.*, **31**, 123 (1961).
195. H. Borsten, *J. Textile Inst.*, **53**, P37 (1962).
196. J. N. Grant and R. S. Orr, *Textile Res. J.*, **36**, 95 (1966).
197. T. F. Cooke, P. B. Roth, J. M. Salsbury, K. G. Switly, and W. J. Van Loo, *Textile Res. J.*, **27**, 150 (1957).
198. J. G. Frick, Jr., B. A. Kottes-Andrews, and D. J. Reid, *Textile Res. J.*, **30**, 495 (1960).

199. J. L. Gardon and R. Steele, *Textile Res. J.,* **31**, 160 (1961).
200. S. J. O'Brien and W. J. Van Loo, *Textile Res. J.,* **31**, 340 (1961).
201. E. I. Valko, B. Bitter, and R. S. Perry, *Textile Res. J.,* **34**, 849 (1964).
202. G. C. Tesoro and S. B. Sello, *Textile Res. J.,* **36**, 158 (1966).
203. J. G. Roberts, *J. Textile Inst.,* **55**, T418 (1964).
204. F. S. Perkerson, W. A. Reeves, and V. W. Tripp, *Textile Res. J.,* **30**, 944 (1960).
205. R. M. Perkins, G. L. Drake, Jr., and L. H. Chance, *Am. Dyestuff Reptr.,* **52**, 521 (1963).
206. W. A. Reeves, R. M. H. Kullman, and J. G. Frick, Jr., *Textile Res. J.,* **33**, 169 (1963).
207. E. I. Valko and K. M. Limdi, *Textile Res. J.,* **32**, 331 (1962).
208. D. D. Gagliardi and A. Wehner, *Textile Res. J.,* **37**, 118 (1967).
209. W. J. Hamburger, *Am. Dyestuff Reptr.,* **35**, 13 (1946).
210. C. J. Gogek and E. I. Valko, *Textilveredlung,* **2**, 423 (1967).
211. Netherland Patent Application No. 6,606,365 (1965) (to Raduner & Co. Horn A. G.).
212. F. B. Shippee and D. D. Gagliardi, *Textile Res. J.,* **36**, 177 (1966).
213. E. R. Cousins, A. L. Bullock, C. H. Mack, and S. P. Rowland, *Textile Res. J.,* **34**, 953 (1964).
214. E. E. Lineken, S. M. Davis, and C. M. Jorgensen, *Textile Res. J.,* **26**, 940 (1956).
215. R. Steele and L. E. Giddings, *Ind. Eng. Chem.,* **48**, 110 (1956).
216. W. B. Achwal and B. B. Pitale, *Textile Res. J.,* **36**, 844 (1966).
217. J. G. Frick, Jr., B. A. Kottes, and J. D. Reid, *Textile Res. J.,* **29**, 314 (1959).
218. R. Steele and R. P. Schiwall, *Textile Res. J.,* **29**, 228 (1959).
219. J. D. Guthrie, *Am. Dyestuff Reptr.,* **51**, 507 (1962).
220. A. L. Bullock, A. W. Post, and S. P. Rowland, *Textile Res. J.,* **36**, 356 (1966).
221. M. L. Rollins, A. T. Moore, and V. W. Tripp, *Textile Res. J.,* **33**, 117 (1963).
222. G. V. Nikonovich, S. A. Leontyeva, and Kh. U. Usmanov, *J. Polymer Sci.,* [C], No. 16, 877 (1967).
223. M. L. Rollins, A. T. Moore, W. R. Goynes, J. H. Carra, and I. V. deGruy, *Am. Dyestuff Reptr.,* **54**, 512 (1965).
224. J. Dlugosz, *Polymer,* **6**, 427 (1965).
225. R. T. O'Connor, E. F. DuPré, and F. R. McCall, *Anal. Chem.,* **29**, 998 (1957).
226. R. Jeffries, *Polymer,* **4**, 375 (1963).
227. S. L. Vail, J. G. Roberts, and R. Jeffries, *Textile Res. J.,* **37**, 708 (1967).
228. E. J. Gonzales, H. M. Ziifle, R. J. Berni, and R. R. Benerito, *Textile Res. J.,* **37**, 726 (1967).
229. H. M. Ziifle, R. J. Berni, A. M. Cannizzaro, and R. R. Benerito, *Textile Res. J.,* **36**, 389 (1966).
230. D. M. Jones, *Textile Res J.,* **34**, 257 (1964).
231. J. J. Willard, R. Turner, and R. F. Schwenker, Jr., *Textile Res. J.,* **36**, 1051 (1966).
232. T. Abend, O. A. Stamm, and H. Zollinger, *Helv. Chim. Acta,* **49**, 1391 (1966).

F. DERIVATIVES WITH UNUSUAL FUNCTIONAL GROUPS

L. S. Gal'braikh and Z. A. Rogovin

This section deals with the synthesis of nonclassical cellulose derivatives—that is, derivatives not resulting from etherification or esterification. Among these are various deoxycelluloses (i.e., derivatives in which the entire hydroxyl group has been replaced), anhydrocelluloses (resulting from the formal loss of 1 mole of water per anhydroglucose unit), and selectively oxidized derivatives (such as "dialdehyde cellulose," which, strictly speaking, is no longer cellulosic since the integrity of the anhydroglucose ring has been destroyed). Other derivatives with unusual functional groups are described in Section G of this Chapter XVII.

1. Nucleophilic Substitution

Nucleophilic-substitution reactions of hydroxyl groups or derivatives thereof (predominantly by the S_N2 mechanism) play at present an important role in carbohydrate chemistry. The application of this type of reaction to cellulose makes it possible to synthesize a variety of new types of cellulose derivatives, the production of which is not possible by conventional methods (esterification, etherification, oxidation). These new derivatives include *C*-substituted cellulose derivatives (deoxycellulose derivatives), anhydro derivatives of cellulose, as well as mixed polysaccharides, whose polymer chains contain, in addition to glucose residues, residues of other sugars or their derivatives. There are two possible directions for nucleophilic substitution:

1. *Intermolecular* nucleophilic substitution: the interaction of cellulose esters or halodeoxycellulose with low-molecular-weight nucleophilic reagents; this reaction permits the synthesis of various *C*-substituted derivatives of deoxycellulose as well as of cellulose ethers and esters.

2. *Intramolecular* nucleophilic substitution: the cleavage of ester groups (sulfonate, nitrate) under the influence of activated (ionized) hydroxyl groups situated in the same unit as the group being split off. This reaction results in the formation of anhydro derivatives; the position of the resulting anhydro ring depends on the relative positions of the attacking and leaving groups.

Although attempts to utilize nucleophilic substitution for the synthesis of new cellulose derivatives were first made as long as 40 years ago [for example, in the synthesis of aminodeoxycellulose (1)], systematic investigation of this attractive and promising method of effecting change in chemical composition, and hence in properties, began only recently. It should be noted that some types of derivatives produced by nucleophilic substitution cannot, strictly speaking, be classified as derivatives of cellulose itself, since one of the principal features of the cellulose macromolecule–the definite steric configuration of the asymmetric carbon atoms of the repeating unit–may change in the course of the reaction.

Nucleophilic-substitution reactions of monosaccharides proceed according to the S_N2 mechanism:

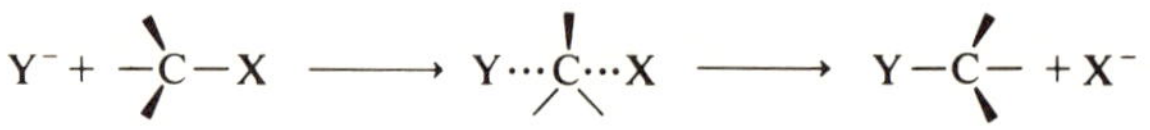

(1)

and are always accompanied by *Walden inversion.* In the case of nucleophilic substitution at the primary carbon atom of a repeating unit of cellulose, the configuration remains, of course, unaltered. If the substituents at the secondary carbon atoms participate in the reaction [which is the case in highly substituted products–for example, the phenyl esters of cellulose (2) or nitrodeoxycellulose (3)], an inversion of the configuration must take place. The resulting derivatives should therefore be regarded as *mixed polysaccharides.*

On the basis of our knowledge of sugar chemistry, *direct* substitution of the hydroxyl group of cellulose should, in principle, be possible. Because of the absence of steric hindrance, substitution reactions proceed most readily at the primary carbon atom (4). Nucleophilic substitution reactions at secondary carbon atoms are more difficult. In this case direct substitution of a hydroxyl group requires the use of sufficiently strong nucleophilic reagents–for example, triphenylphosphite iodomethylate (5). Recently the synthesis of 2(3)-iodo-2(3)-deoxy-6-*O*-tritylcellulose [degree of substitution (DS) of iodine 0.7–1.0] was carried out by reacting 6-*O*-tritylcellulose with this reagent (6).

Nucleophilic-substitution reactions utilizing cellulose derivatives, in which the bond between the carbon atom of the repeating unit and the group being displaced is strongly polarized, proceed much more readily. These derivatives include the cellulose esters of strong organic acids (methanesulfonic, ethanesulfonic, and *p*-toluenesulfonic acids) or mineral acids (nitric, phosphoric, and probably also sulfuric acids), as well as halodeoxycelluloses (iodo and chloro), which are synthesized, in turn, by nucleophilic substitution.

Systematic investigations of the relative reactivity of various types of cellulose derivatives in nucleophilic substitution have been carried out only recently. As shown by Sletkina, Polyakov, and Rogovin (7) in studies of the iodination of cellulose nitrate (DS 2.2) or cellulose tosylate (*p*-toluenesulfonate, DS 2.0) with sodium iodide in cyclohexanone solution at 100 and 120°C, the reactivity of the cellulose tosylate is higher than that of the nitrate; the rate constants of the iodination reaction for cellulose tosylate are 79.0×10^{-4} and 30.5×10^{-3} l/mole-sec at 100 and 120°C, respectively; those for cellulose nitrate are 3.7×10^{-4} and 18.9×10^{-4} l/mole-sec, respectively. Cellulose tosylate is also more reactive than cellulose diphenylphosphate in the reaction with amines of different structure and with sodium nitrite (8). Among the cellulose esters of *p*-substituted benzenesulfonic acids (*p*-toluenesulfonate, *p*-bromobenzenesulfate, *p*-nitrobenzenesulfonate) cellulose tosylate displays the least reactivity in iodination (9); the order is as follows: *p*-nitrobenzenesulfonate > *p*-bromobenzenesulfonate > tosylate.

In the reaction with aminobenzoic acids in aqueous solution (10) cellulose nitrate was found to be much more reactive than the tosylate. The fact that the reaction of cellulose nitrate with *o*-aminobenzoic acid yields *N*-(2-carboxyphenyl)amino derivatives of deoxycellulose with a high DS (i.e., 1.92) points to the nucleophilic substitution of nitrate groups situated not only at the primary but also at the secondary carbon atoms of the repeating unit.

A change in the structure of the group being displaced by nucleophilic attack has a bearing not only on the reaction rate and the DS of the product but also on the stereochemistry of the process. In the interaction of cellulose tosylate or mesylate (methanesulfonate) with potassium acetate in phenol (11) the displacement of secondary tosylate groups proceeds through the intermediate formation of 2,3-anhydro rings, followed by ring opening and formation of *altrose* units, whereas the displacement of secondary mesylate groups takes place as a typical intramolecular nucleophilic substitution; the resulting polysaccharide contains, therefore, *mannose* in its chain.

The synthesis of cellulose derivatives by nucleophilic substitution can be carried out either in organic solvents or in aqueous solution. A series of investigations (7, 12) demonstrate, however, that the rate of reaction in polar aprotic solvents (dimethylformamide, cyclohexanone) is considerably higher than in water or in proton-containing organic solvents (*n*-butanol, ethylene glycol).

The effect of the nature of the nucleophile on the DS of cellulose derivatives has not yet been sufficiently studied. It is known that the order of reactivity of halide ions with cellulose tosylates is as follows: I > Br > Cl. This order is ascribed to decrease in hydration with increase in the radius of the anion (7). If amines of differing structure (e.g., primary or secondary aliphatic, alicyclic, or aromatic amines) are used as the nucleophile, the DS of the aminodeoxycelluloses increases with increasing basicity of the amine (13).

For aromatic nucleophiles the DS of cellulose derivatives will be affected by the position of substituents in the ring. For example, aminobenzoic (10) and aminoarenesulfonic acids (14) can be arranged in the following series, in decreasing order of the rate of reaction with cellulose tosylate or cellulose nitrate: *ortho* > *para* > *meta.*

2. Synthesis by Intermolecular Nucleophilic Substitution

As indicated above, intermolecular nucleophilic substitution allows the synthesis of *C*-substituted deoxycelluloses, such as alkyl-, amino-, hydrazino-, nitro-, halo-, mercapto-, thiocyanato-, or cyanodeoxycelluloses. The use of alkoxides or of the alkali salts of carboxylic acids may result in the formation of cellulose ethers or esters:

$$R_{cell}OTs \xrightarrow{RONa} R_{cell}OR$$
$$R_{cell}OTs \xrightarrow{RCOONa} R_{cell}OOCR \qquad (2)$$

This method has been employed for the preparation of, for example, the phenyl ethers (2), β-naphthyl ethers (15), levulinates (16), propiolates (17), and a number of other cellulose esters. A detailed discussion of the synthesis of these classes of cellulose derivatives by nucleophilic substitution is, however, beyond the scope of this section. Ethers and esters are more fully described in Sections B, C, and D of this Chapter XVII.

It should be noted that, depending on the DS, the deoxycelluloses must be considered either as true cellulose derivatives (in the case of substitution only at the primary carbon atom) or as mixed polysaccharides in which the configuration of the substituents at $C_{(2)}$ and $C_{(3)}$ differs from that in the repeating unit of cellulose. In view of this the term "deoxycellulose derivatives" (amino, nitro, etc.) used here should be regarded simply as trivial terminology, though it is sufficiently descriptive.

(a) DEOXYCELLULOSE

Deoxycellulose is the first representative of the class of cellulose deoxy derivatives. It is the cellulose derivative in which at least a portion of the hydroxyl groups have been substituted by hydrogen atoms. The nucleophilic-substitution reaction cannot be employed for the synthesis of deoxycellulose; however, it is desirable to describe its synthesis first, before the synthesis of the *C*-derivatives.

The possibility of synthesizing deoxycellulose was first demonstrated by Shorygin and Makarova-Zemlyanskaya (Shorygina) during an investigation of the cleavage of cellulose ethers by solutions of metallic sodium in liquid ammonia (18). Later the product resulting from the treatment of methylcellulose (40.8% OCH_3) with a sodium solution in liquid ammonia for 48 hr was separated into three fractions (19): (a) an alcohol-insoluble product with the formula $C_6H_7O_2(OH)(OCH_3)_2$ (34% yield); (b) a product insoluble in an alcohol-ether mixture with the composition $C_6H_7O_2(OH)(H)(OCH_3)$ (4.7% yield); and (c) an ether-insoluble fraction consisting of a mixture of $C_6H_7O_2(H)(OCH_3)_2$ and $C_6H_7O_2(OH)(H)(OCH_3)$. Thus the action of sodium in liquid ammonia results in the partial formation of C–H bonds in the place of C–O bonds.

The cleavage of ether bonds by a solution of sodium in liquid ammonia was employed by Danilov and Lopatenok (20) to study the possibility of synthesizing deoxycellulose from ethylcellulose and cyanoethylcellulose. The treatment of ethycellulose was found to yield a small amount of deoxycellulose; with cyanoethylcellulose (DS 1.5–2.0), the deoxycellulose had a DS of up to 0.42. The composition of deoxycellulose was confirmed by nitration and acetylation (the DS of nitrate and acetyl groups was 2.55). An analogous method was utilized by Carter (21) to synthesize deoxycellulose from cyanoethylcellulose with a DS of 0.98. Based on elementary analysis, acetylation, periodate oxidation, and complete hydrolysis the product contained about 30 repeating units of deoxyglucose per 100 repeating units of the macromolecule. The hydrolyzate contained 2-deoxyglucose as well as a number of other deoxy derivatives of glucose (e.g., methylpentoses accounted for about 10% of the total amount of deoxysugars). The deoxycellulose was found to be insoluble in cuprammonium solution, but it did swell in a 4% sodium hydroxide solution; its rate of hydrolysis was greater than that of cellulose.

Synthesis of deoxycellulose from iododeoxycellulose is also known (22); the reduction of an iododeoxycellulose acetate-tosylate mixed ester (DS of OAc, OTs, and I are 2.33, 0.02, and 0.17, respectively) by zinc-copper alloys in glacial acid produced a mixed polysaccharide whose methanolysis products yielded an isorhamnose (identified as the tetraacetate).

6-Deoxycellulose has been synthesized by Kaverzneva et al. (23) by the reduction of iodo derivatives of cellulose tosylate either with zinc in glacial acetic acid or with metallic sodium in dioxane or in "triethylidenebisglycerol" (the reaction product of trivinylglycerol with glycerol) or, alternatively, by hydrogenation of piperidinodeoxycellulose.

6-Deoxycelluloses have also been recently synthesized by the ionic addition of methanol and glacial acetic acid to the double bond of compound **1**, "5,6-cellulosene" (see subsection 4, "Synthesis of Unsaturated Derivatives of Cellulose"), according to the following equation (shown for the alcohol) (24):

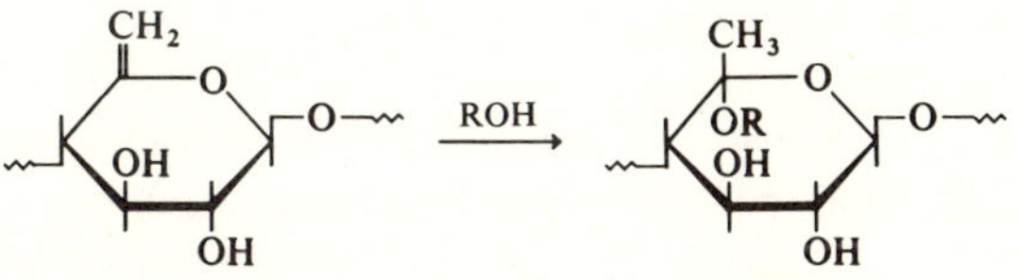

1 **2**

(3)

Based on the analysis of the methanolysis products, the mixed polysaccharide 2 formed in the reaction contained about 45 mole % of 5-*O*-methylisorhamnopyranose or 5-*O*-acetylisorhamnopyranose units.

(b) AMINODEOXYCELLULOSES

The pioneering work on the synthesis of aminodeoxycelluloses was done by Karrer and his co-workers (1). The reaction of tosylates of cellulose (DS 0.1) with aqueous or alcoholic solutions of ammonia or primary, secondary, or tertiary amines (methyl-, ethyl-, dimethyl-, benzyl-, trimethyl-, or α-naphthylamine, aniline, or pyridine) yielded the corresponding derivatives of aminodeoxycellulose.

The reaction proceeded according to the following equations:

 (4)

$$R_{cell}OTs + R_2NH \longrightarrow R_{cell}{-}NR_2{\cdot}TsOH \qquad (5)$$

Later, Sakurada (25, 26) found similar results with cellulose ditosylate and ammonia at 100°C for 20 hr. Hess and Ljubitsch (27) showed that the esterification of cellulose with *p*-toluenesulfonyl chloride in a pyridine medium at elevated temperature also resulted in a side reaction of tosylate groups with pyridine; the reaction product obtained after 4 h at 95°C contained 5.45% nitrogen. Aminodeoxycellulose with a DS close to 1.0 was produced by Scherer and colleagues (28) by the reaction of cellulose nitrate with a solution of potassium or sodium amide in liquid ammonia. The maximum nitrogen content of the aminodeoxycellulose was 8.24%, as compared with 8.6% calculated for a DS of 1.0. Ammonolysis takes place simultaneously with displacement of nitrate groups by the amide ion.

Aminodeoxycellulose is readily soluble in water or in dilute or concentrated mineral acids, but it is insoluble in organic solvents. Scherer and Saul (29) suggested that the nitrate group at $C_{(6)}$ undergoes solvolysis, whereas the nitrate groups at $C_{(2)}$ and $C_{(3)}$ are capable of being replaced by an amino group; this has not yet been supported by sufficiently convincing experimental evidence. Under the reaction conditions involved, an inversion of configuration at secondary carbon atoms may, however, take place. Thus the repeating units of "aminodeoxycellulose" may include residues of altrosamine, mannosamine, etc., and the product should then be regarded as a mixed polysaccharide.

A mixed deoxy derivative of cellulose containing both an amino group and a –C≡CH group has been obtained by the reaction of cellulose nitrate (11.9% N) with a solution of sodium acetylide in liquid ammonia (29). The reaction of ethynyldeoxycellulose with triphenyltin iodide resulted in the formation of a derivative that contained the $(C_6H_5)_3Sn{-}C{\equiv}C{-}$group (DS 1; 25% Sn) and was insoluble in organic solvents.

Subsequently the synthesis of aminodeoxycelluloses derived from heterocyclic bases (pyridine, 3-picoline, isoquinoline) was published. Heating cellulose acetate-tosylate (DS of OAc, 1.35; of OTs, 0.69) at 100°C for 24 hr in a solution of the corresponding amine gave the derivatives with a DS of from 0.42 to 0.49 (30); the reaction involved displacement of only the tosylate groups. A quantitative displacement of tosylate groups in cellulose acetate-tosylate (DS of OAc, 2.18; of OTs, 0.496) has been accomplished with boiling pyrrolidine (31). Saponification of the acetyl groups gave rise to pyrrolidinodeoxycellulose with a DS of 0.51.

Highly substituted aminodeoxycelluloses have been obtained (32) by the reaction of cellulose ditosylate and cellulose tosylate-mesylate (DS of OTs, 1.5; of OMs, 0.89) with pyrrolidine, piperidine, pyridine, and poly(2-methyl-5-vinylpyridine). The piperidine and pyrrolidine derivatives (DS 2.0) were prepared by heating solutions of cellulose ditosylate-mesylate in the corresponding amine at the boiling temperature for 24 hr. The pyridine derivative (DS 1.2) was produced by heating cellulose tosylate-mesylate with pyridine at a temperature of 115°C for 48 hr. It should be pointed out that these workers did not consider the stereochemical implications (32); it may be expected that the reaction is accompanied by inversion of configuration at $C_{(2)}$ and $C_{(3)}$.

Treatment of cellulose acetate-tosylate (DS of OAc, 2.45; of OTs, 0.3) with a solution of poly(2-methyl-5-vinylpyridine) in cyclohexanone at 120°C for 48 hr yielded a product containing 8.73% nitrogen [or a poly(2-methyl-5-vinylpyridine) content of 74.2%] and probably having a crosslinked structure. Its ion-exchange capacity, as determined by titration with 0.1 *N* hydrochloric acid, was approximately 6 meq/g (32). The reaction of cellulose acetate-tosylate (DS of OAc, 2.06; of OTs, 0.94) with isobutylamine or potassium phthalimide proceeds at 68°C for 15 hr to give aminodeoxycellulose derivatives with DS of 0.87 and 0.25, respectively (33).

The formation of aminodeoxycellulose derivatives occurs as a side reaction in the synthesis of cellulose esters with α-amino acids (34). The reaction of cellulose acetate-mesylate (DS of OAc, 2.45; of OMs, 0.55) with the sodium salts of α-amino acids in dimethylformamide at 145°C leads to the alkylation of amino groups:

$$R_{cell}OMs + H_2NCH(R)COONa \longrightarrow R_{cell}{-}NHCH(R)COONa + MsOH \qquad (6)$$

Saponification of the acetate groups yields the *N*-alkylated derivatives of glycine, α-alanine, or α-aminocaproic acid, which may also be regarded as *N*-carboxyalkyl derivatives of aminodeoxycellulose. The DS of these derivatives varies from 0.09 to 0.12.

Aminodeoxycellulose has also been synthesized by the reaction of chlorodeoxycellulose with a 26% aqueous or 16–19% alcoholic solution of ammonia (35). The maximum DS (0.47) was obtained by treating chlorodeoxycellulose (DS 0.63) with a 19% alcoholic solution of ammonia at 160°C for 6 hr. The resulting compound had an ion-exchange capacity of 2.5 meq/g.

(c) HYDRAZINODEOXYCELLULOSE

The synthesis of another type of nitrogen-containing deoxycellulose—namely, hydrazinodeoxycellulose—was first accomplished by Japanese investigators (36), who treated cellulose monotosylate with hydrazine at 100°C for 15 hr. Their publication, however, contains no detailed description of the product obtained. Hydrazinodeoxycellulose was later synthesized (37) by the reaction of chlorodeoxycellulose (DS 0.27–0.63), cellulose tosylate (DS 1.0–2.0), or cellulose mesylate (DS 0.4–1.8) with sodium hydrazide in anhydrous hydrazine or liquid ammonia. The resulting product (maximum DS 0.88) is soluble in water and dilute aqueous solutions of acids and insoluble in alcohol or acetone.

Whistler and Shasha (38) obtained hydrazinodeoxycellulose with a DS of 0.5 by using the reaction of cellulose tosylate (DS 0.64) with hydrazine hydrate. The removal of hydrazine groups by the action of oxidants results in the formation of a mixed polysaccharide containing the elementary units of glucose and 6-deoxyglucose. Polyakov and co-workers (37) showed that it is possible to synthesize phenylhydrazinodeoxycellulose (maximum DS 1.0) by the action of phenylhydrazine on cellulose ditosylate at 100°C for 6 hr. Phenylhydrazinodeoxycellulose is soluble in dimethylformamide and insoluble in acetone, chloroform, and methanol.

(d) NITRODEOXYCELLULOSE

Nitrodeoxycellulose was synthesized several years ago (3, 39). In contrast to cellulose nitrate, this new compound contains C–NO_2 bonds in which the carbon atoms are part of the repeating unit of the cellulose macromolecule; unlike cellulose nitrate, it is therefore a true nitro derivative of cellulose. Nitrodeoxycellulose with a DS of up to 0.4 was synthesized by reacting cellulose tosylate with sodium nitrite in aqueous solution or in dimethylformamide (39). Highly substituted nitrodeoxycellulose (DS 1.64) was obtained by reaction in dimethylformamide at 120°C in the presence of urea and phloroglucinol (3).

(e) HALODEOXYCELLULOSES

The synthesis of *fluorodeoxycellulose* was accomplished by Pacsu and Schwenker (40) by using the reaction of cellulose mesylate with aqueous solutions of sodium fluoride. The DS values obtained were, however, very low (about 0.05). Subsequently low-DS fluorodeoxycellulose was produced by the reaction of cellulose tosylate with sodium fluoride in dimethylformamide (7).

The synthesis of *chlorodeoxycellulose* by the reaction of cellulose with thionyl chloride in pyridine was first described by Carré and Mauclère in 1931 (41). Later this reaction was studied in detail by Boehm (42). Chlorodeoxycellulose with a DS of 1.04 also contained up to DS 0.12 of bound sulfurous acid, probably in the form of a cyclic ester. Hydrolysis of a sample with a DS of 0.29 followed by paper chromatography yielded 6-chloro-6-deoxy-D-glucose. Boehm suggested the following mechanism of chlorination:

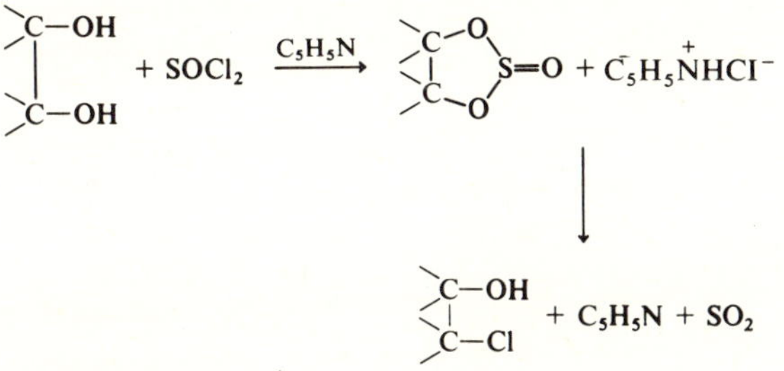

(7)

Hess and Ljubitsch (27) showed, in a study of the possible production of highly substituted cellulose tosylates, that an increase of the reaction time and temperature leads to an increase in the chlorine content of the reaction products. Further experiments on the synthesis of cellulose esters with sulfonic acids (43, 44) showed that the reaction of cellulose with alkanesulfonyl and arenesulfonyl chlorides usually led to the formation of chlorodeoxycelluloses with a DS of 0.4–0.86.

$$R_{cell}OH + RSO_2Cl \xrightarrow{C_5H_5N} R_{cell}OSO_2R + C_5H_5\overset{+}{N}HCl^- \quad (8)$$

$$R_{cell}OSO_2R + C_5H_5\overset{+}{N}HCl^- \longrightarrow R_{cell}Cl + C_5H_5\overset{+}{N}H^-O_3SR \quad (9)$$

Chlorodeoxycellulose has also been synthesized by the reaction of thionyl chloride with cellulose in dimethylformamide (35) and in carbon tetrachloride or chloroform (45). The product thus obtained, even with a relatively high DS (about 1.0), was but slightly colored (the preparations of chlorodeoxycellulose produced in pyridine are, as a rule, deeply colored). It was soluble in cuprammonium solution up to DS 0.6.

Bromodeoxycellulose (DS up to 0.1) was produced in the reaction of cellulose mesylate with an aqueous solution of sodium bromide (40).

The iodination of cellulose mesylates or tosylates, which gives *iododeoxycelluloses*, is employed principally for the quantitative determination of primary [i.e., at $C_{(6)}$] tosylate or mesylate groups. The iodinating agent used is sodium iodide dissolved in ketones (acetone, cyclohexanone). Tosylation followed by iodination permits determination of the distribution of substituents in cellulose derivatives (ethers and esters).

The iodination of cellulose acetate-tosylate (DS of OAc, 2.44; of OTs, 0.45) with sodium iodide in acetone at 100°C for 2 hr gave a DS for iodine of 0.197 (46). An iodo derivative containing no sulfur (DS of I, 0.4) was prepared by the reaction of cellulose acetate-mesylate (DS of OAc, 1.72; of OMs, 1.03) and sodium iodide in acetone or cyclohexanone (43). The reaction of ethylcellulose tosylate (DS of OEt, 2.48; of OTs, 0.517) with sodium iodide in acetonylacetone produces iododeoxycellulose with a DS of 0.126 (47). The iodination of cellulose tosylates (DS up to 2) normally produces iododeoxycelluloses with a maximum DS of 1.0 (23, 48-51); however, Malm and coworkers (50) succeeded in preparing iododeoxycellulose with a DS of 1.21 for iodine (and 0.81 for OTs) by carrying out the reaction in acetonylacetone at 120°C for 360 hr. Timell (52) conducted the iodination of esters of toluenesulfonic, α-naphthalenesulfonic, and ethanesulfonic acids in order to estimate the number of free primary hydroxyl groups in propylcellulose. Iodination was also applied to hydroxyethylcellulose tosylates (53); the products obtained had a DS for iodine of about 1.

Nitrate groups can also be displaced by iodine, the reaction with sodium iodide involving only primary nitrate groups (54). Nitration followed by iodination was used for the quantitative determination of free primary hydroxyl groups in methylcellulose (55).

(f) MERCAPTODEOXYCELLULOSE

The synthesis of mercaptodeoxycellulose was first described in a series of patents (56-58) in 1947. Two years later Izard and Morgan (59) published the results of their investigation into the reaction of cellulose acetate-tosylate (DS of OAc, 2.44 and 1.66; of OTs, 0.3) with sodium thiosulfate, thiourea, and hydrogen sulfide. The products obtained in this reaction are converted by further chemical transformations into mercaptodeoxycellulose derivatives.

The reaction of cellulose tosylate with sodium thiosulfate takes place in a homogeneous medium (a mixture of methyl Cellosolve and water acting as a solvent) at 60-65°C; after 2.5 hr there was produced a derivative with a DS of thiosulfate groups of 0.07:

$$R_{cell}OTs + Na_2S_2O_3 \longrightarrow R_{cell}{-}S{-}\overset{\overset{\displaystyle O}{\|}}{\underset{\underset{\displaystyle O}{\|}}{S}}{-}ONa + NaOTs \qquad (10)$$

Mercaptodeoxycellulose with a DS of 0.28 was prepared by treating cellulose tosylate with a solution of hydrogen sulfide in pyridine (reaction conditions: 8 hr at 35-40°C; then 70 hr at room temperature). The synthesis or isolation of mercaptodeoxycellulose may be accompanied by the formation of disulfide linkages. Treatment of the disulfide for 30 hr at 50°C with a 5% aqueous solution of thioglycolic acid results in conversion to mercapto groups.

The possibility of the formation of disulfide bonds was demonstrated in a study of model compounds, such as methyl 2,3,4-*O*-acetyl-6-*O*-tosyl-α-D-glucopyranoside (60). In another approach, disulfide derivatives of mercaptodeoxycellulose (DS 0.24) were obtained by the alkaline saponification of cellulose thioacetate resulting from the reaction of cellulose tosylate with potassium thioacetate (61).

The synthesis of a polysaccharide containing thiol groups at secondary carbon atoms, by the reaction of 2,3-anhydro derivatives of cellulose (see below) with a solution of hydrogen sulfide in methanol, has been recently reported (62). By analogy with the corresponding reaction with monosaccharides, it may be expected that the mercaptodeoxycellulose is a mixed polysaccharide containing units of glucose and of 3-deoxy-3-thioaltrose.

The reaction of chlorodeoxycellulose with a solution of potassium cyanide in dimethylformamide or methanol at 100–150°C was used by Polyakov and Rogovin (35) to synthesize *cyanodeoxycellulose* (maximum DS 0.41).

$$R_{cell}Cl + KCN \longrightarrow R_{cell}CN + KCl \qquad (11)$$

Cyanodeoxycellulose with a DS of 0.2 is soluble in cuprammonium solution as well as in benzyltriethylammonium hydroxide.

Thiocyanatodeoxycellulose (deoxycellulose thiocyanate) has been synthesized (49) by treating cellulose tosylate (DS 1.77) with a solution of sodium thiocyanate in acetonylacetone at 110–112°C. The maximum DS attained after 11 hr amounted to 1.03. This reaction was proposed for the quantitative determination of primary tosylate groups.

The synthesis of a mixed derivative of cellulose containing both amino and ethynyl groups was described above (28). Proceeding by a similar scheme, the reaction of cellulose tosylate (DS 1.62) with solutions of sodium acetylide and sodium phenylacetylide in liquid ammonia yielded ethynyldeoxycellulose and phenylethynyldeoxycellulose with DS of 1.29 and 1.52, respectively (63).

3. Synthesis by Intramolecular Nucleophilic Substitution

As pointed out above, intramolecular-nucleophilic-substitution reactions give rise to the formation of anhydro derivatives of cellulose. Depending on the position of the leaving sulfonate group in the repeating unit, two types of anhydro rings may be produced: 2,3- and 3,6-anhydro rings.

The formation of anhydro rings in monosaccharides is known to be accompanied by conformational changes (64). Such changes in the repeating unit of cellulose must cause a change in the chemical and physicochemical properties of this essentially novel class of cellulose derivatives (intensity of complex formation; stability of glucoside linkages toward acid hydrolysis; solubility of both the anhydro derivatives and their ethers; energy of intramolecular and intermolecular interaction; solution viscosity, etc.). It should be remembered, however, that conformational transformations of the repeating unit of cellulose are probably hindered by the presence of the acetal linkages.

The synthesis of anhydro derivatives of cellulose was first described by Gardner and Purves (65), who prepared 3,6-anhydrocellulose by the saponification of cellulose acetate–tosylate (DS of OAc, 2.44; of OTs, 0.48) with a solution of sodium hydroxide in methanol. Methanolysis yielded methyl α-3,6-anhydroglucofuranoside and methyl β-3,6-anhydroglucofuranoside. The anhydrocellulose synthesized by Gardner and Purves does not dissolve in 5 and 17% solutions of sodium hydroxide or in a cuprammonium solution. Anhydrocelluloses containing predominantly 3,6- as well as some 2,3-anhydro rings were produced by Danilov and Lopatenok (67) by treating cellulose tosylates (DS 0.25–1.65) with a solution of sodium in liquid ammonia or with piperidine. A synthesis has been reported of 2,3-anhydrocellulose by saponifying 6-*O*-tritylcellulose 2(3)-tosylate with a solution of sodium methoxide in methanol (68) or with a 5–15% aqueous solution of sodium hydroxide (69):

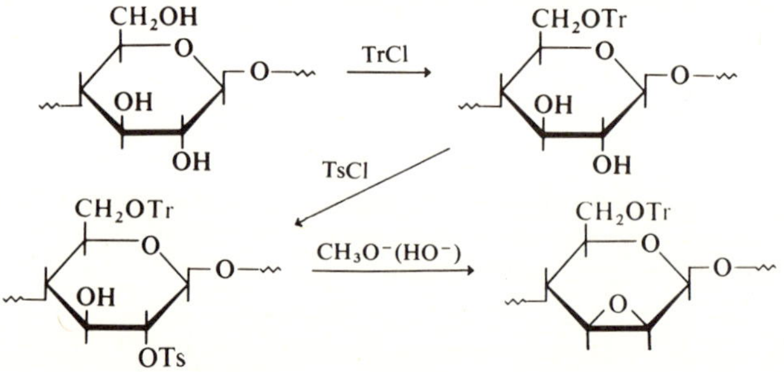

(12)

The intramolecular-nucleophilic-substitution reaction is accompanied by the elimination of tosylate groups, with formation of double bonds and carbonyl groups:

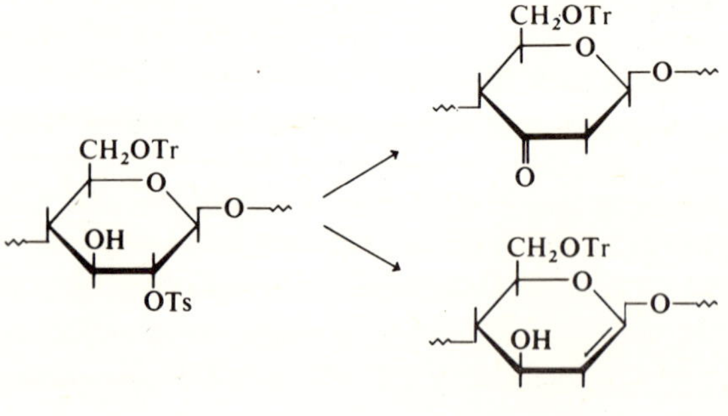

(13)

According to the available evidence (68, 70), the formation of α-oxide rings, C=C bonds, and CO groups involves 45, 30, and 12% of the tosylate groups, respectively. The occurrence of side reactions, accompanying the formation of 2,3-anhydro rings, is probably associated with the difficulties in realizing conformational transformations necessary for the formation of anhydro rings [transformation from the C1 conformation into the 2B or B3 conformation with a *trans* diaxial arrangement of substituents at $C_{(2)}$ and $C_{(3)}$. Such a transformation is likely to cause strains that interfere with the formation of anhydro rings.

Cleavage of the α-oxide ring in 2,3-anhydrocellulose under the influence of acylating reagents (acetic anhydride) or sodium methoxide may proceed with Walden inversion at $C_{(2)}$ or $C_{(3)}$, leading to the formation of a mixed polysaccharide that contains, in addition to glucose units, those of altrose (**3** and **4**) (71):

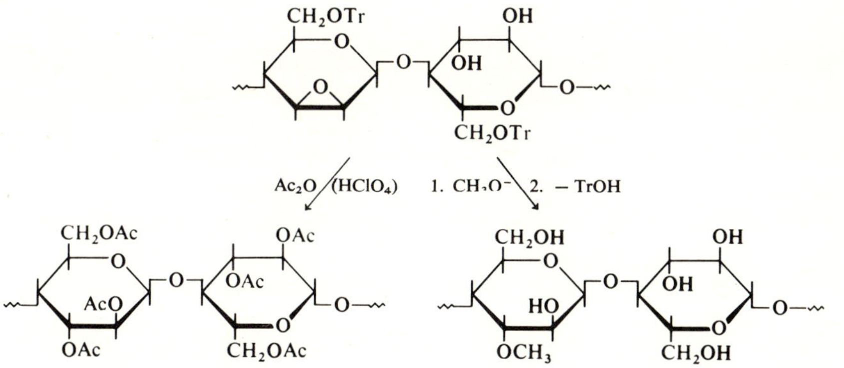

3 **4**

(14)

However, quantitative evaluation of the ratio of altropyanosyl units to glucopyranosyl units in the hydrolyzate of the mixed polysaccharide showed that the opening of rings in anhydrocellulose is caused largely by nucleophilic attack on $C_{(3)}$. 2,3-Anhydrocellulose has been used as a starting material for the synthesis of a mixed polysaccharide containing, in addition to glucose units, units of aminosugars (72). From the hydrolysis of the polysaccharide resulting from the interaction of 2,3-anhydro-6-*O*-tritylcellulose with a solution of ammonia in methanol (100°C, 12 hr), there was identified 3-amino-3-deoxyaltrose, but 2-amino-2-deoxyglucose was absent.

A series of recent publications are devoted to the synthesis and examination of the structure of 3,6-anhydrocellulose synthesized by saponifying cellulose tosylate (**5**, DS 0.95–1.19) or cellulose 2,3-*O*-diacetate-6-tosylate (**6**) with an aqueous or methanolic solution of potassium hydroxide (73–75):

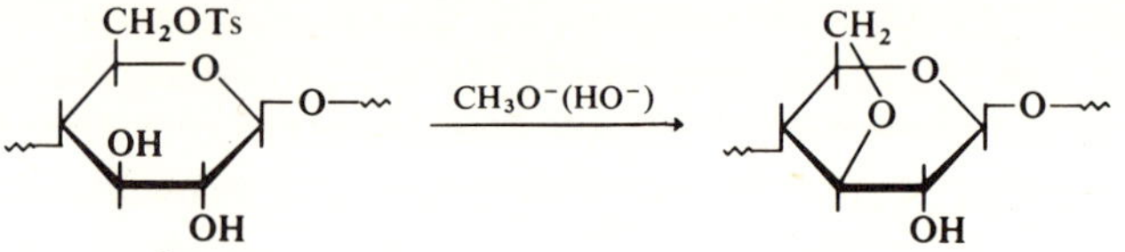

(15)

5

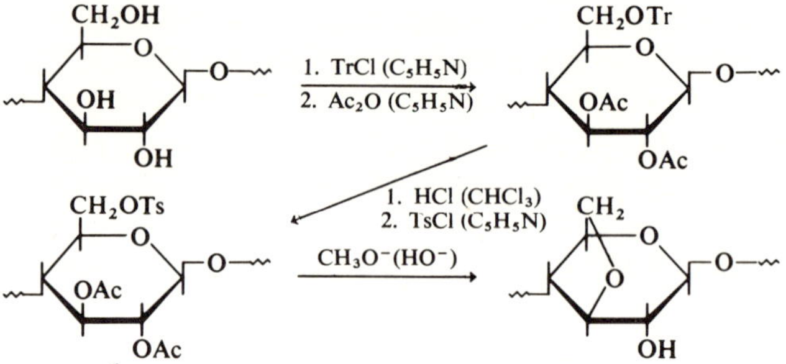

6

(16)

Since the sole synthesis of cellulose 6-*O*-tosylate in the presence of free secondary hydroxyl groups is practically impossible, the cellulose tosylate with the DS indicated also contains a certain amount of secondary tosylate groups whose saponification results, as pointed out above, in the formation of 2,3-anhydro rings (or the products of their cleavage in alkali). The use of compound **6** makes it possible, in principle, to avoid such side reactions.

The infrared spectrum of 3,6-anhydrocellulose showed absorption bands at 810, 850, and 920 cm^{-1}, typical of tetrahydrofuran rings (76). Paper chromatography of sugar mixtures resulting from the hydrolysis of 3,6-anhydrocellulose synthesized according to reactions 15 and 16, yielded 48–66 mole % of 3,6-anhydroglucose and 35–45 mole % of glucose. These results indicate a practically quantitative conversion of 6-*O*-tosylate to 3,6-anhydro rings. The results obtained by periodate oxidation of 3,6-anhydrocellulose are also in agreement with this evidence.

Detailed investigations of the composition of 3,6-anhydrocellulose (73) have made it possible to draw the conclusion that the process of the formation of 3,6-anhydro rings in the repeating units of cellulose involves partial opening of the cyclic acetal 1,5-linkage. The latter is weakened because of the formation of a bicyclic system (tetrahydrofuran–pyranose rings) and possibly also by conformational transformations taking place in the process of formation of 3,6-anhydro rings in the polymer chain (from the C1 conformation to the 1C conformation):

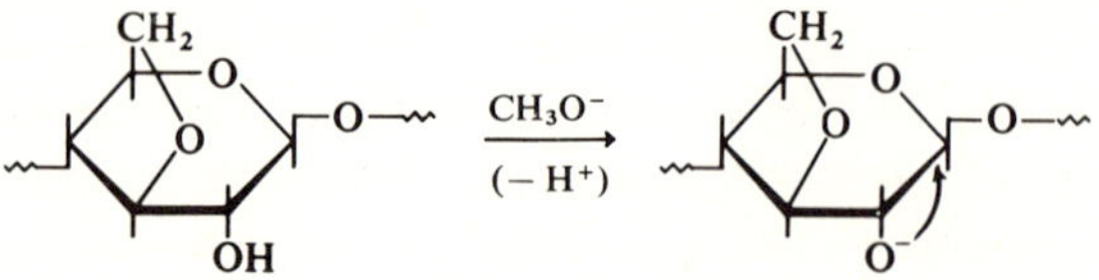

(17)

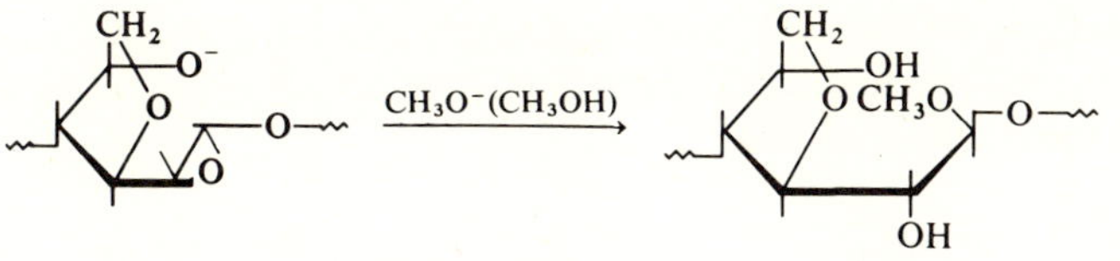

(18)

The investigation of the probable formation of various types of anhydro rings on alkaline saponification of cellulose esters is not only of scientific interest but also possibly of practical value, since it is this process that may account for the chemical and physicochemical peculiarities of cellulose hydrate as compared with native cellulose.

Gal'braikh, Achwal, and Rogovin (77) have recently demonstrated that the formation of 3,6-anhydro rings takes place upon the denitration of cellulose nitrate with an alcoholic solution of ammonium hydrosulfide. The product obtained in this reaction is not cellulose hydrate, it is rather a mixed polysaccharide containing units of glucose and 3,6-anhydroglucose.

4. Synthesis of Unsaturated Derivatives of Cellulose

A variety of methods are known for producing unsaturated derivatives of cellulose containing double bonds between the carbon atoms of the repeating unit. Strictly speaking, these should be regarded as mixed polysaccharides containing the elementary units of mono-, di-, and triglucosene.

Cellulose derivatives whose repeating units contain a system of conjugated double bonds have been synthesized by thermal degradation of cellulose methylxanthate (DS 1.7–19) or cellulose dixanthogenide by Polyakov, Derevitskaya, and Rogovin (78). For the case of a completely substituted cellulose ether, the reaction must proceed as follows:

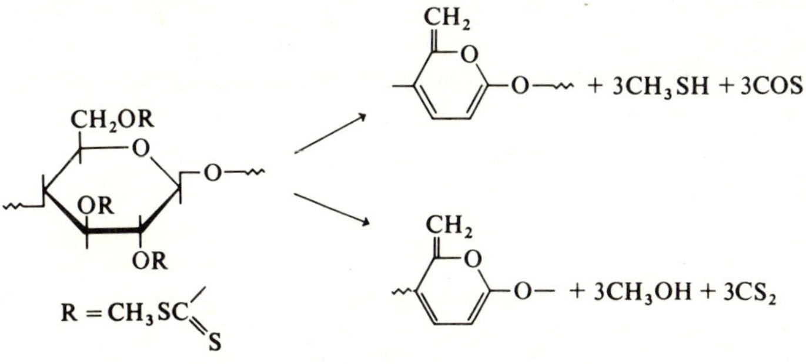

(19)

A major disadvantage of this method is the formation of double bonds in positions that are not fixed. Cellulose derivatives with double bonds in fixed positions—between $C_{(5)}$ and $C_{(6)}$ (which may be termed for convenience

"5,6-cellulosene," **7**)–have been produced by the dehydroiodination of iododeoxycellulose with piperidine (23, 67) or an alcoholic solution of potassium hydroxide (24), as follows:

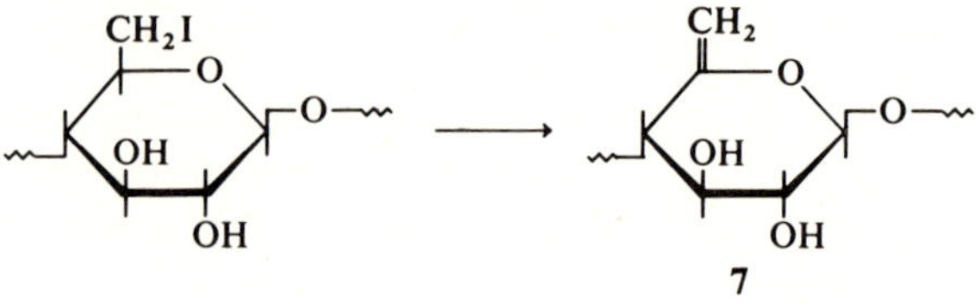

(20)

As indicated above, "5,6-cellulosene" has been used for the synthesis of mixed deoxycellulose polysaccharides containing elementary units of 5-*O*-methylisorhamnose and 5-*O*-acetylisorhamnose.

The reaction of "5,6-cellulosene" with acetic anhydride in the presence of perchloric acid (10% based on the weight of "cellulosene") yields a mixed polysaccharide containing elementary units of glucopyranose and 5,6-di-*O*-acetylgluconopyranose (**7**) (24). The reaction probably goes by way of intermediate 5,6-anhydro rings:

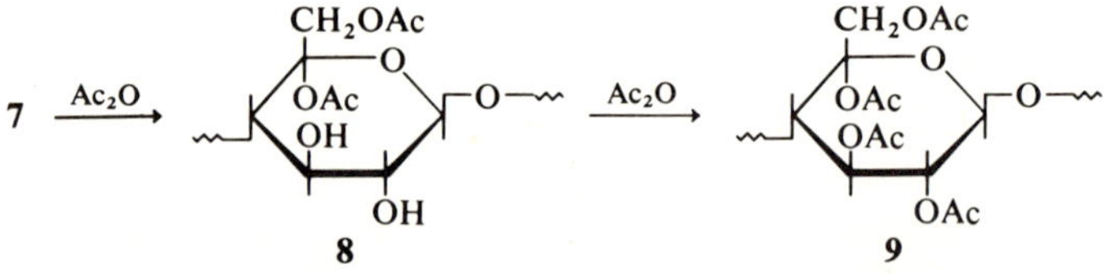

(21)

Subsequent acetylation of the reaction product gives a mixed polysaccharide **9**, which contains four acetylated hydroxyl groups at $C_{(2)}$, $C_{(3)}$, $C_{(5)}$, and $C_{(6)}$.

An interesting method for the synthesis of deoxycellulose derivatives containing various types of functional groups at $C_{(6)}$ is the addition of radicals to the double bond of "5,6-cellulosene" of substances that decompose under the action of ultraviolet radiation or of free-radical initiators. These substances include chloroform, carbon tetrachloride, trichlorosilane, and phosphorus trichloride (79).

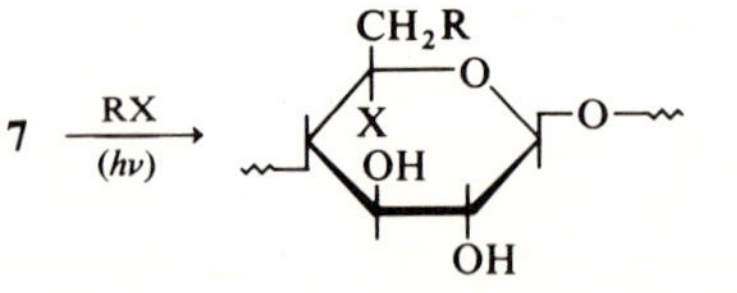

(22)

($R = CCl_3$, $SiCl_3$, PCl_2; $X = H$, Cl)

5. Other Syntheses of Deoxycellulose Derivatives

The reaction of 6-iodo-6-deoxycellulose tosylate (**10**; DS of I, 0.73; of OTs, 0.41) with butyllithium in tetrahydrofuran gave 6-*C*-*n*-butyl-6-deoxycellulose with a DS of butyl groups of 0.84 (80):

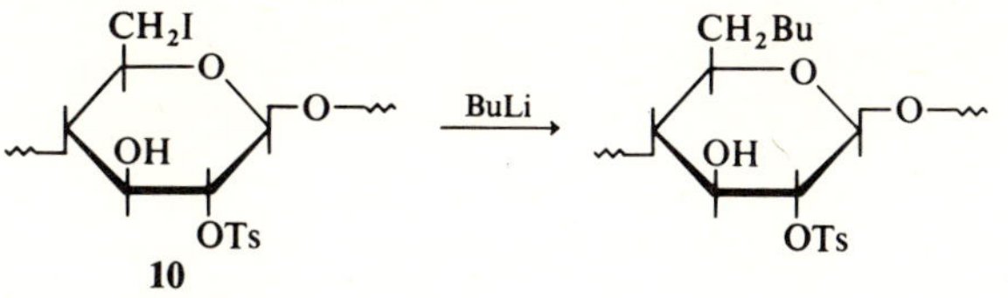

(23)

An analogous scheme has been used for preparing highly substituted phenylcarboranyldeoxycellulose by the reaction of iododeoxycellulose tosylate with phenylcarboranyllithium (81):

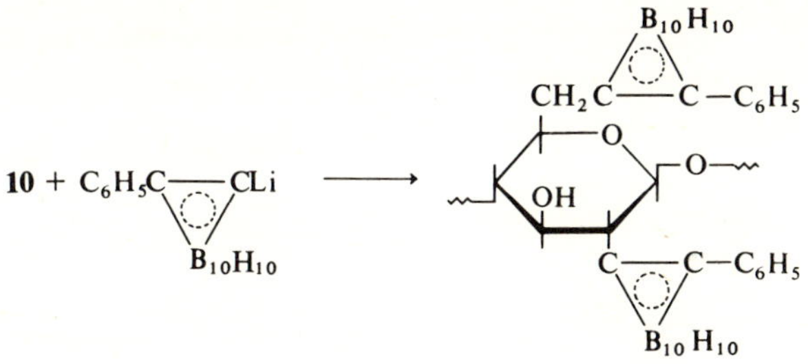

(24)

The DS of the product (1.35) points to the participation in the reaction of secondary tosylate groups. In contrast to the boron-containing cellulose esters described in the literature (82), the carboranyl derivatives of cellulose are stable to hydrolysis.

6. Products of Selective Oxidation

The oxidation of cellulose, by the action of certain types of oxidants (periodic acid and its salts, lead tetraacetate, nitrogen dioxide), may be used to produce derivatives containing aldehyde or carboxylic groups predominantly at definite positions in the repeating unit of the macromolecule. It should be noted, however, that some of these compounds—for example, the so-called dialdehyde cellulose—no longer contain the anhydroglucose ring of cellulose.

The basic features of cellulose oxidation have already been discussed in a number of monographs (see, for example, references 83–85). For this reason we confine ourselves to a consideration of the mechanism and course of the oxidation, with particular attention to selective oxidation.

(a) DIALDEHYDE CELLULOSE AND DERIVATIVES

One of the most selective processes is the oxidation by periodic acid and its salts to form *"dialdehyde cellulose."* The mechanism of this reaction is identical with that for the oxidation of low-molecular-weight 1,2-glycols. According to recent evidence (86), the mechanism consists of the rapid formation of an ester in the form of a singly charged anion (**11**), which rapidly dehydrates into anion (**12**), with subsequent slow, unimolecular rearrangement to the dialdehyde (**13**) (86):

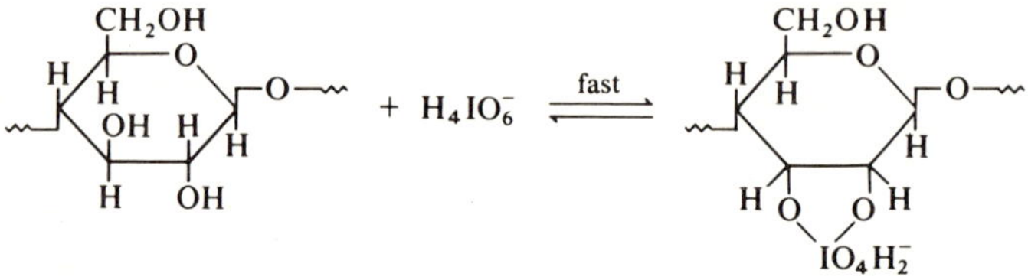

11

(25)

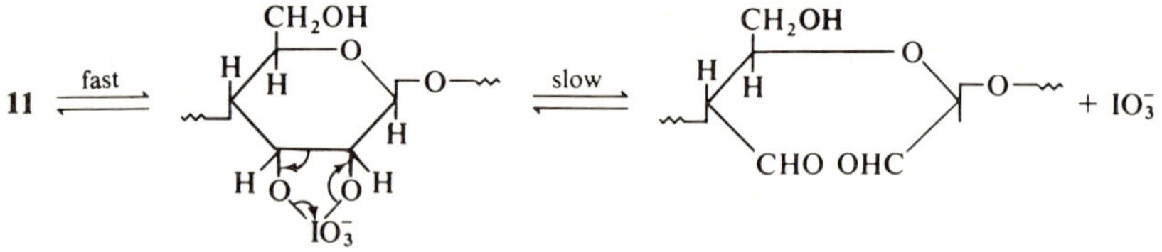

12 **13**

(26)

The oxidation of even a small number of glycol groups to the dialdehyde drastically lowers the stability of the glycosidic linkages toward alkalies or even hot water (87–89) (see Section B in Chapter XVIII for a description of β-alkoxycarbonyl cleavage). The low viscosity of cuprammonium solutions of dialdehyde cellulose is also ascribed to severe degradation in the alkaline medium. A comparative study of cellulose and dialdehyde cellulose (90) has demonstrated that the formation of dialdehyde groups also markedly reduces the stability to acid hydrolysis. Similar evidence has been obtained for low-molecular-weight model compounds (91).

Most chemical transformations of dialdehyde cellulose may be depicted on the basis of structure **13**. However, the infrared spectrum of dialdehyde cellulose shows *no* absorption at 1610 and 1740 cm^{-1}, characteristic of carbonyl groups (92), but a diffuse band at 900 cm^{-1} that can be assigned to hemiacetals (93, 94). This is probably explained by the formation of a cyclic hemiacetal from one of the aldehyde groups with the primary hydroxyl group, accompanied by hydration of the second aldehyde group (**14**):

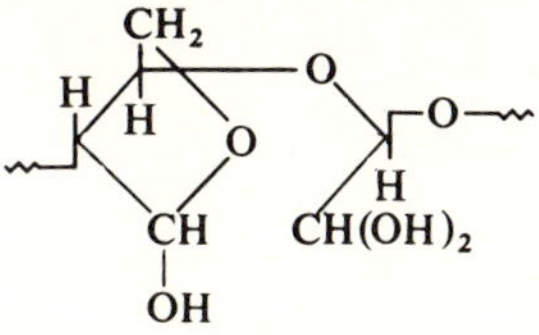

14

The presence of structure **14** in dialdehyde cellulose is confirmed by chemical and spectroscopic evidence. Thus Mester (95) found that, on condensation of dialdehyde cellulose first with phenylhydrazine and then with phenyldiazonium chloride, it is possible to produce a diphenylformazan:

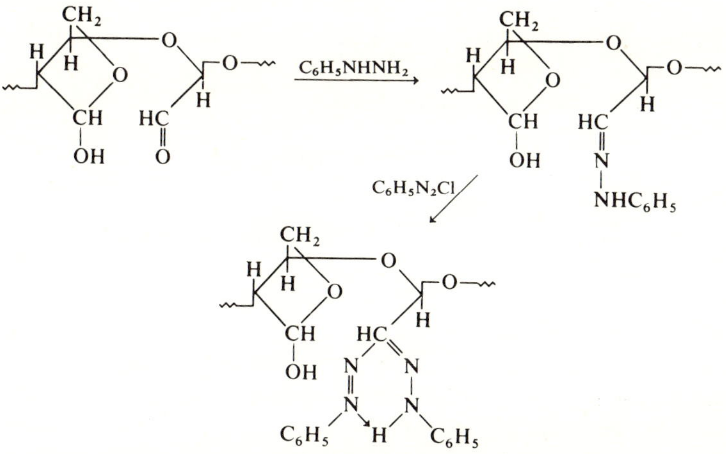

(27)

Head (96) used an ether solution of diazomethane to treat dialdehyde cellulose with a degree of oxidation (average number of oxidized glycol groups per anhydroglucose unit) of up to 0.93 and obtained a product containing up to 18.6% of methoxyl groups (DS about 1.0). The unoxidized cellulose subjected to methylation under similar conditions contained not more than 6% of methoxyl groups. These data were said to confirm structure **14**. The methylation product of dialdehyde cellulose is insoluble in cuprammonium solution and has a copper number of less than unity. The reaction of dialdehyde cellulose (aldehyde content of 18%) and gaseous nitrogen dioxide in the presence of phosphorus pentoxide (97) involves complete conversion of hydroxyl (including hydrated aldehyde) into nitrate groups (calculated as 13.6% N; found to be 12.8% N and 0.73% CHO). These data are convincing evidence for the formation of structure **14**.

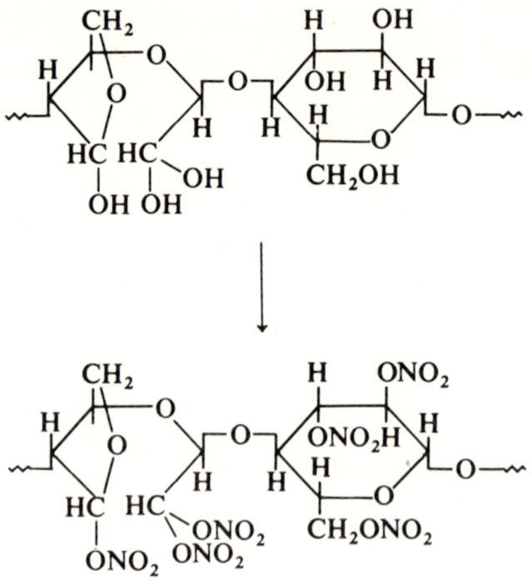

(28)

Reduction of the aldehyde groups in dialdehyde cellulose to hydroxyl groups was first carried out by Meller (98, 99) with aqueous solutions of sodium borohydride. The reduction proceeds as follows:

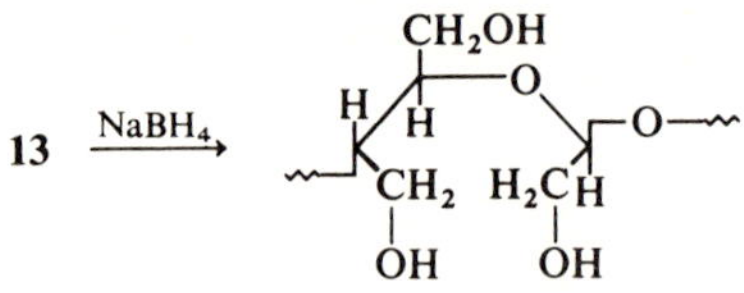

(29)

Lindberg and Theander (100) made use of the reduction with sodium borohydride as a quantitative method for determining carbonyl groups in the oxidation products of cellulose. A detailed study of the conditions of the reduction has been carried out by Head (101), who found that reduction considerably increases the stability of the product toward alkaline reagents. Reduction by sodium borohydride of dialdehyde cellulose with a high degree of oxidation followed by hydrolysis yielded erythritol (102).

Comparative investigations of the rate of acid hydrolysis of cellulose, dialdehyde cellulose, and of the product of the reduction of the latter with sodium borohydride showed that the reduced compound is the least stable to the action of 0.1 and 1.0 *N* solutions of hydrochloric acid at 20°C (103). Similar data have been obtained for model compounds (104). Treatment of the reduced compound with nitrogen dioxide in the presence of phosphorus pentoxide produced a trinitrate that differs from ordinary cellulose trinitrate by the presence of three *primary* nitrate groups in each repeating unit (105).

Dialdehyde cellulose may be used for the synthesis of many new types of derivatives. One of the first publications devoted to this question was by Rogovin, Yashunskaya, and Bogoslovskii (106), who described a synthesis of

Schiff bases from the interaction of dialdehyde cellulose (degree of oxidation of about 0.05) with aromatic amines (e.g., 4-aminophenol). The condensation products were used to prepare chemically colored cellulose fibers by coupling with various diazonium compounds (e.g., those derived from 4-nitroaniline, benzidine, or α-naphthylamine).

Treatment of dialdehyde cellulose (degree of oxidation of 0.44–0.78) with a solution of hydroxylamine hydrochloride (pH 5–6) gives the dioxime derivatives of dialdehyde cellulose (**15**) (107):

13 $\xrightarrow{NH_2OH \cdot HCl}$ 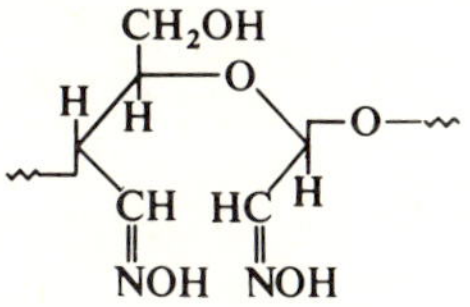 (30)

15

The conversion corresponds to 94–96%. The dioximes have a low resistance to dilute acids and alkalies and to water at elevated temperatures. They, in turn, may be converted to other cellulose derivatives. A product containing 1.12% amino nitrogen was obtained by the reduction of the dioxime of dialdehyde cellulose with a freshly prepared sodium amalgam at 50–55°C for 0.5–5 h (107). Gal'braikh, Derevitskaya, and Rogovin (108) obtained a dinitrile by the dehydration of the dioxime of dialdehyde cellulose with acetic anhydride in the presence of sodium acetate for 6 h at 100°C.

15 $\longrightarrow$ 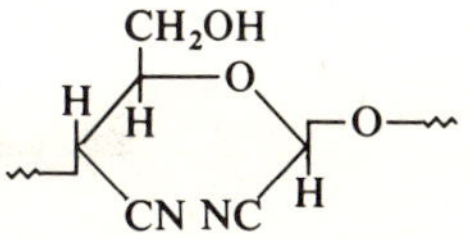 (31)

A few publications have dealt with *reclosure of the ring* that has been broken by oxidation. Some of the reactions result in the formation of a new ring with a number of members greater than six. Kuznetsova, Ivanova, and Shorygina (109) carried out a condensation of dialdehyde cellulose (**13**) with nitromethane; it gave rise to structures of two types, shown in equation 32.

The seven-membered ring (**16**) is formed preferentially on treatment of dialdehyde cellulose with an aqueous solution of nitromethane containing 0.5–1.0 equivalent of sodium hydroxide per aldehyde group (maximum conversion of dialdehyde units into **16** is 60%). At a sodium hydroxide content of 1 equivalent and a nitromethane content of about 15 equivalents per aldehyde group, compound **17** is produced in 85% yield.

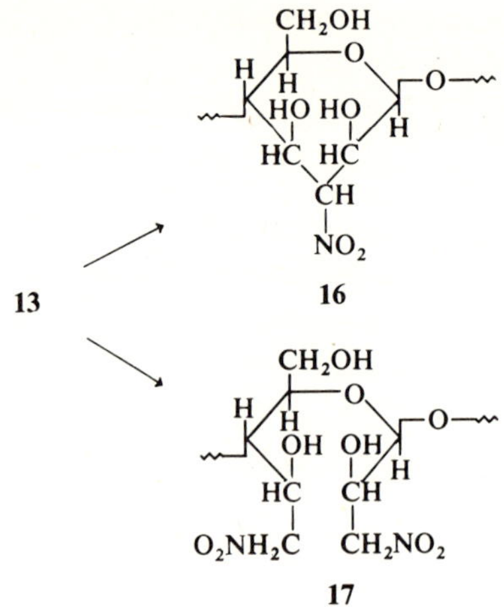

(32)

An investigation of the condensation of dialdehyde cellulose with nitroethane **(18)**, nitroethanol **(19)**, and 2-nitro-1,3-propanediol **(20)** compared with nitromethane **(21)** demonstrated that these compounds may be arranged in the following order of reactivity: **21** > **18** > **19** > **20**(110).

Ring closure to form a six-membered ring occurs on condensation of 2,3-dialdehyde 6-*O*-tritylcellulose with *p*-tolylhydrazine in a mixture of pyridine and water (4 : 1) at room temperature (111, 112). Because the condensation product does not give the formazan reaction, the authors ascribed to it structure **22** (3-deoxy-3-*p*-tolylazo-6-*O*-tritylcellulose).

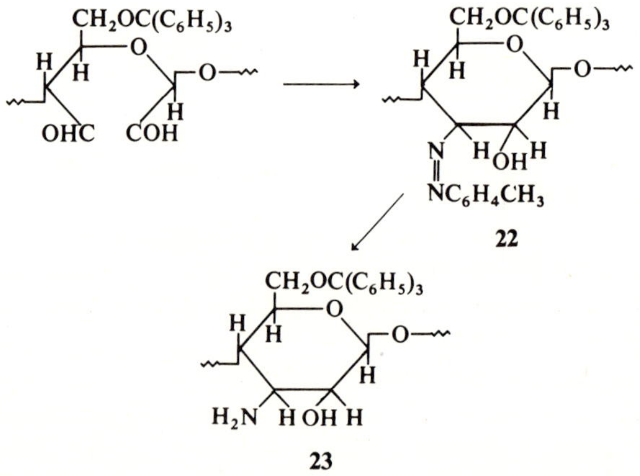

(33)

The reduction of compound **22** with Raney nickel in a dioxane solution leads to the formation of an amino derivative **(23)**. Its infrared spectrum shows no absorption bands at, e.g., 1260, 1520, and 1620 cm^{-1}, which are characteristic of compound **22**. Paper chromatography of its acid hydrolysis products shows the presence in the polymer chains of 3-amino-3-deoxyglucose units (113). This was confirmed by ninhydrin reaction with the mixture of aminosugars (114).

(b) CARBOXYL CELLULOSES

Mild oxidation of dialdehyde cellulose (**13**) with chlorous acid or sodium chlorite yields a derivative containing carboxyl groups at $C_{(2)}$ and $C_{(3)}$ (90, 115). It is the so-called *dicarboxyl cellulose* (**24**). Model compounds react similarly (116).

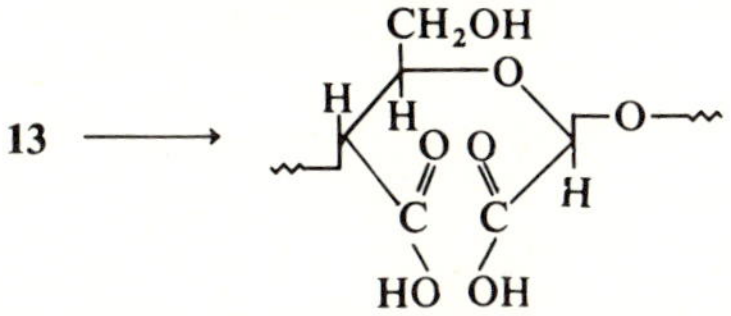

24

(34)

Dicarboxyl cellulose may be used for the recovery of heavy metals, from mixtures of ions, by chelation (117). The tendency to form complexes with the ions of polyvalent metals is much more pronounced with the derivative **25**, obtained by methyl ester formation followed by treatment with a solution of hydroxylamine in methanol, as follows (118):

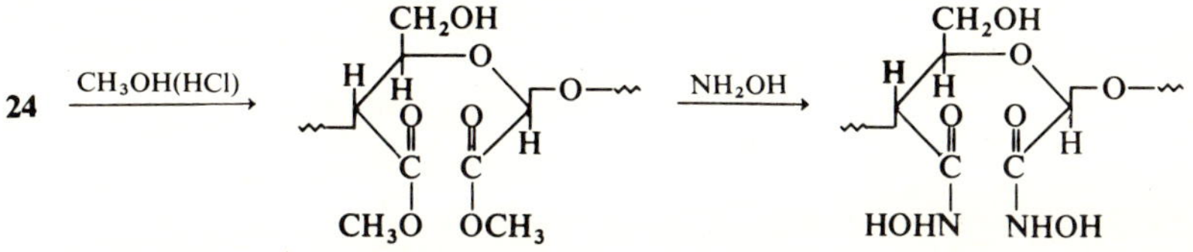

25

(35)

The maximum conversion of the methyl ester to the hydroxamic acid (about 65%) is obtained with a solution of hydroxylamine at pH 9.2. The reaction product forms stable, colored complexes with cupric and ferric ions.

Another type of cellulose derivative containing a carboxyl group in a definite position of the repeating unit may be produced by the oxidation of cellulose with nitrogen dioxide. It is the so-called *monocarboxyl cellulose*. Detailed studies have shown that the oxidation process may be represented as follows:

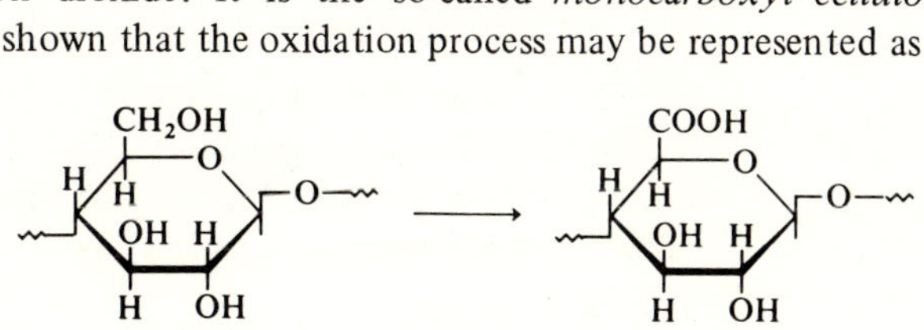

(36)

The mechanism of the oxidation of cellulose with nitrogen dioxide has been examined by a number of authors. A route via the intermediate formation of cellulose nitrate, first proposed by Unruh, Kenyon, and co-workers (119), is not in agreement with later experimental evidence. On the basis of a study of ultraviolet spectra of the products of the oxidation of cellulose with nitrogen dioxide, Pavlyuchenko, Ermolenko, and Kaputskii (120) concluded that cellulose nitrite is first formed, and its decomposition by a radical mechanism, with subsequent stabilization of the radical, yields the oxidation product. Later this scheme was further refined (121).

The oxidation of cellulose with nitrogen dioxide does not proceed with complete selectivity. According to the evidence available in the literature (122), about 15% of the total amount of carboxyl groups in monocarboxyl cellulose is not uronic [at $C_{(6)}$] but is located at $C_{(2)}$ and $C_{(3)}$ of the repeating unit of the macromolecule. Simultaneously with the formation of carboxyl groups there is also taking place (owing to the oxidation of secondary hydroxyl groups) oxidation to keto groups (10–20% of the oxidized groups) (123–126). In some of the repeating units both types of functional group may be present together. This is supported by the fact that a small amount of 3,5-diketotetrahydro-γ-pyrone (**26**) is isolated from the hydrolysis products of monocarboxyl cellulose (127).

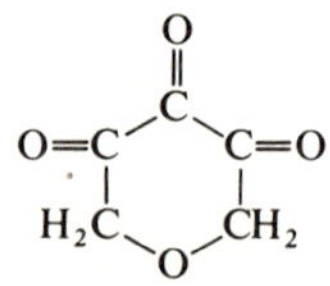

26

Monocarboxyl cellulose is extremely unstable in dilute solutions of alkalies and hot water. When it is dissolved in an aqueous solution of sodium hydroxide, there is observed a gradual reduction of the concentration of sodium hydroxide, after the initial lowering caused by the neutralization of carboxyl groups (128). This phenomenon is probably associated with the presence of keto groups, since polyuronic acids (which have carboxyl groups in the same position) are reasonably stable to alkalies.

According to polarographic evidence, the average length of the fragments formed on alkaline cleavage of the monocarboxyl cellulose chain depends only on the degree of oxidation and is independent of the degree of polymerization (DP) of the starting cellulose. The oxidation of only a few primary hydroxyl groups of cellulose to carboxyl groups does not affect the stability of the glycosidic linkages in the macromolecule to acid hydrolysis (129). An increase in the degree of oxidation (increase in the amount of carboxyl groups from 7.5 to 18%) results in an exponential increase in the rate of hydrolysis (130).

The course of the reaction of cellulose hydroxyl groups with nitrogen dioxide depends on the reaction conditions and, in particular, on the amount of water in the system. The introduction of dehydrating agents (e.g., phosphorus pentoxide) and use of a definite ratio of nitrogen dioxide to phosphorus pentoxide) results in the formation of either monocarboxyl cellulose mononitrate (DS of NO_3, 0.94; of COOH, 0.94) or monocarboxyl cellulose dinitrate (DS of NO_3, 1.82; of COOH, 0.9) (131, 132).

The oxidation of cellulose first with periodate and then with nitrogen dioxide (133, 134) gave a product, called tricarboxyl cellulose, which contained 50.8% of carboxyl groups (about 75% of the theoretical maximum). Tricarboxyl cellulose is soluble in dilute alkali solutions; its ion-exchange capacity is 11.4 meq/g.

References

1. P. Karrer, British Patent 249,842 (1925); P. Karrer, *Helv. Chim. Acta,* 9, 591 (1926).
2. Z. A. Rogovin and T. V. Vladimirova, *Vysokomol. Soedin.,* 2, 342 (1960).
3. H. Kholmuradov, Yu. S. Kozlova, L. S. Gal'braikh, and Z. A. Rogovin, *Izv. Vyssh. Ucheb. Zaved., Khim. i Khim. Tekhnol.,* 9, 470 (1966).
4. N. K. Kochetkov, L. I. Kudryashov, A. I. Usov, and B. A. Dmitriev, *Zh. Obshch. Khim.,* **31**, 3303 (1961).
5. N. K. Kochetkov, and A. I. Usov, *Izv. Akad. Nauk SSSR, Ser. Khim.* 475 (1964).
6. L. S. Gal'braikh, D. G. Dimitrov, and Z. A. Rogovin, *Vysokomol. Soedin.,* **B11**, 911 (1969).
7. L. S. Sletkina, A. I. Polyakov, and Z. A. Rogovin, *Vysokomol. Soedin.,* 7, 199 (1965).
8. N. Kholmuradov, Yu. S. Kozlova, L. S. Gal'braikh, and Z. A. Rogovin, *Vysokomol. Soedin.,* **B9**, 874 (1967).
9. L. G. Nikologorskaya, Yu. S. Kozlova, L. S. Gal'braikh, and Z. A. Rogovin, *Vysokomol. Soedin.,* **A13** 2762 (1970).
10. E. A. Chaikina, L. S. Gal'braikh, Z. A. Rogovin, *Vysokomol. Soedin.,* **B9**, 151 (1967).
11. M. K. Belyakova, L. S. Gal'braikh, and Z. A. Rogovin, *Vysokomol. Soedin.,* **A11**, 577 (1969).
12. N. Kholmuradov, Yu. S. Kozlova, A. I. Polyakov, and Z. A. Rogovin, *Vysokomol. Soedin.,* 7, 439 (1965).
13. N. Kholmuradov, Yu. S. Kozlova, L. S. Gal'braikh, and Z. A. Rogovin, *Vysokomol. Soedin.,* 8, 1089 (1966).
14. E. A. Chaikina, L. S. Gal'braikh, and Z. A. Rogovin, *Cellulose Chem. Technol.,* 1, 625 (1967).
15. T. V. Vladimirova, M. M. Donskaya, L. S. Gal'braikh, and Z. A. Rogovin, *Vysokomol. Soedin., Tselluloza i ee Proizvodnye, Sb. Statei,* 114 (1963).
16. T. V. Vladimirova, Kh. S. Peker, L. S. Gal'braikh, and Z. A. Rogovin, *Vysokomol. Soedin.,* 7, 779 (1965).
17. G. S. Masaidova, L. S. Gal'braikh. V. I. Nikitin, and Z. A. Rogovin, *Vysokomol. Soedin.,* **9A**, 166 (1967).

18. P. P. Shorygin and N. N. Makarova-Zemlyanskaya, *Dokl. Akad. Nauk SSSR,* **14,** 509 (1937).
19. N. N. Shorygina, *Zh. Obshch. Khim.,* **14**, 825 (1944).
20. S. N. Danilov and A. A. Lopatenok, *Zh. Obshch. Khim.,* **28**, 3184 (1958).
21. M. E. Carter, *J. Polymer Sci.,* **51**, 199 (1961).
22. F. B. Cramer and C. B. Purves, *J. Am. Chem. Soc.,* **61**, 3458 (1939).
23. E. D. Kaverzneva, V. I. Ivanov, and A. S. Salova, *Izv. Akad. Nauk SSSR, Otd. Khim. Nauk,* 369 (1949).
24. D. G. Dimitrov, L. S. Antonyuk, W. B. Achwal, L. S. Gal'braikh, and Z. A. Rogovin, *Vysokomol. Soedin.,* **A10**, 1373 (1968).
25. I. Sakurada, *J. Soc. Chem. Ind. Japan,* **32**, 11B (1929).
26. I. Sakurada, *Bull. Inst. Phys. Chem. Research (Tokyo),* **8**, 265 (1929).
27. K. Hess and N. Ljubitsch, *Ann.,* **507**, 62 (1933).
28. P. S. Scherer and J. M. Feild, *Rayon Textile Monthly,* **22**, 607 (1941).
29. P. S. Scherer and G. S. Saul, *Rayon Textile Monthly,* **28**, 474, 537 (1947).
30. F. N. Hayes and C.-H. Lin, *J. Am. Chem. Soc.,* **71**, 3843 (1949).
31. J. F. Haskins and S. G. Sunderwirth, *J. Am. Chem. Soc.,* **79**, 1492 (1957).
32. Yu. G. Kryazhev, A. I. Polyakov, and Z. A. Rogovin, *Vysokomol. Soedin.,* **5**, 48 (1963).
33. J. F. Haskins and A. H. Weinstein, *J. Org. Chem.,* **19**, 67 (1954).
34. A. I. Polyakov, V. A. Derevitskaya, and Z. A. Rogovin, *Vysokomol. Soedin.,* **3**, 1027 (1961).
35. A. I. Polyakov and Z. A. Rogovin, *Vysokomol. Soedin.,* **5**, 11 (1963).
36. T. Nakashima and I. Sakurada, *Bull. Phys. Chem. Res. (Tokyo),* 8, 272 (1929).
37. A. I. Polyakov and Z. A. Rogovin, *Vysokomol. Soedin.,* **5**, 147 (1963).
38. R. L. Whistler and B. Shasha, *J. Org. Chem.,* **29**, 880 (1964).
39. N. Kholmuradov, Yu. S. Kozlova, A. I. Polyakov, and Z. A. Rogovin, *Vysokomol. Soedin.,* **6**, 963 (1964); *ibid.,* **7**, 439 (1965).
40. E. Pacsu and R. F. Schwenker, Jr., *Textile Res. J.,* **27**, 173 (1957).
41. P. Carré and P. Mauclère, *Compt. Rend.,* **192**, 1567 (1931).
42. R. L. Boehm, *J. Org. Chem.,* **23**, 1716 (1958).
43. M. L. Wolfrom, J. C. Sowden, and E. A. Metcalf, *J. Am. Chem. Soc.,* **63**, 1688 (1941).
44. R. W. Roberts, *J. Am. Chem. Soc.,* **79**, 1175 (1957).
45. S. Fumasoni and G. Schippa, *Ann. Chim.* (Rome), **53**, 894 (1963).
46. F. B. Cramer and C. B. Purves, *J. Am. Chem. Soc.,* **61**, 3458 (1939).
47. J. F. Mahoney and C. B. Purves, *J. Am. Chem. Soc.,* **64**, 9 (1942).
48. J. Honeyman, *J. Chem. Soc.,* 168 (1947).
49. J. F. Carson and W. D. Maclay, *J. Am. Chem. Soc.,* **70**, 2220 (1948).
50. C. J. Malm, L. J. Tanghe, and B. C. Laird, *J. Am. Chem. Soc.,* **70**, 2740 (1948).
51. E. Heuser, M. A. Heath, and W. H. Shockley, *J. Am. Chem. Soc.,* **72**, 670 (1950).
52. T. Timell, *Svensk Kem. Tidskr.,* **61**, 146 (1949).
53. C. W. Tasher and C. B. Purves, *J. Am. Chem. Soc.,* **71**, 1023 (1949).
54. G. E. Murray and C. B. Purves, *J. Am. Chem. Soc.,* **62**, 3194 (1940).

55. V. A. Derevitskaya, Yu. S. Kozlova, and Z. A. Rogovin, *Zh. Obshch. Khim.,* **26**, 1466, 3369 (1956).

56. E. F. Izard, U.S. Patents 2,418,938, 2,418,939, and 2,418,940 (1947).

57. E. F. Izard and B. W. Howk, U.S. Patent 2,418,941 (1947).

58. P. W. Morgan, U.S. Patent 2,418,942 (1947).

59. E. F. Izard and P. W. Morgan, *Ind. Eng. Chem.,* **41**, 617 (1949).

60. L. Lifland and E. Pacsu, *Textile Res. J.,* **32**, 170, 797 (1962).

61. R. F. Schwenker and L. Lifland, *Textile Res. J.,* **33**, 107 (1963).

62. S. I. Polukhina, L. S. Gal'braikh, and Z. A. Rogovin, *Vysokomol. Soedin.,* **B10**, 479 (1968).

63. L. S. Gal'braikh, G. S. Masaidova, and Z. A. Rogovin, *Cellulose Chem. Technol.,* **8**, (1969).

64. N. K. Kochetkov et al., *Khimiya Uglevodov*, Izd. Khimya, Moscow, 1967.

65. T. S. Gardner and C. B. Purves, *J. Am. Chem. Soc.,* **65**, 444 (1943).

66. S. N. Danilov, *Uglevody i Uglevodnyi Obmen v Zhivotnykh i Rastitelnykh Organizmakh,* Akad. Nauk SSSR, 1959, p. 48.

67. A. A. Lopatenok, dissertation, Leningrad, 1953.

68. G. N. Smirnova, A. I. Polyakov, and Z. A. Rogovin, *Vysokomol. Soedin.,* **7**, 972 (1965).

69. G. N. Smirnova, L. S. Gal'braikh, A. I. Polyakov, and Z. A. Rogovin, *Vysokomol. Soedin.,* **8**, 1396 (1966).

70. G. N. Smirnova, L. S. Gal'braikh, and Z. A. Rogovin, *Cellulose Chem. Technol.,* **1**, 11 (1967).

71. G. N. Smirnova, L. S. Gal'braikh, and Z. A. Rogovin, *Khim. Prirodn. Soedin.,* **2**, 3 (1966).

72. S. I. Polukhina, L. S. Gal'braikh, and Z. A. Rogovin, *Vysokomol. Soedin.,* **A10**, 2039 (1967).

73. Yu. M. Makhsudov, L. S. Gal'braikh, A. I. Polyakov, and Z. A. Rogovin, *Vysokomol. Soedin.,* **8**, 1289 (1966).

74. Yu. M. Makhsudov, L. G. Krylova, L. S. Gal'braikh, and Z. A. Rogovin, *Khim. Prirodn. Soedin.,* **2**, 372 (1966).

75. Yu. M. Makhsudov, L. S. Gal'braikh, and Z. A. Rogovin, *Vysokomol. Soedin.,* **A9**, 1733 (1967).

76. V. P. Komar, R. G. Zbankov, Yu. M. Makhsudov, L. G. Krylova, L. S. Gal'braikh, and Z. A. Rogovin, *Vysokomol. Soedin.,* **8**, 2012 (1966).

77. L. S. Gal'braikh, W. B. Achwal, and Z. A. Rogovin, *Vysokomol. Soedin.,* **B9**, 863 (1967).

78. A. I. Polyakov, V. A. Derevitskaya, and Z. A. Rogovin, *Vysokomol. Soedin.,* **2**, 386 (1960); *ibid.,* **5**, 161 (1963).

79. D. G. Dimitrov, L. S. Gal'braikh, and Z. A. Rogovin, *Vysokomol. Soedin.,* **7**, 2174 (1965); *ibid.,* **B9**, 685 (1966).

80. E. M. Akovbyan, L. S. Gal'braikh, O. Yu. Okhlobystin, and Z. A. Rogovin, *Vysokomol. Soedin.,* **8**, 959 (1966).

81. E. M. Akovbyan, E. A. Chaikina, L. S. Gal'braikh, V. I. Bregadze, O. Yu. Okhlobystin, and Z. A. Rogovin, *Vysokomol. Soedin.,* **B9**, 566 (1967).

82. Ya. Ya. Makarov-Zemlyanskii and V. V. Gerzev, *Zh. Obshch. Khim.*, **35**, 272 (1965).
83. Z. A. Rogovin and N. N. Shorygina, *Khimiya Tsellulosy i ee Sputnikov*, Goskhimizdat, Moscow, 1953.
84. E. Ott, Ed., *Cellulose and Cellulose Derivatives*, Interscience, New York, 1954.
85. N. I. Nikitin, *Khimiya Tsellulosy i Drevesiny*, Izd. Akad. Nauk SSSR, 1963.
86. T. P. Nevell, *J. Textile Inst.*, **48**, T484 (1957).
87. F. S. H. Head, *J. Textile Inst.*, **38**, T389 (1947).
88. D. O'Meara and G. N. Richards, *Chem. Ind. (London)*, 40 (1958).
89. D. O'Meara and G. N. Richards, *J. Chem. Soc.*, 4504 (1958).
90. V. I. Ivanov, N. Ya. Lenshina, and V. S. Ivanova, *Izv. Akad. Nauk SSSR, Otd. Khim. Nauk*, 1136 (1960).
91. Z. I. Kuznetsova and V. I. Ivanov, *Izv. Akad. Nauk SSSR, Otd. Khim. Nauk*, 2044 (1960).
92. J. W. Rowen, F. H. Forziati, and R. E. Reeves, *J. Chem. Soc.*, **73**, 4484 (1951).
93. B. I. Stepanov, R. G. Zbankov, and I. N. Ermolenko, *Izv. Akad. Nauk. SSSR, Ser. Phys.*, **23**, 1222 (1959).
94. R. G. Zbankov, *Optika i Spektroskopiya*, **4**, 318 (1958).
95. L. Mester, *J. Am. Chem. Soc.*, **77**, 5452 (1955).
96. F. S. H. Head, *J. Textile Inst.*, **43**, T1 (1952).
97. Z. I. Kusnetsova, V. S. Ivanova, and N. N. Shorygina, *Izv. Akad. Nauk SSSR, Ser. Khim.*, 743 (1966).
98. A. Meller, *Chem. Ind. (London)*, 1204 (1953).
99. A. Meller, *Tappi*, **38**, 682 (1955); *ibid.*, **39**, 722 (1956).
100. B. Lindberg and O. Theander, *Svensk Papperstidn.*, **57**, 83 (1955).
101. F. S. H. Head, *J. Textile Inst.*, **46**, T400, T584 (1955).
102. I. J. Goldstein, J. K. Hamilton, R. Montgomery, and F. Smith, *J. Am. Chem. Soc.*, **79**, 6469 (1957).
103. H. Sihtola, *Makromol. Chem.*, **35**, 250 (1960).
104. F. Smith and J. W. Van Cleve, *J. Am. Chem. Soc.*, **77**, 3091 (1955).
105. Z. I. Kuznetsova, V. S. Ivanova, and N. N. Shorygina, *Izv. Akad. Nauk SSSR, Ser. Khim.*, 1682 (1965).
106. Z. A. Rogovin, A. G. Yashunskaya, and B. M. Bogoslovskii, *Zh. Prikl. Khim.*, **23**, 631 (1950).
107. Yu. S. Kozlova, and Z. A. Rogovin, *Vysokomol. Soedin.*, **2**, 614 (1960).
108. L. S. Gal'braikh, V. A. Dervitskaya, and Z. A. Rogovin, *Vysokomol. Soedin*, **3**, 980 (1961).
109. Z. I. Kuznetsova, V. S. Ivanova, and N. N. Shorygina, *Izv. Akad. Nauk SSSR, Otd. Khim. Nauk*, 2087 (1962).
110. Z. I. Kuznetsova, V. S. Ivanova, and N. N. Shorygina, *Izv. Akad. Nauk SSSR, Ser. Khim.*, 2232 (1964).
111. R. G. Krylova, S. N. Ryadovskaya, and O. P. Golova, *Izv. Akad. Nauk SSSR, Ser. Khim.*, 947 (1966).
112. R. G. Krylova, S. N. Ryadovskaya, and O. P. Golova, *Vysokomol. Soedin.*, **A9**, 933 (1967).

113. R. G. Krylova, S. N. Ryadovskaya, and O. P. Golova, *Vysokomol. Soedin.*, **B10**, 703 (1968).

114. R. G. Krylova, S. N. Ryadovskaya, and O. P. Golova, *Vysokomol. Soedin.*, **B11**, 381 (1969).

115. H. Rutherford, F. N. Minor, A. Martin, and M. Harris, *J. Res. Natl. Bur. Std.*, **29**, 131 (1942).

116. V. I. Ivanov and Z. I. Kuznetsova, *Izv. Akad. Nauk SSSR, Otd. Khim. Nauk*, 647 (1958).

117. Z. A. Rogovin, *Khim. i Teknol. Polym.*, N7-8, 174 (1960).

118. L. S. Gal'braikh and Z. A. Rogovin, *Vysokomol. Soedin.*, **5**, 693 (1963).

119. P. A. McGee, W. F. Fowler, Jr., E. Taylor, C. C. Unruh, and W. O. Kenyon, *J. Am. Chem. Soc.*, **69**, 355 (1947).

120. M. M. Pavlyuchenko, I. N. Ermolenko, and F. N. Kaputskii, *Zh. Obshch. Khim.*, **28**, 722 (1958).

121. E. D. Stakheeva-Kaverzneva and A. S. Salova, *Izv. Akad. Nauk SSSR, Otd. Khim. Nauk*, 344 (1959).

122. T. P. Nevell, *J. Textile Inst.*, **42**, T91 (1951).

123. P. A. McGee, W. F. Fowler, Jr., C. C. Unruh, and W. O. Kenyon, *J. Am. Chem. Soc.*, **70**, 2700 (1948).

124. E. D. Stakheeva-Kaverzneva, *Dokl. Akad. Nauk SSSR*, **68**, 865 (1949).

125. E. D. Stakheeva-Kaverzneva, *Izv. Akad. Nauk SSSR, Otd. Khim. Nauk*, 791 (1951).

126. E. D. Kaverzneva, V. I. Ivanov, A. S. Salova, and S. A. Kist, *Izv. Akad. Nauk SSSR, Otd. Khim. Nauk*, 358 (1956).

127. E. Battenberg and A. Berg, *Ber.*, **86**, 640 (1953).

128. I. Rusznak and D. Levai, *Vysokomol. Soedin.*, **6**, 551 (1964).

129. A. A. Konkin, G. S. Smirnova, and Z. A. Rogovin, *Trudy Mosk. Tekst. Inst.*, **13**, 99(1954).

130. I. Rusznak, *Magyar Kem. Folyoirat*, 70, 45 (1964); *Chem. Abstr.*, **61**, 22044 (1964).

131. Z. I. Kuznetsova, V. S. Ivanova, and N. N. Shorygina, *Izv. Akad. Nauk SSSR, Ser. Khim.*, 557 (1965).

132. Z. I. Kuznetsova and V. S. Ivanova, *Vysokomol. Soedin.*, **A9**, 1930 (1967).

133. N. Ya. Lenshima, V. S. Ivanova, and V. I. Ivanov, *Izv. Akad. Nauk SSSR, Otd. Khim. Nauk*, 1559 (1959).

134. N. Ya. Lenshina, V. S. Ivanova, and V. I. Ivanov, *Dokl. Akad. Nauk SSSR*, **129**, 325 (1959).

G. GRAFT COPOLYMERS

V. T. Stannett and H. B. Hopfenberg

Cellulose has the distinction of being among the first polymer substrates to be used for grafting. In 1943 Ushakov (1) synthesized some allyl and vinyl esters of cellulose which he attempted to copolymerize with maleic esters. The insoluble products he obtained were probably the first graft copolymers reported. Since that time a huge literature has developed concerning graft copolymers in general and particularly of cellulose and its derivatives. For example, a bibliography published in 1966 (2) lists 257 references published through 1965. A recent review may be found in reference 3.

The tremendous interest in cellulose grafting was natural since it represented the first novel departure from the traditional means of modifying cellulose through chemical reaction, such as esterification and etherification. The driving force behind this effort was undoubtedly the hope of developing new and commercially interesting cellulose modifications. Graft copolymerizations have been successfully made, for example, to cotton, rayon, paper, cellophane, and to wood itself, as well as to derivatives of cellulose, such as the cellulose esters. Although only minor commercial successes have been achieved in this endeavor so far, there seems little doubt that much larger developments will one day be brought about.

1. Methods of Grafting

A large number of different methods of grafting have been developed (4). By far the greatest effort has been via free-radical vinyl-polymerization routes. This is surprising in view of the large number of functional hydroxyl groups present in cellulose. This is partly due to the difficulty of reacting two polymers together, as would be needed to combine chemically an existing polymer to cellulose via condensation to the pendant hydroxyl groups. Some efforts have been made, however, along these lines.

Most of the reactions discussed here are heterogeneous in nature—that is, the cellulosic component is not in solution—and hence the accessibility of the substrate to the reagents and monomers is of the utmost importance. Accessibility is frequently provided by swelling the cellulose with an appropriate swelling agent, most commonly water. Solvent exchange and other techniques have been used to increase the rate of monomer supply to the growth site. This subject will be discussed further during the treatment of radiation grafting.

(a) FREE-RADICAL-POLYMERIZATION ROUTES

(1) Chain Transfer

In the chain-transfer method the grafting is carried out by polymerizing a vinyl monomer in the presence of a cellulosic material. The occasional chain-transfer steps that occur can lead to grafting in the following manner:

1. $\text{catalyst} \rightarrow 2R\cdot$ 2. $R\cdot + M \rightarrow RM\cdot$	Chain initiation
3. $M_x\cdot + M \rightarrow M_{x+1}\cdot$	Chain growth
4. $M_x\cdot + M_y\cdot \rightarrow M_{x+y}$	Chain termination
5. $M_x\cdot + R_{\text{cell}} \rightarrow M_xH + R_{\text{cell}}\cdot$	Chain transfer to cellulosic material
6. $R_{\text{cell}}\cdot + M \rightarrow R_{\text{cell}}M\cdot$	Graft-copolymer chain initiation
7. $R_{\text{cell}}M\cdot + M \rightarrow R_{\text{cell}}M_2\cdot$, etc.	Grafted-side-chain growth
8. $R_{\text{cell}}M_x\cdot + M_y\cdot \rightarrow R_{\text{cell}}M_{x+y}$	Grafted-side-chain termination
9. $R_{\text{cell}}M_x\cdot + R_{\text{cell}}M_y\cdot \rightarrow R_{\text{cell}}M_{x+y}R_{\text{cell}}$	Crosslinking

Step 9 produces a crosslink, and occasional crosslinks are sometimes obtained in this method. The other termination, step 8, however, apparently dominates under most conditions studied, and this, plus the fact that more than one crosslink is needed to produce insolubility, leads, in general, to low or zero gel content.

This method was first developed by Waltcher, Burroughs, and Jahn (5), who enhanced the chain-transfer reaction by introducing halogen-substituted ester groups into the cellulose molecule. Even with these derivatives only low yields of graft were obtained. More recently Chaudhuri and Hermans (6, 7) reacted cellulose and its derivatives with ethylene sulfide to introduce mercaptyl groups. These groups chain-transfer very readily, and excellent yields of graft copolymer were obtained. Even with this highly favorable situation at least 50% of homopolymer is produced, as can be seen by a consideration of the reaction scheme given above. Nevertheless good yields of graft copolymers are obtained even with unmodified cellulose by special means. These involve the generation of the initiating species in the swollen substrate itself. The concentration of monomer is low under these conditions, and the reaction (equation 1) becomes important compared with the homopolymer-producing step - (equation 2):

$$R\cdot + R_{cell}H \longrightarrow RH + R_{cell}\cdot \quad (1)$$

$$R\cdot + M \longrightarrow RM\cdot \quad \text{(homopolymer radical)} \quad (2)$$

Techniques of carrying out this method include, for example, the saturation of the cellulose substrate with potassium persulfate. A number of details of this reaction and a convincing demonstration that actual grafting does occur have been presented by Sankalia, Chaudhuri, and Hermans (7).

A somewhat similar method that has been developed includes the binding or sorption of one component of an oxidation-reduction initiating system onto the cellulose. This can then be brought into contact with monomer and the oxidation component, again leading to the generation of radicals in the cellulose substrate itself. An example of this method is to absorb ferrous salts into cellulose. If the substrate contains carboxylic acid groups, the ferrous ions are anchored; if not, simple sorption occurs. In either case, when the treated cellulosic material is brought into contact with a solution of monomer and, for example, hydrogen peroxide in water, radicals are produced by the following reaction:

$$Fe^{2+} + H_2O_2 \longrightarrow Fe^{3+} + OH^- + \cdot OH \quad (3)$$

The hydroxyl radicals will then either react with the cellulose, leading to graft copolymer, or with monomer, leading to homopolymer. The extent to which these two reactions occur has been studied by Richards (8) for the case of viscose. A large number of these oxidation-reduction systems have been discussed by Bridgeford (9, 10), who also discusses the deliberate localization of the polymerization reactions at different depths in the substrate and the use of the ion-exchange anchoring of the metallic component. Microscopic and other evidence showed that the deposition of polymer could be controlled at particular levels in the substrate.

(2) Direct Oxidation

The best known example of this technique is the use of the ceric ion, originally developed by Mino and Kaizerman (11). Although the actual reaction is now known to be quite complex, the principles of the reaction can best be illustrated in the following way:

$$R_{cell}H + Ce^{4+} \longrightarrow R_{cell}\cdot + Ce^{3+} + H^+ \quad (4)$$

$$R_{cell}\cdot + M \longrightarrow \text{graft copolymer} \quad (5)$$

Although in principle no homopolymer should be formed, in fact some homopolymer always appears to be produced due to reactions of the ceric ion with the monomers themselves.

The ceric ion method has been extensively investigated with a large number of different cellulosic systems, and many details of the reaction have been elucidated (4). Although the complete details of the reaction are not understood, it seems clear that reversible complex formation with the ceric ion is involved.

$$Ce^{4+} + R_{cell}H \rightleftharpoons \text{complex} \longrightarrow Ce^{3+} + H^+ + R_{cell}\cdot \tag{6}$$

The ceric ion method is particularly rapid and appears to be suited for commercial production. A small pilot plant has been operated successfully for continuous, rapid grafting to paper, cloth, and other substrates. Detailed studies have been reported also on grafting on a laboratory scale onto paper (12), cotton, cloth, and rayon (13–15).

Finally, a number of other oxidizing agents have been developed, although in much less detail. These include manganese and permanganate ion, vanadium salts, and potassium bromate. Most of these studies, however, are described in patents or brief notes. A somewhat similar method employs sodium periodate to oxidize cellulose, breaking the carbon–carbon bonds of the 1,2 dihydric alcohol groups. This method of grafting has been studied by Toda (16).

(3) Cellulosic Initiators

Another general method of grafting to cellulose is to form chemically an initiator, such as a peroxide, on the macromolecule, such as a hydroperoxide (**1**). This can then decompose into radicals and initiate graft copolymerization.

$$\underset{\mathbf{1}}{R_{cell}\text{—OOH}} \longrightarrow R_{cell}\text{—O}\cdot + HO\cdot \tag{7}$$

$$HO\cdot + M \longrightarrow \text{homopolymer} \tag{8}$$

$$R_{cell}\text{—O}\cdot + M \longrightarrow \text{graft copolymer} \tag{9}$$

If a reducing agent is used, the homopolymer formation can be largely eliminated, for example,

$$R_{cell}\text{—OOH} + Fe^{2+} \longrightarrow R_{cell}\text{—O}\cdot + Fe^{3+} + OH^- \tag{10}$$

The peroxides can be formed by ozonization, which has been chiefly studied in the Soviet Union (17, 18), or by radiation of the substrate in air or in hydrogen peroxide solution. Similar reactions have been carried out in cellulose nitrate (19), ethylcellulose, and benzylcellulose (20). There is a danger with these methods that extensive degradation may occur, and parallel studies concerning this effect are badly needed.

In addition to peroxides, other initiating groups can be introduced into the cellulose molecule, including for example, the formation of cellulose peroxy esters. Diazonium salts have also been synthesized and used as grafting initiators. Richards (21) prepared the *p*-aminophenyl acyl ether of cellulose and the *p*-aminophenyl acyl ester of carboxymethylcellulose. These were then diazotized, and the diazonium salts were used as the initiator either by heating or using them as a component of an oxidation–reduction system with ferrous ammonium sulfate. Thus in the case of sodium carboxymethylcellulose the following sequence of reactions occurs:

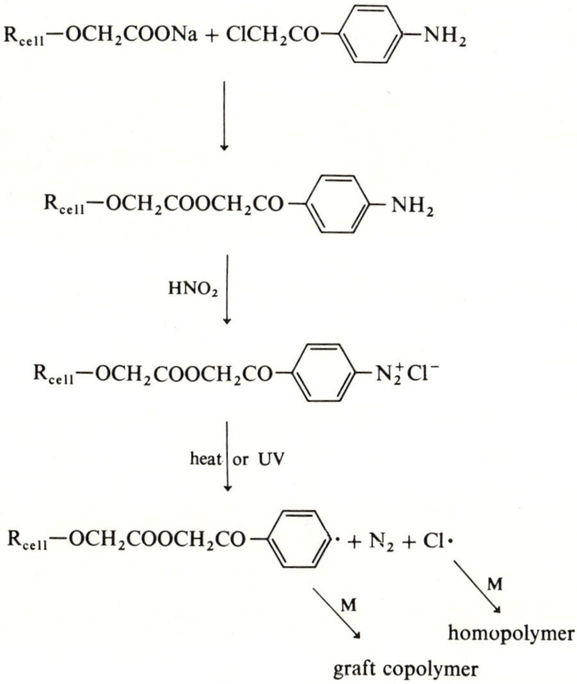

(11)

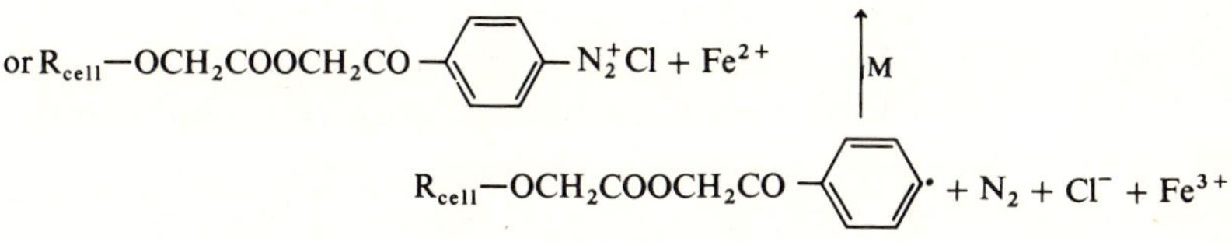

(12)

There are two attractive features to this method. First, no homopolymer is formed when a reducing agent is used. Second, with the carboxymethylcellulose substrate the grafted side chains can be easily and quantitatively removed by mild hydrolysis. This system can therefore be, in principle, very valuable for more precisely characterized grafting studies. Extensive studies have been carried out with the diazonium salts of cellulose and its derivatives by Rogovin and co-workers (22–24).

(4) Cellulose Derivatives as Comonomers

The earliest, and perhaps the most obvious, way to form cellulose graft copolymers is to form vinyl or allyl cellulose derivatives. Among the derivatives prepared have been cellulose allyl esters, cellulose methacrylates, cellulose allyl ethers, and cellulose crotonates and maleates (1, 25, 26). More recently vinyl groups were introduced with cellulose nitrate by reacting with allyl monocarbonate and hexamethylene diisocyanate (27). Grafting to these derivatives can be accomplished by direct free-radical vinyl copolymerization. It can readily be seen that if there are several unsaturated groups on each cellulose molecule, heavy crosslinking will rapidly result. The situation is in many ways comparable to the polymerization of unsaturated polyesters. With low degrees of substitution, however, or properly balanced monomer reactivity ratios graft copolymers that are essentially free from crosslinks can be prepared by this technique.

(5) Radiation Methods

Both ultraviolet light and high-energy radiation have been used to initiate grafting to cellulose and its derivatives. The latter method has, however, been by far the most thoroughly studied. In fact high-energy-radiation grafting has been investigated in more detail than any other method of any kind. Since the chemistry of the process is relatively obscure, radiation initiation has been used for experiments designed to reveal details of the grafting process itself. Since no chemical reagents are used normally for the radiation-grafting techniques, it is essential to make the substrate accessible to the monomer. In fact without a suitable swelling agent only negligible degrees of grafting are possible. All three methods developed for radiation grafting in general have been successfully used for cellulose.

Peroxide Method. In the peroxide method the cellulose substrate is irradiated in air to form peroxide compounds, which can later be used for grafting as described on page 910. This method has been less studied with cellulose than with many other polymers; however, preirradiation grafting in air has been frequently used. Here it seems clear that a combination of peroxide and trapped-radical initiation is operating. Kobayashi (28) increased the grafting yields by first soaking the cellulose, in this case viscose-rayon fibers, in hydrogen peroxide solution.

Preirradiation Method. This method consists of irradiating the cellulosic substrate, preferably in the absence of air, and subsequently contacting the irradiated material with the monomer and swelling agent. Both liquid and vapor-phase grafting has been used, and nearly every kind of cellulosic substrate has been studied. Again, a suitable swelling agent is essential for successful grafting. In general it is better to irradiate dry in order to produce the maximum number of free radicals and then to admit the monomer and swelling agent together. In the case of vapor-phase grafting the swelling agent can also be in the form of vapor, such as water vapor. This method has been studied in considerable detail by Armstrong and Rutherford (29) for cotton and viscose rayon and cellulose acetate fibers. In addition to water vapor, methanol and acetic acid vapors were found to be effective promoters of the grafting process. In liquid-phase preirradiation grafting these additives were also found to be effective. In general the same monomer-swelling agent systems were effective for both the mutual and the preirradiation techniques. These will be discussed further in the section on mutual grafting methods. With the trapped-radical, as opposed to the peroxide, method of preirradiation grafting, very little homopolymer is produced. A disadvantage, however, is that the degradation of the cellulosic backbone is usually greater with the preirradiation method. This is particularly true when the grafting is carried out in the presence of air or oxygen. Less degradation is encountered with the mutual method of irradiation due to the protective action of the vinyl monomers present during the actual irradiation. Preirradiation is often carried out in air, and the combination of peroxides and grafted radicals that results is used to initiate the grafting reaction. In spite of the possibly deleterious effects of the concurrent degradation, the preirradiation method is very attractive economically. It has been used for pilot-plant studies of the grafting of styrene and other monomers to rayon. Some comparative data relating to the preirradiation and the mutual methods of grafting are presented later.

Mutual Method. In this method the cellulosic substrate is irradiated directly with the monomer. The cellulosic material can be actually dissolved in the monomer or monomer–solvent mixture, or simply swollen. Either vapor-phase or liquid-phase monomer can be used. The most usual technique, however, is the irradiation of the swollen cellulosic material in the liquid monomer or monomer solution. It is clear that much homopolymer will also be generated by direct radiolysis of the monomer or monomer–solvent or monomer–swelling agent mixture. The cellulose acetate–styrene system has been studied in considerable detail, and methods developed for separating the homopolymers from true graft copolymer are discussed later. It was apparent from the results of these studies that the degree of swelling of the cellulose substrate has a profound effect on both the yield of graft and the molecular weight of the grafted side chains. The results are illustrated in Figure 1 for cellulose acetate film irradiated in styrene–pyridine mixtures to a total dose of 10 Mrad (30).

At high pyridine contents the cellulose acetate dissolves; only small amounts

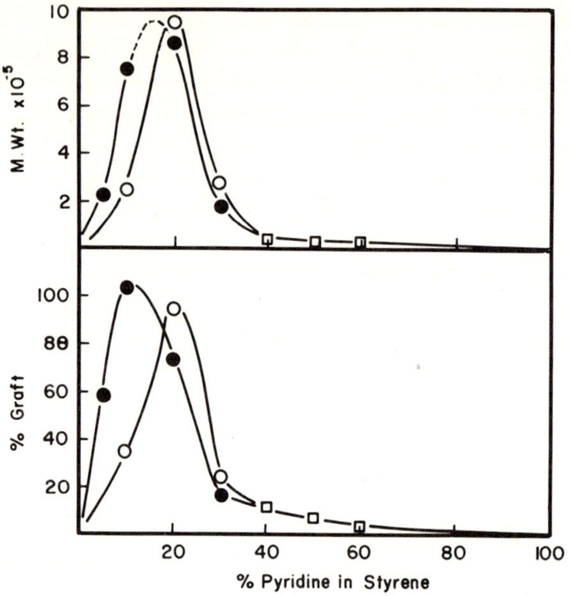

Fig. 1. Effect of swelling on the yields and molecular weights of the grafted side chains for the mutual radiation grafting of styrene to cellulose acetate film (30). Key: ● 0.0025-mm-thick film; ○ 0.025-mm-thick film. Dose of 10 Mrad at 0.34 Mrad/hr at 25°C.

of grafting occur, and the molecular weights of the side chains are rather low. As the styrene content increases, the film remains intact but is highly swollen. The grafting yield increases greatly, as do the molecular weights of the polystyrene side chains. Finally, as the styrene content continues to be increased, the yields and molecular weights reach a maximum and then begin to decrease. In pure styrene the secondary cellulose acetate used was only slightly swollen and essentially no grafting took place. Similar results were obtained with two thicknesses of film, but with the thicker film more swelling agent was needed to achieve the same degree of grafting.

These results make the role of the swelling agent or solvent (pyridine in this case) quite clear. When there is little swelling, as with pure styrene, the monomer cannot reach the bulk of the free radicals created in the film and only some surface grafting can take place. As the pyridine content increases, diffusion of monomer into the film becomes easier and grafting takes place. The termination of the growing chains is highly hindered by the so-called *gel effect*, and the molecular weights increase. Eventually at high pyridine contents ample supplies of monomer are available through diffusion and the process is no longer diffusion controlled. The termination process becomes less hindered and therefore more rapid, and the molecular weights of the side chains diminish; the grafting yields drop correspondingly. In true solution these effects are extended to extremes and only comparatively poor yields of graft are obtained. Table 1 shows some yields of graft copolymer and both homopolymers for a similar system at two thicknesses and dose rates. The diffusion-controlled nature of the process is shown also by the effect of the dose rate. At high dose rates the reaction becomes diffusion controlled to a greater extent and poorer yields per unit dose are obtained.

The need for swelling the cellulosic substrates has been stressed in this section. In the case of cellulosic derivatives it is comparatively easy to find swelling agents that, when mixed with adequate concentrations of monomer, give degrees of swelling up to complete solution. With cellulose itself the choice of swelling agents is more limited. Water is one of the more obvious and effective compounds and has been used alone or mixed with a mutual solvent for the monomer. Dioxane has been found to be an excellent cosolvent with water for styrene grafting, for example, and dissolves the homopolymer also. Methanol, formamide, and acetic acid have also been used alone or mixed with water or each other. An interesting swelling agent used extensively and effectively by Arthur and co-workers (31) is concentrated aqueous zinc chlorine solutions. Some variations of simple monomer-swelling mixtures for radiation grafting have been studied. Huang and Rapson (32) obtained good grafting yields by using water-saturated cellulose in styrene or by simply using wet styrene. A simple solvent exchange with water and styrene was also used effectively. Araki and co-workers (33) used a so-called dry technique in which the cellulose was boiled in water, then in various alcohols, and dried. The treated cellulose was then irradiated directly, or preirradiated, followed by immersion in monomer.

The reader is also referred to Section H of this Chapter XVII.

Ultraviolet-Light Grafting. Very few studies of ultraviolet-light grafting to cellulose and its derivatives have been reported. Oster (34) has reported the grafting of *N*-octylacrylamide to paper, wood, and regenerated-cellulose film by irradiating the cellulosic material in contact with a solution of monomer in ethanol or acetone containing 5–10% of benzophenone as the photosensitizer. Ultraviolet light with a wavelength of 2000–3000 Å from germicidal or carbon arc lamps was used.

More extensive studies using somewhat higher wavelengths have been reported by Geacintov et al. (35, 36). These authors used phototendering dyestuffs, notably the sodium salt of anthracene-2,7-disulfonic acid, as the sensitizer in aqueous solution. For grafting to cellulose acetate and ethylcellulose, 2-methylanthraquinone, which is soluble in organic solvents, was used. The mechanism proposed was the removal of a hydrogen atom from the cellulosic substrate by the light-excited dye molecule. The resulting cellulosic radical yield appeared to be low, but the chain character of the grafting process leads to substantial yields of graft. An interesting feature of this method of

TABLE 1
Yields and Efficiencies of Radiation Grafting of Styrene to Cellulose Acetate[a]

Sample No.	Solvent and its concentration in styrene (%)	Film thickness (mm)	Dose rate (Mrad/hr)	Total dose (Mrad)	Increase in weight% Total	Increase in weight% Due to grafting	Efficiency of grafting Cellulose acetate	Efficiency of grafting Poly-styrene	Molecular weight of grafted side chains (x 10^6)
1	Benzene, 80	–	0.026	0.25	140	49.6	17.4	35.6	–
2	Benzene, 52.5	–	0.026	0.25	211	56.5	16.5	26.8	–
3	Benzene, 34.4	–	0.022	0.48	256	59.0	17.7	23.1	0.29
4	Benzene, 29.5, +DMF, 4.9	–	0.017	0.35	292	52.0	7.1	14.8	0.63
5	DMF, 10	0.025	0.0076	0.45	183	93.6	13.6	51.8	4.2
6	DMF, 10	0.025	0.34	1.0	13.9	6.2	–	–	–
7	DMF, 10	0.0025	0.34	1.0	60.8	51.2	11.1	84.1	1.95
8	Pyridine, 10	0.025	0.34	10.0	53.4	34.0	19.2	63.7	0.25
9	Pyridine, 20	0.025	0.34	10.0	111.9	92.7	17.1	90.0	0.96
10	Pyridine, 30	0.025	0.34	10.0	23.1	23.1	28.3	100.0	0.27
11[b]	Pyridine, 20	0.025	0.34	10.0	72.5	69.0	46.8	94.8	1.63

[a] All samples except No. 11 were grafted by the mutual process.
[b] Grafted by the preirradiation method.

grafting is that enhanced yields are obtained in air, which obviates the need for the deaeration procedures usually involved in free-radical-grafting processes. The nature of the monomer exerts great influence on the grafting rates, acrylonitrile being particularly rapid and styrene and vinyl acetate rather slow. Mixtures of monomers gave maxima in the rates of grafting; in particular, mixtures of styrene and acrylonitrile gave high increases over either monomer alone. Similar results have been reported by Rapson and Kvasnicka (37) with high-energy radiation. As with other heterogeneous grafting methods, the degree of swelling of the substrate also strongly affects the grafting rate.

Guzman, Fatou, and Rosa (38) also studied the grafting of various vinyl monomers to cellulose with ultraviolet light in the presence of a number of photosensitizers. Benzoin was particularly studied. The rates of grafting increased with increasing monomer concentration, tended to decrease with increasing amounts of photosensitizer, and increased with increasing water content of the cellulose. Grafting to cellulose with ultraviolet light was also effected by Tsunooka, Tanaka, and Murata (39) by first reacting with methyl vinyl ketone and then using the carbonyl groups as photosensitizers for acrylonitrile polymerization.

Considerable further work with ultraviolet-initiated grafting is needed since for surface treatments of cellulosic materials the method appears to show considerable promise.

(6) Initiation by Free Radicals Formed by Other Methods

Most of the miscellaneous methods involve the breaking of the main cellulosic backbone chains in the presence of a vinyl monomer. The product forms by the initiation of vinyl polymer chains by the radicals resulting from chain cleavage and hence are probably block copolymers. Ceresa (40) has discussed a number of such methods, including mechanical mastication and vapor-phase swelling. In the case of starch the freezing and thawing of starch–acrylonitrile emulsions brought about the formation of block copolymers. Ultrasonics have been used by Akutin and co-workers (41) to produce copolymers of ethylcellulose in solution with another polymer, such as polysiloxanes. Deters and Huang (42) have used vibratory milling treatments. Kargin and Plate (43) used the initiation of polymerization by the mechanical disruption of inorganic materials in the presence of cellulose to produce graft copolymers. Electrical discharge techniques were used by Akutin and co-workers (44) to graft vinyl chloride to ethylcellulose.

(b) IONIC POLYMERIZATION ROUTES

Ionic methods of grafting, like ionic polymerization itself, have been little studied compared with free-radical processes. The polar nature and frequent need for aqueous systems with cellulosics have even further reduced interest in ionic methods. The alkoxides of cellulosics have, however, been explored in some depth, particularly with acrylonitrile. Feit et al. (45) used this method with acrylonitrile, methacrylonitrile, and methacrylic esters; liquid ammonia, tetrahydrofuran, dimethylformamide, and dimethyl sulfoxide were used as solvents. Considerable homopolymerization was found and attributed to chain-transfer reactions. Schwenker and Pacsu (46) carried out similar studies with sodium cellulosates and acrylonitrile. More recently a detailed study of this system has been made by Avny and Rebenfeld (47). Rayon was used, and tetrahydrofuran was the solvent. Grafts of up to 22% were obtained. It was found that considerable homopolymer was always formed. The side chains of the grafted cellulose were rather short (molecular weight about 15,000), and a rather high degree of substitution (DS) was therefore achieved. The mechanical properties were not improved by the grafting procedure. The reaction scheme for this type of ionic polymerization was presented as follows:

1. Initiation:

$$R_{cell}{-}O^{-}Na^{+} + CH_2{=}CHCN \longrightarrow R_{cell}{-}OCH_2{-}\bar{C}HCN\ Na^{+}$$

2. Propagation:

$$R_{cell}{-}OCH_2{-}\bar{C}HCN + nCH_2{=}CHCN \longrightarrow R_{cell}{-}O(CH_2{-}\underset{\displaystyle CN}{\underset{|}{C}}H)_nCH_2{-}\bar{C}HCN$$

3. Termination (by chain transfer to monomer):

$$R_{cell}{-}O(CH_2{-}\underset{\displaystyle CN}{\underset{|}{C}}H)_nCH_2{-}\bar{C}HCN + CH_2{=}CHCN \longrightarrow R_{cell}{-}O(CH_2{-}\underset{\displaystyle CN}{\underset{|}{C}}H)_n{-}CH_2CH_2CN + CH_2{=}\bar{C}{-}CN$$

4. Termination (by chain transfer to free OH of cellulose):

$$R_{cell}{-}O(CH_2{-}\underset{\displaystyle CN}{\underset{|}{C}}H)_nCH_2{-}\bar{C}HCN + R_{cell}{-}OH \longrightarrow R_{cell}{-}O(CH_2{-}\underset{\displaystyle CN}{\underset{|}{C}}H)_nCH_2CH_2CN + R_{cell}{-}O^{-}$$

5. Homopolymer formation (initiated by highly reactive species $CH_2{=}\bar{C}{-}CN$ formed in step 3):

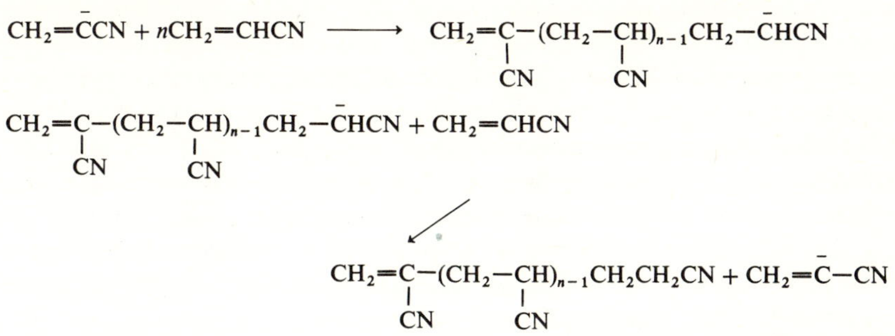

An interesting cationic method was described by Rausing and Sunner (48): boron trifluoride in nitrogen was absorbed into cellulose at low temperatures. The resulting complexed catalyst system was then reacted with isobutylene to give a grafted and partly encapsulated cellulosic material. Excellent water resistance was imparted to paper sheets by this method. Cellulosic materials have been encapsulated, but probably not grafted, by a method developed by the National Lead Company (49, 50). Here one component of a Ziegler-type catalyst, such as titanium tetrachloride, sorbed into cellulose, which is then treated with the second component, such as triisopropylaluminum, in an inert solvent and with a monomer, such as ethylene or propylene. Finally a very interesting series of graft copolymers of methylated xylans were prepared by O'Malley and Marchessault (51), who used the direct addition of "living" polystyryl carbanion to the fully methylated hemicellulose.

(c) CONDENSATION AND RING-ADDITION METHODS

Methods of this type have been little studied compared with the grafting to vinyl polymers. The most extensive studies have been made by Rogovin (52, 53). A distinction can be made between the addition of an already formed polymer to a reactive functional group of the cellulosic material and the utilization of reactive groups on the cellulosic molecule to initiate condensation or ring-opening polymerization.

An example of the former method is the reaction of a low-molecular-weight poly(ethylene adipate) acid chloride with cellulose, using either a solvent-exchange or interfacial polymerization technique. Only preliminary work seems to have been carried out so far with this method (52, 53). A somewhat analogous study (54) involves the reaction of telomers of poly(acrylic acid) having chlorine end groups with β-aminoethylcellulose. Only a small proportion of telomer became attached. Alkali cellulose also did not react extensively with the poly(acrylic acid) telomer.

Rogovin (52, 53, 55) also carried out direct polycondensation grafting of aminoenanthic acid to cellulose and its derivatives. The acid chloride of aminoenanthic acid hydrochloride was reacted with cellulose in the presence of a large quantity of pyridine or triethylamine, using solvent-exchange techniques. Comparatively low DS values and an average degree of polymerization (DP) of the side chains of about 10 were obtained. A similar grafting reaction was carried out by the interfacial polycondensation of a solution of the acid chloride of aminoenanthic acid hydrochloride in dry methylene chloride with a dilute alkaline solution of cellulose xanthate. The xanthate of the product that precipitated at the interface was decomposed with dilute sulfuric acid, and the homopolyamide was removed by extracting with 20% calcium chloride solution in methanol. Again rather low DS and DP of the side chains were obtained. In an earlier study (52, 53, 56) the acid chloride, amide, or methyl ester of carboxymethylcellulose was reacted with the methyl ester of aminoenanthic acid. The reactions were carried out at 100°C for several hours in an inert atmosphere. An average DP of the side chains of 6 was achieved.

The grafting of a number of ring-opening monomers to cellulosic materials has been achieved, starting with the well-known reaction of ethylene oxide with cellulose, forming hydroxyethylcellulose plus some product with a higher DP. Ethylenimine was also reacted (57) with cotton and viscose rayon above 100°C to form graft copolymers, with side chains having an average DP of about 6. β-Propiolactone was also successfully grafted to cotton cellulose with weight gains of up to 60% (58).

Champetier and Lucas (59) used inclusion techniques to react alkali cellulose with ethylene sulfide and propylene sulfide, and with cellulose itself. Graft copolymers were obtained with a relatively high DS and with a low DP of the grafted side chains. The latter was thought to be due to oxidation of the mercaptyl end groups, leading to disulfide crosslinks.

Finally, polyamides were successfully grafted to carboxymethylcellulose (60) by reacting the acid chloride with caprolactam. The acid chloride was prepared by reacting carboxymethylcellulose with thionyl chloride. The polymerization reaction was carried out at 100°C for several hours in the presence of sodium caprolactam. The DP of the grafted side chains was found to be about 6.

2. Characterization of the Grafts

In addition to developing methods of grafting and examining their properties, it is necessary to characterize as closely as possible the pure grafts and to estimate the extent and molecular nature of the resulting of graft. In some respects cellulosic graft copolymers are rather easy to investigate since the solubilities of the two homopolymers and the graft are often quite different from one another. Furthermore, the cellulosic backbone can be destroyed by acid hydrolysis, and the molecular weight and other properties of the isolated side chains can be determined. A number of studies of this kind have been reported (4), and various features of the grafting reaction and of the graft copolymers themselves have been elucidated.

(a) EXTENT OF GRAFTING

Whether the grafting reaction takes place in solution or within a swollen, insoluble cellulosic substrate, such as film or fiber, it seems inevitable that one or more homopolymers are present together with the graft. In rather early work (36) the grafting of acrylamide to cellulose film was carried out by an ultraviolet-light technique. It was found that the graft copolymer and polyacrylamide and the cellulose itself were all soluble in cuprammonium hydroxide. On acidifying, however, only the ungrafted cellulose and the graft copolymer precipitated. By weighing the precipitate the amount of the grafting could be measured, but the amount of ungrafted cellulose could not be determined. This simple procedure was later applied to a number of other grafting methods (61), and the results are shown graphically in Figure 2. It can be seen that each method of grafting gave different efficiencies of grafting with the preirradiation technique being the most efficient.

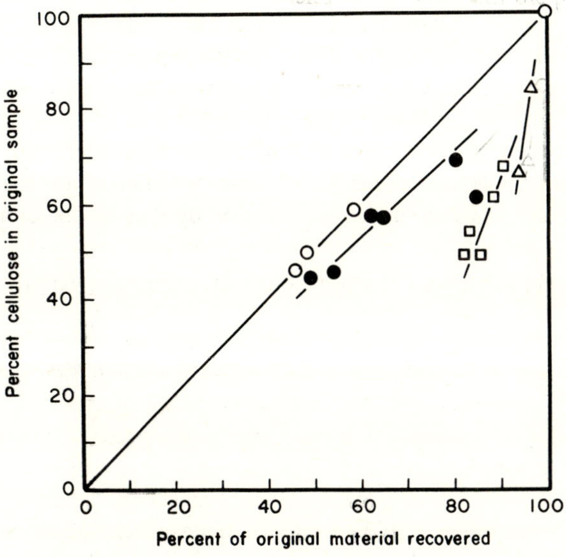

Fig. 2. Proof of grafting and yields for the acrylamide–cellulose film system (36). ○ physical mixture of cellulose and polyacrylamide; ● ceric-ion grafted; □ ultraviolet-light grafted; △ preirradiation (γ-rays) grafted.

Careful work in which the density-gradient ultracentrifuge was used to analyze the completeness of separation of the products showed how difficult it was to separate the homopolymer completely from the grafted material (62). However, a number of efforts have been made, and the results certainly give an indication of the grafting yields. In the case of pure cellulose the extractions are particularly difficult. Kobayashi (28) made one of the best separations with styrene-grafted rayon prepared by a preirradiation method. It was found that after acetylation the homopolystyrene could be removed by extraction with benzene, the ratio of homopolystyrene to grafted polystyrene being 1.25. Again no measure was made of the amount of ungrafted cellulose present.

Perhaps the most complete studies of this kind have been carried out with styrene grafted to secondary cellulose acetate (30, 61–63). It was shown that successive extractions with aqueous acetone and benzene brought about essentially complete separations. Even better separations were made possible by using solution and precipitation or by dissolving and freeze-drying the graft and extracting the resultant finely powdered product. Unfortunately only radiation-grafting techniques have been studied. The results of a number of such studies are summarized in Table 1. All the examples are from heterogeneous preparations and were extracted to constant weight with benzene before the more elaborate separations.

It can be seen that the efficiency of grafting is quite poor, particularly in the case of the mutual radiation method. The maximum amount of the cellulose acetate grafted was only 28.4%. The preirradiation method, on the other hand, grafted nearly 50% in the one example studied. The efficiency of the polystyrene grafting probably reflects more the ease of homopolymer removal than any fundamental fact, and in general thin films and more highly swollen samples gave greater efficiencies. The high efficiency of the preirradiation method was real, however, and was confirmed by separate experiments.

(b) MOLECULAR WEIGHTS OF THE GRAFTED SIDE CHAINS

A large number of samples of grafted cellulosic materials have been acid hydrolyzed, and the molecular weight of the isolated side chains have been determined. However, as shown in the preceding subsection, most of these samples contained occluded homopolymer also. From the point of view of kinetics, the homopolymer should have about the same molecular weight as the side chains whether the grafting be chemical or radiation initiated. The extraction procedures would tend to remove the lower-molecular-weight homopolymer preferentially, leading to higher values for the residual polymer after acid hydrolysis. The values included in Table 1 are for carefully separated grafts and may therefore be considered valid. Regardless of the limitations of the extraction procedures, there is no doubt that under heterogeneous conditions, such as swollen fibers and films, the side chains are disproportionately large, being often on the order of several million. This is undoubtedly due to the so-called gel effect, in which the viscous medium retards the chain-termination step due to diffusion control. In solution grafting, on the other hand, the molecular weights are of the same order of magnitude as polymer found in the free solution since the gel effect is not significant.

The very high molecular weights in heterogeneous grafting are largely responsible for the actual yields of grafts. In fact in a controlled set of radiation-grafting experiments the yields and the molecular weights were found to follow each other methodically. These results are shown graphically in Figure 1. The swelling conditions were varied by adding increasing amounts of a solvent (pyridine) to the monomer (styrene) containing films of secondary cellulose acetate. It can be seen that as the proportion of solvent is progressively added, the yields and molecular weights increase to a maximum and then decrease again, until regular solution grafting takes place. Styrene swells secondary cellulose acetate only to a small extent, and essentially no grafting takes place due to the poor supply of monomer to the free radicals created in the film by the radiation. As more solvent is added, diffusion of monomer into the film is facilitated and grafting takes place. Eventually the film becomes so swollen that the gel effect becomes minimized and the yields and molecular weights begin to decrease again. With thinner film these effects are shifted toward lower amounts of solvent since diffusion is relatively easier into a thinner film.

At least two sets of molecular-weight distributions in grafted side chains have been determined—again, both with the cellulose acetate–styrene systems and radiation grafting (64, 65). Graft copolymers produced by all three radiation methods have been studied. The mutual and peroxide methods gave polystyrene with a broad, but otherwise normal, distribution. The broad distributions are to be expected with such a complex heterogeneous reaction. The preirradiation product gave an unusual bimodal distribution, and this system was subsequently further studied (64, 66). No clear explanation was found, however.

(c) OTHER DETAILS OF CELLULOSIC GRAFTS

There is still no clear picture of the nature of the sites of attachment of the grafted side chains. With certain chemical methods the nature of the connecting links is clear from the method itself, but not the distribution among the three hydroxyl groups. With most methods, however—including the chain-transfer, ceric ion, and radiation techniques—almost nothing is known. Arthur and co-workers (67) used careful electron-spin-resonance (ESR) measurements of the intermediate free radicals and found strong indications that the grafts are probably mainly located in the 2- or 3-, the 6-, and the 5-positions for the ceric ion, ferrous ion-hydrogen peroxide, and the ionizing-radiation procedures, respectively (see also Section I of this Chapter XVII).

Cellulose acetate-styrene grafts have been examined by removing the ester groups through cold alkaline hydrolysis and measuring the loss of grafted polystyrene (63, 68). Again only radiation grafts appear to have been studied in this way. In the samples examined, whether prepared by the preirradiation or the mutual method, about half the polystyrene appeared to have been attached to the ester groups.

Finally, attempts have been made to determine the location of the grafted polymer in the cellulose substrate. Cotton fibers have been particularly well studied in this regard. The location, which appears to be mainly between the inner and outer surface of the fibers, depends on the method of grafting and the monomer system used (31, 67).

3. Properties and Applications of Cellulose Grafts

(a) GENERAL PROPERTIES

The properties of cellulosic materials—pulp and paper, textiles, regenerated cellulose—and cellulose derivatives can be dramatically changed by graft copolymerization. Although data concerning properties and applications are scant in comparison with data concerning synthesis and physicochemical characterization, a sufficient technology has emerged to permit the beginning of a rational tailoring of properties in cellulosics via graft copolymerization.

Graft copolymerization has resulted in improvement in a wide variety of properties, including tensile strength; resistance to microbial degradation, abrasion, and acid; dye receptivity; wet strength of paper; and adhesion. In addition, an entirely new spectrum of properties can be imparted, such as rendering paper and cellophane into ion-exchange membranes by the controlled grafting of anion- and/or cation-exchange groups onto the cellulose. Wellons, Williams, and Stannett (69) and Kesting and Stannett (70) have demonstrated that moisture regain in cellulose and cellulose acetate can be reduced by controlled radiation-induced grafting of styrene. Wellons et al. (69) also demonstrated that the compatibility of dissimilar polymers can be markedly improved by adding small quantities of graft copolymers "constructed" from the two incompatible backbones. The latter observation has vast implications regarding the formation of stable and useful "polymer alloys," or "polyblends."

Since cellulose and cellulose derivatives are often relatively inexpensive, property improvement and/or alteration via graft copolymerization will usually be accompanied by an increase in the cost of the resulting graft in comparison with the unmodified cellulosic. This immediately suggests that applications of end products will tend to be more specialized than those of the starting material.

It is recognized that the grafting process results in the formation not only of true, covalently bonded graft copolymer but also of residual homopolymer of the substrate polymer and newly formed homopolymer corresponding in the chemical repeat unit to the grafted side chains. Many workers (30, 61, 69, 71) have scrupulously attempted to synthesize, isolate, and subsequently characterize the pure graft, whereas others, though acknowledging the presence of residual homopolymer, have proceeded to evaluate the end properties of the graft contaminated by the homopolymer impurities. Although a definitive study on the effect of homopolymer content in the properties of graft mixtures has not been made, the reader is cautioned that graft copolymer containing unextracted homopolymer will have properties that differ from those of a true graft.

(b) CELLULOSIC FIBERS

Saunders and Sovish (72) have studied in detail the effects of acrylonitrile grafting to cotton on the rot resistance, abrasion resistance, solvent resistance, and mechanical stability of cotton sheeting. The grafting was carried out by irradiating cotton in dilute aqueous solutions of acrylonitrile to produce a polyacrylonitrile-modified cotton fabric with improved resistance to rot and mildew, solvents, and abrasion. Their results are presented in Tables 2, 3, and 4.

TABLE 2
Breaking Strength and Elongation of Modified Cotton Sheeting (72)

Modification[a]	Breaking strength (lb)[b]	Ultimate elongation (%)
None (control)	38.0	6.8
Irradiated in H_2O	37.4	6.8
Acrylonitrile:		
2%	38.7	6.9
5%	37.0	6.5
9%	37.3	6.6

[a] Modification with 1.6 Mrad of 2-MeV electrons.
[b] Determined on 1½ by 8-inch samples of cotton fabric.

The data of Table 2 indicate that under the specified grafting conditions the mechanical properties of the ungrafted fabric were retained consequent to modest grafting. The results presented in Table 3 demonstrate that grafting up to 10% acrylonitrile to cotton fabric has little effect on the abrasion resistance of cotton fabric, but that the grafting of 16.8% acrylonitrile results in more than double the abrasion resistance of the control. Significant decreases in moisture regain were encountered as a consequence of grafting.

Kesting and Stannett (70) similarly measured decreased moisture sorption in cellulose as a result of grafting styrene to cellulose fibers. Their results, presented in Table 5, further indicate that sorption, calculated on the basis of pure cellulose, was actually increased as a consequence of grafting. These latter results clearly suggest decreased order, or increased free volume, in the amorphous regions of the cellulosy as a direct consequence of grafting. Similar conclusions were drawn by Yoshitake et al. (73) on methyl acrylate grafting to rayon.

TABLE 3
Moisture Regain and Abrasion Resistance of Cotton Sheeting Modified with Acrylonitrile (72)

Modification[a]	Taber flex abrasion (cycles)[b]	Moisture regain (%)
None (control)	150	6.2
Irradiated (1.6 Mrad)	130	6.7
Acrylonitrile:		
6.5%	165	5.9
8.5%	183	5.2
16.8%	310	4.8

[a] Modification with 1.6 Mrad of 2-MeV electrons.
[b] Determined on 1½ by 8-inch samples of cotton fabric.

Yoshitake et al. envision domains of poly(methyl acrylate) disrupting the amorphous regions of the cellulose, resulting in increased free volume at the same level of crystallinity. Their model, albeit oversimplified, agrees with the domain model discussed by Wellons (69), which will be presented in subsection 3(e), "Cellulose Derivatives."

The soil-burial tests reported in Table 4 reflect the improved resistance to microbial attack as a consequence of acrylonitrile grafting. Clearly acrylonitrile grafting had a dramatic effect on the resistance to a microbial attack even under the extremely stringent test conditions, which resulted in the complete decomposition of the control after 3 weeks. Kulkarni and co-workers (74) reported virtually identical results in a similar set of experiments.

Saunders and Sovish (72) also reported improved dyeability resulting from 10% acrylonitrile graft. The effect, although measured visually, was most striking for acid and basic dyes. Whereas the grafted-cotton fabric dyed uniformly, cotton fabric containing merely occluded homopolymer rather than "true graft" dyed in a spotted, nonuniform manner. These data suggest that the grafting

TABLE 4
Effect of Soil Burial on the Physical Properties of Cotton Sheeting Modified with Acrylonitrile by Radiation Polymerization (72)

	Initial properties		After 3 weeks		After 6 weeks	
Modification	Breaking strength (lb)	Elongation (%)	Breaking strength (lb)	Elongation (%)	Breaking strength (lb)	Elongation (%)
None (control)	31.1	5.0	Decomposed		Decomposed	
Acrylonitrile:						
6%	32.8	4.9	21.5	3.8	[a]	[a]
10%	37.6	5.5	26.8	4.5	[a]	[a]
14%	36.6	5.5	32.9	5.4	23.3	4.2

[a] Sample intact except for a few small areas of degradation that precluded measurement.

process is uniform on an extremely small size scale. These results are to be compared with acrylonitrile treatment by cyanoethylation, discussed in section D of this Chapter XVII.

TABLE 5
Moisture Regain of Cotton Cellulose–Polystyrene Grafts (70)

Percentage of graft[a]	Equilibrium water content (%)	
	Based on total weight	Based on cellulose only
0	10.7	10.7
59	9.0	14.3
98	8.2	16.2
142	7.8	18.9
152	7.4	19.2

[a] Based on 100 parts cellulose.

(c) PULP AND PAPER

Improved tensile strength and bursting strength in both wet and dry conditions have been reported for paper grafted with vinyl monomers. Rather predictably resistance to microbial attack is also dramatically improved as a consequence of vinyl grafting. The data presented in Figures 3 and 4 (75) describe the effect of acrylic grafting on the tensile and bursting strengths of the grafted papers. Ethyl acrylate, acrylonitrile, and acrylamide grafting via mutual irradiation resulted in dramatic increases in tensile and bursting strengths. The

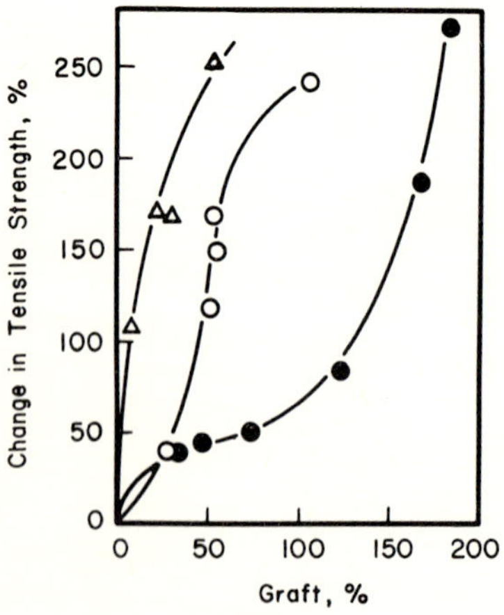

Fig. 3. Change in the tensile strength of paper as a function of the percentage of grafted monomers (75). Key: △ acrylamide; ● ethyl acrylate–acrylonitrile (2–1); ○ acrylonitrile.

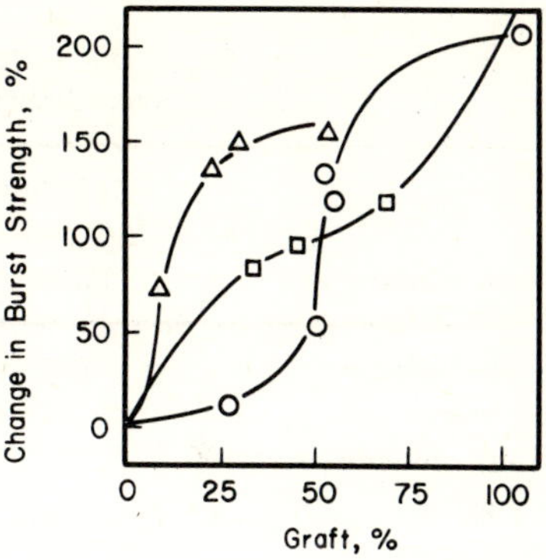

Fig. 4. Change in the burst strength of paper as a function of the percentage of grafted monomers (75). Key: △ acrylamide; ○ acrylonitrile; □ ethyl acrylate–acylonitrile (2–1).

most plausible explanation for the observed increase is that the pendant graft chains aggregate in domains that act as secondary valence crosslinks. This aggregate, or "domain," morphology is further supported by the water-sorption behavior determined on these same grafts.

The authors report that water sorption is substantially decreased in the grafts when the sorption is calculated on the total weight of graft. Only modest decreases are effected when the sorption is calculated on the basis of the initial weight of ungrafted cellulose. Presumably the domains of vinyl polymer are much less hygroscopic than the surrounding cellulose matrix and essentially do not contribute significantly to the equilibrium moisture sorption of the graft "composite." See also the discussion of modified papers in Section D of Chapter XIX.

(d) WOOD–PLASTIC COMPOSITES

A special case of graft copolymerization to cellulosics is the preparation of wood–plastic "combinations," or composites. This application of graft copolymerization is being studied around the world (76). Briefly, the impregnation process generally involves the following steps:

1. Degassing the wood *in vacuo* (~20 mm Hg) for 30 min.
2. Nitrogen purging (~1 atm) for 5 min.
3. Degassing the wood *in vacuo* (~20 mm Hg) for 30 min.
4. Purging the system with degassed monomer (or monomer–solvent mixture) for 15 min.
5. Imposing an overpressure of a few atmospheres for 2–4 hr to impregnate the wood homogeneously.
6. Irradiating the monomer-impregnated wood with ^{60}Co γ-rays at rather low dose rates (~0.01 Mrad/hr to a total dose of ~1 Mrad).

TABLE 6
Effect of Vinyl-Monomer Impregnation on the Bulk Density, Impact Strength, and Bending Strength of Pine (80)

Monomer(s)	Radiation dose (Mrad)	Polymer content (%)	Bulk density (g/cc)	Impact strength (Mkp/cm^2)	Bending strength (kg/cm^2)
None (original pine)	0.0	0.0	0.460	0.79	1019
Vinyl chloride (VCl)	0.4	59	0.844	0.52	1213
Vinyl acetate (VAc)	0.2	40	0.631	1.05	1236
Acrylonitrile	0.3	124	1.045	0.81	1724
VCl–VAc:					
50 : 50	0.4	82	0.744	1.07	1308
80 : 20	0.4	73	0.840	0.84	1393
Styrene–acrylonitrile	0.4	47	0.719	0.80	1504

Much of the pioneering work on wood–plastic composites was done by Karpov et al. (77) in the Soviet Union and by Kenaga (78) in the United States between 1958 and 1960. Research and development rapidly followed in Finland, Sweden, Denmark, and other countries. Presently study of the wood–plastic system is encountered throughout Europe and in Japan as well as in the smaller nations of Asia, including Taiwan, the Philippines, and Thailand (76).

Radiation at the low dose rates cited above has the special advantage of limiting the autocatalytic exotherm generation, which can give rise to temperature increases (~200°C) in the wood–plastic composite—an effect that is deleterious to the ultimate properties of the composite. Wood–plastic composites can be synthesized via chemical initiation quite successfully if careful attention is given to initiator concentration (either azobisisobutyronitrile or oxidation–reduction system) and if suitable retardants are incorporated into the monomer system (79).

The ultimate properties of the composite seem to be more strongly influenced by wood species than by the chemical nature of the impregnating monomer. Monomer systems studied to date include vinyl chloride, vinyl acetate, methyl methacrylate, butyl methacrylate, acrylic acid, acrylonitrile, styrene, various comonomer mixtures, and blends of unsaturated polyester and styrene monomer.

Similar to the results already presented on the effects of vinyl-monomer grafting to cotton, grafting to wood results in reduced moisture sorption and improved resistance to microbial attack. In addition, however, many of the important physical properties of wood that affect its commercial utilization are markedly improved. These properties include dimensional stability, hardness, tensile strength, compressive strength, elastic modulus, shear strength, bending strength, and abrasion resistance. The data presented in Tables 6 and 7 indicate that, for a variety of wood species and monomeric impregnants, dramatic improvements in the various physical properties are observed (80, 81). The hardness of woods is affected even more strongly than the strength properties; increases of up to 1000% in hardness result from grafting about 50% of polyester–styrene mixtures.

Improvement in the dimensional stability of wood has been demonstrated by the extensive work of Stamm (82) via a variety of techniques generally involving the absorption of low-molecular-weight hydrophilic polymers, resins, or inorganic salts (see Section K in Chapter XIX). Kenaga et al. (83) have demonstrated that similar effects result from the radiation-induced graft copolymerization of wood with styrene. Their results indicate that dimensional stability improves monotonically with the percentage of graft and can be significantly affected by the grafting environment.

(e) CELLULOSE DERIVATIVES

Recently extremely careful work has been initiated on the preparation and characterization of scrupulously purified grafts of cellulose acetate (30, 61, 69). This work is instructive because it is virtually unique in combining detailed characterization studies with ultimate-property evaluation. A series of purified graft copolymers of cellulose acetate and polystyrene, carefully prepared and characterized, were used for some initial-property studies (30). Grafts with high-molecular-weight backbones and low-molecular-weight side chains and vice versa, and with both side chains and backbone of high and low molecular weight, were prepared. The graft polymers were found to be insoluble in most solvents but soluble in dimethylformamide and pyridine and in mixtures of solvents for each component. Films were cast from blends of graft plus the homopolymer corresponding to the component of the graft with the highest molecular weight. A graft copolymer with high-molecular-weight backbone and side chains was found to be compatible with both homopolymers separately and with a mixture of both over a wide range of compositions. A blend of the high- and low-molecular-weight grafts was also found to be highly compatible with both homopolymers.

The permeability, diffusion, and sorption of gases and water vapor in the grafts and corresponding homopolymers were also studied. The equilibrium sorption of water vapor in polystyrene is much lower than that in cellulose acetate, whereas the diffusion coefficient for water is much higher in polystyrene than cellulose acetate. Equilibrium water sorption in the grafts was

TABLE 7
Mechanical Properties of Wood–Plastic Composites (76)

Specimen	Polyester content (%)	Compressive strength		Polyester content (%)	Bending strength		Modulus of elasticity		Polyester content (%)	Tensile strength		Abrasion resistance (%)
		kg/cm^2	%		kg/cm^2	%	$kg/cm^2 \times 10^3$	%		kg/cm^2	%	
Birch	0	711	100	0	1427	100	193	100	0	1720	100	100
Polyester	46.2	1212	170	45.7	1940	136	217	112	46.1	2029	118	310
Alder	0	490	100	0	865	100	124	100	0	1170	100	100
Polyester	52.9	1148	240	52.7	1501	174	165	133	54.6	1711	150	602
Aspen	0	468	100	0	939	100	142	100	0	1274	100	100
Polyester	54.1	1043	224	55.9	1623	173	176	124	56.1	1602	126	440
Pine	0	520	100	0	963	100	154	100	0	1097	100	100
Polyester	49.0	962	185	39.1	1512	157	190	123	50.2	1501	137	285

reduced essentially in proportion to the polystyrene content of the graft, but sorption calculated solely on the cellulose acetate content was only modestly reduced. This suggests quite strongly that the polystyrene is present in isolated domains, and it is reasonable to picture the morphology of a graft film as polystyrene domains dispersed in a matrix of essentially unaltered cellulose acetate.

The permeability of grafts to water strongly confirms this model. The values of permeability to water vary monotonically with polystyrene content. A theoretical model describing the effect of dispersed, spherical, nonconducting filler on the apparent permeability of a filled composite is given by Fricke (84):

$$P_{\text{comp}} = P_{\text{cm}} \left(\frac{1 - \phi}{1 + \phi} \right)$$

where P_{comp} is the composite permeability, P_{cm} is the permeability of the conducting matrix, and ϕ is the volume fraction of the dispersed nonconducting filler.

The function expressed by the above equation describes a monotonic decrease of permeability with increasing concentration of dispersed nonconducting filler (convex toward the void-fraction axis). The permeability data of Wellons, Williams, and Stannett (69) conform quite closely to this theoretical model.

This point is extremely important because any judicious tailoring of polymer properties via graft copolymerization requires an understanding of the morphological distribution of the graft components. The overwhelming data to date clearly suggest a domain type of structure, especially when there are relatively few grafted side chains per backbone.

References

1. S. N. Ushakov, *Fiz.-Mat. Nauk (USSR)*, **1**, 35 (1943).
2. L. Roth and J. Weiner, Institute of Paper Chemistry Bibliographic Series, 229, Appleton, Wis., 1966.
3. E. H. Immergut, in N. M. Bikales, Ed., *Encyclopedia of Polymer Science and Technology*, Vol. 3, Interscience, New York, 1965, pp. 242–284.
4. H. A. Krässig and V. Stannett, *Adv. Polymer Sci.*, **4**, 111 (1965).
5. I. Waltcher, R. Burroughs, and E. C. Jahn, *IUPA Abstracts*, Stockholm, 1953.
6. D. K. R. Chaudhuri and J. J. Hermans, *J. Polymer Sci.*, **51**, 373 (1961).
7. S. M. Sankalia, D. K. R. Chaudhuri, and J. J. Hermans, *Can. J. Chem.*, **40**, 2249 (1962).
8. G. N. Richards, *J. Appl. Polymer Sci.*, **5**, *539 (1961)*.
9. D. J. Bridgeford, *Ind. Eng. Chem., Prod Res. Develop.*, **1**, 45 (1962).
10. D. J. Bridgeford, *Tappi*, **46**, 673 (1963).
11. G. Mino and S. Kaizerman, *J. Polymer Sci.*, **31**, 242 (1958).

12. E. Schwab, V. Stannett, D. H. Rakowitz, and J. K. Magrane, *Tappi*, **45**, 390 (1962).
13. T. H. Daniel, S. T. Moore, and N. R. Segro, *Tappi*, **45**, 390 (1962).
14. S. Kaizerman, G. Mino, and L. F. Meinhold, *Textile Res. J.,* **32**, 136, (1962).
15. H. Kamogowa and T. Sekiya, *Textile Res. J.,* **31**, 585 (1961).
16. T. Toda, *J. Polymer Sci.,* **58**, 411 (1962).
17. V. A. Kargin, K. U. Usmanov, and B. I. Aikhodzhaev, *Vysokomol. Soedin.,* **1**, 149 (1959).
18. K. U. Usmanov, B. I. Aikhodzhaev, and U. Azizov, *J. Polymer Sci.,* **53**, 87 (1961).
19. Y. L. Raskia, R. M. Livshits, and A. A. Berlin, *Lakokrasochnye Mat. i ikh Primenenie,* **4**, 6 (1960); *Chem. Abstr.*, **55**, 25341 (1961).
20. E. Gleason and V. Stannett, *J. Polymer Sci.,* **44**, 183 (1960).
21. G. N. Richards, *J. Appl. Polymer Sci.,* **5**, 553 (1961).
22. Z. A. Rogovin, T. Sun, A. D. Virnik, and N. M. Khvostenko, *Vysokomol. Soedin.,* **4**, 571 (1962).
23. A. Y. Kororkova, Y. G. Kryazhev, and Z. A. Rogovin, *Vysokomol. Soedin.,* **6**, 1980 (1964).
24. M. Y. Wu and Z. A. Rogovin, *Vysokomol. Soedin.,* **5**, 706 (1963).
25. A. A. Berlin and T. A. Makarova, *J. Gen. Chem. USSR,* **21**, 1267 (1951).
26. G. Faraone, G. Parasacco, and C. Cogrossi, *J. Appl. Polymer Sci.,* **5**, 16 (1961).
27. A. A. Blagonravova, I. A. Pronina, and I. T. Bondarevskaya, *Lakokrasochnye Mat. i ikh Primenenie* **2**, 4 (1962).
28. Y. Kobayashi, *J. Polymer Sci.,* **51**, 359 (1961).
29. A. A. Armstrong and H. A. Rutherford, *Textile Res. J.,* **33**, 264 (1963).
30. V. Stannett, J. D. Wellons, and H. Yasuda, *J. Polymer Sci.,* [C], No. 4, 551 (1964).
31. J. C. Arthur and F. A. Blouin, *Am. Dyestuff Reptr.,* **51**, 13 (1962).
32. R. Y. M. Huang and W. H. Rapson, *J. Polymer Sci.,* [C], No. 2, 169 (1963).
33. K. Araki, M. Ichino, and T. Hoshino, *Ann. Rep. J.A.R.R.P.,* **2**, 465 (1960).
34. G. Oster, British Patent 856,884 (Dec. 21, 1960).
35. N. Geacintov, V. Stannett, and E. W. Abrahamson, *Makromol. Chem.,* **36**, 53 (1959).
36. N. Geacintov, V. Stannett, E. W. Abrahamson, and J. J. Hermans, *J. Appl. Polymer Sci.,* **3**, 54 (1960).
37. W. H. Rapson and E. Kvasnicka, *Tappi*, **46**, 662 (1963).
38. G. M. Guzman, J. M. G. Fatou, and E. G. Rosa, *Annales Real. Soc. Expan. Fis. Chim. (Madrid),* **B59**, 519 (1963).
39. M. Tsunooka, M. Tanaka, and N. Murata, *Chem. High Polymers (Japan),* **22**, 107 (1965); *Makromol. Chem.,* **85**, 310 (1965).
40. R. J. Ceresa, *Polymer,* **2**, 213 (1961).
41. M. S. Akutin, N. Y. Parlashkevich, I. N. Kogan, S. Pa. Kalinina, and L. I. Menes, *Prim. Ultraakustik. K. Iss. Vesch.,* **10**, 47 (1960).
42. W. Deters and D. Huang, *Faserforsch. u. Textiltech.,* **14**, 183 (1965).
43. V. A. Kargin and N. A. Plate, *J. Polymer Sci.,* **52**, 155 (1961).
44. M. S. Akutin, N. Y. Parlashkevich, I. N. Kogan, V. V. Rubinshtein, and R. N. Gribkova, *Plasticheskie Massy,* **6**, 2 (1960).

45. B. A. Feit, A. Bar-Nun, M. Lahav, and A. J. Zilkha, *J. Appl. Polymer Sci.*, **8**, 1869 (1964).

46. R. F. Schwenker and E. Pacsu, *Tappi*, **46**, 665 (1963).

47. Y. Avny and L. Rebenfeld, *Textile Res. J.*, **38**, 684 (1968).

48. G. Rausing and S. Sunner, *Tappi*, **45**, 203A (1962).

49. D. F. Herman and I. R. Dunlap, *Tappi*, **48**, 418 (1965).

50. D. F. Herman, in N. M. Bikales, Ed., *Encyclopedia of Polymer Science and Technology*, Vol. 8, Interscience, New York, 1968, pp. 736–752.

51. J. J. O'Malley and R. H. Marchessault, *J. Polymer Sci.*, [C], No. 24, 179 (1968).

52. Z. A. Rogovin, *J. Polymer Sci.*, **48**, 443 (1960).

53. Z. A. Rogovin, *Periodica Polytech.*, **5**, 65 (1961).

54. S. A. Volgina, Y. G. Kryazhev, and Z. A. Rogovin, *Vysokomol. Soedin.*, **7**, 1154 (1965).

55. T. Sun and Z. A. Rogovin, *Vysokomol. Soedin.*, **4**, 714 (1962).

56. T. Sun, V. A. Derevitskaya, and Z. A. Rogovin, *Vysokomol. Soedin.*, **1**, 1625 (1959).

57. W. Cooper and R. K. Smith, *Makromol. Chem.*, **40**, 148 (1960).

58. G. C. Daul, R. M. Reinhart and J. D. Reid, *Textile Res. J.*, **24**, 138 (1954).

59. G. Champetier and F. H. Lucas, *Compt. Rend.*, **252**, 2785 (1961).

60. Z. A. Rogovin and J. J. Wu, *Vysokomol. Soedin.*, **1**, 1630 (1959).

61. H. Yasuda, J. A. Wray, and V. Stannett, *J. Polymer Sci.*, [C], No. 2, 387 (1963).

62. H. A. Ende and V. Stannett, *J. Polymer Sci.*, [A], **2**, 4047 (1964).

63. Y. Hachihama and H. Sumitomo, *Tech. Rept., Osaka Univ.*, **12**, 451 (1962).

64. J. D. Wellons, A. Schindler, and V. Stannett, *Polymer*, **5**, 499 (1965).

65. T. Matsuzaki, K. Hayakawa, and K. Kawase, *Chem. High Polymers (Japan)*. **19**, 318 (1962).

66. D. Campbell, J. L. Williams, and V. Stannett, *J. Polymer Sci.*, [A-1], **7**, 429 (1969).

67. J. C. Arthur, *Advances in Chemistry Series*, American Chemical Society, Washington, D.C., in press.

68. J. D. Wellons and V. Stannett, *J. Polymer Sci.*, [A], **3**, 847 (1965).

69. J. D. Wellons, J. L. Williams, and V. Stannett, *J. Polymer Sci.*, [A-1], **5**, 1341 (1967).

70. R. E. Kesting and V. Stannett, *Makromol. Chem.*, **63**, 248 (1963).

71. R. Y. M. Huang, B. Immergut, E. H. Immergut, and W. H. Rapson, *J. Polymer Sci.*, [A-1], **1**, 1257 (1963).

72. F. L. Saunders, and R. C. Sovish, *J. Appl. Polymer Sci.*, **7**, 357 (1963).

73. I. Yoshitake, S. Obuchi, T. Tsuneda, and H. Maeda, *J. Soc. Textile Cellulose Ind. Japan*, **19**, 540 (1963).

74. A. Y. Kulkarni, A. G. Chitale, and B. K. Vaidya, *J. Appl. Polymer Sci.*, **7**, 1581 (1963).

75. E. Schwab, V. Stannett, and J. J. Hermans, *Tappi*, **44**, No. 4, 254 (April 1961).

76. W. E. Mott and V. Stannett, *Isotopes and Radiation Tech.*, **6**, 323 (1969).

77. V. L. Karpov, Yu. M. Molinsky, V. I. Serenkov, R. S. Klimanova, and A. S. Frieden, *Nucleonics*, **18**, 3, 88, (1960).

78. D. L. Kenaga, U.S. Patents 3,077,417, 3,077,418, and 3,077,419 (Feb. 12, 1963).

79. J. F. Siau and J. A. Meyer, *Forest Products J.*, **16**, 47 (1966).

80. T. Czikovsky, *Atomic Energy Review (I.A.E.A. Vienna)*, **6**, 3 (1968).

81. J. K. Miettenen and T. Autio, I.A.E.A. Panel on the Applications of Radio Isotopes in the Pulp and Paper Industry, Helsinki, October 9–13, 1967.

82. A. J. Stamm, *Wood and Cellulose Science*, Ronald Press, New York, 1964.

83. D. L. Kenaga, J. P. Fennessey, and V. Stannett, *Forest Products J.*, 161 (1962).

84. H. Fricke, *Phys. Rev.*, **24**, 575 (1924).

H. REACTIONS INDUCED BY HIGH-ENERGY RADIATION

Jett C. Arthur, Jr.

Since the publication of Parts I, II, and III, the interaction of high-energy radiation with cellulose has become the subject of intensive investigation. There have been two principal interests in using high-energy radiation in processes involving cellulose: (a) radiation sterilization of medical supplies, personal products, and foods; and (b) changing the chemical and physical properties of fibrous cellulose through radiation-initiated reactions (1–12). In the first case the objective has usually been to retain the natural properties of the cellulose with a minimum of postirradiation changes in these properties (5). In the second case the objective has also been to retain the natural properties of fibrous cellulose and at the same time through radiation-initiated reactions to obtain a fibrous cellulosic product with some of the properties of the reacted chemical reagent (1). Generally the reactions of practical interest involve chain copolymerization reactions initiated at free-radical sites formed on the cellulose molecule through the interaction of high-energy radiation with cellulose. Application research in the fields of food sterilization (8), textile fibers (7), paper products (10), and wood products (4) has been reported. Industrial pilot-scale processes involving high-energy radiation have been reported for cellulose-containing textile products and wood products (13).

This discussion of the effects of high-energy radiation on cellulose is concerned principally with natural and regenerated fibrous celluloses. The conclusions concerning the primary effects of radiation on fibrous cellulose are generally applicable to cellulose in food products. There are several advantages in basing the discussion on fibrous cellulose: purified cellulose with a high molecular weight can be prepared by simple extraction procedures; the effects of radiation on the properties of fibrous cellulose can be determined by both chemical and physical techniques; and the effects of radiation-initiated reactions on the macromolecular properties of fibrous cellulose can be determined by the usual testing procedures for fibrous polymers.

1. Interaction of Radiation with Matter

(a) SOURCES OF HIGH-ENERGY RADIATION

The term "high-energy radiation" is used in this discussion to include radiation that interacts with matter in a chemically nonspecific manner; that is, the interaction is nearly independent of chemical structure. The term "ionizing radiation" is also often used to refer collectively to this type of radiation. The energy of this type of radiation is generally many times greater than the energy of a chemical bond. In contrast, the energy of ultraviolet and visible radiation is usually of about the same order of magnitude as a chemical bond. The absorption of ultraviolet and visible radiation is dependent on chemical structure (see Chapter XVIII-D). After the energy is transferred to the cellulose molecule, the localization of energy originating from either of these sources of radiation—and from lower energy sources of radiation, such as heat—initiates similar primary effects in cellulose. In this section primarily the effects of x-rays (14), γ-rays (15–18), electrons, and neutrons on the chemical and physical properties of cellulose are described.

Cobalt-60 is the most commonly used radioisotope as a commercial source of γ-radiation. The excited cobalt-60 atom, which has a half-life of 5.27 years, decays by emitting β-particles (0.306 MeV) and γ-rays (1.17–1.33 MeV).* Sources of high-energy electrons are radioisotopes and high-voltage electrical generators. When radioisotopes decay from an excited state to a ground state, β-particles, that is, electrons may be emitted from the radioactive nucleus during the process. However, the intensity of these sources of electrons is low. On the other hand, electrical machines yield electrons of high energy and intensity; common devices are Van de Graaff generators and linear accelerators. Electron sources with energies as high as 24 MeV and with very high intensity, or flux, are commercially available (14).

The most important and intense source of neutrons is the nuclear fission process found in reactors. Other sources of high-energy particles are available, but these are of little interest as to their effects on cellulose (14, 18).

(b) ENERGY ABSORPTION AND MEASUREMENT

The flux, or intensity, of radiation impinging on a target material is usually indicated by the beam current of electrical generators and by the number of curies of a radioisotope source. One gram of radium undergoes about 3.7×10^{10} disintegrations per second, and this is defined as 1 *curie* of radioactivity. It is now generally accepted that the quantity of any radioisotope that undergoes

*The electron-volt (eV) is a unit of radiation energy corresponding to 1.60×10^{-12} erg.

3.700×10^{10} disintegrations per second is a curie of the radioisotope. The energy absorbed in matter from radiation impinging on a target material is primarily dependent on the intensity, or flux, of the source, the type of radiation, the energy of the radiation, the atomic number of the elements composing the target material, and the density of the target material. The unit of energy absorbed in matter from x- or γ-radiation is the *roentgen*, which is defined as the absorption of 83 ergs per gram of air at 0°C and 760 mmHg. In water 1 roentgen of x- or γ-radiation is equivalent to the absorption of 93 ergs per gram of water; this unit is often referred to as the *roentgen equivalent physical* (rep). A commonly accepted unit is the *rad*, which is independent of the source of radiation and is defined as the quantity of radiation that results in the absorption of 100 ergs per gram of material (14).

A measure of the radiation yield of chemical products is the *G*-value, which is defined as the initial rate of yield of chemical or physical effects per unit of energy absorbed in any material—that is, molecules changed per 100 eV of energy absorbed (19). For many radiation-initiated, nonchain chemical reactions in condensed systems common *G*-values range from 1 to 10. For example, a *G*-value of 7 for the radiation-initiated chain cleavage of the cellulose molecule indicates that initially for each 100 eV of radiation energy absorbed in the cellulose seven chain cleavages occurred (2).

A practical unit for measuring the radiochemical yield of radiation-initiated chain reactions, under given experimental conditions, is the number of molecular actions per 100 eV of radiation energy absorbed in the system. For radiation-initiated graft-copolymerization reactions of cellulose with vinyl monomers, radiochemical yields may be several thousand molecular actions (i.e., molecules of vinyl monomer polymerized) per 100 eV of energy absorbed (2).

The amount of radiation energy absorbed in matter over a period of time is referred to as the *dosage* or *dose*. The rate of deposition of radiation energy in matter is called the dose rate (19). A radiation source is calibrated by determining the dose rate of the source under given experimental conditions. A primary means of determining the dose rate is calorimetry. Usually this is not a convenient means, so that a secondary dosimeter is used. Since the dose rate of a radiation source will be dependent on the characteristics of the absorbing matter, the elemental composition and the density of the dosimeter should be similar to that of the material being irradiated. The dosimeter should be simple, reproducible, and independent of dose rate. When 0.01 *M* ferrous sulfate in 0.5 *N* sulfuric acid is irradiated, the net chemical change is the oxidation of the ferrous ion to the ferric ion. The increase in the concentration of the ferric ion can be determined by the increase in ultraviolet-light absorption of the solution at 305 nm and 25°C. By calorimetric measurements it has been determined that 15.45 atoms of ferrous iron are oxidized per 100 eV of radiation energy absorbed (20). From these data the dose rate in electron volts per unit time per gram of matter can be calculated. This dosimeter is satisfactory for total dosages greater than 5000 roentgens for electrons and x- and γ-radiation. Since the radiation energy is initially absorbed in water, whose elemental composition is similar in atomic number to the elemental composition of cellulose, the ferrous–ferric dosimeter is suitable for investigating the effects of high-energy radiation on cellulose. Another important chemical dosimeter is the ceric–cerous sulfate dosimeter (19).

Electronic instruments—such as ionization chambers, and scintillation, proportional, and Geiger-Müller counters—and photographic film may be used to detect and measure radiation. However, at high dosages the chemical dosimeters are more suitable (14, 19).

(c) ABSORPTION OF HIGH-ENERGY RADIATION

(1) x- and γ-radiation

When low-energy electromagnetic radiation, such as visible or near-ultraviolet radiation, interacts with matter, each photon is absorbed in a single event by a single molecule of the absorbing matter (see Chapter XVIII-D). When high-energy x- or γ-radiation is absorbed by matter, a series of events may occur. Matter composed of elements of low atomic number, such as cellulose, absorbs high-energy electromagnetic radiations in three ways: photoelectric effect, Compton effect, and pair production (14).

For radiations with energies lower than 0.5 MeV the photoelectric effect is important. In this effect the energy of the x- or γ-radiation is transferred to a bound electron, which, with negligible loss of energy, is ejected from its nuclear orbital. The ejected electrons cause additional ionizations, so that many events occur for each absorbed x- or γ-ray.

The Compton effect is an important process in cellulose for the interaction of x- or γ-radiation with energies in the range from 0.5 to 10 MeV. In this effect γ-radiation is scattered on interaction with electrons; that is, an elastic collision between a photon and an electron occurs. Consequently the energy transferred to the electron is directly related to the incident and scattering angles of the γ-radiation.

Pair production is the predominant absorption process for high-energy γ-radiation with energies at least twice the resting mass of an electron, 1.02 MeV. An electron–positron pair is produced. The electron behaves like that produced by the photoelectric effect or the Compton effect. The positron, after losing kinetic energy, combines with an electron to form electromagnetic radiation (14).

The chemical effect of the initial absorption of x- or γ-radiation is very small; the measurable chemical effects are the result of the chain of secondary ionizations caused by the ejected electrons (2, 18). For this reason high-energy γ-radiation and high-energy-particle radiation, such as electrons, cause very similar chemical effects on cellulose.

(2) Particle Radiation

High-energy-electron radiation interacts with the atoms in the absorbing matter to cause ionization. Due to the high rate of absorption in matter, electrons have a maximum range in matter, with the number of electrons decreasing rapidly with increasing depth of penetration (14). Sample size and density are very important factors in determining the uniformity of radiation absorption in the sample. In contrast, the secondary electrons produced when γ-radiation interacts with cellulose are more uniformly distributed throughout the sample.

The effects of densely ionizing particles, such as protons, deutrons, and α-particles, on the properties of cellulose have not been investigated. It would be expected that most of the effects, probably ionization, would be limited entirely to the surface of the sample.

Neutrons, neutral particles with about unit mass, are very penetrating. High-energy, or fast, neutrons may interact with the nucleus to eject a proton. Low-energy, or thermal, neutrons may interact with a nucleus to form a radioactive nucleus that emits β-particles or γ-rays (14, 18). The effects of neutrons on cellulose are the result of the interactions of these secondary radiations with cellulose.

(d) ENERGY-TRANSFER PROCESSES

The state of aggregation of the chemical compound being irradiated will affect the mechanism of energy loss by incident high-energy radiation to the compound, energy transfer in the compound, and energy localization in the compound. The localization of energy in the compound determines the observable chemical and physical effects on the compound (21). In determining radiation effects on cellulose in the solid state the steps in these energy-transfer processes cannot be isolated. However, the effects of radiation on a high polymer, such as cellulose, would likely be (a) random cleavage of the molecular chain, that is, depolymerization; and/or (b) dehydrogenation or dehydroxylation of the molecule, leading to crosslinking or aggregation of the molecules of cellulose (15–17). From the experimental data on the effects of radiation on cellulose, conclusions are drawn as to the nature of these energy-transfer processes.

It can be shown that for many successive secondary events resulting from a common parent

$$N_n = C_1\, e^{-k_1 t} + C_2\, e^{-k_2 t} + \cdots\cdots + C_n\, e^{-k_n t} \tag{1}$$

where N_n is the number of events, t is time, k is a rate constant, and C is a constant (18). For a γ-radiation source, which has a uniform dose rate in the volume being irradiated, the rate constants k should be $k_1 << k_2 \cdots << k_n$;

that is, the secondary electrons should interact with the cellulose at a greater rate than the incident γ-radiation to transfer some of the incident energy to the cellulose molecule. Equation 1 then reduces to approximately

$$\ln N_n = -k_1 t + \ln C_1 \tag{2}$$

If Sippel's equation (22) is extended to the changes in the properties of the cellulose molecule on irradiation, it follows that

$$\frac{dP}{dt} = kP \tag{3}$$

where P is the molecular property. Integrating equation 3 between the limits P_0 to P and t_0 to t, where $t_0 = 0$, and combining with equation 2, we find that

$$\ln P = k' \ln N_n + K' \tag{4}$$

If equation 4 for the proposed energy-transfer processes (23) is graphically solved on a log-log plot, changes in a molecular property versus dosage should be linearly related.

If the common parent is high-energy-particle radiation, such as electrons, the chain of secondary events would follow the steps indicated in equations 1 through 4. However, with increasing thickness of the sample, the number of these secondary events would be fewer than for those originating from γ-radiation.

2. Primary Effects

In order to discuss the effects of high-energy radiation on cellulose it is important to define the starting material as clearly as possible (24). The composition (25, 26), morphology (27), chemical properties (28–35), and chemical and physical modification (24) of the cellulose are important factors in determining the energy-transfer processes in the cellulose molecule. These processes, particularly the localization of energy, determine the observed chemical and physical effects of radiation on cellulose. Purified cotton fibers (25) have been used as the common standard for the comparison of radiation effects on cellulose in this discussion. Experimental data for radiation effects on cellulose from other origins are also indicated where applicable.

(a) FORMATION OF FREE RADICALS

(1) Effect of Lattice Type

γ-Irradiation of purified fibrous cotton cellulose I (after drying at 25°C overnight in vacuum over phosphorus pentoxide) gives products containing long-lived electron paramagnetic species, as shown in the electron spin resonance

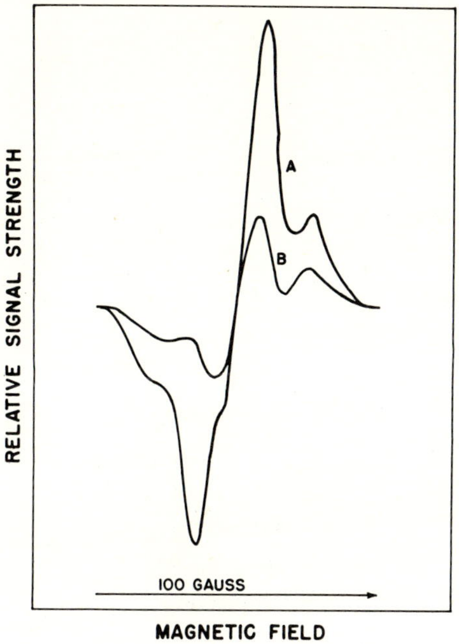

Fig. 1. The ESR spectra of irradiated cellulose I: (*A*) irradiated dry and in vacuum; (*B*) irradiated while containing regain moisture and in air.

(ESR) spectrum in Figure 1*A*. If the cellulose contains regain moisture [about 7% moisture at 21°C and 65% relative humidity (36)] when irradiated, the intensity and shape of the ESR spectrum will change to that shown in Figure 1*B*. This spectrum was obtained either by irradiating cellulose containing regain moisture or by irradiating cellulose dry and then allowing the irradiated cellulose to take up regain moisture. The electron paramagnetic species (shown in Fig. 1*B*) is relatively stable and detectable for long periods (37). (Note: after a storage period as long as 7 years irradiated cotton cellulose still contains detectable concentrations of electron paramagnetic species.)

The ESR spectrum (see Fig. 1*B*) is a triplet with signal strength ratio of about 1 : 1.8 : 1. A free-radical spectrum split by two equivalent hydrogens would ideally give a triplet with a signal strength ratio of about 1 : 2 : 1. Radiation-initiated dehydrogenation to form a free radical at the $C_{(5)}$ atom would give a three-line ESR spectrum, due to the interaction of equivalent hydrogens on $C_{(6)}$, as shown in structure **1**. The localization of energy would be most probably at $C_{(5)}$, due to the delocalizing effect of the oxygen in the glucopyranose ring. However, dehydrogenation of the hydroxyl group on $C_{(6)}$, as shown in structure

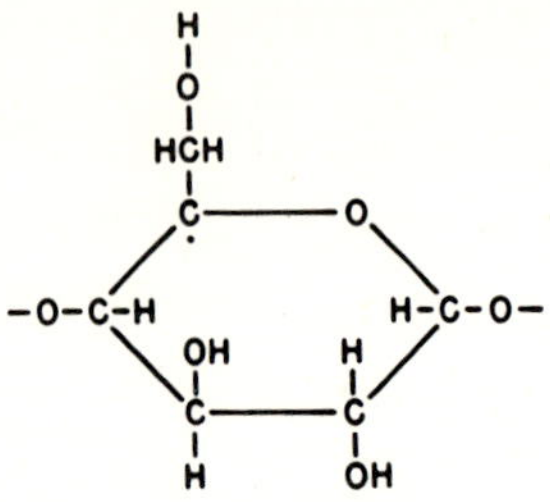

Structure 1

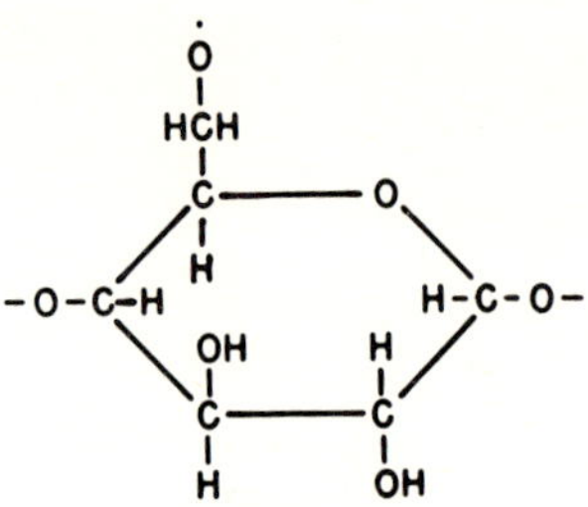

Structure 2

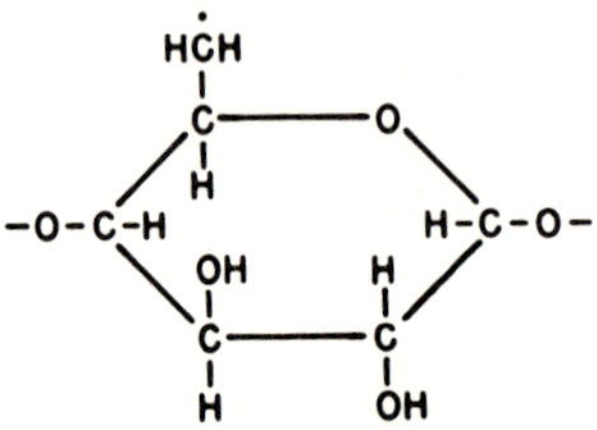

Structure 3

2, or dehydroxylation of $C_{(6)}$, as shown in structure **3**, would also yield free radicals that could interact with the equivalent hydrogens on $C_{(6)}$ to give a triplet spectrum (38).

γ-Irradiation of purified fibrous cotton cellulose II (after drying at 25°C overnight in vacuum over phosphorus pentoxide) also gives products containing long-lived electron paramagnetic species (37–39), as shown in the ESR spectrum in Figure 2*A*. If the irradiated sample is wet out with water, the spectrum will change to that shown in Figure 2*B*. In both cellulose I and cellulose II the line widths of the more intense lines of the ESR spectra are about the same (see Figs. 1*A* and 2*A*). The five-line spectrum of irradiated cellulose II could result from a combination of any two of the three radicals shown in structures **1**, **2**, and **3**. The principal line of the spectra of these radicals should occur at about the same strength of the magnetic field (i.e., *g*-value), and the less intense lines should occur at different magnetic fields, so that two three-line ESR spectra would record as a five-line spectrum (37). One of the radicals would probably occur at the $C_{(5)}$ atom for the reasons given above. The other radical would probably occur at the $C_{(6)}$ atom. The formation of a second long-lived radical in cellulose II, as compared with only one long-lived radical in cellulose I, indicates that the changes in the internal forces in the unit cell on changing the lattice type from I to II increase the delocalizing energy at the $C_{(6)}$ atom. It has been reported that hydrogen bonding through the $-CH_2OH$ [$C_{(6)}$] group is rearranged on changing

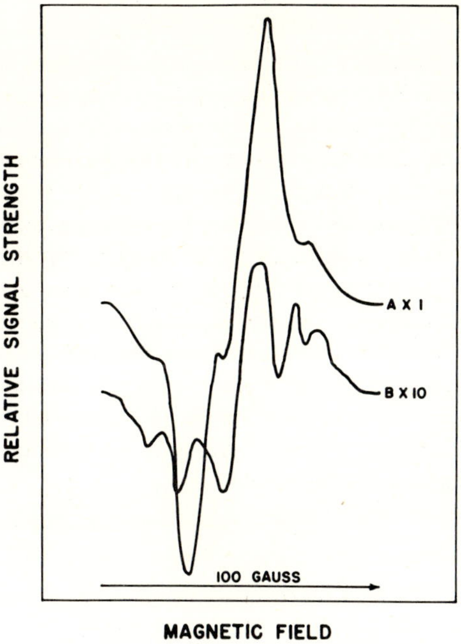

Fig. 2. The ESR spectra of irradiated cellulose II: (*A*) irradiated dry and in vacuum; (*B*) irradiated while containing regain moisture and in air.

the lattice type from I to II (31–33). Also van der Waals forces between the two cellobiose units of the lattice would be greater in lattice type II than in lattice type I, since β is 62° for lattice type II and 84° for lattice type I (28).

When celluloses I and II are irradiated almost dry (see Figs. 1*A* and 2*A*), a short-lived, intense, central line is observed in the ESR spectrum in each case (37, 38). The changes in the ESR spectrum of the irradiated cellulose on addition of moisture (e.g., see Figs. 1*A* and 1*B*) indicate that only a radical that gives a single-line spectrum is reacting. This radical is probably located at $C_{(1)}$ or $C_{(4)}$. Localization of energy at either $C_{(1)}$ or $C_{(4)}$ would lead to the depolymerization of the cellulose molecule and the formation of radicals at $C_{(1)}$ or $C_{(4)}$, which would generate a single-line ESR spectrum. As discussed later, physicochemical evidence indicates that the localization of energy at $C_{(1)}$ or $C_{(4)}$ to initiate the depolymerization of the cellulose molecule is several times greater than the localization of energy to form radicals at $C_{(5)}$ or $C_{(6)}$. This would account for the relative intensities of the spectra shown in Figures 1*A* and 1*B* and 2*A* and 2*B*.

(2) Effect of Oxygen

When cellulose I is irradiated in an atmosphere of oxygen, the intensity of the principal line of the ESR spectra is less than when it is irradiated in an atmosphere of nitrogen (37, 38, 40, 41). Oxygen could react with a free radical to form a peroxy radical, for example, at $C_{(5)}$ (see structure **1**) to form $C_{(5)}O\cdot$. On decomposition of this radical, a carbonyl group would be formed on breaking of the anhydroglucose ring, depolymerization would occur, and an oxygenated radical on $C_{(5)}$ could result. The intensity of the principal line of the ESR spectra for $C_{(5)}O\cdot$ would probably be less than that for $C_{(5)}\cdot$. Also, when cellulose I is irradiated in oxygen, the number of reducing groups should be greater and the molecular weight of the product should be less than when cellulose I is irradiated to the same dosage in nitrogen. This was found to be the case, as discussed below. Comparisons of the ESR spectra obtained on irradiation of celluloses I and II, containing normal regain moisture of about 7%, in different atmospheres are made in Table 1.

TABLE 1
Comparison of the ESR Spectra of Celluloses I and II Containing Regain Moisture and Irradiated in Various Atmospheres (37, 38)

Property of spectrum	Irradiated in atmosphere		
	Nitrogen[a]	Air[b]	Oxygen[a]
Number of lines:			
Cellulose I	3	3	3
Cellulose II	5	5	5
Line width (gauss):			
Cellulose I:			
Major line	19	13	18
Minor line	48	46	48
Cellulose II:			
Major line	21	11	20
Nearest to major line	38	38	38
Other line	57	56	57
Ratio of intensity of lines:			
Cellulose I	1[c]	1 : 1.8 : 1	0.7[c]
Cellulose II	0.5[c]	1 : 1.6 : 1.9 : 1.6 : 1	0.5[c]

[a]Radiation dosage: 1.0×10^{20} eV/g.
[b]Radiation dosage: 2.1×10^{20} eV/g.
[c]Major.

(3) Effect of Moisture

The effects of moisture content at the time of irradiation on the formation of long-lived free radicals in the cellulose I molecule (37, 38, 40) are shown in Figure 3. Prior to irradiation, cellulose I was conditioned for 20 hr at 25°C in a nitrogen atmosphere at the relative humidity indicated to obtain samples with moisture contents ranging from less than 0.5 to about 6%. Then all of the samples were irradiated to a dosage of 2.6×10^{19} eV/g. The concentration of long-lived free radicals decreased with increasing moisture content (37, 40). Below a moisture content of about 2% the concentration of free radicals, at the dosage indicated, reached a maximum of 10^{18} spins/g; at a moisture content of about 6% the concentration of free radicals was about 0.3×10^{18} spins/g (37). Molecules of water are apparently able to diffuse to within the more ordered regions of the fibrous structure to interact with about 70% of the free radicals formed (37). Similar effects were observed when cellulose II of different moisture contents was irradiated (37).

The effects of moisture on the concentration of long-lived free radicals in irradiated cellulose I are smaller if the cellulose is irradiated dry and then moisture is added (37, 40). For example, if cellulose I is equilibrated in a nitrogen atmosphere with 60% relative humidity at 25°C and then irradiated to a dosage of 2.6×10^{19} eV/g, the concentration of free radicals is about 3×10^{17} spins/g. When cellulose I is irradiated dry (about 0.5% moisture) to the same dosage and then equilibrated in a nitrogen atmosphere with 60% relative humidity at 25°C, the concentration of free radicals is about 5×10^{17} spins/g (37).

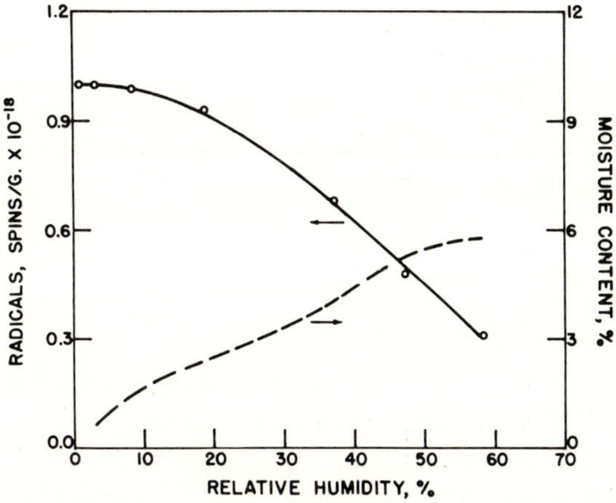

Fig. 3. Effect of moisture content on the formation of trapped radicals in irradiated cellulose I (2.6×10^{19} eV/g; 25°C).

The rate of diffusion of water vapor into the less ordered fibrous structuie is indicated by the initial rate of decay of the free radicals in irradiated cellulose which was directly related to the absolute humidity (37, 39). An equilibrium concentration of free radicals in the irradiated cellulose was obtained in about 15 min. The free-radical decay probably occurs by an abstraction of a hydrogen atom from the water molecule, the hydroxyl radicals formed probably interact to form hydrogen peroxide.

Comparisons of the effects of moisture on the ESR spectra of irradiated celluloses I and II are summarized in Table 2.

TABLE 2
Comparison of the ESR Spectra of Irradiated Celluloses I and II of Different Moisture Contents (37)

Property of spectrum	Moisture content	
	0.5%	6.0%
Number of lines:		
Cellulose I	1 + 3	3
Cellulose II	1 + 5	5
Line width (gauss):		
Cellulose I:		
Major peak[a]	14	17
Minor peak	53	46
New peak	52	–
Cellulose II:		
Major peak[a]	14	12
Nearest to major peak	57	40
Other peak	53	57
New peak	60	–

[a] *g*-Value about 2.0045.

(b) TRAPPED RADICALS

(1) Electron Spin Resonance Evidence

On irradiation at about 25°C trapped-free-radical sites are formed within the crystal lattice of the cellulose molecule and the more ordered regions of the fibrous cellulose. These free radicals have decreased accessibility to moisture vapor and oxygen and, as a result, are long-lived. The trapped free radicals can be detected by ESR spectroscopy (37–48), by initiation of graft-copolymerization reactions with vinyl monomers (49–65), and by postirradiation changes in the properties of the cellulose molecule (66–71).

Cotton cellulose I, containing about 7% moisture and irradiated in air by γ-radiation, produced concentrations of free radicals directly related to the radiation dosages (37). When the irradiated celluloses were immersed in aqueous solutions of zinc chloride and other reagents, the free-radical concentrations decreased with increasing time of reaction at 25°C. However, even after reaction times as long as 71 hr, a large concentration of free radicals in the irradiated cellulose was recorded by ESR spectroscopy (37). This indicates that some of the trapped radicals are probably formed in the crystalline lattice of the irradiated cellulose molecule; the aqueous solution of zinc chloride would be unlikely to penetrate the crystalline lattice completely. For example, when irradiated celluloses were reacted for 71 hr at 25°C with ethanol (15%) in aqueous zinc chloride (75%) the fractions of free radicals that were trapped at various dosages in the lattice or in the more ordered regions of the fibrous cellulose were as follows: 0.26×10^{19} eV/g, 0.33; 0.52×10^{19} eV/g, 0.21; 2.6×10^{19} eV/g, 0.07; and 5.2×10^{19} eV/g, 0.09 (37).

The effects of solutions of aqueous zinc chloride on the fibrous structure of cotton cellulose are dependent on the concentration of the solutions (37). Similarly the effects of these solutions on the reactions of free radicals in the irradiated cellulose with water are dependent on the concentration of the solutions, as shown in Figure 4. As the concentration of the solutions (about 77%) approached saturation at 25°C, the interaction of the irradiated cellulose with aqueous zinc chloride did not decrease the concentration of free radicals. As the concentration of the solutions of zinc chloride decreased, the interaction of the solutions with the irradiated cellulose increased, as demonstrated by a

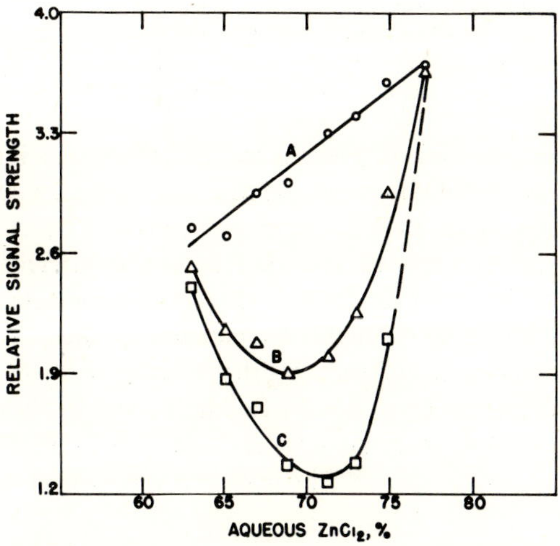

Fig. 4. Effect of aqueous zinc chloride solutions on the stability of trapped radicals in irradiated cellulose I at 25°C: (*A*) after 2 min; (*B*) after 15 min; (*C*) after 25 min.

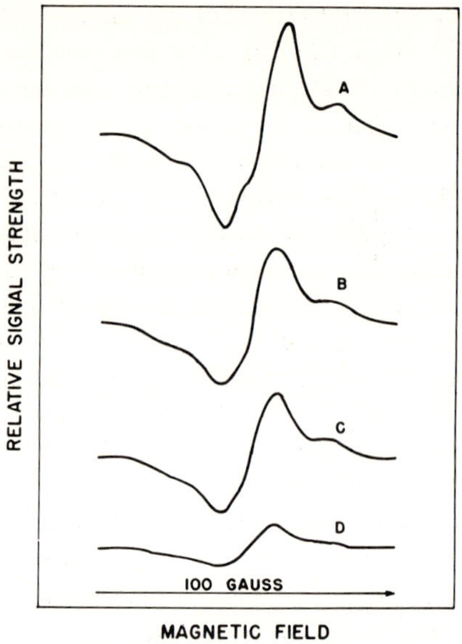

Fig. 5. The ESR spectra of irradiated cellulose I during graft copolymerization with acrylonitrile (15%) in aqueous zinc chloride solution (75%) at 25°C. The spectra were determined at −100°C.: (*A*) irradiated cellulose I; (*B*) reaction time near 0 min, reaction stopped after contact of cellulose with monomer solution by immersion in liquid nitrogen; (*C*) reaction time 34 min; (*D*) reaction time 17 hr.

decrease in the concentration of free radicals. A maximum interaction of the solution with irradiated cellulose was recorded at a concentration ranging from 69 to 75% zinc chloride (37). It has also been observed that solutions of zinc chloride in this range of concentration cause a maximum effect on the swelling and shrinking of fibrous cotton cellulose (59). At lower concentrations of zinc chloride solutions the interaction with irradiated cellulose decreases. Under conditions of maximum interaction of the solutions of zinc chloride with irradiated cellulose, the concentration of free radicals in the irradiated cellulose decreases to about 35% of the initial concentration (37). It is probable that these free radicals are trapped radicals located in the crystalline lattice of the cellulose molecule.

(2) Graft Copolymerization

The trapped radicals, which are probably located in the more ordered regions of the fibrous cellulose but do interact with aqueous solutions of zinc chloride, could also be indicated by graft-copolymerization reactions with vinyl monomers (37). For example, when irradiated cellulose is contacted with an aqueous solution of zinc chloride (75%) containing acrylonitrile (15%), the ESR spectrum of the irradiated cellulose (37) changes during the copolymerization reaction, as shown in Figure 5. The ESR spectrum changes from an apparent combination of a single-line spectrum and a three-line spectrum to a broad single-line spectrum. This would indicate that the acrylonitrile monomer is reacting with the trapped radical, probably at $C_{(5)}$.

Changes in the concentration of trapped radicals in the irradiated cellulose during the copolymerization reaction could also be recorded by ESR spectroscopy (37). The rate of decrease in the concentration of trapped radicals in the irradiated cellulose was greater at the start of the reaction than at the end, although about 30% of the initial concentration of the trapped radicals was apparently not reacted. This probably indicates that a large fraction of the trapped radicals is located within the crystalline lattice. The extent of graft copolymerization, as measured by the percentage of graft copolymer formed, is dependent on radiation dosage (37, 50), as shown in Figure 6. The number of trapped radicals formed in the cellulose molecule is related to radiation dosage (37) (see also Section G of this Chapter XVII).

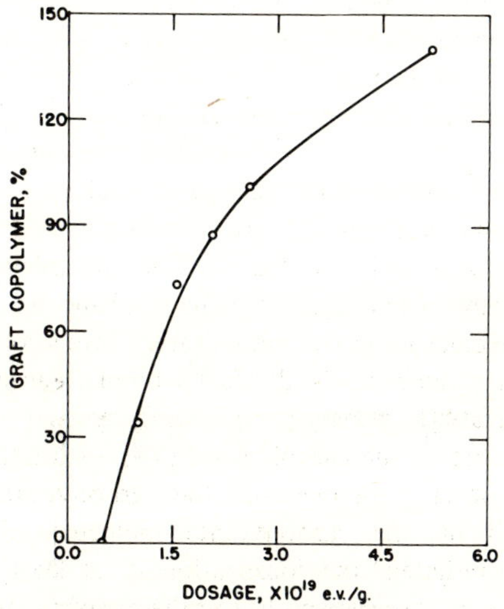

Fig. 6. Formation of trapped radicals in irradiated cellulose I as indicated by post-irradiation graft copolymerization with acrylonitrile. Cellulose was irradiated dry and in vacuum; reacted at 25°C for 48 hr with acrylonitrile (15%) in aqueous zinc chloride solution (75%).

(3) After-Effects

In postirradiation storage trapped free radicals in the irradiated cellulose would react with oxygen and water vapor, as these compounds slowly diffused into the more ordered regions of the fibrous cellulose (40, 69, 70). The most easily detected change in properties is the decrease in the molecular weight of the cellulose molecule as measured by the specific viscosities of solutions (66).

3. Intramolecular Energy Transfer

(a) ENERGY LOCALIZATION

When cellulose is irradiated in the solid state with ionizing radiation, such as γ-radiation, the mechanism of energy loss by the incident radiation to the cellulose molecule is probably an initial random, nonlocalized absorption (and later dissipation) of the energy within the molecule, as energetic electrons in spurs of about 100 Å in diameter (21). Then the rapid localization of the energy from the electron spurs in the molecule results in chemical and physical changes within the molecule. Cellulose belongs to the class of polymers that depolymerize on localization of the energy from these spurs in the molecule (3). If the localization of the energy from these spurs could be directed, depolymerization of the cellulose molecule could be minimized and the radiation-initiated reactions of cellulose could be partially determined.

An early example of controlling the localization of energy from the spurs of electrons was the use of benzene (72). When benzene was added to a hydrocarbon and the mixture was irradiated, a lower yield of gaseous products was obtained than when the hydrocarbon was irradiated separately. It was suggested that the energy was selectively localized within the aromatic group and was dissipated in such a way as to cause less chemical change in the hydrocarbon. In the case of low-molecular-weight carbohydrates in the solid state one of the molecular changes on irradiation is glycosidic bond cleavage (73). It has been shown that the localization of energy for cleaving glycosidic bonds is dependent on the physical structure of the carbohydrate (74), on the formation of certain complexes between other molecules and the carbohydrate (75), and on the chemical substitution on the carbohydrate molecule (75). Similarly, when fibrous cellulose is irradiated, the localization of at least part of the energy results in cellulosic chain cleavage and loss in the breaking strength of the fibers. It is considered probable that the substitution of an aromatic group on the carbohydrate molecule can cause localization of energy to be effected through intramolecular energy transfer to the aromatic group and/or through selective energy absorption by the aromatic group. Then, considering the π-electron system of the aromatic ring, the energy could be dissipated as heat or light without causing bond cleavage in the carbohydrate molecule (76).

(b) LOW-MOLECULAR-WEIGHT CARBOHYDRATES

When methyl α-D-glucoside is exposed in the solid state to γ-radiation, the glycosidic bond is cleaved with an initial G_{glucose} value of 2.33. The localization of energy at the glycosidic bond is decreased on the substitution of aromatic groups on the glucoside (77, 78). When the aromatic group is phenyl, the initial G_{glucose} value is reduced from 2.33 to 0.8. The phenyl group is also cleaved from the glucoside with an initial G_{phenol} value of 0.8. When the aromatic group is substituted as a benzoyl group, no detectable glucose, as measured by reducing power, is found at dosages as high as 4.5×10^{21} eV/g. Other aromatic groups (79–81) that also protect the glycosidic bond from radiation cleavage include 2,3,4,6-tetra-*O*-(phenylcarbamoyl), 2,3,4,6-tetra-*O*-(*p*-toluenesulfonyl), 4-chloro-2,3,6-tri-*O*-(*p*-toluenesulfonyl), 2,3,4,6-tetra-*O*-(*p*-methoxybenzoyl), 2, 3,4,6-tetra-*O*-(*p*-nitrobenzoyl), 2,3,4,6-tetra-*O*-(*p*-ethoxycarbonylbenzoyl, 2,3,4, 6-tetra-*O*-(*o*-chlorobenzoyl), and 2,3,4,6-tetra-*O*-nicotinoyl. In all of these examples, except for the phenyl group, the bond linking the aromatic group to the glucoside is also very stable to radiation cleavage. In the case of the toluenesulfonyl esters irradiated samples have a new infrared-absorption band at 1725 cm^{-1}, probably indicating a carbonyl group (radiation damage) (79, 80).

This localization of energy, which minimizes radiation damage to the glycosidic bond, could be effected through intramolecular transfer of energy to the aromatic group or selective absorption of energy by the aromatic group, or both. One simple way to deactivate the excited state of the aromatic group is by fluorescence; all of these glucosides, substituted by aromatic groups, exhibit fluorescence (79, 80). Generally the presence of substituents on the aromatic groups of the glucosides does not interfere with the availability of the effective number of π-electrons present in the ring, as evidenced by the resistance of the glycosidic bond to radiation damage. In the case of the toluenesulfonyl esters the apparently decreased resistance of the glycosidic bond to radiation damage could be attributed to the *d*-orbital participation of the sulfur atom present in these compounds (79, 80).

(c) FIBROUS CELLULOSE

The effects of γ-radiation on the breaking strengths (82, 83) of purified, mercerized, and modified fibrous cotton celluloses, in the form of yarns, are shown in Table 3. The conversion of the type I crystalline lattice (purified cellulose) to type II (mercerized cellulose) slightly decreased the localization of energy, which resulted in chain depolymerization and loss in the breaking strength of the fibrous cellulose. However, the substitution of aromatic groups on the cellulose molecule significantly increased the resistance of fibrous cellulose to radiation damage (84–89). The dosage, which was used to evaluate

TABLE 3
High-Energy Localization in Fibrous Cellulose as Effected by Substituted Aromatic Groups (84–87)

Type of cotton cellulose	Lattice type	DS	Breaking strength (lb) Control	Breaking strength (lb) Irradiated sample[a]	Retention of strength (%)
Purified	I	0.00	10.4	1.8 (1.5)	17
Mercerized	II	0.00	11.3	4.0 (1.5)	35
Modified derivative:					
Phenylcarbamoyl	I	0.00	10.3	2.1 (1.5)	20
		1.00	10.4	9.6 (1.5)	92
		1.60	10.4	9.6 (1.5)	92
Benzoyl	I	0.00	10.5	2.7 (1.3)	26
		0.80	9.8	8.0 (1.3)	82
		1.10	12.6	7.9 (1.3)	63
		1.30	12.8	10.3 (1.3)	80
Naphthoyl	I	0.00	10.5	2.8 (1.2)	27
		1.44	11.6	5.7 (1.2)	49
Cinnamoyl	I	0.00	10.0	3.0 (1.7)	30
		1.52	9.0	6.9 (1.7)	77
Benzyl	II	0.00	13.2	3.4 (1.6)	26
		0.52	9.1	3.2 (1.6)	35
		0.00	10.4	1.8 (1.5)	17
		0.74	8.7	2.5 (1.5)	29
Benzhydryl	II	0.00	12.5	3.1 (1.6)	25
		0.31	11.8	8.6 (1.6)	73
		0.46	10.9	8.7 (1.6)	80
		0.87	10.0	8.1 (1.6)	81
		1.12	8.3	6.8 (1.6)	82
		1.22	9.2	6.3 (1.6)	68
Trityl	II	0.00	11.3	3.2 (1.6)	28
		0.30	10.8	8.7 (1.6)	81
		0.53	8.4	7.4 (1.6)	88

[a]Numbers in parentheses show radiation dosage in 10^{21} electron-volts per gram.

the radioprotective effects of aromatic group substitution on the cellulose molecule, was about 10 times that required for sterilization and about 30 times that required for radiation-initiated graft-copolymerization reactions.

All of the substituted aromatic groups, with the exception of the benzyl group, decreased the localization of energy, which resulted in decreased breaking strength (86). On postirradiation analysis of the irradiated substituted celluloses the loss in aromatic groups was less than 5% for all of the groups listed in Table 3, with the exception of the benzyl group, in which case the loss was about 55–60%. The loss of benzyl groups from the cellulose is similar to the radiation-initiated cleavage of phenyl groups from substituted glucosides (77, 78).

As shown in Table 3 the radioprotection of the aromatic groups for the cellulose molecular chain was effective at different degrees of substitution (DS). Assuming that the groups are randomly substituted on the cellulose molecule and that each hydroxyl group is equally accessible for reaction, the maximum distances between the groups along the molecule were calculated as a function of the DS by means of equations 5 or 6, assuming a binomial distribution with a probability of reaction $\geqslant 0.95$ (90).

$$p(x) = \binom{n}{x} p^x (1-p)^{n-x} = 0.95 \tag{5}$$

where x is the number of successes, p is DS/3, and n is the number of hydroxyl groups.

For the case of $x = 0$ successes equation 5 would reduce to

$$p(0) = \binom{n}{0} p^0 (1-p)^{n-0} = 0.05 \tag{6a}$$

$$(1-p)^n = 0.05 \tag{6b}$$

For example, for DS = 0.30

$$\left(1 - \frac{0.30}{3}\right)^n = 0.05; \quad n = 28.4$$

That is, for DS = 0.30 and probability of reaction $\geqslant 0.95$ the maximum spacing of groups would be at least one group for about 28 hydroxyl groups or about 5 cellobiose units. If 10.3 Å is taken as the length of one cellobiose unit (28, 29), the maximum spacing of the groups would be about 50 Å. In the experimental case the assumptions concerning equal accessibility of the hydroxyl groups will not be exactly true; therefore maximum actual experimental distances between the substituted groups will tend to be greater than those calculated. The calculated maximum spacings for the groups and the average number of aromatic rings per cellobiose unit for a number of derivatives of cellulose (84, 86, 87) are tabulated in Table 4. The effect of a substituted group on the localization of energy apparently extends over more than one cellobiose unit. The effects of benzoyl groups on the localization of energy are detectable at DS values as low as 0.20, equivalent to a maximum spacing of 70–80 Å (84). The nature of the trapped radicals formed in irradiated substituted celluloses is not changed (86), as shown in their typical trapped-radical spectra.

As already discussed, the mechanism of energy loss by high-energy irradiation of the cellulose molecule would probably include random, nonlocalized formation of high-energy electrons in spurs of about 100 Å in diameter (21, 76). Apparently the substitution of aromatic groups on the cellulose molecule leads to the selective absorption of the energy from these spurs and then the dissipation of the energy as light or heat. This would minimize the localization of energy, which would cause depolymerization and loss in the strength of the fibrous cellulose.

TABLE 4
Calculated Maximum Distance between Substituted Groups on the Cellulose Molecule as a Function of Degree of Substitution (84, 86, 87)[a]

Derivative	DS[b]	Average number of aromatic rings per cellobiose unit	Maximum distance between groups (Å)
Phenylcarbamoyl	1.00	2.0	10–20
	1.60	3.2	10
Benzoyl	0.80	1.6	10–20
Naphthoyl	1.44	2.9	10
Cinnamoyl	1.52	3.0	10
Benzyl	0.74	1.5	10–20
Benzhydryl	0.46	1.8	30–40
	0.87	3.5	10–20
Trityl	0.30	1.8	40–50
	0.79	4.7	10–20

[a]Calculated as a binomial distribution of aromatic groups with a probability of reaction ⩾0.95.

[b]Assuming that the groups are randomly substituted on the cellulose molecule and that each hydroxyl group is equally accessible for reaction.

(d) POSTIRRADIATION STABILITY

Postirradiation reactions of trapped radicals with oxygen and water apparently do not cause significant depolymerization of cotton cellulose in the solid state or serious loss in breaking strength (66, 91). Postirradiation decreases in the viscosities of solutions of irradiated celluloses have been observed; these changes probably result from the reactions of trapped radicals with oxygen and water–reactions that increase the alkali sensitivity of the anhydroglucose rings to cleavage (92, 93). The postirradiation stability of the breaking strength of fibrous cellulose, under ambient conditions, indicates the feasibility of using high-energy radiation for sterilizing cellulosic products and for initiating chemical reactions to change the properties of fibrous cellulose.

4. Macromolecular Properties

(a) DEPOLYMERIZATION

The depolymerization of the cellulose molecule is one of the reactions initiated when cellulose is irradiated (3, 5, 94–104), as shown in Table 5. Irradiation of cellulose in an atmosphere of nitrogen or oxygen indicates that a reactive free radical which generates a strong singlet ESR spectrum is formed at the $C_{(1)}$ or $C_{(4)}$ atoms (37, 38). The depolymerization of the irradiated cellulose molecule probably results principally from the localization of energy at these carbon atoms. When cellulose is irradiated in an atmosphere of oxygen,

TABLE 5
Effect of Moisture Content and Atmosphere on the Radiation-Initiated Depolymerization of Cellulose (3, 95)

Dosage (10^{19} eV/g)[a]	Degree of polymerization: Irradiated in oxygen, 6% H_2O	Irradiated in nitrogen, Moisture content %: 1	6	70
0.00	4400	4400	4400	4400
0.52	3100	–	3100	–
5.2	800	790	880	1200
26	290	–	320	–
52	180	230	210	260
260	72	–	79	–
275	–	89	75	92
520	39	–	56	–

[a] Cobalt-60 γ-radiation at ambient temperature.

depolymerization is greater than when cellulose is irradiated in an atmosphere of nitrogen. Oxygen apparently reacts with the free radical formed at the $C_{(5)}$ atom, causing anhydroglucose-ring cleavage; radiation-initiated depolymerization is therefore slightly greater in an atmosphere of oxygen.

At a high radiation dosage cellulose is sensitized to alkali cleavage (3, 95, 99), as shown in Table 6. At the lowest dosage (0.52×10^{19} eV/g) there was a significant change in solubility of the irradiated cellulosic residue, which had a viscosity-average degree of polymerization (DP) of 3100 in 1% sodium hydroxide solution. The solubility of the irradiated cellulosic residue in water was significant only at dosages greater than 52×10^{19} eV/g, when the viscosity-average DP was 70–80. The high solubility of the irradiated cellulosic residue in

TABLE 6
Radiation-Initiated Formation of Soluble Degradation Products from Cellulose (3, 95)

Dosage (10^{19} eV/g)[a]	Atmosphere of irradiation	Solubility (% residue) Solvent: Water	Solubility (% residue) Solvent: 1% NaOH
0.00	–	0.1	0.8
0.52	O_2	0.0	1.5
	N_2	0.0	1.0
5.2	O_2	0.0	4.0
	N_2	0.0	4.9
26	O_2	0.2	16
	N_2	0.1	16
52	O_2	0.4	30
	N_2	0.5	24
260	O_2	4.5	61
	N_2	3.3	58
520	O_2	9.7	74
	N_2	10	71

[a]Cobalt-60 γ-radiation at ambient temperature.

1% sodium hydroxide solution probably indicates cleavage of anhydroglucose rings and additional lowering of the molecular weight of the irradiated cellulosic residue.

The presence of water in the irradiated cellulose decreases the formation of free radicals (37, 38, 40). When cellulose samples of various water contents were irradiated in nitrogen at a dosage of 5.2×10^{19} eV/g, water apparently reacted with free radicals, as they were formed, and decreased intramolecular transfer and localization of energy, which would result in depolymerization (3), as shown in Table 5. As the dosage was increased, the DP of the cellulosic residue was decreased, so that degradation products, which were soluble in water and sodium hydroxide solutions, influenced the results.

(b) BREAKING STRENGTH

Radiation-initiated depolymerization of the cellulose molecule also results in a decrease in the breaking strength of fibrous cellulose (3, 95, 105–115). The average number of chain breaks per molecule at a given dosage was found to be slightly less for the cellulose irradiated in an atmosphere of nitrogen than that for cellulose irradiated in an atmosphere of oxygen. However, the breaking strengths of the irradiated fibrous cellulose at a given dosage were about the same even when 15–20 chain breaks per molecule had occurred. No significant strength remained when the irradiated cellulose had 50–60 chain breaks per molecule.

(c) CRYSTALLINITY

Natural cellulose with lattice type I is highly crystalline (24). The effects of high-energy radiation on the crystallinity have been determined by examination of the infrared spectra (116), x-ray spectrometry (30), rate of acid hydrolysis (117), and sorption of iodine (118). The irradiated cellulose retained all of the infrared-absorption bands of the purified unirradiated cellulose, without lessening of intensity (3). It had been reported that when cellulose is partially decrystallized, the series of sharp bands in the 1050–1430 cm^{-1} region broadens and lessens in intensity. At dosages lower than 52×10^{19} eV/g there was apparently no change in the crystallinity of cellulose on irradiation. At higher dosages the x-ray determination indicated no change in crystallinity. The acid hydrolysis and iodine-sorption methods, however, seemed to indicate changes in the cellulose molecule; however, at the dosages at which these changes were observed the cellulose molecule is depolymerized to a great extent (see Table 5). The reactivities of the degradation products were probably being measured by the chemical methods, rather than the true crystallinity of the cellulose itself (3, 119–121).

The inaccessibility of trapped radicals to water vapor and liquid and to aqueous zinc chloride solution (37, 40) also indicates the high degree of crystallinity in irradiated cellulose (see Figures 3 and 4). At dosages lower than 5.2×10^{19} eV/g there is no indication of any decrease in the crystallinity of cellulose. It seems to be highly probable that at higher dosages no decrease in the crystallinity of cellulose occurs.

(d) INTERMOLECULAR BONDING

It seems probable that the radiation-initiated dehydrogenation of the cellulose molecule would lead to intermolecular bonding. Owing to depolymerization, the usual physical testing procedures do not conclusively show whether intermolecular bonding does occur. Usually at low dosages a small increase in the breaking strength of irradiated cellulose over that of unirradiated cellulose is observed (3, 95); this could be interpreted as being due to increased intermolecular bonding.

The lower absorption of water in the vapor or liquid states by cellulose irradiated at low dosages (3, 122–125), as compared with that by unirradiated cellulose, probably indicates intermolecular bonding in the irradiated cellulose. Similarly, when cellulose is irradiated dry and then allowed to regain moisture under controlled conditions, the concentration of trapped radicals formed at a given dosage is greater than if the cellulose, containing regain moisture, is irradiated to the same dosage. This could be interpreted to mean that radiation-initiated intermolecular bonding decreases the accessibility of the trapped radicals to moisture vapor (37, 40). Reaction of water with the trapped radicals would yield hydrogen peroxide and a decreased concentration of trapped radicals. However, it appears that the radiation yield for the depolymerization of cellulose is much greater than the yield for the intermolecular bonding of cellulose (15–17).

Methylcellulose in solution is apparently crosslinked to some extent on exposure to radiation. The increased intermolecular bonding is shown by an increase in the viscosity of irradiated solutions (126–130).

5. Chemical Properties

(a) GASEOUS PRODUCTS

The primary chemical effects of energy localization in the cellulose molecule are dehydrogenation and destruction of anhydroglucose rings to yield the gaseous products hydrogen, carbon monoxide, and carbon dioxide (2, 8, 131–134). When cellulose is irradiated in an atmosphere of nitrogen, under constant volume–constant temperature conditions, the pressure caused by the gaseous products evolved increases linearly with dosage (3). When cellulose is irradiated in air or oxygen, the absolute pressure decreases until most of the oxygen is consumed in the radiation-initiated reactions and then increases with dosage (3).

Typical compositions of gases evolved on the irradiation of cellulose in various atmospheres (2) at 25°C are shown in Table 7.

(b) OXIDATIVE DEGRADATION

The interaction of radiation with the cellulose molecule, to cause depolymerization and evolution of gaseous products, leads to an oxidative degradation of the solid residue of irradiated cellulose (8, 97, 132–159). With increasing dosage there is an increasing formation of reducing groups (160, 161) and acid groups (162, 163), as shown in Table 8.

TABLE 7
Typical Composition of Gases Evolved on the Irradiation of Cellulose (2)

Atmosphere of irradiation	Dosage (10^{20} eV/g)[a]	Gaseous products (micromoles per gram of cellulose)	Composition (%)		
			H_2	CO	CO_2
Vacuum	33	205	35	22	43
Oxygen	42	430	19	27	54
Air	14	127	13	59	28
Nitrogen	22	163	22	56	22
	32	217	24	51	25
	38	247	24	51	25

[a]Cobalt-60 γ-radiation at ambient temperature.

TABLE 8
Radiation-Initiated Formation of Reducing and Acid Groups in Cellulose (2, 3, 95)

Dosage	Irradiated in nitrogen		Irradiated in oxygen	
(10^{19} eV/g)[a]	Reducing groups (mmoles/g)	Acid groups (mmoles/g)	Reducing groups (mmoles/g)	Acid groups (mmoles/g)
0.00	0.00	0.002	0.00	0.002
0.52	0.01	0.002	0.01	0.004
5.2	0.05	0.005	0.06	0.007
26	0.22	0.015	0.29	0.022
52	0.36	0.023	0.58	0.039
260	1.30	0.070	1.92	0.107
520	2.68	0.139	3.04	0.155

[a]Cobalt-60 γ-radiation at ambient temperature.

When the cellulose is irradiated in an atmosphere of oxygen, the yields of these groups are slightly greater than those obtained in an atmosphere of nitrogen (2, 3, 95). The reaction of oxygen with the free radicals formed in the amorphous or more accessible regions of the fibrous cellulose could lead to the formation of reducing groups, as follows:

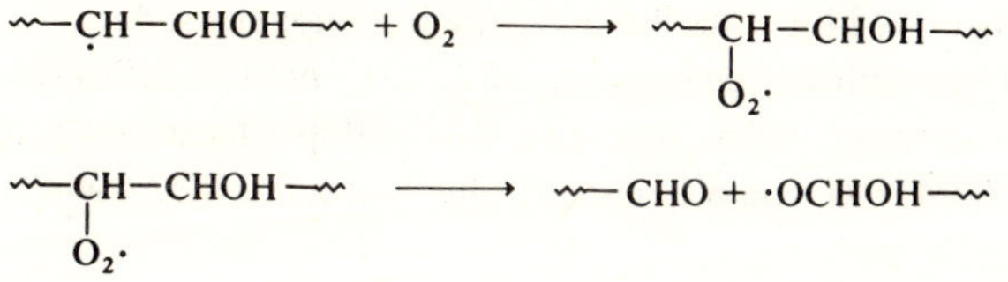

The formation of ·OCHOH—~~ would also probably cause additional cleavage of the anhydroglucose ring, further depolymerization of the cellulose molecule, and formation of acid groups.

The infrared-absorption spectrum of irradiated cellulose indicates the formation of a component absorbing at 1725 cm^{-1}. The intensity of this band increases as the dosage is increased (2, 3, 95, 122, 154); it is probably due to carboxyl groups. Since the concentration of reducing groups in the irradiated cellulose was found to be several times greater than the concentration of acid groups, the reducing groups are probably hemiacetals (122). The data also indicate the presence of an ionizable chromophore that is probably 2-ketogluconic acid (122).

(c) BASE EXCHANGE

Irradiated cotton cellulose exhibits base-exchange properties, probably due to the acid groups formed (164). By the use of radioisotopes, the exchange of H^+ by Ca^{2+} has been shown to be proportional to the radiation dosage and to be dependent on the pH of the solution. The Ca^{2+} absorbed in the irradiated cellulose increased to a maximum as the solution was made alkaline and decreased to a minimum as the solution was made acidic. The relative affinity of irradiated cellulose for metal ions examined was $Ca^{2+} > Co^{2+} > Na^+ > Mg^{2+}$.

(d) DEGRADATION PRODUCTS

The nonvolatile degradation products of irradiated cellulose were found to consist of water-soluble and water-insoluble fractions (122, 132, 133, 145). The water-soluble fraction was then divided into ethanol-insoluble and ethanol-soluble fractions. About 4% of the solid residue of cellulose, irradiated in an atmosphere of oxygen to a dosage of about 5.2×10^{21} eV/g, was soluble in ethanol. The components of the ethanol-soluble fraction were examined by paper chromatography (165); the products that were identified are listed in Table 9.

Further evidence that a homologous series of degradation products are produced on the irradiation of cellulose is the proportional relationships between the DP and R_f values of the phenylhydrazine derivatives of the products (166, 167), as follows:

$$0.345 \text{ DP} = \log(1/R_f - 1) + 0.692 \tag{7}$$

If an R_f value for glucose of 0.69–0.71 is used, the observed R_f value, the DP, and the DP calculated by equation 7 for the homologous series of products given in Table 9, are as follows: cellobiose, 0.52, 2.0, and 1.9; cellotriaose, 0.31, 3.0, and 3.0; cellotetraose, 0.16, 4.0, and 4.1; cellopentaose, 0.08, 5.0, and 5.1; cellohexaose, 0.04, 6.0, and 6.0; and celloheptaose, 0.02, 7.0, and 6.9.

TABLE 9

Some of the Ethanol-Soluble Degradation Products of Irradiated Cellulose (122)

Product[a]	Phenylhydrazine derivative of product[a]
Cellobiose	Cellobiose
Glucose	Glucose
2-Ketogluconic acid	(2-Ketogluconic acid)
Arabinose	Arabinose
Glyoxal	(Erythrose)
(2-Ketocellobionic acid)	Cellotriose
(2-Ketogluconic acid lactone)	Cellotetraose
(Six other components)	Cellopentaose
(Four acids)	Cellohexaose
	(Celloheptaose)
	(Acid)
	(Two oligosaccharides)

[a]Parentheses indicate tentative identification.

6. Morphological Properties

(a) RADIATION EFFECTS

When fibrous cotton cellulose is irradiated, the morphology of the fiber is changed markedly (168–170), beginning at a dosage of about 5×10^{19} eV/g. At a dosage of about 2.6×10^{20} eV/g the typical fragment shown in the electron microphotograph in Figure 7*A* has changed to that shown in Figure 7*B*. Similarly the long fibers, shown by light microscopy in Figure 8*A*, on irradiation tend to break up to form short fibers (Fig. 8*B*). The fibrils, which spiral around the center axis and along the length of the fiber, are weakened when cellulose is irradiated. When irradiated fibers are swollen in solution, the ballooning effect obtained with unirradiated fibers is reduced because the structure has been weakened.

(b) EFFECTS OF RADIATION-INITIATED REACTION

The effects of the radiation-initiated chain reactions of vinyl monomers with cellulose on the morphology of fibrous cotton cellulose are dependent on the following experimental variables: simultaneous reactions—that is, contact of reagent with cellulose during irradiation; postirradiation reactions – that is, irradiation followed by contact with chemical reagent; chemical modification of

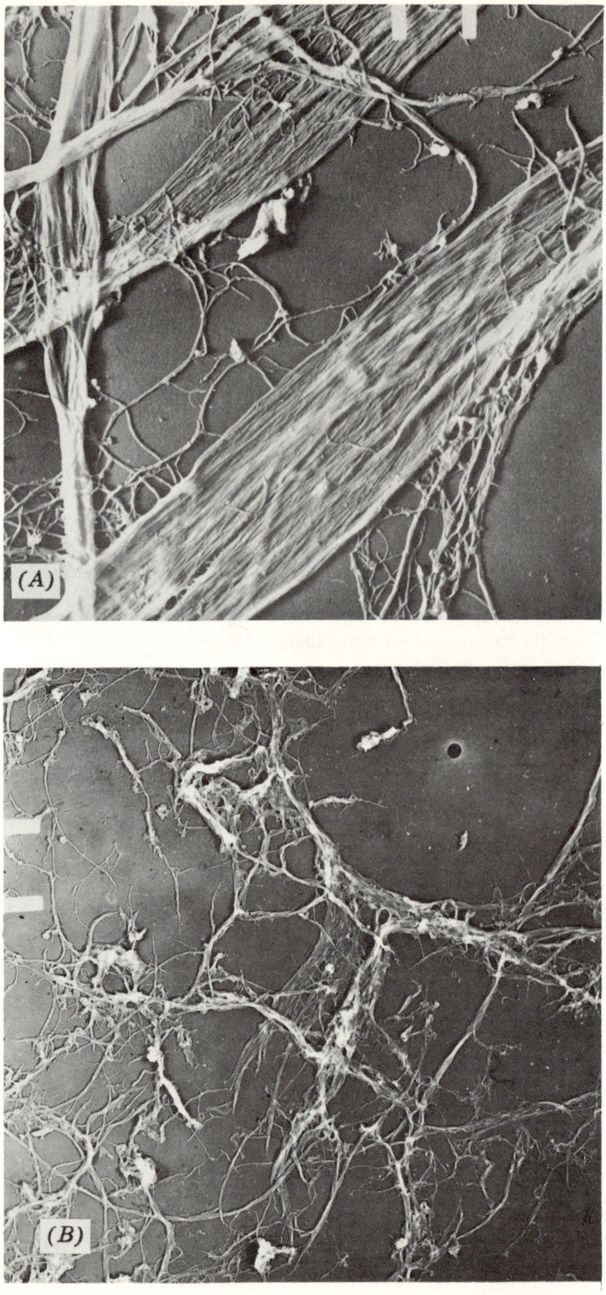

Fig. 7. Fragments of cotton fibers: (*A*) control; (*B*) irradiated at a dosage of about 2.6×10^{20} eV/g.

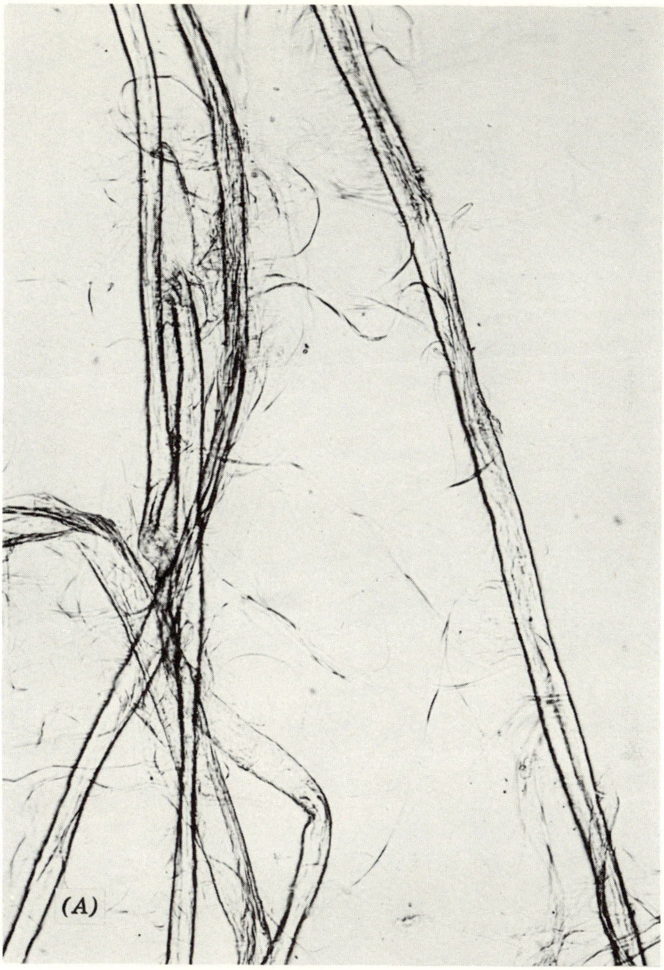

Fig. 8. Longitudinal view of cotton fibers (magnification 200 ×): (*A*) control.

cellulose prior to irradiation and reaction; interaction of the solvent for the chemical reagent with the cellulose, particularly any interaction that causes swelling of the fibrous structure; and, of course, the properties of the chemical reagent (50, 59–61, 63, 171–175).

The postirradiation reaction of cellulose with acrylonitrile monomer dissolved in concentrated aqueous zinc chloride solution (see Fig. 9*A*) shows little effect on the natural, cross-sectional shape of the fibrous cotton cellulose (60). However, when the acrylonitrile monomer is dissolved in dimethylformamide and after irradiation reacted with cellulose, the natural fibrous shape collapses (Fig. 9*B*) (61). When fibrous cellulose is swollen in concentrated

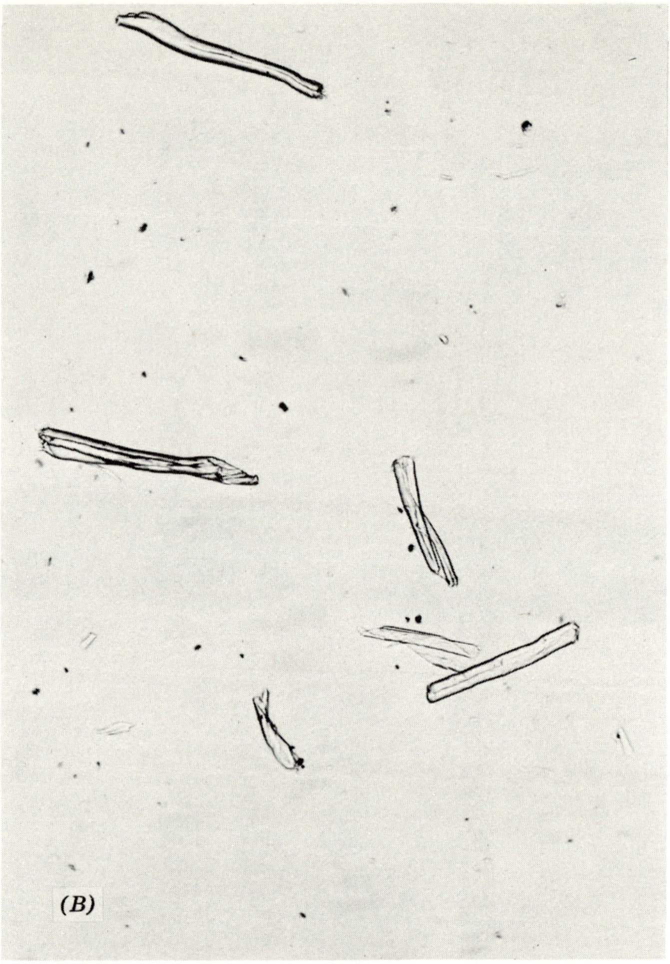

Fig. 8. Longitudinal view of cotton fibers (magnification 200 x): (*B*) irradiated at a dosage of about 5×10^{21} eV/g.

aqueous zinc chloride, containing acrylonitrile monomer, and then irradiated (simultaneous reaction), the fibrous structure is swollen and rounded (172); however, there is no layering, as shown in the swollen cross section of unirradiated fibers. The outer layers of the reacted fibers are soluble in cupriethylenediamine dihydroxide; the polyacrylonitrile is distributed within the fibrous structure. Similar results were obtained with fibrous cellulose that was cyanoethylated (DS 0.7) and then irradiated with acrylonitrile monomer dissolved in concentrated aqueous zinc chloride (simultaneous reaction) (171). This is shown in Figure 10*A*.

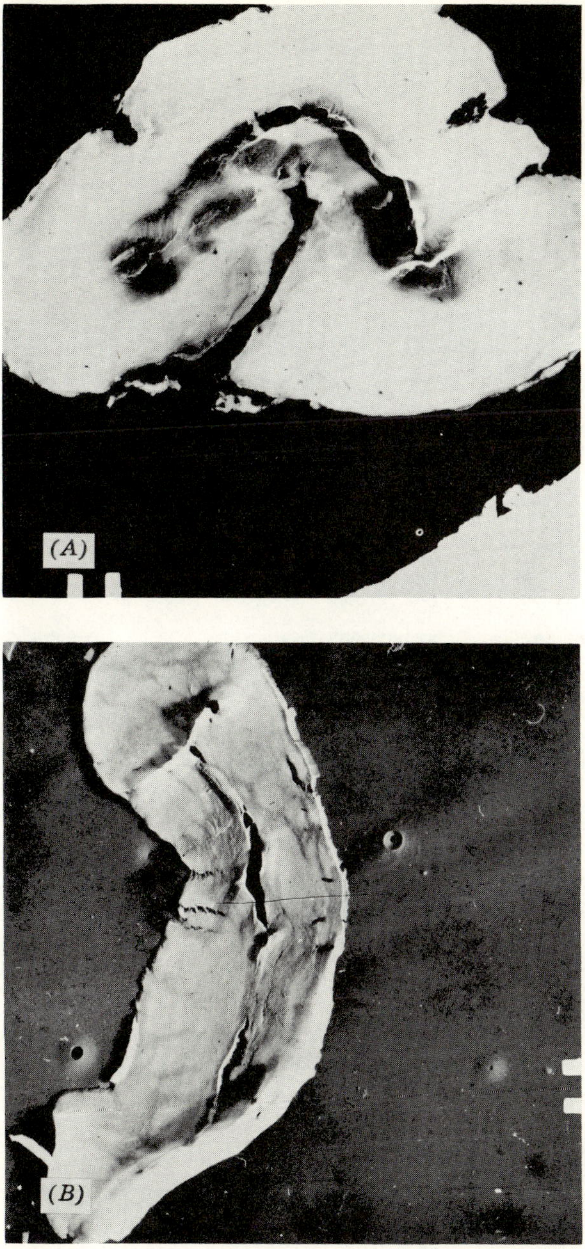

Fig. 9. Cross-sectional view of cotton fiber copolymerized with acrylonitrile (distance between markers is 1 μ.): (*A*) postirradiation, 32% monomer in 80% aqueous $ZnCl_2$; (*B*) postirradiation, 32% monomer in dimethylformamide.

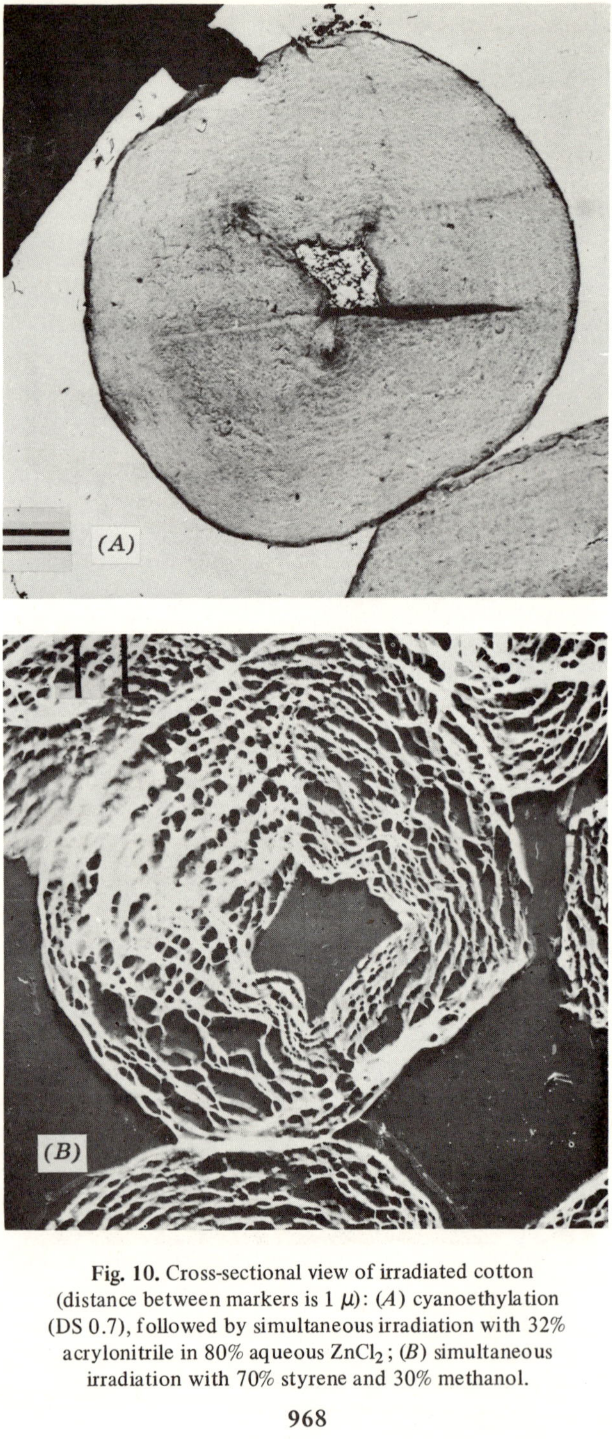

Fig. 10. Cross-sectional view of irradiated cotton (distance between markers is 1 μ): (*A*) cyanoethylation (DS 0.7), followed by simultaneous irradiation with 32% acrylonitrile in 80% aqueous $ZnCl_2$; (*B*) simultaneous irradiation with 70% styrene and 30% methanol.

When fibrous cellulose is irradiated with styrene monomer dissolved in methanol (simultaneous reaction), the fibrous structure is rounded, but a layering effect is obtained (50, 63), as shown in Figure 10*B*. Similarly, when fibrous cellulose is irradiated with alkyl methacrylate monomer dissolved in methanol (simultaneous reaction), the fibrous structure is rounded and a layering effect is obtained. The extent of this layering effect appears to be directly related to the molecular weight of the monomer (60).

The postirradiation reaction of cotton with vinyl acetate monomer, dissolved in concentrated aqueous zinc chloride solution, causes a rounding of the fibrous structure, and the poly(vinyl acetate) is apparently distributed throughout the structure.

7. Conclusions

The reactions induced in cellulose by high-energy radiation can be classified as oxidative-degradation reactions of the cellulose molecule. On irradiation of cellulose, dehydrogenation, destruction of anhydroglucose units to yield carbon monoxide and carbon dioxide, and random depolymerization of the molecule to yield an homologous series of products are observed. Trapped free radicals are

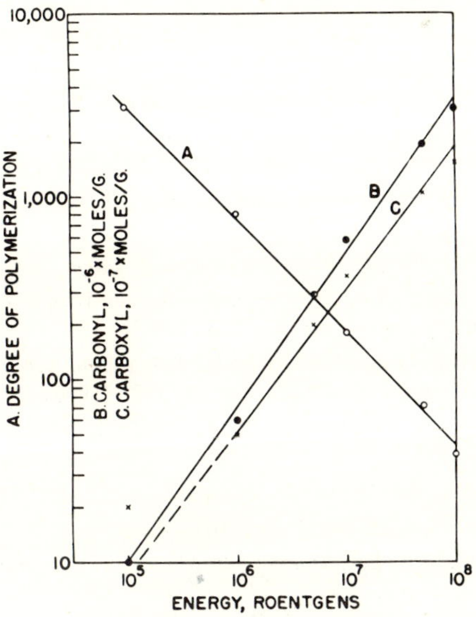

Fig. 11. Effect of energy transfer and localization processes in irradiated cellulose I on the molecular properties of the products.

formed in the crystalline lattice of the cellulose molecule, probably on the $C_{(5)}$ and/or $C_{(6)}$ atoms. The *G*-values for major reactions induced in the cellulose molecule are estimated to be as follows: for molecular depolymerization in nitrogen $G = 6.5$; for acid-group formation in nitrogen $G = 3.5$; for reducing-group formation in nitrogen $G = 58$; for hydrogen in vacuum $G = 2.6$; for carbon monoxide in vacuum $G = 0.80$; for carbon dioxide in vacuum $G = 1.6$; for free radicals, dry and in vacuum $G = 4.0$; and for trapped radicals in oxygen $G = 1.6$. The energy-transfer processes and localization of high-energy radiation in cellulose apparently results from a series of successive secondary events from a common parent of incident radiation, as indicated by equation 4. When changes in molecular properties (depolymerization, reducing-group formation, and acid-group formation) are plotted as a function of radiation dosage, as shown in Figure 11, a linear plot is obtained, as predicted from theoretical considerations (23, 176).

References

1. J. C. Arthur, Jr., *Cotton Trade J., Intern. Ed.*, 41 (1964–1965).
2. J. C. Arthur, Jr., in G. O. Phillips, Ed., *Energetics and Mechanisms in Radiation Biology*, Academic Press, London, 1968, p.153.
3. J. C. Arthur, Jr., F. A. Blouin, and R. J. Demint, *Am. Dyestuff Reptr.*, **49**, 383 (1960).
4. G. Becker and A. Burmester, *Materialpruefung*, **4**, 416 (1962); through *Chem. Abstr.*, **58**, 7031 (1963).
5. F. A. Blouin and J. C. Arthur, Jr., *Am. Dyestuff Reptr.*, **53**, 951 (1964).
6. A. S. Freĭdin and Y. M. Malinskiĭ, *Usp. Khim. i Tekhnol. Polimerov, Sbornik*, 1960, 130; through *Chem. Abstr.*, **55**, 11320 (1961).
7. A. B. Gupta, *Bombay Technologist*, **15**, 63 (1965); through *Nucl. Sci. Abstr.*, **19**, 44063 (1965).
8. H. Lueck and F. Dell, *Chimia*, **17**, 1 (1963); through *Chem. Abstr.*, **59**, 5441 (1963).
9. F. Muenzel and J. Schenkel, *Schriftenreihe Bundesmin. Wiss. Forsch., Radionuklide*, 141 (1964); through *J. Textile Inst.*, **57**, 133 (1966).
10. W. Ruttinger and H. D. Heinze, *Schriftenreihe Bundesmin. Wiss. Forsch., Radionuklide*, 120 (1964); through *J. Textile Inst.*, **57**, 133 (1966).
11. A. Pande, *Mundo Textil Argentino*, No 31, 24 (1965); through *J. Soc. Dyers Colourists*, **82**, 163 (1966).
12. A. Pande, *Textile Rec.*, **82**, 60 (1964); through *J. Textile Inst.*, **55**, 772 (1964).
13. Anon., *Chem. Eng. News*, **45**, No. 8, 48 (1967).
14. L. F. Curtis, in D. E. Gray, Ed., *American Institute of Physics Handbook*, McGraw-Hill, New York, 1963, Section 8, p. 1.
15. F. A. Bovey, *The Effects of Ionizing Radiation on Natural and Synthetic High Polymers*, Interscience, New York, 1958.
16. A. Chapiro, *High Polymers*, Vol. XV, *Radiation Chemistry of Polymeric Systems*, Interscience, New York, 1962.
17. A. Charlesby, *Atomic Radiation and Polymers*, Pergamon, London, 1960.

18. G. Friedlander and J. W. Kennedy, *Nuclear and Radiochemistry,* Wiley, New York, 1955.
19. B. M. Tolbert and W. E. Siri, in A. Weissberger, Ed., *Physical Methods of Organic Chemistry,* Vol. I, *Technique of Organic Chemistry,* 3rd ed., Part IV, Interscience, New York, 1960, p. 3335.
20. R. H. Schuler and A. O. Allen, *J. Chem. Phys.,* **24**, 56 (1956).
21. M. Burton, *Discussions Faraday Soc.,* No. 36, 7 (1963).
22. A. Sippel, *Kolloid-Z.,* **112**, 80 (1949).
23. J. C. Arthur, Jr., *Textile Res. J.,* **28**, 204 (1958).
24. J. C. Arthur, Jr., in H. F. Mark, N. G. Gaylord, and N. M. Bikales, Eds., *Encyclopedia of Polymer Science and Technology,* Vol. 4, Interscience, New York, 1966, p. 244.
25. J. H. Kettering and C. M. Conrad, *Ind. Eng. Chem. Anal. Ed.,* **14**, 432 (1942).
26. J. D. Guthrie, C. L. Hoffpauir, M. F. Stansbury, and W. A. Reeves, *Survey of the Chemical Composition of Cotton Fibers, Cottonseed, Peanuts, and Sweet Potatoes,* US Department of Agriculture, AIC-61, 1949.
27. V. W. Tripp, A. T. Moore, I. V. deGruy, and M. L. Rollins, *Textile Res. J.,* **30**, 140 (1960).
28. E. G. V. Percival, *Structural Carbohydrate Chemistry,* Miller, London, 1962.
29. K. Ward, Jr., *Chemistry and Chemical Technology of Cotton,* Interscience, New York, 1955.
30. L. Segal, L. Loeb, and J. J. Creely, *J. Polymer Sci.,* **13**, 193 (1954).
31. C. Y. Liang and R. H. Marchessault, *J. Polymer Sci.,* **37**, 385 (1959).
32. R. H. Marchessault and C. Y. Liang, *J. Polymer Sci.,* **43**, 71 (1960).
33. M. L. Nelson and R. T. O'Connor, *J. Appl. Polymer Sci.,* **8**, 1311 (1964).
34. R. T. O'Connor, E. F. Dupré, and D. Mitcham, *Textile Res. J.,* **28**, 382 (1958).
35. R. H. Mauersberger, Ed., *Mathews' Textile Fibers, Their Physical, Microscopic, and Chemical Properties,* 6th ed., Wiley, New York, 1954.
36. *ASTM Standards*, American Society for Testing and Materials, Philadelphia, ASTM D629-55T.
37. P. J. Baugh, O. Hinojosa, and J. C. Arthur, Jr., *J. Appl. Polymer Sci.* **11**, 1139 (1967).
38. J. C. Arthur, Jr., T. Mares, and O. Hinojosa, *Textile Res. J.,* **36**, 630 (1966).
39. G. S. Park and J. C. Ward, *Nature,* **202**, 389 (1964).
40. R. E. Florin and L. A. Wall, *J. Polymer Sci.,* A, **1**, 1163 (1963).
41. Z. Kuri, Y. Fujiwara, H. Ueda, and S. Shida, *J. Chem. Phys.,* **33**, 1184 (1960).
42. R. J. Abraham and D. H. Whiffen, *Trans. Faraday Soc.,* **54**, 1291 (1958).
43. S. Dilli, I. T Ernest, and J. L. Garnett, *Australian J. Chem.,* **20**, 911 (1967).
44. S. Dilli and J. L. Garnett, *Nature,* **198**, 984 (1963).
45. R. E. Kesting and V. Stannett, *Makromol. Chem.,* **55**, 1 (1962).
46. V. N. Makatun, A. K. Potapovich, and I. N. Ermolenko, *Vysokomol. Soedin.,* **5**, 467. (1963); through *Chem. Abstr.,* **59**, 1831 (1963).
47. J. Ubersfeld, *Compt. Rend.,* **238**, 1397 (1954); see reference 43.
48. J. Ubersfeld, *Ann. Phys.,* **1**, 395 (1956); see reference 43.
49. J. C. Arthur, Jr., and F. A. Blouin, *Am. Dyestuff Reptr.,* **51**, 1024 (1962).
50. J. C. Arthur, Jr., and F. A. Blouin, *J. Appl. Polymer Sci.,* **8**, 2813 (1964).

51. J. C. Arthur, Jr., and F. A. Blouin, US Atomic Energy Commission Document TID-7643, 319 (1962).
52. J. C. Arthur, Jr., and D. J. Daigle, *Textile Res. J.,* **34**, 653 (1964).
53. J. C. Arthur, Jr., and R. J. Demint, *Textile Res. J.,* **31**, 988 (1961).
54. J. C. Arthur, Jr., and R. J. Demint, US Patent 3,109,798 (Nov. 5, 1963).
55. J. C. Arthur, Jr., and R. J. Demint, US Patent 3,115,460 (Nov. 17, 1964).
56. J. C. Arthur, Jr., R. J. Demint, W. F. McSherry, and J. F. Jurgens, *Textile Res. J.,* **29**, 759 (1959).
57. O. R. Bernard, D. Gagnaire, and P. Servoz-Gavin, *J. Chim. Phys.,* **60**, 1348 (1963); through *Nucl. Sci. Abstr.,* **18**, 17900 (1964).
58. F. A. Blouin, and J. C. Arthur,, Jr., *Polymer Preprints,* **6**, 359 (1965).
59. F. A. Blouin and J. C. Arthur, Jr., *Textile Res. J.,* **33**, 727 (1963).
60. F. A. Blouin, A. M. Cannizzaro, J. C. Arthur, Jr., and M. L. Rollins, *Textile Res. J.,* **38**, 811 (1968).
61. F. A. Blouin, N. J. Morris, and J. C. Arthur, Jr., *Textile Res. J.,* **36**, 309 (1966).
62. F. A. Blouin, N. J. Morris, and J. C. Arthur, Jr., *Textile Res. J.,* **38**, 710 (1968).
63. R. J. Demint, J. C. Arthur, Jr., A. R. Markezich, and W. F. McSherry, *Textile Res. J.,* **32**, 918 (1962).
64. R. J. Demint, J. C. Arthur, Jr., and W. F. McSherry, *Textile Res. J.,* **31**, 821 (1961).
65. N. J. Morris, F. A. Blouin, and J. C. Arthur, Jr., *J. Appl. Polymer Sci.,* **12**, 373 (1968).
66. J. C. Arthur, Jr., and F. A. Blouin, *Textile Res. J.,* **34**, 733 (1964).
67. P. K. Christensen and B. M. Tolbert, *Norsk Skogind.,* **19**, 12 (1965); through *J. Textile Inst.,* **56**, 502 (1965).
68. E. F. Degering, G. J. Caldarella, and M. Mancini, ASTM Special Tech. Pub. No. 276, 1960, p. 244.
69. R. E. Glegg and Z. I. Kertesz, *Radiation Res.,* **6**, 469 (1957).
70. R. E. Glegg and Z. I. Kertesz, *J. Polymer Sci.,* **26**, 289 (1957).
71. Z. Kuri and H. Ueda, *J. Polymer Sci.,* **50**, 349 (1961).
72. J. Lamborn and A. J. Swallow, *J. Phys. Chem.,* **65**, 920 (1961).
73. G. O. Phillips, in M. L. Wolfrom and R. S. Tipson, Eds., *Advances in Carbohydrate Chemistry,* Vol. 16, Interscience, New York, 1961, p. 13.
74. G. O. Phillips and P. Baugh, *Nature,* **198**, 282 (1963).
75. G. O. Phillips, *Discussions Faraday Soc.,* No. 36, 281 (1963).
76. J. C. Arthur, Jr., in G. O. Phillips, Ed., *Energy Transfer in Radiation Processes,* Elsevier, Amsterdam, 1966, p. 29.
77. G. O. Phillips, F. A. Blouin, and J. C. Arthur, Jr., *Nature,* **202**, 1328 (1964).
78. G. O. Phillips, F. A. Blouin, and J. C. Arthur, Jr., *Radiation Res.,* **23**, 527 (1964).
79. I. M. Sarkar and J. C. Arthur, Jr., *Carbohydrate Res.,* **6**, 207 (1968).
80. I. M. Sarkar and J. C. Arthur, Jr., *Chem. Ind.* (*London*), 2085 (1967).
81. I. M. Sarkar, J. C. Arthur, Jr., and O. Hinojosa, *Carbohydrates Res.,* **6**, 333 (1968).
82. *ASTM Standards,* American Society for Testing and Materials, Philadelphia, ASTM D76-53.
83. *ASTM Standards,* American Society for Testing and Materials, Philadelphia, ASTM D1445-53T.
84. J. C. Arthur, Jr., and T. Mares, *J. Appl. Polymer Sci.,* 9, 2581 (1965).
85. J. C. Arthur, Jr., T. Mares, and M. George, *Textile Res. J.,* **35**, 1116 (1965).

86. J. C. Arthur, Jr., D. J. Stanonis, T. Mares, and O. Hinojosa, *J. Appl. Polymer Sci.*, **11**, 1129 (1967).
87. I. M. Sarkar, J. C. Arthur, Jr., and M. George, *Textile Res. J.*, **38**, 1145 (1968).
88. C. P. J. Glaudemans, E. Passaglia, and E. A. Wielicki, *Tappi*, **45**, 542 (1962).
89. C. J. Westberry and W. L. Hyden, *Am. Dyestuff Reptr.*, **56**, 13 (1967).
90. W. J. Dixon and F. J. Massey, Jr., *Introduction to Statistical Analysis*, McGraw-Hill, New York, 1957, p. 228.
91. F. A. Blouin and J. C. Arthur, Jr., *Am. Dyestuff Reptr.*, **53**, 951 (1964).
92. *ASTM Standards*, American Society for Testing Materials, Philadelphia, ASTM D539-53.
93. C. M. Conrad, V. W. Tripp, and T. Mares, *J. Phys. Colloid Chem.*, **55**, 1474 (1951).
94. J. C. Arthur, Jr., and R. J. Demint, *Textile Res. J.*, **32**,108 (1962).
95. F. A. Blouin and J. C. Arthur, Jr.,*Textile Res. J.*, **28**, 198 (1958).
96. F. A. Blouin, J. C. Arthur, Jr., R. S. Orr, and V. J. Ott, *Textile Res. J.*, **31**, 597 (1961).
97. R. E. Glegg and Z. I. Kertesz, *Science*, **124**, 893 (1956).
98. J. L. Neal and H. A. Kraessig,*Tappi*, **46**, 70 (1963).
99. T. S. A. Padmanabhan, L. R. Sud, P. F. E. Mann, S. K. Ranganathan, and T. S. Subramanian, *Current Sci. (India)*, **18**, 47 (1949); through *Chem. Abstr.*, **43**, 6407 (1949).
100. J. F. Saeman, M. A. Millett, and E. J. Lawton, *Ind. Eng. Chem.* **44**, 2848 (1952).
101. A. Sugimatsu, S. Senda, and Y. Harada, *Kogyo Kagaku Zasshi*, **62**, 576 (1959); through *Chem. Abstr.*, **57**, 15391 (1962).
102. O. Teszler and H. Hefti, *Textil-Rundschau*, **13**, 61 (1958).
103. O. Teszler and H. A. Rutherford, *Proc. UN Intern. Conf. Peaceful Uses Atomic Energy, 2nd, Geneva, 1958*, **29**, 228 (1959); through *Chem. Abstr.*, **54**, 23344 (1960).
104. J. Zurakowska-Orszagh and J. Pieniazek, *Polish Acad. Sci., Inst. Nucl. Res., Rept. No. 206* (1960); through *Chem. Abstr.*, **56**, 15706 (1962).
105. C. Doree and J. W. W. Dyer, *J. Soc. Dyers Colourists*, **33**, 17 (1917).
106. E. S. Gilfillan and L. Linden, *Textile Res. J.*, **25**, 773 (1955).
107. E. S. Gilfillan and L. Linden, *Textile Res. J.*, **27**, 87 (1957).
108. N. Kishi, *Sen-i Gakkaishi*, **17**, 1116 (1961); through *Chem. Abstr.*, **56**, 4280 (1962).
109. G. A. Klein, M. P. Tikhomolova, E. L. Aizenshtat, and M. Sultanova, *Izv. Akad. Nauk Uz. SSSR, Ser. Fiz.-Mat. Nauk*, **7**, 61 (1963); through *Chem. Abstr.*, **59**, 14149 (1963).
110. Huo-Ping Pan, B. E. Proctor, S. A. Goldblith, H. M. Morgan, and R. Z. Naar, *Textile Res. J.*, **29**, 415 (1959).
111. Huo-Ping Pan, B. E. Proctor, S. A. Goldblith, H. M. Morgan, and R. Z. Naar, *Textile Res. J.*, **29**, 422 (1959).
112. A. E. Raubenheimer and R. L. Galbraith, *Am. Dyestuff Reptr.*, **53**, 1059 (1964).
113. O. Schwab, K. Nital,and H. Klingelhoffer, *Kolloid-Z.*, **199**, 108 (1964).
114. K. Seifert, *Holz. Roh-Werkstoff*, **22**, 267 (1964); through *Nucl. Sci. Abstr.*, **19**, 30259 (1965).
115. O. Teszler, L. H. Kiser, P. W. Campbell, and H. A. Rutherford, *Textile Res. J.*, **28**, 456 (1958).
116. R. T. O'Connor, E. F. Dupré, and E. R. McCall, *Anal. Chem.*, **29**, 998 (1957).
117. M. L. Nelson and C. M. Conrad, *Textile Res. J.*, **18**, 149 (1948).

118. A. V. Bailey, E. Honold, and E. L. Skau, *Textile Res. J.*, **28**, 861 (1958).
119. M. Isozaki, N. Ishikawa, and Y. Minoshima, *Tokyo Kogyo Shikendo Hokoku*, **59**, 358 (1964); through *Chem. Abstr.*, **66**, 11286 (1967).
120. H. Sobue and Y. Saito, *Bull. Chem. Soc. Japan*, **34**, 1343 (1961); through *Chem. Abstr.*, **56**, 3053 (1962).
121. S. Takamuru, Y. Miyamoto, and Y. Hachihama, *Kogyo Kagaku Zasshi*, **64**, 2027 (1961); through *Chem. Abstr.*, **57**, 3663 (1962).
122. F. A. Blouin and J. C. Arthur, Jr., *J. Chem. Eng. Data*, **5** 470 (1960).
123. G. S. Park, *J. Polymer Sci.*, B1, 617 (1963).
124. T. Mares and J. C. Arthur, Jr., *J. Polymer Sci.*, B, **7**, 419 (1969).
125. H. Sulser, H. Mohler, and H. Luethy, *Mitt. Gebiete Lebensm. Hyg.*, **55**, 264 (1964); through *J. Textile Inst.*, **56**, 416 (1965).
126. W. J. Hillend and H. A. Swenson, *J. Polymer Sci., A2, 4921 (1964).*
127. F. C. Leavitt, *J. Polymer Sci.*, **45**, 536 (1960).
128. F. C. Leavitt, *J. Polymer Sci.*, **51**, 349 (1961).
129. A. A. Miller, US Patent 2,895,891 (July 21, 1959).
130. F. Muenzel and J. Schenkel , *Andwendung Radioaktiven Isotope in der Textilforschung und Technik*, Karl Thiemig, Munich, 1964, pp. 141–164; through *Nucl. Sci. Abstr.*, **20**, 4034 (1966).
131. F. L. Dalton, M. R. Houlton, and J. A. Sykes, *Nature*, **200**, 862 (1963).
132. S. Dilli and J. L. Garnett, *Chem. Ind. (London)*, No. 10, 409 (1963).
133. M. Horio, R. Imamura, and H. Mizukami, *Bull. Inst. Chem. Res., Kyoto Univ.*, **41**, 17 (1963); through *Chem. Abstr.*, **60**, 9468 (1964).
134. C. S. Schoepfle and L. H. Connell, *Ind. Eng. Chem.*, **21**, 529 (1929).
135. A. Aziz and A. Husain, *Pakistan Atomic Energy Comm. Chem.*, **105** (1966).
136. F. A. Blouin, V. J. Ott, T. Mares, and J. C. Arthur, Jr., *Textile Res. J.*, **34**, 153 (1964).
137. R. C. L. Bosworth, I. Ernst, and J. L. Garnett, *Internl. J. Appl. Radiation and Isotopes*, **11**, 152 (1961).
138. A. S. Freĭdin, Yu. M. Malinskiĭ, and V. L. Karpov, *Gidroliz. i Lesokhim. Prom.*, **12**, 4 (1959); through *Chem. Abstr.*, **53**, 22927 (1959).
139. A. S. Freĭdin, Yu. M. Malinskiĭ, and V. L. Karpov, *Tr. Tashkentsk. Konf. po Mirnomu Ispol'z. At. Energii, Akad. Nauk Uz. SSR*, **1**, 401 (1961); through *Chem. Abstr.*, **57**, 1785 (1962).
140. A. S. Freĭdin, Yu. M. Malinskiĭ, and V. L. Karpov, *Vysokomol. Soedin.*, **1**, 784 (1959); through *Chem. Abstr.*, **54**, 17876 (1960).
141. J. L. Garnett and J. W. T. Merewether, *Chem. Ind. (London)*, No. 39, 1215 (1959).
142. J. L. Garnett and J. W. T. Merewether, *Technol. Use Radiation, Proc. Conf., Sydney, 1960*, 76 (1961); through *Chem. Abstr.*, **59**, 1835 (1963).
143. Y. Hachihama and S. Takamuku, *Technol. Repts. Osaka Univ.*, **10**, 871 (1960); through *Chem. Abstr.*, **55**, 14906 (1961).
144. R. Imamura and H. Minagami, *Kyoto Daigaku Nippon Kagakusemi Kenkyusho Koenshu*, **18**, 73 (1961); through *Chem. Abstr.*, **62**, 4204 (1965).
145. R. Imamura, *Genshiryoku Kogyo*, **9**, 62 (1963); through *Chem. Abstr.*, **60**, 7018 (1964).
146. A. I. Krasinov, *J. Exptl. Theoret. Phys. USSR*, **9**, 1343 (1939); through *Chem. Abstr.*, **35**, 1698 (1941).

147. E. J. Lawton, W. D. Bellamy, R. E. Hungate, M. P. Bryant, and E. Hall, *Science,* **113**, 380 (1951).
148. E. J. Lawton, W. D. Bellamy, R. E. Hungate, M. P. Bryant, and E. Hall, *Tappi,* **34**, No. 12, 113A (1951).
149. E. J. Lawton, A. M. Bueche, and J. S. Balwit, *Nature,* **172**, 76 (1953).
150. S. K. Majumdar and W. H. Rapson, *Textile Res. J.,* **33**, 1 (1963).
151. T. Mares and J. C. Arthur, Jr., *Textile Res. J.,* **33**, 661 (1963).
152. I. Polcin, *Jad. Energ.,* **13**, 145 (1967); through *Nucl. Sci. Abstr.,* **21**, 30466 (1967).
153. Huo-Ping Pan, B. E. Proctor, and S. A. Goldblith, *Textile Res. J.,* **29**, 425 (1959).
154. J. Polcin and J. Kapustova, *Bull. Vyskum. Ustavu Papieru Celulozy,* **5**, (1962); through *Chem. Abstr.,* **58**, 3595 (1963).
155. J. Polcin and M. Karhanek, *Faserforsch. u. Textiltech.,* **14**, 357 (1963); through *J. Textile Inst.,* **54**, 808 (1964).
156. T. Sasaki and K. Hayashi, *Kobunshi Kagaku,* **21**, 637 (1964); through *Nucl. Sci. Abstr.,* **19**, 9203 (1965).
157. C. Trogus, H. Halberschadt, and K. Hess, *Naturwissenshaften,* **18**, 846 (1930); through *Chem. Abstr.,* **25**, 876 (1931).
158. Kh. U. Usmanov, E. I. Kalabanovskaya, and R. B. Damovskii, *Vysokomol. Soedin.,* **3**, 233 (1961); through *Chem. Abstr.,* **56**, 2607 (1962).
159. N. N. Winogradoff, *Nature,* **165**, 72 (1950).
160. D. A. Clibbens and A. Geake, *J. Textile Inst.,* **15**, 27 (1924).
161. M. Somogyi, *J. Biol. Chem.,* **195**, 19 (1952).
162. G. F. Davidson and T. P. Nevell, *J. Textile Inst.,* **39**, 102 (1948).
163. G. F. Davidson and T. P. Nevell, *Shirley Inst. Mem.,* **21**, 85 (1947).
164. R. J. Demint and J. C. Arthur, Jr., *Textile Res. J.,* **29**, 276 (1959).
165. E. Lederer and M. Lederer, *Chromatography–A Review of Principles and Applications*, Elsevier, Amsterdam, 1957.
166. T. A. Scott, Jr., and F. R. Senti, *J. Am. Chem. Soc.,* **77**, 3816 (1955).
167. W. J. Whalen, J. M. Bailey, and P. J. P. Roberts, *J. Chem. Soc.,* 1293 (1953).
168. M. L. Rollins and V. W. Tripp, *Textile Res. J.,* **24**, 345 (1954).
169. F. I. Korchemkin, Yu. M. Malinskii, and G. V. Sukhov, *Tr. Leningrad. Lesotekh. Akad. im. S. M. Kirova*, 101 (1960); through *Chem. Abstr.,* **56**, 2607 (1962).
170. B. R. Porter, V. W. Tripp, I. V. deGruy, and M. L. Rollins, *Textile Res. J.,* **30**, 510 (1960).
171. J. C. Arthur, Jr., and J. N. Grant, *Textile Res. J.,* **36**, 934 (1966).
172. J. C. Arthur, Jr., and R. J. Demint, *Textile Res. J.,* **30**, 505 (1960).
173. J. C. Arthur, Jr., A. R. Markezich, and W. F. McSherry, *Textile Res. J.,* **33**, 896 (1963).
174. M. L. Rollins, A. M. Cannizzaro, F. A. Blouin, and J. C. Arthur, Jr., *J. Appl. Polymer Sci.,* **12**, 71 (1968).
175. M. L. Rollins, A. T. Moore, W. R. Goynes, J. H. Carra, and I. V. deGruy, *Am. Dyestuff Reptr.,* **54**, P512 (1965).
176. A. Charlesby, *J. Polymer Sci.,* **15**, 263 (1955).

I. ELECTRON SPIN RESONANCE STUDIES

Jett C. Arthur, Jr.

The importance of free-radical intermediates in reactions initiated by heat, light, high-energy radiation, oxidation-reduction systems, and mechanical action has often been suggested by cellulose chemists. The detection of these intermediates has usually been by indirect methods. For example, the decoloration of iodine, the initiation of polymerization of vinyl monomers, and the reaction of oxygen leading to oxidative degradation have been used as methods to detect free-radical intermediates in chemical reactions (1).

Free radicals are reactive intermediates that have one unpaired electron. The tendency of electrons to pair in reactions results in that most intermediates, having unpaired electrons, occur in very low concentrations and are highly reactive. The half-lives of these intermediates are usually very short (2). However, many free radicals, formed in crystalline or highly ordered chemical systems, are stable for months and even years. Diphenylpicrylhydrazyl (DPPH) is a stable free radical and is often used as a standard for electron spin resonance (ESR) spectroscopy (3, 4).

An electron has spin and, as a consequence, magnetic moment. The net magnetic moment of paired electrons with opposed spins is zero. A free radical, because of the presence of its unpaired electron, has a net magnetic moment. Developments in microwave spectroscopy, usually called electron spin resonance or, alternatively, electron paramagnetic resonance (EPR), have made possible the measurement of the net magnetic moment of an unpaired electron. The resonance frequency is associated with the energy of the moment of the electron in an applied direct-current magnetic field; therefore the resonance frequency and magnetic field, where any free-radical spectra will be generated, are known. If a sample containing unpaired electrons is placed in a direct-current magnetic field, the electrons will align themselves with their spins and moments parallel or antiparallel to the applied field. If radiation of frequency ν is now applied to the sample, there will be a net absorption of energy provided that

$$h\nu = g\beta H$$

where ν = frequency of radiation, typically held constant at about 9500 MHz
β = Bohr magneton
H = applied magnetic field, typically about 3400 gauss
h = Planck's constant
g = spectroscopic splitting factor

An ESR spectrum is recorded by varying the applied magnetic field a few gauss above and below the point at which maximum absorption of energy occurs; the frequency of the applied radiation is held constant (2).

The number of lines in an ESR spectrum of a free radical depends on the interaction of the unpaired electron with the nuclei composing the molecule of which it is a part. A free radical, containing unpaired electrons not coupled to any nuclei, will generate a singlet ESR spectrum with a *g*-value of 2.0023. Interaction of an unpaired electron of a free radical with *n* identical hydrogen atoms generates an ESR spectrum with $n + 1$ lines and with *g*-values differing from the free-electron value. For example, a methyl radical $CH_3\cdot$ would generate a quartet of lines with relative intensities of 1 : 3 : 3 : 1, based on quantum energy levels (2–4). Consequently for cellulose and cellulose derivatives whose organic chemical structure is known the ESR spectra generated in these compounds by ionizing radiation, light, heat, oxidation–reduction systems, or mechanical action can be analyzed to determine the structure of the free radicals formed. Reactions of these free radicals can also be readily followed by observing the changes in the ESR spectra.

Another application of ESR spectroscopy is in the investigation of the type of bonding in complex ions and coordination compounds. The spin-orbit coupling of an unpaired electron in a complex ion causes anisotropic splitting of the ESR spectral lines (2, 4). Interactions of cellulose with complex ions, such as cupriethylenediamine, can be investigated by this technique.

1. Free-Radical Reactions

(a) REACTIONS INITIATED BY HEAT

When dextrose, sucrose, cellobiose, other carbohydrates of low molecular weight, or cellulose are charred, products are obtained that exhibit paramagnetic resonance (5–13). Hoskins and Pastor (9) suggested the use of dextrose, charred at 560°C, as a standard for ESR spectroscopy. Charred dextrose generates a singlet ESR spectrum with a constant line width, *g*-value, and spin density over a temperature range from about 4 to 300°C. Typical ESR spectra for charred cellobiose and for charred and heated cellulose (5) are shown in Figure 1. The singlet ESR spectra for charred cellulose and cellobiose indicate that unpaired

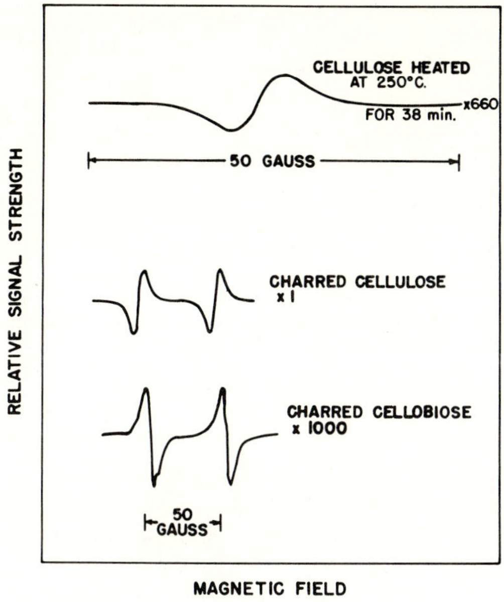

Fig. 1. ESR spectra generated on heating cellulose and cellobiose.

electrons located on free carbons probably generated them. In the case of heated cellulose the symmetrical singlet ESR spectrum also indicates that the thermally initiated decomposition of cellulose probably resulted in the formation of free carbon with unpaired electrons. If the unpaired electrons were located on carbon atoms in the anhydroglucose ring, hyperfine splitting of the spectrum would have occurred. The effects of temperature and time of heating on cellulose on the generation of free radicals on the cellulose molecules, as determined by ESR spectroscopy at the temperature of heating, are shown in Figure 2. The rate of change in the concentration of free radicals in heated cellulose increases rapidly at temperatures above 225°C. From an Arrhenius interpretation of these rate data the apparent activation energy for the thermal initiation of free radicals on the cellulose molecule was calculated to be about 33 kcal/mole (5). Potential applications of ESR spectroscopy in studying pyrolytic and flame-proofing mechanisms for cellulose degradation are obvious from this type of finding (see also Chapters XVIII-C and XIX-I).

(b) REACTIONS INITIATED BY LIGHT

Applications of ESR spectroscopy to study the effects of ultraviolet irradiation on cellulose in the presence of hydrogen peroxide were reported by Ichikawa and Hirota (14), who suggested that hydrogen peroxide is decomposed

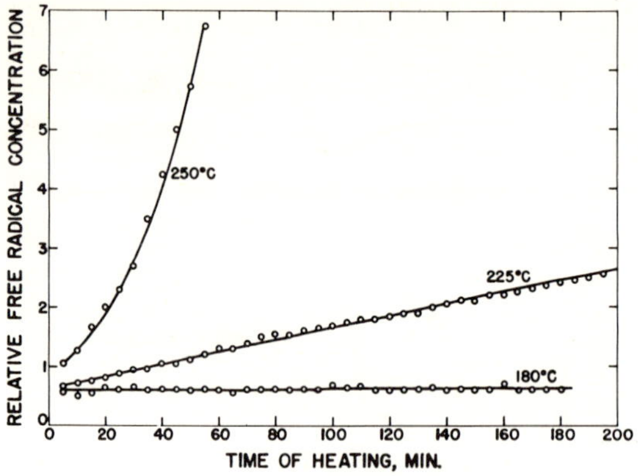

Fig. 2. Effects of temperature and time of heating on rate of formation of free radicals in cellulose.

by ultraviolet light to yield ·OH radicals. These free radicals react with cellulose to generate a complex ESR spectrum, probably composed of a broad singlet and a triplet spectrum. For linear polyols, such as sorbitol or erythritol, broad singlet ESR spectra are generated.

Kleinert (15, 16) used ESR spectroscopy to record free-radical reactions initiated in 95% α-cellulose on ultraviolet irradiation. He irradiated cellulose in air, nitrogen, oxygen, and vacuum, and suggested that molecular chain cleavage results in the formation of free radicals, which generate the ESR spectrum. He reported a slow rate of decay of the free radicals after irradiation (15). Kleinert (17) also recorded the presence of stable free radicals, which generated a singlet ESR spectrum, in various lignins extracted from paper and pulp preparations.

There has been considerable uncertainty about the wavelength of light required to photolyze cellulose directly. Generally it has been concluded that light of wavelength greater than 3400 Å could not photolyze pure cellulose directly. To explain the photolysis of pure cellulose it has been proposed that the primary absorption of light is at about 2650 Å, due to the presence of the hemiacetal group at the $C_{(1)}$ atom of the anhydroglucose unit of the cellulose molecule (18, 19). Later ESR spectroscopy was used to detect the absorption of light by the cellulose molecule (15, 20). This sensitive method detected the free radicals formed in cellulose during photolysis.

Typical ESR spectra of purified cellulose and dyed cellulose, generated on exposure to light (20), are shown in Figure 3. For purified cellulose a narrow singlet ESR spectrum with a line width of about 8–10 gauss and a *g*-value of 2.0045 is generated. Using a high-pressure mercury-vapor lamp as a light source and filters, it was observed that light in the region from 3250 to 4000 Å is absorbed

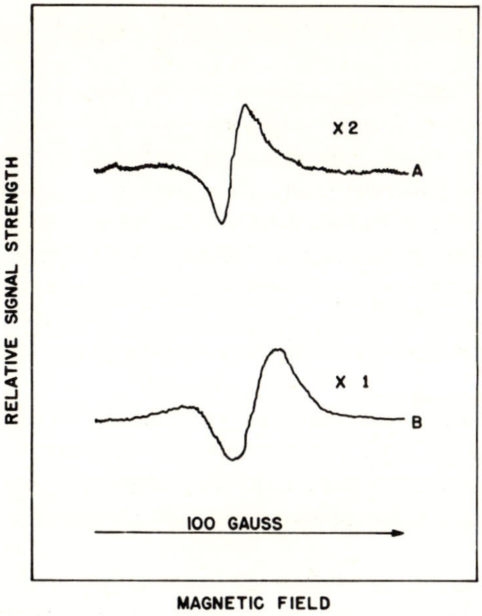

Fig. 3. ESR spectra generated on irradiation with light at 25°C: (*A*) purified cellulose; (*B*) cellulose dyed with Indanthrene Golden Yellow RKA.

and results in the formation of free radicals in cellulose. Light of wavelengths below 3250 Å is responsible for 15–20% of the total concentration of free radicals; light of wavelengths above 4300 Å is responsible for 10–15% of the free radicals. Apparently absorption of light of about 3600 Å results in the most efficient formation of free radicals on the cellulose molecule (20).

When irradiated with light, cellulose dyed with direct, vat, sulfur, or naphthol dyestuffs gives narrow singlet ESR spectra, indicating the formation of free radicals on the cellulose molecule. The line widths of the singlet spectra were 6–11 gauss and similar to those observed for charred samples. There is no obvious correlation between the absorption of light by dyed cellulose, as measured by the formation of free radicals, and loss of breaking strength of weathered, dyed cellulosic fibers. To account for the strong ESR signals it can be concluded that some of the energy absorbed is localized to form free radicals on the cellulose molecule without depolymerization and loss of fiber strength. Photolysis of dyes on cellulose, as observed by the fading of the dyes on irradiation, generates only singlet ESR spectra, indicating that localization of energy is on the cellulose molecule (21). In sensitized photolysis of model compounds, such as alcohols, more complex ESR spectra were obtained (22).

(c) REACTIONS INITIATED BY HIGH-ENERGY RADIATION

The effects of high-energy radiation on the formation of free radicals in carbohydrates of low molecular weight have been investigated by ESR spectroscopy (23–34). The effects of molecular aggregation (27), substitution of groups on the carbohydrate molecular (particularly aromatic groups) (28), and water on the concentration of free radicals detected (26, 30) were of value in interpreting the effects of high-energy radiation on cellulose. After energy absorption, energy transfer and localization occur in the irradiated carbohydrates, causing chemical changes and the formation of free radicals (24, 27, 32). As the temperatures of the irradiated samples are raised to their melting points, the free radicals disappear (24, 30).

Similarly it has been shown that the type of the crystal lattice, oxygen, water, and temperature are factors in determining the concentration of free radicals detected in irradiated cellulose by ESR spectroscopy (35–48). The substitution of aromatic groups on the cellulose molecule influences energy transfer and localization.

Free radicals in irradiated cellulose and starch are very unreactive to oxygen and sulfur dioxide, probably because of intramolecular and intermolecular hydrogen bonds (49, 50). ESR spectroscopy has shown that the free radicals formed in irradiated cellulose acetate react with sulfur dioxide (49); however, the free radicals in irradiated ethylcellulose apparently react to form crosslinks and do not react with sulfur dioxide (50). ESR spectroscopy has also detected stable free radicals in irradiated starch (51), dextran (52), and flour (53). The presence of moisture in the carbohydrates during irradiation decreases the concentration of free radicals detected, as in the case of cellulose. See Section H of this Chapter XVII for additional discussion concerning free radicals in cellulose.

(d) REACTIONS INITIATED BY OXIDATION–REDUCTION SYSTEMS

The presence of free-radical intermediates formed on the interaction of the ceric ion with cellulose has been demonstrated by ESR spectroscopy (54–56). A typical ESR spectrum of fibrous cellulose [degree of polymerization (DP) about 4400], saturated with ceric ammonium nitrate (0.05 *M*) in nitric acid (0.05 *M*), is shown in Figure 4*A*. Ceric ions probably form a chelate through the $C_{(2)}$ and $C_{(3)}$ hydroxyls; a transfer of a single electron from the cellulose molecule to Ce^{4+} would then result in reduction to Ce^{3+}. Formation of Ce^{3+} and cleavage of the $C_{(2)}$–$C_{(3)}$ bond would result in the formation of a free radical on one of the carbon atoms. The rate of decay of these free radicals at 25°C, as measured by ESR, is shown in Figure 5. The free radical could be terminated by reaction with Ce^{4+} (55).

The rates of formation and decay of free radicals on hydrolyzed cellulose

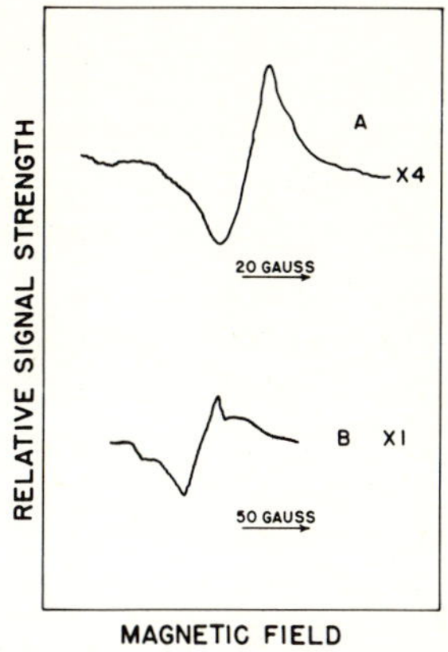

Fig. 4. ESR spectra generated on reaction of the ceric ion with cellulose: (*A*) saturated with ceric ammonium nitrate in nitric acid for 20 sec at 25°C, then frozen at − 100°C to record spectrum; (*B*) saturated as in (*A*), followed by reaction with acrylonitrile for 6 min at 25°C, frozen at − 100°C to record spectrum.

(microcrystalline, DP of about 220), measured by ESR spectroscopy while Ce^{4+} solution is continuously flowing through the sample at 25°C, is shown in Figure 6. The half-lives of the free radicals in the various atmospheres are about as follows: nitrogen, 6 sec; argon 13 sec; air, 88 sec; and oxygen, 74 sec (55).

In oxidation-reduction systems, such as, Fe^{2+}-H_2O_2, Fe^{3+}-H_2O_2, Ce^{4+}-H_2O_2, and Ti^{3+}-H_2O_2, $\cdot OH$ and $HO_2\cdot$ radicals are generated and react readily with carbohydrates, particularly cellulose (57-59). Two useful applications of these reactions are (a) oxidative degradation of carbohydrate molecules to shorten the molecular chain length and (b) initiation of graft copolymerization of vinyl monomers with cellulose (see Chapter XVII-G). In both applications the initial reactions of the $\cdot OH$ and $HO_2\cdot$ radicals with carbohydrates abstract hydrogen atoms and leave highly reactive free-radical sites on the

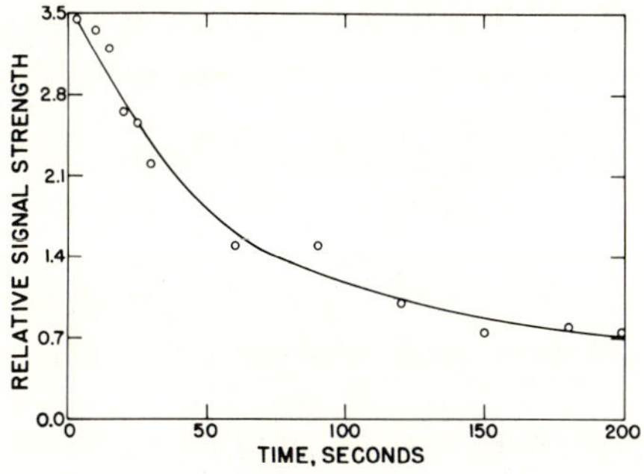

Fig. 5. Rate of decay of free radicals formed in cellulose by reaction with the ceric ion at 25°C. (Time measured from saturation at 25°C to freezing at −100°C to record spectrum).

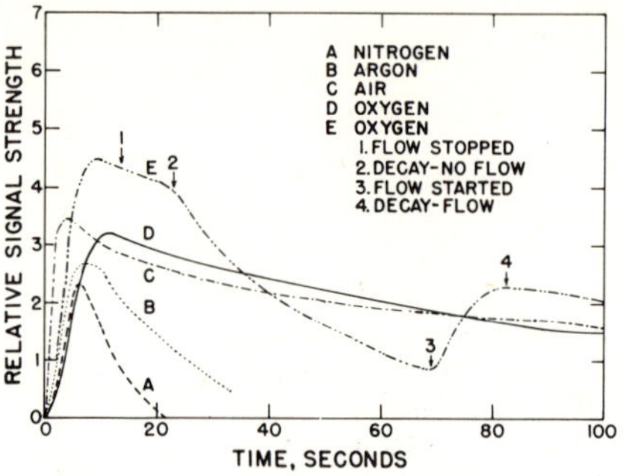

Fig. 6. Rates of formation and decay of free radicals on the reaction of the ceric ion with microcrystalline cellulose (DP 220) under various atmospheres under flow conditions at 25°C.

carbohydrate molecule (57, 59). By using flow systems, the reactions of ·OH and $HO_2\cdot$ radicals with carbohydrates have been found to generate ESR spectra. When fibrous cellulose is placed in an aqueous sample cell in the resonant cavity of an ESR spectrometer and Fe^{2+} and H_2O_2 solutions are mixed while flowing through the cellulose at 25°C, no detectable ESR spectrum is generated. On the other hand, when the fibrous cellulose is immersed in a solution of ferrous ammonium sulfate to wet the sample thoroughly and then dried in a flowing stream of nitrogen, and hydrogen peroxide solution is drawn onto the sample, which is then immediately immersed in liquid nitrogen, an ESR spectrum (measured at −110°C) is generated, as shown in Figure 7. The intensity of the signal is very weak. The three most intense lines appear to form a triplet spectrum (about 1 : 2 : 1, hyperfine splitting about 31 gauss). This type of ESR spectrum would be generated if the ·OH radical abstracted a hydrogen atom from $C_{(5)}$ or from the hydroxyl group on $C_{(6)}$ (57).

(e) REACTIONS INITIATED BY MECHANICAL ACTION

Butyagin (60) has reported the detection (by ESR spectroscopy) of active intermediate states during the mechanical degradation of carboxymethylcellulose. Kleinert and Morton (61) used ESR spectroscopy to detect free radicals in mechanically damaged wood. They found that the concentration of free radicals increases, as the particle size decreases. Urbanski (62) recorded the ESR spectrum of cellulose ground in an agate mortar. The free radicals formed were easily quenched by water. A high concentration of free radicals was obtained in alkaline wood pulps; oxygen or sulfur were found to react with the free radicals to quench them (62).

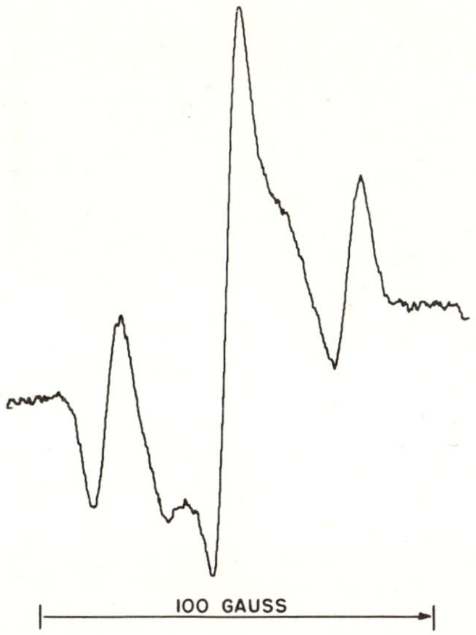

Fig. 7. ESR spectrum generated on the reaction of the $Fe^{2+}-H_2O_2$ solution with cellulose; recorded at $-110°C$.

2. Chain Copolymerization Reactions

The kinetics and mechanism of chain copolymerization reactions of vinyl monomers have also been studied with ESR spectroscopy. Fibrous cellulose, reacted with Ce^{4+} ions and then allowed to react with acrylonitrile generated the ESR spectrum shown in Figure 4*B*. On initiation of the copolymerization reaction there was found to be a change in the number of lines and a broadening of the spectrum. Apparently a second radical on the end of the growing polymer chain was being recorded. During the chain copolymerization reaction the concentration of free radicals increased initially and then decreased after about 15 sec of reaction time. The initial free radical formed on the cellulose molecule was maintained on the end of the growing polymer chain; then apparently additional ceric ions formed chelates with the cellulose molecule, leading to the formation of additional free radicals. However, as indicated, the termination reaction for the free radicals became the predominant reaction after about 150 sec (54–56).

Fibrous cellulose, reacted with ferrous ions in solution, then dried in a stream of nitrogen, and contacted with a solution of hydrogen peroxide and acrylonitrile generated the ESR spectrum shown in Figure 8. The intensity of this spectrum is greater than that of the spectrum generated in Fe^{2+} - cellulose by hydrogen peroxide solution alone (see Fig.7). The triplet spectrum disappears

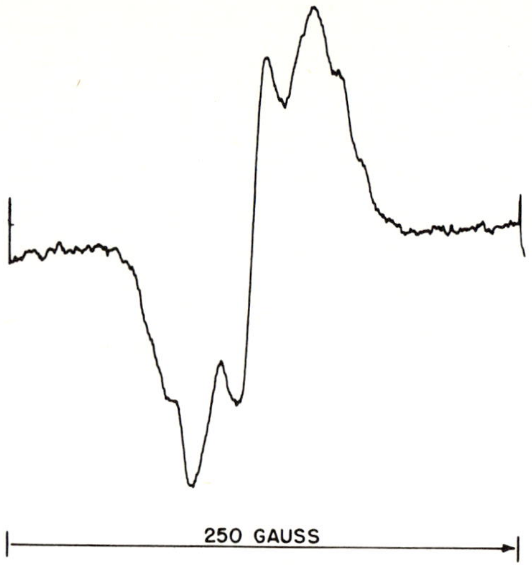

Fig. 8. ESR spectrum generated on the reaction of cellulose with Fe^{2+}–H_2O_2 plus acrylonitrile solution; recorded at –110°C.

during the reaction, and an ESR spectrum, which appears to be a combination of two singlet spectra, is recorded (54,57).

ESR studies of chain copolymerization reactions of vinyl monomers with irradiated fibrous cellulose (63–67), cellophane (66), and wood (68) have been reported. Free radicals on irradiated cellulose, which were not readily accessible to reaction with water, were shown by ESR spectroscopy to be the site of the copolymerization reactions (63, 64) (see also Sections G and H of this Chapter XVII).

3. Cellulose–Copper Complexes

Cupriethylenediamine dihydroxide (cuen) and cupriammonium dihydroxide (cuam) interact with cellulose in aqueous solutions to yield soluble complexed products. The determination of the intrinsic viscosity of solution of these complexes with cellulose is commonly used to estimate the molecular weight of cellulose. ESR spectroscopy has been used to observe directly the nature and extent of definition of the chemical bonds involved. Reaction of either of the complex ions with cellulose causes anisotropic splitting of their ESR spectra.

In cuen itself at 25°C, due to the interaction with the copper nuclear spin of 3/2, four hyperfine components can be resolved with $g_{center} = 2.1000$. When the ESR spectrum is recorded at –100°C, the four hyperfine lines are reduced to a single line (69). Cuen (0.0005 *M*) adsorbed on cotton cellulose (crystalline lattice type I) generates the ESR spectrum shown in Figure 9; it consists of possibly two sets of components: one set centered at $H_{\perp}$ and the other at $H_{\parallel}$. From the spectrum $g_{\parallel}$ and $A_{\parallel}$ were calculated to be 2.1923 and 177 gauss, respectively; the value of $g_{\perp}$ was estimated to be 2.0522. Cellobiose (0.1 *M*) dissolved in cuen (0.0005 *M*) generates an ESR spectrum at −100°C that is similar to that generated by cellulose–cuen and has a *g*-value of 2.1781 and an *A* value of 182 gauss (69).

When fibrous cellulose I–cuen is aligned parallel to the external magnetic field ($B_{\parallel}$), there is a maximum contribution of the components centered at $H_{\parallel}$. If the external magnetic field is rotated until the fibers are perpendicular across the field ($B_{\perp}$), then a minimum contribution of these components is found. These effects are dependent on the cellulose lattice type and also the origin of the fibrous cellulose. When cuen–ramie is used, the changes in the intensity of the spectrum, as the magnetic field is rotated, are greater and sharper than those when cuen–cotton cellulose (crystalline lattice type II) is used. There is little difference between the changes in the spectra of cotton cellulose I–cuen and cotton cellulose II–cuen as the magnetic field is rotated. Similar results are obtained with cuam. These changes in spectra with orientation of the fibrous structure complexes in the external magnetic field indicate the relative degree of molecular orientation (69). A thorough discussion of cellulose–metal complexes can be found in Section A of Chapter XIV.

When purified cellulose is treated with solutions of sodium hydroxide, an ESR spectrum is generated that most likely indicates the presence of trace

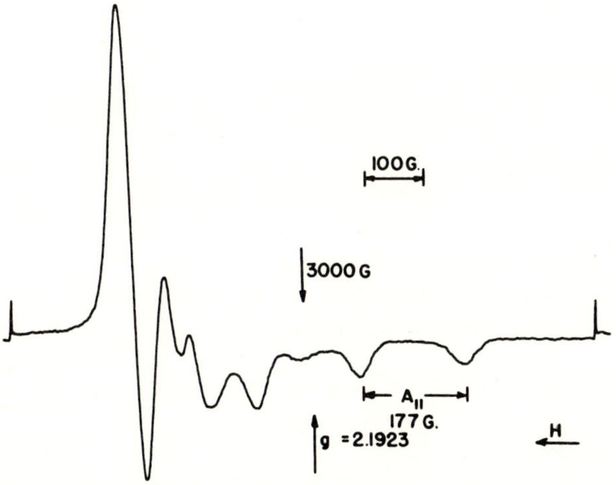

Fig. 9. ESR spectrum of cuen (0.0005 *M*) adsorbed on cellulose at 25°C.

quantities of copper (70). Natural lignin (71) and humic acid (72) have been reported to generate an ESR spectrum. Adsorbed or complexed paramagnetic metal on cellulose has been used to determine carboxyl groups (73). Similarly starch–iodine complexes have been studied by ESR (74). Autoxidation of alkali cellulose did not generate an ESR spectrum (75).

References

1. W. A. Pryor, *Chem. Eng. News,* **46,** No. 3, 70 (1968).
2. G. K. Fraenkel, in A. Weissberger, Ed., *Physical Methods of Organic Chemistry,* Vol. 1, Technique of Organic Chemistry, 3rd ed., Part IV, Interscience, New York, 1960, p. 2801.
3. S. A. Al'tshuler and B. M. Kozyrev, *Electron Paramagnetic Resonance,* C. Poole, Jr., transl. Ed., Academic Press, New York, 1964.
4. M. C. R. Symons, in V. Gold, Ed., *Advances in Physical Organic Chemistry,* Vol. I, Academic Press, New York, 1963, p. 283.
5. J. C. Arthur, Jr., and O. Hinjosa, *Textile Res. J.,* **36**, 389 (1966).
6. R. M. Elofson and K. F. Schulz, *Am. Chem. Soc., Div. Fuel Chem. Preprints,* Part 2, **11,** 513 (1967).
7. J. J. Krebs, *J. Chem. Phys.,* **34,** 326 (1961).
8. J. J. Krebs and J. K. Thomson, *J. Chem. Phys.,* **36,** 2509 (1962).
9. R. H. Hoskins and R. C. Pastor, *J. Appl. Phys.,* **31,** 1506 (1960).
10. R. C. Pastor and R. H. Hoskins, *J. Chem. Phys.,* **32,** 264 (1960).
11. R. C. Pastor, J. A. Weil, T. H. Brown, and J. Turkevich, *Phys. Rev.,* **102,** 918 (1956).
12. L. S. Singer and W. J. Spry, *Bull. Am. Phys. Soc.,* **1,** 214 (1956).
13. J. Uebersfeld and E. Erb, *Compt. Rend.,* **243**, 2043 (1956); through *Chem. Abstr.,* **51,** 6331*g* (1957).
14. T. Ichikawa and K. Hirota, *Nippon Kagaku Zasshi,* **83,** 652 (1962); through *Chem. Abstr.,* **58,** 9773*g* (1963).
15. T. N. Kleinert, *Holzforschung,* **18,** 24 (1964); through *Chem. Abstr.,* **61,** 10859a (1964).
16. T. N. Kleinert, *Monatsh. Chem.,* **96,** 1925 (1965); through *Chem. Abstr.,* **64,** 12940*d* (1964).
17. T. N. Kleinert, *Tappi,* **50,** 120 (1967).
18. G. O. Phillips and J. C. Arthur, Jr., *Textile Res. J.,* **34,** 497 (1964).
19. G. O. Phillips and J. C. Arthur, Jr., *Textile Res. J.,* **34,** 572 (1964).
20. G. O. Phillips, O. Hinojosa, J. C. Arthur, Jr., and T. Mares, *Textile Res. J.,* **36,** 822 (1966).
21. P. J. Baugh, O. Hinojosa, T. Mares, M. J. Hoffman, and J. C. Arthur, Jr., *Textile Res. J.,* **37,** 942 (1967).
22. P. J. Baugh, G. O. Phillips, and J. C. Arthur, Jr., *J. Phys. Chem.,* **70,** 3061 (1966).
23. A. J. Bailey, S. A. Barker, J. A. Brimacombe, D. Pooley, and D. H. Spence, *Nature,* **190,** 259 (1961).
24. M. A. Collins, *Nature,* **193,** 1061 (1962).

25. S. Dilli and J. L. Garnett, *Nature,* **198,** 984 (1963).
26. G. Lofroth, L. Ehrenberg, and A. Ehrenberg, *Preprints Papers Intern. Symp. Free Radicals, 5th Uppsala, 1961,* 38–1; through *Chem. Abstr.,* **60,** 145a (1964).
27. G. O. Phillips and P. J. Baugh, *J. Chem. Soc.,* A, 370 (1966).
28. I. M. Sarkar, J. C. Arthur, Jr., and O. Hinojosa, *Carbohydrate Res.,* **6,** 633 (1968).
29. H. Shields and P. Hamrick, *J. Chem. Phys.,* **37,** 202 (1962).
30. F. K. Truby and W. H. Storey, Jr., *J. Chem. Phys.,* **31,** 857 (1959).
31. H. Ueda, *J. Phys. Chem.,* **67,** 966 (1963).
32. H. Ueda, *J. Phys. Chem.,* **67,** 2185 (1963).
33. H. Ueda, Z. Kuri, and S. Shida, *J. Chem. Phys.,* **35,** 2145 (1961).
34. D. Williams, J. E. Geusic, M. L. Wolfrom, and L. J. McCabe, *Proc. Natl. Acad. Sci. U.S.,* **44,** 1128 (1958).
35. R. J. Abraham and D. H. Whiffen, *Trans. Faraday Soc.,* **54,** 1291 (1958).
36. J. C. Arthur, Jr., T. Mares, and O. Hinojosa, *Textile Res. J.,* **36,** 630 (1966).
37. S. Dilli, I. T. Ernst, and J. L. Garnett, *Australian J. Chem.,* **20,** 911 (1967).
38. I. N. Ermolenko, A. K. Potapovich, and V. N. Makatun, *Vesti Akad. Navuk Belarusk. SSR, Ser. Fiz.-Tekhn. Nevuk,* 65, (1963); through *Chem. Abstr.,* **59,** 4710*h* (1963).
39. R. E. Florin and L. A. Wall, *J. Polymer Sci., A 1, 1163 (1963).*
40. R. E. Glegg, *J. Polymer Sci.,* **26,** 289 (1957).
41. R. E. Kesting and V. Stannett, *Makromol. Chem.,* **55,** 1 (1962).
42. V. N. Makatun, A. K. Potapovich, and I. N. Ermolenko, *Vysokomol. Soedin.,* **5,** 467 (1963); through *Chem. Abstr.,* **59,** 1832*a* (1963).
43. G. S. Park and J. C. Ward, *Nature,* **202,** 389 (1964).
44. H. Shields, W. B. Ard, Jr., and W. Gordy, *Bull. Am. Phys. Soc.,* **1,** 200 (1956).
45. J. Uebersfeld, *Ann. Phys.,* **1,** 395 (1956); through reference 37.
46. J. Uebersfeld, *Compt. Rend.,* **238,** 1397 (1954); through reference 37.
47. H. Ueda, Z. Kuri, and S. Shida, *Nippon Kagaku Zasshi,* **82,** 8 (1961); through *Chem. Abstr.,* **55,** 17232*h* (1961).
48. M. P. Votinov, M. A. Khenokh, E. A. Kuzicheva, V. F. Evodokimov, and V. V. Antuf'ev, *Tr. 2-go(Vtorogo) Vses. Soveshch. po Radiats. Khim., Akad. Nauk SSSR, Otd. Khim. Nauk, Moscow, 1960,* p. 335 (published in 1962); through *Chem. Abstr.,* **58,** 5172*g* (1963).
49. Z. Kuri and H. Ueda, *J. Polymer Sci.,* **50,** 349 (1961).
50. B. R. Loy, *Soc. Plastics Engrs. Trans.,* **2,** 157 (1962); through *Chem. Abstr.,* **57,** 4218*a* (1962).
51. K. Adamic, P. Cevc, and K. A. Korotchenko, *Staerke,* **19,** 336 (1967).
52. K. A. Granath and P. O. Kinell, *Acta Chem. Scand.,* **15,** 141 (1961).
53. C. C. Lee, *Cereal Chem.,* **39,** 147 (1962).
54. J. C. Arthur, Jr., *Cellulose Graft Copolymers, Advances in Chemistry Series No. 91,* Am. Chem. Soc., Washington, 1969, p. 574.
55. J. C. Arthur, Jr., P. J. Baugh, and O. Hinojosa, *J. Appl. Polymer Sci.,* **10,** 1591 (1966).
56. J. C. Arthur, Jr., P. J. Baugh, and O. Hinojosa, *Preprints of Scientific Papers, IUPAC, Internatl. Symposium on Macromol. Chem.,* Tokyo and Kyoto, 1966, pp. IV–28.
57. J. C. Arthur, Jr., O. Hinojosa, and M. S. Bains, *J. Appl. Polymer Sci.,* **12,** 1411 (1968).

58. M. S. Bains, J. C. Arthur, Jr., and O. Hinojosa, *J. Phys. Chem.*, **72**, 2250 (1968).

59. P. J. Baugh, O. Hinojosa, and J. C. Arthur, Jr., *J. Phys. Chem.*, **71**, 1135 (1967).

60. P. Yu. Butyagin, *Dokl. Akad. Nauk SSSR.*, **140**, 145 (1961); through *Chem. Abstr.*, **56**, 10367*c* (1962).

61. T. N. Kleinert and J. R. Morton, *Nature*, **196**, 334 (1962).

62. T. Urbanski, *Nature*, **216**, 577 (1967).

63. O. R. Bernard, D. Gagnaire, and P. Servoz-Gavin, *J. Chim. Phys.*, **60**, 1348 (1963); through *Chem. Abstr.*, **60**, 12220*h* (1964).

64. O. R. Bernard and D. Gagnaire, *Proc. Colloq. AMPERE (Atomes Mol. Etudes Radio Elec.)*, **12**, 269 (1963) (published in 1964); through *Chem. Abstr.*, **62**, 9340*g* (1965).

65. P. J. Baugh, O. Hinojosa, and J. C. Arthur, Jr., *J. Appl. Polymer Sci.*, **11**, 1139 (1967).

66. H. A. Kraessig, *Tappi*, **46**, 654 (1963).

67. J. Kroh and M. Walicki, *Bull. Acad. Polon. Sci., Ser. Sci. Chim.*, **13**, 635 (1965).

68. K. V. Ramalingam, G. N. Werezak, and J. W. Hodgins, *J. Polymer Sci.*, C, No. 2, 153 (1963).

69. P. J. Baugh, O. Hinojosa, J. C. Arthur, Jr., and T. Mares, *J. Appl. Polymer Sci.*, **12**, 249 (1968).

70. M. S. Bains, O. Hinojosa, and J. C. Arthur, Jr., *Carbohydrate Res.*, **6**, 233 (1968).

71. M. I. Chudakov and A. P. Samsonova, *Sb. Tr. Gos. Nauch.-Issled. Inst. Gidroliz. Sul'fitne-Spirit. Prom.*, **15**, 285 (1966); through *Chem. Abstr.*, **67**, 91821*R* (1967).

72. C. Steelink, in N. M. Bikales, Ed., *Encyclopedia of Polymer Science and Technology*, Vol. 7, Interscience, New York, 1967, pp. 530–539.

73. I. N. Ermolenko, A. K. Potapovich, and R. N. Sviridova, USSR Patent 174,423 (1965); through *Chem. Abstr.*, **64**, P911*d* (1966).

74. R. Bersohn and I. Isenberg, *J. Chem. Phys.*, **35**, 1640 (1961).

75. J. A. Mattor, *Tappi*, **46**, 586 (1963).

Chapter XVIII

DEGRADATION OF CELLULOSE AND ITS DERIVATIVES

A. ACID HYDROLYSIS AND ALCOHOLYSIS

A. Sharples

Hydrolytic degradation of cellulose has been described in Section C–2 of Chapter III (pp. 130–140) of Part I.

Chemically, the hydrolysis of cellulose is similar to that of the simpler and more easily studied disaccharides (1). It is an acid-catalyzed reaction involving the rapid formation of an intermediate complex between the glycosidic oxygen and a proton; this is followed by the slow, rate-determining scission of the glycosidic bond adjacent to $C_{(1)}$(2). In the presence of excess water the reaction is first order, although in many cases of cellulose hydrolysis only small extents of reaction are studied, with the result that the data can be fitted equally well to a zero-order plot.

The method most commonly used for following the hydrolytic degradation of cellulose involves the determination of molecular weight, or degree of polymerization, P, as a function of time. Because of its sensitivity this is potentially a very accurate method. Thus for a chain of length one thousand units, the breaking of one bond corresponds to an extent of reaction of only 0.1%, and yet this leads to a decrease in the measured value of P from 1000 to 500. As the reaction is first order

$$\frac{dn}{dt} = -k_1 n$$

n is the total number of bonds available in all the cellulose molecules in the system, and this is related to P by

$$n = A - \frac{A}{P}$$

where A is the total number of monomeric units in the system (a constant quantity) and P, the degree of polymerization (DP), is the number-average value. This equation represents the fact that each chain has one less bond in it than there are monomeric units. Therefore

$$\frac{d\,[1-(1/P)]}{dt} = -k_1\left(1-\frac{1}{P}\right)$$

and

$$\ln\left(1-\frac{1}{P_0}\right) - \ln\left(1-\frac{1}{P_t}\right) = k_1 t$$

where P_t is the degree of polymerization after time t. This is the exact first-order relation, but if P_0 and P_t are large (i.e., if the extent of reaction is small), then

$$\frac{1}{P_t} - \frac{1}{P_0} \simeq k_1 t$$

This approximation is identical with the expression obtained by assuming a zero-order reaction for the starting point, that is,

$$\frac{dn}{dt} = k_0$$

The important point is that a rectilinear plot is obtained if the *reciprocal* of P is plotted as a function of the time of hydrolysis. This apparently obvious fact has been overlooked by some workers in the past, who have used the direct plot of P as a function of time. This, in fact, takes the form of a rectangular hyperbola ($y = a/x$) and is obviously misleading.

An alternative approach to determining the rate and extent of hydrolysis in cellulose is to measure the formation of soluble fragments, either directly by analysis of the reducing power of the hydrolyzing medium or indirectly from the loss in the weight of the cellulose. This method can be informative in certain cases, but it is essential to realize that it gives only an indirect assessment of the number of bonds broken. Of the degraded fragments formed during the course of the reaction, only a fraction is in fact soluble; and the value of this fraction itself depends on the extent of degradation and on the nature of the hydrolyzing medium. An analysis of the difficulties arising from this approach has been made by Stewart and Williams (3), and although their treatment has been questioned by Meller (4), the objection still remains that the determination is an indirect one.

Interest in the hydrolysis of cellulose has extended over a considerable number of years and relates to analytical and preparative problems. The analytical interest is two-fold. First, a study of the hydrolysis kinetics can in principle be used to provide information on whether there are present in the cellulose chain anomalous sugars or other monomeric units that differ from the normal (1→4)-β-linked α-D-glucose. Second, the course of hydrolysis in dilute acids is always complicated by the fact that cellulose is insoluble in the hydrolyzing medium. The structure is, however, swollen, and the regions penetrated by water (to a first approximation, the "amorphous" regions) are in principle available for degradation by dissolved acid. The method can thus be used to provide information on the physical structure of the various types of cellulose.

On the preparative side, the possibility has long been considered that the reaction might be used to produce reducing sugars from wood pulp, suitable for subsequent conversion to alcohol.

1. Effect of Chemical Structure on Hydrolysis

If the physical structure of cellulose can be destroyed by dissolving it in a suitable hydrolyzing medium, any anomalies in the subsequent kinetics should be related solely to abnormalities in the chemical structure of the sample; that is, deviations from the normal (1→4)-β-glucosidic chain. Freudenberg and co-workers (5) in 1930 established that, for cellulose below a chain length of about 100, the rate constant of hydrolysis in dissolving strength sulfuric acid is consistent with the figures for cellobiose, cellotriose, and cellotetraose, thereby demonstrating that the majority of the bonds are identical. Subsequent work has been concerned with determining whether other sources of cellulose may show anomalous behavior particularly in the earlier stages of degradation ($P > 100$).

Only units that gave rise to abnormally *high* rate constants will be detected by this method, and even then it must be appreciated that an unexpectedly high rate of chain scission measured in practice from the average decrease in DP will depend not only on the rate constant but also on the fractional number of the foreign bonds. Thus chains containing one bond in a hundred that hydrolyze 10 times as fast as the normal rate will give rise to an overall rate that is only 10% greater than that for pure cellulose.

In practice there has been observed in the past a consistent difference between the hydrolytic behavior of cellulose from wood and from cotton, under homogeneous degradation conditions (Fig. 1). Immergut, Rånby, and Mark (6), for example, showed in 1955 that the overall rate of hydrolysis in phosphoric acid solution is roughly twice as great as it is for cotton, and this observation has subsequently been confirmed (7–10). The possibility that foreign units are present in the chain structure of some types of cellulose does thus seem to be real. Early suggestions concerning the chemical nature of these units were based on hemiacetal structures (11) and also on the possible inclusion of sugars, such as xylose (12). The former has subsequently been discounted (13), but the well-established presence of xylan and mannans in wood pulp does raise the

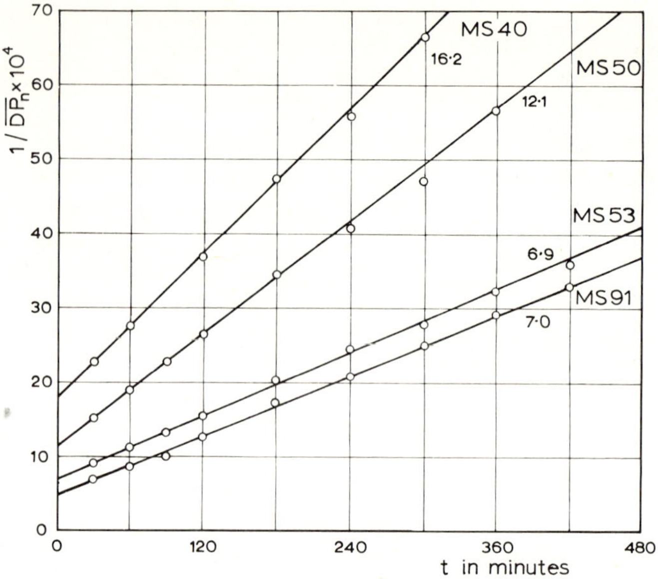

Fig. 1. Homogeneous hydrolysis in phosphoric acid, showing enhanced rate for wood-pulp samples (MS40 and MS50) compared with that for cotton samples (MS53 and MS91) (6).

possibility that some of the component sugars may also occur in chemical combination with cellulose. Early data on the hydrolysis of the xylose–xylose link (14) indicated that it was hydrolyzed more readily than the normal glucosidic link in cellulose, to an extent of about thousandfold. It was subsequently realized, however, that this large difference was due mainly to the physical structural differences between the xylan and cellulose samples studied, and recent work has shown that the increase in rate constant is in fact only about five-fold, both for the xylose link (15) and for mannose (16). Thus, although these and other foreign sugar residues may be present to a small extent in the cellulose chain, it is unlikely that they would cause a measureable increase in the hydrolysis rate constant. An explanation for the observed increase in the case of wood pulps must thus be sought along other lines.

Much attention has been given in recent years to the possibility that oxidized groups may be present along the chain, capable of modifying by an inductive effect (17–19) the rate-determining step in hydrolysis, which involves electron transfer to the glycosidic oxygen:

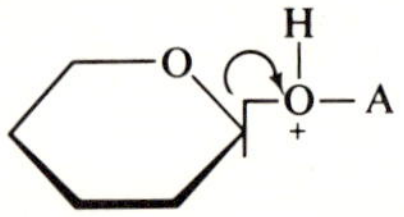

An aldehyde or carboxyl group near to the glycosidic oxygen, and in the glycosyl moiety, is predicted to reduce the rate; a similarly placed group in the aglycon moiety A will increase it. Such oxidized groupings are more likely to be present in wood cellulose than in cotton, as a result of the pulping processes to which the former is subjected. Marchessault and Rånby (8) have shown that there is a correlation between the initial high rate of hydrolysis in wood pulps and the content of oxidized groupings, and also that reduction of the oxidized groups reduces the rate to that for cotton (7, 9). The various possibilities include carboxyl and aldehyde groups at the $C_{(6)}$ position in the anhydroglucose unit and aldehyde groups in the $C_{(2)}$ and $C_{(3)}$ positions. A considerable amount of effort has been made to determine which of these, and of related modified groupings, is the one responsible for the observed increase in rate.

It would appear to be well established now that a carboxyl group at the $C_{(6)}$ position is excluded as a possibility (10, 20–24), although there are two reservations to this general finding. Rusznak (20) found an increase in rate when the carboxyl content at $C_{(6)}$ exceeded 7.5%; and Nakano and Rånby (21), using model compounds, found an increased rate, but only when concentrated phosphoric acid was used as the hydrolyzing medium.

In contrast to this negative result, there is considerable evidence to show that oxidation and cleavage at the $C_{(2)}$ and $C_{(3)}$ positions to either the dialdehyde or the dicarboxyl derivative does lead to a considerable increase in rate (10, 19, 25). The increase is retained even when the aldehyde groups are reduced to give the corresponding dialcohol (10, 26). In addition, Lindberg (24), who used model compounds, has obtained evidence to show that an aldehyde group at $C_{(6)}$ also increases the normal rate by as much as 70 times. As this latter effect can also be removed by reduction [unlike the effect with aldehydes at $C_{(2)}$ and $C_{(3)}$], the oxidation product involving an aldehyde at $C_{(6)}$ appears to be the most satisfactory explanation at present for the increased susceptibility of wood celluloses to acid degradation. The preferred tendency of this effect to be observed in wood rather than in cotton is consistent with oxidation during pulping.

The alternative approach to using hydrolysis diagnostically in the determination of chemical structure involves the qualitative analysis of the partial or complete degradation products from cellulose. Homogeneous degradation can be effected in dissolving-strength acid, and the resulting sugars can be estimated by paper chromatography, after dilution, neutralization, and deionization of the hydrolysis mixture. Various sugars, particularly xylose and mannose, have been detected by various workers who used this approach (27–29). The possible objection here is that the foreign sugars may arise from occluded hemicelluloses, and it is essential with this method that careful attempts be made to isolate the cellulosic component, either by extraction of impurities or by fractional precipitation from solution (30–33). Even with fractional precipitation of a suitable derivative from solution there is the danger that coprecipitation of mannan and xylan may occur, and hence the detection of a foreign sugar in the hydrolyzate from a cellulose fraction does not constitute conclusive proof that the unit was originally linked to glucose in the cellulose chain. The isolation of the products of partial hydrolysis and the detection of disaccharides containing glucose-xylose or glucose-mannose linkages are somewhat more convincing in this respect, and observations of the latter have been made in the degradation products from slash-pine α-cellulose (34). On balance, however, there is no conclusive evidence to show that the true cellulosic microfibrils from cotton or wood contain anything other than (1→4)-β-linked glucose, in the natural state.

2. Effect of Physical Structure on Hydrolysis

(a) EARLY STAGES

The swelling of cellulose by water, unlike the swelling of most semicrystalline polymers by liquids, is exceptional in that the penetrated regions are well defined and relatively little affected by such external factors as temperature. Results obtained from deuteration studies [see, for example, Jeffries (35)] have established that these accessible regions correspond, at least to a first approximation, to the amorphous component of the polymer. Aqueous reactions in cellulose are thus dependent, both in rate and extent, on the amount of amorphous material present, and hydrolysis by dilute acids is no exception.

It was observed many years ago that the *extent* of hydrolysis is limited and that the weight loss associated with the formation of soluble oligosaccharides, although initially rapid, is subsequently very much slower (36). This readily degraded fraction can be equated to the amorphous content (37) and provides an experimental means of determining the latter. The approximate correlation between this and other methods for determining amorphous content is now well established, and this approach has been used to make estimations with a range of natural and regenerated celluloses (38-45).

Several objections have been raised to the use of this method if it is to be considered as a precise quantitative method of characterizing crystalline structure. One that is certainly valid is that the process of hydrolysis may itself increase crystallinity, by allowing broken chains to reorient to more highly ordered structures. This point was established by Hermans and Weidinger (46) in 1949 and has since been confirmed (45, 47-49). It is now known that this recrystallization takes place at a very early stage of hydrolysis (50), so that estimates of amorphous content based on determinations of the initial rapid weight loss in cellulose are likely to be lower than the true values.

The way to circumvent this difficulty is to determine the rate of *bond scission* under standard conditions, from measurements of the decrease in DP during hydrolysis, as this can be carried out at a very early stage and without disturbing the structure significantly. This rate of hydrolysis is less than that for the completely accessible (1→4)-β-glucosidic link, by a factor α, which represents the fraction of bonds available for hydrolysis in a practical cellulose sample (51). The experimental approach involves the determination of DP as a function of time of hydrolysis, t, when

$$\frac{1}{P_t} - \frac{1}{P_0} = k\alpha t$$

The rate constant of hydrolysis for the completely accessible bonds, k, can be determined from experiments on soluble oligosaccharides. Amorphous contents, α, determined by this method (51) are higher than those from the weight-loss approach (37, 52) owing to the elimination of recrystallization.

A further objection to the hydrolytic method that has been raised in the past is that although it places cellulose samples of varying crystallinity in the same order as do other techniques, there is disagreement on the quantitative values. Thus the amorphous content determined hydrolytically for cotton is in the region of 10%, whereas the figure determined by the x-ray method is nearer 30%. This overlooks the fact that even in a simple so-called two-phase picture of polymer structure there are in fact three distinct regions: the amorphous regions the crystalline regions, and the surface regions of the crystallites. For the small crystallite sizes involved in cellulose the surface regions may constitute as much as 30% of the total, and the amorphous component determined by x-rays includes a contribution from this fraction. Different methods for determining amorphous content are thus likely to give varying results, simply because the crystallite surfaces are included to an arbitrary extent. See also Section G in Chapter XIII.

Another aspect of this problem is that when the course of hydrolysis is followed in a sample of cellulose with a fairly high degree of crystallinity, the crystalline fraction as measured by x-rays does not appear to increase appreciably, even though amorphous material is being removed (52–54). McKeown and Lyness (55) in fact claim that there is no increase at all, within experimental error (± 4%). The explanation is probably that the amorphous content, which in any case is initially small, is reduced even further during hydrolysis, by recrystallization. The fraction of amorphous material that is removed during the early stages of hydrolysis from, for example, cotton cellulose, is thus only about 5%. The remaining crystalline material, however, still contains an apparent amorphous content of about 30% by the x-ray method, owing to the contribution from the disordered surface component. The overall change in crystallinity is thus barely detectable by the x-ray method.

(b) LATER STAGES

After the readily degradable amorphous component has been removed by hydrolysis, there is nevertheless a further attack on the residual crystalline regions, involving a rate of weight loss that is roughly 100 times as slow (39, 56,

57). Much can be learned from a study of this process concerning the mechanism of hydrolysis and its relation to cellulose morphology.

The rate of weight loss during this later stage is accurately semilogarithmic: Figure 2 shows some data for a range of celluloses in which the logarithm of the residual weight is plotted as a function of time (58). The extrapolated intercepts at zero time correspond, of course, to the amorphous content determined by the weight-loss method. The hydrocelluloses undergoing degradation take the form of rod-shaped particles, whose length depends on the nature of the original cellulose and is, for example, about 400 Å for the case of cotton (53); the width is about 100 Å. In view of the current controversy over the exact nature of the morphology of cellulose, it is important to recognize that such discrete particles are formed during hydrolysis and that they are likely to be related to the original crystalline regions in the undegraded material. The lengths of these particles, as determined from electron microscopy, have been correlated with the lengths of the component chain molecules, determined from the measurement of solution properties (Fig. 3) (59), and this further indicates that the cellulose chains are oriented parallel to the long axes of the crystallites. A similar orientation of the chains is known to be present in natural cellulose microfibrils (60), so that it is reasonable to deduce that the crystallites alternate with the amorphous regions along the lengths of the microfibrils.

The slow degradation of the crystallites during the later stages of hydrolysis

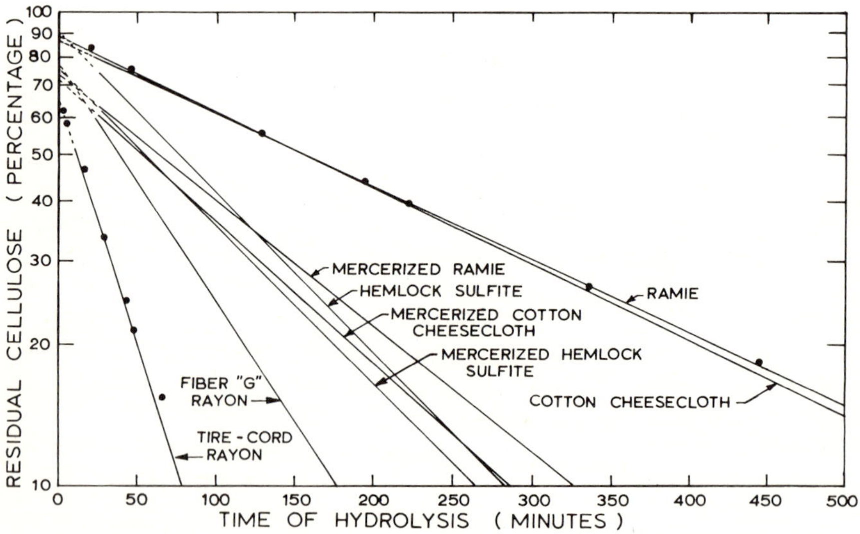

Fig. 2. Weight loss occurring during the heterogeneous hydrolysis of cellulose (58). The semilogarithmic relation to time is illustrated for a range of natural and regenerated celluloses.

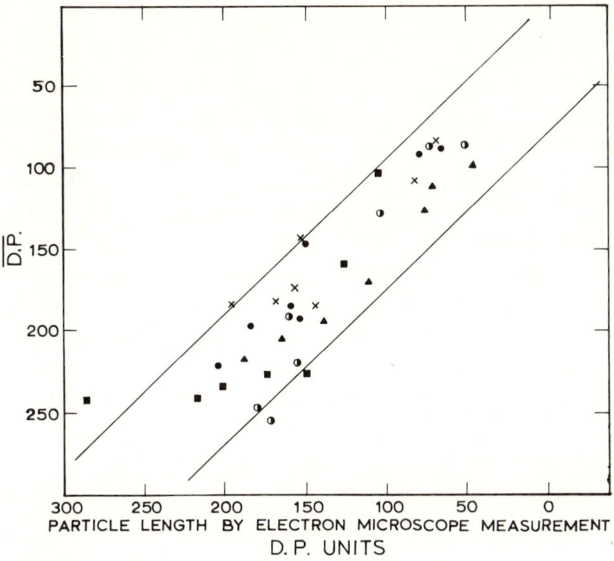

Fig. 3. Relation between particle length determined by electron microscopy and chain length (DP) of the component chain molecules; for crystallites isolated from various cellulose samples by heterogeneous acid hydrolysis (59).

(Fig. 2) is characterized by the fact that several of their properties remain constant, even during the removal of as much as 80% of the material. Thus the length of the particles determined by electron microscopy (53), the length of the component chain molecules (57, 58, 61, 62), and water sorption (58, 62) remain unchanged. The crystallinity of cotton crystallites, which has already been noted to be less than 100% owing to the disorder present at the surface, is also constant (53).

The simple explanation for the invariant properties of a given hydrocellulose during hydrolysis, and also of the logarithmic decrease in weight with time, is that the crystallite lengths are exponentially distributed and that the acid attack is on the ends of these particles (63). It is a feature of the exponential distribution that it is unchanged by displacement of the abscissa, so that all aspects of the particle-size distribution, including the average length, remain constant, even though the mass of material is decreasing. It follows from this picture that the rate of hydrolysis during the later stage, which is known to vary from one sample of cellulose to another (Fig. 2), will be determined by the number of particle ends present in the system. The expected correlation between reciprocal particle length and the rate of hydrolysis has been observed (64).

(c) OVERALL PROCESS

The overall process of hydrolytic degradation and its relation to physical structure thus involves attack initially on the glucosidic bonds in the amorphous regions. The bonds in the surfaces of the crystallites, although accessible to attack by aqueous acids, are in fact remarkably resistant. Degradation during this stage can be followed either by observing the attendant decrease in molecular weight or by measuring the weight loss associated with the formation of soluble oligosaccharides. The latter approach suffers from the disadvantage that it is indirect and can be carried out only after recrystallization has occurred, so that the results are likely to be low. Values for amorphous content determined by any hydrolytic method will also be lower than those obtained by such techniques as deuterium exchange and x-ray diffraction, which do contain contributions from the crystallite surfaces.

The subsequent attack is on the ends of the crystallites, at a much reduced rate. The exponential distribution of particle lengths necessary to account for the characteristic features of this second stage is to be expected if the amorphous regions in the component microfibrils are sited at random along the lengths of the latter (63). This picture has a precise significance in the case of natural celluloses, in which the existence of well-defined microfibrils is well established, and is probably correct also for regenerated material, where the existence of shorter, connected microfibrillar regions has been observed (65).

(d) FOLDED-CHAIN STRUCTURES

In this section so far the accumulated data from a range of studies on the hydrolysis of cellulose have been interpreted on the basis of a simple two-phase structure, with crystallites and amorphous regions constituting the essential components, as envisaged in the fringed-micelle concept developed 30 years ago. The only exception is that it has been necessary to stress the importance of the crystallite surfaces as a third component, providing an additional contribution to the amorphous content as measured by some methods, for example, x-ray diffraction.

It is of course now accepted that this picture is incomplete when applied to most melt-crystallized polymer structures, where the organization of physical structure takes place on a much larger size level and is based on the spherulite as the prime unit of growth during crystallization (66). In the case of cellulose, however, which in the regenerated form is crystallized by precipitation from solution during the spinning process, the position is probably that the number of nuclei is so large that the incipient spherulites develop only to a very limited extent, owing to rapid impingement on other growing centers. As the early stage of development of a spherulite is rodlike in nature, the approximation to the fringed-micelle-network picture is probably not only justified in this case but even to be preferred. The small size of the growth units in a structure that is still nevertheless quite crystalline is consistent with the exceptionally high transparency to be found in cellulose (e.g., in the large-diameter monofilaments) compared with, say, poly(ethylene oxide), where the growth units are much larger and give rise to appreciable scatter.

The above reasoning suggests that cellulose is different in degree from other crystallizable polymers (in possessing a higher nucleation density and hence a smaller growth-unit size), but not necessarily in kind. This difference may arise either from the basic nature of the cellulose molecule or more simply from the way in which its structure is normally laid down—that is, by crystallization from solution. The similarity to other polymers is indicated by the fact that spherulites can be formed under the appropriate conditions in cellulose derivatives (67). Recently Manley (68) has shown that cellulose triacetate can be induced to crystallize in the form of single-crystal lamellae, exhibiting characteristic chain folding, in much the same way as polyethylene and other synthetic polymers. Attempts to produce cellulose single crystals by hydrolysis of the triacetate, however, failed (69), owing to disruption of the structure and subsequent recrystallization in the aqueous hydrolysis medium. Nevertheless the demonstration of chain folding as a possibility in cellulosic materials has led recent workers to inquire whether supercrystalline ordering, perhaps involving chain folding, should be invoked to explain some of the phenomena of cellulose hydrolysis. Measurements on the accessibility determined by deuterium exchange have shown that this parameter remains unchanged during the course of degradation and have led to the suggestion that there is in fact no intercrystalline amorphous component, but that the diffuse scatter observed in the x-ray diffraction diagram is the result of lattice imperfections (70). This latter point in effect is the explanation given above in connection with the observed invariance of other hydrocellulose properties, where the surfaces of the crystallites were found to constitute the dominant contribution to the disordered, amorphous component. The specific proposal that has been made along these lines—namely, that the properties of natural (71) and regenerated (72) cellulose microfibrils can be accounted for in terms of single-crystal folded-chain lamellar structures with helical ordering—would thus seem to have no advantage over the established concept of alternating crystalline and amorphous regions (with an additional amorphous component in the crystallite surfaces). It suffers, moreover, from the strong objection that it provides no explanation for the formation of discrete crystallites on hydrolysis, whose widths are similar to those of the original microfibrils and which can only be accounted for by by the presence of an acid-degradable, *intercrystalline,* amorphous component. A similar objection has been raised on the basis of degradation observed under alkaline conditions (73).

The objections to this highly specific picture do not, of course, rule out the possibility of points of chain folding occurring generally in natural or regenerated celluloses (74). It is not unreasonable to expect that folds may be located in longitudinally sited amorphous regions in a microfibril and that they may even be the cause of such regions. There is, however, as yet no direct evidence to substantiate such a possibility. See also the detailed discussions in Section C.2 of Chapter XIII and Section B of Chapter XVI.

(e) WEAK BONDS

For many years the literature has contained references (75) (76) to the possibility that cellulose may contain abnormally acid-sensitive linkages. The *chemical* basis for such a concept, involving oxidized sites, is now well established and is discussed in subsection 1 of this section. There is also evidence, however, that such bonds may arise from *physical* abnormalities in the structure and that the rate constant may be increased by as much as 10,000 times (77), that is, very much greater than in the case of the chemical effect.

The basis for conclusions of this nature needs careful scrutiny, as the observations are necessarily made under heterogeneous conditions, where the presence of the crystalline component can complicate the interpretation. Nevertheless, it has been shown that the initial rate of hydrolysis, particularly in regenerated celluloses, is appreciably greater than that for the normal and completely accessible $(1\rightarrow4)$-β-glucosidic bonds, as determined from experiments with soluble oligosaccharides (77). This result has subsequently been confirmed (78). In native celluloses the effect, although significant, is small. If the cellulose structure is dissolved and regenerated, however, the number of these weak bonds is increased appreciably, and when account is taken of their concentration, the associated rate constant is roughly four orders of magnitude higher than that for the normal glucosidic linkages. This, together with the fact that they are formed in a variety of regenerated celluloses, regardless of the chemical nature of the regeneration process (79), has led to the conclusion that they must involve physical, rather than chemical, structural abnormalities. One obvious possibility is that abnormal strains may be set up in the solid state at a small number of localized points (79, 80) and an association with chain folds has been proposed in this respect (see also the discussion of the effect of morphology on reactivity, Chapter XVII-A).

3. Alcoholysis

The alcoholysis of cellulose in general shows similar characteristics to hydrolysis, although for the lower-molecular-weight alcohols, such as methanol and ethanol, the rate of reaction is increased. According to Valley (81), this is because the catalytic species, the hydrogen ion, is preferentially absorbed by the cellulose, thereby causing an increase in the intermediate complex. If, however, the rate is measured by following weight loss arising from the formation of alcohol–soluble fragments, a slight reduction is observed for the lower alcohols (82) either because the higher-molecular-weight oligosaccharides are not soluble or because the removal of the degradation products becomes diffusion controlled. With the higher alcohols, such as butanol, penetration of the liquid into the cellulose becomes the limiting factor (83, 84), so that the rate is no longer determined directly by the amount of amorphous material.

4. Saccharification

At various times consideration has been given to the possibility of exploiting the hydrolytic degradation of cellulose commercially, in order to provide sugar solutions suitable for fermentation to industrial alcohol. Under normal conditions the economics of the process are unfavorable, but in isolated areas, and in times of national emergency, the process may become viable.

Two main practical difficulties are encountered. First, the presence of relatively unreactive crystalline regions results in a considerable reduction in reaction rate compared with that for other polysaccharides, such as starch, if the process is carried out with dilute acids and under heterogeneous conditions. On the other hand, if concentrated, dissolving-strength acids are used, the neutralization process becomes an expensive one. The second problem is that the drastic conditions necessary to produce significant saccharification of cellulose also result in decomposition of the sugars that are formed.

Process based on the two extremes of conditions have been developed, although emphasis has been placed on the low-temperature, concentrated-acid method. Saeman (85) has nevertheless shown that temperatures as high as 180°C, with 0.8% sulfuric acid, are potentially feasible and that under these conditions the cellulose hydrolysis is in fact favored relative to the sugar degradation. The sugar yield can be increased further by prior irradiation of the cellulose with high-energy radiation, which disrupts the crystalline regions, but this is unlikely to improve the economic viability of the process (86).

More recently a considerable amount of information has been published, mainly by Soviet workers who used the concentrated-acid, low-temperature method. Typical conditions involve concentrated hydrochloric acid and temperatures in the range of 40–60°C. The research effort is primarily concerned with minimizing the acid-to-cellulose ratio and obtaining maximum extraction of acid from the unconverted residues, while still maintaining a high degree of saccharification. The acid concentration is quite critical in that the rate increases rapidly above about 40% (87). In order to obtain uniform reaction it is desirable to allow diffusion of the acid to take place at low temperatures and then subsequently to initiate reaction by raising the temperature (88, 89). Under these conditions reducing sugar yields of up to 90% are claimed.

References

1. W. Z. Hassid and C. E. Ballou in W. Pigman, Ed., *The Carbohydrates, Chemistry, Biochemistry, Physiology,* Academic Press, New York, 1957, p. 489.
2. C. A. Bunton, T. A. Lewis, D. R. Llewellyn, and C. A. Vernon, *J. Chem. Soc.*, 4419 (1955).
3. C. M. Stewart and E. J. Williams, *Chem. Ind. (London)*, 1350 (1955).
4. A. Meller, *Chem. Ind. (London)*, 139 (1956).
5. K. Freudenberg, W. Kuhn, W. Dürr, F. Bolz, and G. Steinbrunn, *Ber.*, **63**, 1510 (1930).
6. E. H. Immergut, B. G. Rånby, and H. Mark, *Ricerca Sci.*, **25**, 308 (1955).
7. B. G. Rånby and R. H. Marchessault, *J. Polymer Sci.*, **36**, 561 (1959).
8. R. H. Marchessault and B. G. Rånby, *Svensk Papperstidn.*, **62**, 230 (1959).
9. R. H. Marchessault and B. G. Rånby, *Proc. 2nd Cellulose Conf., Syracuse, 1959,* 34.
10. M. Usuda, O. Suzuki, J. Nakano, and N. Migita, *Kogyo Kagaku Zasshi,* **70** (3), 349 (1967); through *Chem. Abstr.*, **67**, 1268z (1967).
11. E. Pacsu, *Textile Res. J.*, **17**, 405 (1947).
12. G. V. Schulz, E. Husemann, and H. J. Löhmann, *Z. Physik. Chem.*, **B52**, 23 (1942).
13. A. Meller, *Holzforschung,* **9**, 149 (1955).
14. G. V. Schulz and E. Husemann, *Z. Physik. Chem.*, **B52**, 42 (1942).
15. J. Hollo and J. Szejtli, *Acid Hydrolysis of Starch,* Akademie Verlag, Berlin, 1966.
16. L. M. Vinogradova, A. A. Konkin, and Z. A. Rogovin, *Nauch.-Issled. Tr. Moskov. Tekstil. Inst.*, **13**, 87, (1954); through *Chem. Abstr.*, **50**, 14221 (1956).
17. G. N. Richards, *Chem. Ind. (London)*, 228 (1955).
18. B. G. Rånby, *J. Polymer Sci.*, **53**, 131 (1961).
19. N. Ya. Lenshina and V. I. Ivanov, *Vysokomol. Soedin.*, **4**, (11), 1647 (1962); through *Chem. Abstr.*, **59**, 4157 (1963).
20. I. Rusznak, *Magyar Kem. Folyoirat,* **70** (1), 45 (1964); through *Chem Abstr.*, **61**, 2044 (1964).
21. J. Nakano and B. G. Rånby, *Svensk Papperstidn.*, **65**, 29 (1962).
22. A. A. Konkin, G. S. Smirnova, and Z. A. Rogovin, *Nauch.-Issled. Tr. Moskov. Tekstil. Inst.*, **13**, 99 (1954); through *Chem. Abstr.*, **50**, 14221 (1956).
23. R. L. Whistler and G. N. Richards, *J. Am. Chem. Soc.*, **80**, 4888 (1958).
24. B. Lindberg, *Chim. Biochim. Lignine, Cellulose Hemicelluloses, Actes Symp. Intern., Grenoble, France, 1964,* 303; through *Chem. Abstr.*, **64**, 19742 (1966).
25. V. I. Ivanov, Z. I. Kuznetsova, N. Ya. Lenshina, and V. S. Ivanova, *Tr. Leningrad. Lesotekh. Akad. in S. M. Kirova,* No. 91, part 1, 33 (1960); through *Chem. Abstr.*, **56**, 1639 (1962).
26. N. E. Virkola, *Paperi Puu,* **41**, 551 (1959); through *Chem. Abstr.*, **54**, 16822 (1960).
27. A. Bourman, *Textile Res. J.*, **23**, 888 (1953).
28. A. Hatano and H. Sobue, *Bull. Chem. Soc. Japan,* **26**, 403 (1953).
29. D. B. Das, M. K. Mitra, and J. F. Wareham, *Nature,* **174**, 228 (1954).
30. G. Centola, *Ricerca Sci.*, **23**, 1780 (1953).
31. E. Dymling, W. Giertz, and B. G. Rånby, *Svensk Papperstidn.*, **58**, 10 (1955).
32. G. A. Adams and C. T. Bishop, *Tappi,* **38**, 672 (1955).

33. R. L. Mitchell, M. A. Millet, W. E. Moore, and J. F. Saeman, *Tappi,* **39,** 571 (1956).
34. A. Anthis, *Tappi,* **39,** 40 (1956).
35. R. Jeffries, *Polymer,* **4,** 375 (1963).
36. G. F. Davidson, *J. Textile Inst.,* **34,** T87 (1943).
37. R. G. Nickerson and J. A. Habrle, *Ind. Eng. Chem.,* **39,** 1507 (1947).
38. M. L. Nelson and C. M. Conrad, *Textile Res. J.,* **18,** 149 (1948).
39. A. Meller, *J. Polymer Sci.,* **10,** 213 (1953).
40. J. R. Modi, S. S. Trivedi, and P. C. Mehta, *J. Appl Polymer Sci.,* **7,** 15 (1963).
41. F. Krejci, *Sb. Ved. Praci, Vysoka Skola Chem.-Technol., Pardubice,* Part 2, 131 (1961).
42. V. I. Sharkov and V. P. Levanova, *Vysokomol. Soedin.,* **5** (5), 729 (1963); through *Chem. Abstr.,* **59,** 4157 (1963).
43. L. Kudlacek and J. Ruzicka, *Vysokomol. Soedin.,* **6** (4), 587 (1964); through *Chem. Abstr.,* **61,** 2043 (1964).
44. L. Kuniak and B. Alince, *Faserforsch. u. Textiltech.,* **16** (3), 155 (1965).
45. V. I. Sharkov, *Sb. Tr., Gos. Nauch.-Issled. Inst. Gidrolizn. i Sul'fitno-Spirt. Prom.,* **10,** 7 (1962); through *Chem. Abstr.,* **64,** 12948 (1966).
46. P. H. Hermans and A. Weidinger, *J. Polymer Sci.,* **4,** 317 (1949).
47. J. Linderot, *Sevensk Papperstidn.,* **59,** 37 (1956).
48. E. H. Daruwalla and G. M. Nabar, *J. Polymer Sci.,* **20,** 205 (1956).
49. Yu. T. Tashpulatov, T. Saidaliev, and Kl. U. Usmanov, *Uzbek. Khim. Zh.,* **10** (1), 39 (1966); through *Chem. Abstr.,* **64,** 19865 (1966).
50. V. I. Sharkov and V. P. Levanova, *Dokl. Akad. Nauk* SSSR, **130,** 109 (1960); through *Chem Abstr.,* **54,** 10312 (1960).
51. A. Sharples, *J. Polymer Sci.,* **13,** 393 (1954).
52. P. H. Hermans, *Makromol. Chem.,* **6,** 25 (1951).
53. E. H. Immergut and B. G. Ranby, *Ind. Eng. Chem.,* **48,** 1183 (1956).
54. W. B. Achwal, E. H. Daruwalla, G. M. Navar, and P. Subramanian, *J. Polymer Sci.,* **35,** 93 (1959).
55. J. J. McKeown and W. I. Lyness, *J. Polymer Sci.,* **47,** 9 (1960).
56. H. J. Phillipp, M. L. Nelson, and H. M. Ziifle, *Textile Res. J.,* **17,** 585 (1947).
57. E. H. Daruwalla and R. T. Shet, *Textile Res. J.,* **32,** 942 (1962).
58. M. A. Millett, W. E. Moore, and J. F. Saeman, *Ind. Eng. Chem.,* **46,** 1493 (1954).
59. O. A. Battistà, S. Coppick, J. A. Howsmon, F. F. Morehead, and W. A. Sisson, *Ind. Eng. Chem.,* **48,** 333 (1956).
60. R. D. Preston and G. W. Ripley, *Nature,* **174,** 76 (1954).
61. O. A. Dmitrieva, N. P. Potapova, and V. I. Sharkov, *Zh. Publ. Khim.,* **37** (7), 1583 (1964); through *Chem. Abstr.,* **61,** 9620 (1964).
62. W. B. Achwal and G. M. Nabar, *Textile Res. J.,* **30,** 872 (1960).
63. A. Sharples, *Trans. Faraday Soc.,* **53,** 1003 (1957).
64. A. Sharples, *Trans. Faraday Soc.,* **54,** 913 (1958).
65. J. Dlugosz and R. I. C. Michie, *Polymer,* **1,** 41 (1960).
66. A. Sharples, *Introduction to Polymer Crystallization,* Edward Arnold Ltd., London, 1966, pp. 23–25.
67. D. J. Plazek and J. D. Ferry, *J. Phys. Chem.,* **60,** 289 (1956).

68. R. St. J. Manley, *J. Polymer Sci.,* [A], **1,** 1875 (1963).
69. R. St. J. Manley, *J. Polymer Sci.,* [A], **1,** 1893 (1963).
70. I. L. Wadehra and R. St. J. Manley, *Makromol. Chem.,* **94,** 42 (1966).
71. R. St. J. Manley, *Nature,* **204,** 1155 (1964).
72. R. St. J. Manley, *J. Polymer Sci.,* [B], **3,** 691 (1965).
73. D. W. Haas, B. F. Hrutfiord, and K. V. Sarkanen, *J. Appl. Polymer Sci.,* **11,** 587 (1967).
74. H. Dolmetsch and H. Dolmetsch, *Kolloid-Z.,* **185,** 106 (1962).
75. E. Pacsu, *J. Polymer Sci.,* **2,** 565 (1947).
76. G. V. Schulz, *J. Polymer Sci.,* **3,** 365 (1948).
77. A. Sharples, *J. Polymer Sci.,* **14,** 95 (1954).
78. E. H. Daruwalla and M. G. Narsian, *Tappi,* **49** (3), 106 (1966).
79. R. I. C. Michie, A. Sharples, and A. A. Walter, *J. Polymer Sci.,* **51,** 85 (1961).
80. A. Sippel, *Papier (Darmstadt),* **13,** 413 (1959).
81. R. B. Valley, *Textile Res. J.,* **25,** 930 (1955).
82. V. I. Orlov and V. I. Sharkov, *Izv. Vysshikh Uchebn. Zavedenii, Lesn. Zh.,* 8 (3), 141 (1965); through *Chem. Abstr.,* **63,** 11848 (1965).
83. N. I. Klenkova, O. M. Kulakova. and L. A. Volkova, *Zh. Prikl. Khim.,* **36,** 166 (1963); through *Chem. Abstr.,* **59,** 1793 (1963).
84. H. Nishimura, O. Nakao, and N. Mibita, *Sen-i Gakkaishi,* **18,**(12), 1076 (1962); through *Chem. Abstr.,* **63,** 3152 (1965).
85. J. F. Saeman, *Ind. Eng. Chem.,* **37,** 43 (1945).
86. J. F. Saeman, M. A. Millett, and E. J. Lawton, *Ind. Eng. Chem.,* **44,** 2848 (1952).
87. N. V. Lebedev and A. A. Banikova, *Sb. Tr., Gos. Nauch-Issled. Inst. Gidrolizn. i Sul'fitno-Spirt. Prom.,* 9, 7 (1961); through *Chem. Abstr.,* **60,** 13430 (1964).
88. N. V. Korotkov and N. V. Chalov, *Sb. Tr., Gos. Nauch.-Issled. Inst. Gidrolizn. i Sulfitno-Spirt. Prom.,* **14,** 180 (1965); through *Chem. Abstr.,* **66,** 96335 (1967).
89. J. Kusama and T. Ishii, *Kogyo Kaguku Zasshi,* **69**(3), 476 (1966); through *Chem. Abstr.,* **65,** 15641 (1966).

B. ALKALINE DEGRADATION

G. N. Richards

The reactions with which we are concerned in this section are those that occur between alkali or the hydroxyl ion and cellulose or an oxidized cellulose. Oxidation by molecular oxygen in alkaline conditions is not considered here, but it should be emphasized that such autoxidation may as a first step produce carbonyl-containing oxidized celluloses, which are subsequently degraded by alkali as described below.

1. Technological Importance of Alkaline Degradation*

The technological importance of alkaline degradation is most evident in wood-pulp manufacture (both cooking and refining) and in the degradation of all types of carbonyl-containing oxidized celluloses (see also Chapter XIX-L). Alkaline degradation during the pulping of wood is of course most marked in the soda process; but it almost certainly occurs extensively also in the kraft process and, in fact, in any process in which cellulose is exposed to a significant concentration of hydroxyl ions at elevated temperatures. Under such conditions cellulose is always subject to "erosion" (or "unzipping") of the molecule from the reducing end group. The mechanism of this stepwise depolymerization process is described in detail below. Its effect is to remove one glucose unit at a time, mainly by conversion to soluble D-glucoisosaccharinic acid. This liberates the next glucose unit, which is again subject to alkaline degradation in a repeating process. This process normally continues until it is interrupted in a particular molecule by an irreversible stabilization reaction that causes the cellulose molecule to become completely resistant to any further erosion. The mechanism of the stabilization reaction is also described below. This type of alkaline degradation can cause the conversion of cellulose (particularly a hydrocellulose) to soluble products and hence a reduced pulp yield without at the same time having any marked effect on the degree of polymerization (DP) (2). This situation, in which a considerable weight loss (e.g., 50%) by an end-erosion process leads to

*For a detailed review see reference 1.

only a very slight decrease, or even an increase in DP, is analogous to the effect of heterogeneous acid hydrolysis on leveling-off DP (LODP) hydrocellulose. In the latter system also the attack on the cellulose molecules occurs effectively only at the ends of the molecules. Sharples (3) has discussed this situation in detail and has shown that the lack of any marked effect of such a process on the DP is an inevitable result of the particular exponential molecular-weight distribution in LODP hydrocelluloses (see also Section A of this Chapter XVIII).

The alkaline depolymerization process can be prevented by chemical modification of the reducing end group or, alternatively, the extent of the process may be reduced by adopting conditions that favor the stabilization reaction. Attempts to reduce degradation by such methods are described below.

High temperatures and high hydroxyl-ion concentration, such as occur in the soda and kraft processes, also lead to a significant amount of molecular scission, whose probable mechanism is described below. This scission does not appear to depend on the presence of molecular oxygen and is probably followed by erosion from any new reducing end group produced by the scission process. The scission, of course, results in a reduction in molecular weight, but this process does not appear to occur very extensively under the normal conditions of hot alkali refining.

Any oxidized cellulose that contains carbonyl groups at any position other than end groups is very labile to alkali, and under very mild conditions (e.g., 0.05 *N* sodium hydroxide at room temperature for 1 hr), almost complete scission may be anticipated of the cellulose molecule at any unit that contains a carbonyl group. The mechanism of such a process is described below. Such oxidized celluloses may be produced by oxidative bleaching treatments, such as chlorination, and also by autoxidation, especially in the presence of sensitizers such as anthraquinones (which occur in many common vat dyes). The oxidation itself may not cause any significant chain scission and hence no mechanical weakening (e.g., of a woven fabric), but any subsequent mild alkaline treatment, such as scouring or laundering, will cause alkaline scission of cellulose molecules at the point of oxidation. It is at this stage that the effects on mechanical properties, such as strength, become apparent.

2. Mechanism of the Alkaline Erosion Reactions

A cellulose molecule with a normal reducing end group may be considered as a glucose molecule substituted in the $C_{(4)}$ position with the remainder of the cellulose chain. On this basis the alkaline erosion reaction is entirely in accordance with the theory of β-alkoxycarbonyl elimination, originally postulated by Isbell (4) and subsequently verified (5, 6) by investigations of the mechanism of alkaline degradation of other 4-*O*-substituted glucose derivatives, such as 4-*O*-methyl-D-glucose (7) and maltose (8). The mechanism of the

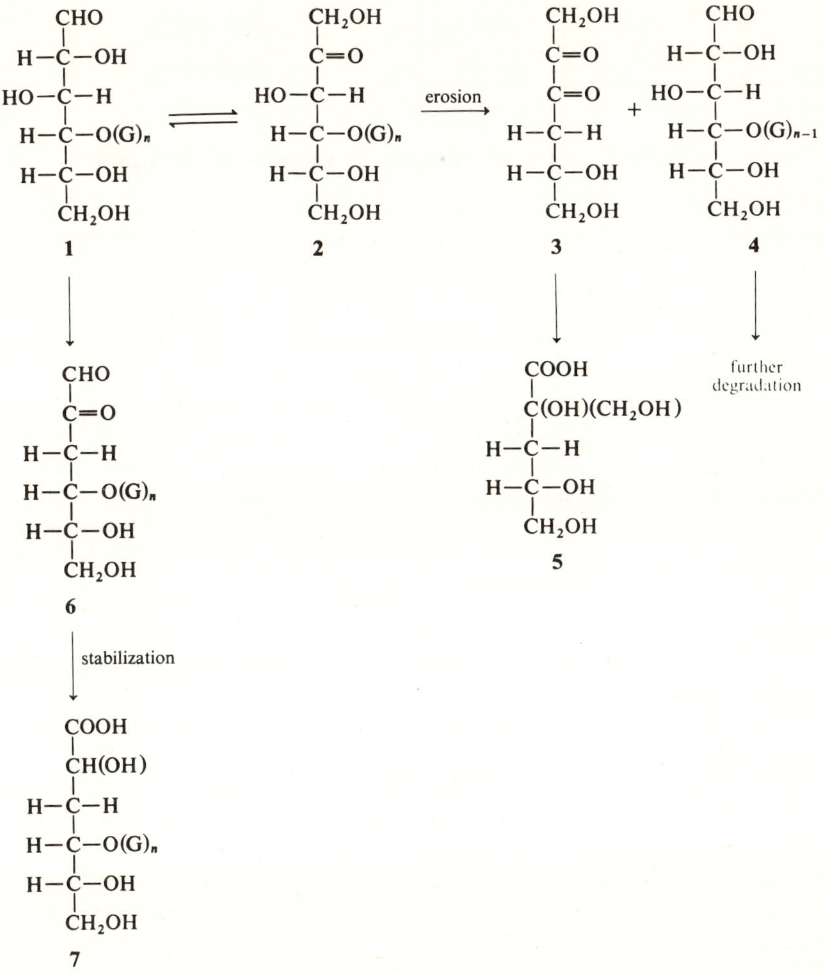

Fig. 1. Alkaline erosion and stabilization of cellulose.

cellulose erosion reaction was first definitely elucidated (9, 10) after a general study of such simple model compounds. The degradation proceeds predominantly by the mechanism shown and has been verified (11) by isolation of the intermediate compound **3** in model systems (see Fig. 1). The reducing end group thus becomes converted predominantly to D-glucoisosaccharinic acid, but at higher temperatures the soluble products become quite complex, as increasing amounts of other acids, and especially formic acid, are also produced (9, 10, 12). The mechanism of formic acid formation is not understood at present, but it is probable that it arises at least in part from an alternative degradation of the intermediate (compound **3**) (11). The mixture of reaction products is simplified in the presence of calcium ions, which appear to catalyze the rearrangement of compound **3** to compound **5** (11).

3. Mechanism of the Alkaline Stabilization Reactions

The alkaline erosion of a cellulose molecule would continue until the cellulose were completely degraded to soluble products in the absence of branch points or any other heterogeneities in the cellulose molecule, or if there is no competing reaction that produces an alkali-stable cellulose molecule. Neither branch points nor any other linkage irregularities have ever been detected in cellulose, but it is known that cellulose dissolves partly, but not completely, in hot alkali. Since this process is accompanied by an increase in the carboxyl content of the cellulose (13), it had long been suspected that a stabilization reaction was operative. Richtzenhain and Abrahamsson (14) first suggested tentatively that stabilization might arise by a saccharinic acid rearrangement of the reducing end group in cellulose. This was finally established as the mechanism of the stabilization reaction by Machell and Richards (15), who treated cotton hydrocellulose with hot, oxygen-free sodium hydroxide to obtain alkali-stable hydrocellulose and, by a selective acid hydrolysis of the end groups, were able to isolate D-glucometasaccharinic acid, which could only have arisen by alkaline rearrangement of a reducing end group that was still attached to the cellulose molecule. The stabilization reaction therefore proceeds as shown in Figure 1 (**1** → **7**) and competes with the erosion reaction. In any given cellulose molecule, once the elimination reaction (**1** → **6**) has occurred, the rearrangement very rapidly proceeds to a cellulose molecule (compound **7**) with D-glucometasaccharinic acid in place of the normal reducing end group, and this product is not subject to further alkaline degradation. The erosion process, however, normally occurs at a considerably greater rate than the stabilization reaction, and commonly several hundred glucose units may be removed from the average cellulose molecule before stabilization.

The occurrence of the same type of stabilization reaction during sulfate pulping has subsequently been demonstrated by the isolation of D-glucometasaccharinic acid from a pine sulfate pulp (16).

4. Alkaline Scission of Cellulose

The nonoxidative scission of cellulose has been less extensively studied than other types of alkaline degradation. This is no doubt partly due to the difficulties in ensuring freedom from traces of oxygen in such reactions under conditions in which alkaline autoxidation would proceed very rapidly. Furthermore, under the conditions in which alkaline scission probably occurs (e.g., 1.0 *N* sodium hydroxide, 170°C) the erosion reactions also occur very extensively

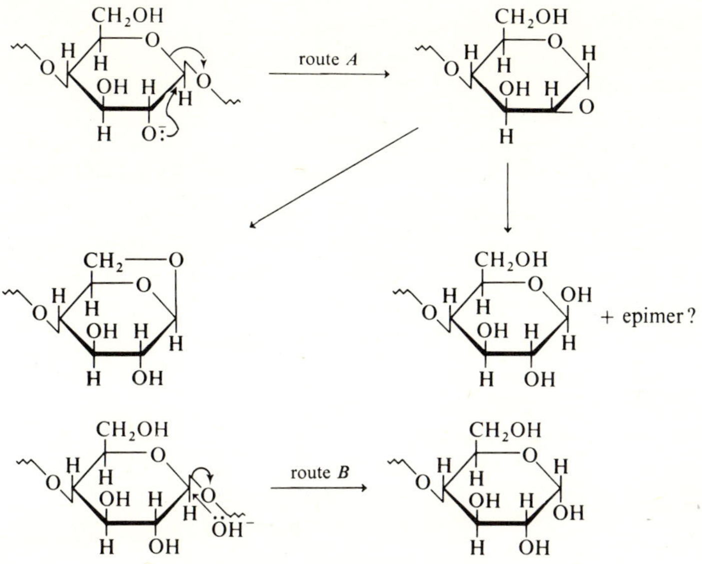

Fig. 2. Alkaline scission of cellulose.

and rapidly, and, although the latter reactions have very much less effect on molecular weight than the scission reactions, the two effects are not easy to separate.

Samuelson and co-workers (17) have reported evidence for the alkaline scission of cellulose by 1.0 *N* sodium hydroxide at 170°C, and in a study in which exhaustive attempts were made to remove all traces of oxygen in such conditions Corbett and Richards (12) have concluded that such a reaction can occur nonoxidatively. By analogy with alkaline degradative studies on model glycosides, the scission reaction may be postulated as occurring either by intramolecular displacement [route *A*, compare Lindberg (18, 19)] or by addition of the hydroxyl ion [route *B*, compare Overend and co-workers (20)]. This is shown in Figure 2. Either mechanism could yield a new reducing end group at the point of scission and would be followed by alkali erosion.

There is some possibility that the "xylan retake" observed in alkaline pulping processes may be associated with a subsequent reaction of xylose or xylans with cellulose molecules that have undergone alkaline scission [for a review see Meller (21)].

5. Attempts To Minimize Alkaline Degradation of Cellulose

The commercial incentives to reduce the alkaline degradation of cellulose are considerable. For example, alkaline erosion is responsible for the majority of the losses of cellulose that occur in the alkaline cooking and refining of wood pulps.

Attempts to minimize such losses have followed two different approaches, which depend either on the modification of the reducing end group to a grouping that is relatively stable to alkali or on influencing the relative rates of erosion and stabilization reactions so as to favor the latter.

The reducing end group has been modified by reduction to glucitol (22–24), oxidation to gluconic acid (25, 26), by conversion to the methyl glucoside (27), and by conversion to the anhydride form (28). Of these approaches the only one that appears to hold much possibility of economic advantage is the reduction with sodium borohydride. It should be remembered, however, that the hemicelluloses also contain reducing end groups and are subject in most cases to the same type of erosion reactions as cellulose. One of the objects of the refining process, and to a smaller extent the cooking process, is to reduce the content of hemicelluloses in the pulp, and yet it is probable that the mechanism of removal of hemicelluloses in these processes depends at least in part on alkaline-erosion degradation. Any attempt to stabilize the cellulose molecule by end-group modification therefore will inevitably stabilize the hemicelluloses in the same way, and the latter will then sometimes become rather resistant to removal by conventional refining processes. Meller (23, 24), however, has studied the effect of sodium borohydride on pulp yield and purity in kraft cooking, and his results appear to indicate a preferential reductive stabilization of hexosans in comparison with xylans under such conditions.

Attempts to favor the stabilization reaction with respect to the erosion reaction were at first rather arbitrary, although the successful use of added calcium ion had some basis in model-compound work (27). It has been evident for some time, however, that the reaction temperature was important in determining the relative rates of the reactions, and this situation has now been placed on a quantitative basis by the determination (29) of activation energies for the erosion and stabilization reactions. The former has an E value of 24 kcal/mole and the latter has an E value of 32 kcal/mole for cellulose in 5% sodium hydroxide solution. The rate of the latter reaction therefore increases the more rapidly with temperature.

6. Alkaline Degradation of Carbonyl-Containing Oxidized Celluloses

A considerable advance in understanding of the mechanism of alkaline degradation of carbonyl-containing oxidized celluloses has resulted from studies of alkaline degradation of model compounds, such as the methyl ethers of reducing sugars and disaccharides (5, 6). In such reactions and in confirmation of Isbell's original hypothesis (4) it has been established that, in any situation where a β-alkoxycarbonyl group is present in alkali, a rapid elimination of the alkoxyl group will occur as follows:

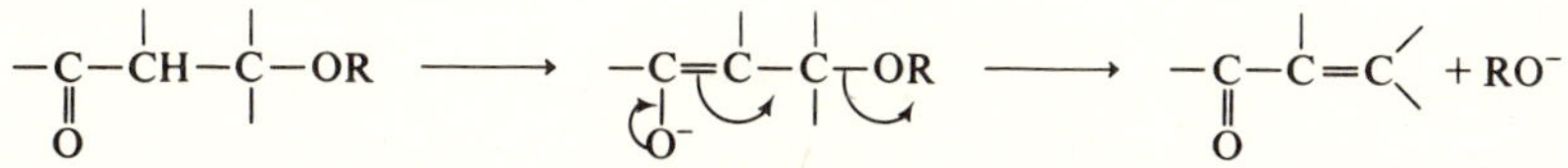

The *β*-alkoxycarbonyl grouping occurs in all possible situations where a carbonyl group is introduced into a cellulose molecule (see Fig.3), and in all such circumstances it may be assumed that rapid scission will occur by the above mechanism in dilute alkali. The only type of carbonyl-containing oxidized cellulose that has been isolated in the pure state is periodate-oxidized cellulose, and the alkaline degradation of this material itself (30) and also of model compounds (31) has been shown to proceed predominantly as shown in Figure 4. This type of degradation occurs rapidly at room temperature in very dilute alkali (e.g., 0.05 *N*), and if the cellulose sample has been extensively oxidized, it will be completely dissolved by the alkali.

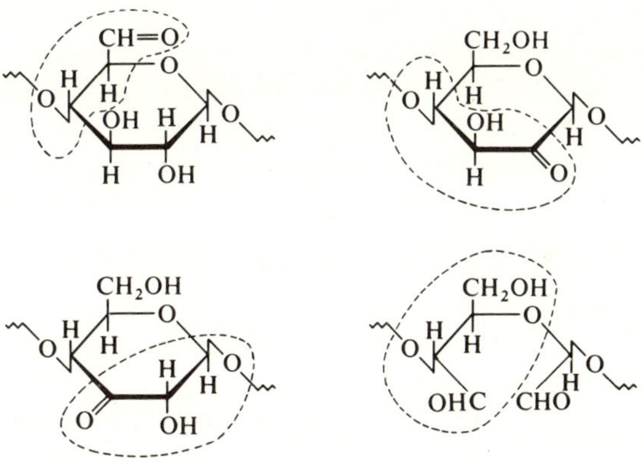

Fig. 3. *β*- Alkoxycarbonyl groups in carbonyl-containing oxidized cellulose.

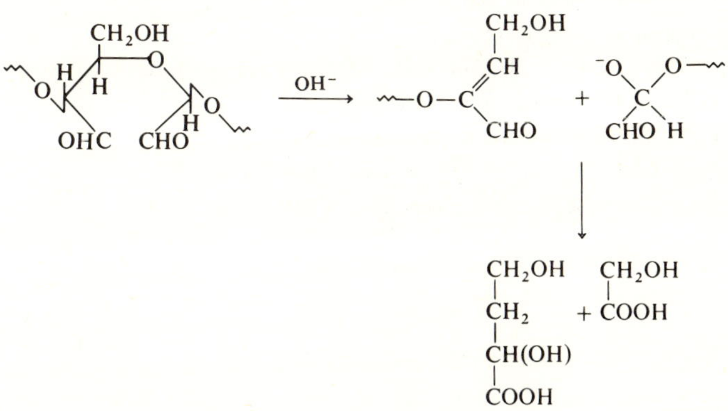

Fig. 4. Alkaline degradation of carbonyl-containing oxidized cellulose.

References

1. A. Meller, *Holzforschung,* **14**, 78, 129 (1960).
2. G. Machell and G. N. Richards, *J. Chem. Soc.,* 4500 (1957).
3. A. Sharples, *Trans. Faraday Soc.,* **53**, 1003 (1957).
4. H. S. Isbell, *J. Res. Natl. Bur. Std.,* **32**, 45 (1944).
5. G. N. Richards, *Pulp Paper Mag. Can.,* **59**, 122 (1958).
6. R. L. Whistler and J. BeMiller, in *Advances in Carbohydrate Chemistry,* Vol. 13, Academic Press, New York, 1958, p. 289.
7. J. Kenner and G. N. Richards, *J. Chem. Soc.,* 1810 (1955).
8. W. M. Corbett and J. Kenner, *J. Chem. Soc.,* 2245 (1953).
9. G. Machell, G. N. Richards, and H. H. Sephton, *Chem. Ind. (London),* 467 (1957).
10. G. N. Richards and H. H. Sephton, *J. Chem. Soc.,* 4492 (1957).
11. G. Machell and G. N. Richards, *J. Chem. Soc.,* 1932 (1960).
12. W. M. Corbett and G. N. Richards, *Svensk Papperstidn.,* **60**, 791 (1957).
13. O. Samuelson and A. Wennerblom, *Svensk Papperstidn.,* **57**, 827 (1954).
14. H. Richtzenhain and B. Abrahamsson, *Svensk Papperstidn.,* **57**, 538 (1954).
15. G. Machell and G. N. Richards, *J. Chem. Soc.,* 4500 (1957).
16. B. Alfredsson, O. Samuelson, and B. Sandstig, *Svensk Papperstidn.,* **66**, 703 (1963).
17. O. Samuelson, G. Grangord, K. Jönsson, and K. Schramm, *Svensk Papperstidn.,* **56**, 779 (1953).
18. B. Lindberg, *Svensk Papperstidn.,* **59**, 531 (1956).
19. J. Janson and B. Lindberg, *Acta Chem. Scand.,* **13**, 138 (1959).
20. R. J. Ferrier, W. G. Overend, and A. E. Ryan, *J. Chem. Soc.,* 3484 (1965).
21. A. Meller, *Tappi Third International Pulp Bleaching Conference Preprints,* 1964, p. 23.
22. H. Richtzenhain, B. O. Lindgren, B. Abrahamsson, and K. Holmberg, *Svensk Papperstidn.,* **57**, 363 (1954).
23. A. Meller, *Tappi,* **46**, 317 (1963).
24. A. Meller and E. L. Ritman, *Tappi,* **47**, 55 (1964).
25. A. Meller, *Svensk Papperstidn.,* **65**, 307 (1962).
26. A. Meller, *Tappi,* **45**, 481 (1962).
27. G. Machell and G. N. Richards, *Tappi,* **41**, 12 (1958) and earlier references therein.
28. A. Meller, *Svensk Papperstidn.,* **65**, 629 (1962).
29. D. W. Haas, B. F. Hrutfiord, and K. V. Sarkarnen, *J. Appl. Polymer Sci.,* **11**, 587 (1967).
30. D. O'Meara and G. N. Richards, *J. Chem. Soc.,* 4504 (1958).
31. D. O'Meara and G. N. Richards, *J. Chem. Soc.,* 1204 (1958).

C. THERMAL DEGRADATION

Frank J. Kilzer

At temperatures above about 120°C the thermal instability of cellulose and many of its derivatives becomes increasingly apparent. These materials appear to undergo internal changes accompanied by decomposition into smaller, volatile molecules. Such "thermal" reactions are of great importance. At temperatures only slightly above ambient (such as occur during processing and use) fabrics, fiberboard, tire cord, and the like experience heat-accelerated loss of desirable properties. At higher temperatures cotton and wood products can be decomposed into useful small molecules. Above about 300°C cellulose and its derivatives can be converted into graphitized or carbonized materials.

Early controlled studies of the thermal degradation of cellulose were carried out toward the goal of analyzing the structure of cellulose. Thermal degradation is a time-honored tool used by organic chemists to help unravel the structure of complicated molecules. In principle a knowledge of the structure of the resulting smaller molecules helps to define the structure of the parent molecule.

Pictet and Sarasin (1) dry-distilled cotton cellulose under vacuum. In addition to gaseous and volatile products, they obtained a viscous oil, from which they isolated a solid corresponding to about 45% by weight of the starting cellulose. Because of its optical rotation and ready hydrolysis to glucose, they named it levoglucosan (**1**):

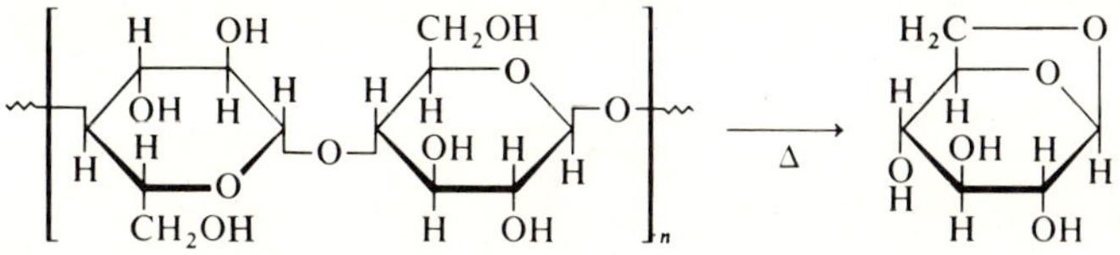

1

Efforts were made to determine the structure of this major product and to use this knowledge to solve the riddle of the primary structure of cellulose (2). Josephson (3) proposed the now-accepted structure for levoglucosan in 1929 as 1,6-anhydro-β-D-glucopyranose (see Fig. 1). However, this knowledge confused the search for the cellulose structure: the 1,6-anhydride linkage was taken as

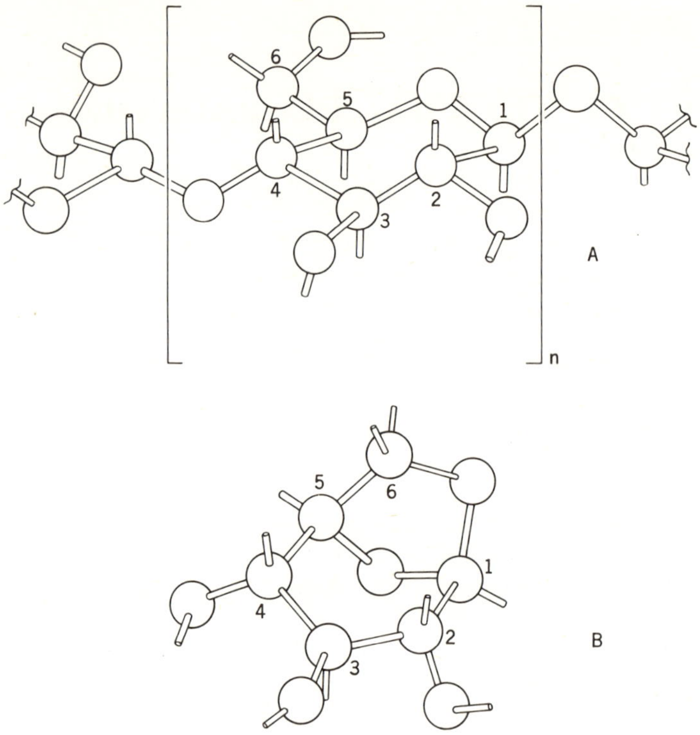

Fig. 1. Models of the (1→4)-β-linked glucosyl repeating unit of cellulose (*A*) and of levoglucosan (*B*). The steric arrangement at $C_{(1)}$ of the cellulose prevents the formation of levoglucosan by a direct displacement reaction.

evidence that the glucose rings in cellulose are linked together through their 1- and 6-positions instead of through their 1- and 4-positions, as demonstrated by subsequent evidence.

In the following decades other reasons for studying the thermal-degradation processes in cellulose became apparent (5–16). One of the major incentives has been the increased awareness of the importance of pyrolytic processes in providing the volatile molecules that are the immediate fuel in the flaming combustion of cellulosic materials. Another is to relate cellulose pyrolysis to the flameproofing of cotton textiles (4) (see Section I in Chapter XIX).

This section deals particularly with studies made in the absence of oxygen. The analytical tools used and the major findings and conclusions are reported. Special emphasis is placed on those investigations that provide the foundation for a mechanistic interpretation of the degradation processes.

1. Experimental Studies on Cellulose and Its Derivatives

Physicochemical techniques used to probe thermal behavior can be divided broadly into two classes:

1. Methods that yield information about the *solid sample* itself: thermogravimetric analysis, differential thermal analysis, etc.
2. Various analyses of the *volatile products* from the reaction.

These methods are discussed first, together with some of their findings. The mechanistic conclusions to be drawn from these studies are presented subsequently.

(a) THE SOLID SAMPLE

By far the majority of investigations have depended on techniques that yield information about changes in the solid sample—that is, those that measure such parameters as weight, tensile strength, and color. A few investigators have performed wet chemical analyses of the residues.

(1) Thermogravimetry

The study of the relation of loss of mass to temperature on samples that volatilize when heated, called thermogravimetric analysis (TGA) (17), can provide insight into the underlying reactions. Thermogravimetric analysis can be performed by recording changes in sample weight at a constant temperature (static TGA) or as the sample temperature is increased (dynamic TGA).

The TGA results of various investigators are difficult to correlate because of differences in experimental conditions; however, some representative findings are presented here.

Static TGA. Madorsky, Hart, and Straus (18) studied weight loss by static TGA for samples of celluloses ("pure" or with added inorganic salts) heated between 245 and 305°C in a vacuum of 10^{-6} mmHg. For the temperature range 285–295°C for cotton, a hydrocellulose, and a viscose rayon they calculated activation energies of 50, 47, and 49 kcal/mole, respectively, for the overall weight-loss process. To perform the calculations they selected arbitrary "carbonization end points," but this did not imply the presence of only a single reaction. In fact a consideration of weight data together with analysis of volatile products suggested the participation of at least two major reactions. To obtain apparent initial rates they extrapolated the rate curves to 0% volatilization, since only at the beginning of the degradation did all the samples have the same composition.

Golova, Krylova, and Nikolaeva (19) measured the weight loss at 305°C in 10^{-5} mmHg vacuum of a cotton cellulose and a regenerated cellulose (a "cellulose hydrate"). The regenerated cellulose volatilized faster initially. Its weight loss leveled off at about 77% loss (23% residue); that of the cotton cellulose leveled off at about 84% loss. The weight-loss curves eventually crossed. The authors suggest the participation of several reactions in the overall process. For undetermined reasons the regenerated cellulose undergoes greater initial conversion to volatile products but ultimately retains a greater weight of residue (see also Chapter XVII-A). These results appear to confirm those of Madorsky et al. (18).

Chatterjee and Conrad (20) followed the weight loss from absorbent cotton and ball-milled cotton between 270 and 310°C, and noted that rapid heating of the sample was required to avoid obscuring early decomposition reactions. They defined two reactions; the first, up to about 40% weight loss, has energies of activation of 54.3 and 75.0 kcal/mole for cotton and ball-milled cellulose, respectively. For the second reaction, a first-order process, the activation energies calculated were 33.0 and 37.1 kcal/mole, respectively. The latter values are in good agreement with those derived for related processes by other authors: 30.5 kcal/mole by Stamm (21), 33.0 kcal/mole by Esteve et al. (22), and 39.4 kcal/mole by Murphy (23). Other studies (24, 25) involved measurements of the glucosan content after TGA.

Kato and Takahashi (26) studied microcrystalline cellulose under a nitrogen atmosphere at temperatures between 240 and 300°C, recording an induction time (1.4 min), a volatilization rate during the linear weight loss (50% per minute), a "time to 50% weight loss" (2.1 min), and a residual weight (23%) at 300°C. For the linear weight-loss process they calculated an activation energy of 22.0 kcal/mole.

Static TGA has also been applied to cellulose derivatives, such as cellulose acetate, and extensively to cellulose nitrate. Madorsky et al. (27) studied cellulose triacetate *in vacuo* at 297°C and estimated the apparent initial-weight-loss rate to be 0.37% per minute, with a calculated activation energy of 45 kcal/mole. Von Gizycki (28) determined the weight loss of cellulose nitrate at 100°C. The rate of volatilization was retarded by diphenylamine and other free-radical scavengers, whereas the decrease in the degree of polymerization (viscosity) was not affected by them. Abe et al. (29) have attributed the weight loss from cellulose nitrate below 160°C to the loss of water. Andreev and co-workers (30–32) studied the evolution of two gas fractions as well as the weight loss of cellulose nitrate in 5×10^{-4} mmHg vacuum at temperatures between 135 and 160°C. In their view the less volatile gases arise at the lower temperatures along with a residue said to be an "intermediate" of low stability. The lower temperature reactions comprise a first-order process. At higher temperatures the intermediate decomposes into noncondensable gases in self-accelerating reactions (32). Ettre and Varadi (33) found that reactions at temperatures between 175 and 250°C produce a residue that does not decom-

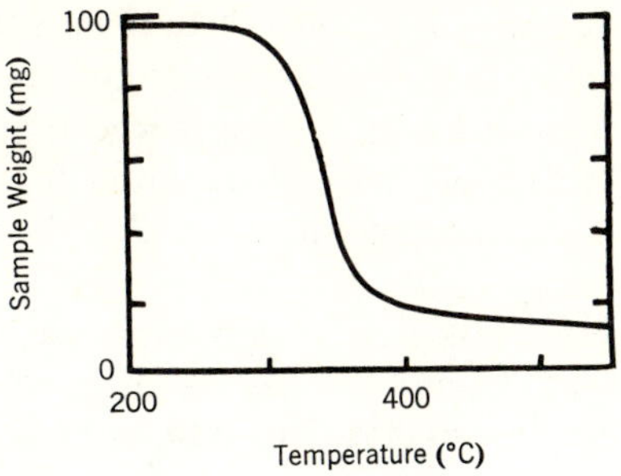

Fig. 2. Dynamic thermogravimetric trace of α-cellulose paper in He (36).

pose at higher temperatures, but samples heated above 300°C leave little residue, and pyrolysis at 500°C leaves no measurable residue.

Gas chromatography (33) and infrared spectroscopy (34) are useful tools in conjunction with TGA.

Dynamic TGA. Dynamic TGA studies of cellulose and its derivatives are far more common than static studies. Often they are performed to compare the properties of materials before and after various chemical and physical treatments—for example, in conjunction with flameproofing investigations (35–40).

A typical TGA plot for cellulose is shown in Figure 2. α-Cellulose powder was heated at an increment of 3°C/min under a vacuum of 0.3 mmHg. Observable weight loss begins somewhere above 270°C, accelerated until an inflection point near 335°C and levels off somewhat above 350°C. Similar curves have been published by other authors (38–40 and references therein). Purification by bleaching and scouring raises the temperature of decomposition of cotton fabrics (37). On the other hand, chemical modification (such as benzhydrylation, tritylation, tosylation, cyanoethylation, thioacetylation, and disulfide crosslinking) tends to lower thermal stability (38, 40–43).

The derivation of kinetic parameters from TGA data has been criticized by several authors (41–44); the complexity of the processes makes such data valuable mainly for empirical correlations.

(2) Differential Thermal Analysis and Differential Scanning Calorimetry

Many organic substances undergo relatively few and simple energy exchanges when heated. Melting and boiling endotherms are the most common (17). In contrast most carbohydrates, including cellulose, exhibit a series of endotherms and exotherms. These can be used to characterize the substance or for comparison purposes (45, 46). The relative energy released or absorbed during a reaction can provide a qualitative thermodynamic parameter for proposed mechanisms (38, 47).

During the 1950s pioneering differential thermal analyses (DTA) were performed on cellulose and other carbohydrates by Morita and Rice (45, 46), Costa and Costa (48), and Berkowitz (49). As knowledge of the usefulness of this sensitive technique spread, more authors employed it in their studies of the underlying reactions or for comparative purposes (36–51), and new references to its results constantly appear.

Figure 3 shows three variants of such curves obtained in an inert atmosphere for various pure celluloses. Figure 4 shows the range of complexity for some cellulose derivatives. Regardless of the source of the cellulose, its curves show some common features: a slight endotherm above 100°C, attributed to the volatilization of adsorbed water, and a strong endotherm between 310 and 350°C, related in some way to the formation and volatilization of levoglucosan. The curves for cellulose are relatively plain and show few other (and variable) features. Both Berkowitz (49) and Broido (42), working with cellulose filter paper in nitrogen observed a very slight exotherm at about 220°C. Schwenker et al. (50) noted a slight exotherm at about 310°C, just before the major endotherm, for cotton fabric in nitrogen. A number of workers have recorded extensive exothermic

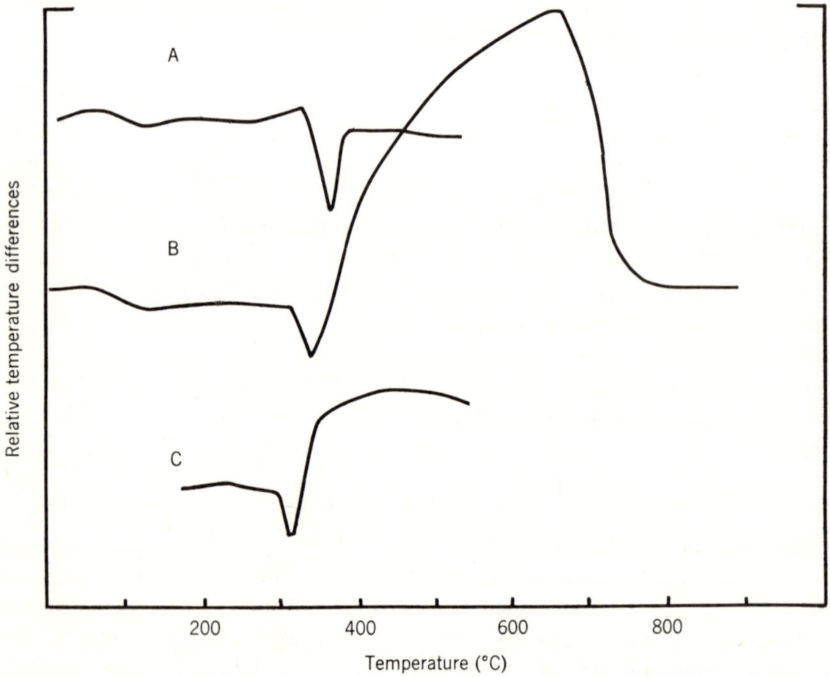

Fig. 3. Differential thermal analyses of celluloses: (*A*) alkali-scoured cotton fabric (50); (*B*) Whatman cellulose powder (45); (*C*) Whatman cellulose (49).

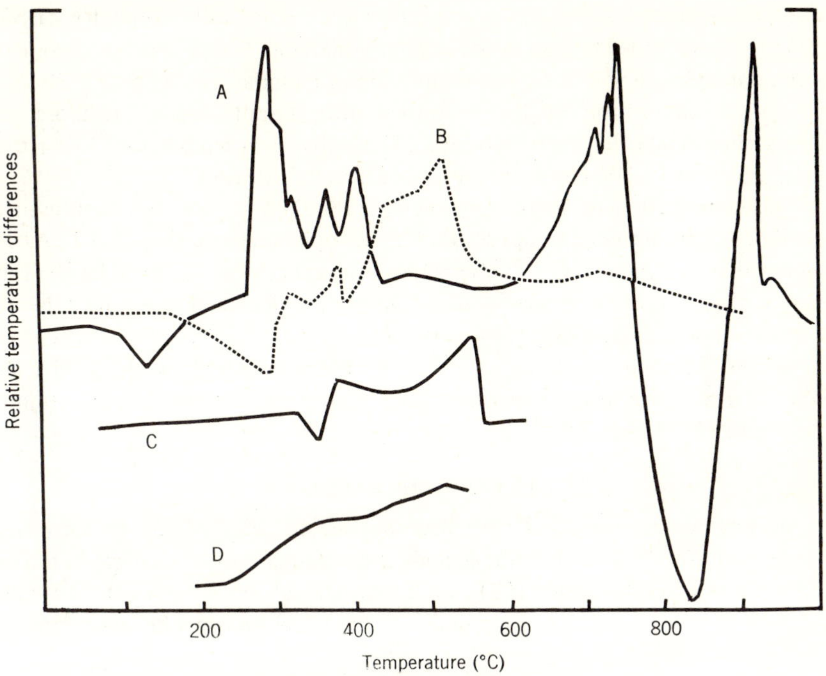

Fig. 4. Differential thermal analyses of cellulose derivatives: (*A*) cellulose triacetate (45); (*B*) carboxymethylcellulose (45); (*C*) benzhydrylated cotton cellulose, DS 0.31 (41); (*D*) "partially carbonized" cellulose (49).

behavior following the major endotherm (41, 42, 45, 49, 51). Running the thermograms under air introduces a large exotherm at 320°C, as well as a major exotherm above 360°C, following the endotherm that peaks at 360°C (38). (Or the endotherm may be described as overcoming or interrupting a single major exotherm beginning at about 270°C and extending upward.) Stepwise addition of potassium bicarbonate "ash" to ash-free filter paper (42) eliminates the major endotherm but does not affect the exotherm above 350°C. Thus neither the ash content of the sample nor the atmosphere during the analyses seems to explain the differences in these curves.

Berkowitz (49) partially carbonized Whatman's cellulose at temperatures between 200 and 250°C in a sealed tube for 1 or 2 h. The thermogram of this material (Fig. 4D) lacks any hint of the cellulose endotherm near 320°C as well as the subsequent sharp exotherm. Instead it shows a stepwise series of exotherms. The DTA curves of Morita and Rice (45, 46) for cellulose acetate and a carboxymethylcellulose with a degree of substitution (DS) of 0.5 are recorded in Figure 4. The authors made no attempt to interpret these complicated traces. Mack and Hobart (51) studied the thermal behavior of phenylcarbanilated celluloses with differential scanning calorimetry (DSC), using the unmodified cotton as the reference material. Concurrent gas-chromatography analysis showed the single endotherm of material with a DS of 1.50 to be due to decomposition of the derivative into its antecedents, cellulose and phenyl isocyanate, between 226 and 252°C—that is, a temperature lower than that needed to initiate observable reactions in cellulose itself.

The combination of thermogravimetry with DTA or DSC has been used to clarify mechanisms of decomposition. For example, matched DTA and TGA have shown that at 220–335°C, decomposition of cyanoethylated cotton involves the loss of the cyanoethyl group; this is followed, at higher temperatures, by the typical cellulose degradation (38).

Because of their sensitivity to variables, thermal methods, such as DTA and DSC, demand care in the experimentation and caution in the interpretation of the results (38, 45, 46, 49).

(3) Mechanical Parameters

Mechanical measurements have been applied to cellulose and its derivatives during and after heat treatment, including determinations of changes in dimensions during graphitization (52), of papermaking properties (53), softening behavior (54–56), modulus of elasticity (57–61), and torsional behavior (56).

(4) x-Ray Analysis

x-Ray analysis has been used to correlate accessibility parameters (62–65) and crystallinity with thermal behavior (65–69). For example, it was found that the crystallinity of cellulose triacetate and of cyanoethylcellulose was increased by heat treatments below their decomposition temperatures (67, 68).

(5) Infrared Spectroscopy

Studies of heated solid cellulosic materials by infrared techniques have met with mixed success.

Tang and Bacon (70) heated regenerated-cellulose films under argon at 40°C/hr increments to the specified temperature, at which the samples were held for 12 hr (see Fig. 5). Up to 100°C only water desorption occurs. At 200 and 240°C weak bands appear for carbonyl stretching (5.75 μ) and ethylene vibration (6.15 μ), and the O–H and C–H stretching bands (2.95 and 3.46 μ) decrease slightly. Rapid and major changes occur at temperatures between 240 and 400°C. The bands for O–H, C–H, C–OH, and C–O–C weaken and all but disappear. The carbonyl stretching band (5.87 μ) predominates at 280°C but decreases with increasing temperature. A conjugated ethylene band (6.2 μ) appears at 280°C and becomes dominant. Extraction of the residues with a variety of solvents changed the carbonyl band slightly and did not affect the

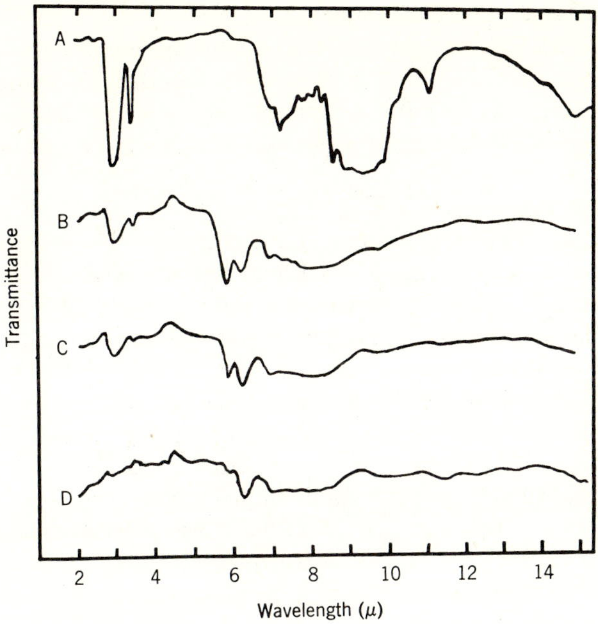

Fig. 5. Infrared spectra of regenerated-cellulose films held at various temperatures: (*A*) 100°C; (*B*) 280°C; (*C*) 360°C; (*D*) 450°C (70).

ethylene band. Other investigators, likewise, confirmed the appearance of strong carbonyl absorption at about 250°C, and decrease in O–H and C–H at about 300°C (71–74). On the other hand, brief heating at 300°C (< 1 min) produces no change in absorption (61).

(6) Electron Spin Resonance

The appearance of free radicals in cellulose during pyrolysis has been noted and studied. Ermolenko et al. (75) heated cellulose at temperatures between 200 and 700°C in a 10^{-3} mmHg vacuum, cooled it to room temperature, and observed a single symmetrical line in the electron spin resonance (ESR) spectrum. The radical demonstrated unexpected stability: the ESR line was not affected by exposing the samples to oxygen.

The stability of these radicals was further documented by Arthur and Hinojosa (76). In one sample repeatedly heated to 180°C and cooled, the radical concentration decreased on cooling to about two-thirds the amount at 180°C but then remained unchanged indefinitely. Three heating–cooling cycles generated a radical concentration about equal to that obtained by heating a sample continuously for that duration. Apparently, below a "decomposition temperature," the radical concentration reaches an equilibrium value. Above 180°C "decomposition" sets in. By 225°C the free-radical concentration increases linearly with time, and by 250°C the increase is exponential: an autocatalytic or chain-branching process was inferred (see also Section I in Chapter XVII).

(7) Molecular-Weight Determination

Golova et al. (19) heated cellulose at 305°C under 10^{-5} mmHg vacuum and examined the molecular weights of the cellulose nitrate obtained by nitrating the solid residues. The degree of polymerization (DP) of cotton fell rapidly to about 200 within the first 2–4% loss in weight, then remained constant through about 80% weight loss. The DP of regenerated cellulose fell steadily through about 40% weight loss. After even brief pyrolysis the derived nitrates contained some acetone-insoluble material. To prepare completely soluble nitrates specimens had to be nitrated two to four times. As weight loss increased, more nitrations were required; at around 40% weight loss the nitrates could not be dissolved despite repeated nitrations. This presence of insoluble nitrate indicates that reactions other than simple chain cleavage were occurring.

A similar breakdown of cellulose to a "leveling-off DP" of 200 was observed by Dmitrieva et al. (77), who heated samples at 200°C under carbon dioxide or in a high-boiling hydrocarbon. In this respect the thermal degradation of cellulose resembles acid hydrolysis, where a "leveling-off DP" is also encountered (78, 79) (see Section A in Chapter XVIII).

(8) Chemical Composition

A variety of chemical analyses have been performed on the solid residues from thermal degradation.

Suga (80), utilized a combination of acetylation, methylation with diazomethane, hydroxamation, and Grignard reactions to analyze residues carbonized to various carbon contents. In chars of 70–90% carbon, the hydroxyl, carbonyl, and acetal oxygens decreased to about the same extent. At this level resemblance to carbohydrate structure was slight. At 90–96% carbon the acetal oxygens decreased more rapidly than the hydroxyl and free carbonyl groups, which would be part of aromatic quinonoid structures. The acetal oxygens were the last vestiges of carbohydrate structure.

Other studies involved a correlation of weight loss with glucosan (24) and carbonyl and carboxyl (26) contents of residues. Elemental analyses of cellulosic residues, especially with respect to the atomic ratio of hydrogen to carbon and oxygen to carbon, indicate that dehydration is the main reaction at temperatures between 200 and 280°C; between 280 and 400°C dehydration continues but is combined with decarbonylation and decarboxylation; above 400°C dehydrogenation and demethanation can be inferred (70, 72).

(b) THE VOLATILE PRODUCTS

The nature and amount of volatile thermal products from cellulose and its derivatives determine the course of their combustion. Thus the desire to reduce or modify the flammability of cellulosic materials has provided the major incentive for the analysis of the volatile products.

Since levoglucosan was first isolated from the tarry products of cellulose pyrolysis, increasingly sophisticated techniques have disclosed more compounds among the volatile products. The appearance of such a large number of distinct products creates the suspicion that more than one reaction may be at work. However, much past research has been concerned with the effects of additives or structural modifications on the burning characteristics of the product, with little attempt to discern how the volatile products arise. Only for a few major products have the kinetics of their formation been investigated.

(1) Products from Cellulose

Effects of Variables. The effect of the type and rate of heating on the course of thermal degradation is especially noticeable in the production of volatile products. Thus if cellulose samples are "preheated" at or below 200°C for several hours before heating between 250 and 400°C, the weight of resulting residue ("char") increases and the amount of volatiles decreases correspondingly (24). Heating cellulose quickly and holding above 300°C produces usable amounts of levoglucosan (81–86). However, heating rapidly to above 400°C results in cracking of starting material and initial breakdown products into water, carbon dioxide, carbon monoxide, and other small molecules (87–90). "Hot-wire" pyrolysis followed by gas chromatography may provide fingerprint chromatograms of polysaccharides and smaller carbohydrates (91–93).

The explosion of cellulose into small molecules when exposed to high radiant energy flux has been studied to determine the effects of flame-retardant treatments or added salts on the production of volatiles and on ignition and flaming behavior (87, 88, 94–96). The amount and composition of volatiles are also affected strongly by inorganic impurities in the sample. Many studies had been performed before the extent of this influence was recognized, and experimental results on samples from different sources should only be compared judiciously.

Product Characterization. The volatile products generated from cellulose range from small molecules (hydrogen, methane, carbon dioxide) through intermediate-size hydrocarbons, alcohols, and carbonyl compounds to levoglucosan and other anhydrides equal in weight to the cellulose monomer. The total number isolated and identified (if only tentatively) is well over 50 and includes many compounds of carbon, hydrogen, and oxygen.

Separations of the less volatile products (loosely classified as the "tar" fraction) have been achieved through paper chromatography (71, 80), thin-layer chromatography of carbonyl derivatives (97), and column chromatography on alumina (98). Among the larger molecules isolated are levoglucosan in yields of 40% and higher (99); 1,6-anhydro-D-glucofuranose (1.5%), 5-hydroxymethyl-furfuraldehyde (1.3%), and 1,4 : 3,6-dianhydro-D-glucopyranose (0.7%) (100); and at least 10 polynuclear hydrocarbons, such as acenaphthylene, in trace amounts (98).

The more volatile components (the "gas" fraction) are separated by gas chromatography (33, 72, 91–93, 100–102). Under suitable conditions major products include water, carbon monoxide, carbon dioxide, formic acid, and glyoxal (18, 102). Minor products include hydrogen, methane, and other small hydrocarbons (87, 103). Schwenker and Beck (102) published one of the most extensive tabulations of volatile products. From purified cotton heated at 15°C/min to 370°C they isolated 37 peaks by gas chromatography.

Identification of isolated substances depends on a variety of techniques, such as determination of the melting point of derivatives (104) and mass spectroscopy (18). The combination of gas chromatography with mass spectroscopy promises to be especially powerful (88).

Kinetic Studies. Cerniani (105) measured volumes of gases generated from cellulose filter paper, glucose, sucrose, lactose, and potato starch during stepwise heating. Any one of five products (CO_2, CO, H_2, CH_4, and "other hydrocarbons") was evolved at about the same rate from all substrates. Carbon dioxide evolution peaked at 250°C. Carbon monoxide evolution from cellulose peaked near 340°C, but evolution from other carbohydrates peaked near 400°C. Methane production began at about 350°C, rose to about 445°C, and then leveled off. The total gas evolution from cellulose peaked at about 280°C. Madorsky, Hart, and Straus (18) studied the rate of evolution of water, carbon dioxide, and carbon monoxide; above 240°C the mass spectra were too complicated for detailed analysis. Golova et al. (19) used the increase in reducing content of the tars after hydrolysis to measure levoglucosan yield from cotton at 305°C. This yield peaks at about 20% decomposition and slopes off through 80% decomposition. Murphy (23) analyzed water, carbon dioxide, and carbon monoxide produced from α-cellulose at temperatures between 100 and 250°C under 10^{-4} mmHg vacuum. The data fitted two parallel but slightly displaced straight lines—evidence that two reactions were occurring. The first (activation energy of 34 kcal/mole) produces water, carbon dioxide, and carbon monoxide in a volume ratio close to 3 : 1 : 1 and is said to be decomposition of a secondary "intermicellar" structure. The second (activation energy of 39.4 kcal/mole) reportedly involves decomposition in the "interior of the micelle." giving water, carbon dioxide, and carbon monoxide in a ratio near 10 : 1 : 1.

(2) Products from Cellulose Derivatives

A few typical examples of studies carried out on cellulose derivatives are cited.

Ethers. Reilly (106) and Pakhomov et al. (107) examined the volatile products of trimethylcellulose but characterized only 2,3,6-tri-*O*-methyl-1,4-anhydro-1,5-α-glucopyranose, which could be isolated in 42% yield. Gas chromatography appears to be particularly appropriate to this problem since many of the larger products would already be in a suitable "derivatized" form [cf. Neely et al. (108)].

Acetates. Madorsky et al. (27) heated cellulose triacetate under nitrogen or in vacuum and collected 45–48% acetic acid and 45% of a tar that had an average molecular weight of 236 (versus 288 for the monomer). Its infrared spectrum differed from that of cellulose triacetate and bore little relation to that of 2,3,4-triacetyllevoglucosan. Other products included acetaldehyde, methanol, carbon dioxide, carbon monoxide, and acetone. No water was detected. Gokcen and Cates (93) and Kammermaier (92) pyrolyzed cellulose triacetate and identified among the products acetic acid, water, carbon monoxide, carbon dioxide, methane, and hydrogen; the latter appeared only above 700°C.

Nitrates. Wolfrom et al. (109–115) investigated the formation of volatile products from cellulose nitrate ignited in a gentle stream of nitrogen. The pressure affected the nature and amount of products. At a pressure of 2–3 mmHg a solid nitrated oxidized cellulose formed. At higher pressures there appeared a liquid mixture that consisted largely of water, formic acid, formaldehyde, and glyoxal but included many other carbonyl compounds in lesser quantities. By igniting nitrate labeled with ^{14}C at positions $C_{(1)}$ and $C_{(6)}$ they determined that $C_{(1)}$ converts predominantly to carbon dioxide, but also to formic acid and glyoxal (112, 113). Carbon-6 forms mostly formaldehyde, plus smaller amounts of formic acid and carbon dioxide. The $C_{(2)}$ and $C_{(5)}$ positions contribute some carbon dioxide and formic acid, but little formaldehyde.

Gelernter et al. (116) decomposed cellulose nitrate labeled at position 6 with ^{15}N. The percentage of ^{15}N in the gas was lower than in the residue but increased with time, indicating that the nitrate is more stable at position 6 than at positions 2 or 3. The infrared spectrum of the residue showed the appearance of hydroxyl groups (2.9 μ) and a gradual buildup of carbonyl groups (5.75 μ). Gas chromatography may be used to analyze the volatiles (32, 117).

(c) THE EFFECTS OF SAMPLE COMPOSITION

(1) Supermolecular Structure

There have been several attempts to relate supermolecular structure (118, 119) to cellulose pyrolysis (65, 99, 120). Complications arising from different ash contents of the cellulose samples have obscured the true meaning of the findings (see also Section A of Chapter XVII).

(2) Inorganic Contamination: Ash Content

The content of inorganic materials – "ash content"–in cellulose samples exerts a major effect on the course of their thermal degradation. Venn (121) observed in 1924 that removing inorganic contaminants before dry-distilling cellulose dramatically raised the levoglucosan yield. The relationship between levoglucosan yield and flammability of cellulosics encouraged study of the effect of inorganic salts on this yield.

Addition of 2% by weight of various flame-retarding salts lowers the active weight-loss temperature and increases the amount of residue for α-cellulose (36). Inorganic ash content softens the abrupt weight loss and eliminates the endotherm exhibited by pure cellulose above 300°C (42). Added sodium chloride (8%) or sodium carbonate (7%) decreases the yield of tar (supposedly levoglucosan) and increases the yields of water, carbon dioxide, and carbon monoxide (18). The general effect of salts seems to be to decrease the tar and increase the residue, in some cases dramatically. The yields of carbon dioxide and carbon monoxide do not change greatly (103).

Golova and Krylova (99) measured the rate of weight loss and the levoglucosan yield from celluloses containing from less than 0.001% up to 0.26% ash. As much as 70% levoglucosan arose from the "ash-free" material. With increasing ash content the initial weight loss was accelerated, the levoglucosan yield was decreased, and the products were shifted away from those formed by C–O–C bond cleavage toward those produced by "deep-seated" C–C bond cleavage–that is, breakup *within* the monomers to molecules smaller than the anhydroglucoses. Golova, Epstein, and Durynina (122) completely degraded cellulose in superheated steam at temperatures between 260 and 360°C in the presence of 0.04–3.0 wt % of various salts and used the difference between actual and theoretical yields of levoglucosan to estimate "deep decomposition." The order of effectiveness in promoting "deep decomposition" was as follows: monovalent chlorides > sulfates > divalent chlorides.

The addition of 1.5% potassium bicarbonate increases the yields of all volatile products and decreases the ratio of flammable volatiles to nonflammable volatiles (carbon dioxide and water) from α-cellulose exposed in helium to high-intensity thermal radiation (123). As a result the treated samples did not support sustained flaming combustion (see also Section I in Chapter XIX).

2. Behavior of Model Compounds

Many investigators have looked to molecules that differ monotonically from cellulose for models to provide parameters for mechanisms describing the reactions of cellulose and its derivatives.

Monosaccharides, such as glucose, chiefly react above their melting points by condensing to a host of oligomers and polymers (124–128). Inorganic materials catalyze such polymerization and subsequent decompositions (125, 129). The condensation products are felt to be the major source of the many volatile products formed during subsequent degradation. Glucose degrades at 300°C to at least 56 compounds, but chiefly to 1,4 : 3,6-dianhydro-D-glucopyranose (130). However, the proposed intermediacy of glucose in the thermal formation of levoglucosan from cellulose was discounted by the finding of Golova et al. (131) that the addition of glucose actually depresses the yield of levoglucosan.

Levoglucosan in a sealed system was observed by Pictet (132) to polymerize when heated. Later studies of this polymerization showed the products to range from disaccharides (133) to moderately large polymers (134). The reactions are catalyzed by Lewis acids (135–137). On the basis of such behavior, Laible (12), Schwenker and Beck (102), and others proposed the "levoglucosan mechanism" whereby levoglucosan—the initial thermal product—both polymerizes to generate the residue and decomposes to provide all the observed smaller volatile products. Esteve and Wright (138) suggested that levoglucosan first rearranges to 5-hydroxymethylfurfural, which then polymerizes or decomposes. Support for this theory comes from the observation that inorganic substances that catalyze the polymerization of levoglucosan also act as flame retardants (4) and the similarity between the products from levoglucosan or cellulose heated at 170–360°C in a sealed oven (139).

Madorsky, Hart, and Straus (18) compared the behavior of levoglucosan and celluloses under 10^{-3}–10^{-5} mmHg vacuum. Levoglucosan evaporated at near 130°C and gave peaks only for water, carbon dioxide, and carbon monoxide—peaks that, however, were quite small compared with those from the celluloses. Honeyman (7) reported that hot-wire pyrolysis at 350°C (and gas chromatography) of levoglucosan produces small amounts of acetaldehyde and dimethoxymethane, but little of other volatiles. Such results contrast sharply with the results of sealed-system pyrolysis.

Amylose, a linear starch fraction, differs from cellulose only by stereochemistry at the point of attachment of the monomers to one another, making it a plausible model for comparing with cellulose. Bryce and Greenwood (140–143) heated each in vacuum. The production of carbon dioxide and carbon monoxide was the same from both up to 300°C; above this temperature amylose produced twice as much carbon dioxide and 5–10 times as much carbon monoxide. The residue after 20 min at 291°C from amylose was 9% and from cellulose 57%; at 326°C amylose left 8% residue and cellulose 9%. At these two temperatures the yields of tar from amylose were 14 and 53%; from cellulose, 0.2 and 2.0%, respectively. On the basis of TGA results, Carroll and Liskowitz (144) concluded that the nature of the impurities present in various starch preparations plays a more important role in the decomposition than do the structural differences.

3. Mechanisms of Thermal Degradation

(a) DEVELOPING A MODEL

Whereas degradations, such as hydrolysis, produce smaller and usually simpler substances, thermal degradation of cellulose and its derivatives leads to new chemical and physical structures in the solid and a bewildering array of volatile products. Mechanistic descriptions of the thermal reactions are complicated by commonly encountered confusion between what was experimentally measured and what can reasonably be concluded from the data. The difficulties inherent in developing rate expressions and extending them to molecular terms is compounded by the need to define carefully such common words as "concentration," "rate," and even "molecule" when they are applied to reactions of high polymers (44).

In an early review of the process (145), it was proposed that cellulose initially decomposes to levoglucosan, which then reacts to provide all the observed products: char, tar, and volatiles. Variations on this theme were composed as described above. A more recent schematic depicts different monomers reacting in one of two alternative sequences to give two intermediates: a monomer that can rearrange into levoglucosan or other anhydrides, and a modified residue ("anhydrocellulose") that further reacts to form char and volatiles. Since the descriptions differ formally only slightly, rate expressions derived from either model accommodate the same sets of data resonably well (20, 24, 25). In this sense no rate expression "proves" a molecular model, it merely narrows down the possibilities.

At present the weight of evidence described in the preceding discussion seems better accommodated by a set of reactions that compete for the native polymer. To reiterate briefly: temperatures between 150 and 250°C lead to the production of a substance that is different from cellulose in DTA behavior and in the production of more char and of little or no levoglucosan at higher temperatures (24, 49). Different temperatures lead to different ratios of volatile products (18, 23) and different anhydrosugar-to-residue ratios (140, 143) (typically a single reaction would produce only changes in rate with temperature). Additionally different rate expressions and different activation energies apply at different temperatures and different stages of decomposition (23, 24, 99). Finally, results of studies on simple molecules, such as glucose (131) and levoglucosan (7, 18) contradict their exclusive intermediacy in all the observed behavior of cellulose.

The present status of the mechanistic model for cellulose can be summed up as follows:

1. During heating cellulose undergoes several distinct transformations. These may be sequentially interrelated but, more likely, include two processes that compete for the monomer units: char formation and depolymerization.
2. The residue, or char, arises from the decomposition of the monomers *in situ*, possibly preceded by coupling between monomers.
3. Anhydrides (levoglucosan, etc.) arise through depolymerization: uncoupling of monomers with subsequent rearrangements.
4. Smaller volatile products arise from the decomposing residue.

Variations of this pattern describe the thermal behavior of cellulose derivatives:

1. *Cellulose nitrate* reacts differently, depending on the temperature used; at lower temperatures much nitrogen-containing residue arises, which is stable at higher temperatures; at higher temperatures the samples volatilize more or less completely into small molecules—no "monomers" are formed.
2. *Cellulose acetate* undergoes several phase transitions in addition to at least two reactions: the elimination of acetic acid, leaving a reactive substance that becomes the char, and some form of depolymerization.
3. *Cellulose ethers* vary in behavior. Methyl and ethyl ethers of high DS appear largely to depolymerize; larger derivatives (such as triphenyl and benzhydryl) and those of lower DS appear to eliminate hydrocarbon or alcohol fragments, leaving an unsaturated substance that forms the char.
4. *Cellulose carbanilates* revert to the isocyanate and cellulose at moderate temperatures.
5. *Other esters*, such as tosylates, mesylates, and phosphates, appear to eliminate the parent acid, leaving reactive unsaturated residues that form char.

In general, compared with smaller carbohydrates, many other polysaccharides, and even most of its derivatives, cellulose is remarkably stable toward thermal degradation.

(b) TRANSFORMATIONS OF CELLULOSE AND ITS DERIVATIVES

Thermally induced chemical transformations of cellulose are known to include the appearance of free radicals; the formation of carbonyl, ethylene, and aromatic groups; and the loss of acetal, hydroxyl, and C–H groups; that is, the ultimate conversion of each group in the repeating unit into some other functional group. However, there is still uncertainty concerning the specific fate of each functional group, the rates of specific transformations, and their detailed mechanisms.

One transformation must involve cleavage of the acetal bonds as a prerequisite for depolymerization. Madorsky et al. (18) suggested that this was a nonrandom process. Pakhomov (146) argued by analogy with other heteroatom-containing polymers that the acetal linkage was the weakest bond in cellulosic polymers. A homolytic type of cleavage was suggested. The significant concentrations of free radicals in heated cellulose (75, 76) could originate from such a reaction and support the proposed mechanism.

Detailed analysis of this process has been hindered by lack of knowledge of subsequent reactions and resulting products. It has been suggested that "active sites" (whether free radical or ionic in nature) could disproportionate, open to form carbonyl groups, or interact at other sites to form anhydride bridges, crosslinks, and so on, in addition to simply recombining. On the assumption that at certain temperatures the overwhelming reaction of these sites produces anhydride "end groups" that rapidly form volatile monomers, Chatterjee and Conrad (20, 25) developed rate expressions to accommodate static TGA data and, by assuming a variety of subsequent reaction situations, dynamic TGA data as well (44). Their first-order expressions fit the data, and activation energies calculated over the early decomposition are compatible with similar results from other workers (18, 24). However, their derivations assume initial reaction conditions; when the derivations are applied to later stages of decomposition, discrepancies in activation energies and data plots are noted (44).

Ermolenko et al. (75) explained the stability of the radicals toward oxygen by assuming that they were inaccessible to the oxygen, "caged," so that the oxygen could not diffuse to reach them. Since they observed the radicals after cooling the samples to room temperature, radical recombination must have been in some way hindered. The autocatalytic formation of radicals at 250°C and higher temperatures (76) may be due to "release" of caged radicals by the increased mobility of the polymer segments and the newly formed molecular ends. Once initiated, these radicals could partake in a rapid "unzipping" type of depolymerization.

Other transformations of cellulose appear to involve random cleavage of carbon–carbon bonds, elimination of water, and disproportionations of hydroxyl and carbonyl groups. The plausibility of such reactions is based on infrared studies (7). Mechanisms attempting to explain some of these observations are described below in connection with char formation. In general they lack kinetic support.

Madorsky et al. (27) interpreted the decomposition of cellulose triacetate as being comprised also of at least two processes. Although they estimated an activation energy for the overall weight loss, they did not study the kinetics of the separate processes in detail.

At lower temperatures the reactions of cellulose nitrate appear to involve the concerted loss of nitrate groups and acetal linkages: the latter appear to revert to carbonyl groups. Phillips et al. (34) calculated the activation energies to be between 40.1 and 45.3 kcal/mole for the reactions, which apparently lead to the char obtained under these conditions (35). Higher temperatures induce an explosive decomposition directly to smaller molecules (35). Possible mechanistic sequences for this were outlined by Wolfrom and co-workers (109–115) (see below under the discussion of volatile products).

The initial transformation of trimethylcellulose is felt to involve cleavage of the acetal linkage. Pakhomov (146) points out that this reaction is most likely followed by depolymerization. Other transformations of cellulose ethers involve eliminations of hydrocarbons, alcohols, or vinyl compounds depending on the nature of the ether. Unsaturated residues are formed in some cases whose further reactions have not been described in detail.

Schwenker et al. (34) sketched the eliminations of acids from cellulose tosylates, mesylates, etc., which resulted in new covalent bonds. They did not investigate the kinetics of the various processes.

(c) ANHYDROSUGAR FORMATION

Cellulose, trimethylcellulose, and cellulose triacetate generate large amounts of monomeric or dimeric anhydrosugars when heated under suitable conditions. Weight-loss and glucosan-loss measurements have been fitted by numerous workers into two distinct rate expressions. The earlier decomposition follows pseudo-zero-order kinetics; later decomposition follows pseudo-first-order laws (20, 24, 25, 44). The fitting of degradation-rate data into zero-order rate expressions is compatible with a chain-reaction depolymerization process (147, 148). According to this model, the radical, or other active center, after initiation moves down the length of the polymer. At each transfer a "monomer" breaks away and a fresh active center is generated. The number of reactive species thus remains constant despite an overall loss of sample. [Chatterjee's explanation differs from this slightly in that he combines the original "initiation" process with each new scission of a glucosidic bond (25); his "propagation" step is simultaneous with this second type of "initiation.] After the average length of the molecules drops to some value, this rejuvenation decreases in relative importance and the observed rate will depend on the initiation process. The number of reactive species thus corresponds more closely to the number of molecules and hence to the weight of the sample, and the kinetics shift to first order.

In addition to bond cleavage, which separates monomers from the chain, the production of anhydrosugars requires the formation of new bonds. Since the attachment of the extrapyranose oxygen is the same in levoglucosan and cellulose (see Fig. 1), the formation of the new anhydride ring in levoglucosan by direct displacement is unlikely; an inversion of configuration would be

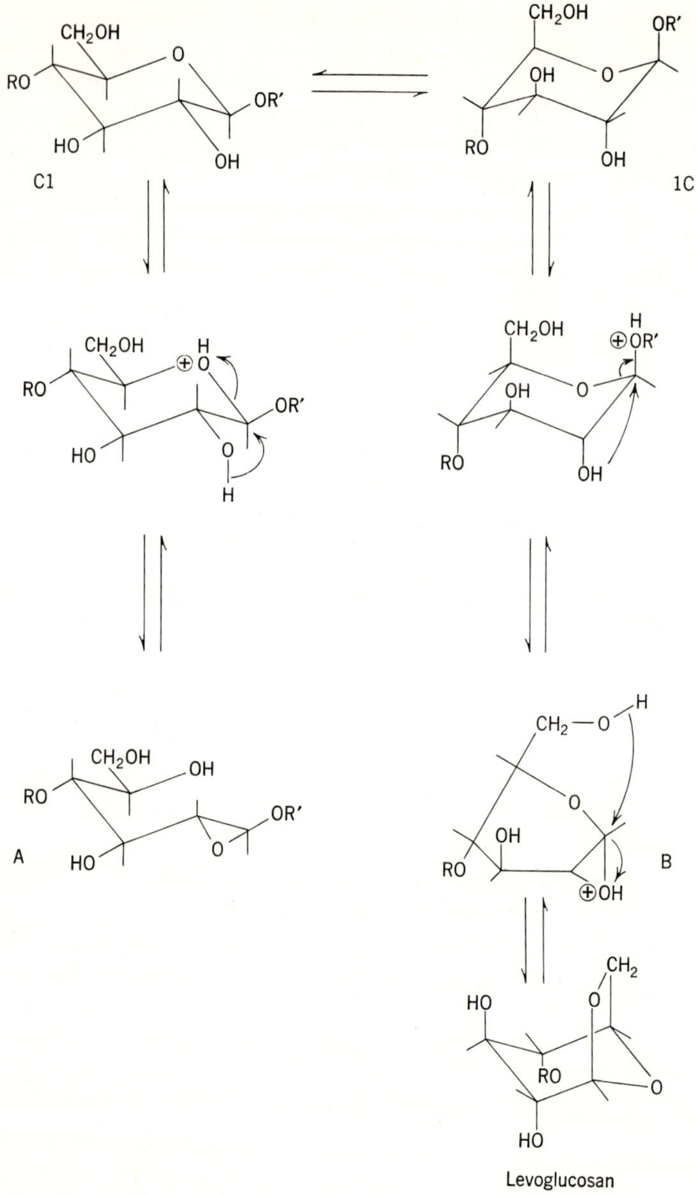

Fig. 6. Proposed rearrangements of cellulose monomer involving 1,2-anhydride formation (7). The ring-opened anhydride (*A*) can rearrange to give furan derivatives (see Fig. 8). The ring-closed pyranosic anhydride (*B*) can rearrange into the bicyclic levoglucosan structure.

expected (149). Madorsky et al. (18) circumvent this by proposing an intermediate 1,2-anhydrosugar (see Fig. 6) similar to that used by McCloskey and Coleman (150) to describe the formation of levoglucosan from β-glucosides.

Kilzer and Broido (47) describe a concerted chain-reaction displacement of each $C_{(1)}$-attached group by the monomer's own oxygen or hydroxyl group, which generates a novel 1,4-α-anhydride and a new reactive group at $C_{(4)}$ simultaneously. This anhydride then rearranges to give levoglucosan or other anhydrides, since it is relatively unstable (Fig. 7). Chatterjee and Conrad (20) suggest that the new 1,6-anhydride bridge is formed before the second bond holding the monomer to the chain is cleaved.

Honeyman (7) and Gardiner (100) provide detailed pictures of the formation of levoglucosan, other anhydrides, and many smaller molecules. Some of their proposals are illustrated in Figures 6 and 8. Gardiner refers to Madorsky's (18) suggested 1,2-epoxide to rationalize the formation of the levoglucosan 1,6-β-anhydride link from the β-oriented group at $C_{(1)}$ in cellulose. After an unexplained rupturing of the $C_{(4)}$ attachment, the levoglucosan is free to vaporize. Gardiner suggests that under pyrolytic conditions concerted displacement reactions in the solid state are more favorable than reactions requiring solvation or hydration: the charge dispersal in the transition state of a concerted displacement provides an alternative type of stabilization.

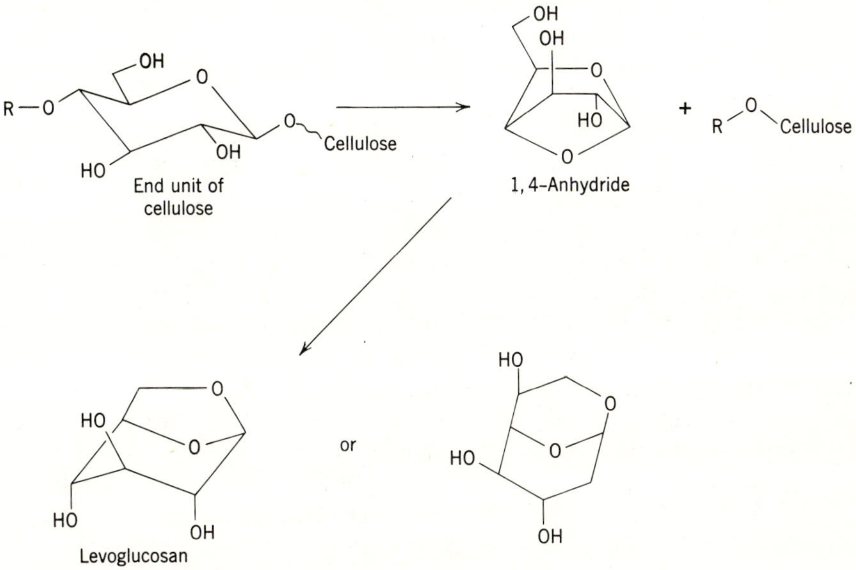

Fig. 7. Proposed "unzipping"-type of mechanism for depolymerization of cellulose or trimethylcellulose. The nature of the R is not specified and may be free radical, anion, or protonated (47).

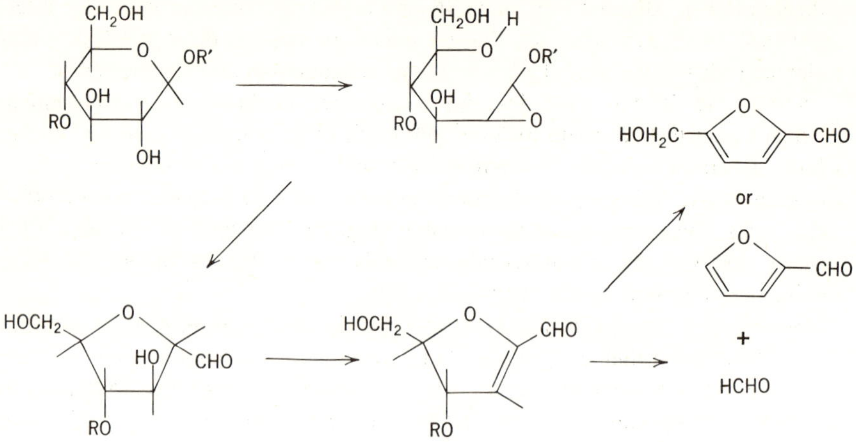

Fig. 8. Proposed mechanisms for the decomposition of cellulose into furans and formaldehyde through a ring-opened 1,2-anhydride (7).

The simplest description of the formation of the 1,4-anhydride from trimethylcellulose invokes a simultaneous displacement-depolymerization comparable to that applied to cellulose (47). The presence of the methyl groups on the substituent oxygens effectively blocks any other action as well as any other mechanism for anhydride formation. Thermal decomposition of a triethylcellulose conceivably follows a similar course. In either case lower degrees of substitution free hydroxyl groups for other reactions and decrease the amount of 1,4-anhydride produced. Other ethers do not appear to generate anhydride-type products.

The thermal products of cellulose triacetate depend on the ambient pressure: in vacuum a tar and acetic acid each account for 45% of the starting material; under nitrogen the decrease in tars nearly equals the increase in acetic acid. In vacuum the fully acetylated derivatives of fairly large carbohydrates or their anhydrides—as large as disaccharides, according to Bishop (151, 152)—are volatile, whereas under nitrogen they would remain in the hot zone and decompose into acetic acid and nonvolatile residue. The tar could then consist of a mixture of acetylated anhydrosugars, including the monomer (molecular weight 288) and a dianhydride (molecular weight 228). (The average molecular weight of the tar was 236.) The tar was not characterized further, nor were kinetic studies of this process published.

For a further discussion of anhydro derivatives of cellulose see Section F in Chapter XVII.

(d) RESIDUE OR CHAR FORMATION

The reactions leading to the solid residue have not been as throughly studied as those leading to anhydrides, but the growing importance of carbonized materials is changing this balance. Since the known small products of thermal degradation appear to be too volatile under most conditions to play important roles in char formation, the burden of investigation rests on reactions of the polymer.

Madorsky et al. (18) have proposed a random dehydration along the cellulose chain to compete with depolymerization for the original monomeric units. Scission of C–C bonds would lead to carbon dioxide, carbon monoxide, and the carbon residue. Kilzer and Broido (47) suggest a preliminary displacement at $C_{(4)}$ that frees the attached carbonyl group; this open carbonyl group then rearranges to a tetrahydro-5-hydroxymethylfurfural group with the loss of water (see Fig. 9). This scheme rationalizes many observations about the course of char formation, but the proposed structure of their "anhydrocellulose" remains speculative.

The participation of crosslinking reactions in the formation of the residue has been proposed by Back and co-workers (59–61). Operationally a crosslinking reaction can be defined as one that leads to decreased solubility of the polymer or, in fibrous materials, to an increase in dimensional stability at right angles to the fiber direction. By this definition Back's results support crosslinking, as do the observations of Golova et al. (19) that heating cellulose near 300°C generates a substance that repeated nitrations fail to render acetone-soluble. Similarly Lipska and Parker (24) found increasing amounts of nonhydrolyzable residue in cellulose samples heated at 250–300°C.

Bacon and Tang (52) have described an alternative mechanism since they observed an increase in the length of carbonized cellulose filaments with increasing stretch during the treatment. They feel that crosslinking should decrease the length with increased stretch. As additional adverse evidence, they cite a decrease in the rigidity of rayon filaments pyrolyzed above 240°C. Their mechanism for carbonization (70) involves the following four steps:

1. An initial physical desorption of water.
2. This is followed by an intramolecular dehydration with the formation of conjugated carbonyl and ethylenic groups at temperatures between 150 and 240°C.
3. Next the glycosidic linkages between rings rupture along with other C–O and some C–C bonds at temperatures between 240 and 400°C. This is described as the major reactive stage preceding the formation of tar, water, carbon monoxide, and carbon dioxide.

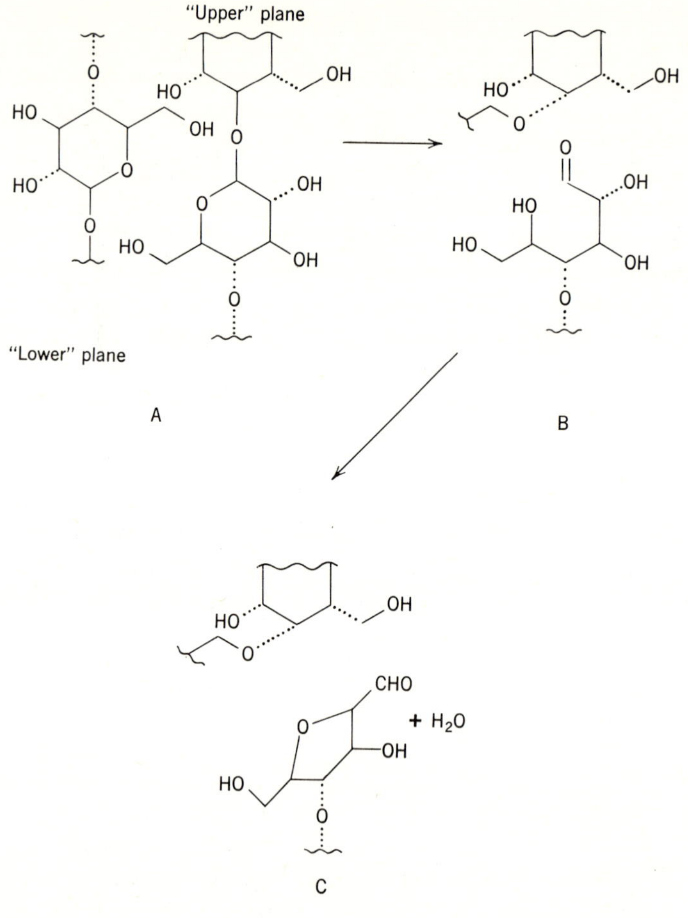

Fig. 9. Proposed mechanism for thermal crosslinking in cellulose (47). The $C_{(6)}$ hydroxyl group displaces the oxygen attached to $C_{(4)}$ of another glucosyl ring in a higher plane (*A*) to produce a new $C_{(4)} - C_{(6)}$ ether linkage and a ring-opened glucosyl unit (*B*). The hydroxyaldehyde (*B*) rearranges to give a tetrahydro-5-hydroxymethylfurfural unit plus water (*C*).

4. Without suggesting a distinct chemical intermediate, they propose an aromatization stage above 400°C, consisting of the polymerization of four-carbon residues from each pyranose ring to form graphite layers.

Steps 3 and 4 are described as chain reactions involving free radicals. The carbon "chains" of the resulting graphite are formed along the axes of the original cellulose chains.

Formation of residues from cellulose esters appears to proceed by an initial elimination of an acid (or nitrogen dioxide from nitrates) to leave an unsaturated reactive substance. The loss of acetic acid from cellulose triacetate (27) may be used as an example. Extensive studies of acetate pyrolysis [reviewed by DePuy and King (153)] support a *cis*-elimination mechanism. Two such reactions are depicted in Figure 10 for cellulose triacetate. Elimination from $C_{(6)}$ gives a methylene group; elimination from $C_{(2)}$ or $C_{(3)}$ gives a double bond inside the ring.

The first-order decomposition of cellulose nitrate at temperatures below 250°C to a residue was interpreted by Gelernter et al. (116) as shown in Figure 11. An initial homolytic loss of nitrogen dioxide is followed by C–C cleavage between the vicinal 2,3-dinitrate group and elimination of another molecule of nitrogen dioxide. The resulting dialdehyde reacts with the nitrogen dioxide to give carbon dioxide, formic acid, etc. In the third stage an accelerated decomposition

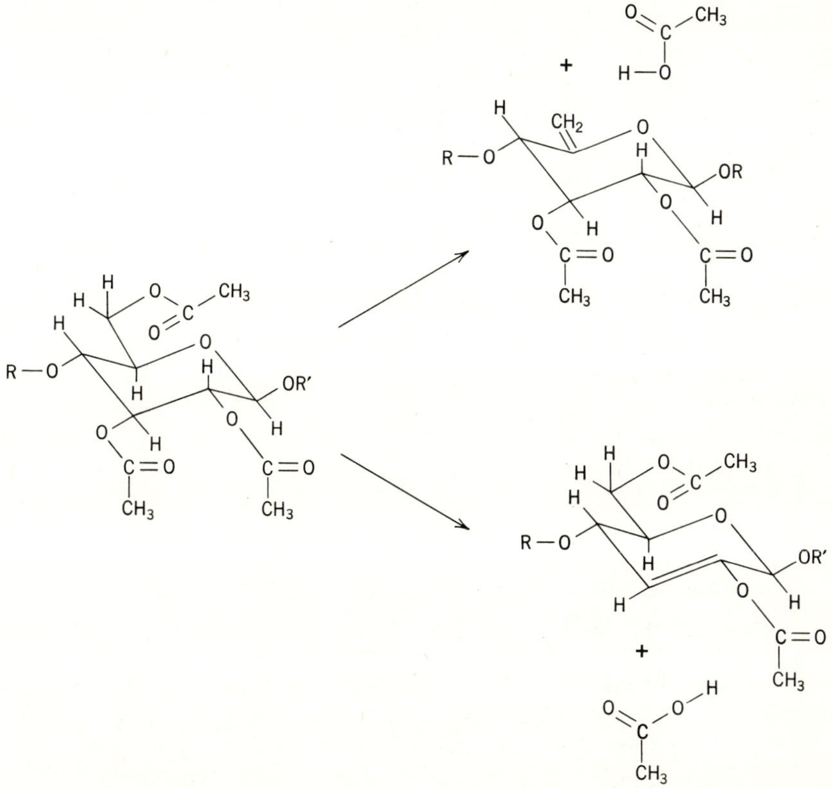

Fig. 10. Proposed mechanisms for pyrolytic *cis*-eliminations of acetic acid from cellulose triacetate (153). Depolymerization would involve other monomeric units.

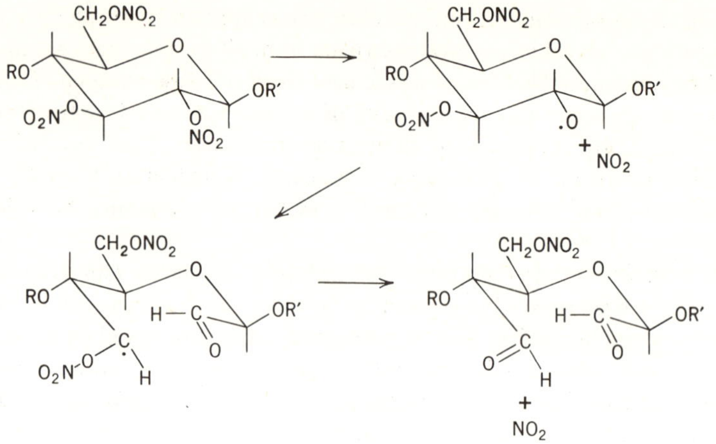

Fig. 11. Proposed mechanisms for the decomposition of cellulose nitrate (116).

of the nitrate groups at $C_{(6)}$ into nitrogen dioxide and formaldehyde is accompanied by the decomposition of the cellulose chain into smaller fragments. At lower temperatures the dialdehyde could react to form the thermally resistant residue observed by several workers.

Reactions similar to acetate pyrolysis may account for the behavior of mesylates, tosylates, phosphates, etc., since these functions are known to be good "leaving groups" of wide synthetic application (149). Schwenker et al. (38) cite such reactions as forerunners of reactive ethylene groups, which may then undergo a variety of polymerizations catalyzed by the newly formed acids.

A detailed description of residue formation from cellulose ethers has not been attempted.

See also Section F of Chapter XVII for a description of unsaturated derivatives of cellulose.

(e) OTHER VOLATILE PRODUCTS

The nature and amount of smaller volatile molecules generated from cellulose and its derivatives are largely dictated by the nature of the groups attached to the pendant oxygens. Production of these molecules seems more straightforward in the case of the derivatives than from cellulose itself. For example, isocyanates are liberated from carbanilates (51), acids from esters, etc. The unsaturated polymeric substance that may be formed concurrently undergoes further decomposition into volatile molecules and a char. This latter stage of decomposition has received scant attention. Cellulose itself and cellulose nitrate are two exceptions to this rule. In fact the formation of these small molecules from cellulose nitrate is of prime importance.

A minor proportion of the volatiles generated from cellulose is felt to arise from the decomposition of levoglucosan and similarly sized intermediates. The major quantity appears to arise from decomposition of the solid residue (7, 47, 96, 146).

At high temperatures and low pressures cellulose nitrate decomposes more or less completely into small volatiles. Formaldehyde and some formic acid and carbon dioxide arise, from $C_{(6)}$. Most of the formic acid and carbon dioxide originate from $C_{(1)}$. Glyoxal comes from $C_{(1)}$ and $C_{(2)}$. Wolfrom et al. attribute large differences in the relative yields of products to competing fragmentation processes and subsequent secondary reactions. From studies of other polysaccharide nitrates, it can be concluded that all such compounds react by similar mechanisms during low-pressure decomposition (113). Wolfrom et al. (110) noted that ignition at nitrogen pressures below about 5 mmHg allows a polymeric oxidized cellulose of low DP to form. Andreev (31) and others have pointed out that the gaseous decomposition products can react further with the solid residue. Their presence seems important to the explosive decomposition at higher temperatures; their absence allows the polymeric residue to arise.

The reader is also referred to Section I of Chapter XIX for a related discussion.

References

1. A. Pictet and J. Sarasin, *Compt. Rend.*, **166**, 38 (1918).
2. J. C. Irvine and J. W. H. Oldham, *J. Chem. Soc.*, **119**, 1744 (1921).
3. K. Josephson, *Ber.*, **62B**, 313 (1929).
4. H. A. Schuyten, J. W. Weaver, and J. D. Reid, *Fire Retardant Paints, Advances in Chemistry Series*, No. 9, American Chemical Society, Washington, D.C., 1954, p. 7.
5. W. Sandermann and H. Augustin, *Holz Roh- und Werkstoff*, **21**, 256 (1963).
6. G. N. Rozmarin, *Usp. Khim.*, **34**, 1965 (1965).
7. J. Honeyman, *A Fundamental Study of the Pyrolysis of Cotton Cellulose*, Final Report, Project No. UR-E29-(20)-9, the Cotton Silk and Man-Made Fibres Research Association, Shirley Institute, Manchester, England, 1964.
8. A. Broido and F. J. Kilzer, *Fire Research Abstr. Rev.* **5**, 157 (1963).
9. G. D. M. MacKay, in *Wood: Fire Behavior and Fire Retardant Treatment*, Canadian Wood Council, Ottawa, Canada, November 1966.
10. A. E. Lipska, *The Pyrolysis of Cellulosic and Synthetic Materials in a Fire Environment*, USNRDL-TR-1113, U.S. Naval Radiological Defense Laboratory, San Francisco, December 14, 1966.
11. S. Martin, *The Mechanisms of Ignition of Cellulosic Materials by Intense Thermal Radiation*, USNRDL-TR-102- NS 081-001, U.S. Naval Radiological Defense Laboratory, San Francisco, February 10, 1956.
12. R. C. Laible, *American Dyestuff Reptr.* **47**, 173 (1958).
13. F. L. Browne, *Theories of the Combustion of Wood and Its Control*, Report No. 2136, U.S. Department of Agriculture, Forest Products Laboratory, Forest Service, Madison, Wis., December 1958.

14. J. M. Church, *Fundamental Studies of Chemical Retardants for the Fire Resistant Treatment of Textiles,* Research and Development Report, Textile Series Report No. 38, Department of the Army, Office of the Quartermaster General, Columbia University, New York, 1952.
15. I. M. Gottlieb and L. R. Beck, Jr., *Ann. N.Y. Acad. Sci.,* **82**, 724 (1959).
16. I. S. Walker, *Bushfire Combustion Studies–Fuel Pyrolysis,* Division of Physical Chemistry, Commonwealth Scientific and Industrial Research Organization, Chemical Research Laboratories, Melborne, Australia, March 1963.
17. W. W. Wendlandt, *Thermal Methods of Analysis,* Interscience, New York, 1964; J. Chiu, *Appl. Polymer Symposia,* No. 2, 25 (1966).
18. S. L. Madorsky, V. E. Hart, and S. Straus, *J. Res. Natl. Bur. Std.,* **56**, 343 (1956).
19. O. P. Golova, R. G. Krylova, and I. I. Nikolaeva, *Vysokomol. Soedin.,* **1**, 1295 (1959).
20. P. K. Chatterjee and C. M. Conrad, *Textile Res. J.,* **36**, 487 (1966).
21. A. J. Stamm, *Ind. Eng. Chem.,* **48**, 413 (1956).
22. R. M. Esteve, Jr., M. H. Gollis, and R. Garcia, Progress Reports No. 12 and 13, Q. M. Contract No. DA 44-109-QM-1414, University of Rhode Island, cited in reference 20.
23. E. J. Murphy, *J. Polymer Sci.,* **58**, 649 (1962).
24. A. E. Lipska and W. J. Parker, *J. Appl. Polymer Sci.,* **10**, 1439 (1966).
25. P. K. Chatterjee, *J. Appl Polymer Sci.,* **12**, 1859 (1968).
26. K. Kato and N. Takahashi, *Agr. Biol. Chem.,* **31**, 519 (1967).
27. S. L. Madorsky, V. E. Hart, and S. Straus, *J. Res. Natl. Bur. Std.* **60**, 343 (1958).
28. J. F. von Gizycki, *Chem. Ztg.,* **74**, 649 (1950).
29. T. Abe, S. Tobisawa, and Y. Nomura, *Bull. Chem. Soc. Japan,* **31**, 1041 (1958).
30. K. K. Andreev and B. S. Samsonov, *Dokl. Akad. Nauk SSSR,* **114**, 815 (1957).
31. K. Andreev, *Explosivstoffe,* **8**, 275 (1960); through *Chem. Abstr.,* **55**, 10891 (1961).
32. K. K. Andreev, *Ind. Chim. Belge, Suppl.,* **2**, 222 (1959); through *Chem Abstr.,* **55**, 22824 (1961).
33. K. Ettre and P. F. Varadi, *Anal. Chem.,* **35**, 69 (1963).
34. R. W. Phillips, C. A. Orlick, and R. Steinberger, *J. Phys. Chem.,* **59**, 1034 (1955).
35. F. L. Browne and W. K. Tang, *Effects of Various Chemicals on Thermogravimetric Analysis of Ponderosa Pine,* FPL 6, U.S. Department of Agriculture, Forest Products Laboratory, Forest Service, Madison, Wis., June 1963.
36. W. K. Tang and W. K. Neill, *J. Polymer Sci.,* [C], No. 6, 65 (1964).
37. R. M. Perkins, G. L. Drake, Jr., and W. A. Reeves, *J. Appl. Polymer Sci.,* **10**, 1041 (1966).
38. R. F. Schwenker, Jr., L. R. Beck, Jr., and R. K. Zuccarello, *Am. Dyestuff Reptr.,* **53**, 30 (1964).
39. K. Akita and M. Kase, *J. Polymer Sci.,* [A-1], **5**, 833 (1967).
40. C. M. Conrad, D. J. Stanonis, W. D. King, and P. Harbrink, *J. Appl. Polymer Sci.,* **12**, 1219 (1968).
41. C. M. Conrad and D. J. Stanonis, *Appl. Polymer Symposia,* No. 2, 121 (1966).
42. A. Broido, *Pyrodynamics,* **4**, 243 (1966).
43. D. J. Stanonis, W. D. King, P. Harbrink, and C. M. Conrad, *J. Appl. Polymer Sci.,* **11**, 817 (1967).

44. P. K. Chatterjee and C. M. Conrad, *J. Polymer Sci.,* [A-1], **6**, 3217 (1968).
45. H. Morita and H. M. Rice, *Anal. Chem.,* **27**, 336 (1955).
46. H. Morita, *Anal. Chem.,* **29**, 1095 (1957); *ibid.,* **28**, 64 (1956).
47. F. J. Kilzer and A. Broido, *Pyrodynamics,* **2**, 151 (1965).
48. D. Costa and G. Costa, *Chim. Ind. (Milan),* **33**, 71 (1951).
49. N. Berkowitz, *Fuel,* **36**, 355 (1957).
50. R. F. Schwenker, Jr., L. R. Beck, Jr., and W. J. Kauzmann, *Thermal Characteristics of Coated Textile Fabrics,* Final Report Contract N140-(132)-57442B, Textile Research Institute, Princeton, N.J., February 15, 1960.
51. C. H. Mack and S. R. Hobart, *Appl. Polymer Symposia,* No. 2, 133 (1966).
52. R. Bacon and M. M. Tang, *Carbon,* **2**, 221 (1964).
53. V. P. Alikin, N. F. Gracheva, Ya. V. Kurylev, and N. P. Perekal'skii, *Tr. Leningr. Tekhnol. Inst. Tsellyulozu Bunazhn. Prom. No.,* **18**, 7 (1956); through *Chem. Abstr.* **65**, 5657 (1966).
54. D. A. I. Goring, *Consolidation of the Paper Web,* Transactions of the Cambridge Symposium 1965, British Paper and Board Makers' Association, p. 555.
55. D. A. I. Goring, *Pulp Paper Mag. Canada,* **64**, T517 (1963).
56. J. K. Gillham and R. F. Schwenker, *Appl. Polymer Symposia,* No. 2, 59 (1966).
57. E. L. Back and L. O. Klinga, *Svensk Papperstidn.,* **66**, 745 (1963).
58. F. Johanson and E. L. Back, *Svensk Papperstidn.,* **69**, 199 (1966).
59. E. L. Back, *Pulp Paper Mag. Canada,* **68**, T165 (1967).
60. E. L. Back, M. T. Htun, M. Jackson, and F. Johanson, *Textile Res. J.,* **37**, 432 (1967).
61. E. L. Back, M. T. Htun, M. Jackson, and F. Johanson, *Tappi,* **50**, 542 (1967).
62. C. M. Conrad and J. J. Creely, *J. Polymer Sci.,* **58**, 781 (1962).
63. C. Houtz and D. McLean, *J. Phys. Chem.,* **45**, 111 (1941).
64. T. V. Gatovskaya, O. F. Golova, R. G. Krylova, and V. A. Kargin, *Zh. Fiz. Khim.,* **33**, 1418 (1959).
65. B. Philipp, J. Baudisch, and A. Gaudig, *Faserforschung u. Textiltech.,* **18**, 9 (1967); through *Chem. Abstr.,* **66**, 56,447*f* (1967).
66. K. A. Murty and P. L. Blackshear, *Pyrodynamics,* **4**, 285 (1966).
67. J. J. Creely and C. M. Conrad, *Textile Res. J.,* **32**, 184 (1962).
68. N. N. Iznairskaya, P. A. Il'chenko, M. A. Martynov, and N. V. Shueninova, *Plasticheskie Massy,* 67 (1964); in serial translation *Soviet Plastics,* of Rubber and Plastics Research Association of Great Britain.
69. V. B. Ryzhov, N. D. Burkhanova, and P. V. Kozlov, *Vysokomol. Soedin.,* **6**, 1471 (1964).
70. M. M. Tang and R. Bacon, *Carbon,* **2**, 211 (1964).
71. F. H. Holmes and C. J. G. Shaw, *J. Appl. Chem.,* **11**, 210 (1961).
72. G. A. Byrne, D. Gardiner, and F. H. Holmes, *J. Appl. Chem.,* **16**, 81 (1966).
73. T. Urbanski, W. Hofman, T. Ostrowski, and M. Witanowski, *Bull. Acad. Polon. Sci., Ser. Sci., Chim., Geol., Geograph.,* **7**, 851 (1959); through *Chem. Abstr.,* **55**, 23049 (1961).
74. H. W. Davidson and H. W. W. Losty, *Conf. Ind. Carbon Graphite, Papers, 2nd, London, 1965,* 20 (1966); through *Chem. Abstr.,* **66**, 3964 (1967).

75. I. N. Ermolenko, R. N. Sviridova, and A. K. Potapovich, *Vesti Akad. Navuk Belarus. SSSR, Ser. Khim. Navak,* 111 (1966); through *Chem. Abstr.,* **66**, 24303 (1967).
76. J. C. Arthur and O. Hinojosa, *Textile Res. J.,* **36**, 385 (1966).
77. O. M. Dmitrieva, N. P. Potapova, and V. I. Sharkov, *Zh. Prikl, Khim.,* **37**, 1583 (1964).
78. L. F. McBurney, in E. Ott, H. M. Spurlin, and M. W. Grifflin, Eds., *Cellulose and Its Derivatives,* Part I, Interscience, New York, 1954, p. 130.
79. O. A. Battista, *Ind. Eng. Chem.,* **42**, 502 (1950).
80. S. Suga, *Nenryo Kyokaishi,* **44**, 515 (1965); through *Chem. Abstr.,* **63**, 14601 (1965).
81. Ya. A. Epstein, O. P. Golova, and L. I. Durynina, *Izv. Akad. Nauk SSSR, Otd. Khim. Nauk,* 1126 (1959).
82. A. K. Esterer, US Patent 3,298,928 (Jan. 17, 1967); through *Chem. Abstr.,* **66**, 56886 (1967).
83. V. N. Piyalkin, A. K. Slavyanskii, and E. A. Isyganov, *Mater. Nauch.-Tekh. Konf. Leningr. Lesotekh, Akad.,* No. 4, 168 (1966); through *Chem. Abstr.,* **67**, 6201 (1967).
84. L. J. Carlson, U.S. Patent 3,235,541 (Feb. 15, 1966); through *Chem, Abstr.* **64**, 16122 (1966).
85. D. V. Tischenko and I. S. Sorokin, USSR Patent 136,387 (March 14, 1961); through *Chem. Abstr.,* **55**, 25250 (1961).
86. V. N. Sergeeva, A. Kaluius, and S. Biseniece, *Tr. Inst. Lesokhoz. Problem, Akad. Nauk Latv. SSR,* No. 17, 77 (1959); through *Chem. Abstr.,* **54**, 10314 (1960).
87. S. Martin, *Tenth Symposium (International) on Combustion,* The Combustion Institute, 1965. p. 877.
88. K. A. Lincoln, *Pyrodynamics,* **2**, 133 (1965).
89. I. F. Bogdanov, M. L. Mischenko, and I. L. Farberov, *Teoriyai Tekhnol. Protsessov Pererabotki Topliv., Inst. Goryuch, Iskop.,* 188 (1966); through *Chem. Abstr.,* **65**, 8613 (1966).
90. E. W. Stern, A. S. Logiudice, and H. Heinemann, *Ind, Eng. Chem., Process Des. Develop.,* **4**, 171 (1965); through *Chem. Abstr.,* **62**, 12009 (1965).
91. C. T. Greenwood, J. H. Knox, and E. Milne, *Chem. Ind. (London),* 1878 (1961).
92. J. Kammermaier, *Kolloid-Z., Z. Polymere,* **209**, 20 (1966); through *Chem. Abstr.,* **65**, 5649 (1966).
93. U. Gokcen and D. M. Cates, *Appl. Polymer Symposia,* No.2, 15 (1966).
94. D. L. Simms, *Combustion and Flame,* **5**, 369 (1961).
95. D. L. Simms, *Combustion and Flame,* **6**, 303 (1962).
96. J. B. Berkowitz-Mattuck and T. Noguchi, *J. Appl. Polymer Sci.,* **7**, 709 (1963).
97. G. A. Byrne, *J. Chromatog.,* **20**, 528 (1965).
98. J. A. S. Gilbert and A. J. Lindsey, *Brit J. Cancer,* **11**, 398 (1957).
99. O. P. Golova and R. G. Krylova, *Dokl. Akad. Nauk SSSR,* **135**, 1391 (1960).
100. D. Gardiner, *J. Chem. Soc.,* C, 1473 (1966).
101. K. Ettre and P. F. Varadi, *Anal. Chem.,* **34**, 752 (1962).
102. R. F. Schwenker, Jr., and L. R. Beck, Jr., *J. Polymer Sci,* [C], No. 2, 331 (1963).
103. K. Tamaru, *Bull. Chem. Soc. Japan,* **24**, 164 (1951).
104. R. F. Schwenker and E. Pacsu, *Chem. Eng. Data Series,* **2**, 83 (1957).
105. A. Cerniani, *Ann. Chim. (Rome),* **41**, 455 (1951).
106. J. Reilly, *Helv. Chim. Acta,* **4**, 616 (1921).

107. A. M. Pakhomov, O. P. Golova, and I. I. Nikolaeva, *Izv. Akad. Nauk SSSR, Otd. Khim. Nauk,* 521 (1957).
108. W. B. Neely, J. Nott, and C. B. Roberts, *Anal. Chem.,* **34**, 1423 (1962).
109. M. L. Wolfrom, *J. Am. Chem. Soc.,* 77, 6573 (1955).
110. M. L. Wolfrom, *J. Am. Chem. Soc.,* 78, 4695 (1956).
111. M. L. Wolfrom, A. Chaney, and P. McWain, *J. Am. Chem. Soc.,* **80**, 946 (1958).
112. F. Shafizadeh and M. L. Wolfrom, *J. Am. Chem. Soc.,* **80**, 1675 (1958).
113. F. Shafizadeh, M. L. Wolfrom, and P. McWain, *J. Am. Chem. Soc.,* **81**, 1221 (1959).
114. M. L. Wolfrom, A. Cheney, and K. S. Ennor, *J. Am. Chem. Soc.,* **81**, 3469 (1959).
115. M. L. Wolfrom and G. P. Arsenault, *J. Am. Chem. Soc.,* **82**, 2819 (1960).
116. G. Gelernter, L. C. Browning, S. R. Harris, and C. N. Mason, *J. Phys. Chem.,* **60**, 1260 (1956).
117. H. Schubert and F. Volk, *Explosivstoffe,* **14**, 1 (1966); through *Chem. Abstr.,* **64**, 15667 (1966).
118. M. Lahau and G. M. J. Schmidt, *J. Chem. Soc.,* B, 312, (1967).
119. T. I. Sogolova, *Mekhan. Polimerov, Akad. Nauk Latv. SSR,* 5 (1965); through *Chem. Abstr.,* **63**, 3062 (1965).
120. O. P. Golova, A. M. Pakhomov, and E. A. Andrievskaya, *Dokl. Akad. Nauk SSSR,* **112**, 430 (1957).
121. H. J. P. Venn, *J. Textile Inst.,* **15**, T414 (1924).
122. O. P. Golova, Ya. E. Epstein, and L. I. Durynina, *Vysokomol. Soedin.,* **3**, 536 (1961).
123. A. Broido and S. B. Martin, *Fire Res. Abstr. Rev.,* **3**, 193 (1961).
124. C. D. Hurd and O. E. Edwards, *J. Org. Chem.,* **14**, 680 (1949).
125. H. W. Durand, M. F. Dull, and R. S. Tipson, *J. Am. Chem. Soc.,* **80**, 3691 (1958).
126. P. S. O'Colla and E. E. Lee, *J. Chem. Soc.,* 2351 (1964).
127. H. Sugisawa and H. Edo, *Chem. Ind. (London),* 892 (1964).
128. H. Sugisawa and H. Edo, *J. Food Sci.,* **31**, 561 (1966).
129. L. Segal and D. J. Stanonis, *Anal. Chem.,* **35**, 1750 (1963).
130. K. Heyns, R. Stute, and H. Paulsen, *Carbohydrate Res.,* **2**, 132 (1966).
131. O. P. Golova, R. G. Krylova, and I. I. Nikolaeva, *Vysokomol. Soedin.,* **1**, 1305 (1959).
132. A. Pictet, *Helv. Chim. Acta,* **1**, 226 (1918).
133. M. L. Wolfrom, A. Thompson, and R. B. Ward, *J. Am. Chem. Soc.,* **81**, 4623 (1959).
134. H. Abe and W. Prins, *Makromol. Chem.,* **42**, 216 (1961).
135. V. V. Korshak, V. A. Sergeev, Ya. A. Surua, and R. Ya. Pernikis, *Vysokomol. Soedin.,* 9, 1593 (1963).
136. J. daSilva Carvalho, W. Prins, and C. Schuerch, *J. Am. Chem. Soc.,* **81**, 4054 (1959).
137. E. R. Ruckel and C. Schuerch, *J. Org. Chem.,* **31**, 2233 (1966).
138. R. M. Esteve and G. E. Wright, *Abstracts of the 138th Meeting,* American Chemical Society, September 1960, p. 17E, No. 43.
139. S. Glassner and A. R. Pierce, *Anal. Chem.,* **37**, 525 (1965).
140. D. J. Bryce and C. T. Greenwood, *Stärke,* **15**, 166 (1963); *ibid.,* 285 (1963); *ibid.,* 359 (1963).
141. D. J. Bryce and C. T. Greenwood, *Stärke,* **17**, 275 (1965).
142. D. J. Bryce and C. T. Greenwood, *Appl. Polymer Symposia,* No. 2, 149 (1966).

143. D. J. Bryce and C. T. Greenwood, *Appl. Polymer Symposia,* No. 2, 159 (1966).

144. B. Carroll and J. W. Liskowitz, *Agric. Food Chem.,* **15**, 701 (1967).

145. W. G. Parks, *Abstracts of the 127th Meeting,* American Chemical Society, p. 6E, April 1955.

146. A. M. Pakhomov, *Izv. Akad. Nauk SSSR, Otd. Khim, Nauk,* 1497 (1957).

147. J. H. Perry, Ed., *Chemical Engineers' Handbook,* McGraw-Hill, New York, 1950, pp. 321–328.

148. W. J. Moore, *Physical Chemistry,* 2nd ed., Prentice-Hall, Englewood Cliffs, N.J., 1955 pp. 528–543.

149. J. Hine, *Physical Organic Chemistry,* 2nd ed., McGraw-Hill, New York, 1962.

150. C. M. McCloskey and G. H. Coleman, *J. Org. Chem.,* **10**, 184 (1945).

151. C. T. Bishop F. P. Cooper, and R. K. Murray, *Can. J. Chem.,* **41**, 2245 (1963).

152. C. T. Bishop, *Methods Biochem. Anal.,* **10**, 1 (1962).

153. C. H. DePuy and R. W. King, *Chem. Rev.,* **60**, 431 (1960).

D. PHOTOCHEMICAL DEGRADATION

P. J. Baugh and Glyn O. Phillips

It has been known for over 80 years that sunlight degrades cellulose and cellulosic materials. In 1883 Witz (1) first established that chemical changes in cellulose could be induced by light. Later Harrison (2) showed that the degradation of cellulose by light gave oxidized cellulose ("oxycellulose") and strongly reducing organic materials. Prolonged irradiation with a quartz mercury-vapor lamp converted cellulose into a fine powder (3). Vignon (4) noted the role played by moisture and oxygen in the degradation of cellulose by light. Loss in the tensile strength (5) of yarn or fabric and the production of carboxyl and carbonyl groups in the cellulose chain were subsequently reported (6–10).

Until 1940 little was known about the nature of the degradation products or the mechanism of photodegradation. A more fundamental approach was initiated by Stillings and Van Nostrand (11) and Launer and Wilson (12). Race (13) has subsequently shown that outdoor weathering of cellulosic materials is the result of photochemical and acidic degradation.

Although much of the work has been technological in character, it is possible to distinguish two broad mechanisms of photodegradation in cellulose. These are here designated direct-photolysis and photosensitized-degradation processes. The proceedings of a symposium, *Photochemistry in Relation to Textiles* (14), held in 1949, adequately summarizes the information available up to that time. Therefore attention here is devoted mainly to reviewing the current position by reference to the information obtained by studying model systems. These have yielded valuable mechanistic data that are directly relevant to cellulose photodegradation. When dyed cellulose is exposed to light, fading of the dye is frequently observed. This process is not necessarily related to the chemical changes induced in the cellulose. A full account of the factors leading to dye fading on textiles has been given (15). This section focuses attention on the chemical changes and possible mechanisms that lead to the deterioration of cotton cellulose.

1. Primary Process in Model Systems

(a) ACTIVATION PROCESSES

The starting point in any photochemical interpretation is the long-established principle that only light that is absorbed by a molecule is effective in producing chemical change. Thus the first step in any study of a photochemical process is to obtain the absorption spectrum of the compound to be photolyzed. This basic information has unfortunately not been obtained for cellulose, so that considerable speculation about possible types of primary absorption processes that can lead to degradation has arisen.

When light is absorbed by a molecule, the energy of the molecule is raised by a quantity equal to the absorbed photon energy, which is

$$E = h\nu = \frac{hc}{\lambda} \tag{1}$$

where h is Planck's constant, c is the velocity of light, and ν and λ are the frequency and wavelength, respectively, of light. Changes may occur in the electronic, vibrational, or rotational energy of the molecule. Electronic changes can be effected only by visible or ultraviolet light, and they will be accompanied by simultaneous changes in vibrational and rotational energy. During the formation of electronically excited states an electron undergoes a transition from a lower-energy to a high-energy level in the molecule. Such changes in electronic energy are on the order of 100 kcal/mole and give rise to spectra in the visible or ultraviolet regions at frequencies of about 10^{15} sec^{-1}. The associated changes in vibrational energy are about 5 kcal/mole, and the rotational changes are about 0.02 kcal/mole; they give rise to coarse vibrational and finer rotational structures in the electronic spectral band. Electronic transitions of this type are governed by well-defined selection rules. One such rule (Franck-Condon) states that, as the transition from the vibrational levels of one state to the vibrational levels of another state occur within about 10^{-15} sec, the nuclei of the molecule can be considered as stationary. Since electronic absorption spectra arise from transitions of this kind, the maximum absorption will correspond to the Franck-Condon (vertical) transition.

(b) TRIPLET STATE

Generally molecules in the ground state have two electrons in each orbital, and according to the Pauli principle these electrons must be spin-paired. Such a state is the *singlet state.* When one of the paired electrons is excited, the spins can be paired or unpaired. When electrons have parallel spins, they contribute angular momentum to the system and confer paramagnetism on the molecule.

Such a paramagnetic, metastable state is the *triplet state,* which is a diradical (16). Triplet excited states have been postulated to account for the mechanistic pathway of certain photochemical processes. Singlet excited states are diamagnetic in nature, since then the electron spins are in opposition.

The natural lifetime of an excited state is inversely proportional to the strength of absorption to that state, usually expressed as the molar absorptivity coefficient ϵ, which is defined as

$$\epsilon = \frac{1}{cl} \log \frac{I_0}{I} \tag{2}$$

where c is the concentration in moles per liter, l is the length of the absorption path in centimeters, and I/I_0 is the fraction of the incident light that is transmitted. In the near-ultraviolet region the following expression can be used for estimating the lifetime of excited states:

$$\text{natural lifetime} = \frac{10^{-4}}{\epsilon_{\text{max}}} \tag{3}$$

It can be estimated that the probability of a singlet–triplet transition is about one million times less than that for the corresponding single–singlet transition.

(c) TYPES OF ORBITAL TRANSITION

By using formaldehyde as a reference for the various types of valence electrons, it is possible to illustrate the transitions that can take place (17). The electrons form single bonds (σ) or double bonds (π) and are located as unshared or nonbonding electrons (n).

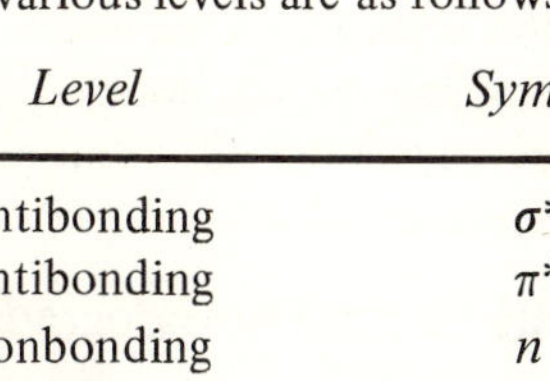

(4)

The symbols used for the various levels are as follows:

	Level	*Symbol*
↑	Antibonding	σ^*
	Antibonding	π^*
E	Nonbonding	n
	Bonding	π
	Bonding	σ

After light absorption the order of increasing energy for the possible orbital promotions (19) is $n \to \pi^*$, $n \to \sigma^*$, $\pi \to \pi^*$, and $\sigma \to \sigma^*$. The energy required to raise these electrons to antibonding orbitals varies with the type of bond. The n-electrons are the least stongly bound in a molecule, and in bonding levels π-electrons have higher energies than the corresponding σ-electrons. For the latter two in the antibonding levels the order is reversed (18). The latter transitions (e.g., as for paraffins) are observed only in the far-ultraviolet region (< 1700 Å). Transitions involving bonding π-orbitals ($\pi \rightarrow \pi^*$) require less energy, and although some are observed in the far-ultraviolet region (as for ethylene), many are found in the near-ultraviolet region. The $n \rightarrow \sigma^*$ transition is normally exhibited by saturated molecules in the far-ultraviolet region (e.g., COH, CNH_2, and CCl). The $n \rightarrow \pi^*$ transitions are found in molecules having a heteroatom multiply bound to another atom (C=O, C=S, N=N) and occur generally in the near-ultraviolet region; for example, acetone shows the $n \rightarrow \pi^*$ transition at 2770 Å and the $n \rightarrow \sigma^*$ transition at 1900 Å.

(d) LOSS OF EXCITATION ENERGY

The excitation of a molecule by light does not necessarily lead to chemical change. Such processes as fluorescence, internal conversion of energy, and nonradiative transitions involving the triplet state and energy transfer are all methods by which the excited state may lose its energy without chemical change occurring (20). Permanent chemical changes may arise as a result of two broad general types of process. Where transitions lead to a repulsive state with lifetimes less than one vibration, dissociation of the excited molecules can occur. Alternatively the molecule can possess sufficient vibrational energy in either its upper or its lower state to cause it to dissociate. Generally the bond that breaks is the bond responsible for the absorption of light. Chemical reaction can also be the result of intermolecular quenching of the excited state by a foreign molecule. The process may be *primary*—that is, one in which there is reaction between the quencher molecule and the excited molecule. Alternatively a *secondary reaction* can lead to the quencher molecule's acquiring sufficient energy from the excited molecule by an inelastic collision to cause it to react. The excited molecule that becomes quenched is frequently referred to as a photosensitizer. Truly sensitized reactions are those of the secondary type discussed above, where the absorbing substance is unchanged and acts only as a catalyst. The term "photosensitizer" is, however, loosely used to cover both the primary and the secondary quenching effects outlined here.

2. Processes Leading to the Photodegradation of Carbohydrates

Any attempt to systematize the literature on photodegradation of cellulose is complicated by the variety of conditions under which numerous investigations have been conducted. It has been pointed out that one or more of several processes may initiate photodegradation when a carbohydrate is irradiated in aqueous solution or where there are water molecules associated with the fibers (21). These processes are as follows:

1. The photolysis of water by radiation of wavelength lower than 2000 Å and subsequent attack on the carbohydrate by the free radicals formed.
2. Sensitized decomposition of water to produce radicals that may attack the carbohydrate.
3. Direct photolysis of the carbohydrate.
4. Photosensitized decomposition of the carbohydrate by a hydrogen atom or by an electron-transfer mechanism.

Through the study of model systems considerable information has been collected recently about the nature of such processes and the requirements for their propagation. Consideration is given to these, for all such processes may be operative when cotton cellulose is exposed to light.

(a) PHOTOLYSIS OF WATER

The absorption spectrum of water vapor displays a continuum from 1850 to 1450 Å, with a maximum at about 1650 Å. Below 1450 Å there is a series of diffuse bands superimposed on a second continuum, which extends below 1200 Å. The molar absorptivity ϵ at the first maximum is 1480. The first broad continuum (22–25) is to be expected if the primary absorption results in a decomposition of a water molecule according to the following equation:

$$H_2O \xrightarrow{h\nu} H\cdot + \cdot OH \tag{5}$$

Chen and Taylor (22) confirmed that such photolysis occurs when water vapor is irradiated in a flow system with 1650 Å radiation from a hydrogen discharge. There is chemical evidence that in water in the liquid state photolysis occurs by this path also (26–28). During the photolysis of deaerated aqueous solutions of methanol (10^{-2} *M*) with 1849-Å radiation from a low-pressure mercury lamp, the formation of hydrogen and ethylene glycol is attributed to the following processes:

$$H_2O \longrightarrow H\cdot + \cdot OH \tag{6}$$

$$H\cdot + CH_3OH \longrightarrow H_2 + \cdot CH_2OH \tag{7}$$

$$\cdot OH + CH_3OH \longrightarrow H_2O + \cdot CH_2OH \tag{8}$$

$$2\cdot CH_2OH \longrightarrow (CH_2OH)_2 \tag{9}$$

This demonstrates the role played by H· and ·OH species in aqueous solution.

(b) PHOTOSENSITIZED DECOMPOSITION OF WATER

It has been suggested that in the presence of compounds that can act as photosensitizers the water molecule can be cleaved, although the energy of the radiation may be insufficient to induce the direct photolysis of water. It has long been recognized that water can be decomposed by excited mercury (29, 30), but in more recent years other types of photosensitizer have been recognized. Riboflavin is an example, and the mechanism has been studied in several laboratories (31–36). In deoxygenated aqueous solution riboflavin is reduced to dihydroriboflavin on exposure to visible light. Addition of "activators," such as tertiary amines or thioethers and other compounds containing electronegative nitrogen or sulfur atoms, increases the rate and permits photoreduction of riboflavin in oxygen. The mechanism of photoreduction of riboflavin (R) in the presence of an activator (A) has been represented as follows:

$$R + A + H_2O \longrightarrow [R{-}H_2O{-}A] \xrightarrow{h\nu} RH_2 + AO \tag{10}$$

The water molecule may be regarded as "suspended" between riboflavin and the activator and "pulled apart" as the complex dissociates on absorption of radiant energy. In deoxygenated solution, through the formation of a riboflavin–water complex, the bond strength of water can be lowered enough to facilitate cleavage by light of 4400-Å wavelength when the energy absorbed is 65 kcal/mole, a quantity that would clearly be incapable of initiating direct photolysis of water.

Mooney and Stonehill (37) have suggested that anthraquinone derivatives, such as sodium anthraquinone-2-sulfonate, can also photosensitize the decomposition of water by light of wavelength greater than 4000 Å. The complex consisting of a molecule of hydration water hydrogen bonded to the carbonyl-oxygen atom of the salt is supposedly cleaved during the photoexcitation of the photosensitizer to give a hydroxyl radical, which then attacks the anthraquinone or anthraquinol ring. Although the energy of the light used is only about one-half that of the H–OH bond, the validity of the mechanism is claimed on the basis of calculations. If correct, this mechanism may have an important bearing on the photodegradation of carbohydrates sensitized by vat dyes. The hydroxyl radical produced might initiate abstraction processes in a molecule like cellulose and could lead to considerable chain scission and other oxidative processes, particularly in the presence of oxygen.

(c) DIRECT PHOTOLYSIS

Many diverse views here have been expressed about the conditions necessary for the direct photolysis of carbohydrates (21). For example, in the case of D-glucose one body of opinion is that no chemical degradation occurs unless photosensitizers, such as iron or uranium salts, are present (38–40). Another group, conversely, claims to have observed changes in optical rotation, reducing power, and liberation of carbon dioxide and hydrogen on ultraviolet irradiation in the absence of a photosensitizer (41, 42).

Many investigations on the photolysis of carbohydrates have been conducted without reference to the basic concepts of photochemistry. Little attention has been given to the active wavelength of the light and such factors as pH, degree of oxygenation, and others that might influence the nature of the primary absorbing species.

It is necessary to ascertain (a) whether sugars absorb visible or ultraviolet light directly and (b) whether the absorbed light is sufficiently energetic to rupture bonds in the molecules. Unfortunately, owing to their relative transparency, little attention has been given to the detailed absorption spectra of sugars in neutral solution. An early claim (43) that there was a selective absorption band at about 2800 Å has not been substantiated (44, 45). When acid is added to neutral solutions of D-glucose, an absorption maximum at 2840 Å is observed, apparently due to the conversion of nonabsorbing cyclic forms into open-chain carbonyl forms (46).

It is possible to predict the nature of the absorption spectra of sugars by considering the nature of the spectra of simple aliphatic alcohols and cyclic ethers. Aliphatic alcohols are transparent in the near-ultraviolet region; in the 2000 Å region (47, 48) a continuum sets in, with a maximum occurring at about 1800 Å (49–52).

Tetrahydrofuran, tetrahydropyran, and *p*-dioxane, three compounds that contain one or more oxygen atoms, show bands in the vacuum ultraviolet between 2000 and 1667 Å (52, 53). These compounds are practically transparent at wavelengths above 2550 Å. There is a slight increase in absorption below this point. If the *p*-methylene group of tetrahydropyran is replaced by an oxygen atom to give *p*-dioxane, the band moves toward the visible region.

The spectrum of D-glucitol, measured in aqueous solution, accords with the above observations. Measurements show that the solutions are almost transparent at wavelengths up to 3000 Å; thereafter a continuum sets in, extending into the vacuum ultraviolet. After 2127 Å the absorbency rises sharply. In the region 2000–2537 Å the solutions obey Beer's law, and molar absorptivities have been calculated (21). Since the straight-chain hydrocarbon corresponding to D-glucitol absorbs in the vacuum ultraviolet, the above spectrum can be attributed to electrons associated with the oxygen atoms, with the strong probability that the nonbonding electrons are involved ($n \rightarrow \sigma^*$).

On the basis of this information Phillips and Criddle (54) examined the chemical effects of filtered and unfiltered light from a Hanovia medium-pressure mercury lamp on oxygenated aqueous solutions of D-glucitol. Radioactive-tracer methods showed that the reaction proceeds in a stepwise fashion. By the use of filter solutions it was demonstrated that absorption of radiation in the spectral region 2300–2400 Å is responsible for the major portion of the photodegradation. These results demonstrate that D-glucitol absorbs the radiation directly, and although the amount of light absorption is small, there is excitation of the nonbonding electrons to antibonding levels ($n \rightarrow \sigma^*$ transition), which initiates dissociation. Light of 2300–2537 Å, corresponding to an energy of 124–113 kcal/mole, is absorbed by the D-glucitol molecule, and it is sufficient to rupture C—C or C—0 bonds (80–90 kcal/mole) or to remove a hydrogen atom (100 kcal/mole).

It has been shown that fragmentation takes place when aqueous D-glucose solutions are irradiated with unfiltered light from a medium-pressure mercury lamp, and prolonged irradiation reduces the initial substrate completely to carbon dioxide (55).

Beelik and Hamilton (56) studied the ultraviolet (2200–4000 Å) irradiation of cellobiose, cellopentaose, and other model compounds related to cellulose. The effects of irradiation were very similar. It was suggested that the degradation is initiated as a result of primary light absorption near 2650 Å by an "acetal chromophore" that arises from the presence of the lactol oxygen atom. Subsequent results on the photolysis of D-glucose in aqueous solution support this view (57). The initial product of photodegradation is D-gluconic acid and is followed by the production of D-arabinose. Photochemical reaction therefore occurs at the lactol carbon atom, which is consistent with the view that this is the primary absorbing site.

(d) PHOTOSENSITIZED DEGRADATION

Considerable information about the probable nature of the processes operative during the fading and tendering of dyes adsorbed on cellulosic fibers has been obtained by the study of the photosensitized oxidation of alcohols. Sensitizers initiating photosensitized oxidation include benzophenone (58–60), benzoquinone (61), hydrogen peroxide (62), and azo compounds (63). For each of the sensitizers the pattern of chemical change is similar. In the absence of oxygen primary and secondary alcohols are converted into the corresponding carbonyl compounds and the sensitizer is hydrogenated. In oxygenated solutions primary alcohols yield hydrogen peroxide, the corresponding aldehyde, and carboxylic acid. Secondary alcohols yield almost exclusively the ketone and hydrogen peroxide. The mechanism has been examined in detail when sensitizers are derivatives of anthraquinone because of their close structural relationship to the anthroquinonoid vat dyes (64, 65).

During the autoxidation of aliphatic alcohols and carbohydrates in aqueous solutions photosensitized by sodium anthraquinone-2-sulfonate, the energy of the photoexcited quinonoid sensitizer (A*) is degraded by the following processes:

$$A^* + RH \xrightarrow{k_1} \cdot AH + R\cdot \qquad (11)$$

$$A^* \xrightarrow{k_0} A \qquad (12)$$

The radicals R• produced in the hydrogen-transfer reaction are oxidized to peroxy radicals

$$R\cdot + O_2 \xrightarrow{k_3} RO_2\cdot \qquad (13)$$

with $k_3 >> k_1$, and the semiquinone reverts rapidly to the quinone form, as follows:

$$\cdot AH + O_2 \xrightarrow{k_2} A + HO_2\cdot \qquad (14)$$

with $k_2 >> k_1$. Reaction (12) is a first-order, adiabatic, radiationless transition from the photoexcited state to its ground state. Wells (65) has shown for a series of carbohydrates that the rate of oxygen absorption, which may be measured experimentally, varies linearly with the reciprocal of the solute concentration, and from the slope can be calculated k_1/k_0; the intercept is constant for a given light intensity. Thus, if one sensitizer is used, the ratio constitutes a measure of its relative reactivity.

Barber and Phillips (66, 67) examined the validity of the above approach with respect to polyhydric alcohols and other hexitols, using sodium anthraquinone-2-sulfonate and sodium anthraquinone-2,6-disulfonate as sensitizers. The rate of disappearance of hexitol (estimated by measuring the rate of formation of the corresponding hexose) was shown to vary linearly with the reciprocal of hexitol concentration. The slope and intercept are constants at a given light intensity. These results fully support the conclusions of Wells. The general mechanism for the photosensitized degradation of sugar alcohols may therefore be represented as follows:

$$A + h\nu \longrightarrow A^* \qquad (15)$$

$$RCH_2OH + A^* \longrightarrow \cdot AH + RCHOH \qquad (16)$$

$$\cdot AH + O_2 \longrightarrow A + HO_2\cdot \qquad (17)$$

$$R\underset{\bullet}{C}HOH + O_2 \longrightarrow R\underset{\underset{\dot{O}_2}{|}}{C}HOH \qquad (18)$$

$$R\underset{\underset{\dot{O}_2}{|}}{C}HOH + HO_2\cdot \longrightarrow RCHO + H_2O_2 + O_2 \qquad (19)$$

Reactions of the type $2RO_2\cdot \rightarrow RCOOH + H_2O_2$ also occur to a smaller extent.

During the photosensitized oxidation of ethers and glycosides Lock (68) observed that cleavage of the glycosidic link occurred also. When RCH_2OR' represents the glycoside benzyl β-D-glucopyranoside (R = phenyl and R′ = β-D-glucopyranosyl), the following reactions occur:

$$RCH_2OR' \longrightarrow \underset{\displaystyle \dot{O}_2}{\overset{}{RCHOR'}} \longrightarrow RCHO + R'OH + H_2O_2 \quad (20)$$

$$\underset{\displaystyle \dot{O}_2}{RCHOR'} \longrightarrow RCO_2R' + H_2O_2 \quad (21)$$

The photodegradation of *N*-alkyl amides (as simple models for the nylon molecule) indicates that there is a close similarity in their behavior to that of alcohols, ethers, and glycosides (51).

More recently electron spin resonance has been used to study the nature of radical intermediates produced during the degradation of alcohols and carbohydrates photosensitized by anthraquinonesulfonates (69). It was shown that an excited or ground-state anthrasemiquinone radical (AH·) is produced, which is present as the radical ion ($A^{\overline{\cdot}}$) in alkaline solution. In this work it was suggested that the radical species observed in neutral solution was ·AH and that this could abstract a hydrogen atom from a carbohydrate. Subsequent investigations have shown that additional radical species are produced when anthraquinonesulfonate concentrations of 5×10^{-3} *M* are used. This observation invalidates the above proposal. These newly observed species have an important bearing on the "colored products" that have been observed by many investigators during reactions photosensitized by anthraquinone salts and that act as internal filters in reducing reaction rates. Kinetic and flash studies on anthraquinonesulfonates (A) alone in solution indicate a reaction between excited (A*) and ground-state anthraquinonesulfonate, as follows:

$$A^* + A \longrightarrow A^{\overline{\cdot}} + A^{\overset{+}{\cdot}} \quad (22)$$

where $A^{\overset{+}{\cdot}}$ and $A^{\overline{\cdot}}$ are the semioxidized and semireduced forms (70). In oxygen unchanged A can be produced by the reaction

$$A^{\overline{\cdot}} + O_2 \longrightarrow A + O_2^{\overline{\cdot}} \quad (23)$$

and hydroxylated anthraquinones, which have been identified as the "colored products", can be produced from $A^{\overset{+}{\cdot}}$ by the reaction

$$A^{\overset{+}{\cdot}} + OH^- \longrightarrow A + \cdot OH \quad (24)$$

with subsequent hydroxylation by the ·OH radicals produced. Accurate rate-constant data have shown that the excited sensitizer (A*) will react about 100 times faster with a carbohydrate substrate than with A in the commonly observed side reaction that leads to destruction of the sensitizer.

From this brief consideration of the factors that control the photodegradation of model systems it is clear that if a fundamental understanding of the processes that occur during the deterioration of cotton cellulose under the influence of light is to be obtained, it is necessary to consider the following factors:

1. The wavelength of the radiation used.
2. The proportion of water associated with the textile.
3. The absorption characteristics of any reacted chemical or dye.
4. The atmosphere surrounding the textile during exposure.

One substantial difficulty in assessing previous work in this field in relation to the mechanism is the lack of consideration given to these vital parameters in many practical studies that have been conducted on cellulose photodegradation. Generally, in the textile studies, the practical outcome has understandably been the prime consideration. Nevertheless, indications of two broad types of photodegradation can be obtained. These are considered below.

3. Direct Photolysis of Cellulose

(a) CHEMICAL CHANGES

This type of direct photolysis leads to direct rupture of the chemical bonds by the absorption of light. It is therefore necessary that the energy of the absorbed quanta be sufficient to achieve bond scission. For cellulose the cleavage of either carbon–carbon or carbon–oxygen bonds will require an energy of 80–90 kcal/mole. About 100 kcal/mole is required to break a carbon–hydrogen bond (13, 71).

During irradiation with monochromatic light of frequency ν, quanta of known energy ($h\nu$) are delivered. Generally the photochemist refers to a mole of quanta ($Nh\nu$, where N is Avogadro's number), and this is called an einstein of radiation. The energy E per mole of quanta is given in kilocalories and can be evaluated from the following derived expression:

$$E = \frac{2.858 \times 10^5 \text{ kcal/einstein}}{\lambda} \tag{25}$$

where λ is in angstrom units. Thus it can be seen that light of 4000-Å wavelength is equivalent to 71 kcal, 3000-Å to 95 kcal, and 2000-Å to 143 kcal. Therefore, provided that light is absorbed by the cellulose molecule (this is an essential prerequisite for any photochemical reaction), the energy requirement for the degradation of cellulose by direct photolysis is met by ultraviolet light with a wavelength of 3400-Å or shorter.

Many investigators have reported chemical changes after the ultraviolet irradiation of cotton, (1, 6–10, 11, 14, 72–84) and wood cellulose (12, 56, 71, 85–89).

Considerable effort has been applied toward establishing whether the presence of a photosensitizer is in fact necessary to promote the degradation of cellulose by light. Normally, techniques to establish this involved the comparison of the rates of degradation in the presence and absence of sunlight. It had been thought by a number of workers that oxygen influences the rate of degradation (9, 11, 75), but it has been established that degradation by 2537 Å light is independent of oxygen (12, 71, 72–74, 77).

The initial disagreement on this important point probably arose because of variation in the output of lamps at the critical region around 2537 Å and in types of samples used (77). Stillings and Van Nostrand (11) irradiated cellulose in nitrogen, taking stringent precautions to exclude oxygen during irradiation. Degradation was observed chiefly by a decrease in the degree of polymerization (DP) and α-cellulose content, an increase in copper number, and liberation of carbon monoxide and carbon dioxide. When the oxygen content of the atmosphere was increased, an enhanced degradation ensued. The mercury lamp used by these workers, according to Egerton (77), emits radiation at wavelengths from 2483 to 5780 Å and of the total light produced, 58% is of wavelength greater than 3500 Å and only 22% of wavelength less than 3000 Å; 18% is of 3654-Å wavelength. On this evidence it is most unlikely that these investigators were observing the direct-photolysis process, for it has been demonstrated that no significant degradation occurs when pure cotton is exposed through glass (which cuts off radiation at wavelengths below 3400 Å) in the absence of oxygen.

With a quartz mercury-vapor lamp, which emits 90% of its radiation at a wavelength of 2537 Å, loss in the strength of cotton in atmospheres of carbon dioxide, nitrogen, or helium is about the same as in oxygen (12, 71). For this type of photolytic breakdown it is significant that degradation, whatever the assay method that is used, is greatly diminished by the presence of vat dyes. The desensitizing effect of vat dyes was obtained in dry oxygen and atmospheric air. This effect of dyes constrasts markedly with their acceleration of degradation of tensile materials by light of wavelengths greater than 3400 Å (81, 90–92).

Launer and Wilson (12, 71) report that water vapor exerts an inhibitory effect on the degradation initiated by light of 2537 Å wavelength and take this to be an indication that water molecules are in fundamental chemical combination with the cellulose. It is believed that, when water molecules enter the interchain spaces, a decreased freedom of displacement occurs for the carbon atoms in the vicinity of an –H–O–H– bond. Thus, when a far-ultraviolet photon is absorbed, the deformation of a given glucose unit is rendered less likely. Such an effect of deformability within the cellulose macromolecule is indicated by the results of Mason and Rosevear (93). With a silica "pile of plates" used to produce polarized ultraviolet light, parallel ramie fibers were irradiated in a cell containing metallic sodium to eliminate air or moisture, which might otherwise cause secondary nondirectional reactions. With light vibrating at right angles to the fibers, degradation, as measured by the copper number, was greater by about 25% than it was with the vibrations parallel. By analogy with simpler compounds, it is concluded that photochemical degradation of cellulose chains is greatest if they are oriented so that the light energy vibrates in the direction in which they are most deformable.

The rates of photochemical degradation of undyed, bleached cotton yarn in oxygen, nitrogen, and air at 0 and 100% relative humidity (RH) were determined by using decreases in tensile strength as a measure of degradation. The rate of degradation was dependent on the presence of oxygen and was greater in the atmospheres having a higher relative humidity (9,91). These effects of moisture and oxygen on the degradation of cellulose did not confirm the observations of Launer and Wilson in the far-ultraviolet region, where they used increases in the fluidity of cellulosic solutions as a measure of degradation in cellulose (12, 71). They reported that in the near-ultraviolet region the effect of moisture and oxygen on the degradation of cellulose was the reverse of that in the far-ultraviolet region. These apparent differences in the effects of moisture and oxygen on the photochemical degradation of cellulose are attributed to the relationships between bond energies in the cellulosic chain and energies of the photons involved. In the far-ultraviolet region most photons would have sufficient energy to rupture bonds. In the near-ultraviolet region the photon energy would not be sufficient to rupture bonds and would require the participation of oxygen (12, 71). Hydrogen peroxide formed during irradiation could also participate.

Egerton and Shah (94) and Egerton, Attle, and Rathor (95) have since investigated the effects of temperature, humidity, and atmosphere on the photochemical breakdown of cellulose in the far-ultraviolet region. The degradation assessed by tensile strength, fluidity, reducing power, and carbonyl content increased with temperature in the presence of oxygen. In the absence of oxygen this was independent of oxygen at temperatures of up to 50°C. In the temperature range 15–70°C the loss in tensile strength was greater in the absence of water vapor. The difference compared with a high humidity was greatest for short exposures. For short periods the reducing power in dry air was only slightly greater than it was in humid air, but the reverse was true for longer periods. The change in fluidity was greater in dry air, but for short periods the differences were negligible. From these results it was deduced that irradiation in dry air produces a large loss in tensile strength for a small change in fluidity or degree of polymerization. This indicated a highly localized degradation near the chain end. In the presence of moisture the degradation is not so localized and the course of reaction is different.

Generally, in evaluating the effect of ultraviolet light, consideration is given to modifications of the polymer system, including increase in solubility and reducing power, formation of carbonyl groups, or the reduction in degree of polymerization (71, 77, 96–101). Gaseous degradation products were also identified, notably carbon dioxide (11, 80, 96), carbon monoxide (11, 80), and hydrogen (80, 102). The latter was identified during the irradiation of dry, purified cotton *in vacuo* at 40°C with 2537 Å radiation. It was suggested that the rigorous degassing procedure might have removed certain inhibiting factors that account for the previous nondetection of hydrogen. The rate of hydrogen evolution follows a parabolic rate law:

$$\frac{dx}{dt} = \frac{k}{(1 + ax)} \tag{26}$$

where t is the time, x is the number of moles of hydrogen formed, k is the initial rate of evolution, and a is a constant. A plot of log k against log I_a (intensity of light) for any particular cellulose gives a straight line of unit slope. Thus $k = I_a \Phi_H$, where Φ_H is the initial quantum yield for the production of hydrogen, with a value of 10^{-2} moles/$Nh\nu$, and I_a is the intensity of light absorbed. The rate of evolution of carbon monoxide and carbon dioxide remains relatively constant, and the zero-order rate equation shows a combined quantum yield of 10.9×10^{-4} moles/$Nh\nu$ for the first series of celluloses and $7\text{–}8 \times 10^{-4}$ moles/$Nh\nu$ for the second series. Jortner (103) quotes a quantum-yield value of 2.5×10^{-3} moles Nh/ν for cellulose irradiated with 2537 Å radiation.

The quantum yield calculated by Sippel (104–106) is nearly unity for degradation by light of the middle-ultraviolet region (2537 Å). The inefficiency of light in the near-ultraviolet region in the direct-photolysis process is further confirmed by the low quantum yield (10^{-4}–10^{-8}) moles/$Nh\nu$ calculated for degradation by light in this region (71, 104–106).

A postirradiation process, observed by Launer and Wilson (12, 71) initially, was confirmed by Stillings and Van Nostrand (11) and Egerton (77). Degradation of cotton that had been irradiated in nitrogen or in carbon dioxide was increased by storage in the dark for several weeks. Cellulose irradiated in oxygen also showed a postirradiation effect. However, the effect was smaller and became almost negligible when the irradiation time in oxygen was increased from 4 to 100 days. The thermal nature of the postirradiation process was indicated by the enhancing effect that followed raising the temperature to 70°C.

The exact mechanism of degradation is difficult to ascertain, and most investigators have drawn some analogy with acid hydrolysis to assume a cleavage of the glycosidic bond. Launer and Wilson (12, 71) mention the possibility of rupture in the vicinity of $C_{(1)}$, $C_{(2)}$, or $C_{(3)}$ of the glucose residues, rather than at the glycosidic linkage. If, for example, the bond between $C_{(1)}$ and $C_{(2)}$ was disrupted, this would not necessarily cause immediate rupture of the cellulose chain. But $C_{(1)}$, attached in this state to possibly three electronegative groups, could be sensitive to future thermal oxidation, thus disrupting the chain according to the postirradiation process already mentioned.

To account for the production of hydrogen Flynn and co-workers (80, 107) suggested that photolysis of the alcohol group occurs by the reaction

$$\gt C(OH)H + h\nu \longrightarrow \gt C{=}O + H_2 \tag{27}$$

The increasing carbonyl concentration from this reaction at the surface might act as a nonvolatile filtering layer that would decrease the rate of photolysis of the alcohol groups and so explain the parabolic rate law that governs the evolution of hydrogen during irradiation.

Flynn and Morrow (102) have subsequently shown that if the hydrogen in the hydroxyl groups is replaced by deuterium and the cellulose is irradiated *in vacuo* at 77, 313, and 365°K with 2537 Å light, 50–70% of the hydrogen evolved is deuterium hydride and little deuterium is formed. The total amount of deuterium hydride that is produced is independent of temperature.

The changes in the quantum yields of hydrogen, carbon monoxide, carbon dioxide, hydrocarbons, aldehyde groups, carboxyl groups and chain scissions, with respect to the quanta of light absorbed and the initial concentration of functional group, indicate that the photolysis of alcohol groups plays a dominant role in the complex of interdependent primary and secondary reactions taking place (82). The dominant mechanism of hydrogen production is disproportionation, rather than atom abstraction or atom recombination. Much of the aldehyde formation takes place during the photolysis and probably results from secondary photolysis or disproportionation of radicals. It is suggested that chain scission and carbonyl group formation result to a large extent during the postirradiation oxidation of long-lived radicals. The yields of hydrogen, carbon dioxide, and carbon monoxide increase with temperature.

An encouraging systematic attempt to establish a pattern of degradation was undertaken by Beelik and Hamilton (82), who examined the smaller, water-soluble fragments produced during the initial stages of the photodegradation of wood cellulose by ultraviolet light. The light source was a mercury lamp, with 60% of its energy output being in the ultraviolet region, with emission lines at 2537, 2652, 3022, 3131, and 3654 Å. The ultraviolet-absorption spectrum of the cellulose-water extract after irradiation had a weak maximum at 2600 Å, which was intensified and shifted to 2700 Å, at pH 12 but disappeared at pH 2 (88, 100). A product with similar absorption characteristics is produced during the irradiation of aqueous solution of D-glucose (108–111). Volatile and nonvolatile acids are formed, but 60% of the water-soluble portion is composed of neutral sugars. The following sugars have been isolated and identified: xylose;

xylobiose, xylotriose, xylotetraose, and xylopentaose; D-glucose, cellobiose, and cellotriose; D-arabinose, 3-*O*-β-D-arabinose, 3-*O*-β-cellobiosyl-D-arabinose; and D-mannose.

In the ethanol extract of ultraviolet-irradiated cellulose from hardwood and softwood Peng and collaborators (89) found xylose, arabinose, glucose, galactose, mannose, and rhamnose. Xylose was more abundant in irradiated hemicellulose from hardwood than in that from softwood.

Beelik and Hamilton (56) observed that model compounds related to cellulose such as cellobiitol and cellopentaitol, when irradiated with ultraviolet light, both gave cellobiose, and the latter gave cellopentaose also. These workers suggested that the carbon-oxygen bonds either in the ring or in the chain of the cellulose molecule act as chromophores and undergo scission (88). Bera et al. (112, 113) claim that the photochemical degradation of cellulose involves both hydrolysis and oxidation, occurring either concurrently or alternately in random order. Bera suggests that this theory explains the formation of cellobiose and cellopentaose from cellopentaitol, but that Beelik and Hamilton's theory (88) does not. Nevertheless, both groups of workers agree broadly on the identification of degradation products. Bera and co-workers (114) isolated in addition to the above sugars, an unknown trisaccharide, consisting of glucose, arabinose, and xylose. Its formation was explained by assuming that the cellotriose formed by cellulose hydrolysis subsequently underwent dehydrogenation, oxidation, and finally decarboxylation. In 1965 Beelik and Hamilton (115) disputed the validity of this product and claimed that the unknown trisaccharide was actually glucosyl arabinose. Native cotton normally contains a xylan in small quantities. This contaminant can largely be removed by repeated extraction of the cotton with cold aqueous sodium hydroxide of mercerizing strength. Beelik and Hamilton have irradiated with ultraviolet light linters extracted with 10% sodium hydroxide, unextracted linters, and wood cellulose, containing 0.03, 0.5, and 2.7% xylan, respectively. The amounts of xylose and xylose oligosaccharides produced by photodegradation were barely detectable in the sodium hydroxide-extracted linters and were greatest in the wood cellulose. The amounts of other products were similar in all three samples. It was suggested that xylose observed by Bera and co-workers (113, 114) arose largely from the photodegradation of a xylan contaminant present in their cotton sample. The evidence for Bera's unknown trisaccharide's being glucosyl arabinose is based only on the chromatographic analysis of the products. The mobility of the trisaccharide was greater than that of cellobiose but less than that of glucose. All trisaccharides observed by Beelik and Hamilton had mobilities smaller than cellobiose, whereas glucosyl-arabinose, one of the disaccharides produced, had a mobility similar to the unknown trisaccharide. The xylose found in its hydrolyzate originated from a water-soluble xylan extracted from the chromatography paper with the oligosaccharide. According to Beelik and Hamilton, oligosaccharides recovered from paper chromatograms by extraction with water must be treated with hot alcohol to free them from the alcohol-insoluble xylan contaminant.

Although Bera (116) has cited other work in defense of his theory, it appears that it is not yet supported by experimental evidence. Moreover, Daruwalla and Da Costa (84), in their study of the influence of environmental conditions on the products of the photochemical degradation of cellulose, could not detect any trisaccharide, as claimed by Bera. They found that in the presence of nitrogen at 0% relative humidity (RH) only glucose was formed as a water-soluble product. At 90% RH glucose, arabinose, and a disaccharide were detected, but at 0% RH xylose was formed in addition to the first two. With continuing irradiation at 0% RH in oxygen, increasing amounts of uronic acids could be detected, but these were not present at 90% RH in an oxygen or nitrogen atmosphere. They suggested that xylose could be formed by decarboxylation of the uronic acid during irradiation at 0% RH in oxygen.

(b) ELECTRON SPIN RESONANCE MEASUREMENTS

Little information is available about the nature of the free-radical species produced in cellulose on ultraviolet irradiation. This is mainly due to the very small concentrations that are formed as a result of low light absorption. Kleinert (117, 118) produced a measurable quantity of free radicals by irradiating dissolving-grade bleached pulp (94% α-cellulose) *in vacuo,* and in nitrogen and oxygen atmospheres directly in the microwave cavity and outside, and subsequently measuring the radical yield. In each case the electron spin resonance (ESR) signal area was linearly related to chain splitting (expressed as $1/\eta$). It was therefore concluded that chain cleavage induced by the photon energy led to free-radical formation and appeared to be the primary reaction irrespective of whether or not oxygen was present. The actual ESR spectrum exhibited hyperfine structure similar to that of γ-irradiated cellulose I.

Phillips et al (119) studied the radiation wavelengths that were most efficient in promoting free-radical formation, they used various filters interposed between a mercury high-pressure point arc and cellulose I yarns located within the cavity of a microwave spectrometer. The most active wavelengths were in the region 3000–4000 Å. The ESR spectrum appears as a single resonance line with a *g*-value close to that calculated for the free electron and is unlikely to be the same species as that obtained by Kleinert (117, 118). In the region 3000–4000 Å the photon energy is hardly sufficient to induce bond scission. It is more probable that the signal is associated with trapped conduction-band electrons (see also Section I of Chapter XVII).

(c) ABSORPTION PROCESSES IN CELLULOSE

Kujirai (120) has shown that cellulose films have a strong absorption at wavelengths below 2000 Å but appear to be completely transparent at longer wavelengths. Earlier he had suggested that a weak absorption existed in the region 2000–4000 Å (97). Treiber (121, 122), in a review of the absorption properties of cellulose and its derivatives, concludes that at about 2780 Å there is probably a selective absorption that is noticeable but weak. This absorption is attributed to the lactol ring and glycosidic atoms.

Clearly, in cellulose films any weak absorption would be difficult to observe. Transmission spectroscopy demands that the sample be reasonably transparent in the wavelength region being investigated. Therefore, to observe an absorption with a very low absorptivity, the thickness of the film has to be increased, and this poses considerable transmission difficulties. Multiple transmission of the light through the cellulose film could be one method of overcoming this difficulty. Recently it has been found (123), by using *diffuse-reflectance spectroscopy,* that cellulosic yarns and fabrics consistently have a weak, broad absorption band with a maximum in the region 2600–2700 Å. With the diffuse-reflectance method a strongly absorbing and a weakly absorbing species can both be observed simultaneously.

Thus for cellulose, in the solid state, a weak absorption at 2600–2700 Å can be detected by diffuse-reflectance spectroscopy despite the presence of the more dominant end absorption that is present at wavelengths of 2000 Å and below. Such a discrete absorption is not detectable in D-glucose when examined in aqueous solution even at high concentrations and could well be a property of the cellulose aggregate, which is extensively hydrogen bonded. There is always the possibility that any such weak absorption is due to the presence of impurity, and since the method is so sensitive, in any such examination it is essential that the cellulose be purified extensively prior to any spectral measurement. After such purification and in a wide range of samples the absorption was persistently present.

It has been demonstrated that radiation of wavelength greater than 3400 Å is unable to photolyze pure cellulose (56, 88, 99). For example, when Pyrex glass (transmitting only 1% of 2900 Å light and only 10% of 3100 Å light) was used as a filter, irradiated wood cellulose showed no change in solubility, viscosity, or content of carboxyl and aldehyde groups. The formation of water-soluble acids was halved, and only traces of D-arabinose were detected by paper chromatography.

Kujirai (99) has sought to distinguish the mechanisms of photolysis by 1850 and 2537 Å radiation on the basis of an oxygen effect. It is suggested that with 1850 Å radiation the degradation is of a dissociative type, whereas with 2537 Å radiation a different photochemical activation occurs (124). Degradation by 2537 Å radiation is accelerated by temperature and is independent of oxygen pressure. With 1850 Å radiation the decrease in rate with increasing oxygen pressure is due only to light absorption by oxygen. After irradiation the ultraviolet-absorption spectra of cellulose films show a strong continuum below 2400 Å when 2537 Å light is used (125). Earlier Egerton and Roach (126) had shown that cellulosic films irradiated in a nitrogen atmosphere with 2537 Å light exhibited a greater absorption at 2600 Å. Changes in absorption during the irradiation of cellulose acetate films with 2537 Å radiation were strongly dependent on temperature (127). Kujirai (128) claims that 1850 Å radiation selectively produces aldehyde groups, whereas 2537 Å radiation leads to the production of carboxyl groups in the cellulose.

Cellulose peroxide is produced in films treated with 2537 Å radiation (129). Initially it was thought that the primary dissociation due to 1850 Å irradiation leads to chain scission. More prolonged irradiation leads to the formation of alkali-sensitive linkages as a result of intramolecular-bond scissions. Such alkali-sensitive linkages are not found after treatment with 2537 Å radiation (130).

The reflectance spectra of cellulose yarn and fabric irradiated with the full spectrum of a Hanovia medium-pressure lamp show a pronounced absorption in the region of 2600–2700 Å under normal- and high-humidity conditions (123).

Although there can be little doubt that the wavelength of light used in various direct photolyses is sufficiently energetic to rupture carbon–hydrogen, carbon–carbon, and carbon–oxygen bonds in cellulose, another essential prerequisite is that light be absorbed by the molecule. At this time it cannot be unequivocally stated that a true, identifiable absorption at wavelengths above 2000 Å is present in cellulose. Nevertheless, as has been pointed out, strong indications have been obtained for such an absorption, which is a solid-state property of the cellulose, as distinct from the absorption present in any single D-glucose unit. Until this concept can be rigorously demonstrated, there will remain an uncertainty about the nature of the light-absorption process at 2537 Å that leads to direct photolysis of cellulose. There remains the postulate of the acetal chromophore to account for the weak absorption near 2650 Å arising from the acetal group at $C_{(1)}$ of the D-glucose residue (56, 88). In D-sorbitol (54) exposure to 2537 Å radiation produces degradation as a result of $n \rightarrow \sigma^*$ absorptions, and this possibility exists for cellulose, particularly in view of the photolysis of alcohol groups observed by Flynn and Morrow (102).

In conclusion, therefore, it can be stated that there is little doubt that direct photolytic breakdown of the cellulose takes place under the influence of wavelengths in the region of 2536 Å or lower, although the mechanism of energy absorption cannot as yet be clearly envisaged.

4. Photosensitized Degradation of Cellulose

Light of wavelengths longer than 3400 Å cannot induce degradation of cellulose directly. However, certain dyes and compounds, such as zinc oxide and titanium dioxide, are capable of absorbing light in the near-ultraviolet or visible part of the spectrum and utilize this energy to induce the degradation of cellulose. The processes may be collectively classified as photosensitized degradations of cellulose, although it is highly unlikely that all sensitized reactions proceed by a common mechanism. The more general problem of photosensitized processes in polymer chemistry has been exhaustively treated in a recent review (131).

The photoinitiated tendering of cellulose and synthetic fabrics treated with vat dyes has been recognized for many years (81, 91, 132–140). The technological aspects have been fully reviewed (14), and more attention has been given to practical considerations than to fundamental studies of the mechanism of initiation of the tendering process. A conspicuous exception, however, is the series of studies into the mechanism of the autoxidation of cellulosic materials photosensitized by anthraquinonoid systems related to vat dyes. Here studies on model compounds reported in a review by Phillips and Arthur (141) allow a possible extrapolation to the mechanism of tendering of cellulosic materials (64, 65, 68, 77, 142), although recently the reliability of such extrapolations has been questioned (143).

(a) TENDERING BY VAT DYES

In marked contrast to direct photolysis, the photosensitized deterioration of cellulose is dependent on the presence of oxygen (81, 91, 136–139). During this process dyes act as initiators and not inhibitors. Moisture enhances the degradation, not only in the dyed fabric, but also by catalyzing the breakdown of adjacent, undyed films (91, 136–139). A similar behavior is observed when zinc oxide and titanium dioxide are employed as sensitizers.

Several views have been expressed about the detailed path of the reaction. Egerton (91, 136–139) has suggested that light energy absorbed by the dye is transferred to oxygen in the surrounding atmosphere, with the production of activated oxygen. In the presence of water vapor the activated oxygen is supposed to react with the water vapor, bringing about the formation of hydrogen peroxide. The degradation of the textile material could thus be promoted by the activated oxygen and/or hydrogen peroxide, depending on the particular textile fiber and the experimental conditions.

Although associating photodegradation with the presence of hydrogen peroxide, Bamford and Dewar (142) do not concur with Egerton (91, 136–139) about its mode of formation. These workers consider that the primary step leading to the formation of hydrogen peroxide by the excited dye (D*) is

$$D^* + OH^- \longrightarrow D^- + \cdot OH \tag{28}$$

On this basis, hydrogen peroxide would be formed subsequently by the combination of two hydroxyl radicals (·OH) or by the reaction of two peroxy radicals ($HO_2\cdot$) formed according to the equation:

$$D^- + H^+ \longrightarrow \cdot DH \xrightarrow{O_2} D + HO_2\cdot \tag{29}$$

Neither of these views, however, can be considered as entirely satisfactory if H· and ·OH radicals can be produced by the sensitized photodegradation of water considered in the preceding subsections. The views of Bamford and Dewar (142, 144) on the initiation of the photosensitized oxidation appear to be closer to current thought. These authors envisage the initial stage of reaction in the absence of moisture as occurring between the cellulose ($R_{cell}H$) and the excited dye (D*):

$$D^* + R_{cell}H \longrightarrow R_{cell}\cdot + \cdot DH \tag{30}$$

Tendering may then take place as follows:

$$R_{cell}\cdot + O_2 \longrightarrow R_{cell}O_2\cdot$$

$$R_{cell}O_2\cdot + \cdot DH \longrightarrow R_{cell}O_2H + D \tag{31}$$

This explanation leaves the sensitizer unchanged. When moisture is present, initiation may also take place by the abstraction of hydrogen by hydroxyl radicals that can arise by photosensitized rupture of the water molecule (145):

$$R_{cell}H + \cdot OH \longrightarrow R_{cell}\cdot + H_2O \tag{32}$$

The hydrogen atom formed could then lead to the production of hydrogen peroxide under these conditions:

$$2H\cdot + 2O_2 \longrightarrow 2HO_2\cdot \longrightarrow H_2O_2 + O_2 \tag{33}$$

More recently Lock (68) and others (64, 65) in investigations on model compounds, such as alcohols, glucosides, and disaccharides, have substantiated the view that the initial step in the tendering process is the abstraction of a hydrogen atom from the cellulose by the photoexcited sensitizer molecule (instead of any kind of electron-transfer process), followed by addition of oxygen to give a peroxy type of radical. At least two reactions can follow, depending on the site of initial attack. In one case decomposition of the peroxy radical severs the cellulose chain and leads to carbonyl formation, or carboxyl group formation results within the chain without complete disruption. Alternatively, a carbonyl group can be formed.

Preliminary investigations seeking to apply these mechanisms to amylose (146) show that changes in the viscosity of amylose in different solvents during photodegradation correlate with the number of glycosidic bonds broken. Oxygen absorbed during photodegradation correlates with glycosidic bonds as required by the mechanism, which therefore seems to explain adequately photooxidation in the absence of moisture. When moisture is present, such processes as have been described in relation to the sensitized photodecomposition of water can further contribute to the initial abstractions.

Indirect confirmation of the above mechanism of phototendering comes from experiments in which ultraviolet light was used to initiate grafting on to cellulose (147–149). After the abstraction of a hydrogen atom from the cellulose by photoexcited dye molecule, a free radical should be produced on the cellulose backbone. Evidence for the formation of such intermediate radicals is given by their ability to initiate vinyl polymerization, and, although not certain, it is probable that a copolymer is grafted to the cellulose backbone. The following mechanism is envisaged for grafting and the simultaneous production of homopolymer:

absence of oxygen

$$D + h\nu \longrightarrow D^* \tag{34}$$

$$D^* + R_{cell}-\overset{\large OH}{\underset{\large H}{C}}-H \longrightarrow \cdot DH + R_{cell}-\overset{\large OH}{\underset{\large H}{C}}\cdot \tag{35}$$

$$R_{cell}-\overset{\large OH}{\underset{\large H}{C}}\cdot + MH \longrightarrow R_{cell}-\overset{\large OH}{\underset{\large H}{C}}-MH\cdot \longrightarrow \text{graft} \tag{36}$$

$$R_{cell}-\overset{\large OH}{\underset{\large H}{C}}\cdot + MH\ (\text{or } XH) \longrightarrow R_{cell}-\overset{\large OH}{\underset{\large H}{C}}-H + M\cdot(\text{or } X\cdot) \longrightarrow \text{homopolymer} \tag{37}$$

$$R_{cell}-\overset{\large OH}{\underset{\large H}{C}}-(MH)_n^{\cdot} + XH \longrightarrow R_{cell}-\overset{\large OH}{\underset{\large H}{C}}-(MH)_nH + X\cdot \tag{38}$$

$$D^* + MH \longrightarrow \cdot DH + M\cdot \longrightarrow \text{homopolymer} \tag{39}$$

Where MH represents the vinyl monomer, XH, the solvent (a possible hydrogen donor) and $R_{cell} - CH_2OH$ a cellulose molecule (only primary hydroxyl groups shown for simplicity). The process is initiated by the excited dye (reaction 34), and grafting occurs by reactions 35 and 36. Homopolymerization predominates under these conditions by reactions 37, 38, and 39.

presence of oxygen

$$\cdot DH + O_2 \longrightarrow HO_2\cdot + D \qquad (40)$$

$$R_{cell}-\overset{\displaystyle OH}{\underset{\displaystyle H}{\overset{|}{\underset{|}{C}}}}\cdot + O_2 \longrightarrow R_{cell}-\overset{\displaystyle OH}{\underset{\displaystyle H}{\overset{|}{\underset{|}{C}}}}-O_2\cdot \qquad (41)$$

$$R_{cell}-\overset{\displaystyle OH}{\underset{\displaystyle H}{\overset{|}{\underset{|}{C}}}}-O_2\cdot + \cdot HO_2 \longrightarrow R_{cell}-C{=}O + H_2O_2 + O_2 \qquad (42)$$

$$R_{cell}-\overset{\displaystyle OH}{\underset{\displaystyle H}{\overset{|}{\underset{|}{C}}}}-H + \cdot OH \longrightarrow R_{cell}-\overset{\displaystyle OH}{\underset{\displaystyle H}{\overset{|}{\underset{|}{C}}}}\cdot + H_2O \qquad (43)$$

$$\underset{(\text{or } \cdot OH)}{MH + \cdot HO_2} \longrightarrow \underset{(\text{or } H_2O)}{M\cdot + H_2O_2} \longrightarrow \text{homopolymer} \qquad (44)$$

Both grafting and homopolymerization are inhibited in oxygen, indicating that under these conditions reactions 41 and 42, which occur during phototendering, predominate (see also Section G in Chapter XVII).

Although the sensitizer operates mainly by a cyclic mechanism, there is evidence that side reactions lead also to degradation of the sensitizer, with formation of oxidation products (150–152). It has now been conclusively demonstrated by Broadbent (153) that the colored products are hydroxylated derivatives of the anthraquinone sensitizer. The previous suggestions of Mooney and Stonehill (37) are thus substantiated. More recently Broadbent (153) and Clarke and Stonehill (154) have shown that the exact positions (α or β) of hydroxylation are dependent on the pH of the medium. At pH $>$ 10 substitution occurs exclusively in the β-positions, whereas at lower pH values α- substitution occurs. Further anomalies are found when the reaction is carried out in a sulfuric acid concentration greater than 70%. In dyed cellulose, therefore, it is to be expected that chemical changes will be found in the dye as well as in the cellulose, and it should not be supposed that these are always interrelated processes.

Recently reflectance spectroscopy and viscosity measurements have been used to show that parallel reactions in the sensitizer and the substrate occur during the photodegradation of cellulose fabric and yarn sensitized by anthraquinonesulfonates (155). By using sodium anthraquinone-2- and -1-sulfonates as examples of good and bad sensitizers, it has been shown that the formation of the corresponding anthrahydroquinone (AH_2) and the degradation of cellulose are in this instance associated reactions that proceed at the same rate for any particular sensitizer. The good sensitizer, the 2-sulfonate, promotes photodegradation and AH_2 formation 15 times faster than the poor sensitizer (1-sulfonate); this is reasonable agreement with Wilkinson's results (156) for the ratio of quantum yields of AH_2 for these sensitizers in deoxygenated solution. The results indicate that the mechanism is similar in the solid state to that which has been proposed for model systems in aqueous solution, which were outlined in subsection 1.

Egerton and Assaad (157) have investigated the role of dye semiquinones in cellulose films dyed with 16, 17-dimethoxydibenzanthrone irradiated with 2537 and 3650–5790 Å light. The leuco compound is detected only in the film irradiated with 2537 Å light. They conclude that the mechanism cannot be compared with that of photooxidation of alcohols sensitized by anthraquinone-sulfonates.

Egerton, in a series of papers, has repeatedly pointed out that the mechanism of formation of the reduced or leuco form of the anthraquinonoid dye in solution during irradiation does not necessarily bear any relation to the mechanisms of reactions that occur in the solid state. Although it can be spectrally demonstrated that the reduced form is produced when 1,5-diamino-anthraquinone is photolyzed in ethanol, cellulose acetate and nylon (158) treated with the same sensitizer show no significant formation of this product. The production of the corresponding anthrahydroquinone and anthrone when aminoanthraquinones are irradiated in ethanol, *n*-hexane, ethyl acetate, and carbon tetrachloride indicate that hydrogen abstraction is not the only path of degradation of the dye that results in the fading of the color (159). The situation in the solid-state dyed-polymer system is even more complex. When six aminoanthraquinones were examined, only the 1- and 1,5-amino derivatives on *N*-methoxymethyl nylon gave the reduced form on irradiation in a nitrogen atmosphere. In contrast, cellulose acetate and nylon similarly treated showed no appreciable formation of the leuco compound. A similar behavior was found when *N*-methylaminoanthraquinones (160), piperidinoanthraquinones (161), and acylaminoanthraquinones were used as sensitizers (143).

Feichtmayr and Schlag (162) have studied the phototropy of flavanthrone and pyranthrone dyes on cellulose by ultraviolet, visible, and ESR spectroscopy. Stable free radicals were produced during the phototropic reaction; this suggests an interaction between the excited dye and the cellulose substrate. The color change is indicative of the formation of the reduced quinone, since a similar change in absorption spectrum occurs when the dye is reduced in alkaline dithionite solution. Phillips and co-workers (119) found that cotton vat-dyed with anthraquinonoid dyes exhibited a free spin center prior to photolysis. The increase during irradiation was marked and independent of temperature in the range -140 to 25°C. The effects of light, as measured by the relative production of free radicals, showed that vat dyes sensitize cellulose, whereas hydrogen-bonded and reactive dyes protect it. Cellulose yarn dyed with 24 different dyestuffs selected from eight broad classifications of dyes was investigated by ESR spectroscopy both before and after photolysis, and also for weather resistance (163). No obvious correlation was found between the increase in free spin centers during photolysis and weather resistance as measured by retention in tensile strength. It is not yet clear whether the formation of the free-radical species and the leuco compound generally involves reaction between the photoexcited dye and the cellulose substrate. An important observation is that the extent of free-radical production is greater when the cellulose is present than when the dye is irradiated alone in the solid state. The formation of charge-transfer complexes between the dye and the cellulose that assist in free-radical production cannot be excluded at present. When the polymer is anionic, this type of interaction leading to free-radical production on irradiation has been demonstrated. No further consideration of the reactive states that promote dye fading is made here because this subject has been adequately dealt with in a review (15).

(b) PHOTOSENSITIZATION BY METALLIC OXIDES

Although it has long been recognized that metallic oxides can sensitize the photodegradation of cellulose, little attempt has been made to elucidate the mechanism of the process (132, 164). It has been shown, for example, that tungsten and molybdenum oxides (165), when finely dispersed on dry cellulose (filter paper) or other polysaccharides, assume a blue color when exposed to light of 2300–3900 Å. Reduction to colored oxides of lower valency is presumed to be responsible for the change. The presence of moisture influences the process. In the case of tungsten oxide irradiation leads to considerable degradation of the cellulose.

Consideration was given by Schurz and Kienzl (166) to the rate of titanium dioxide as a catalyst in the light damaging of cellulose. It is postulated that the photooxidation with titanium dioxide produces first an oxidation of the glucose ring without splitting the glucose bonds, followed by secondary splitting of these bonds.

Egerton (140) has studied the degradation of undyed cotton yarn exposed to sunlight when in contact with solid films composed of various sensitizers, such as zinc sulfide, zinc oxide, and titanium dioxide. A peroxide vapor that is produced by these sensitizers exerts a marked effect on the cellulose in moist air. Furthermore, cellulose yarn irradiated in proximity to yarn containing these sensitizers shows an increased degradation in the presence of oxygen and moisture. No volatile hydrogen peroxide was formed in the absence of water vapor. When zinc sulfide was precipitated by double decomposition to obtain optimum contact, the decomposition was observed even in dry air, where normally there was no effect. Egerton has compared photosensitization by zinc sulfide in oxygen with the photosensitization by the dye Caledon Yellow 5G. A common mechanism is advanced, involving the species O_2^-, as follows:

Pigments:

$$ZnS + O_2 \longrightarrow ZnS^+ + O_2^- \quad (45)$$

$$H^+ + O_2^- \longrightarrow HO_2\cdot \quad (46)$$

$$2HO_2\cdot \longrightarrow H_2O_2 + O_2 \quad (47)$$

Dyes:

$$D\text{ (triplet)} + O_2 \longrightarrow D^+ + O_2^- \quad (48)$$

$$R_{cell} + O_2^- \longrightarrow \text{products} \quad (49)$$

However, additional mechanisms cannot at this stage be ruled out. Sensitized decomposition of water by the metal oxide or sulfide sensitizers, followed by direct abstraction by free radicals from the cellulose, presents one possibility. The observation of reactive, abnormal valence states during pulse radiolysis (167) as a result of electron-transfer points also to the possible participation of such species in these reactions. General indications can be found in the literature for participation of electron-transfer reactions.

When Baur and Neuweiler (168) exposed suspensions of zinc oxide in water in the presence of air to sunlight for 10–15 hr, it was observed that the solutions liberated iodine from potassium iodide solution. When glucose was present during a similar exposure, hydrogen peroxide was formed, which led these and later investigators (169–172) to conclude that hydrogen peroxide was the agent responsible for the liberation of iodine from iodide and also for the photodegradation of cellulose. However, it is highly probable from recent radiation studies of aqueous solutions that the processes are more satisfactorily explained in terms of the precursors of the hydrogen peroxide in irradiations in the presence of zinc oxide.

When light is absorbed by the zinc oxide, the initial reaction may be a transfer of an electron from zinc to oxygen absorbed on the surface, and the resulting species is a radical ion (O_2^-). It has been shown that organic alcohols (173, 174) can be oxidized by way of this initiation step, followed either by transfer of electrons to the photopositive surface of the zinc oxide (after loss of a photoexcited electron by the crystal to 0_2) or hydrogen abstraction by the peroxide radical to form HO_2^- . The peroxide formed in the absence of organic material (175) may occur through a reaction of the type

$$O_2^- + H_2O \longrightarrow HO_2^- + \cdot OH \quad (50)$$

The ·OH radical has been shown to be a powerful oxidizing entity and would find an ideal oxidizable substrate in cellulose. Similarly the liberation of iodine by sunlight from iodide solutions containing zinc oxide is probably due to such processes as

$$\cdot OH + I^- \longrightarrow I\cdot + OH^- \qquad (51)$$

$$\cdot I + I\cdot \longrightarrow I_2 \qquad (52)$$

For photodegradation of cellulose to occur, it is not, therefore, necessary to postulate the prior formation of hydrogen peroxide.

However, it is clear that the photodegradation of cellulose in the presence of metallic oxides needs a thorough reappraisal in the light of recent knowledge. There is little doubt that zinc oxides and possibly certain dyes may induce photodissociation of water with light that at first sight is energetically unfavorable. The role of water in the photosensitized decomposition of cellulose seems clearly to be more complex than has hitherto been imagined.

References

1. G. Witz, *Bull. Soc. Ind. Rouen,* **11,** 188 (1883).
2. W. Harrison, *J. Soc. Dyers Colourists,* **30,** 206 (1914).
3. C. Doree and J. W. W. Dyer, *J. Soc. Dyers Colourists,* **33,** 17 (1917).
4. L. Vignon, *Compt. Rend.,* **170,** 1322 (1920).
5. P. Heerman and H. Sommer, *Leipzig Monatsh. Textile Ind.,* **40,** 207 (1925).
6. G. Barr and I. H. Hadfield, *J. Textile Inst.,* **18,** T490 (1927).
7. A. J. Turner, *J. Soc. Dyers Colourists,* **36,** 165 (1920).
8. W. Scharwin and A. Pakschwer, *Angew. Chem.,* **40,** 1008 (1927).
9. P. W. Cunliffe and F. D. Farrow, *Shirley Inst. Mem.,* **7,** 1 (1928); *J. Textile Inst.,* **19,** T169 (1928).
10. R. Haller and L. Wyszewianski, *Melliand Textilber.,* **17,** 325 (1936).
11. R. A. Stillings and R. J. Van Nostrand, *J. Am. Chem. Soc.,* **66,** 753 (1944).
12. H. F. Launer and W. K. Wilson, *J. Res. Natl. Bur. St.,* **30,** 55 (1943).
13. E. Race, *J. Soc. Dyers Colourists,* **65,** 56 (1949).
14. Symposium published in *J. Soc. Dyers Colourists,* **65,** 585 (1949).
15. C. H. Giles and R. B. McKay, *Textile Res. J.,* **33,** 527 (1963).
16. C. Reid, *Quart. Rev. (London),* **12,** 205, (1958).
17. M. Kasha, in M. Burton, J. S. Kirby-Smith, and J. L. Magee, Eds., *Comparative Effects of Radiation,* Wiley, New York, 1960; *Radiation Res. Suppl.,* **2,** 246 (1960).
18. S. F. Mason, *Quart. Rev. (London),* **15,** 287 (1961).
19. A. B. F. Duncan and F. A. Matsen, in A. Weissberger, Ed., Vol. 9, *Chemical Applications of Spectroscopy,* W. West, Ed., Interscience New York,1956, pp. 581, 629.
20. J. P. Simons, *Quart. Rev. (London),* **13,** 3 (1959).
21. G. O. Phillips, in *Advances in Carbohydrate Chemistry,* Vol. 18, Academic Press, New York, 1963, pp. 9–59.
22. M. C. Chen and J. A. Taylor, *J. Chem. phys.,* **27,** 857 (1957).

23. C. J. Giles and R. B. McKay, *Textile Res. J.*, **33**, 527 (1963).
24. W. C. Price, *J. Chem. Phys.*, **4**, 147 (1937).
25. M. Zelikoff and K. Watanabe, *J. Opt. Soc. Am.*, **43**, 753 (1953).
26. J. Barrett and J. H. Baxendale, *Trans. Faraday Soc.*, **56**, 37 (1960).
27. H. Fricke and E. J. Hart, *J. Chem. Phys.*, **4**, 418 (1936).
28. M. S. Matheson, *Proc. Intern. Conf. Peaceful Uses At. Energy*, Geneva, **29**, 387 (1958).
29. H. W. Melville, *Proc. Roy. Soc. (London)*, **A157**, 621 (1936).
30. H. Seinflieben and I. Rehren, *Z. Physik*, **37**, 529 (1926).
31. D. Manzerall, *J. Am. Chem. Soc.*, **82**, 1832 (1960).
32. W. J. Nickerson, *Biochim. Biophys. Acta*, **14**, 303 (1954).
33. W. J. Nickerson and J. R. Merkel, *Proc. Natl. Acad. Sci. U.S.*, **39**, 901 (1952).
34. W. J. Nickerson and G. Strauss, *J. Am. Chem. Soc.*, **82**, 5007 (1960); ibid., **83**, 3187 (1961).
35. G. K. Oster, G. Oster, and G. Prati, *J. Am. Chem. Soc.*, **79**, 4836 (1957).
36. G. Oster and N. Wotherspoon, *J. Am. Chem. Soc.*, **79**, 4836 (1957).
37. B. Mooney and H. I. Stonehill, *Chem. Ind. (London)*, 1309 (1961).
38. D. Berthelot and H. Gaudechon, *Compt. Rend.*, **155**, 831 (1912).
39. L. M. Dillman, *J. Lab. Clin. Med.*, **17**, 236 (1931).
40. C. Neuberg, *Biochem. Z.*, **39**, 158 (1912).
41. A. L. Bernouilli and R. Cantieni, *Helv. Chim. Acta*, **15**, 119 (1932).
42. H. von Euler and E. Linderg, *Biochem. Z.*, **394**, 410 (1912); *J. Chem. Soc.*, **102**, 4107 (1912).
43. J. E. Purvis, *J. Chem. Soc.*, **123**, 2519 (1923).
44. F. Micheel, in *Chemie der Zucker and Polysaccharide*, Leipzig, Akademisches Verlags Ges. 1939, p. 218.
45. B. Tollens and H. Elsner, *Kurzes Handbuch der Kohlenhydrate*, 4th ed. Barth, Leipzig, 1935, p. 86.
46. E. Pacsu and L. A. Hiller, *J. Am. Chem. Soc.*, **70**, 523 (1948).
47. G. Herzberg and G. Schiebe, *Z. Physik. Chem. (Leipzig)*, **B7**, 390 (1930).
48. W. C. Price, *J. Chem. Phys.*, **3**, 256 (1935).
49. G. Fleming, M. M. Anderson, A. J. Harrison, and L. W. Pickett, *J. Chem. Phys.*, **30**, 351 (1959).
50. A. J. Harrison, B. J. Cederholm, and M. A. Terwilliger, *J. Chem. Phys.*, **30**, 355 (1959).
51. J. S. Lake and A. J. Harrison, *J. Chem. Phys.*, **30**, 361 (1959).
52. L. W. Pickett, N. J. Hoeflich, and T.-C. Liu, *J. Am. Chem. Soc.*, **73**, 4862 (1951).
53. L. W. Pickett, M. Muntz, and E. M. McPherson, *J. Am. Chem. Soc.*, **73**, 4862 (1951).
54. G. O. Phillips and W. J. Criddle, *J. Chem. Soc.*, 3984 (1963).
55. G. O. Phillips and G. J. Moody, *J. Chem. Soc.*, 3398 (1960).
56. A. Beelik and J. K. Hamilton, *J. Org. Chem.*, **26**, 5074 (1961).
57. G. O. Phillips and T. Rickards, unpublished results in T. Rickards, Ph.D. Thesis, University of Wales, Cardiff, 1965.
58. H. L. J. Backstrom, *The Svedberg Memorial Volume*, Uppsala, Sweden, 1944, p. 5.

59. A. Berthoud, *Helv. Chem. Acta,* **16,** 592 (1933).
60. W. D. Cohen, *Rec. Trav. Chim.,* **40,** 433 (1920).
61. A. Berthoud and D. Porret, *Helv. Chim. Acta,* **17,** 494 (1934).
62. H. S. Taylor and A. J. Gould, *J. Phys. Chem.,* **37,** 367 (1933).
63. B. E. Blaisdel, *J. Soc. Dyers Colourists,* **65,** 618 (1949).
64. J. L. Bolland and H. R. Cooper, *Proc. Roy. Soc. (London),* **A225,** 405 (1954).
65. C. F. Wells, *Discussions Faraday Soc.,* **29,** 219 (1960); *Trans. Faraday Soc.,* **57,** 1703, 1719 (1961).
66. P. Barber, Ph. D. Thesis, University of Wales, Cardiff, 1962.
67. G. O. Phillips and P. Barber, *J. Chem. Soc.,* 3990 (1963).
68. M. V. Lock, personal communication; *Abstract Papers Am. Chem. Soc.,* **131,** 6D (1957); *Proc. Chem. Soc.,* 358 (1960).
69. P. J. Baugh, G. O. Phillips, and J. C. Arthur, Jr., *J. Phys. Chem.,* **70,** 3061 (1966).
70. G. O. Phillips, N. W. Worthington, J. McKellar, and R. R. Sharpe, *Chem. Comm. (London),* 835 (1968); *J. Chem. Soc.* (in press).
71. H. F. Launer and W. K. Wilson, *J. Am. Chem. Soc.,* **71,** 958 (1949).
72. H. Kauffman, *Melliand Textilber.,* **7,** 617 (1916).
73. P. Waentig, *Angew. Chem.,* **36,** 357 (1923).
74. P. Heerman and H. Sommers, *Leipzig. Monatschr. Textil-Ind.,* **40,** 161, 207 (1925).
75. S. Ogyri and T. Yamaguchi, *J. Soc. Chem. Ind. Japan,* **40,** Suppl. 300B, 356B (1937).
76. E. Heuser and G. N. Chamberlain, *J. Am. Chem. Soc.,* **68,** 79 (1946).
77. G. S. Egerton, *J. Soc. Dyers Colourists,* **65,** 764 (1949).
78. A. Agster, *Textilber.,* **35,** 129 (1954); *ibid.,* **36,** 1 (1955).
79. A. Agster and O. Holzinger, *Textil-Praxis,* **11,** 825 (1956).
80. J. H. Flynn, W. K. Wilson, and W. L. Morrow, *J. Res. Natl. Bur. Std.,* **60,** 229 (1958).
81. G. S. Egerton, E. Attle, F. Guirguis, and M. A. Rathor, *J. Soc. Dyers Colourists,* **79,** 49 (1963).
82. J. H. Flynn and W. L. Morrow, *J. Polymer Sci.,* [A], **2,** 81 (1964).
83. Cr. Simionescu, G. L. Rhozmarin, and V. Bulacovschi, *Celuloza Hirte (Bucharest), 14* 319 (1965).
84. E. H. Daruwalla and N. M. da Costa, *Chem. Ind. (London),* 597 (1966).
85. G. Centola, *Assoc. Tech. Ind. Papetiere Bull.,* **5,** No. 3, 11 (1951).
86. L. Chadeyron, *Assoc. Tech. Ind. Papetiere Bull.,* **7,** No. 3, 21 (1954).
87. F. A. Abdie-Maumert, *Papeterie,* **77,** 593 (1955).
88. A. Beelik and J. K. Hamilton, *Papier,* **13,** 77 (1959).
89. J. Peng, K. Minami, and T. Yoshimoto, *Nippon Mokuzai Gakkaishi,* **13** (4), 155 (1967).
90. G. S. Egerton, *J. Soc. Dyers Colourists,* **64,** 336 (1948).
91. G. S. Egerton *J. Textile Inst.,* **39,** T293 (1948).
92. G. S. Egerton, *Textile Res. J.,* **18,** 659 (1948).
93. C. W. Mason and F. B. Rosevear, *J. Am. Chem. Soc.,* **61,** 2995 (1939).
94. G. S. Egerton and K. M. Shah, *Nature,* **202,** 81 (1964).

95. G. S. Egerton, E. Attle, and M. Rathor, *Nature,* **194,** 968 (1962).
96. V. L. Frampton, L. P. Foley, and H. H. Webber, *Arch. Biochem.,* **18,** 345 (1948).
97. C. Kujirai, *Bull. Inst. Chem. Res. Kyoto Univ.,* **23,** 35 (1950).
98. C. Kujirai, *Bull. Inst. Chem. Res. Kyoto Univ.,* **24,** 42 (1951).
99. C. Kujirai, *Bull. Inst. Chem. Res. Kyoto Univ.,* **31,** 228 (1953).
100 J. Schurz, *Svensk Papperstidn,* **59,** 98 (1955).
101. H. Sihtola and B. C. Fogelberg, *Paperi Puu,* **36,** 430 (1954).
102. J. H. Flynn and W. L. Morrow, *J. Polymer Sci.,* [A], **2,** 91 (1964).
103. J. Jortner, *J. Polymer Sci.,* **37,** 199 (1959).
104. A. Sippel, *Faserforsch. u. Textiltech.,* **3,** 211 (1952).
105. A. Sippel, *Kolloid-Z.,* **127,** 79 (1952).
106. A. Sippel, *Melliand Textilber.,* **33,** 645 (1952).
107. J. H. Flynn, *J. Polymer Sci.,* **27,** 83 (1958).
108. T. C. Laurent, *J. Am. Chem. Soc.,* **78,** 1875 (1956).
109. T. C. Laurent and E. M. Wertheim, *Acta Chem. Scand.,* **6,** 678 (1952).
110. G. O. Phillips, *Nature,* **173,** 1044 (1954).
111. G. O. Phillips, G. J. Moody, and G. L. Mattock, *J. Chem. Soc.,* 3522 (1958).
112. B. C. Bera, S. D. Kapila, and D. A. L. Rao, *J. Sci. Industr. Res,* **21D,** 444 (1962).
113. B. C. Bera and P. N. Agarwal, *L. J. Sci. Tech.,* **2** (2), 1 (1964).
114. P. N. Agarwal, B. C. Bera, and L. K. Chaturvedi, *Indian J. Technol.,* **1,** 356 (1963).
115. A. Beelik and J. K. Hamilton, *Chem. Ind. (London),* 1341 (1965).
116. B. C. Bera, *Chem. Ind. (London),* 2068 (1965).
117. T. N. Kleinert, *Monats. Chem.,* **95** (2), 387 (1964).
118. T. N. Kleinert, *Holzforschung,* **18** (1–2), 24 (1964).
119. G. O. Phillips, O. Hinojosa, J. C. Arthur and T. Mares, *Textile Res. J.,* **36,** 822 (1966).
120. C. Kujirai, *Sen-i Gakkaishi.* **21** (4), 183 (1965).
121. E. Treiber, *Kolloid-Z,* **130,** 39 (1953),
122. E. Treiber *Svensk Papperstidn.,* **58,** 185 (1955).
123. G. O. Phillips and co-workers, unpublished results.
124. C. Kujirai, *Sen-i Gakkaishi,* **21** (4), 191 (1965).
125. C. Kujirai, *Sen-i Gakkaishi,* **21** (5), 260 (1965).
126. G. S. Egerton and A. G. Roach, *Nature,* **180,** 189 (1957).
127. G. S. Egerton and S. L. Fitton, *Nature,* **178,** 41 (1956).
128. C. Kujirai, *Sen-i Gakkaishi,* **21** (12), 626 (1965).
129. C. Kujirai, *Sen-i Cakkaishi,* **21** (12), 631 (1965).
130. C. Kujirai, *Sen-i Gakkaishi,* **22** (1), 20 (1966).
131. J. F. Rabek, *Photochem. Photobiol.,* **7,** 5 (1968).
132. D. K. Appleby, *Am. Dyestuff Reptr.,* **38,** 149, 189 (1949).
133. D. Ashton and M. E. Probert, *J. Textile Inst.,* **44,** Tl (1953).
134. A. D. Baskin and A. M. Kaplan, *Textile Res. J.,* **28,** 554 (1958).
135. W. N. Berard, S. G. Gremillion, Jr., and C. F. Goldthwait, *Textile Res. J.,* **26,** 81 (1956).

136. G. S. Egerton, *Am. Dystuff Reptr.*, **36**, 561 (1947).
137. G. S. Egerton, *J. Soc. Dyers Colourists,* **63**, 161 (1947).
138. G. S. Egerton, *J. Soc. Dyers Colourists,* **64**, 659 (1948).
139. G. S. Egerton, *J. Textile Inst.*, **39**, T305 (1948).
140. G. S. Egerton, *Nature,* **204**, 1153 (1964).
141. G. O. Phillips and J. C. Arthur, Jr., *Textile Res. J.*, **34**, 497 (1964).
142. C. H. Bamford and M. J. S. Dewar, *J. Soc. Dyers Colourists,* **65**, 674 (1949).
143. G. S. Egerton, N. E. N. Assaad, and N. D. Uffindell, *J. Soc. Dyers Colourists,* **84**, 220 (1968).
144. C. H. Bamford and M. J. S. Dewar, *Proc. Roy. Soc. (London)*, **A,198**, 252 (1949).
145. A. O. Allen, *The Radiation Chemistry of Water and Aqueous Solutions,* Van Nostrand, New York, 1961.
146. G. A. Gilbert, H. Baum, and M. V. Lock, reported in *Chem. Eng. News,* **35**, 26 (1957).
147. N. Geacintov, V. Stannett, and E. W. Abrahamson. *Makromol. Chem.*, **36**, 52 (1959).
148. N. Geacintov, V. Stannett, E. W. Abrahamson, and J. J. Hermans, *J. Appl. Polymer Sci.*, **3**, 54 (1960).
149. V. Stannett, J. J. Hermans, and N. Geacintov, paper presented at the 138th Meeting of the American Chemical Society, New York, 1960.
150. R. Haller and G. Ziersch, *Melliand Textilber,* **10**, 951 (1929).
151. E. Hibbert, *J. Soc. Dyers Colourists,* **43**, 261 (1927).
152. I. Rusznak and M. Fehervari, *Magy. Textiltech.*, 127 (1954).
153. A. Broadbent, *Chem. Comm.*, 382 (1967).
154. K. Clarke and H. I. Stonehill, private communication.
155. P. J. Baugh, G. O. Phillips, and N. W. Worthington, unpublished results.
156. F. Wilkinson, *J. Phys. Chem.*, **66**, 2569 (1962).
157. G. S. Egerton and N. E. N. Assaad, *Chem. Ind. (London)*, 2112 (1967).
158. G. S. Egerton, J. M. Gleadle, and A. G. Roach, *Nature,* **202**, 345 (1964).
159. G. S. Egerton, J. M. Gleadle, and A. G. Roach, *J. Soc. Dyers Colourists,* **82**, 369 (1966).
160. G. S. Egerton, N. E. N. Assaad, and N. D. Uffindell, *J. Soc. Dyers Colourists,* **83**, 45 (1967).
161. G. S. Egerton, N. E. N. Assaad, and N. D. Uffindell, *J. Soc. Dyers Colourist,* **83**, 409 (1967).
162. F. Feichtmayr and J. Schlag, *Melliand Textilber.*, **45**, (5), 526 (1964).
163. P. J. Baugh, O. Hinojosa, T. Mares, M. J. Hoffman, and J. C. Arthur, Jr., *Textile Res. J.*, **37**, 942 (1967).
164. H. M. Robinson and W. A. Reeves, *Am. Dyestuff Reptr.*, **50**, 17 (1961).
165. A. Bollinger and N. T. Hinks, *Australian J. Exptl. Biol. Med. Sci.*, **27**, 569 (1949).
166. J. Schurz and E. Kienzl, *Svensk Papperstidn.*, **61**, 844 (1958).
167. G. E. Adams, J. H. Baxendale, and J. W. Boag, *Proc. Chem. Soc.*, 241 (1963).
168. E. Baur and C. Neuweiler, *Helv. Chim. Acta,* **10**, 901 (1927).
169. F. Scholefield and E. H. Goodyear, *Melliand Textilber.*, **10**, 867 (1929).
170. C. F. Goodeve, *Trans. Faraday. Soc.*, **33**, 340 (1937).

171. K. Yamafuji, N. Nishioeda, and H. Imagawa, *Biochem. Z.*, **301**, 404 (1939).

172. C. N. Chari and M. Qureshi, *J. Indian Chem. Soc.*, **21**, 97 (1944).

173. M. C. Markham, M. C. Hannan, and W. S. Evans, *J. Am. Chem. Soc.*, **76**, 820 (1954).

174. M. C. Markham, M. C. Hannan, R. M. Paternosto, and C. B. Rose, *J. Am. Chem. Soc.*, **80**, 5394 (1958).

175. M. C. Markham and K. J. Laidler, *J. Phys. Chem.*, **57**, 363 (1953).

E. ENZYMATIC DEGRADATION

E. T. Reese and Mary Mandels

A number of reviews (1-5) and symposia (6-11) have dealt with cellulose decomposition, since the review by Siu in Part I of this work (12). Earlier emphasis was on microbial action on cellulose and on the prevention of deterioration. This section stresses the action of the enzymes themselves. Limitation of space prevents our covering such interesting areas as the following:

1. Inhibition of cellulase (13, 14).
2. The role of cellulase in the rumen (15-17), in soils (18), in wood rotting (5, 19), and in plant pathology (14).
3. *Trichoderma viride* as the source of cellulase (20-22), a subject of increasing interest.

1. Production, Secretion, and Purification of Cellulolytic Enzymes

Cellulolytic enzymes are produced chiefly by certain protozoa (23), bacteria (23), fungi (23), nematodes (23, 24), mollusks (23, 25, 26), and insects (23, 27). For these organisms cellulose is a major food, and the enzyme is required for its digestion. It has also been reported that cellulase is produced by higher plants (23), where its role is probably that of softening cell walls prior to cell enlargement. Cellulose is not a food for higher plants nor for most higher (nonruminant) animals.

Since cellulose is insoluble, it must be digested *outside* the cell membrane; that is, either at the fungal cell surface or at some distance from it. Fungal hyphae bore holes in fibers and other cell walls at the point of contact, but dissolution of cell walls at some distance from penetrating hyphae has also been observed (28). In culture, fungi commonly secrete cellulolytic enzymes into the culture media. The secretion appears to be from living cells, rather than from dead cells, but nothing is known of the secretory mechanism.

Cellulases in bacteria may be constitutive (i.e., produced by the organism growing on *any* substrate) (29), but in fungi they are usually adaptive, that is, induced or increased when the organism is growing on cellulose (30-32) (Table 1). Cellulases are produced when fungi are grown on cellulose or on related glucans containing the (1→4)-β linkage (e.g., lichenin and the β-glucans of oat and barley). They are also induced by soluble cellodextrins, including cellobiose, and by certain glucose trimers that contain the cellobiose moiety, such as 3^2-*O*-β-glucosylcellobiose and 2^1-*O*-α-glucosylcellobiose (30). 4^2-*O*-β-Glucosyl laminaribiose is not an inducer (30); the cellobiose moiety is present in this trimer, but at the nonreducing end of the molecule. Lactose (4-*O*-β-D-galactosyl-glucose), an analog of cellobiose, is a good inducer for some fungi. Sophorose (2-*O*-β-D-glucosylglucose) is a powerful inducer, as can be seen from Table 1, but only for *T. viride* (30). Glucose is not an inducer (30). The apparent inducing effect of *reagent* glucose has been traced to the sophorose present in it as a minor (0.006%) impurity.

TABLE 1

Induction of Cellulase in *Trichoderma viride*

Inducer	Units produced per milligram of inducer	
	C_1[a]	(1→4)-β-Glucanase
Glucans:		
Cotton sliver	9.8	5.4
Cotton duck	10.0	8.3
Solka Floc (spruce-wood cellulose)	8.8	10.0
Filter paper	7.0	8.5
Carboxymethylcellulose (DS 0.5)	–	0.6
Cellulose sulfate (DS 0.4)	–	0.8
Lichenin	–	7.8
Oat glucan	–	0.8
Barley glucan	–	6.6
Oligosaccharides:		
Cellobiose	2.5	3.8
Cellotriose	–	0.6
Cellohexaose	–	0.4
Cellobiose octaacetate (plus esterase)	–	1.5
2^1-α-Glucosylcellobiose	–	2.8
6^2-β-Glucosylcellobiose	–	18.0
Lactose	1.0	4.8
Sophorose	1200.0	2800.0
Noninducers (examples):		
Glucose (purified on charcoal)	0	0
Starch	0.03	0.04
Glycerol	0.02	0.03

[a]For description see discussion of the cellulase complex.

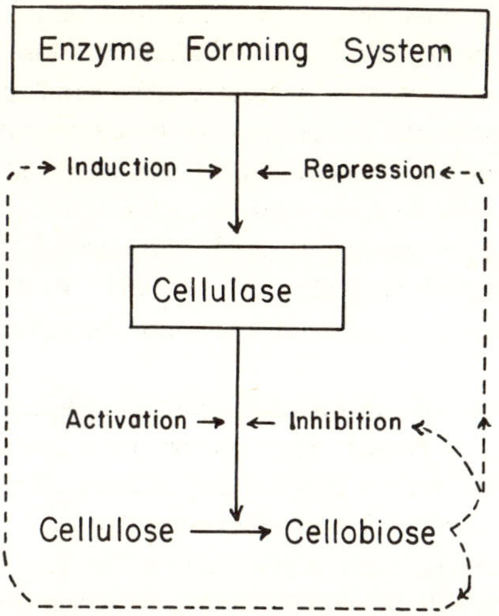

Fig. 1. Regulation of cellulase production and activity. Adapted from Witkop (35).

The true inducer of cellulase in fungi growing on cellulose is not the cellulose but the soluble hydrolysis products (32), especially cellobiose. High concentrations of cellobiose (32) or of other readily metabolizable carbon sources (33, 34) in the growth medium strongly repress cellulase production by fungi (Fig. 1). However, by controlling the rate at which cellobiose is consumed, one can obtain as much cellulase as when cellulose is used as the substrate. This creates the interesting situation in which cellobiose is at once the product of enzymatic action, the inducer and repressor of enzyme production, and an inhibitor of enzymatic action. When fungi are grown on cellulose, a low level of enzyme (produced constitutively) breaks down a little of the cellulose, producing cellobiose, which induces the formation of more enzyme. The cellobiose is consumed immediately, so that it never builds up to repressing levels. Induction on cellobiose can be demonstrated only when the rate of metabolism of cellobiose is reduced, as by means of suboptimum temperature, reduced aeration, or excesses or deficiencies of mineral nutrients. Alternatively, cellobiose can be supplied continuously at low levels, by using cellobiose octaacetate as an inducer (32). Many fungi possess an esterase that hydrolyzes the insoluble octaacetate, slowly releasing cellobiose. This process is quite similar to the slow release of cellobiose during the growth of microorganisms on cellulose.

Mineral nutrition, pH, and aeration (31) influence cellulase production. Suitable media for cellulase production by various fungi in submerged culture have been described (36), and a dry koji process for producing *T. viride* cellulase commercially on wheat bran (22) has been developed in Japan.

During growth the enzymes are secreted into liquid media. They can be readily separated from mycelium and residual cellulose by centrifugation or filtration. Water extracts of koji cultures can be handled in the same manner. Purification can be achieved by dialysis, or by precipitation with ammonium sulfate, or with solvents (acetone, ethanol, 2-propanol, etc.). Separation of the various members of the enzyme complex from each other, and from other proteins, is achieved by passing through columns of Sephadex gel (37–40), diethylaminoethyl(DEAE)-Sephadex (21, 41–43), ion-exchange resins (41, 43, 44), and various forms of cellulose (20); or it may be achieved by electrophoresis (21).

Occasionally reports of the crystallization of cellulase (45) or of one of its components (46) have appeared. Such preparations contain not only cellulase but also other enzymes, such as amylase and xylanase. No one has yet succeeded in obtaining cellulase crystals free of other activities.

The fungal enzymes are most stable under acidic conditions. At a pH higher than 8 they are quickly destroyed; maximum activity is found at pH 4.5 (± 1.0). A temperature of 50°C gives best results in tests lasting less than 4 hr.

2. The Cellulase Complex

Cellulase is a group of enzymes acting together and in sequence (Fig. 2). The existence of the several components was first postulated because of the fact that while cellulolytic fungi grow readily on all cellulosic substrates and consume all the cellulose, the action of cell-free enzymes from such cultures is usually

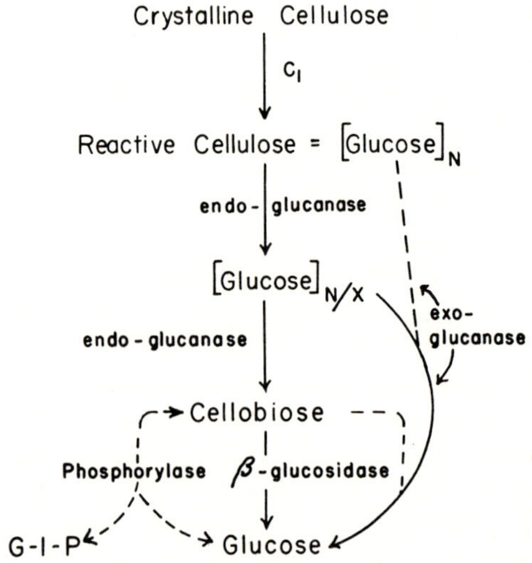

Fig. 2. The cellulase complex.

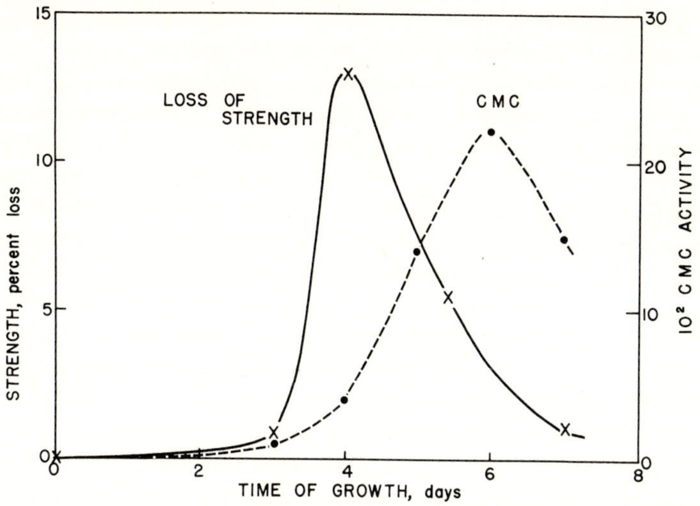

Fig. 3. The cellulolytic enzyme that produces strength loss in cotton yarn appears earlier during the growth of *Myrothecium verrucaria* than does the β-glucanase that acts on the soluble carboxymethylcellulose (CMC) (47).

limited to hydrolysis of cellulose derivatives or of swollen celluloses. Such enzyme preparations are low in the component C_1 required for the attack on *crystalline* cellulose. The existence of components was first demonstrated by differences in activity on various cellulosic substrates for the cellulases of different organisms or for cellulases of the same organism grown under various conditions (Table 1, Fig. 3). Later the components were separated in many laboratories, and differences in their action were demonstrated (21, 39, 46, 48, 49). These components are the following:

1. C_1 – the enzyme required for action on crystalline cellulose
2. *endo*-(1→4)-β-glucanases } C_x
3. *exo*-(1→4)-β-glucanases } C_x
4. β-glucosidases

[The designations used here conform to cellulose and carbohydrate nomenclature. In enzyme nomenclature, the Greek letter usually precedes the numbers, i.e., β-(1→4)].

The enzyme C_1 acts on crystalline cellulose (e.g., cotton fiber) in such a way that subsequent action by the C_x enzymes becomes possible. Enzyme C_1 has been found in large amounts in cell-free preparations of *T. viride* (21, 48–50) and of *Trametes sanguinea* (46). The nature of its action is not understood. Indeed it is questionable whether all C_1 assay procedures measure exactly the same thing. The "hydrocellulase" method (20, 51–53) is perhaps the closest known approximation to C_1. It measures a change in the turbidity of a microcrystalline (Avicel) preparation. The concurrent appearance of cellobiose may indicate contamination with glucanase.

The purest "hydrocellulase" probably produces no sugars in the complete absence of *endo*-glucanase (39). This preparation has a very interesting effect (52, 53). It fragments microcrystalline particles (6-μ diameter) into numerous (800–1500) smaller fragments. In this action it is related to "fragility" factors observed by others (50, 54).

Several reported components resemble C_1. The "A" enzyme (47) is responsible for tensile-strength loss in cotton. A similar component (22) reduces the strength of thread. Enzyme C_1 has been further subdivided by some workers who used various methods for its detection. The use of "C_1" for action on *noncrystalline* cellulose is questionable. Thus maceration of filter-paper strips (22) is a complex action that probably involves more than just C_1.

The manner by which C_1 converts crystalline cellulose to a form susceptible to the action of other enzymes is quite unknown. The need for β-glucanase in the C_1 assay suggests that C_1 is continuously transforming small amounts of crystalline cellulose to the reactive form and that this must be removed by the glucanase before there is further C_1 action. If true, this would explain why it has been so difficult to detect the changes mediated by C_1. [It has been suggested that C_1 may not be an enzyme in the usual sense, but a protein that hydrogen bonds to cellulose more tightly than cellulose hydrogen bonds to itself and thus opens up the micellar structure (53).]

The C_x complex of enzymes hydrolyzes (1→4)-β bonds in the cellulose molecule. The designation covers the random-acting *endo*-(1→)-β-glucanases and the endwise-acting *exo*-(1→4)-β-glucanases. The recognition of the components of C_x makes it desirable to reduce the use of the term "C_x" since it is synonymous with (1→4)-β-glucanase. With the assistance of C_1, the (1→4)-β-glucanases hydrolyze crystalline celluloses; in the absence of C_1, they hydrolyze only *noncrystalline* celluloses, such as those produced by swelling, grinding, or reprecipitation from solution.

The *endo*-enzymes are the characteristic random-acting glucanases responsible for hydrolysis of the high-molecular-weight glucans. (In such material there are too few ends to obtain appreciable rates of hydrolysis by *exo*-enzymes.) The *endo*-(1→4)-β-glucanases (E.C.3.2.1.4, as defined by the International Commission on Enzyme Nomenclature) act preferentially on longer chains. [The situation is somewhat confused since above a degree of polymerization (DP) of 6 the solubility of the dextrins falls rapidly.] Acting on the soluble cellulose derivatives, the random-acting enzymes cause a rapid increase in fluidity, with a relatively slow increase in reducing end groups. The action is not entirely random. Both end linkages appear to be somewhat less affected than the internal linkages. Another limitation is the lower limit of size for susceptibility. The rate of hydrolysis increases with DP—within the limits of substrate solubility. Cellobiose is not hydrolyzed by *endo*-glucanase; cellotriose is highly, or completely, resistant; cellotetraose is hydrolyzed by most *endo*-glucanases, but not by all. The major products of *endo*-(1→4)-β-glucanase action are cellobiose and cellotriose (20). Like most other *endo*-glucanases, the cellulose *endo*-enzyme acts with retention of configuration.

The effect of chemical modification of cellulose on its susceptibility to *endo*-glucanase action is complex. Anything done to increase the solubility of the polymer in water, or to open up the crystalline structure, will increase the *rate* of hydrolysis. At the same time the presence of a substituent effectively prevents hydrolysis of linkages adjacent to it. In other words, the *extent* of hydrolysis is reduced as the number of substituents increases.

At one time it was believed that a substituent on every anhydroglucose unit of the chain was required in order to obtain complete resistance of cellulose to enzymatic hydrolysis. Now it appears that substitution of every other unit is sufficient, provided that substitution can be uniformly accomplished (55). Investigation of the hydrolysis of an arabinoxylan (56) showed that the minimum requirement for xylanase action is two adjacent unsubstituted xylose units. Results on cellulose derivatives (57) revealed a similar relationship for cellulase. In the latter case, however, a substituent at $C_{(6)}$ of the "*aglucone*" unit [but not at $C_{(2)}$ or $C_{(3)}$] can be tolerated.

Data from other *endo*-glucanases tend to support the general requirement for adjacent unsubstituted units (58). As a result both end units of the branched *products* of hydrolysis are usually unsubstituted.

The *exo*-(1→4)-β-glucanases (E.C.3.2.1–, not yet assigned a definite number by the Enzyme Commission) successively remove glucose units from the nonreducing end of the glucan. These enzymes have not been as well investigated as the *endo*-glucanases. They have been reported from *T. viride* (20), and the "cellobiase" of *Stachybotrys atra* (59) may also belong here. They act most rapidly on the soluble chains of four to seven units produced by the action of the *endo*-enzymes and as such play an important part in the overall digestion of cellulose to glucose. Because of their high specificity, *exo*-(1→4)-β-glucanases have very limited action on cellulose derivatives and on glucans having mixed linkages, or branches. Although they can remove one or two unsubstituted terminal units, progress down the chain is halted by any deviation from the regular structure.

β-Glucosidase (E.C.3.2.1.21) hydrolyzes cellobiose and other very-short-chain cellodextrins to glucose. These enzymes do not act on cellulose. Both *exo*-glucanases and β-glucosidases act on cellobiose, cellotriose, and cellotetraose. The *exo*-enzyme acts preferentially on longer oligomers, and acts by inversion. β-Glucosidase hydrolyzes dimer and trimer very rapidly, but the rate decreases markedly as the DP increases. The β-configuration is retained by the product

(60). Furthermore, β-glucosidases are strongly inhibited by gluconolactone, and they hydrolyze (1→1)-β, (1→2)-β, (1→3)-β, and (1→6)-β, as well as (1→4)-β linkages (60). *Exo*-glucanases have greater linkage specificity.

Actual separation and definition of the roles of all these components have not been easy. The number of workers who still consider a single homogeneous enzyme preparation to be responsible for all activities is rapidly diminishing.

The molecular weights of the various enzyme components have been estimated (Table 2). Enzyme C_1 from *T. viride* appears to be a glycoprotein with a molecular weight of about 60,000 (37). The *endo*-(1→4)-β-glucanases show

TABLE 2
Estimated Molecular Weights of Cellulase Components of Selected Organisms

Organism	C_1	*Endo*-(1→4)-β-glucanase	*Exo*-(1→4)-β-glucanase	β-Glucosidase	Reference
Irpex lacteus		10,000			42
		50,000			42
Myrothecium verrucaria		5,300			39
		30,000			39
		55,000			39
		49,000			61
Penicillium chrysogenum		32,000			62
Penicillium notatum		35,000			63
Polyporus versicolor		11,000			64
		51,000			64
Trichoderma koningii		26,000			65
		50,000			65
Trichoderma viride	60,000	52,000	76,000[a]		20
	61,000	12,600		50,000[a]	66
		50,000			66

[a] It is not clear whether *exo*-glucanase or β-glucosidase has been measured.

considerable variation in size, with molecular weights ranging from 5300 to 55,000. It is possible that the smaller particles represent subunits of the "normal" enzyme. *Exo*-β-glucanases and β-glucosidases are reported to have molecular weights of 50,000–76,000. It has been suggested that glycosidases are generally larger molecules than *endo*-glycanases (62).

Amino acid analyses have been made on a number of *endo*-(1→4)-β-glucanases (44, 64, 65). In some cases the basic acids predominate, in others the acidic amino acids predominate. The amount of sulfur-containing amino acids is usually low (less than 2% of the total). Polymeric carbohydrate is usually associated with the various enzyme fractions, the amounts varying from 1 to 50% (64, 65). This association of carbohydrate with enzyme may be responsible, in part, for the multiplicity of (1→4)-β-glucanase components that have been reported. As yet no relationship between structure and function is apparent.

3. Effects of Enzymatic Attack on the Properties of Cellulose

Microbial attack on cellulose fibers involves penetration of the fiber wall, growth of the hyphae within the lumen, and digestion from within. Under favorable conditions the attack is rapid, and the cellulosic material loses its strength. Fabric pulls apart or is easily punctured; wood crumbles. Loss of functionality precedes extensive digestion. The action of the organism on the cellulose is by means of digestive enzymes secreted by the fungus. It is largely a localized effect, and most of the cellulose retains its original properties. The differences between microbial effects and the changes obtained by immersing the entire sample in enzyme can be attributed to the effect of localized, as contrasted to a general, enzyme action.

The action of these enzymes on cellulose results in the following changes (66):

1. *Alkali swelling* provides the most rapid and sensitive measure of the activity of cellulase. Cotton fibers treated with very dilute culture filtrates for as little as 30 min (at 50°C) show increased swelling in 18% sodium hydroxide. No other changes are detectable. The effect is probably a hydrolysis of the cellulose network of the primary wall. The weakening of the outer wall removes a barrier to the swelling in alkali of the underlying secondary wall.

2. *Transverse cracking* occurs on short exposure (for 2 hr) of fibers to enzyme and *spiral fissures* later develop (within 5 days). Similar changes occur in the cell walls of wood. The losses in tensile strength that follow may be due in part to these transverse and spiral cracks. These "fragility" effects seem to be related to the activities of Selby's "A" enzyme, King's "hydrocellulase" fragmentation factor, and to one of Halliwell's components.

3. *The DP is lowered.* This is in contrast to attack by the organism where the DP of the *undigested* residue remains constant because of localized action and the rapid consumption of short-chain fragments. An increase in *alkali* (10% NaOH) solubility also occurs, apparently due to formation of chains of intermediate lengths.

4. *Loss of weight* and production of *reducing sugars* (Fig. 4) occur much more slowly than the preceding changes, especially if the enzyme preparation is low in C_1. The residue shows a decrease in moisture regain and an increased resistance to enzymatic attack. This is probably due to removal of the more amorphous, readily accessible regions of the cellulose.

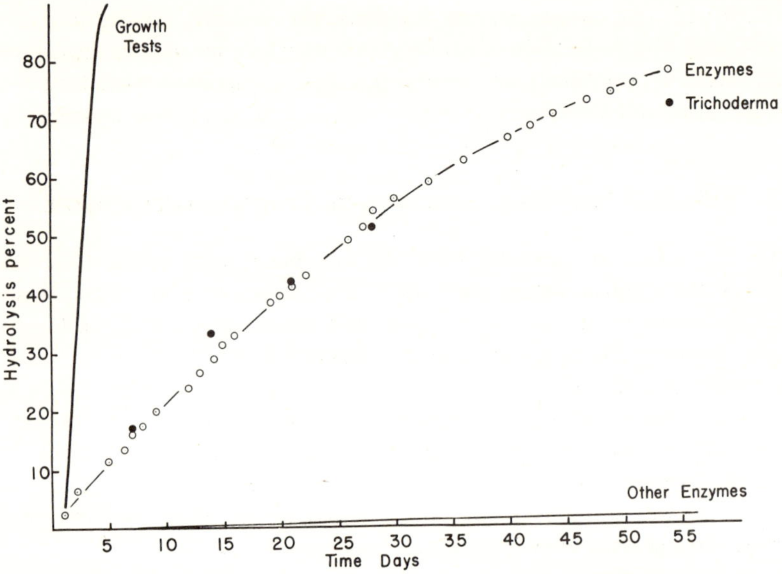

Fig. 4. Hydrolysis of cotton by organisms and by cell-free enzymes. Growth tests: consumption of cellulose by active cellulolytic fungi; *Trichoderma* enzymes: open circles, sugar production – closed circles, weight loss of cellulose (21): other enzymes: cell-free cellulases from organisms other than *Trichoderma*.

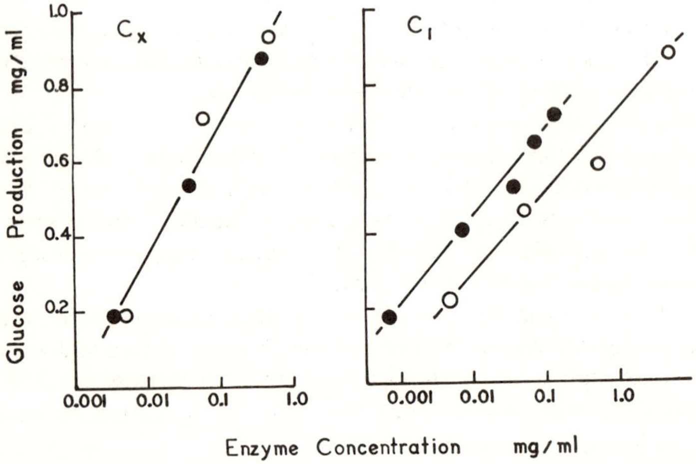

Fig. 5. Effect of enzyme concentration on rate of hydrolysis: C_x – *endo-β-*glucanase, action on carboxymethylcellulose, 1 hr, 50°C; C_1 – action on cotton, 24 hr, 40°C; open circles – crude cellulase of *Trametes sanguinea;* closed circles – crystalline C_1 component of *Tr. sanguinea.* Enzymes are described in reference 46 and were supplied to authors for these tests.

The changes caused by cellulase action as discussed above are largely hydrolytic ones due to (1→4)-β-glucanase action. The resistance of cellulose to enzymatic attack is in large measure due to the inaccessibility of the crystalline, hydrogen-bonded regions of the cellulose fiber to the large enzyme molecules. The role of C_1 is in opening up, or hydrating, the cellulose and so making the glucosidic bonds in these areas accessible to β-glucanase. For the latter to act it must bind to the substrate. The evidence indicates that for *endo*-(1→ 4)-β-glucanase (E) this binding requires a minimum of two adjacent unaltered glucosyl units (G) linked (1→4)-β (linkage *b*):

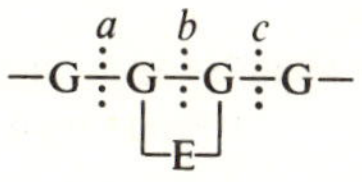

The linkage broken, however, is the adjacent one (linkage *c*). Since the glucosidic linkages in cellulose are also (1→4)-β, the effect cannot be detected by analysis of the hydrolysis products. When glucans of mixed linkage are attacked, the enzyme binds to a (1→4)-β-dimer but may break a (1→3)-β or a (1→4)-β bond. Thus the specificity of *endo*-β-glucanase is determined more by the nature of the fragment (of the polymer) bound to the enzyme than by the nature of the particular linkage hydrolyzed.

The kinetics of cellulose decomposition by enzymes have not received much attention because of difficulties in working with an insoluble substrate and with a complex of enzymes (Fig.5). If cellulose modification is used to achieve solubility, the substituent interferes. Results with soluble oligosaccharides (20, 67) (Table 3) give an approximation of what the activity would be if cellulose were soluble. Attempts to explain the action of enzymes on an insoluble substrate (20, 68) have led only to the obvious conclusion that the rate depends on the amount of available surface. The rate of cellulose decomposition is proportional to the log of enzyme concentration (Fig. 5).

TABLE 3
Michaelis Constants of *Trichoderma viride* Cellulases (20)

	K_m (*M*)	
Substrate	*Exo*-β-glucanase ($\times 10^{-5}$)	*Endo*-β-glucanase ($\times 10^{-4}$)
Cellobiose	230.0	190.0
Cellotriose	18.0	31.0
Cellotetraose	6.5	24.0
Cellopentaose	6.0	7.0
Cellohexaose	16.0	1.0

The number of bonds cleaved per enzyme molecule per second has been calculated in a few cases (48, 62, 67, 69). The reported values, based on soluble oligosaccharides and on cellulose derivatives, fall in the range of 5–29 bonds per second for both the *endo*-and the *exo*-(1→4)-β-glucanases. The rates are much lower for solid cellulose. Such values are very low compared with those reported for the analogous starch-hydrolyzing enzymes, 300 for α-amylase and 4000 for β-amylase.

4. Uses of Cellulase and Related Enzymes

Cellulases and related enzymes can be very useful in the isolation and analysis of polysaccharide materials because of their highly specific nature and the mild conditions under which they act. In contrast, chemical procedures are much less specific and usually involve much more drastic procedures. Thus in the isolation of lignin from spruce wood, cellulase removes the cellulose with little degradation of the lignin (70). Similarly the β-glucan contaminant of xylan (commercial) is removed by the action of an enzyme from *Rhizopus arrhizus* (71).

A material can be characterized by the products it yields on enzyme hydrolysis. A material that is completely converted to cellobiose and cellotriose by *endo*-(1→4)-β-glucanases can be only cellulose. In a similar manner a complex mixture can be shown to contain cellulose if it yields cellobiose on hydrolysis by purified cellulase. Negative results do not always signify the absence of cellulose. In many structures the cellulose may be imbedded in a matrix that prevents the enzyme from reaching it. In these cases fine grinding may be required. Cellulase has been used to identify cellulose in the fruiting body of slime molds (72) and in cell walls of phycomycetes (73). Related enzymes have proved useful in identifying other glucans in fungi and are showing promise in elucidating the structure of complex polysaccharides (71).

An interesting application of cellulase has been in the removal of cell walls to form protoplasts of higher plant cells (74, 75) and of some phycomycetes (73). Other hydrolytic enzymes have been used to remove cell walls from pathogenic fungi. The resulting protoplasts are now popular for physiological investigations including the problem of cell-wall synthesis.

Cellulosic materials have many commercial uses, and in these applications they are susceptible to degradation by cellulolytic enzymes. Cotton, wood, paper, cellophane film, and viscose rayons are readily degraded, but acetate film or fiber—because of a high degree of substitution—resist degradation. Modifications intended to improve functionality may also increase rot resistance. For example, resin treatment to impart crease resistance increases rot resistance. Sometimes unexpected problems develop. The development of wet strength in disposable paper diapers brought resistance to decomposition in sewage systems. The new water-based paints contain carboxymethylcellulose or methylcellulose as thickening agents; very little fungal or bacterial growth on these is sufficient to degrade them rapidly so that the paint loses its viscosity and is ruined.

For many years the Japanese have used microbial fermentation to upgrade grains and soybeans for use as human food (76). Today they are exploiting cellulases in a like manner (77, 78). It is claimed that the nutritive value of food for animals is enhanced by such treatment (79–81). Toyama, who is preeminent in this field, has shown that cellulase and related enzymes can be applied to many foodstuffs with beneficial results. Recently he has found that treatment of rice with enzyme increases soaking efficiency and water absorption. An "instant rice" may be attainable in this way (82).

Valuable contents of the plant cell (22) can be released more effectively by use of enzymes. Cellulases have been proposed as a means of removing crude fiber from oil-seed press-cakes (83, 84), as a means of processing microbial cells for human food, and for extracting valuable cell contents from plant materials, such as protein from grass or soybean (85), agar from red algae (86, 87), and essential oils and flavors from various sources (76).

Cellulases have been used in the clarification of citrus juice and in the production of stable juice concentrates. They have been incorporated into products promoting breakdown in septic tanks, and they have been employed in papermaking to decrease the beating time of pulp.

The dream of cellulase investigators is to develop a commercially feasible process for converting waste cellulose to glucose. The recent successful commercial practice of enzymatically converting *starch* to glucose gives new hope to the dream. The chief problem is obtaining sufficiently active enzymes and highly reactive substrates so that relatively high concentrations of sugar can be obtained in a reasonable time. The sugars produced can be used as a source of glucose or they can be converted to protein, or to fat, by feeding them to the appropriate organism. The sugars could also be used for the production of alcohol. It has been suggested that internal combustion engines using alcohol would not pollute the atmosphere. After many years of combating microbial degradation of cellulose, investigators are now trying to accelerate the process.

At a recent symposium (88) advances in understanding of basic problems, such as C_1, were discussed. It was suggested that C_1 (or "affinity factor") may act by virtue of its ability to hold the hydrolytic factor ($= C_x =$ *endo*-(1→4)-β-glucanase) on solid cellulose in such a way as to promote hydrolysis. Progress is being made in relating the pore size of solid materials to enzyme diffusability. Renewed enthusiasm was shown for applications of cellulase in upgrading animal feeds in the converison of waste cellulose to microbial protein by *Cellulomonas* and to glucose by enzymes. Recently glucose concentrations as high as 30% were obtained in small-scale enzymatic digests. This required very high enzyme concentrations and pretreatment of the cellulose by heating and fine grinding (89). Glucose concentrations of up to 10% were obtained in pilot-plant continuous digesters using heated and milled cellulose and a culture filtrate of *T. viride* as the enzyme (88). A broad program on the disposal of cellulose solid wastes is being pursued by the Department of Health, Education, and Welfare.

References

1. J. Gascoigne and M. Gascoigne, *Biological Degradation of Cellulose*, Butterworths, London, 1960.
2. K. W. King, *Virginia Polytech. Inst. Tech. Bull.*, **154**, (1961).
3. B. Norkrans, *Ann. Rev. Phytopath.*, **1**, 325 (1963).
4. D. L. Ray, Ed., *Marine Boring and Fouling Organisms,* University of Washington Press, Seattle, 1959.
5. B. Norkrans, *Adv. Appl. Microbiol.*, **9**, 191 (1967).
6. *Proc. 1st Symposium on Cellulase and Related Enzymes*, Cellulase Association, Osaka University, Japan, 1961.
7. *Proc. 2nd and 3rd symposia on Cellulase and Related Enzymes*, Cellulase Association, Osaka University, Japan, 1963.
8. *Proc. 4th Symposium on Cellulase and Related Enzymes,* Cellulase Association, Osaka University, Japan, 1964.
9. *Proc. 5th Symposium on Cellulase and Related Enzymes*, Cellulase Association, Osaka University, Japan, 1965.
10. *Proc. 6th Symposium on Cellulase and Related Enzymes*, Cellulase Association, Osaka University, Japan, 1966.
11. E. T. Reese, Ed., *Advances in Enzymic Hydrolysis of Cellulose and Related Materials*, Pergamon Press, London, 1963.
12. R. G. H. Siu, in E. Ott, H. M. Spurlin, and M. W. Grafflin, Eds., *High Polymers*, Vol. V, *Cellulose*, 2nd ed., Interscience, New York, 1954, pp. 183–191.
13. M. Mandels and E. T. Reese, in E. T. Reese, Ed., *Advances in Enzymic Hydrolysis of Cellulose and Related Materials*, Pergamon Press, London, 1963, pp. 115–158.
14. M. Mandels and E. T. Reese, *Ann. Rev. Phytopath.*, **3**, 85 (1965).
15. G. Halliwell and M. P. Bryant, *J. Gen. Microbiol.*, **32**, 441 (1963).
16. K. W. King, *Virginia Agr. Exp. Sta. Tech. Bull.*, 127 (1956).
17. R. E. Hungate, *The Rumen and Its Microbes,* Academic Press, New York, 1966.
18. M. V. Tracey, in D. L. Ray, Ed., *Marine Boring and Fouling Organisms*, University of Washington Press, Seattle, 1959, pp. 253–264.
19. E. Cowling, *USDA Forest Service Rpt. No. 2116, 1958.*
20. K. W. King, *J. Fermentation Technol. (Japan)*, **43**, 79 (1965).
21. M. Mandels and E. T. Reese, *Develop. Ind. Microbiol.*, **5**, 5 (1964).
22. N. Toyama, in E. T. Reese, Ed., *Advances in Enzymic Hydrolysis of Cellulose and Related Materials,* Pergamon Press, London, 1963, pp. 235–256.
23. E. T. Reese and M. Mandels, in E. T. Reese, *Advances in Enzymic Hydrolysis of Cellulose and Related Materials*, Pergamon Press, London, 1963, pp. 197–234.
24. V. Dropkin, *Nematologica,* **9**, 444 (1963).
25. F. L. Myers and D. H. Northcote, *Biochem. J.*, **71**, 749 (1959).

26. G. Okada, T. Nisizawa, and K. Nisizawa, *Biochem. J.*, **99**, 214 (1966).
27. J. B. Adams and M. E. Drew, *Can. J. Zool.*, **41**, 1205 (1963).
28. E. B. Cowling, in E. T. Reese, Ed., *Advances in Enzymic Hydrolysis of Cellulose and Related Materials*, Pergamon Press, London, 1963, pp. 1–32.
29. R. A. Hammerstrom, K. D. Claus, J. W. Coghlan, and R. H. McBee, *Arch. Biochem. Biophys.*, **56**, 123 (1955).
30. M. Mandels, F. W. Parrish, and E, T. Reese, *J. Bacteriol.*, **83**, 400 (1962).
31. M. Mandels and E. T. Reese, *J. Bacteriol.*, **73**, 269 (1957).
32. M. Mandels and E. T. Reese, *J. Bacteriol.*, **79**, 816 (1960).
33. N. T. Keen and J. C. Horton, *Can. J. Microbiol.*, **12**, 443 (1966).
34. J. C. Horton and N. T. Keen, *Can. J. Microbiol.*, **12**, 209 (1966).
35. B. Witkop, *Am. Scientist*, **55**, 114 (June 1967).
36. E. T. Reese and M. Mandels, in R. L. Whistler, Ed., *Cellulose. Methods of Carbohydrate Chemistry*, Vol. III, Academic Press, New York, 1963, pp. 139–143.
37. G. Petterson, *Biochem. Biophys. Acta*, **77**, 665 (1963).
38. G. Petterson, E. B. Cowling, and J. Porath, *Biochem. Biophys. Acta*, **67**, 1 (1963).
39. K. Selby and C. C. Maitland, *Biochem. J.*, **94**, 578 (1965); *ibid.*, **104**, 716 (1967).
40. D. R. Whitaker, K. R. Hanson, and P. K. Datta, *Can. J. Biochem. Physiol.*, **41**, 671 (1963).
41. T. Niwa, K. Kawamura, and K. Nisizawa, *J. Fermentation Technol. (Japan)*, **43**, 286 (1965).
42, K. Wakabayashi, T. Kanda, and K. Nisizawa, *J. Fermentation Technol. (Japan)*, **44**, 669 (1966).
43. T. Iwasaki, K. Hayashi, and M. Funatsu, *J. Biochem. (Japan), 209 (1964).*
44. G. Okada, T. Niwa, H. Suzuki, and K. Nisizawa, *J. Fermentation Technol. (Japan)*, **44**, 682 (1966).
45. H. H. Higa, R. D. O'Neill, and M. W. Jennison, *J. Bacteriol.*, **71**, 382 (1964).
46. K. Nara, T. Fugano, and H. Yoshino, in *Proc. 5th Symposium on Cellulase and Related Enzymes*, Cellulase Association, Osaka University, Japan, 1965, pp. 77–84.
47. K. Selby, in E. T. Reese, Ed., *Advances in Enzymic Hydrolysis of Cellulose and Related Materials*, Pergamon Press, London, 1963, pp. 33–49.
48. L. H. Li, R. M. Flora, and K. W. King, *Arch. Biochem. Biophys.*, **11**, 439 (1965).
49. K. Selby and C. C. Maitland, *Biochem. J.*, **104**, 716 (1967).
50. G. Halliwell, *Biochem. J.*, **100**, 315 (1966).
51. T. Iwasaki, R. Ikeda, K. Hayashi, and M. Funatsu, *J. Biochem. (Tokyo)*, **57**, 467 (1965).
52. K. W. King, *Biochem. Biophys. Res. Commun.*, **24**, 295 (1966).
53. T. H. Liu and K. W. King, *Arch. Biochem. Biophys.*, **120**, 462 (1967).
54. T. Kitamokado and N. Toyama, *Hakko Kogaku Zasshi*, **40**, 85 (1962).
55. T. Cayle, *Wallerstein Lab. Comm.*, **25**, 349 (1962).
56. A. Perlin and E. T. Reese, *Can. J. Biochem. Physiol.*, **41**, 1842 (1963).
57. W. Klop and P. Kooiman, *Biochem. Biophys. Acta*, **99**, 102 (1965).
58. D. H. Hutson and H. Weigl, *Biochem. J.*, **88**, 588 (1963).
59. M. A. Jermyn, *Australian J. Biol. Sci.*, **20**, 193 (1967); personal communication, 1967.
60. E. T. Reese, A. Maguire, and F. W. Parrish, *Can. J. Biochem.*, **46**, 25 (1968).

61. P. K. Datta, K. R. Hanson, and D. R. Whitaker, *Can. J. Biochem. Physiol.*, **41**, 697 (1963).
62. E. Ahlgren and K. E. Ericksson, *Acta CHem. Scand.*, **21**, 1193 (1967).
63. K. A. Almin and K. E. Eriksson, *Biochem. Biophys. Acta*, **139**, 238 (1967).
64. G. Petterson and J. Porath, *Biochem. Biophys. Acta*, **67**, 9 (1963); T. Iwasaki, R. Ikeda, K. Hayashi and M. Funatsu, *J. Biochem. (Tokyo)*, **57**, 478 (1965).
65. K. Selby and C. C. Maitland, *Biochem. Biophys.*, **118**, 254 (1967).
66. E. T. Reese, in D. L. Ray, Ed., *Marine Boring and Fouling Organisms*, University of Washington Press, Seattle, 1959, pp. 265–290.
67. D. R. Whitaker, *Arch. Biochem. Biophys.*, **53**, 439 (1954).
68. G. Terui, in *Proc. 5th Symposium on Cellulase and Related Enzymes*, Cellulase Association, Osaka University, Japan, 1965, p. 9.
69. M. Katz, personal communication, 1967.
70. J. C. Pew and P. Weyna, *Tappi*, **45**, 247 (1962).
71. E. T. Reese and M. Mandels, *J. Am. Chem. Soc.*, **80**, 4625 (1958).
72. C. Ward and B. Wright, *Biochemistry*, **4**, 2021 (1965).
73. S. Bartnicki-Garcia and E. Lippman, *Biochem. Biophys. Acta*, **136**, 533 (1967).
74. E. C. Cocking, *Nature*, **187**, 962 (1960).
75. A. W. Ruesink and K. V. Thimann, *Science*, **154**, 280 (1966).
76. L. A. Underkofler, in E. T. Reese, Ed., *Advances in Enzymic Hydrolysis of Cellulose and Related Materials*, Pergamon Press, London, 1963, p. 255.
77. K. Yasumatsu, S. Bichu, S. Hori, and H. Shimazono, *Nippon Shokuhin Kogyo Gakkaishi*, **13**, 291 (1966).
78. N. Fujii and N. Toyama, *Hakko Kogaku Zasshi*, **42**, 105 (1964).
79. E. T. Moran, *Dissertation Abstr.*, **26**, 2951 (1965).
80. H. Morimoto, D. Kubota, and N. Akai, *Nippon Kakin Gakkaishi*, **3**, 166 (1966).
81. H. Morimoto, D. Kubota, and H. Maeshima, *Nippon Kakin Gakkaishi*, **3**, 171 (1966).
82. N. Fujii and N. Toyama, *J. Fermentation Technol. (Japan)*, **45** (7), 681 (1967).
83. R. Tazaki and K. Ouye, *Hakko Kogaku Zasshi*, **40**, 195 (1962).
84. K. M. Abdo, Ph.D Thesis, Virginia Polytechnic Institute, Blacksburg, Va. 1966.
85. K. Ramamott and D. S. Johat, *Nature*, **198**, 481 (1963).
86. N. Toyama, *Hakko Kogaku Zasshi*, **40**, 199 (1962).
87. M. Hachiga and K. Hayashi, *Hakko Kogaku Zasshi*, **42**, 207 (1964).
88. G. Hajny and E. T. Reese, Ed., *Cellulase and Its Applications*, Advances in Chemistry Series Monograph 95, American Chemical Society, Washington, D.C., 1969.
89. M. Katz and E. T. Reese, *Appl. Microbiol.*, **16**, 419 (1968).

Chapter XIX

NEW DEVELOPMENTS IN THE TECHNOLOGY OF CELLULOSE AND ITS DERIVATIVES

A. DURABLE-PRESS TREATMENTS

Herman B. Goldstein

Patents issued in 1928 to Foulds, Marsh, and Wood (1, 2) can be considered to be the first disclosure of the far-reaching discovery that cellulosic fabrics can be rendered shrink and wrinkle resistant through the application of precondensates of urea-formaldehyde or phenol-formaldehyde, followed by curing the fabric. Although Foulds and his co-inventors were able to demonstrate useful effects on all types of cellulosic fabrics, only the application to regenerated cellulose became commercially important initially. Application of these processes to linen and to cotton had to wait for further developments to make such treatments practical and attractive.

During the period 1945–1955 a succession of synthetic fibers were introduced in commercial quantities; garments manufactured from them were advertised for their "easy-care" characteristics (3–7); that is, freedom from wrinkling while the garment is being worn and rapid-drying and smooth-drying characteristics after laundering. The synthetic fabrics therefore competed in what were the traditional outlets for cotton.

Largely on the basis of earlier experience with treatments for rayon, the cotton industry attempted to upgrade its products by the use of urea–formaldehyde and melamine–formaldehyde resins. However, these treatments enjoyed only limited success because of two serious shortcomings—namely, poor washfastness and poor chlorine resistance (see subsection 1, "Definitions and Test Methods"). Fortunately in the late 1940s dimethylolethyleneurea (DMEU) was made available on a commercial scale, and the advantages of this compound were quickly recognized (see also Section E in Chapter XVII). With the introduction of DMEU, the cotton industry had, for the first time, an agent that provided many of the desirable characteristics necessary for a successful wrinkle-resistant finish, with improved laundry fastness and reduced sensitivity to chlorine bleaches. According to current terminology, a "wrinkle-resistant" finish is not expected to provide a cotton fabric with a high level of wrinkle resistance; it does, however, provide an improvement in wrinkle *recovery*, both wet and dry. Typically, when less than 2% resin solids is applied, the treated fabric shows a total wrinkle recovery (warp plus fill) of about 200–230° and less than 20% loss in tensile strength as compared with the untreated cloth (wrinkle recovery about 150°). From the consumer's standpoint this property proves beneficial primarily because of easier ironing characteristics.

By 1955 improved reactants had been introduced on a commercial scale, most notably the dimethyloltriazone-type reactants (8–11). For best performance it was found necessary to apply up to 4% resin solids on the weight of the fabric. This caused tensile-strength losses of up to approximately 35%, as compared with untreated fabric; however, such treatments provided wrinkle recoveries (total warp plus fill) of 270–290°. Garments made from such fabrics were often called "wash-and-wear". More conservative and considerably more accurate descriptive terms would include the following: "easy care"; "minimum care"; "minimum ironing"; "drip dry"; and the like. Such easy-care garments required touch-up ironing because of seam puckering and poor pleat retention, but, as compared with ordinary untreated cottons, they offered the consumer considerable advantages because of more rapid drying and easier ironing properties. In addition, the garments remained neat and muss-free while being worn. As a result they were commercially highly successful (12).

The high wrinkle recovery of easy-care cottons, which is so beneficial in the ways indicated above, itself introduces a serious problem. This is the problem associated with the deliberate creases and pleats that are pressed into a garment, as called for by styling; for example, the crease in men's trousers. With fabrics composed wholly or in part of synthetic fibers, the problems of imparting durable creases and pleats are somewhat smaller because of the thermoplastic nature of the synthetic fibers. In such fabrics creases and pleats are made to withstand laundering by hot pressing.

In the ordinary method of producing easy-care or wash-and-wear cotton fabrics the textile is treated in the mill and cured while it is in a flat condition. Any wrinkles or creases imparted in use will tend to uncrease, thereby restoring the fabric to the flat condition it had during the curing. The higher the wrinkle recovery, the higher the tendency for the cloth to return to the flat condition. A garment made from this type of fabric will lose almost completely any deliberately imparted creases or pleats when the garment is subjected to laundering. Pressing after laundering is therefore required to restore the desirable creases and pleats. Also, with ordinary easy-care cotton fabrics, it is virtually impossible to sew seams that will not pucker excessively when subsequently laundered. It was the poor crease-and-pleat retention and bad seam-puckering properties of 100% cotton easy-care or wash-and-wear fabrics that prevented their widespread use in men's slacks, suits, walking shorts, and work pants; in ladies' pleated dresses and skirts, as well as slacks; and in pleated children's clothes.

Since approximately 1956 many workers have attempted to improve the crease-and-pleat retention and seam-smoothness properties of chemically treated cotton fabrics, but these efforts were not commercially successful until 1963. Garments made as a result of these developments were designated "permanent press" or "durable press." They use fabrics of blends of cotton with synthetic fibers because of generally superior wearing characteristics as compared with fabrics consisting solely of cotton. When the durable-press process is applied to cotton alone, the necessary high level of crosslinking causes a 40–70% loss in tensile strength and a 60–90% loss in abrasion resistance in the fabrics; such severe strength losses result in poor service wear. Blending synthetic fibers (15–20% nylon or, more usually, 50–80% polyester) with cotton produces a reinforcing effect that greatly improves the wearing characteristics. Also blends of polyester, acrylic fiber, and/or wool are blended with rayon to produce high-style fabrics that are similar in appearance and hand to worsted fabrics but have durable-press properties.

Among the various durable-press procedures that have been described, the most important are resin treatment of garments (13–17); recuring (18); deferring or postcuring (19–24); precuring or re-forming (25–27); two-stage curing (28, 29); wet fixation (30–34); and cold curing (35, 36). Of these, only the post-curing and precuring approaches have achieved widespread commercial success through 1970.

These methods have made possible the manufacture of truly wash-and-wear garments—that is, garments that resist wrinkling while being worn; may be dry-cleaned or laundered; and, without pressing, will dry to a smooth surface, with good crease-and-pleat retention and excellent seam smoothness. From the standpoint of the processes involved there are two critical features that distinguish all durable-press processes from earlier wrinkle-resistant or easy-care processes. These critical features are the following:

1. Application of a minimum of 6% resin solids on the weight of fabric.
2. Performing special pressing and/or curing operations on the completed garment.

1. Definitions and Test Methods

Considering the differences in terminology and interpretations, it is well to define some of the terms used in this section.

Crease (pleat): a desirable fold in the fabric required by the styling of the garment.

Crease retention: the retention of desirable creases (and pleats) through wearing and washing. Crease-retention properties are evaluated and rated by a newly developed variation (37a) of the durable-press test (37b) or, alternatively, by the crease-angle-retention test, proposed by Goldstein and May (21).

Wrinkle: the undesirable ridges, furrows, corrugations, or puckers imparted to a fabric by wearing or washing.

Wrinkle resistance (recovery): the result of a fabric treatment that reduces the severity of these distortions. Krais (38) and Hall (39) suggested the use of crease angle to measure wrinkle recovery. A variety of tests that use this approach are now in use throughout the world, including the AATCC method (37c), the TBL method (40), the British standard method (41), and the TEFO new method (42). Details of these and other wrinkle-recovery tests have been reviewed by Bostwick (43). In the AATCC method tests are usually performed in both the warp and filling directions, and it is common to report the results as the total of both warp plus filling (W + F) values. Perfect recovery would be represented by 180° in each direction, or a total of 360°. Untreated cotton fabrics usually show a total wrinkle-recovery value of approximately 150°; chemically modified wrinkle-resistant cottons, values of 200–300°; untreated intimate blends of polyester and cellulosic fibers, values of 250–280°; and such blends treated for durable press, wrinkle-recovery values of 290–325°. If the test is performed on the conditioned, but otherwise dry fabric, the results are designated "dry wrinkle recovery"; if the fabric specimen is wet or saturated with water, the results are designated "wet wrinkle recovery" (44).

Wash-and-wear ("easy-care"): a term applied to garments or household textile products that require only a slight ironing after washing and little or no ironing after wearing.

Durable (permanent) press: A term applied to garments or household products that require *no* ironing after wearing or washing. The smoothness properties of fabrics or garments after laundering may be evaluated by the overhead-lighting method using the newly developed three-dimensional replicas (37b, 45). Seam puckering in garments is evaluated by the overhead-lighting method (37d). It should be noted that the appearance characteristics of a given fabric or garment may be considerably affected by the specific drying conditions used after laundering (35), such as drip-drying, or spin-tumble-drying.

In the *postcure* (also known as the deferred cure) durable-press process the fabric is impregnated with an appropriate reactant, catalyst, and other auxiliaries and is then dried in such a manner that there is little or no reaction with the fiber. The dried fabric (which is said to be in a "sensitized" condition) may then be stored or shipped. The garment manufacturer then cuts, sews, presses, and eventually cures the garments to cause crosslinking and to permanently set the garment in the final pressed condition.

In the *precure* process the fabric is impregnated with the chemicals and then dried and cured (as in conventional wrinkle-resistant treatments); optionally the fabric may be afterwashed to eliminate odor and unreacted chemicals. At this point a second catalyst may also be applied. The fabric is then shipped to the garment manufacturer, who cuts, sews, and then sets the garment into its final shape with high-temperature, high-pressure presses. The presence of thermoplastic fibers in the fabric is highly desirable in the precure process.

Chlorine resistance: this term refers to the stability of the chemical finish or the textile toward chlorine-containing bleaches, such as hypochlorite. The American Association of Textile Chemists and Colorists' Research Committee RA-35 test (37e) consists of immersing the fabric in a specified solution of sodium hypochlorite for 15 min, rinsing, and air-drying; strips of fabric are then placed between two heated plates at 185°C; after specified conditioning, the tensile strength of the strips is determined and compared with the strength of similarly bleached, but not heated, strips. The percentage of loss in strength due to heating is recorded as "damage due to chlorine." In addition to the strength loss, it is also of interest to note whether there is any yellowing and any discoloration due to scorching.

Soil release: this term means that stains and spots can be readily removed by laundering or dry cleaning. "Anti-wet soiling" means that soil redeposition during laundering has been reduced or eliminated; this is particularly important for fabrics dyed in pastel shades (46).

2. Reagents and Their Function

It is beyond the scope of this section to enter the controversy as to whether durable-press properties are due to resin formation (polymerization) or to crosslinking. However, the preponderance of evidence already at hand, seems to indicate that, if resin formation plays any role at all, it is a minor one in which the resin acts, itself, as the crosslinking agent through the formation of covalent or strong hydrogen bonds with the cellulose (47, 48). Furthermore, the hypothesis that the crosslinking of cellulose is the most important factor in the changes observed is supported by very substantial evidence in its own right (49–55) (see also Section E of Chapter XVII). Thus as the essential component of any wrinkle-resistant or durable-press process is an appropriate *crosslinking agent* (reactant).

Almost invariably some type of *catalyst* is also needed to induce the reaction between the crosslinker and the fiber. Sometimes two catalysts with different activities are used to permit two discrete steps of reaction (28–30). The role of the catalyst is to increase the rate of curing to permit more economical processing of the textiles. Additionally, it allows the use of a lower reaction temperature, which is frequently desirable. There are some facts (55) and considerable conjecture concerning the mechanism by which the catalyst functions. At least with the metal-salt types discussed below, formation of a complex between the catalyst and the crosslinker, to reduce the activation energy, appears to be the most logical explanation.

In addition to the reactant and the catalyst, most durable-press treatments include a softener or lubricant to improve the hand of the treated cloth and, equally important, to preserve adequate sewing properties and tearing strength. Durable-press treatments may also make use of wetting agents, bodying agents, abrasion improvers, fluorescent brighteners, and soil-releasing agents, depending on specific requirements.

The most important types of products in current use for durable-press processes are, with very few exceptions, suitable in principle for all types of cellulose (linen, cotton and regenerated cellulose); however, the concentrations applied and the reaction conditions must be varied for different fiber substrates. Furthermore, the same general formulations (with small adaptations) can be used for 100% cellulosic fabrics as well as for blends of cellulosics with thermoplastic fibers (such as polyester, acrylic, and nylon) or wool, or for ternary blends of cellulosic, woolen and thermoplastic fibers.

(a) REACTANTS

Among the important reactants for durable-press fabrics, the following should be noted:

Dimethylolethyleneurea (50, 56–59) (see p. 839 of Chapter XVII-E) is useful for precure durable-press processes because it cures and recures under mild conditions. It is, however, unsuitable for postcure because sensitized cloth cures spontaneously (20). Its maximum performance level is fair, and there are occasional problems with odor, unless the treated fabric is afterwashed. Its chlorine resistance is poor, unless the fabric is afterwashed (60, 61), and it exerts an adverse effect on the lightfastness of dyed goods (62).

Bis(methoxymethyl)uron (63) (see p. 840 of Chapter XVII-E) is sold in most commercial products as a blend of bis(methoxymethyl)uron with a buffering-type resin, such as a triazone (64) or a melamine–formaldehyde (65). It is generally unsuitable for postcure work because of excessive free formaldehyde (bad odor), but it is satisfactory for precure processes. It provides a good initial white color but has only marginal chlorine resistance, especially after multiple commercial washes. The hand of cured cloth is somewhat firm and/or full.

Dimethylolpropyleneurea (66–69) (see p. 839 of Chapter XVII-E) is generally unsuitable for postcure processes because a substantial degree of crosslinking occurs merely on drying, even at low temperatures. It is suitable for precure work, but the maximum level of performance is only fair. It is widely used in Europe because of excellent initial white color. Its washfastness is, however, relatively poor in commercial laundering. The reactant is sometimes used in conjunction with the slower curing monoureins to produce a "partial postcure" treatment.

Dimethylol monocarbamates (70–74) (see p. 841 of Chapter XVII-E) are generally unsuitable for postcure processes because of excessive free formaldehyde or volatility of the active agent. They are well suited for precure processes, especially where maximum washfastness and chlorine resistance are desired. They provide an excellent initial white color and whiteness retention.

Glyoxal monourein derivatives (75–79), such as dimethyloldihydroxyethyleneurea (DMDHEU, see p. 839 of Chapter XVII-E), are widely used for postcure processes (20, 21, 24); they provide for excellent storage stability of the sensitized cloth, have essentially no odor, give excellent initial white color and whiteness retention, satisfactory chlorine resistance even after commercial laundering, and have no adverse effect on the lightfastness of dyestuffs (62, 81). They have a very high level of performance (fabric smoothness, crease retention, and seam smoothness), excellent washfastness, and may also be used for precure processes.

Miscellaneous types. The five major types of reactants described above represent the major classes in use at the present time throughout the world. In addition to these, there are a number of other types of lesser importance, listed below:

Polymethylolurea is available in formaldehyde-to-urea molar ratios of 1.3–2.0 : 1, the lower ratio being used primarily on rayon and the higher ratio on cotton. It produces an excellent, resilient, full hand.

Bis(methoxy)trimethylolmelamine (82–85) is used as a hand modifier where stiffness with resilience or wrinkle-recovery properties are desirable.

"Triazine" (86) is a commercial product consisting of a blend of hexa(methoxymethyl)melamine with dimethylolethyleneurea. The finish tends to yellow on repeated exposure to chlorine bleach. It is used primarily on cottons for easy-care-type treatments.

"Triazone" (8–11) is widely used as a low-cost, chlorine-resistant, easy-care finish. It is often used as a blend of dimethyloltriazone with dimethylolurea for reduced amine odor (87).

Acetals (88–91) are nonnitrogenous shrinkage-control agents for cellulose. They do not provide much wrinkle resistance or resilience unless blended with a nitrogenous type, such as dimethylolethyleneurea.

(b) CATALYSTS

The most commonly used catalysts are listed in Table 1.

In general the main advantage of the ammonium salts is their low cost. On the other hand, they may cause development of odor and, with resin-forming

TABLE 1
Catalysts for Durable-Press Reactants

Ammonium salts (80)	Amine salts	Organic acids	Metal salts
Monoammonium phosphate	Hydrochloride of: 2-amino-2-methyl-1-propanol (92);	Tartaric acid	Zinc chloride
Diammonium phosphate	Monoethanolamine	Maleic acid	Zinc nitrate
Ammonium chloride		Diglycolic acid	Zinc fluoroborate
Ammonium sulfate		Citric acid	Magnesium chloride
			Aluminum chloride
			Magnesium phosphate (93)
			Metal salt–organic acid combination (94)

compounds, also tend to shorten the bath life. Strong catalysts, such as ammonium chloride, should be used only in treating regenerated cellulose; they are too acidic to be used on cotton.

The amine salts generally provide superior bath stability and are less likely to cause problems with odor; however, in general, they are comparatively high in cost and may cause discoloration under high-temperature or severe curing conditions.

The organic acids can be used only with reactant compounds that do not polymerize in the bath. Furthermore, these acidic compounds can be used only with reactants that have some buffering power to prevent excessive tendering of the cloth due to the acidity.

The metal salts are generally moderate in cost and usually quite effective, with little or no tendency to cause odor. As a matter of fact, with the use of magnesium chloride the tendency for formaldehyde odor development is sometimes actually reduced. Aluminum chloride is extremely acidic; as such it is a very fast catalyst but tends to cause severe strength losses. For this reason aluminum chloride is often used as a minor component in mixed-catalyst systems.

The use of magnesium phosphate as a catalyst was patented by Tootal Broadhurst Lee (93) for use in their "Fixaform" process for postcure durable press. Although it came into widespread vogue only since about 1960, the use of metal salts has greatly increased, and at this time metal salts are used practically exclusively in all durable-press processes. Of the many metal salts available, the three most widely used compounds are zinc nitrate, zinc chloride, and magnesium chloride. For most resins the zinc salts are much faster catalysts than magnesium chloride, and they often prove to be the more economical from the standpoint of efficient utilization of the crosslinker. The curing efficiency of metal-salt catalysts can be still further improved by the addition of an organic acid (94).

However, it is important to point out that, aside from efficient curing of the resin, there are several other significant factors in selecting the catalyst (35). Of these, the most important are the possible effect on the shade of the dyed fabric and on the effectiveness of brighteners; possible diffusion of dispersed colors from polyester fibers; the possible development of discoloration during prolonged storage of sensitized cloth; and interference with efficiency of soil-release treatments. Based on his own experience, the author can make the following generalizations:

1. For postcure durable-press treatment with monourein-type reactants zinc nitrate catalysts should be used without exception.
2. For precure durable-press treatment with monourein reactants magnesium chloride catalysts may be used if chlorine resistance is not needed; otherwise, zinc chloride is preferred.
3. For precure durable-press treatment with carbamates zinc chloride or modified magnesium chloride types are recommended.
4. For precure durable-press treatment with dimethylolethyleneurea, dimethylolpropyleneurea, and bis(methoxymethyl)uron magnesium chloride is generally satisfactory; zinc chloride types can be used for improved chlorine resistance.

(c) SOFTENERS AND LUBRICANTS

A softener is almost always required in easy-care and durable-press treatments to improve the hand and drape of crosslinked-cellulose fabrics. A lubricant is also highly desirable to improve fiber and yarn mobility so as to preserve tearing strength and sewability. It should be noted that, although most lubricants seem to improve abrasion resistance as measured in laboratory abrading devices (such as the Stoll Flex Tester), they have only marginal beneficial effect (if any) as measured by actual wear-and-wash tests. Some compounds are primarily softeners, whereas others are primarily lubricants; still others combine both properties to some extent. Some agents withstand laundering and/or dry cleaning, but others are completely removed during the first cleaning. Until recently there was a general trend to use durable softeners and lubricants; however, with the recent emphasis on soil-release properties in permanent-press garments, it has been found that many of the durable compounds interfere with soil-release treatments or contribute significantly to wet soiling (95, 96). Consequently there is now a widespread return to the nondurable types, especially where soil-release or anti-wet-soiling characteristics are desired.

The more important softeners (S) and lubricants (L) that are now in widespread use are the following:

Durable types: polyethylene emulsions (L); polypropylene emulsions (L); silicone emulsions (S, L); stearamide emulsions (S); methylolstearamide (S); cationic fatty compounds (S); and soft acrylic latexes (S, L).

Nondurable types: glycerol monostearate (S); mono fatty esters of polyglycols (S); sulfated fatty esters (S); wax emulsions (L).

As a nondurable softener is removed during laundering, the surface resin (which is almost inevitably present to some extent) is also being removed, thereby actually leaving the cloth with an undiminished soft hand.

(d) MISCELLANEOUS ADDITIVES

In addition to the resins, catalysts, softeners, and lubricants, durable-press processes also usually require wetting agents, firming, and bodying agents [such as poly(vinyl acetate) or acrylic emulsions, hydroxyethylstarch, poly(vinyl alcohol), or thermosetting syrups], soil-release agents (such as flurochemicals, acrylic copolymers, copolymers of maleic anhydride with methyl vinyl ether), anti-wet-soiling agents (hydrophilic polyesters), fluorescent brighteners, etc. These various materials must be judiciously chosen to produce the best overall effect.

3. Mill Processing

As with ordinary resin finishing, the fabric to be processed for durable press will previously have been singed, desized, scoured, bleached, mercerized, dyed, printed, etc., as required. With fabrics containing synthetic fibers it is customary to heat-set as well. Open-width processing throughout these preparative steps is highly desirable with blends containing cotton and is mandatory with blends containing rayon if optimum fabric smoothness is to be achieved. The fabric to be finished must be absorbent and free of impurities. Alkalinity from previous processing steps must be removed to prevent interference with the catalyst (97). The chemical treatment is applied to the fabric by padding, which consists of running cloth in open width one or more times through the bath and then passing it between the nip of rollers to squeeze out surplus liquor.

In the *postcure process* the impregnated fabric is then dried on a tenter frame. Usually a weft straightener is inserted at the front of the drying range (98), and it is customary to use a predryer in front of the frame to minimize buildup of resin on the clips. The drying range is operated at a speed-and-temperature combination that will dry the cloth to a residual moisture content of 6–10% (based on cellulose). On exiting from the drying frame, the cloth is passed over one or more cooling cans to bring it to room temperature and then batched on a roll. As far as the textile mill is concerned, this is essentially the end of the postcure process, except for inspection and packing.

In the *precure process* the fabric is impregnated and dried as already indicated above, but it is then passed into the curing range, where the temperature is regulated to provide most efficient use of the crosslinking agent. After curing, depending on the chemical treatment used, the cured fabric may be afterwashed (to eliminate undesirable odors or to improve the hand), dried, and batched. Alternatively, with certain types of crosslinkers, it is possible to eliminate the afterwash and batch directly after cooling the cloth.

4. Garment Manufacture

It is beyond the scope of this work to dwell on the details involved in garment manufacture. However, one of the critical features of durable-press treatments lies in the fact that some special pressing and/or curing operations are performed on the completed garments. Thus to complete this description a brief outline of the major operations involved is included.

(a) POSTCURE PROCESSES (19–21, 24, 26, 27)

The sensitized cloth, on rolls, is delivered to the cutting plant. Usually the patterns are cut somewhat larger than normal because of the tendency of the sensitized fabric to shrink slightly during pressing and curing. Because the crosslinker has not yet produced any appreciable effect on the fiber, the sensitized cloth can be cut and sewn according to procedures used on untreated cloth; it can be cut at high speed and sewn without danger of needle cutting. After the sewing operation, the garment is pressed so as to impart the desirable creases and pleats, to flatten the seams, and, in general, to put the garment into its final shape. Then the garment is hung and cured to effect crosslinking. Curing may be performed in suitable closed rooms where the garments may be hung on racks; or preferably, the garments may be cured by hanging them on a moving chain that passes through a heated tunnel. After the curing step the garments are hung at room temperature, to allow them to cool before shipment.

For optimal performance the garments should be constructed with linings and pocketings that have also been treated by this deferred-curing technique; similarly zippers, sewing thread, and buttons should be treated and/or designed for deferred curing.

(b) PRECURE PROCESSES (25–27)

In the precure process, the fabric is impregnated, dried, and cured in the piece; that is, while it is in a smooth and flat condition. As a consequence it would normally be difficult to permanently shape a garment from such a fabric. However, by using fabrics that contain thermoplastic fibers blended with cellulose and by employing high-pressure, high-temperature presses, it is possible to obtain good durable-press performance.

It should be noted that the resin-treated cellulose component (which was set in the smooth and flat condition) tends to oppose garment shaping (such as inserting creases and pleats); consequently the precure process tends to be limited to light-weight fabrics and to applications in which very sharp creases and pleats are not required (e.g., men's shirts, ladies' blouses and dresses, nurses' uniforms, bed sheets, tablecloths, and the like). Even with light-weight fabrics, it is difficult to obtain permanently smooth seams when using the precure process. Consequently special care must be exercised to minimize puckering. Two excellent reports from the British Clothing Institute (99, 100) give details of proper sewing techniques.

5. Current Developments

(a) COTTON FABRICS

Although the original commercial offerings in durable-press textiles were based on 100% cotton fabrics, there was a rapid shift to the use of blended fabrics as soon as the inadequate wear life of highly resinated cotton fabrics was recognized. At present durable-press textiles, with very few exceptions, involve blended fabrics. In the search for processes suitable for 100% cotton one promising approach involves replacing part of the conventional crosslinker with a soft polymer, such as a polyurethane, polyacrylate, polyether, or silicone (101, 102). Another involves crosslinking the cotton while it is in a swollen, rather than in a collapsed, condition, which is typical of normal crosslinking processes. There are several different procedures utilizing this principle under study, and Table 2 shows some of the most important features.

(b) RADIATION PROCESSES

For a number of years the use of radiation to activate the crosslinking of cellulose has been the subject of many extensive research programs. The work at Heberlein in Switzerland (105) and of Walsh and co-workers at the University of North Carolina (106) has probably been the most noteworthy. In this approach the crosslinker that is used contains at least one functional group that can be

TABLE 2
Experimental Crosslinking Processes for Cotton

Process	Type	Proponent	Resin type	Catalyst type	References
Wet fixation two-step	Postcure	National Cotton Council	Polymer former and crosslinker	Mineral acid in first step (polymerization); metal salt in second step (crosslinking)	30–34
Damp curing	Precure	Many (European origin)	Pyrimidinones carbamates	Mineral acid (mild cure)	103
Polyset two-step	Postcure	Southern Regional Research Laboratory, U.S. Department of Agriculture	*N*-methylol crosslinkers in both steps	Weak metal salt in first step (polymerization); strong metal salt in second step (crosslinking)	104
Cold curing	Precure	Sun Chemical Corp.	Triazones	Mineral acid	35, 36

reacted with cellulose under the influence of high-energy radiation (such as ultraviolet, β, or γ-radiation). One of the prime objectives is to eliminate or reduce the degradation of the cotton that normally occurs when crosslinking is carried out at elevated temperatures with an acidic catalyst.

Through the use of an unsymmetrical crosslinker in which one functional group can be radiation activated and another activated with heat and a conventional catalyst, it is possible to achieve postcure durable-press effects. Specifically, the compound may be monofunctionally attached to the cellulose at the mill, and then, after the garment has been manufactured, the other functional group may be activated to complete the crosslink and provide shape retention to the garment.

Considerable commercial interest was generated in 1966 when the Deering Milliken Corporation announced that its VISA finish is produced with radiation, using an electron accelerator (107). It is widely believed that the chemical being used commercially at present in *N*-methylolacrylamide. If this compound is applied to the cloth and exposed to radiation, grafting on the cellulose occurs through the vinyl group. If the fabric is then washed to remove unreacted agent and an acidic-type catalyst is applied, the dried fabric will be in a "sensitized" state. After garments are made from this cloth, they can be postcured with heat by completing the crosslink through reaction of the pendant methylol group.

An alternative sequence of functional-group activation has been proposed by Walsh and Rutherford (108). The unsymmetrical crosslinker is first reacted through the use of heat and an acidic catalyst; after washing and drying steps, garments are manufactured and then "cured" by being exposed to radiation. Thus in the first of these competitive radiation processes the radiation step is carried out only at the textile mill, whereas in the second case the radiation treatment is at the garment plant (but no heat is required).

It must be remembered, however, that high-energy radiation can cause fiber damage, and the level of radiation must be carefully controlled to prevent excessive strength losses (see also Section H in Chapter XVII). To date there has not been sufficient advantage demonstrated in the use of radiation processes to result in widespread use; however, with continued research, there is a good likelihood that increased usefulness will develop.

(c) VAPOR-PHASE TREATMENTS

Largely as an outgrowth of work by scientists of the U.S. Department of Agriculture (109) and subsequent work by Gagliardi (110), there has been widespread interest in the possible merits of applying crosslinker, catalyst, and other auxiliaries from the vapor phase instead of by normal aqueous impregnation. Because of its volatility as well as effectiveness as a crosslinker, it is not surprising that formaldehyde is utilized in many of the vapor-phase studies. For fast crosslinking, a strong catalyst is desirable; for this reason, and because it is volatile, hydrogen chloride has been intensively studied. Other volatile acids, such as formic and acetic, have also been evaluated.

It is possible to obtain multifunctional effects with certain compounds. For example, through the use of methylchlorosilanes, which are volatile, one can obtain a catalytic effect and at the same time render the treated cloth water-repellent. Not all components of the treatment need to be volatile. For example, hand modifiers, sewing lubricants, and the like can be applied in a typical aqueous padding application prior to vapor-phase treatment with the crosslinker and catalyst. Similarly a nonvolatile catalyst could be applied to the cloth by padding or spraying prior to vapor-phase contact with the crosslinker. Alternatively even part of the crosslinking agent can be applied by aqueous padding before vapor-phase treatment. For example, cyclic ethyleneurea can be applied to the fabric by padding. After having been dried, the cloth is exposed to formaldehyde vapor and a volatile catalyst; the formaldehyde reacts first with the ethyleneurea, forming dimethylolethyleneurea, which in turn reacts with the fiber.

Among the advantages of the vapor-phase process, one of the most important is an improvement in the strength retention of 100% cotton fabrics. In this connection an extremely critical factor is the moisture content of the cloth at the time of crosslinking. Since the vapor-phase process can be applied either at the textile mill of by the garment maker, the equivalent of either precure of postcure procedures may be affected.

At present, although there is some indication that the use of vapor-phase application will provide a better relationship between strength retention and performance on 100% cotton fabrics, there are some extremely difficult technological problems that have held back commercial-scale utilization of this approach. For example, some of the more promising crosslinker–catalyst systems are so corrosive that only very selected materials of construction may be used for the vapor-phase chamber, making the cost prohibitively high. Furthermore, the rate of vapor diffusion into (to effect crosslinking) and out of the cloth (to eliminate odor and other undesirable side effects) is so slow that extremely long processing times are required, especially for garments. There has been, therefore, no commercially successful use of vapor-phase processes up to this time. It should be noted, however, that there is considerable interest at present in the vapor-phase process developed at Stanford Research Institute (111) and licensed to American Laundry Machine Co. In this process, garments are exposed to formaldehyde, sulfur dioxide, and water vapor.

The reader is also referred to Section E of Chapter XVII for a description of the preparation and properties of crosslinked cellulose.

References

1. R. P. Foulds, J. T. Marsh, F. C. Wood, H. Boffey, and J. Tankard, British Patent 291,473 (1928) (to Tootal Broadhurst Lee Co., Ltd.).
2. R. P. Foulds, J. T. Marsh, and F. C. Wood, British Patent 291,474 (1928) (to Tootal Broadhurst Lee).
3. G. K. Lake, *Textile Res. J.,* **22**, 138 (1952).
4. L. G. Ray, Jr., *Textile Res. J.,* **22**, 144 (1952).
5. J. H. Andreen, *Textile Res. J.,* **23**, 11 (1953).
6. G. Hazlewood, *Am. Dyestuff Reptr.,* **37**, P295 (1948).
7. J. B. Quig, *Am. Dyestuff Reptr.,* **39**, P329 (1950).
8. W. J. Burke, U.S. Patent 2,321,989 (1943). (to E. I. du Pont de Nemours & Co.).
9. A. M. Paquin, *Angew. Chem.,* **A60**, 267 (1948).
10. H. B. Goldstein, *Am. Dyestuff Reptr.,* **48**, 44 (1959).
11. H. B. Goldstein, British Patent 845,468 (1960) (to Sun Chemical Corp.); U.S. Patent 3,004,870.
12. U.S. Bureau of Census, *Facts for Industry* (M22S and M22T) Aug. 4, 1969 and Aug. 16, 1968).
13. J. D. Reid, L. W. Mazzeno, R. M. Reinhardt, and A. R. Markezich, *Textile Res. J.,* **27**, 252 (1957).
14. A. B. Hilton, U.S. Patent 3,025,622 (1962) (to Courtaulds, Inc.).
15. M. D. Hurwitz, *Textile Res. J.,* **27**, 267 (1957).
16. R. T. Graham, *Textile Res. J.,* **27**, 270 (1957).
17. L. R. Mizell, U.S. Patent 3,096,524 (1963) (to Williamson-Dickie Manf.).
18. N. F. Getchell, U.S. Patent 3,138,802 (1964) (to Cotton Producer's Institute of the American Cotton Council of America).
19. W. K. Warnock and F. G. Hubener, U.S. Patent 2,974,432 (1961) (to Koret of California).
20. H. B. Goldstein and J. M. May, *Textile Research J.,* **34**, 325 (1964).
21. H. B. Goldstein and J. M. May, *Am. Dyestuff Reptr.,* **54**, 738 (1965).
22. J. Margeson, *Textile Res. J.,* **34**, 3 (1964).
23. M. D. Hurwitz, U.S. Patent 2,950,553 (1960) (to Rohm & Haas Co.).
24. H. B. Goldstein and M. A. Silvestri, British Patent 1,049,683 (1964), French Patent 1,402,238 (1964), and Belgian Patent 649,620 (1964) (all assigned to Sun Chemical).
25. C. M. Pyke and F. S. Pyke, U.S. Patent 3,341,955 (1967) (to Everprest Inc.).
26. *American Fabric Magazine*, Part I, issue 66, (1964–1965); Part II, issue 67 (1965); Part III, issue 68 (1965).
27. S. M. Suchecki, *Textile Ind.*, January 1965.
28. G. C. Tesoro and I. Pensa, *Textile Res. J.,* **34**, 960 (1964).
29. P. H. Egrie, I. E. Pensa, R. O. Rau, and G. C. Tesoro, *Textile Res. J.,* **36**, 279, (1966).
30. N. R. S. Hollies and N. F. Getchell, *Textile Res. J.,* **37**, 70 (1967).
31. S. R. Chafitz, K. A. Farquhar, and N. R. S. Hollies, *Textile Res. J.,* **39**, 497 (1969).
32. D. V. Parikh, S. L. Vail, G. B. Verburg, and A. H. P. Young, *Textile Res. J.,* **39**, 505 (1969).

33. R. L. Colbran, C. C. Maitland, and W. J. Roff, *Textile Res. J.*, **39**, 512 (1969).
34. R. L. Colbran, M. G. Corless, and C. C. Maitland, *Textile Res. J.*, **39**, 521 (1969).
35. H. B. Goldstein, *Textilveredlung,* **2**, 6, 384 (1967).
36. H. B. Goldstein, Canadian Patent 793,186 (1968) (to Sun Chemical).
37. Technical Manual of the American Association of Textile Chemists and Colorists, Vol. 44, 1968: (a) Test 88C–1964T, p. B116; (b) Test 124–1967, p. B109; (c) Test 66–1968, p. B173; (d) Test 88B–1964, p. B115; (e) Test 92–1967, p. B122.
38. P. Krais, *Tex. Forsch.*, **1**, 71 (1919).
39. A. J. Hall, *Rayon Record,* 689 (1920).
40. J. R. Marsh, *An Introduction to Textile Finishing*, Chapman and Hall, London, 1947, p. 388.
41. British Standards Institute, B.S. 3086:/1959.
42. C. O. Bostwick, *Analyzing the Wrinkling of Textile Materials by a New Technique*, No. 23 (1961).
43. C. O. Bostwick, *Am. Dyestuff Reptr.*, **51**, 386 (1962).
44. C. R. Williams, *Am. Dyestuff Reptr.*, **45**, 472 (1956).
45. E. J. Stavrakas, *Am. Dyestuff Reptr.*, **56**, 708 (1967).
46. J. V. Beninate, E. L. Kelly, G. L. Drake, and W. A. Reeves, *Am. Dyestuff Reptr.*, **55**, 37 (1966).
47. E. E. Lineken, S. M. Davis, and C. M. Jorgensen, *Textile Res. J.*, **26**, 940 (1956).
48. E. Elod, H. M. Holtz, and W. Graeser, *Reyon, Zellwolle, Chemiefasern,* **33**, 7 (1955).
49. D. D. Gagliardi and F. B. Shippee, *Textile Res. J.*, **31**, 316 (1961).
50. H. C. Walter, J. K. Buxbaum, and L. Q. Green, *Textile Res. J.*, **27**, 146 (1957).
51. W. G. Cameron and T. H. Morton, *J. Soc. Dyers Colourists,* **64**, 329 (1948).
52. R. Steele, *Am. Dyestuff Reptr.*, **54**, 6 (1965).
53. H. Rath, *Melliand Textilber.*, **40**, 526 (1959).
54. W. A. Reeves, G. L. Drake, O. J. McMillan, and J. D. Guthrie, *Textile Res. J.*, **25**, 41 (1955).
55. H. M. Ziifle, R. J. Berni, and R. R. Benerito, *Textile Res. J.*, **31**, 349 (1961).
56. F. W. Hoover and G. T. Vaala, U.S. Patent 2,373,136 (1945) (to Du Pont).
57. W. J. Burke and F. W. Hoover, U.S. Patent 2,374,647 (1945) (to Du Pont).
58. A. T. Larson and H. R. Dittmar, U.S. Patent 2,436,311 (1948) (to Du Pont).
59. R. L. Wayland, U.S. Patent 2,825,732 (1958) (to Dan River Mills).
60. H. H. Mosher, *Textile Res. J.*, **29**, 88 (1959).
61. A. R. Smith, *Am. Dyestuff Reptr.*, **48**, 49 (1959).
62. H. B. Goldstein and D. N. Koenig, *Textile Res. J.*, **29**, 66 (1959).
63. H. Kadowaki, *Bull. Chem. Soc. Japan,* **11**, 248 (1936).
64. D. H. Abrahams and L. Goldberg, U.S. Patent 3,264,241 (1966) (to Dexter Chemical Corp.).
65. P. B. Roth, U.S. Patent 3,063,869 (1962), French Patent 1,272,869 (1961) (to American Cyanamid Co.).
66. Sun Chemical, British Patent 931,559 (1963).
67. Sun Chemical, British Patent 931,560 (1963).
68. J. G. Frick, B. A. Kottes, and J. D. Reid, *Am. Dyestuff Reptr.*, **48**, 23 (1959).

69. R. L. Wayland, U.S. Patent 3,158,501 (1964) (to Dan River Mills).
70. R. L. Arceneaux, J. G. Frick, J. D. Reid, and G. A. Gautreaux, *Am. Dyestuff Reptr.*, **50**, 849 (1961).
71. R. L. Arceneaux and J. D. Reid, *Am. Dyestuff Reptr.*, **51**, 800 (1962).
72. J. D. Reid, R. M. Reinhardt, and J. S. Bruno, *Am. Dyestuff Reptr.*, **54**, 485 (1965).
73. J. G. Frick, Jr., French Patent 1,327,420 (1963).
74. Sun Chemical, French Patent 1,498,147 (1967).
75. B. Reibnitz, W. Ruemens, and K. Beideck, U.S. Patent 2,731,364 (1956) (to Badische Anilin- und Soda-Fabrik AG.).
76. B. Reibnitz, A. Woerner, and H. Scheuermann, 2,764,573 (1956) (to Badische Anilin).
77. Badische Anilin- und Soda-Fabrik AG., British Patent 717,287 (1954).
78. H. B. Goldstein and M. A. Silvestri, French Patent 1,286,054 (1962), British Patent 979,974 (1965), and U.S. Patent 3,049,446 (1962) (all assigned to Sun Chemical).
79. S. L. Vail, P. J. Murphy, J. G. Frick, and J. D. Reid, *Am. Dyestuff Reptr.*, **50**, 550 (1961).
80. N. Goetz, W. Ruettiger, W. Ruemens, and R. Zeidler, *Textil Rundschau*, **16**, 761 (1961).
81. H. B. Goldstein and M. A. Silvestri, French Patent 1,278,589 (1961), British Patent 975,359 (1961), and U.S. Patent 3,355,242 (1967) (all assigned to Sun Chemical).
82. F. C. McGrew, U.S. Patent 2,454,078 (1948) (to Du Pont).
83. F. A. Bonzagni, U.S. Patent 2,645,625 (1953) (to Monsanto).
84. R. F. Nickerson, U.S. Patent 2,684,347 (1954) (to Monsanto).
85. W. J. Van Loo and J. M. Salsbury, U. S. Patent 3,087,837, (1963) (to American Cyanamid).
86. American Cyanamid Co., French Patent 1,286,956 (1962).
87. G. S. Y. Poon, U.S. Patent 3,324,062 (1967) (to Dan River Mills).
88. E. Abrams, U.S. Patent 2,785,948 (1957) (to Quaker Chemical Products).
89. B. H. Kress and E. Abrams, U. S. Patent 2,785,947 (1957); B. H. Kress, U.S. Patents 2,885,443 (1959), 2,903,328 (1959), 2,785,996 (1957), and 2,786,081 (1957) (all assigned to Quaker Chemical).
90. Quaker Chemical Products, British Patent 776,468 (1957).
91. J. B. Irvine and B. H. Kress, *Textile Res. J.*, **28**, 148 (1958).
92. M. J. Scott, U.S. Patent 2,467,160 (1949) (to Monsanto).
93. J. J. Ryan and P. J. Taylor, U.S. Patent 3,006,784 (1961) (to Tootal Broadhurst Lee).
94. A. G. Pierce and J. G. Frick, *Am. Dyestuff Reptr.*, **57**, 864 (1968).
95. L. W. Mazzeno, R. M. H. Kullman, R. M. Reinhardt, H. B. Moore, and J. D. Reid, *Am. Dyestuff Reptr.*, **47**, 299 (1958).
96. A. C. Nuessle, L. M. Nageley, and E. O. J. Heiges, *Textile Res. J.*, **33**, 146 (1963).
97. H. B. Goldstein, *Teintex*, **21**, 975 (1956).
98. W. M. Gallahue, *Am. Dyestuff Reptr.*, **56**, 73 (1967).
99. C. M. C. Dorkin and N. H. Chamberlain, Clothing Institute (London) Technological Report No. 10, 1961.
100. T. Townsend and N. H. Chamberlain, Clothing Institute (London) Technological Report No. 16.
101. R. J. Harper, E. J. Blanchard, J. T. Lofton, and G. A. Gautreaux, *Textile Res. J.*, **37**, 233 (1967).

102. R. J. Harper, E. J. Blanchard, and J. D. Reid, *Am. Dyestuff Reptr.*, **55**, 369 (1966).

103. H. Ender and G. Pusch, *Textile Res. J.*, **36**, 322 (1966).

104. W. A. Reeves, C. H. Hamalainen, H. H. St. Mard, and A. S. Cooper, *Textile Res. J.*, **37**, 76 (1967).

105. G. Heberlein, F. Munzel, U.S. Patent 3,252,750 (1963); G. Heberlein, F. Munzel, and W. Kunz, U.S. Patent 3,140,197 (1963); F. Munzel, U.S. Patent 3,206,273 (1963) (all assigned to Heberlein & Co.).

106. W. K. Walsh, C. R. Jin, and A. A. Armstrong, *Textile Res. J.*, **35**, 648 (1965).

107. F. W. Marco, U.S. Patent 3,377,249 (1968) (to Deering Milliken).

108. W. K. Walsh and H. A. Rutherford, *Textile Res. J.*, **37**, 89 (1967).

109. J. D. Guthrie, *Am. Dyestuff Reptr.*, **51**, 507 (1962).

110. F. B. Shippee and D. D. Gagliardi, *Textile Res. J.*, **36**, 177 (1966).

111. K. Wilson, J. Gamarra, and R. Swidler, *Textile Res. J.*, **38**, 401 (1968); Canadian Patent 836,132 (1970).

B. HIGH-WET-MODULUS AND OTHER HIGH-TENACITY RAYONS

J. W. Schappel and G. C. Bockno

High-performance rayons, such as the high-wet-modulus (HWM) and other high-tenacity types, are relatively new when one considers the history of rayon fibers. The development of these improved-type rayons can be traced back to the period around 1940. Prior to this time rayon production had shown enormous growth, but this could not be attributed to major physical property improvements of the fiber.

The modern development of improved rayons began in 1940 with the substitution of rayon for cotton as reinforcement in the tires required for the Allied war effort. Improvements came rapidly in these early years and, in 1944, the Truman Commission established that rayon was superior to cotton as a tire-reinforcement textile. Since then rayon has taken a prime position in the tire-cord market as a result of continued research efforts to improve the properties and performance of the product. The development of high-tenacity-rayon tire cords led to adaptation of the technology to staple fiber, and as a consequence high-tenacity-rayon staples are commodity products today.

Sometime later high-wet-modulus, high-tenacity rayons were developed, although some of the precursors of these families of fibers can be traced back to 1925.

1. History

Production of the improved rayons described here may be classified according to two spinning systems: (a) the zinc-based processes (current tire yarns and modified HWM) and (b) the processes not based on zinc (Polynosic). The chronological development of these processes is presented in Tables 1 and 2.

TABLE 1
Evolution of Zinc-Based Processes (1)

Date	Inventor	Viscose	Spin bath	Stretch bath	Stretch (%)	Conditioned tenacity (g/den)
1906	Müller	Normal	$H_2SO_4 + Na_2SO_4$	None	Nil	1.2
1912	Napper	Normal	$H_2SO_4 + Na_2SO_4 + ZnSO_4$ (1%)	None	Nil	1.2
1928	Courtaulds	Normal	As above	Air	25	2.2
1937	Givens (tire yarn)	High γ-value[a]	$H_2SO_4 + Na_2SO_4 + ZnSO_4$ (4%)	Hot water	75	3.5
1950	Cox (Super tire yarn)	Modifiers; high CS_2; very underripe; higher cellulose DP	As above	Hot water	100	4.0
1955–1965	Many inventors (Super 2, 3, 4, tire yarn)	As above; combination of modifiers	As above	Hot water	110	5.0
1955–1965	Many inventors (modified HWM)	As above	As above	Hot water	125–150	5.0

[a] The number of moles of carbon disulfide bound as xanthate per 100 moles of $C_6H_{10}O_5$. The theoretical maximum, 300, is never attained.

TABLE 2

Evolution of Processes not Based on Zinc (Polynosic Type) (1)

Date	Inventor	Viscose	Spin bath	Stretch bath	Stretch (%)	Conditioned tenacity (g/den)
1906	Müller	Normal	$H_2SO_4 + Na_2SO_4$	None	Nil	1.2
1925	Lilienfeld	Underripe	64% H_2SO_4	Air	200	7.0
1943	Tachikawa (Toramomen)	Very low cellulose; very low NaOH; high CS_2; very underripe; high DP; high viscosity	Weak H_2SO_4 + weak Na_2SO_4	Same as spin bath	150	3.5
1955	Cramer et al. (Fiber G)	High NaOH	Phosphate (weakly acidic)	Same as spin bath	250	5.0[a]
1950–1960	Tachikawa et al. (Polynosic)	As for Tachikawa above	Weak H_2SO_4 + weak Na_2SO_4; low temperature[b]	Same as spin bath	250	4.5
1950–1965	Tachikawa et al. (Super Polynosic)	As above	As above[c]	Same as spin bath	300	7.5
1964	Courtaulds	As above	As above, plus formaldehyde	Hot water	600	10.0

[a] Very low elongation.
[b] Plus a very small percentage of zinc sulfate.
[c] Plus formaldehyde, in some cases.

(a) ZINC-BASED PROCESSES

Over 95% of today's rayon production is based on systems using zinc-containing spinning baths and includes tire yarn, modified HWM, regular, and crimped rayons. The following are some of the most significant developments on which the improved rayons of today are based:

1. The use of zinc sulfate in the spinning bath to influence coagulation was patented by Napper in 1912 (2), but the modification was not used commercially until later. The influence of zinc as a regeneration retardant was not recognized until 1934.
2. The two-godet stretch process was commercialized in 1928. This system, which used driven rolls or godets, applied about 25% stretch to freshly coagulated filaments.
3. Givens et al. (3) in 1937 patented the use of hot water or steam to plasticize the coagulated thread during stretching. Combining this treatment with 3-4% of zinc sulfate in the spinning bath resulted in threefold stretch increases.
4. Addition of modifiers to the spinning bath was patented by Cox (4) in 1950. This process technology resulted in significant improvements in the properties of rayon tire yarn and staple.
5. Since 1955 developments of improved viscose modifiers and use of synergistic combinations of modifiers and regeneration systems have resulted in high-tenacity, fatigue-resistant tire yarns and modified HWM rayon.

(b) PROCESSES NOT BASED ON ZINC (POLYNOSIC TYPE)

In contrast to the many rayon products that are based on the zinc-spinning-bath system, only a few rayon products are based on the spinning-bath system not incorporating zinc. The HWM polynosic rayon is the outstanding product derived from this system, although recent developments related to this type of rayon show the use of spinning baths containing low concentrations of zinc. The following are the most significant developments that have contributed to Polynosic-rayon technology:

1. Viscose rayon became commercially practical and profitable around 1900 with the advent of the Müller acid-salt bath (5).
2. The first rayon exhibiting high-tenacity, polynosiclike properties was patented by Lilienfeld in 1928 (6). This rayon was spun from an unripened, high-DP (degree of polymerization) viscose into a 64% sulfuric acid bath that served to coagulate and swell the filaments. The swollen filaments were stretched 200% and hardened in an aqueous sodium sulfate bath. A degree of crystallinity of about 50% and high orientation were obtained. This process yielded fibers

with tenacities of up to 7.0 g/den and wet modulus of 30 g/den (see definition in subsection 3, "High-Wet-Modulus Rayons").

3. Although Tachikawa's low-acid Toramomen process was patented in 1943, interest in HWM rayons did not develop until the late 1950s, when commercialization of this process occurred. However, extensive research on the Toramomen process was carried out since 1950, as is shown by the abundant patent literature.

4. Fiber G, a polynosic fiber, was developed in the United States during the mid-1950s and was based on a phosphate-bath spinning system. The fiber never achieved a position of commercial importance because the manufacturer terminated rayon-staple production.

5. The first reference to the use of formaldehyde as a spinning-bath component or viscose additive was in 1948 (7). Courtaulds commercialized a fiber of this type in 1964 (Ten x) and reported a tenacity of 10 g/den.

6. In the last few years, various workers (8) have shown that a very small amount of zinc sulfate (about 1 g/l) in polynosic-type spinning baths permits an increase in bath temperature and spinning speed and results in fibers with higher elongations.

2. Market Position of High-Tenacity and HWM Rayons

The position of high-tenacity rayon yarns in the tire-yarn market reached its peak during the 1955–1960 era. Since that period competition from synthetic fibers, such as nylon and polyester, has resulted in a market regression. In the United States rayon tire cord still holds a prominent position in the market, but its future is uncertain and may depend in large part on the development of radial tires, in which rayon shows excellent performance. Industrial uses of high-tenacity rayon yarns (e.g., reinforcement yarns for beltings and hoses) continue to grow.

High-tenacity rayon staples are finding application in the apparel-textile market and in the industrial field for tarpaulins and tentage.

The future for HWM-rayon staple is extremely bright. In the United States the production of this type of rayon was insignificant before 1961 and by 1969 had reached 170 million lb. It is estimated that production of this fiber will reach 300 million lb in 1975. The significant increases in production of the HWM rayon are directly related to its use in fiber blends, in which it is substituted for cotton. This is due to both a distinct economic advantage and to some superior properties of HWM rayon, such as softer hand and higher tensile strengths in plain and resin-finished fabrics.

3. High-Wet-Modulus Rayons

The wet, or secant, modulus for rayon is arbitrarily defined by the stress required to produce a strain of 5% when the fiber is loaded in the wet state. The wet modulus of the fiber is reported in grams per denier which is the product of the stress (g/den) at 5% extension x 20. Rayon fibers characterized as HWM fibers have moduli of 10 g/den or greater.

The property term "high wet modulus" describes one of the more significant properties of this new class of rayons–a property that explains the improved performance of fabrics containing the fiber. Under moderate forces, HWM rayons, like cotton, resist deformation in the wet state. Fabrics produced from HWM rayon may be handled on continuous-wet-processing equipment, are resistant to caustic in mercerization processes, accept compressive shrinkage (Sanforizing), and exhibit good dimensional stability.

Regular rayons that are characterized by their low-wet-modulus property are easily deformed by the moderate forces that are applied to the fibers in wet finishing operations and laundering. These deformations and the resultant unrelieved stresses account in large part for the relatively poor dimensional stability of regular rayon fabrics. Furthermore, the low wet modulus of regular rayon makes shrinkage control by compressive shrinkage processes impossible.

(a) TYPES OF HWM RAYON

Two distinctly different types of HWM rayon are produced in the world: (a) the Polynosic type and (b) the modified type. In France and certain other foreign countries the term "polynosic fiber" is a generic name for regenerated rayons with a homogeneous fibrillar structure. The term "modified HWM rayon" refers to fibers produced via spinning systems that employ viscose and spinning-bath modifiers or additives (or both) for the control of the regeneration.

The two types of HWM rayon fibers differ significantly in some physical, mechanical, and fine-structural properties as well as in manufacturing schemes. An understanding of the relationship of the two fiber types to each other can be had by comparing the fiber properties.

(b) FIBER PROPERTIES

(1) Physical Properties

In Figure 1, typical conditioned and wet stress-strain curves are given for regular rayon, Polynosic and modified HWM rayons, and cotton. The hatched area between the stress-strain curves shows the limits of tensile properties for the two families of HWM fibers, and although these two types of fiber overlap each other in many of the properties, the Polynosic fibers usually have lower elongations and higher moduli of elasticity, both conditioned and wet. Comparison of the properties of HWM rayons with those of regular rayon shows large differences in tenacity and moduli, both conditioned and wet. The wet moduli of cotton and the modified rayons are nearly identical.

Table 3 presents the physical properties of Polynosic and modified HWM

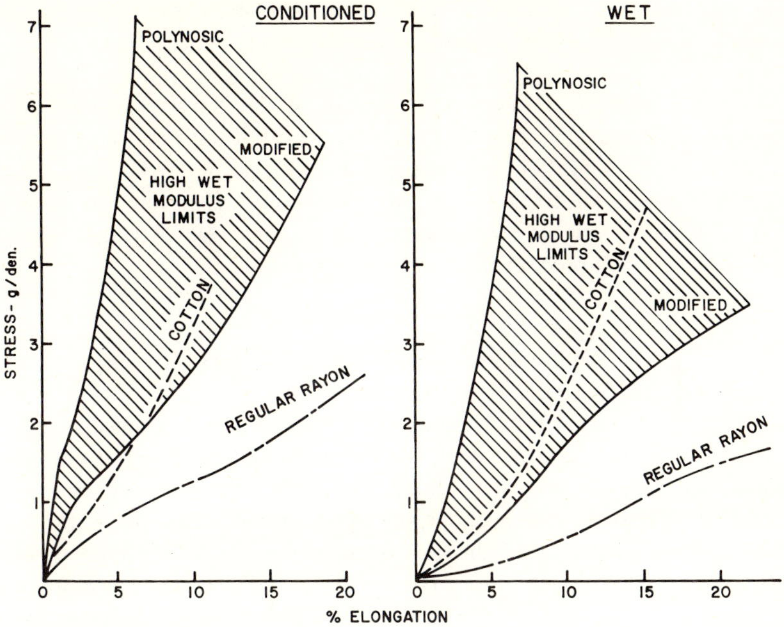

Fig. 1. Relationship among the single-fiber stress–strain properties, conditioned and wet, of the HWM rayons, regular rayon, and cotton.

rayons, regular rayon, and cotton. The initial wet moduli of the HWM rayons are 3–20 times as high as that of regular rayon and are at least equivalent to that of cotton. Measurements have shown that rayon fibers with an initial wet modulus of about 10 g/den permit the manufacture of fabrics that are dimensionally

TABLE 3
Physical Properties of Cellulose Fibers (9)

Property	Standard rayon	Modified HWM rayon	Polynosic HWM rayon	Cotton
Conditioned tenacity (g/den)	2.2–3.5	3.5–6.0	3.5–6.5	2.0–6.0
Conditioned elongation (%)	10–30	8–18	6–12	7.0–12.0
Wet tenacity (g/den)	0.5–2.2	2.5–4.6	2.7–4.5	3.0–6.5
Wet elongation (%)	22–35	15–22	17–14	13
Wet modulus (g/den)	3–4	9–25	20–70	7–15
Loop test (g/den)	0.3–1.0	0.7–3.0	0.7–1.2	1–2
Moisture regain (%)	11–13	10–12	9–11	7–7.5
Water retention (%)	90–115	60–80	55–75	35–45
Conditioned modulus (g/den)	60–90	80–110	120–180	

stable and respond to compressional shrinkage processes. As the wet modulus is increased beyond 10 g/den, marginal improvements in dimensional stability are found.

The loop strength of fibers is a measure of the transverse properties or toughness, reflected in carding performance and fabric's resistance to abrasion. The decrease in elongation and increase in modulus of HWM rayons is not accomplished without some loss in fiber flexibility. Therefore the modified HWM rayons with higher loop strengths show advantage over the Polynosic HWM rayon fibers in processability and abrasion resistance.

(2) Fiber Cross Sections

Figure 2 shows photomicrographs of thin (<5-μ), stained cross sections of Polynosic and modified HWM rayons. The procedure employed for producing the stained cross section is based on work by Preston (10), Morehead and Sisson (11), and Tyler and Wooding (12). Regular rayon is marked by the irregular, serrated, and skin-and-core cross section. The HWM rayons are characterized by round or nearly round cross sections, but significant differences are noted in the amount of skin. Differences in the skin-to-core relationship of these fibers reflect differences in the fine structure of the fibers. Voids are also observed in certain HWM rayons, most notably the polynosics. Zinc and other modifiers in the spinning bath or in the viscose are recognized as a factor that controls the size of the cellulose crystallite. Skin observed in the stained cross section of rayon is indicative of many small-sized crystals, whereas the core section of the fiber is indicative of a coarser crystalline structure. The modified HWM rayons, which are spun with zinc-containing baths, show this characteristic skin structure; the Polynosic fibers, on the other hand, are generally characterized by the all-core type of cross section.

Compared with regular rayons, the HWM rayons exhibit reduced water retention and, in some instances, approach the swelling characteristics of cotton.* The water-retention measurement serves to define the fine structure of the fiber, particularly with respect to the intercrystalline bonds. An increasing number of these bonds develop as the fiber structure is shifted from regular rayon to the modified and Polynosic HWM rayons.

(3) Fine Structure of HWM Rayons

Data presented in Table 4 show the fine-structure comparison for various rayon fibers. The HWM rayons, both modified and Polynosic, have higher degrees of crystallinity than the regular rayons. Mark (9) notes the largest differences in crystallinity among the rayon-fiber types are observed on the basis

*Water retention is a measure of water held by the rayon fiber after the initial gel fiber is dried, reswollen in water, and centrifuged under standardized conditions.

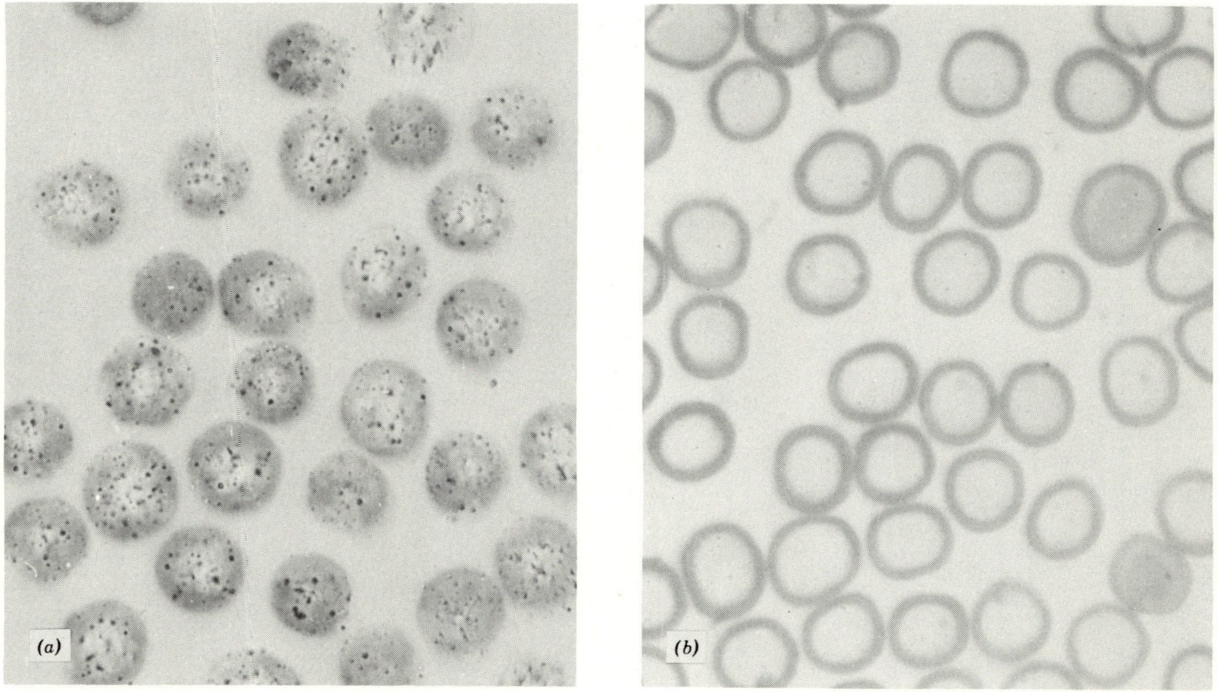

Fig. 2. Cross sections of HWM rayon fibers: (*a*) Polynosic, 1.5 den/filament (1000 ×); (*b*) modified HWM, 1.5 den/filament (1000 ×).

TABLE 4
Comparison of the Fine Structure of Various Rayon Fibers (9)

Fine-structure properties	Regular rayon	Modified HWM rayon	Polynosic HWM rayon
Degree of crystallinity (%):			
Based on x-ray diffraction	30–36	38–44	40–47
Based on infrared–deuterium oxide	35–37	35–40	39–50
Based on density	28–34	38–42	36–45
Length of the crystallite (DP)			
Based on x-ray diffraction	80	85	110–130
Based on leveling-off DP	60–80	80–95	100–140
Based on electron microscopy		100	150
Thickness of crystallites (Å):			
Based on wide-angle x-ray diffraction	50–70	70–100	80–100
Based on small-angle x-ray diffraction	80	100	100–110
Based on electron microscopy		200	200
Accessibility of OH groups:			
Based on H/D exchange	60–70	55–60	50–55
Based on moisture absorption	65	60	50
Chain orientation (%)			
Based on x-rays	70–80	75–85	80–90
Based on birefringence	60–70	70	85
Based on infrared dichroism		80	85
Degree of polymerization	350–450	500–550	600–800

of density determinations, which show the density of the noncrystalline regions to be greater for the new types of fibers than for the regular rayon. Furthermore, the well-oriented Polynosic fibers have no more than a 50% crystalline structure. The comparison parameters for the degree of crystallinity indicate the HWM fibers to be lower than ramie and to approach cotton in this structural property.

Mark (9) suggests that there are large differences in the length of the crystalline regions for various rayons. Calculations based on the DP, the average length of the crystallite, and the degree of crystallinity permit an assessment of the average number of monomer units from each cellulose chain that are present in the crystalline state. In the case of regular rayon about 140 monomer units, or less than half of each cellulose chain, are in the crystalline state; this explains the low molecular bonding of these fibers, expecially in the wet state, when the amorphous region is plasticized.

In the case of modified HWM rayon about 220 anhydroglucose units of each cellulose chain lie in the highly ordered regions, which suggests the participation of each chain in two crystallites. Similarly in Polynosic HWM rayons about 300 anhydroglucose units of each cellulose chain lie in the crystalline regions, and therefore each cellulose chain, on the average, participates in about 2.5 crystallites. The strength of the supermolecular lattice increases progressively from the regular rayon structure to that of the modified and to that of Polynosic HWM rayon structures, corresponding to the greater percentage of the monomer units in the crystalline regions. The data on accessibility suggest that the modified and Polynosic HWM rayon fibers have smaller percentages of amorphous regions; hence these fibers have a stronger internal structure.

The orientation of macromolecules increases significantly as the fiber structure is shifted from the regular to modified HWM and to Polynosic HWM rayons. This factor, together with the average length of the crystallite and the higher DP for Polynosic fibers, results in a higher modulus, higher dry and wet strength, and a greater resistance to swelling.

The average DP of the HWM rayons, modified and Polynosic, is about 35–60% higher than that of regular rayons and accounts in part for the improvement in the structure and physical properties of these newer rayons. An increase in the average DP of the modified HWM rayon from 500 to 600 shows only a small effect on the conditioned and wet tensile properties, but a significant reduction in alkali solubility occurs.

The properties of Polynosic rayons are more responsive to a shift in the average DP of the fiber. As the DP of the viscose is increased, it is possible to achieve higher spinning stretches and hence higher degrees of orientation, which result in higher tensile strengths and higher moduli, both conditioned and wet.

The relationship of fine structure to the physical properties of rayon has also been reported by Bingham (13) whose studies suggest that the strength of the fiber is more dependent on the intercrystallite linkages or the amorphous regions than on the crystallites. Therefore, in order to improve fiber strength, it is necessary to orient the cellulose chains in the amorphous regions, permitting a growth of the crystallites and an improvement in the lateral order. In addition, it is postulated that orientation of the amorphous areas will increase the number of chains connecting the crystallites and thereby strengthen the intercrystallite links.

Bingham's hypothesis is supported by the relationship of the wet and dry strength of rayon. When water penetrates the rayon structure, the amorphous regions are weakened as a result of the rupture of accessible hydrogen bonds, whereas the crystalline regions are essentially unaffected. Bingham observed that a sharply increased ratio of wet to dry strength (T_w/T_d) occurs as the rayon structure is shifted to one of higher orientation and order in the amorphous regions, resulting in a reduction of accessible hydrogen bonds.

According to Bingham, the parameter $1-(T_w/T_d)$ may represent the proportion of accessible hydrogen bonds in the amorphous regions. When this parameter was plotted against corrected birefringence, the points described a straight line, as shown in Figure 3. Furthermore, employing deuteration techniques to describe the accessible regions of the fiber, Charles (1) noted that the deuterium

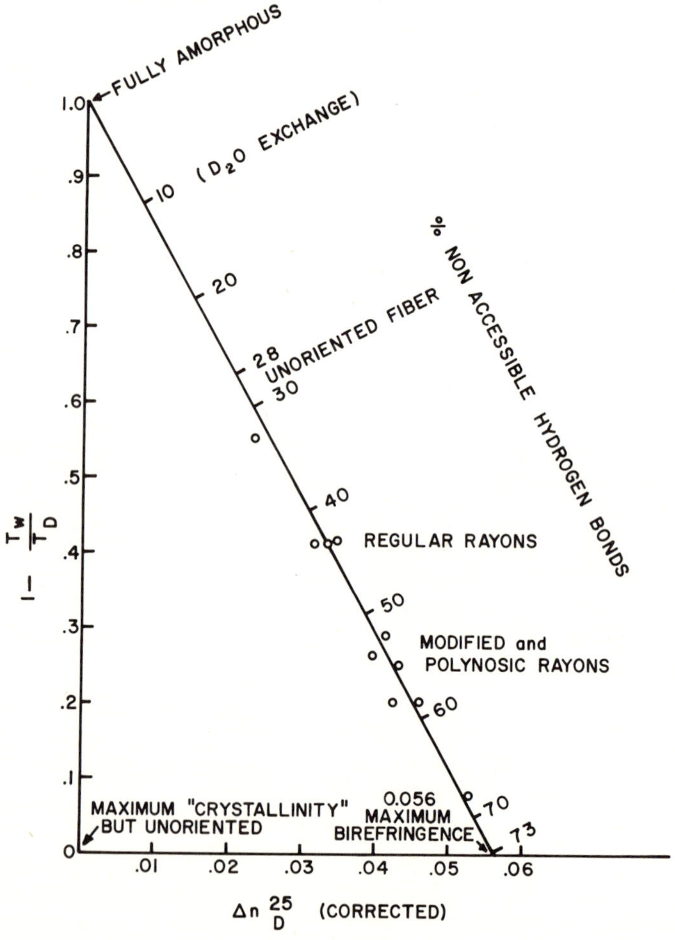

Fig. 3. Proportion of accessible hydrogen bonds in the amorphous regions as determined by the plot of $1-(T_w/T_d)$ versus corrected birefringence (1, 13).

exchange shows a fair degree of linearity with birefringence. The relative deuterium-exchange values shown on the plot of Figure 3 furnish additional support for Bingham's hypothesis. See also Section G of Chapter XIII.

(4) Fibrillar Structure

Highly oriented rayon fibers are characterized by a fibrillar macrostructure. Distinct differences in the fibrillar structure of the Polynosic and modified HWM rayons are observed. Drisch and Priou (14) and Herrbach (15) report that the Polynosic HWM fibers have a structure consisting of an agglomerate of fine fibrils similar to those of cotton. This structure is observed when the fibers are subjected to swelling in nitric acid and mechanical grinding. It is further noted that rayon fibers produced via the modifier system possess a macrostructure based on aggregates of thick fibrils.

Improved macrostructures for the Polynosic fibers are reported by Drisch (16). Since the earlier Polynosic fibers were based on structures of well-organized fibrils that were straight and strictly parallel to the fiber axis, the lateral cohesion of the fiber was relatively weak and had a strong tendency to fibrillate. In the case of the newer Polynosic fibers (type 8) the macrostructure consists of fibrils that have a highly organized internal structure and are angularly disposed with respect to the axis of the fiber. Such a macrostructure causes an entanglement of the fibrils, which considerably increases the cohesion and reduces the tendency to fibrillate. The differences in fibrillar morphology between Polynosic and modified HWM rayons have also been demonstrated by an ultrasonic decomposition of the fibers in alkaline media (17).

Fiber brittleness and inferior resistance to abrasion may be related to the wet fibrillation properties of the HWM fibers. The fibrillar splitting of various cellulosic fibers was demonstrated by Uchida (17), who tumbled fibers in a 3% sodium hydroxide solution in a mixer rotating at 3000 rpm for 50 min.

Splitting of the HWM fibers has also been explored by methods that parallel the action of laundering. Schappel (18) reports fibrillation to occur by flexing or working the fibers in water. In these studies the modified HWM rayons showed no evidence of fibrillation, whereas the Polynosic fibers exhibited fibrillation that varied as a direct function of their wet-modulus property. Figure 4 illustrates the fibrillation characteristics of HWM fibers subjected to wet-working.

(5) Alkali Resistance of HWM Rayons

Regular rayon fibers are sensitive to the alkaline treatments of bleaching and mercerization. This factor limits their use in certain areas, such as in blends with cotton.

Herrbach (15) has studied the physical property changes of cellulosic fibers subjected to relaxation in 5% sodium hydroxide solutions. The results of this study are shown in Table 5 and give evidence for the greater alkali resistance of

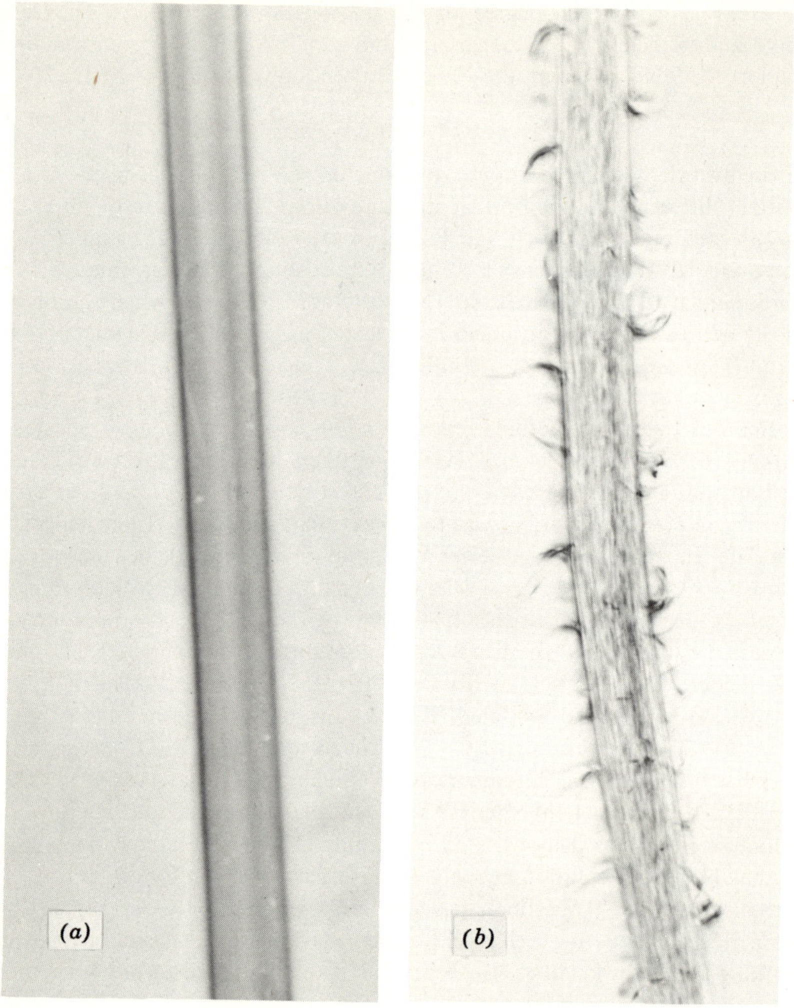

Fig. 4. Comparison of fibrillar characteristics of (*a*) modified and (*b*) Polynosic HWM rayons as detemined by mechanical working of wet fibers (750 ×) (18).

the Polynosic HWM fibers compared with that of modified HWM and regular rayons. In addition, the Polynosic HWM fibers are shown to be less soluble in sodium hydroxide than the modified HWM rayons.

Schappel (18) reports the alkaline solubility of the modified HWM rayon fibers to decrease as the DP is increased from 500 to 700. This improvement can be attributed to an increase in the ordered crystalline regions of the fiber and a decrease in accessibility.

TABLE 5
Alkali Treatment of Cellulosic Fibers (15)

Property	Cotton	Rayon				
		Polynosic I	Polynosic II	HWM (modified)	Regular	High tenacity
Wet strength (g/den):						
Control	3.8	2.8	5.0	3.2	1.5	2.8
Treated with 5% NaOH	3.2	2.1	4.9	1.8	1.0	1.1
Wet elongation at break (%):						
Control	14	14.5	12.1	17.9	23	37
Treated with 5% NaOH	15	26.6	14.3	48.6	29	59
Wet elongation at 0.5 g/den (%):						
Control	4	2.7	1.9	3.6	9	10
Treated with 5% NaOH	4	6.6	2.3	13.3	16	28.5
Solubility in 10% NaOH at 20°C (%)	2	25	9	42	51	52

(c) GENERAL PROCESS TECHNOLOGY

Two systems are employed in the spinning of HWM rayons, and the processes are designated by the fiber description names, modified HWM and Polynosic HWM. Various modifications are applied to the two general process systems, but the basic characteristics of each system are readily identified.

Modified HWM rayons, as the name implies, are based on the addition of chemical modifiers to the viscose or the spinning bath (or both), a system used in rayon-tire-yarn spinning. The modified viscose is spun into sulfuric acid, zinc sulfate, and sodium sulfate spin baths of moderate concentrations at relatively low temperatures and under conditions that yield fibers of high degrees of orientation and a slightly greater degree of crystallinity than that found in tire yarns.

Polynosic HWM rayons are based on viscoses of low alkali content that are spun into low-temperature, low-concentration spinning baths of sulfuric acid and sodium sulfate under conditions yielding a high degree of orientation and very high degrees of crystallinity.

Both the modified and the Polynosic HWM rayons are produced as staple fibers in deniers ranging from 1.0 to 3.0 den/fil. A modified HWM continuous-filament yarn is also produced in the United States.

(1) Viscose Compositions

High-wet-modulus fibers, either of the modified or Polynosic type, are generally produced from the premium-type dissolving pulps that are characterized by high α-cellulose levels (92 to over 98%), high DP, and high purity. Pulps are chosen to yield excellent viscose quality with respect to filtration and spinnability. Both prehydrolyzed sulfate pulps and special sulfite pulps are used. Treiber (19) has shown the range of properties for various special pulps available worldwide for the production of HWM fibers. The properties of these pulps are

TABLE 6
Properties of Pulps Used for the Production of HWM Rayon Fibers

Property	Range
α-Cellulose (%)	96–97.5
R 18%[a]	92.5–98.9
Viscosity (cp)	19–45
Ash (%)	0.03–0.08
Extractables (%)	0.01–0.25

[a] The weight percentage of pulp that is insoluble in 18% sodium hydroxide as determined by TAPPI Specification T 235 m.

TABLE 7
Compositions and Properties of Viscose Used in HWM Rayon Processing

Constituent or Property	Modified HWM	Polynosic HWM
Cellulose (%)	5–8	4–7
Sodium hydroxide (%)	5–10	2.5–5.5
Carbon disulfide (%)	30–38	45–55
Modifier (%, based on the weight of cellulose)	0.2–5	0–1.0
DP of alkali cellulose	450–600	600–800
Viscosity (poises)	100–150	150–600
γ-Value[a]	35–45	55–100
Salt index (NaCl)[b]	8–12	10–20

[a] The number of moles of carbon disulfide bound as xanthate per 100 moles of $C_6H_{10}O_5$.

[b] A ripeness measurement—namely, the minimum concentration of aqueous sodium chloride required to coagulate viscose completely under standard test conditions.

given in Table 6. The usual compositions of viscoses employed in the processing of modified and Polynosic HWM rayons are listed in Table 7.

A comparison of the viscose compositions employed for the modified and Polynosic HWM processes reveals distinct differences. The modified HWM viscose generally has a higher cellulose content, a lower carbon disulfide concentration (lower γ-value), lower viscosity, and lower cellulose DP than the Polynosic HWM viscose. In addition, viscose modifiers are employed in the modified HWM system as in the spinning of high-strength, high-elongation yarns. Although Polynosic HWM fibers may be spun without viscose modifiers, the most recent technology describes the use of the modifier (20).

(2) Spinning Conditions

Modified HWM Rayon. As in the case of tire-yarn spinning, the modified HWM fiber is spun into coagulation baths that are relatively rich in zinc (2–5% zinc sulfate). The zinc ions, along with the viscose modifiers, serve to control the course and rate of cellulose regeneration. The sodium sulfate concentration of the coagulation bath is 10–14%, lower than that used in tire yarn baths. The temperature of the coagulation bath is in the range of 20–40°C and is adjusted as an inverse function of acid concentration, 6–9%, to control the degree of regeneration prior to stretching the fibers.

The lower bath temperatures employed in this process produce a highly deformable gelled fiber that requires relatively low spinning speeds, 20–40 m/min. The gel filaments leaving the coagulation bath are very plastic and are stretched 120–160% in an early stage of gel dehydration and decomposition. Stretching is accomplished in a high-temperature (90–98°C), low-acid regeneration bath. Orientation of the molecules occurs by the time the cellulose xanthate decomposition has proceeded far enough for crystallization to occur. The high-temperature regeneration bath completes the decomposition of the cellulose xanthate while the fiber is held in a condition of stress, thus increasing the orientation of the chain molecules.

A more detailed discussion of the technical aspects of the "high-zinc" spinning process and the role of viscose modifiers is presented in subsection 4, "High-Strength, High-Elongation Rayons."

Polynosic HWM Rayon. The present technology for the spinning of Polynosic rayon is based on the principles set forth by Tachikawa in 1952 (21). The process uses a high-viscosity viscose (about 300 poises) that has a low alkali content and a high γ-value and is spun into a low-temperature (5–35°C), low-acid coagulation bath. The coagulation bath may consist of 2–4% sulfuric acid, 4–6% sodium sulfate, and 0–0.1% zinc sulfate (i.e., it is weak in regeneration power). Spinning speed is in the range of 10–30 m/min, with stretching of the fibers being accomplished incrementally by means of stepped godets in the coagulation bath or in a second bath of about the same composition as the coagulation bath. Stretches of 90–200% are employed in Polynosic-rayon-spinning processes. The regeneration of the fiber is completed in a third bath of dilute sulfuric acid (0.5–1.5%) at elevated temperature (95°C).

The regenerated Polynosic fibers may be chemically purified by the procedures used for regular rayon; these include desulfurizing, washing, soft-finishing, and drying. New process technology (22) suggests the treatment of the regenerated gel fibers with solutions of sodium hydroxide (4%), which improve the knot strengths or transverse physical properties of the fibers.

Since the spinning of Polynosic fibers takes place in weak acid baths in which coagulation is very slow, the highly swollen xanthate gel filaments are formed with relatively few cellulosic crystal nuclei. These nuclei have an opportunity to grow into long crystallite formations during the slow decomposition of the cellulose xanthate. The coagulated filaments are then stretched at low temperatures, which ensures a very high orientation of the structure. The highly oriented filaments are finally regenerated in high-temperature baths to produce a network structure of long, crystalline fibrils having internal disordered regions that are partially oriented and relatively inaccessible.

Formaldehyde Spinning Processes. The reaction of formaldehyde with the xanthic acid derivative of cellulose is the basis for modifying the regeneration rate of cellulose xanthate and the structure of HWM and high-tenacity rayons. Several procedures may be used for introducing formaldehyde into the spinning system. Formaldehyde may be added directly to the viscose in concentrations of 0.05–1.0% based on the viscose weight, or it may be introduced into the coagulation–regeneration baths in concentrations of up to 1%.

As the viscose solution is passed into the acid spinning bath, the cellulose xanthate filaments are coagulated and converted to the cellulose xanthic acid state. If formaldehyde is present in the viscose or spinning bath, it will react with the cellulose xanthic acid, forming the *S*-hydroxymethyl cellulose xanthate, as shown in Figure 5. Phifer and Ticknor have shown the formaldehyde derivative to be relatively stable in acid solutions at moderate temperatures.

Due to the chemical stability of the *S*-hydroxymethyl cellulose xanthate in acid solutions containing formaldehyde, the freshly spun filaments are preserved in a metastable state and exhibit good toughness, plasticity, and high stretchability. These filaments will allow stretch as high as 600% since the bulky methylol derivative groups on the xanthate inhibit the crystallization of the cellulose and offer potential for toughening the fiber through transient methylene crosslinks between xanthate or hydroxyl groups on the cellulose molecule.

Extremely-high-tenacity fibers (about 10 g/den) with very high wet-modulus properties are spun from modifier-free viscoses with high γ-values ($\gamma = 80$). Spinning can be carried out in zinc-free or zinc-containing formaldehyde-modified baths at relatively low temperatures (about 25°C). Very high stretching

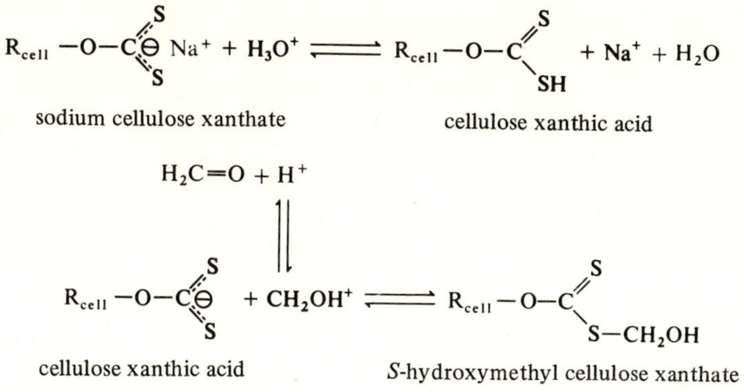

Fig. 5. Reaction of cellulose xanthate with formaldehyde in acid media.

of the fibers begins before a significant decomposition of the *S*-hydroxymethyl cellulose xanthate occurs. As the filaments are passed through hot (95°C) dilute-acid regeneration baths, the derivative is decomposed in stages. The extremely high tensile properties of these fibers depend on a unique structure that results from crystallizing the well-ordered cellulose molecules. This structure is characterized by long crystallites, high orientation, and well-ordered noncrystalline regions.

Tough, thick-skin to all-skin rayon fibers, with either high or low wet-modulus properties can also be obtained with formaldehyde spinning systems (24). Although formaldehyde can be considered as a substitute for viscose modifiers and the zinc-type spinning baths, processes dependent on employing viscose modifiers with zinc–formaldehyde or acid–formaldehyde spinning baths are useful in producing a variety of high-strength, high-wet-modulus fibers exhibiting good toughness properties (25).

Rayons spun with formaldehyde spinning systems show some evidence of methylene crosslinks in the finished fiber. The DP of these fibers shows an increase of 50–100 units in converting the viscose to fiber. The higher DP of these fibers is believed to be due to the coupling of adjacent molecular chains as a result of decomposition of *S*-hydroxymethyl cellulose xanthate and release of formaldehyde, which can react with the cellulose.

Commercialization of the formaldehyde spinning processes has been hindered by various problems, such as contamination of reclaimed sodium sulfate and the reaction of formaldehyde with hydrogen sulfide to form an insoluble trithiane (1), which deposits in bath troughs, pipe conduits, and reclaim systems.

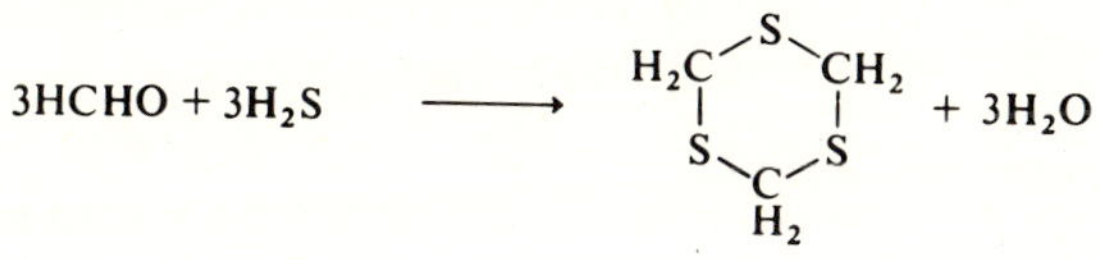

4. High-Strength, High-Elongation Rayons

By arbitrary definition, high-tenacity rayons are those with tenacities exceeding about 3.0 g/den when conditioned at 58% relative humidity (RH) and 24°C. As a group, they have high toughness values and find wide applications where high strength, toughness, improved tearing strength, and durability are required. The discussion that follows refers to this group as high-strength, high-elongation (HSHE) rayons.

(a) TYPES OF HSHE RAYON

(1) Tire Cord

Since the development of improved rayons for use as tire cord in about 1940 (26), successive improvements in tire-cord strength and durability have been achieved through the continuing research-and-development effort of the rayon producer as well as the industry's suppliers. Today's modern rayon tire yarn has a strength level over twice that of the early material. It is representative of the best HSHE rayons that the industry's technology can produce.

(2) Reinforcing Yarns

Closely related to tire yarn are those rayons that are employed as reinforcement in a variety of industrial rubber goods, such as conveyor belting, V-belts, and pressure hoses. As with tire cord, the requirements for these materials are product toughness, strength, durability or fatigue resistance, and stability. For most of these applications the yarns are used in lower twist constructions than the high twist levels normally employed for tire reinforcement. Much the same process technology is used to produce this class of yarns, although the individual product requirements—such as shrinkage resistance, initial modulus, and breaking elongation—require individual process specifications to achieve the best balance of physical properties.

(3) Industrial Sewing Threads

High-strength, high-elongation continuous rayon yarns with specially engineered physical properties are used as heavy-duty sewing threads in the manufacture of burlap and multiwall paper bags. Special finishes on the yarns ensure good sewing efficiency and abrasion resistance. The inherent high elongation of the yarn enables it to absorb the high seam stresses often encountered during shipment.

(4) Textile Yarns

Light-denier continuous-filament HSHE rayons find applications in tape reinforcement and other specialty products, where the combination of strength, light weight, and ease of fabrication is required. They are not used extensively in strictly textile applications, such as apparel fabrics and home furnishings, where regular rayon-type yarns perform adequately. When higher tensile and tearing strengths are required in special applications—for example, tent fabrics—the HSHE yarns can then be utilized to advantage.

(5) Staple

The high strength and toughness of the HSHE staple fibers make them particularly attractive in a variety of applications. Some important uses are mechanical rubber goods; fabric-reinforced vinyl sheeting; 100% rayon fabrics; blended-yarn apparel fabrics and sheetings (e.g., combinations with synthetic fibers or cotton); and rayon–cotton combination fabrics, such as those used in apparel and sheetings (cotton warp and rayon filling).

(b) FIBER PROPERTIES

(1) Physical Properties

Single fibers of HSHE rayon have a characteristic stress–strain diagram. Continuous-spun, tension-dried rayon yarns, however, have high moduli (i.e., lower elongations and higher tensile strength resulting from drying under tension). On rewetting and relaxed drying, however, these high-conditioned-moduli fibers revert to their normal, low-modulus state.

Characteristic stress–strain diagrams for HSHE tire yarns dried while relaxed and dried under tension are shown in Figure 6.

(2) Flex-Fatigue Properties

High-strength, high-elongation rayons exhibit a significantly higher degree of flex-fatigue resistance as compared with either the regular or the HWM types of rayon. Single-fiber flex-fatigue data for three rayon types (1.5 den/fil) and a domestic cotton are shown in Table 8.

A sharp difference in both primary swelling and secondary swelling (e.g., swelling of the fully regenerated fiber, never-dried and dried, respectively) exists between the high-tenacity rayons and the regular-rayon fibers. As noted by Mieck (28), reduced swelling can be correlated with an increased fatigue resistance for HSHE type rayons.

(3) Fiber Cross Sections

The fiber cross sections of various HSHE rayons fibers have circular, oval, and kidney-bean shapes. As a rule, the surfaces are rather smooth, showing few, if any, crenulations characteristic of the regular or crimped rayons. Particular process conditions employed in the fiber's manufacture—such as viscose regeneration modifiers, the degree of viscose ripeness, spinning bath composition, and spinning conditions—are the important factors influencing a fiber's cross section. Morehead and Sisson (11) showed a relationship between increasing zinc sulfate

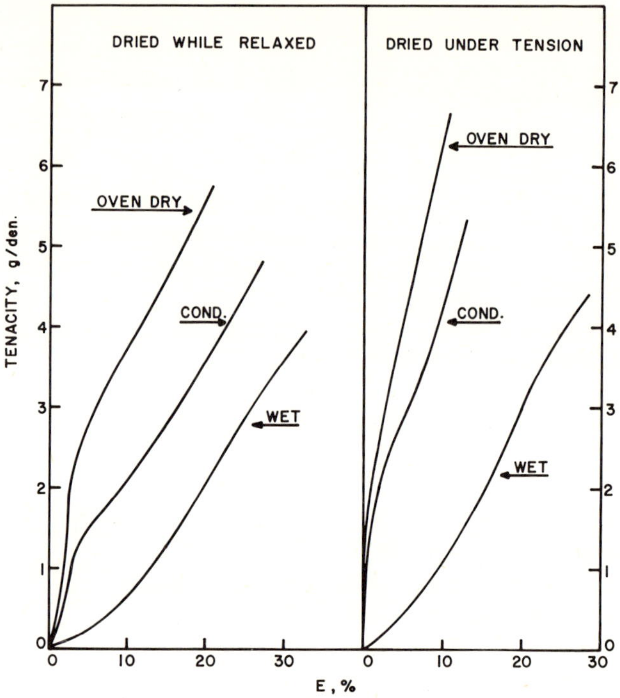

Fig. 6. Stress–strain curves – 1650/1500 denier tire yarn. Comparison of relaxed and tension drying.

TABLE 8
Single-Fiber Flex-Fatigue Data for Three Types of Rayon

Fiber	Single-fiber flex cycles (x 10^{-3})[a] (0.005-in. bar radius, 0.57 g weight)
High-tenacity rayons	400–1500
Regular rayon	150–250
High-wet-modulus rayon	80–150
Cotton	70–150

[a] Determined with a modification of the procedure developed by Lefferdink and Briar (27).

concentration in the spinning bath and the proportion of skin in the fiber's cross section. See also Section E in Chapter XIII.

According to Klare (29), the basic viscose-regeneration modifiers, such as amines and polyamides, influence the cross sections of fibers, depending on whether the viscose ripening is rapid or slow. He notes that less ripening promotes the circular or kidney-bean shape, whereas extended ripening tends to produce an increased irregularity in the fiber's cross section. Accompanying this loss in cross-sectional regularity there is a progressive deterioration of the fiber's physical properties; that is, tenacity, elongation, and flex fatigue decrease while the water swelling increases. These losses of desirable properties reflect a major change in the fiber's fine structure.

Cross-sectional photomicrographs of various types of HSHE rayon are illustrated in Figure 7.

There is good evidence that the "all-skin" type of structure is not uniform throughout the cross section. Ohara (30) and Horio (31) have noted the presence of an outer layer referred to as the cuticle. Cumberbirch and Ford (32) in their work also have shown the presence of an unstained, thin, outer cuticle and a heterogeneous pitted structure through the fiber's cross sections, having more voids near the cuticle and fewer toward the center of the fiber.

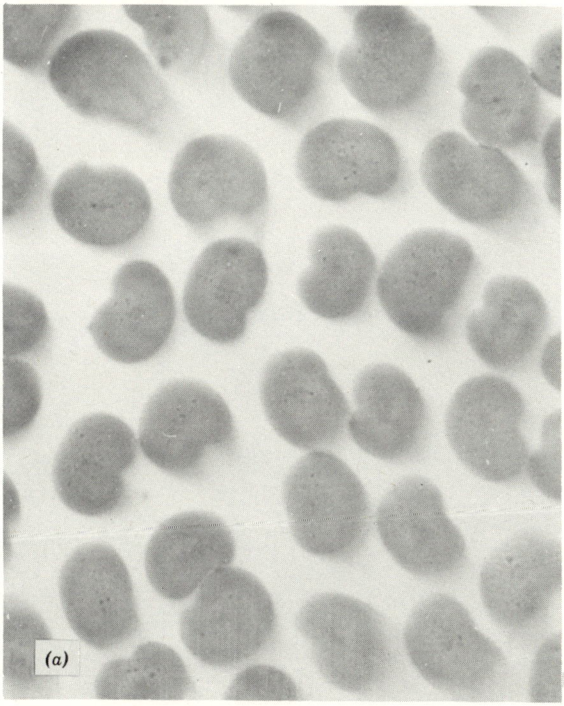

Fig. 7. Cross sections (1000 x) of HSHE rayons: (*a*) 1.5-den/fil staple; (*b*) 1.5-den/fil tire yarn; (*c*) 2.2-den/fil tire yarn.

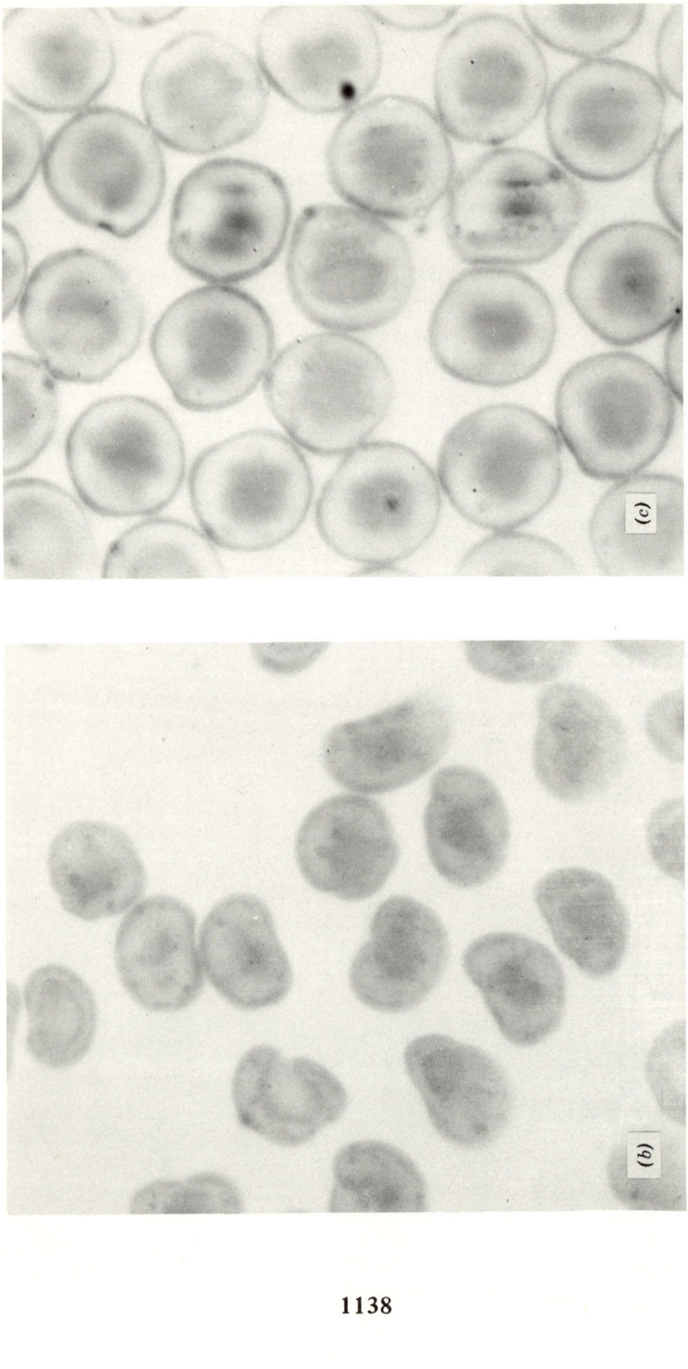

Fig. 7 (*cont.*)

Morehead's (33) fine-structure examinations of the skin and core portions of fibers sampled near the spinning nozzle and preserved in the gel state show a difference in structure in the skin and core zones. These electron-microscope studies show that, for normal rayons, the skin portion consists of an open lacy type of structure, whereas the core section is relatively dense and voidfree. A gel state, "all-skin" tire yarn, on the other hand, has a porous, open network near the outer portion, diminishing gradually to a somewhat more dense structure toward the interior of the fiber.

In contrast, electron micrographs of fully processed, dried, all-skin, HSHE rayons show a fairly uniform, dense, granular structure, viewed either in cross or in longitudinal section. Severe reswelling, as with 8% caustic soda, reopens the fiber structure, revealing the original porous network in the outer portion of the fiber.

Electron cross-sectional micrographs of fully processed normal and HSHE rayons are shown in Figure 8.

(4) Fine Structure

Compared with regular-rayon fibers, the HSHE fibers have smaller and more numerous crystallites. As noted by Götze and Hilgers (34), these small crystallites contain fewer molecules, but because of the higher DP of the cellulose, these molecules are more extended and the crystallites are more elongated than in the regular fibers. These larger molecules provide a greater

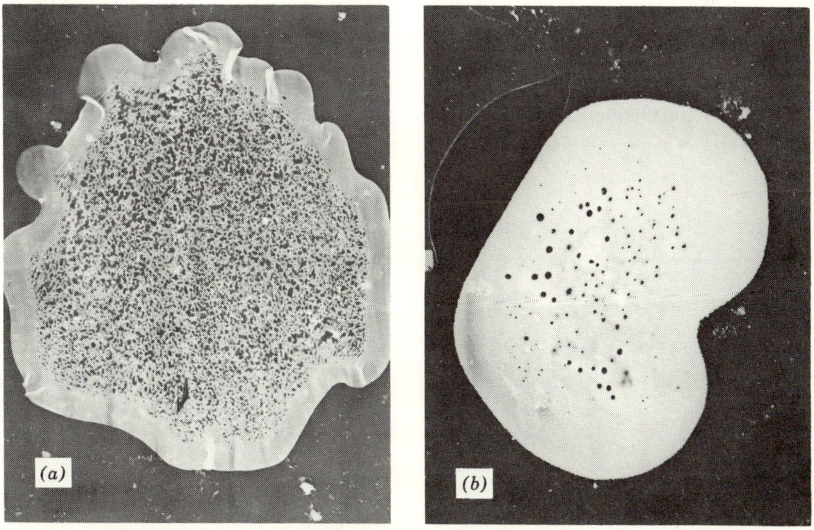

Fig. 8. Electron photomicrographs (5000 ×) showing fine structures of (*a*) regular-type rayon and (*b*) HSHE rayon.

number of molecular-chain contact points in the amorphous regions of the regenerated-cellulose structure. The greater number of amorphous-zone tie points and the higher degree of crystallite orientation, particularly in the interior regions, in large measure contribute to the improved tenacity, superior flex-fatigue resistance, and lower swelling values of the HSHE rayons.

Most rayons have about the same degree of crystallinity; that is, about 40% of the total material is crystalline. Koblitz (35), by determining the "nonsolvent water," has established that the degree of crystallinity of the HSHE rayons is about the same as that of the ordinary type rayons; this further indicates that the difference in physical properties is due to the nature, size, and distribution of the crystalline material.

Longitudinal electron-microscope examination and longitudinal birefringence measurements of various rayons as conducted by Felton and Morehead (36) show that, for regular rayon and tire-yarn-type HSHE rayon the birefringence values (Δn) are different in the exterior and interior portions of the fibers. Table 9 shows the values of Δn (Senarmont method) found in these regions. These data show that the tire yarn has a different relative distribution of the two birefringence zones, and that the interior birefringence is significantly higher than that of the regular rayon.

TABLE 9

Longitudinal Birefringence Values of Regular Rayon and Tire-Yarn Rayon

Sample	Distribution of Δn across fiber	Relative proportion of each Δn (%)		Birefringence Δn of longitudinal sections	
		A	B	A	B
Regular rayon	A–B–A	40 to 20	60 to 80	0.033	0.0175
Tire yarn	A–B–A	75	25	0.366	0.0264

(5) Swelling Properties

The HSHE rayons characteristically have significantly lower degrees of water retention and lower cross-sectional swelling than the normal rayons. Moisture-regain values, however, are characteristically higher for HSHE rayons. Typical values are shown in Table 10.

(6) Chemical Resistance

The high-tenacity rayons generally are similar in their response to chemical action and other degradative environments as are the other rayon types. An exception to this is a greater susceptibility to the action of caustic soda as compared to the new high-wet-modulus rayon types.

TABLE 10
Water-Retention and Moisture-Regain Data

Sample	Water retention (%)[a]	Moisture regain (%)[b]
Normal rayon	95–115	11.0–12.0
High-tenacity rayon	75–85	12.5–13.5

[a]Determined by the centrifugal method.
[b]Determined at 58% relative humidity and 24°C.

(c) GENERAL PROCESS TECHNOLOGY

Most high-strength, high-elongation rayons, whether they are produced as yarns or as staple, employ basically similar process technology. Commercially, they are produced as fine denier filaments, ranging from about 0.5 den/fil to 2.5 den/fil for the continuous yarns and between 1.0 den/fil and 3.0 den/fil for the staple varieties. The total denier of the continuous yarns range from about 300 to over 3000.

(1) Spinnerets

Yarns are spun with small spinnerets having face diameters ranging from 3/8 to 3/4 in., or more, with the number of holes ranging from less than 100 to 3000 or more. Hole sizes are usually within a narrow range, from about 50 to 100μ for most yarn types.

Staple (tow) spinnerets are primarily designed to provide a large output per unit. For these the number of holes ranges from about 5000 to over 10,000, with face diameters of 2–3 in. As with the smaller yarn jets, the hole diameters are in the 50-to-100μ range.

(2) Viscose Compositions

Premium pulps with α-cellulose levels of 95 to over 98% (bone-dry basis) are commonly employed. Compared with regular-rayon-type viscoses, higher ratios of soda to cellulose and higher carbon disulfide levels are the rule.

Viscose solutions for HSHE rayons are spun at moderate or low ripeness levels (high salt index); that is, where the level of sulfur present as cellulose xanthate is relatively high in proportion to the total sulfur present. Viscose-regeneration retardants (modifiers) at a level of about 0.2–4% (based on cellulose content) are universally employed (39). These modifiers play a key role in the spinning process by modifying the coagulation and retarding the regeneration of the viscose filaments, thus allowing the technologist greater latitude in controlling fiber properties and achieving the optimum degree of fiber orientation during the spinning–stretching process.

A tabulated comparison of regular rayon and HSHE-type viscose is shown in Table 11.

High-strength rayon fibers are spun from relatively dilute viscose solutions where the cellulose has a relatively high degree of polymerization (450–500). The longer cellulose chains, as noted by Götze and Hilgers (34) and Rimaboschi (40), participate in the formation of more micelles. These micelles are increasingly interconnected, with the result that a stronger gel structure is formed. Increased chain length also leads to the formation of longer micelles with many points of adhesion between them. In contrast, the micelles formed in a regular rayon are shorter and thicker and have fewer points of contact within the structure. The viscosity of the spinning solution increases sharply with increase in cellulose DP so that, in normal practice, a lower concentration of cellulose must be employed.

TABLE 11
Viscose Compositions and Properties of Regular and HSHE Rayons

Constituent or property	Regular rayon	HSHE rayon
Cellulose (%)	8.0–9.0	6.0–8.0
Sodium hydroxide (%)	5.5–6.5	5.0–7.0
Carbon disulfide (%)	30–34	36–42
Salt index[a]	4.0–6.0	7.0–12.0
Modifier(s) (%, based on the weight of cellulose)	Nil	0.2–4
γ-Value[a]	30–40	45–60
DP of alkali cellulose	350–400	500–600

[a] See Table 7 for definition.

Compared with the viscose composition used for regular rayon, the ratio of alkali to cellulose in the viscose of HSHE rayon is significantly higher, ranging from about 0.85 to 1.0 or more. Klare's (29) work on the effect of an increasing ratio of alkali to cellulose with unmodified viscoses spun into a high-zinc bath shows a profound effect on the strength of HSHE yarn (See Figure 9).

In making high-tenacity rayon the viscose is spun in an unripened state—that is, when a high concentration of cellulose xanthate is still present. γ-Values (i.e., number of xanthate groups per 100 glucose units) of about 45–50 or more are typical for these viscoses. These higher γ-values are attained through a combination of increased carbon disulfide (up to about 44%, based on cellulose) and a low degree of ripeness, where by-product formation is at or near a minimum.

Though the viscose is spun in the relatively unripened state, a certain degree of ripening is necessary to exceed the minimum point of the viscosity curve as established by Heuser and Schuster (41).

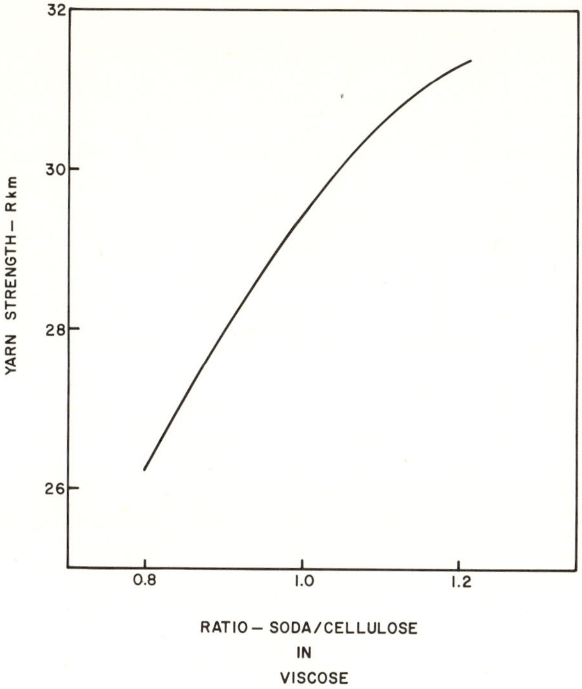

Fig. 9. Effect of soda-to-cellulose ratio in unmodified viscose on the strength of HSHE yarn (29).

(3) Effect of Pulps

Common to all viscose solutions used in making HSHE rayon is the use of high-quality purified pulps. Klare's (29) work with various grades of wood dissolving pulps, which were processed into viscose solutions under similar conditions, shows an excellent relationship between the pulp used and the properties of the rayon yarn. These data are depicted in Figure 10.

The beneficial effect of eliminating the low-molecular-weight fractions in the mercerized cellulose is clearly indicated when curves 2 and 4 are compared.

On the other hand, Croon et al (42) indicate that the hemicellulose in spruce sulfite pulps has a marginal influence on strength properties. They suggest that higher quantities of hemicellulose than presently used could be accepted in rayon products if impurities, such as metal ions and resins, were kept at extremely low levels.

(4) Spinning Conditions

Spin-Bath Compositions. The HSHE rayons are spun in baths that contain an appreciable zinc content in addition to the sulfuric acid and sodium sulfate

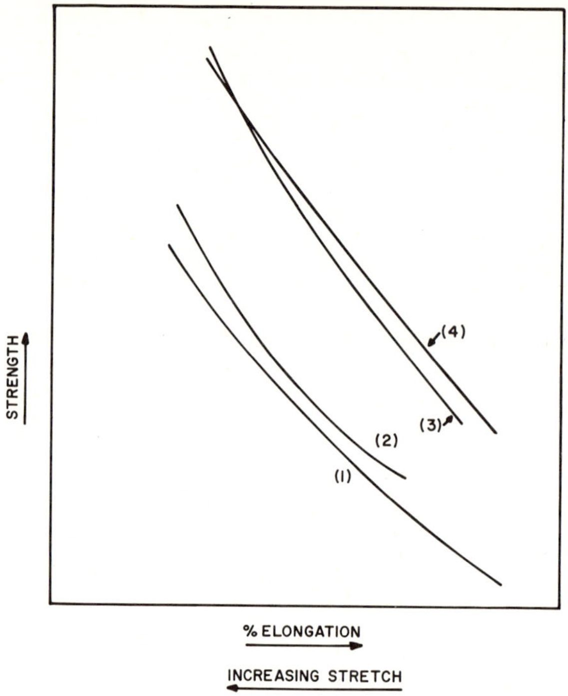

Fig. 10. Effect of pulp type on the physical properties of HSHE rayon (29): (1) spruce sulfite, unpurified, 89% α-cellulose; (2) spruce sulfite, purified, 95.1% α-cellulose, high proportion of low-molecular-weight fractions after mercerization; (3) prehydrolyzed sulfate, 97.7% α-cellulose; (4) spruce sulfite, purified, 95.1% α-cellulose, lower proportion of low-molecular-weight fractions after mercerization.

normally present. Zinc ions play a key role in the regeneration process in that they react with the sodium cellulose xanthate to form zinc cellulose xanthate, a derivative that is more resistant to acid decomposition (43).

Spinning Process. The typical spinning process for HSHE rayon consists of the following:

1. Viscose-regeneration modifiers are employed, either in the viscose, the spin bath, or in combination.
2. Fairly-high-temperature spin baths ranging from about 50 to 60°C.
3. A low sulfuric acid concentration, at a level comparable to the alkali content of the viscose solution—that is, in the range of about 70–100 g/l.
4. A high zinc sulfate concentration, ranging generally from about 70 to 110 g/l, and a sodium sulfate concentration in the range of 130–200 g/l.

5. Spinning speed (takeup speed) is low by normal rayon standards, the commercial range being about 35–55 m/min.

(5) Filament Formation

Role of Sulfuric Acid. Filament formation is initiated by the hydrogen ions diffusing into the fine streams of viscose. During this inward diffusion the pH of the system varies from one of alkalinity to one of acidity, and the reaction of the acid with the caustic soda in the viscose produces sodium sulfate and water in relatively large amounts.

Sulfuric acid reacts also with the viscose by-products, such as sodium trithiocarbonate and sodium sulfide, producing additional sodium sulfate. Elemental sulfur is also formed as the result of other reactions.

Sodium cellulose xanthate groups or the newly formed zinc cellulose xanthate groups are decomposed by the action of the acid, with the addition of hydrogen to form hydroxyl groups.

The concentration of sulfuric acid is maintained at a relatively low level to minimize the decomposition of the sodium cellulose xanthate before it can be converted to the more desired zinc cellulose xanthate form. Sufficient acid is needed for the neutralization of the caustic soda and for reactions with the by-products. Additionally, a minimum level of acidity is required to preclude the precipitation of zinc as the hydrate.

Effect of Sodium Sulfate. Coagulation of the viscose streams is largely due to the presence of the sodium sulfate in the spin bath. This coagulation is induced by the dehydration from the interior of the viscose stream.

Normal rayon spinning with high sodium sulfate concentrations in the spin bath produces fibers with a crenulated surface. This results from the rapid collapse of the gel structure arising from the severe dehydration of the filament body. Spinning baths for HSHE rayons avoid collapse of the gel structure because the sodium sulfate concentration is maintained at a level that limits the salt-concentration gradient between the spin bath and the interior of the filament. The prevention of a structural collapse produces a more desirable, uniform, porous, and open gel structure.

Effect of Zinc and Modifiers. The high-zinc bath (43) and the use of viscose modifiers (39) are key features of the modern HSHE rayon process. These developments, along with the technique of imparting stretch to freshly spun filaments (44), individually and collectively are the important historical milestones of the HSHE rayon processes.

The presence of zinc ions in the spin bath results in a retardation of the xanthate-decomposition rate, with the result that improved-quality, more elastic filaments are formed. The investigations of Poznanski (45) established that the zinc ions substitute for the sodium ions of the xanthate groups, forming zinc cellulose xanthate. The zinc cellulose xanthate has been shown to be more stable to acid decomposition than the sodium cellulose xanthate that it replaces. As a consequence the rate of decomposition of the xanthate by the acid entering the structure is reduced. This slower xanthate-decomposition rate has a favorable effect on the structural properties of the filament.

Zinc, being divalent, can combine two xanthate groups as a result of its covalency. This zinc-crosslinked xanthate is capable of more stretching than the uncrosslinked sodium xanthate in which slippage of macromolecules probably occurs at fairly low stretch levels.

The advantage derived from the formation of zinc cellulose xanthate has resulted in a considerable effort to optimize and/or enhance its formation. Klare and Gröbe (46) have shown that the extent and degree of zinc penetration into the filament depend on many factors. It was established that penetration is enhanced by an increased concentration of zinc in the bath and increased residence time of the filament in the bath. Zinc migration may also be influenced by the amount of water passing from the filament's interior outward into the bath. As the amount and velocity of the liquid issuing from the filaments increase, the diffusion rate of zinc into the filament is appreciably retarded. As a consequence the zinc-diffusion velocity is influenced by the concentration of sodium sulfate. Other factors that influence or limit zinc penetration have been noted by other investigators (34, 47).

The spinning technology for HSHE rayon thus depends on promoting conditions that convert most or all of the sodium cellulose xanthate to the preferred zinc cellulose xanthate. Since the number of xanthate groups in an unripened, high-γ-value viscose is much greater, high zinc levels in the spin baths are necessary to allow for a maximum degree of conversion of the sodium cellulose xanthate to the zinc form.

Recent studies (48, 49) indicate that the conversion of sodium cellulose xanthate to the zinc cellulose xanthate, as determined by analytical techniques, proceeds to only a limited extent. This work indicates that the zinc content of the fibers at the neutral point is about 1–1.5%, based on cellulose, as compared with over 5% content with complete conversion. Also zinc content has been found to decrease with increasing residence time in the spin bath. This is in contrast to earlier reports (46).

Investigations as to where zinc reacts in the fiber cross sections were made by using radioactive zinc ions in the spin bath. Radioautographs led to the conclusion that zinc penetrates the fiber to only a limited extent and is essentially confined in the narrow outer boundary (48, 49).

(6) Viscose Modifiers

As noted earlier, sodium trithiocarbonate, a by-product of the viscose formation and aging process, plays an important role in the spinning of HSHE fibers. Investigations have demonstrated that a definite minimum amount of by-products is generally required in the viscose for it to show a tendency to molecular aggregation and for the subsequent formation of micelles (50). It has also been shown that HSHE fibers can be produced when a sufficient amount of sodium trithiocarbonate is added to a modifier-free viscose spun under the proper conditions (51). On the other hand, trithiocarbonate-free viscose, with modifiers, fails to produce the HSHE fibers. This evidence indicates that the viscose by-products are essential to the HSHE-fiber-spinning process and that the presence of modifiers complements and enhances the effect of the by-products (52).

Since their introduction into the viscose process (39), many compounds of an organic nature have been found to function as viscose-regeneration modifiers. Over the years, however, the practical considerations of efficiency, economics, compatibility, and toxicity, among others, have reduced this list to a few types of compounds. The search for more effective modifiers continues, however, in view of their considerable importance to the modern HSHE-rayon-spinning process.

Modifiers, when added to the viscose, are used in concentrations ranging from about 1.0 to 3.0 g per kilogram of viscose. It is common practice in some processes to maintain a 500-to 1000-ppm concentration of these materials in the spin bath.

Types of Modifiers. Modifying compounds currently used commercially include aliphatic amines, ethoxylated amines, poly(oxyethylene) glycols, aromatic amines, and quaternary ammonium salts. Of this group, the first three are perhaps the most widely used. Some processes involve the use of the combination of an amine and a polyglycol (53).

Role of Modifiers. Some of the more popular theories that have been advanced to explain the action of modifiers are as follows:

1. Modifiers produce on the fiber's surface a very thin film that serves to retard the penetration of acid into the interior of the extruded filaments, thus retarding the decomposition of the cellulose xanthate. [Evidence for this is the extended time for the filament to lose its alkalinity (52, 54).]

2. To the extent of their dissociation, modifiers form ions, specifically organic cations and hydroxyl ions. In the acid medium these form salts in which the organic cation has a positive charge while the acid radical is negative. Modifiers, therefore, are in the form of salts in acidic regions of the freshly spun filaments and in the surrounding spin bath as well (34).

3. Amines react with the caustic soda and carbon disulfide to form sodium dithiocarbamate, which in turn reacts with zinc to form the relatively insoluble zinc dithiocarbamate, a material that acts as a barrier against the rapid diffusion of hydrogen ions into the interior of the filament (54).

4. The fine pores of the skin that forms on the fiber's surface when modifiers are present become rapidly clogged, with mixed precipitates of zinc sulfide and zinc trithiocarbonate acting as a barrier to the diffusion of the acid to the fiber's

interior. A gradual breakdown of the zinc sulfide in the pores then permits acid diffusion (and xanthate decomposition) to accelerate gradually until it is finally completed very rapidly in the succeeding hot stretching–plasticizing bath (49).

(7) Stretching Process

Normally the HSHE-fiber-spinning process involves only a limited degree of initial regeneration (i.e., xanthate decomposition) in the spin bath since the bath acts mainly as a coagulating medium. Compared with the normal process, the xanthate sulfur content of the HSHE yarns leaving the spin bath is significantly higher.

The regeneration–stretch bath that immediately follows the spin bath is maintained at about 90–100°C and consists of dilute sulfuric acid as well as spin-bath salts carried in by the yarn. Very rapid and essentially complete xanthate decomposition is effected in this bath coincident with the applied yarn stretch of 100% or more.

During the stretching process orientation of the micelles in the direction of the fiber axis is developed throughout the fiber's cross section, this being permitted by the relatively uniform state of coagulation in the interior of the structure. Stretching expresses the liquid surrounding the micelles as well as gaseous reaction products, leaving an open, porous structure.

Subsequent drying involves a collapse of the open structure and the transformation of the micelles to "crystallites" of regenerated cellulose; however, no further change in the basic structure takes place.

References

1. F. R. Charles, *Can. Textile J.*, **83** (16), 37 (1966).
2. S. S. Napper, Samuel Courtaulds & Co., Ltd. U.S. Patent 1,045,731 (1912).
3. J. H. Givens, H. W. Biddulph, and L. Rose, British Patent 467,500 (1937).
4. N. L. Cox, U.S. Patent 2,535,045 (1950).
5. M. Müller, U.S. Patent 836,452 (1906).
6. L. Lilienfeld, U.S. Patent 1,683,199 (1928).
7. G. M. H. Kayser, U. S. Patent 2,452,130 (1948), (to American Enka Co.).
8. Tachikawa Research Institute, Belgian Patent 593,672 (1960).
9. H. Mark, *Chemiefasern*, **15**, 422 (1965).
10. J. M. Preston, *J. Soc. Chem. Ind.*, **50**, T199 (1931).
11. F. F. Morehead and W. A. Sisson, *Textile Res. J.*, **15**, 443 (1945).
12. D. N. Tyler and N. S. Wooding, *J. Soc. Dyers Colourists*, **74**, 283 (1958).
13. B. E. M. Bingham, *Makromol. Chem.*, **77**, 139 (1964).
14. N. Drisch and R. Priou, *Bull. Inst. Textile France*, **101**, 667 (1962).
15. P. Herrbach, *Lenzinger Ber.* (17), 13 (1964).
16. N. Drisch, *Pure Appl. Chem.*, **14**, 317 (1967).

17. Y. Uchida, *Pure Appl. Chem.*, **14**, 461 (1967).
18. J. W. Schappel, *Tappi*, **46**, (10), 18A (1963).
19. E. Treiber, *Lenzinger Ber.* (18), 15 (1965).
20. N. Drisch, P. Herrbach, and H. Rodier, U.S. Patent 3,337,671 (1967) (to Chimiotex S.A.).
21. S. Tachikawa, U.S. Patents 2,592,355 (1952) and 2,732,279 (1956).
22. N. Drisch and P. Herrbach, U.S. Patent 3,352,957 (1967) (to Chimiotex S.A.).
23. L. H. Phifer and L. B. Ticknor, *J. Appl. Polymer Sci.*, **9**, 1055 (1965).
24. Rayonier, Inc., Belgian Patent 582,089 (1959).
25. R. L. Mitchell, J. W. Berry, and W. H. Wadman, U.S. Patent 3,018,158 (1962) (to Rayonier, Inc.).
26. F. P. Alles, U.S. Patent 2,053,123 (1936) (to Du Pont Rayon Co.).
27. T. Lefferdink and H. Briar, *Textile Res. J.*, **29**, 477 (1959).
28. K. Mieck, *Textil-Praxis,* **16**, 7, 684 (1961).
29. H. Klare, *Faserforsch. u. Textiltech.*, **11** (12), 582 (1960).
30. K. Ohara, *Sci. Papers Inst. Phys. Chem. Res. (Tokyo)*, **25**, 152 (1934).
31. M. Horio, *Textile Res. J.*, **24**, 971 (1954).
32. R. J. E. Cumberbirch and J. E. Ford, *J. Textile Inst.*, **51**, 132 (1960).
33. F. F. Morehead, unpublished work, 1965.
34. K. Götze and F. Hilgers, *Tinctoria,* **57**, No. 3, 89 (1960).
35. W. Koblitz and H. Kiessig, *Das Papier,* **14**, (5), 179 (1960).
36. C. D. Felton and F. F. Morehead, unpublished work, 1959.
37. F. F. Morehead, unpublished work, 1957.
38. J. A. Howsman, paper presented at 120th meeting. American Chemical Society, New York, Sept. 1951.
39. N. L. Cox, U.S. Patent 2,536,014 (1950) (to E. I. du Pont de Nemours & Co.).
40. V. Rimaboschi, *Riv. Tiss.*, **37** (11), 483 (1961).
41. E. Heuser and K. Schuster, *Cellulosechem.*, **7**, 17 (1926).
42. I. Croon, H. Jonsen, and H. G. Olafsson, *Svensk Papperstidn.*, **71** (2), 40 (1968).
43. I. P. Davis, U.S. Patent 2,312,152 (1943) (to American Viscose Corp.).
44. H. C. Stuhlmann, U.S. Patent 1,901,007 (1933) (to I.G. Farbenindustries A.K.G.).
45. S. Poznanski, *Przem. Chem.*, **22**, 463 (1938).
46. H. Klare and A. Gröbe, *Faserforsch. u. Textiltech.*, 8, 310 (1957); *ibid.*, 9 (1958).
47. P. Kassenbeck, *Melliand Textilber.*, **39**, 55 (1958).
48. H. Klare, *Lenzinger Ber.* (19), 21 (1965).
49. H. Klare and A. Gröbe, *Österr. Chemiker-Ztg.*, **65** (7), 218 (1964).
50. B. Phillip. *Faserforsch. u. Textiltech.*, 8, 21 (1957).
51. N. L. Cox, U.S. Patent 2,581,835, (1951) (to Du Pont).
52. D. K. Smith, *Textile Res. J.*, **29**, 32 (1959).
53. R. L. Mitchell, J. W. Berry, and W. H. Wadman, U.S. Patent 2,942,931 (1960) (to Rayonier).
54. H. Klare and A. Gröbe, *Faserforsch. u. Textiltech.*, **9**, 262 (1958).

C. CELLULOSE TRIACETATE FIBERS

C. L. Smart and Carl N. Zellner

Cellulose triacetate is a relative newcomer to the realm of commercially important fibers, offering hydrophobicity and high crystallinity after heat setting. Thus it imparts to fabrics such desirable characteristics as ease of care, resistance to wrinkling, dimensional stability, heat stability, and rapid drying. The literature on cellulose triacetate was sparse until the publications on the crystallinity of cellulose esters by Baker, Fuller, and Pape and by Work (see Part I, pp. 338–346).

Intense investigation of cellulose triacetate and its fibers started in 1949. Beginning in 1954, commercial fibers were announced in rapid succession in the United States, Canada, England, France, and Germany. Triacetate fiber has also been produced in the Soviet Union since 1965. In this section we attempt to summarize the state of knowledge concerning polymer properties, fiber formation, and fiber properties—and especially to review significant papers published since 1954, that bear on these areas.* (See also Section B in Chapter XVII).

1. Polymer Properties

(a) COMPOSITION

The U.S. Federal Trade Commission has ruled that if not less than 92% of the hydroxyl groups of cellulose are acetylated, the term "triacetate" may be used as a generic description of the fiber. Although an ester showing an acetyl value of 42.7% [degree of substitution (DS) of 2.75] could be termed a triacetate, in actual practice the acetyl value of polymer converted to commercial fiber is appreciably higher (44.3–44.8%). Assuming that there are few, if any, monoacetate or unsubstituted glucose residues at an acetyl content of 40.5% (slightly above the commercial cellulose acetate level) there is, on the average,

* For a general review of the preparation and the physical and chemical properties of cellulose triacetate fibers, see reference 1.

one triacetate residue for each diacetate residue; this ratio is not propitious for crystal formation. However, at an acetyl content of 44.3% there are approximately 15 triacetate residues for each diacetate residue. This predominance of triacetylated glucose residues facilitates good crystal formation and explains the special, attractive properties of cellulose triacetate fibers.

(b) SOLUBILITY

The solubility of a cellulose triacetate depends on chain length and distribution, degree and spread of substitution, degree of crystallinity, and hence on the mode and severity of the polymer's coagulation and drying. Cellulose acetate hydrolyzed to an acetyl value of about 14% is water soluble. As the degree of substitution is increased, water solubility is decreased and the polymer at the diacetate state, 32–38% acetyl, is soluble in such solvents as acetone, ethyl acetate, or phenol. In this range of substitution the polymer might be considered both a Lewis base and a Lewis acid and consequently is soluble in both acidic and basic solvents. Fully acetylated triacetate (44.8% acetyl) is soluble only in acidic solvents. Solvents that yield a 10% (wt/wt) solution of cellulose triacetate were listed in Part II on page 794 as including chloroform, *sym*-tetrachloroethane, formic acid, and methylene chloride–methanol (4 : 1). As an example of the influence of cosolvents, a 10% solids solution of a 44.3%-acetyl triacetate shows a viscosity of 9900 cp at 20°C in methylene chloride and 8000 cp in methylene chloride–methanol (9 : 1) (2).

Smooth, clear 20% (wt/wt) solutions of a commercial, fiber-forming triacetate (44.3% acetyl) may also be prepared in acetic acid, *m*-cresol, cyclopentanone, ethylene chlorohydrin, propylene chlorohydrin, 85% aqueous phenol, pyrrole, 1,1,2-trichloroethane, trichloroethanol, and mixtures of dioxane with alcohol and water, with nitromethane, with pyrrole, and with sulfolane. At elevated temperatures good solutions are obtained in dimethylformamide, dimethylacetamide, glycerol formal, sulfolane, and trioxane. In view of the known solubility parameters of these various liquids, it is apparent that the solubility parameter for this particular triacetate approximates 10.5–11.5. However, knowledge of the solubility parameter of a liquid is not sufficient to ensure its utility as a solvent. For cellulose acetates two additional parameters—dipole moment and hydrogen-bonding value—seem to define solvent capability more completely (3).

Swelling, but not solution, at the 20% concentration level is observed at room temperature in chlorobenzene, dioxane, formamide, difluorodichloromethane, furfural, methyl formate, methyl acetate, nitromethane, pyrrolidone, and 1,2-propylene glycol. Acetone would be a most desirable solvent, and in fact smooth, 15-wt. % solutions are obtained by cooling a slurry to −40°C and then allowing the mass to warm to room temperature. Alternatively solutions are obtained at room temperature by rapidly agitating amorphous fibers in acetone. However, reproducibility is poor, the solutions are not stable, and fibers from such solutions are brittle.

(c) MOLECULAR-WEIGHT DISTRIBUTION

The literature cites numerous, rather unsuccessful efforts to fractionate cellulose triacetate with regard to molecular weight. In spite of careful control of fractionation conditions and choice of solvent and precipitant only slight fractionation takes place. Howard and Parikh (4) have examined the problem in great detail and concluded that the familiar method of obtaining fractions of wide molecular-weight range—to reacetylate fractions of lower-acetyl-content polymer—is still best. They suggest that cellulose triacetate dissolves in such solvents as chloroform, *sym*-tetrachloroethane, and acetic acid by forming hydrogen bonds with these liquids, the extent of such polymer-solvent interaction accounting for the poor fractionation observed. They found *sym*-tetrachloroethane to be the best solvent of a series examined and argued that the effectiveness of this particular solvent is due to the availability of two hydrogen atoms per molecule for hydrogen bonding with carbonyl groups on the polymer. Furthermore, Howard and Parikh noted the solubility of very-low-molecular-weight (DP ~ 23) cellulose triacetate in both proton-accepting and aprotic solvents at room temperature. Here improved solubility could be attributed to an increase in the entropy of mixing arising from the greater randomization of a polymer-solvent mixture containing short chains.

Tanghe and co-workers (5) attacked the problem in a most ingenious manner. They determined the molecular-weight distribution of cellulose triacetate by converting the triester to the tricarbanilate, a derivative that is more amenable to separation. Thus high-acetyl-content acetates were saponified with sodium methoxide and the regenerated cellulose was washed and then reacted in pyridine with 50% excess of the theoretical phenyl isocyanate. Viscosity measurements suggested that very little chain scission had occurred during the entire procedure. Precipitation fractionation of the tricarbanilate from acetone solution was accomplished by incremental additions of acetone-aqueous sodium chloride. The last low-molecular-weight fraction was precipitated by addition of distilled water to give a total recovery of about 95%. Intrinsic-viscosity values for each fraction were measured in dioxane and ranged from 0.3 to 3.3 dl/g, whereas the original sample showed a value of 1.7 dl/g. Gel permeation chromatography was then applied to each fraction. Weight-average to number-average molecular-weight ratios were about 2.1 for the total unfractionated sample and 1.3-1.6 for representative fractions. The combined techniques of fractional precipitation and gel permeation chromatography indicate the presence of very-high- and very-low-molecular-weight material not otherwise detected (see Figure 1). There

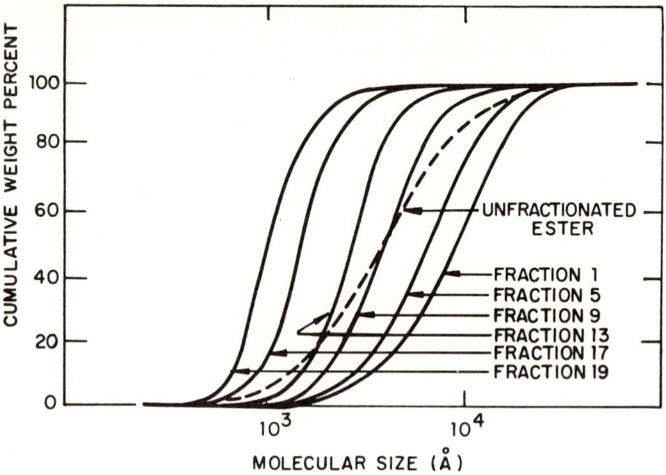

Fig. 1. Sequential fractionation of cellulose carbanilate (prepared from cellulose triacetate) by gel permeation chromatography and precipitation methods.

appears, for these trisubstituted celluloses, to be a maximum prevalance at an intrinsic viscosity of 1.1 dl/g, corresponding to a 6000-Å overall molecular length. A surprising result of the work of Tanghe is the lack of any appreciable difference in the curves whether the chain lengths of the original triacetates had been lowered heterogeneously or homogeneously. On the other hand, Bischoff and Philipp (6) claim that the molecular-weight distribution of a triacetate depends on the mode of preparation. After preparing triacetates by homogeneous and heterogeneous acetylation, the acetyl content was reduced in aqueous acetone to the level of the secondary acetates, and the latter esters were separated by fractional precipitation. The triacetate prepared homogeneously showed wider chain-length range and a greater fraction of long chains. The conclusions are not a direct contradiction of those of Tanghe, since esterification conditions and characteristics of the solutions differed widely (see also Section E in Chapter XIV).

In view of the breadth of the molecular-weight-distribution curve for a cellulose triacetate preparation, any attributions of number-, weight- or viscosity-average molecular weight should consider acetylation conditions, polymer-purification procedures, polymer characterization, solvent system, and temperature of measurement. To illustrate some of the complexity, Flory, Spurr, and Carpenter (7) measured the intrinsic viscosities of an unfractionated cellulose triacetate (DS 2.86) and of a series of five triacetates (DS 3.00) covering a range of molecular weights. A plot of intrinsic viscosity versus temperature shows striking dependence on temperature and on solvent.

A typical fiber-forming cellulose triacetate (unfractionated) shows (8) a number-average molecular weight of 57,300 (DP_n about 200) and a second virial coefficient of 1.28 x 10^3 (mole-ml/g^2) (values obtained on 0.1-0.5% solutions in *m*-cresol at 65°C). The intrinsic viscosity in 0.5 *M* cupriethylenediamine at 25°C of the cellulose regenerated from this commercial polymer is 2.0 dl/g, to be compared with 1.6 dl/g for cellulose regenerated from commercial fiber-forming cellulose acetate (i.e., secondary acetate). For a crude triacetate fraction showing an intrinsic viscosity of 2.0 dl/g in cuprammonium hydroxide solution, Cumberbirch and Harland (9) assign a viscosity-average degree of polymerization (DP_v) of about 550. They calculate the exponent in the Mark-Houwink equation, $[\eta] = K M_v^a$, to be 0.834 for triacetate in methylene chloride-ethanol (80 : 20, v/v). This value may be compared with 0.803 for secondary acetate in acetone-water (80 : 20). Both esters are more coiled than cellulose trinitrate, which shows an *a*-value of 0.933 in acetone. The viscosity-average molecular weight of triacetate, $\overline{M}_v$, has been calculated most often by attempting to prepare narrow fractions, determining number-average or weight-average molecular weights by absolute methods, and then making the assumption, undoubtedly erroneous, that the fractions are sufficiently monodisperse to justify writing $\overline{M}_v = \overline{M}_n = \overline{M}_w$. Sharples and Major (10) suggest an alternative approach that uses the fact that for a distribution arising from random degradation $\overline{M}_v = \text{constant} \cdot \overline{M}_n$. The exponent *a* is then calculated to be 1.02 and *K* to equal 2.51 x 10^{-4}. The *K*-value is consistent with that published in the comprehensive investigation by Howlett et al. (11), 2.39 x 10^{-4}, with *a* assumed to equal unity.

Extremely-high-molecular-weight triacetate, prepared (12) via amine pretreatment of cotton and subsequent low-temperature heterogeneous acetylation forms clear, homogeneous solutions in (90 : 10) methylene chloride-methanol. The authors claim that films of 1200 DP triacetate show the same level of tensile strength as films of a more conventional DP level but they measure a resistance to double folding of 20,000 cycles as compared with 1700 cycles for conventional films. Triacetate oligomers with terminal hydroxyl groups and suitable as components of hard segments of elastomeric fibers are prepared (13) by the acid degradation of a full triacetate dissolved in an acetic acid-water mixture. The level of the reacted water is the most critical variable affecting the retention of acetyl content.

(d) LIGHT AND HEAT STABILITY

Cellulose triacetate has good light and heat stability. At a wavelength of 313 nm, little effect is noted (14). In general triacetate fibers, properly stabilized, show good resistance to outdoor weathering and to exposure to sun under glass (15).

The stability of the triacetate polymer to heat is excellent, as exemplified by

TABLE 1
Volatiles Produced on Heating Cellulose Triacetate (18)

Compound	Atmosphere	Volatile (wt %)						Comments
		CO	CO_2	H_2O	CH_3CHO	CH_3COOH	Unknown component	
Triacetate	Air	1.2	34.8	5.9	0.3	56.4	1.3	CO_2 first product
Acetate	Air	4.4	48.9	5.5	0.3	38.9	1.9	CO_2 first product
Triacetate	Helium	3.2	24.2	4.5	0.5	64.3	3.2	CH_3COOH first product; traces of CH_3OH, CH_2O
Acetate	Helium	1.4	19.8	7.2	0.6	66.3	4.7	CH_3COOH first product; traces of CH_3OH, CH_2O

the retention of fiber properties at elevated temperatures. Differential scanning calorimetry on triacetate polymer (44.3% acetyl) heated at 10°C per minute under nitrogen indicates (16) the onset of melting at approximately 290°C, a value that is in agreement with the literature. Light-microscopy studies indicate a melting point at 305–307°C for crystalline, oriented fiber from such polymer.

Madorsky (17) determined the percentage of volatilization at various temperatures and analyzed the volatiles. Thus a triacetate held at 250°C in air for 30 min lost only 2% of its total weight. The same polymer heated from 25 to 465°C over a 5.4-hr period produced 90% of volatile materials. Of the volatiles, equal amounts (45%) of tar and acetic acid were produced, with the remaining 10% consisting mainly of carbon dioxide, carbon monoxide, and methane. The monomer levoglucosan acetate was not detected in the tar, evidently because of the blocking of the $C_{(6)}$ by the acetyl group. Zitomer and DiEdwardo (18) have compared the products of thermal breakdown of both triacetate polymer (44.3% acetyl) and a commercial secondary acetate polymer. Thermogravimetry–mass spectrometry analysis, a technique disclosed recently (19), was used for determining the products. The results of quantitative analysis of the volatiles generated above 300°C are shown in Table 1.

Torsional braid analysis of a composite specimen with glass braid indicates transitions at about 190 and at about 290°C, accompanied by drastic decreases in rigidity, prominent maxima in damping, and significant untwisting of the specimen (20). It should be noted that the glass-transition temperature of triacetate fiber, based on work recovery at 1% extension, is between 160 and 180°C.

2. Fiber Formation

Fibers of cellulose triacetate can be formed by melt spinning, or by dry, or wet solution-spinning techniques, but to date the first two are of the greatest commercial importance. In addition, the heterogeneous triacetylation of cotton and viscose-rayon fibers has been explored (see, for example, reference 21). Polymer to be converted to fiber must be as uniform as possible. Typically such polymer shows about 2% moisture, an acidity expressed as an acetic acid content of not more than 0.003%, 0.03% sulfate, a bulk density of 0.2 g/cm^3, and gives solutions that are virtually gel-free and colorless.

(a) MELT SPINNING

Melt spinning of cellulose triacetate is of interest because it avoids the need for solvents and their recovery. Because the polymer is somewhat unstable above its melting point, early experimental work on melt extrusion met with difficulties. However, melt spinning can be carried out if contact time in the melt is reduced to seconds (22).

The process can produce relatively-large-diameter (high-denier) fibers, exploited commercially in carpeting. Such fibers are difficult to produce by solution-spinning techniques. The two general spinning methods—melt and solution (dry)—thus complement each other for producing a variety of yarns for different end uses.

(b) SOLUTION SPINNING

(1) Solution Preparation

As already mentioned, solvents for cellulose triacetate include acetic acid, chlorinated hydrocarbons in admixture with alcohols, and mixtures of dioxane, ethanol, and water. Most frequently described in the literature are mixtures of methylene chloride and an alcohol in a ratio of about 90 : 10.

Solution preparation is accomplished in turbine-blade mixers where polymer flake is dissolved in the solvent system to produce a solution of the proper viscosity. Filter aid and a delustrant, such as titanium dioxide, may be added to thc mixers. The spinning solution proceeds to storage tanks and through a series of filtration steps. Filtrations tests provide a measure of the gel and foreign-particle content in terms of filterability of the spinning solutions. Another approach to preparing a smooth, clear solution involves acetylation of cellulose in methylene chloride, countercurrent extraction of the acetic acid with water, and then removal of residual water from the cloudy solution by contact with a regenerated hydrophilic fabric or film, such as a regenerated cellulose (23).

(2) Dry Spinning

The filtered solution is conveyed to a number of vertical, air-heated spinning columns. Precise quantities of solution are pumped and forced through spinnerets at the top of the columns. As the filaments descend through the spinning columns, the solvent is evaporated by means of hot air. The term "dry spinning" refers to this method, in which solvent is removed from the filaments by evaporation. This is to be contrasted to wet spinning, in which fine streams of the viscous solution are passed into a bath for solvent removal in various nonsolvents. At the bottom of the dry-spinning columns the filaments are lubricated and collected uniformly on bobbins of constant yardage. The yarn may be twisted to commercial levels and put on cones or beams as may be desired. In the manufacture of staple the untwisted yarn is combined into a tow, which is dried, crimped, cut, and baled.

Radial heterogeneity or the presence of a skin has been demonstrated in many fibers. Cellulose triacetate is particularly advantageous as a model polymer for such studies since a wide variety of coagulation conditios can be utilized for fiber preparation. Singleton, Sieminski, and Sprague (24) saponified the surfaces of dry-spun (rapid coagulation) and wet-spun (slow coagulation) triacetate fibers. On dissolution of the cellulose, the mechanical and dyeing properties of the rapidly coagulated, dry-spun fiber were improved but remained unchanged for the slowly coagulated wet-spun fiber. These data indicate that skin formation is associated with rapid coagulation.

(3) Wet Spinning

Wet spinning of cellulose triacetate is not practiced commercially in the United States, although procedures are described in the literature. Rogovin (25) investigated ethylene glycol as the coagulating agent and later used 2-propanol because of its lower viscosity and boiling point. The spinning solution consisted of the polymer in an acetylation mixture of methylene chloride and acetic acid. The sulfuric acid catalyst had been neutralized with sodium acetate. Under certain conditions it was shown by Rogovin, and by others, that for the same polymer, fiber spun by the wet process had a greater strength than fiber spun by the dry process. Further increase in strength was achieved by spinning polymer fractions of higher DP from which low-molecular-weight fractions had been removed. From a practical standpoint dilute acetic acid was suggested as a coagulant in place of 2-propanol.

The effect of tension within the spin bath on the pore structure of wet-spun cellulose triacetate was studied by Quynn (26). Measurements were made with a high-pressure mercury porosimeter. The application of tension drastically reduced the void volume and improved the luster of the resulting fiber. Fibers made at low tension had much greater dye takeup, due mainly to increased internal surface area, as indicated by poor dye washfastness.

Riley (27) describes wet spinning triacetate from a solution of 18–26% concentration in a halogenated lower alkane containing up to 15% of a lower alkanol (e.g., methylene chloride plus methanol). Fibers issuing from the jet pass into a spin bath consisting of halogenated alkane with a sufficiently high proportion of alkanol to precipitate the triacetate. The individual filaments are generally circular in cross section and show a higher modulus of elasticity and resistance to deformation than do dry-spun filaments. In the course of preparation there was a tendency for the filaments to coalesce or adhere to one another. To overcome this tendency Soehngen (28) passed the freshly formed swollen filaments through a hot-air jet to remove adherent spin-bath liquid. The treatment introduces crimp to the filaments with no physical damage. When the resulting tow is cut into staple fiber, much less breakage of filaments occurs.

An alternative method (29) of avoiding coalescence of wet-spun filaments is to extrude the polymer solution into a spin bath containing hydrocarbon oils, silicone oils, or polyoxyalkylene glycols. A hydrocarbon, such as a white mineral oil, with a molecular-weight range of 200–300 is effective in retarding the rate of coagulation of the filaments and in improving spinning stability. In one example a 22% solution of triacetate (44.4% acetyl) in methylene chloride–methanol (91 : 9) was extruded as a monofilament into a bath of methylene chloride–methanol (41.5 : 58.5) containing 0.5% by weight of mineral oil. The tenacity of the monofilament was 2.1 g/den and its elongation was 26.2%, both values being several percent higher than those of a filament produced in the absence of mineral oil.

(c) HEAT TREATMENT

The high regularity of triacetate molecules makes possible association into a more perfect order. As discussed in subsection 1(a), "Composition," secondary acetate, in which acetyl and hydroxyl groups are irregularly distributed along the chains, has a much lower tendency for crystallization. Work (pp. 338–346, Part I) has shown that on raising the temperature of the triacetate filament to slightly below its melting point a more regular or crystalline structure is obtained.

The effect of heat treatment is apparent in x-ray diffraction patterns of triacetate yarn heated for various lengths of time in a hot oven or treated under steam pressure. With increase in the rigor of heat treatment the diffraction arcs begin to separate and become sharper and more intense. This is the result of an increase in the size and perfection of ordered regions (i.e., of crystallites). The changes induced in the crystalline structure by heat treatment can also be shown by radial x-ray diffractometer curves (30). During heat treatment, nucleation and crystallization are initiated, leading to a stabilization of the fiber. As a result of the heat treatment, the softening point in terms of safe ironing temperature is raised from about 180°C before heat treatment to 240–250°C. Resistance to glazing, reduced shrinkage, resistance to wrinkling, dye washfastness, and low water absorption for quick-drying characteristics are realized.

The heat treatment must be carefully controlled so that mechanical properties are not adversely affected by heat degradation and by the formation of large crystals, which decrease the fiber's flexibility. Temperature and exposure times depend on equipment design and fabric weight. The temperature level required may be lowered by small amounts of plasticizers, such as water, which accounts for the effectiveness of steam treatment in the pressure range of 1.4–2.8 kg/cm^2. Temperatures from 180 to 245°C are required, depending on fabric type and operating conditions. For example, 2 sec at about 220°C may be sufficient. Shrinkage and distortion must be avoided by proper restraint of the fabric.

3. Properties of Fibers and Yarns

(a) CRYSTALLINE STRUCTURE

Early work on the crystalline structure of cellulose triacetate indicated the presence of at least two polymorphic crystalline forms, related to the cellulose source and method of esterification. It was not until the effect of heat on the crystallization of cellulose triacetate was discovered (Part I, pp. 338–346) that well-defined x-ray diagrams for both structures could be obtained. Sprague, Riley, and Noether (31) examined the effects of cellulose source and conditions of acetylation on the crystalline structure of cellulose triacetate. Cellulose in the form of wood pulp, cotton, and ramie (cellulose I structure) was acetylated both heterogeneously and homogeneously. The heterogeneous procedure involved pretreatment of the cellulose in 80% aqueous acetic acid followed by acetylation in a 50 : 50 mixture of acetic anhydride–amyl acetate with a perchloric acid catalyst. Homogeneous acetylation involved acetic acid pretreatment and acetylation in acetic anhydride with sulfuric acid as the catalyst. Heat treatment of the products for development of crystallinity was carried out in hot air at temperatures of 210–280°C for 1–5 min.

Heterogeneous acetylation of cellulose I results in a structure whose first two equatorial diffractions are at approximately 11.5 and 5.4 Å. This structure was called cellulose triacetate I. Homogeneous acetylation of cellulose I, followed by fiber spinning and heat treatment, led to a much different x-ray pattern, and such a structure was named triacetate II. Dissolving triacetate I in a solvent and reprecipitation also led to patterns with spacings equivalent to those of triacetate II. Structure II is probably the preferred (low-energy) structural form. x-Ray patterns of cellulose triacetate are shown in Figure 2. It was further noted that homogeneous or heterogeneous acetylation of cellulose II (saponification product from wet-spun, stretched cellulose acetate fiber) yielded triacetate II. Rather unexpectedly, the heterogeneous acetylation of cellulose IV (high-tenacity-rayon tire yarn) also resulted in the triacetate II pattern. In general treatment with swelling agents does not transform the crystalline structure even at concentrations sufficient to swell the fibers drastically. Recently it was

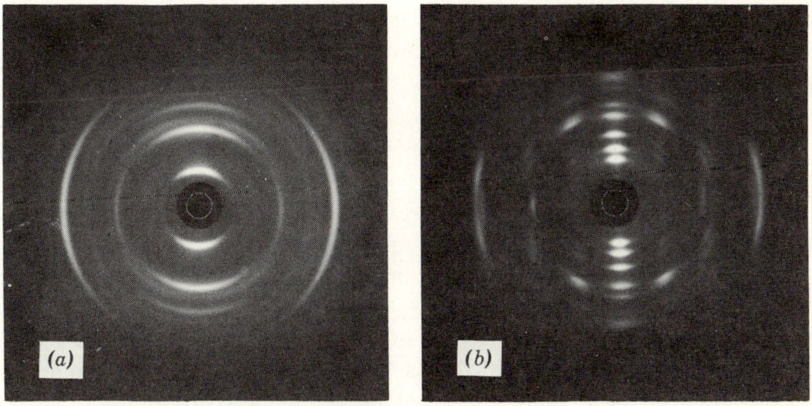

Fig. 2. x-Ray diffraction patterns of (*a*) cellulose triacetate I and (*b*) cellulose triacetate II.

reported by Creely and Conrad (32) that triacetate I is, however, converted to triacetate II by treatment with formic acid.

Observation of the fiber diagrams of triacetates I and II revealed that the unit-cell length in the fiber direction is the same for both structures and has a value of 10.5 Å, which is similar to that of native and regenerated celluloses and corresponds to two glucose units per repeat length. Unit-cell dimensions for triacetate II have been proposed by Dulmage (33). Manley (34) has obtained crystals of cellulose triacetate (from nitromethane solution) in which the molecules appear to be oriented perpendicularly to thin crystalline layers. It was suggested that chain folding, similar to that of some other polymer crystals, may also occur in cellulose triacetate, in spite of the relative stiffness of the molecular chains. The crystallization of amorphous cellulose triacetate has been followed by proton magnetic resonance (35). The wide-line nuclear magnetic resonance technique was found to corroborate the increase in crystalline order of triacetate on heat treatment as observed by x-ray measurements. Heat treatment was shown to result in a higher glass-transition temperature and in an increase in the quantity of protons in the rigid state at any temperature, relative to the sample that was not heat treated.

(b) OTHER PHYSICAL PROPERTIES

Both bright and dull cellulose triacetate fibers and yarns are manufactured. Dull yarns contain 1–2% of a delustrant, such as titanium dioxide. Light reflected from such yarn is nearly all diffuse. Yarns of excellent whiteness can be produced and can be dyed to clean, pure colors in a range of shades and depths. Colored organic or inorganic pigments can be incorporated into the spinning solution prior to spinning. Exceptional colorfastness is obtained in this way, where the pigment is an integral part of the fiber.

Absorption of moisture by triacetate fibers depends on the relative humidity (RH) and temperature, and on whether equilibrium is approached from the dry or wet side, as illustrated in Table 2 (36).

Of importance for easy-care and quick-drying textiles is the percentage of water imbibition. This is determined by measuring the moisture remaining in the fiber after equilibrium is established between the fiber and air at 100% relative humidity while centrifuging the fiber at forces of up to 1000 *g*. Average recorded values are as follows: triacetate without heat treatment, 16%; heat-treated triacetate, 10%; and secondary acetate fibers, 24%.

Sticking and softening temperatures of fibers depend on yarn diameter, fabric construction and geometry, and on the test procedure used. Other important variables are polymer composition and degree of crystallization. As already mentioned, properly heat-treated triacetate fabrics can be ironed at temperatures as high as 240°C, as compared with about 180°C for acetate and triacetate without heat treatment. Mechanical properties of triacetate fibers are not

TABLE 2
Moisture Regain of Triacetate Fiber on Absorption and Desorption at 22°C (36)

Relative humidity (%)	Absorption (%)	Desorption (%)
33	1.8	2.2
53	3.3	4.3
65	4.2	5.4
86	6.8	8.4
94	8.8	9.7
100	12.5	–

significantly altered within the range of normal climatic temperatures. At an elevated temperature (130°C) in air triacetate fibers retain 88% of their initial strength after 180 hr and 70% after 500 hr.

The stress–strain properties of commercial triacetate yarns are similar to those of acetate (see Chapter XI in Part III). The initial modulus of such fibers is generally in the range of 35–40 g/den. The wet modulus at various temperatures is of significance from a textile point of view because of its bearing on the degree of creasing and mussiness resulting from laundering. Figure 3 shows the wet moduli of several fibers as a function of temperature. A summary of the physical properties of cellulose triacetate fiber is presented in Table 3 (37).

(c) CHEMICAL AND ENVIRONMENTAL STABILITY

Cellulose triacetate fiber is resistant to many common solvents. It is significantly more resistant to the alkalies encountered in textile operations than is acetate. Triacetate is unaffected by dilute solutions of weak acids and also has unusually high resistance to chlorine at the concentrations normally used in home washing. It is attacked by stronger solutions of sodium hypochlorite (5%), acetic acid, phenol, mineral acids, and sodium and ammonium hydroxides.

Triacetate is resistant to attack by moths and carpet beetles and is substantially unchanged after 12 weeks of soil burial. Strength retention on exposure to sunlight is slightly better than that of acetate, rayon, and cotton.

Cellulose triacetate fibers are not self-extinguishing. Flame retardants for such fibers impart resistance to burning or retard flame propagation of the burning material. The flame-retardant additive should also be relatively resistant to laundering and dry cleaning, and should not appreciably alter the hand, physical properties, and dyeing behavior of the fiber or textile. One such additive is tris(2,3-dibromopropyl) phosphate (38), which is generally made by fully brominating triallyl phosphate (see also Section I of this Chapter XIX). The light and heat stability has been discussed above, under subsection 1(d).

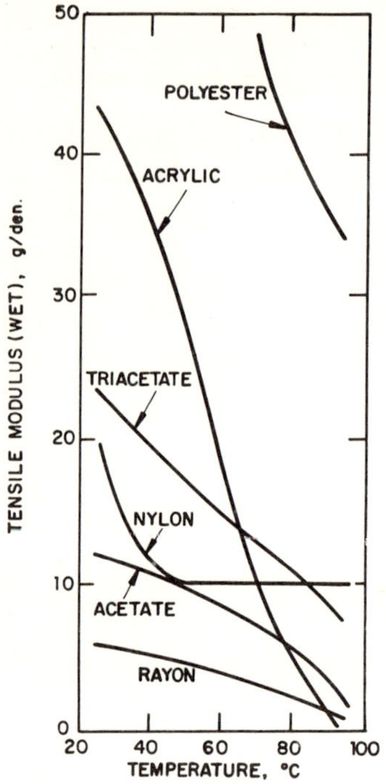

Fig. 3. Effect of water temperature on the wet modulus of fibers.

(d) DYEING AND FINISHING

Yarns and fabrics of cellulose triacetate can be dyed in a full range of shades with disperse dyes. Because of its hydrophobicity, triacetate has a slower dyeing rate than does acetate. Some of the accelerants that have been used to increase the rate of dyeing of triacetate are tributyl phosphate and other phosphate esters, dialkyl phthalates, trialkyl citrates, and pine-oil alcohols. Deep shades can be achieved with selected disperse dyes at 96°C in the presence of accelerants. Because triacetate does not display the initial "strike" characteristics of acetate and rayon at 75–80°C, the leveling problems experienced with these rapidly dyeing fibers are not encountered. Heat treatment after dyeing with selected dyes will give good washfastness. Detailed information on dyeing procedures is available from manufacturers.

Triacetate can be screen or roller printed with disperse dyes to give prints with good definition, levelness, and colorfastness equivalent to that of dyed fabric. The printed fabric does not stain on further processing or washing if properly heat-treated during printing.

The effect of nitrogen dioxide and other components in the atmosphere on dyed fabrics, a phenomenon termed gas fading, has been studied extensively.

TABLE 3
Physical Properties of Cellulose Triacetate Fibers (37)

Property	Range
Tenacity, (g/den):	
Standard conditions (65% RH, 21°C)	1.2–1.4
Wet	0.8–1.0
Bone dry	1.4–1.6
Knot, standard conditions	1.0–1.2
Loop, standard conditions	1.0–1.2
Elongation at break (%)	
Standard conditions	22–28[a]
Wet	30–40
Initial modulus (g/den)	35–40
Tensile recovery (%):	
1% stretch	100
3% stretch	88
5% stretch	75
10% stretch	43
Specific gravity (H_2O at 25°C/25°C)	1.33
Refractive index:	
Parallel to fiber axis	1.474
Perpendicular to fiber axis	1.471
Standard moisture regain (%)	3.2
Water absorptivity:[b]	
Without heat treatment	16–17
Heat-treated	10–11
Area cross-sectional swelling in H_2O (%):[c,d]	
Without heat treatment	4.0
Heat-treated	1.5

[a] For staple fibers elongation at break is 34–40%.
[b] At 100% RH, centrifuging method.
[c] Determined by immersion at 22°C from an initial conditioning at 65% RH, 22°C.
[d] For comparison, acetate = 6–8%; cotton = 40–42%; wool = 25%.

Heat treatment improves resistance to fading in part because of redistribution of dye away from the outer surfaces and in part because a denser structure develops. Use of selected gas-fast dyes and of codyeing inhibitors provides a practical level of protection (39).

For a relatively nonreactive polymer like cellulose triacetate, finishing processes generally involve insolubilization of materials on the surface of the individual fibers. For example, silicone resins can be applied during heat treatment to impart durable water repellency and spot resistance. A combination of silicone resin with dispersions of fine-particle-size silica increases fiber-to-fiber friction and minimizes yarn slippage. Fluorinated polymers and antistatic finishes must be applied, if at all, after heat treatment.

A finishing treatment unique to cellulose triacetate is saponification of the fiber surface with alkali to yield a film of regenerated cellulose. This surface film results in antistatic properties, reduced soilage, higher resistance to fading, better hand, increased safe ironing temperature, greater tearing strength, and versatility in choice of resin finishings.

References

1. K. C. Laughlin, *Man-Made Fibers, Sci. Technol.*, **2**, 103 (1968).
2. F. S. Sherman and P. V. Kozlov, *Zh. Prikl. Khim.*, **26**, 524 (1953); through *Chem. Abstr.*, **47**, 10839 (1953).
3. J. D. Crowley, G. S. Teague, Jr., and J. W. Lowe, Jr., *J. Paint Technol.*, **39**, No. 504, 19 (1967).
4. P. Howard and R. S. Parikh, *J. Polymer Sci.*, [A–1], **4**, 407 (1966); *ibid.*, **6**, 537 (1968).
5. R. J. Brewer, L. J. Tanghe, S. Bailey, and J. T. Burns, *J. Polymer Sci.*, [A–1], **6**, 1967 (1968).
6. K. H. Bischoff and B. Philipp, *Faserforsch. u. Textiltech.*, **17**, (3), 115 (1966).
7. P. J. Flory, O. K. Spurr, Jr., and D. K. Carpenter, *J. Polymer Sci.*, **27**, 231 (1958).
8. H. J. Notarius and C. P. Driscoll, Celanese Research Co., unpublished results.
9. R. J. E. Cumberbirch and W. G. Harland, *J. Textile Inst.*, **49**, T279 (1958).
10. A. Sharples and H. M. Major, *J. Polymer Sci.*, **27**, 433 (1958).
11. F. Howlett, E. Minshall, and A. R. Urquhart, *J. Textile Inst.*, **35**, T133 (1944).
12. N. I. Klenkova, O. M. Kulakova, N. D. Tzimara, N. A. Matvennya, and E. N. Khlebosolova, *J. Appl. Chem. USSR*, **41** (3), 620 (1968).
13. H. W. Steinmann, U.S. Patent 3,386,932 (1968) (to Celanese Corp.).
14. C. S. H. Chen, S. Jankowski, and A. Brother, in *Symposium on Irradiation of Polymers, Advances in Chemistry Series*, 66, American Chemical Society, Washington, D.C., 1967.
15. R. W. Singleton, R. K. Kunkel, and B. S. Sprague, *Textile Res. J.*, **35**, 228 (1965).
16. C. Driscoll, Celanese Research Co., unpublished results.
17. S. L. Madorsky, *Thermal Degradation of Organic Polymers*, Interscience, New York, 1964, pp. 244–261.
18. F. Zitomer and A. H. DiEdwardo, Celanese Research Co., unpublished results.
19. F. Zitomer, *Anal. Chem.*, **40**, 1091 (1968).
20. J. K. Gillham, *Appl. Polymer Symp.*, No. 2, 45 (1966).
21. C. M. Conrad and P. Harbrink, *Textile Res. J.*, **33**, 784 (1963).
22. *Chem. Eng.*, **72**, 106 (Nov. 8, 1965).
23. H. W. Steinmann, U.S. Patent 2,849,444 (1958) (to Celanese Corp.).
24. R. W. Singleton, M. A. Sieminski, and B. S. Sprague, *Textile Res. J.*, **31**, 917 (1961).
25. Z. A. Rogovin, *Khim. Volokna*, No. 1, 48 (1961).
26. R. G. Quynn, *Textile Res. J.*, **33**, 21 (1963).

27. J. L. Riley, U.S. Patent 3,057,039 (1962) (to Celanese Corp.).
28. J. W. Soehngen, U.S. Patent 3,057,038 (1962) (to Celanese Corp.).
29. C. Cipriani and J. W. Soehngen, U.S. Patent 3,109,697, (1963) (to Celanese Corp.).
30. R. G. Stoll, *Textile Res. J.,* **25**, 650 (1955).
31. B. S. Sprague, J. L. Riley, and H. D. Noether, *Textile Res. J.,* **28**, 275 (1958).
32. J. J. Creely and C. M. Conrad, *Textile Res. J.,* **35**, 184 (1965).
33. W. J. Dulmage, *Nature,* **172**, 1053 (1953); *J. Polymer Sci.,* **26**, 277 (1957).
34. R. St. J. Manley, *J. Polymer Sci.,* **47**, 509 (1960).
35. R. L. Foley, Celanese Research Co., unpublished results.
36. D. K. Beever and L. Valentine, *J. Textile Inst.,* **49**, T95 (1958).
37. F. Fortess, *Am. Dyestuff Rep.,* **44**, 524 (1955).
38. W. D. Paist and N. Van Gorder, U.S. Patent 2,662,834, (1953) (to Celanese Corp.).
39. A. F. Tesi, *Am. Dyestuff Rep.,* **45**, 512 (1956).

D. PAPER FROM CHEMICALLY MODIFIED CELLULOSE

Kyle Ward, Jr., and Leo J. Bernardin

In 1969 the world production of paper was 139,896,000 short tons (1). This is the greatest single use of cellulose in the purified or semipurified state, and it is somewhat surprising that the paper industry was affected so slightly by the widespread investigation of chemically modified cellulose in the first half of the twentieth century. In the first three parts of the second edition of this work there are only two places (2, 3) where chemical modification (methylation and acetylation) is mentioned in regard to paper, at least in the sense used in this chapter. In the last 15 years, however, more attention has been paid to this field, and chemically modified cellulosic fibers are now used industrially. They are, of course, more expensive to produce and so find their present utility only in certain specialty papers where their properties justify the cost. If one can judge from the great research activity, this type of product will attain considerably more commercial importance in the future.

This section represents an effort at thorough coverage of the research on, and utilization of, partially esterified or etherified, crosslinked, or grafted cellulosic fibers in the fields of paper and board. We are limiting our definition of the term "chemically modified" to the types just mentioned, and are excluding hydrolytic and oxidative treatments. We are also eliminating such essentially swelling treatments as mercerization, parchmentizing, and the production of vulcanized fiber. Neither do we discuss the use of nonfibrous cellulose derivatives, such as methylcellulose and carboxymethylcellulose in paper. Finally, we do not try to cover man-made fibers from cellulose. These are used in certain specialty papers and especially in nonwoven fabrics, which are structures usually held together by adhesives rather than by fiber-to-fiber bonds; we deal with the subject only where the emphasis seems to fall on the chemical modification of the cellulose in the fiber.

1. Fiber Esterification and Etherification

(a) IMPROVEMENT OF BONDING AND STRENGTH PROPERTIES

Jayme and Froundjian (4) demonstrated that slight methylation of pulp appreciably increases paper strength. A few years later Bletzinger (5) and Aiken (6) showed a similar effect with acetylation. Papers from acetylated pulp are indeed in use today, but the main interest is not in the strength properties. Walecka (7) showed that substitution by hydrophilic groups greatly magnifies the increases in sheet strength.

(1) Acetylation

Acetylation may be carried out either on the pulp or on the sheet [see subsection 1(b)(*1*)]. Acetylation of the sheet improves wet strength, but the effect on dry strength is not so marked. Klinga and Back (8) found that dry tensile strength increases at low levels of paper esterification and then decreases (Fig. 1); Klinga and Tarkow (9) found a similar increase with hardboard, but without the decrease at higher acetyl contents. Bletzinger (5), Aiken (6), and Higgins and co-workers (10–13) found that the strength of paper from beaten acetylated pulps is also increased only at low substitutions and that beyond this point it decreases sharply (Fig. 2); unbeaten pulps (5, 14) behave differently.

The usual explanation for the effect of slight acetylation on handsheet properties is that the introduction of a few acetyl groups opens up the fiber structure and prevents hydrogen bonding between cellulose hydroxyls in the vicinity of the acetyl groups, thus facilitating moisture sorption and swelling. At higher degrees of substitution (DS) the decrease in hydroxyl content outweighs

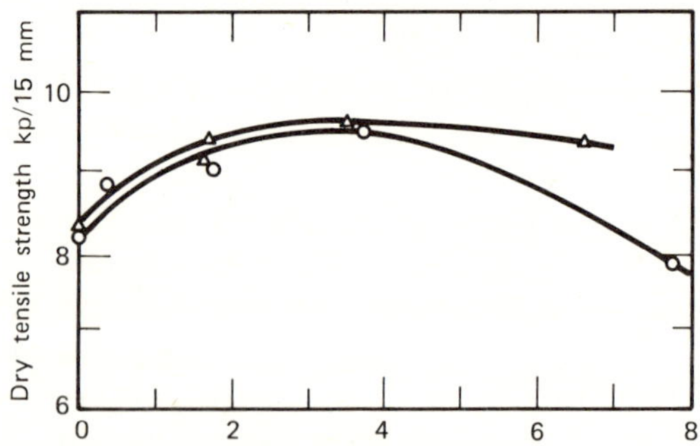

Fig. 1. Dry (65% relative humidity) tensile strength of two sulfate papers as a function of their apparent acetyl content (8).

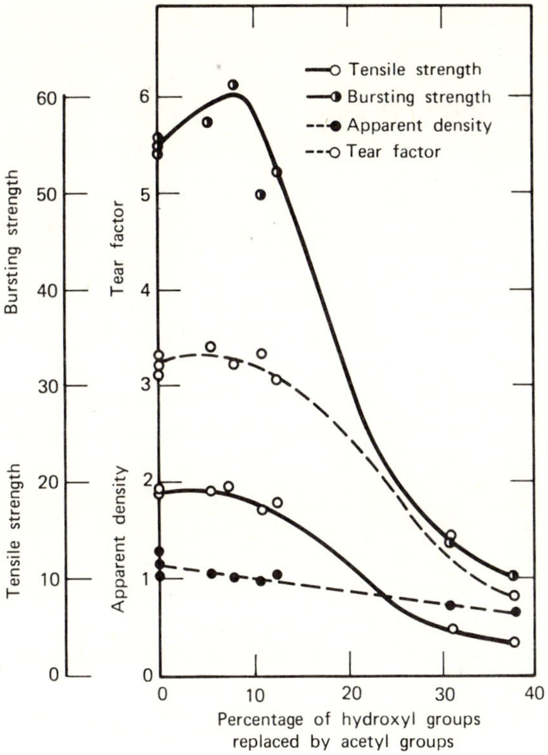

Fig. 2. Effect of acetylation on the strength and apparent density of beaten rag stock (5). (Reproduced by permission of the American Chemical Society, the copyright owner.)

this. This explanation is not sufficient for the case where fiber-to-fiber bonds are formed before acetylation, as when the paper sheet itself is acetylated. Klinga et al (8, 9, 15) explain the increased strength in this case by the formation of crosslinkages of the type proposed by Back and Klinga (16) to explain the action of heat on paper; that is, the formation of hemiacetals between carbonyl and hydroxyl groups. This would also be a major factor, if accepted, in explaining the increased wet strength.

Schulwitz (17) improved the strength properties of paper made from acetylated pulp of 15–30% acetyl content (DS 0.7–1.6) by moistening the sheet with an organic liquid such as methylene chloride, pressing, and drying. This procedure has been modified by others to improve opacity by incorporating titanium dioxide into the impregnating liquid (18) or by using an emulsion (19) or a solution (20) of a cellulose ester as the impregnant or coating. Sizing of various types has also been used to increase the strength of paper from acetylated fibers (21). Herdle and Griggs (22) have described the properties and potential uses of partially acetylated fibers (DS ~ 1.0) in some detail. At this DS interfiber bonding is low, causing lower folding endurance and bursting strength. However, tearing and wet tensile strengths, opacity, and stiffness are higher.

(2) Cyanoethylation

The explanation advanced to explain the handsheet properties of acetylated pulps predicts that practically any substituent in small amounts will improve sheet strengths, provided it does not cause degradation or crosslinking. One instance of this is cyanoethylation.

Cyanoethylated paper is used largely when heat resistance and electrical resistance are important requirements. Certain strength properties are also improved, and that aspect is mentioned here, although others will be covered in more detail in subsection 1(b)(*2*).

The data of Harpham et al. (23) show that cyanoethylation of linters produces increases in bursting, tearing, and folding strengths at a lower level of substitution (DS 0.06). Spadaro and co-workers (24) obtained increases in strength at a DS value of about 0.2 with decreases as substitution increased further, but these were considered to be insignificant.

Morton and Bikales (25) used kraft pulps at higher substitutions (DS 0.2–0.6). They found a slight increase in tensile strength at a DS of about 0.19 with a regular decrease at higher substitutions. They are doubtful about the

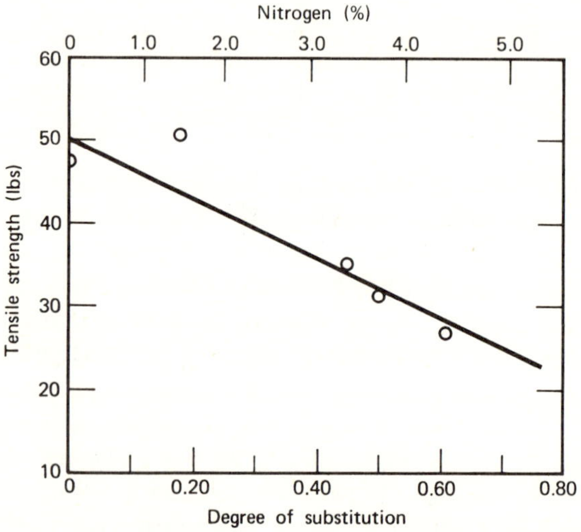

Fig. 3. Tensile strength of cyanoethylated handsheets as a function of the degree of substitution (25).

significance of this maximum (Fig. 3), for in the case of bursting strength there was no such maximum. They pointed out, however, that Spadaro found optimal properties, especially folding endurance, for cyanoethylated linters at about this substitution. McKay and Bikales (26) later found a similar effect. This maximum is reasonable; cellulose etherified with other hydrophobic constituents (like methyl or acetyl groups) shows just this type of behavior: increased affinity for water at low DS values and a subsequent decrease. This increased moisture sorption results in swelling during beating and higher strengths in the sheet.

Mizutani et al. (27) obtained increased folding strengths and elongations with a cyanoethylated kraft pulp of somewhat higher DS (0.4). They found, however, slight decreases in tearing and tensile strengths. Morton and Bikales (25) found a decrease in bursting strength that is linear with nitrogen content. With blends as well, of untreated kraft plus cyanoethylated pulps, they found a decrease; the plots are concave upward (Fig. 4), indicating that the blends are weaker than

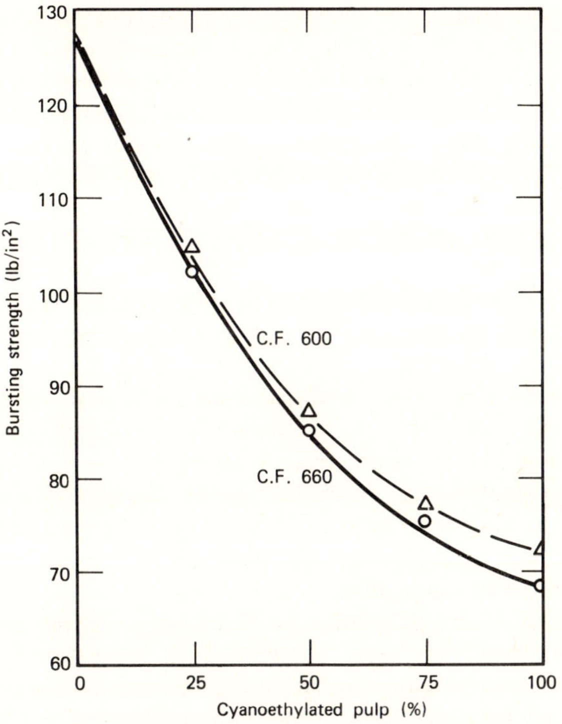

Fig. 4. Bursting strength (Mullen) of handsheets made from mixtures of bleached sulfate, at two Canadian freeness values, and cyanoethylated bleached sulfate (4.9% N) (25).

unblended cyanoethylated pulps at the same nitrogen level. Such blends have been patented (28) for making extensible paper. The bulk of the furnish is untreated wood pulp, but 1-10% of cyanoethylated, hydroxyethylated, or carboxyethylated pulp is also used. Specially regenerated cyanoethylated cellulose may replace the cyanoethylated pulp (29). Jayne (30) has shown that the strength of the web from cyanoethylated fibers can be enhanced by heat bonding while it still contains at least 10% moisture. Wink et al. (31) found a similar effect on the dry sheet to play a part in the improvement of thermal stability. This will be discussed in subsection 1(b), "Improvement of Heat Resistance and Electrical Properties" (see also Section D in Chapter XVII).

(3) Hydroxyethylation

The use of hydroxyethylated fibers had been proposed by Richter (32) in 1936 and by Schorger (33) in 1950, but there was no evidence of commercial use until announcements in 1957 and 1958 of the availability of hydroxyethylated linter pulp (34, 35). The level of hydroxyethylation used for papermaking pulps is very slight (23, 36, 37); although the improvement in properties increases with substitution, the increasing solubility and swelling in water cause drainage on the machine to slow down and eventually prevents sheetmaking. The molar substitution (MS) of the hydroxyethylated pulps is generally from 0.15 to 0.2; at these low substitutions the difference between MS and DS is probably negligible.

Hydroxyethylation can be carried out on an alkali cellulose prepared from the pulp by steeping in aqueous sodium hydroxide and then pressing or centrifuging to remove excess solution. This alkali cellulose is then reacted with ethylene oxide (23, 36, 37) or ethylene chlorohydrin (37) at room temperature. The pulp is then neutralized and washed to remove the by-products of the reaction. Superior results (36, 38-40) are obtained by using raw linters instead of purified linters and carrying out conventional alkaline digestion and bleaching on the hydroxyethylated or otherwise modified pulps. In this case the slight change in crystallinity that occurs in purifying untreated linters is absent and the x-ray pattern at the low substitutions in question is the same as that of linters treated with alkali alone (36, 38, 41). Mitchell (42) has proposed a continuous process of hydroxyethylating fibers.

The effect of hydroxyethylation on the sheet properties is very marked, especially for pulps that form poorly bonded sheets, such as linters or mercerized wood pulps. Not only are tensile strength, folding endurance, and bursting strength increased but so is the tear factor (see Fig. 5) (36, 37, 39, 40). This has been shown to be the result of greatly enhanced bonding (43). In bonding strength hydroxyethylated linters surpass not only ordinary linters but also rag stock (44). Part of the increase, but not all of it, is due to the presence

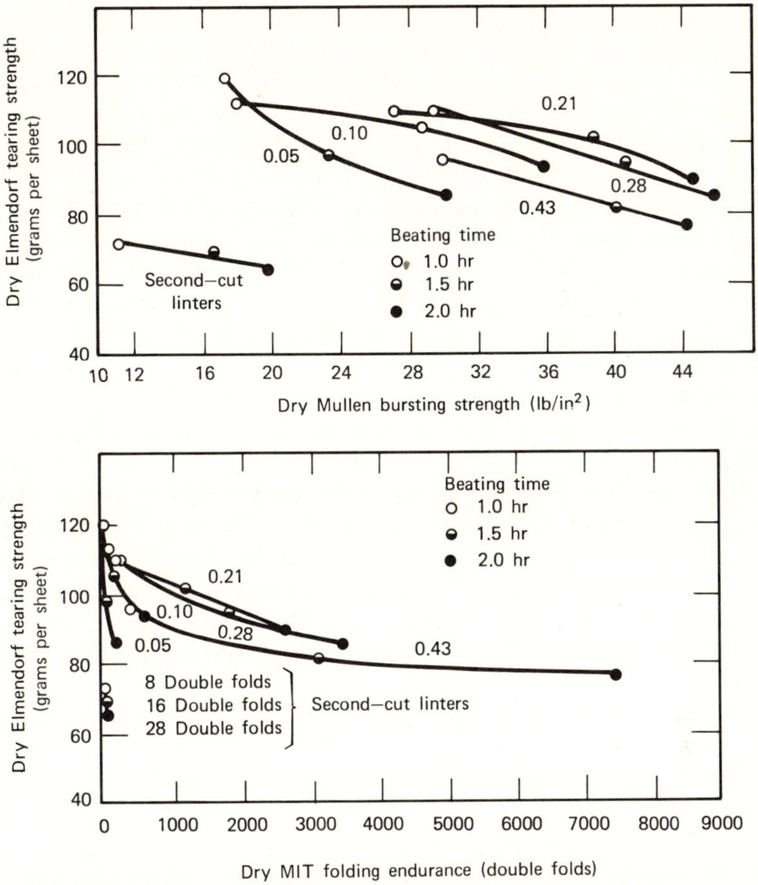

Fig. 5. Valley beater curves of hydroxyethylated second-cut cotton linters. Raw linters hydroxyethylated prior to purification. Numbers on curves denote DS; handsheet basis weight = 40 lb (24 x 36–500 sheet ream) (36).

of fines (44, 45). The development of strength during beating is more rapid than with untreated fibers, and the rate increases with the degree of etherification (36). Sheet formation is as good (43) or better (36) than in the case of the untreated material. Resistance to thermal breakdown is also excellent (36); this is discussed in subsection 1(b). Both Harpham (36) and Ward (37) have shown that tearing strength increases to a maximum and then falls off with increasing DS. Harpham also reported that this applies to bursting strength as well.

The moisture content of handsheets made from hydroxyethylated linters is unchanged according to Harpham et al. (36), although Klenkova (46) found a very considerable increase in the affinity for moisture on hydroxyethylation. This is probably due to the concentration of alkali used in etherification, for Ward and co-workers (47) have shown that this is the most important factor governing the moisture content of hydroxyethylated pulps. Harpham et al. (36) also found that tearing strength is increased by increasing the relative humidity. Bursting strength is affected very little, and the results for folding endurance are inconclusive. The pulp sizability is improved by hydroxyethylation (39). The ink resistance of handsheets is greatly increased even with a reduction in the amount of rosin used for sizing (36). Hydroxyethylated linters are stronger than untreated linters at the same porosity level and more porous at the same strength. It is possible to increase both properties at once by hydroxyethylation (36).

Harpham (36) found no change in the refractive index on hydroxyethylation. The refractive index of the hydroxyethylated fiber was 1.555 and of the sheet, 1.561. However, one very important optical property of the sheet is changed by hydroxyethylation. Both Spadaro (24) and Ward (37) observed that the opacity drops off markedly; at higher MS the sheets are translucent. This is another result of the increased bonding mentioned earlier. Plunguian (48) has indicated how this effect can be increased to produce transparent papers. Harpham (36) found little change in density with hydroxyethylation. Spadaro (24) found a decrease except at high (0.34) MS. Ward (37) found considerable scatter but, in general, an increase (Fig. 6). Thomas (49), working with kraft pulps instead of linters, also found increased density.

High-DS hydroxyethylated cellulose can be added to untreated fibers as a beater adhesive to give an improved product (48, 50). A similar effect occurs in blending less highly hydroxyethylated fibers with linters (51, 52); sheets so produced are stronger than those from uniformly hydroxyethylated fibers of the same MS (53). Beyer et al. (52) found the best blend with linters to be that using fibers hydroxyethylated after steeping in 14% alkali. The Buckeye Cellulose hydroxyethylated product was particularly recommended for cotton–wood pulp blends (54). Didwania (45) has shown that the increased bonding with hydroxyethylation is a result partly of an increase in relative bonded area, but much more of an increase in specific bonding strength (increased conformability of the fiber, increasing the number of actual bonds per unit of bonded area). Gillespie et al. (41) showed that, in linters swollen with aqueous alkali, hydroxyethylation is not limited to the surface; in fact, when mercerizing strengths of alkali are used, the hydroxyethylating agent penetrates the crystal lattice. This may not be necessarily true if nonswelling media are used; for instance, Spadaro (24) used carbon tetrachloride solutions of ethylene oxide, but even in this case the linters were first swollen with 12.5% sodium hydroxide. Akim and Nikitin (55) found no change in the x-ray pattern for hydroxyethylation to low substitutions of both mercerized and nonmercerized pulps.

Both Morak (37, 43), using alcohol, and Gillespie (41), using alkali, were able

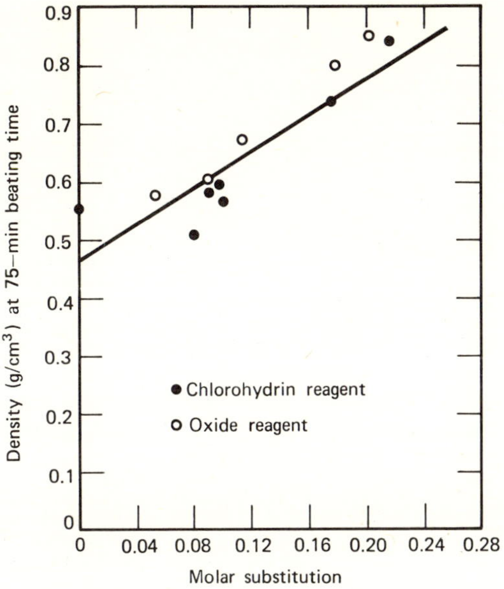

Fig. 6. Relationship of density to molar substitution in hydroxyethylcellulose (37).

to fractionate pulps into a soluble material of higher MS and an insoluble residue of lower MS. Gillespie showed that surface phenomena are important, even though the reaction does occur within the fiber as well. He omitted the final neutralization step and washed the alkaline pulp directly. Much alkali-soluble material was removed, but the product still had a high MS. However, its strength properties were only slightly better than those of the untreated linters and were greatly inferior to those of a hydroxyethylated sample of the same MS that had been neutralized in the usual way. This seems to indicate that the concept of a "built-in" beater additive is valid. Hydroxyethylation occurs throughout the fiber, but the major improvement in bonding is due to alkali-soluble material adsorbed on the fiber during the neutralization step.

Ward and co-workers (47) have studied the effect of the concentration of the alkali steep on the properties of the product. If the concentration is nonswelling below mercerization strength, the bonding as reflected in the handsheet properties is less at the same substitution than it is when the concentration is higher (see Fig. 7; the breaking length refers to the free-hanging length that will cause a tensile failure). In fact there is, in general, an increase in strength with increased concentration of alkali, but there is a more marked difference between nonmercerized and mercerized fibers. If mercerized fibers are hydroxyethylated at lower concentrations of alkali in a two-stage process, the various physical

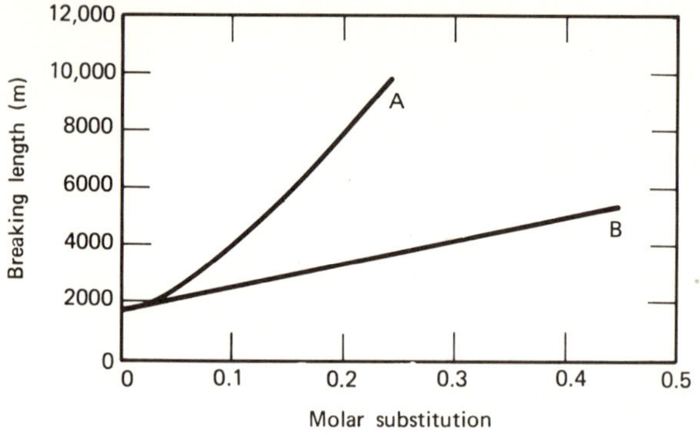

Fig. 7. Effect of hydroxyethylation on breaking length; 75-min. beating (47). Curve *A*: 10–30% NaOH; curve *B*: 1–5% NaOH.

properties respond differently, and in this way one may achieve a balance between, for example, folding endurance and opacity. However, hydroxyethylation with resulting improvement in papermaking properties has been effected with concentrations of sodium hydroxide as low as 1% (47, 55) and as high as 30% (47).

Battista, Cruz, and Reichel (56) have patented papers from regenerated fibers of low-MS hydroxyethylcellulose produced by spinning the alkali solution into a suitable bath. Data are given for handsheets of the treated fibers alone, but most of the examples are for blends with wood pulp.

As already mentioned, hydroxyethylation of mercerized wood pulp improves the bonding strength. Thomas (49, 57) has brought out that this makes it feasible to take advantage of the improved purity and absorbency of mercerized pulps and still obtain the requisite strength that such pulps lack. Since Thomas' product is recommended particularly for thermal stability, it will be discussed in more detail under that heading. It is only necessary to say here that hydroxyethylation of kraft pulps (49, 58–60) brings about the same changes that have been discussed for linters pulps. Thomas did find one difference—namely, that freeness is developed more slowly than it is in the original or in the mercerized pulp. This is probably due to the strong alkali used (18% sodium hydroxide); Ward (58), using 10% sodium hydroxide, found the same rapid development of freeness with kraft as with linters, but a single experiment with 20% alkali showed much slower development. Thomas' sheets are more porous to air at low levels of beating, but at 300-ml freeness the air resistance has increased considerably. Photomicrographs indicate separation of the primary wall on beating, but Thomas attributes much of the bonding to compaction of the conformable secondary wall. This agrees with the views of Ward (47) and Didwania (45) on hydroxyethylated linters (see also Section C in Chapter XVII).

(4) Carboxymethylation

Walecka (7) used carboxymethylated rag pulp to test the hypothesis that hydrophilic substitution of pulps should increase the strength of handsheets and found this to be true. The etherification can be carried out in a number of media. Walecka (7) and Ward et al. (61) found that the reaction of alkali cellulose from aqueous alkali with sodium chloroacetate is not suitable because only at relatively high DS does one overcome the deleterious effect of alkali swelling on sheet strength. Harpham and co-workers (23, 36), however, using a less alkaline solution of sodium chloroacetate and sodium hydroxide, were able to get improved sheet properties at low DS values (0.06). Walecka (7) steeped his pulp in an 2-propanol solution of chloroacetic acid and then reacted it in methanolic 2-propanol containing a very slight excess of alkali. The pulp was separated, neutralized with dilute aqueous acetic acid and converted to the sodium-salt form with sodium bicarbonate solution. Later work (61, 62) has shown that dimethyl sulfoxide as the reaction medium gives much higher substitutions than 2-propanol. The DS rather than the reaction medium seems to govern the level of strength improvement.

The effects of increasing DS up to 0.06 are to increase the rate of beating, the bursting strength, the tensile strength, and the zero-span tensile strength (TAPPI Standards T–481 SM 60 and T–231 SM 60; an index of individual fiber strength), and to decrease the opacity (7). The tearing strength at constant freeness goes through a maximum. Both bonding and moisture adsorption are increased. Carboxymethylated fibers fibrillate well on beating and have high bonding strength. Walecka suggests that the latter is due not only to fibrillation but also to increased affinity for water and to improved plasticity (even unbeaten pulps are stronger than the beaten untreated pulps). The increased zero-span values and increased fiber transparency indicate that intrafiber bonding has also occurred. Talwar (63), however, could not find any increase in single-fiber breaking loads.

Talwar (63) showed the conversion to the salt form to be a very important step in the process. The free-acid form is highly flocculated, whereas the sodium-salt form has much better dispersion, lower freeness, higher filtration resistance, and increased strength. These differences relate to the affinity for water. The sodium salt of carboxymethylcellulose is water soluble at a DS at which the free acid is insoluble, and it was found that improvement in sheet strength could be entirely lost if the liquid in wet sheets (sodium-salt forms) was displaced with acid or alum solutions before drying (Table 1). The effect of pH is reversible. Small amounts of material of higher DS can be separated by extraction with water at pH 8.5, but the DS of the residue is changed very little. If 3% sodium hydroxide is used for the extraction, slightly more material is extracted, but the residue contains more carboxyl groups, probably due to atmospheric oxidation during the extraction.

TABLE 1
Properties of Handsheets of Carboxymethylcellulose Treated with Various Solutions (63)

DS of pulp	Solution	Apparent density[a]	Mullen bursting strength[b]	Tensile strength[c]	Opacity (%)
Untreated control	–	9.12	35.1	8.80	82.7
Treated control	–	9.37	40.3	9.98	82.0
0.017	Alkali (pH 9.0)	9.60	48.6	12.35	79.5
0.017	0.001% alum	8.74	37.7	9.60	81.0
0.017	Acid (pH 3.5)	9.18	41.7	9.85	81.2
0.044	Alkali (pH 9.0)	10.53	78.1	16.76	75.5
0.044	0.001% alum	9.16	39.0	9.65	81.2
0.044	Acid (pH 3.5)	9.25	41.4	9.80	81.3

[a]Apparent density $= \frac{\text{basis weight}}{\text{thickness}} = \frac{\text{lb/ream}}{\text{mils}} = \frac{\text{lb/500,000 in.}^2}{\text{mils}}$; (1 mil = 0.001 in.).

[b]Bursting strength in pounds per square inch measured according to TAPPI Standard Method T–403 TS–63; values in table are for psi/100 lb of basis weight.

[c]Tensile-strength units: pounds per inch of width for a standard-weight sheet. It is generally given for paper as the ratio of mass to width because of the difficulties inherent in measuring the cross-sectional area of a sheet of paper.

Talwar found the compressibility and the specific volume to increase with substitution for the unbeaten pulps, although the specific surface is little affected. The filtration resistance is lowered if tap water is used instead of distilled water or if an excess of alum is used. Freeness is lowered by alum at low concentrations, but when the concentration reaches about 3×10^{-5} moles/l, the trend is reversed and freeness rises again.

Some information on the causes of these strength increases was also obtained by Talwar (63). Formation, bonding strength, and relative bonded area are all increased by carboxymethylation. In the acid-displaced sheets, however, only formation is increased. Bonding strength and relative bonded area, like bursting strength, are no different from the untreated control if the sheets are in the acid form.

Carboxymethylation of linters pulps is carried out before purification, according to Turner (40) and Harpham (59). Carboxymethylation of wood pulps for papermaking has also been considered. In one series of patents pine soda pulp was carboxymethylated in an aqueous medium (64), but usually variants of Walecka's procedures are used (65-67).

Eucalyptus kraft pulp was treated (65) in methanolic 2-propanol (aqueous solutions were also used, but the DS so reached was very low). Bursting, tensile, and tearing strengths were all improved in the unbeaten samples, but the advantage of the carboxymethylated samples was very slight after beating. Blends with untreated pulp were slightly better than an unblended stock of the same DS. Nelson and Kalkipsakis (66) studied a number of salts of the carboxymethylated eucalyptus kraft pulp. In the unbeaten state the acid form and all 16 salts studied gave sheets that were stronger than the untreated pulp, except the two trivalent salts (chromium and iron). All the other salts except for barium were stronger than the acid form. Only the divalent salts responded to beating, however; in fact the monovalent-salt forms were slightly weaker than the acid form. These statements apply to the relatively high DS of 0.11. A few salts were also prepared at lower DS, in which cases sodium, potassium, and ammonium all responded well to beating and gave the highest strengths when beaten. The authors have pointed out that the strengths of unbeaten sheets in the salt form are in the same order as the degree of hydration of the cations in water, in the series of monovalent cations.

Ruffini (67) has corroborated the improved strengths obtained by the carboxymethylation of both sulfite and sulfate pulp. His data indicate that some esterification of hydroxyl and carboxyl groups may occur on drying the sheet. Such esterification may give stronger bonding, improved wet strength, and some degree of water repellency.

An interesting product that has recently appeared on the market consists of soluble or readily disintegrating paper. There is an early patent by Cady and Steinberg (68) covering a disposable diaper in carboxymethylated fibers, but this seems to have been making use only of the increased absorbency of the fibers; in fact insolubility in water is stressed. The later material seems more of the nature of that described by the Mishima Paper Manufacturing Company (69). Carboxymethylated fiber in the acid form is made into a sheet and given a heat treatment that increases the wet strength enough for the sheet to be impregnated with alkali (sodium bicarbonate, carbonate, or hydroxide) to convert it into the salt form. The paper is then squeezed and dried. The speed of solution or disintegration can be controlled by the nature of the alkaline treatment as well as by the stock, which may include other fibers as well as the carboxymethylcellulose. The Dissolvo paper manufactured by the Gilbreth Company (70) has these same characteristics, but the details of its manufacture have not been disclosed. A later patent describes a high-wet-strength paper that disintegrates in alkali and is prepared from ordinary pulp blended with a small amount of water-insoluble, alkali-soluble carboxymethylated fiber (71).

(5) Miscellaneous Modifications

Many other chemical modifications have been proposed as possibilities for bringing about higher strengths in papermaking, some of which are related to those already discussed. Harpham has shown that hydroxypropylated (23, 36, 59) and hydroxybutylated linters (39, 40) are effective, as well as the hydroxyethylated linters described above. It appears from the data that both the extent of etherification and the improvement in properties drop off as the length of the chain increases. The porosity of hydroxypropylated wood pulp is improved at least as much as that of hydroxyethylated samples. Battista, Cruz, and Reichel (56) have also given examples of paper handsheets made of blends of wood pulp and low-MS hydroxypropylcellulose. The latter was in the form of fibers spun from an alkaline solution. Ward et al. (50) have experimented with polyglycol chlorides, obtaining similar or possibly superior products to those of hydroxyethylation except that for a given DS, the MS was larger.

Homologs of carboxymethyl ethers have also been proposed. Data on carboxyethylated linters (prepared from acrylonitrile with a subsequent caustic treatment) show high strength properties even at a DS as low as 0.01 (40, 59). Sheets have been made from blends of viscose rayon and regenerated carboxyethylcellulose or from the carboxyethyl fibers alone (56, 72). The carboxyethylation of various types of pulps has been most thoroughly studied by Baker and Sepall (73). It can be achieved by a secondary hydrolysis after reaction with acrylamide, acrylonitrile, or an acrylic ester. Baker and Sepall used acrylamide as the reagent and chose such reaction conditions that the secondary hydrolysis step was included in the preparation. The product contained both carbamoylethyl and carboxyethyl groups, and the ratio of these groups depended strongly on caustic concentration (Fig. 8). The pulps were used in the sodium-salt form (see also Section D in Chapter XVII).

The bursting, tensile, and folding strengths of the unbeaten carboxyethylated pulps are increased with increasing DS, resembling the results obtained by beating unsubstituted pulp, but tearing strength decreases with increasing DS, and there is also a loss in opacity. Gains in porosity, equilibrium moisture content, and stretch were observed. All these changes are similar to those observed with carboxymethylation and can be similarly explained. The carboxyethylated pulps have remarkable resistance to the deleterious effects of drying, and air-drying of pulps produces a slight improvement in beating response, which would be of great value in connection with the reuse of broke. It would be interesting to know if these pulps also exhibit the ester formation observed by Ruffini (67) on drying of carboxymethylated pulps. Miller (74) has recently patented a process for preparing, by shear precipitation, a regenerated-cellulose fiber that contains cyanoethylated groups and can be converted to paper, either with or without the addition of other fibers. Andersson (75) investigated the effect of methylation on a sulfite pulp. He found no increase in the flocculation tendency but corroborated the earlier observation of Jayme and Froundjian (4) that dry strength increases.

Cationic pulps prepared by McKenzie (76) showed no loss in strength; they are discussed below in regard to ion-exchange properties. Ward et al. (50) tested several modified linters pulps containing hydrophilic groups. Carbamoylethylation was carried out with acrylamide under conditions that cause little carboxyl-group formation (low alkali concentration and brief heating). Aminoethylation was achieved with aminoethylsulfuric acid. Both of these gave marked increases

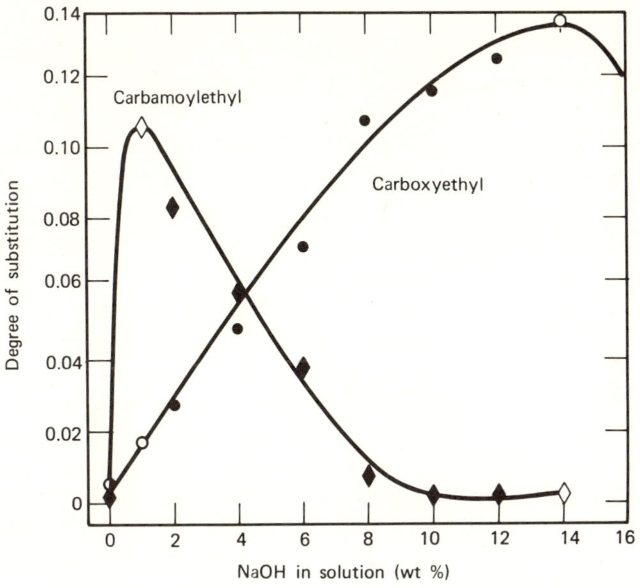

Fig. 8. Effect of caustic concentration on DS of bleached kraft pulps prepared by reaction with acrylamide at 93°C (73). (Reproduced by permission from the Canadian Pulp and Paper Association, Technical Section.)

in folding endurance and tensile strength, but not as striking as those obtained with the reaction product from β-propiolactone. Bikales and Stephens (77) have also suggested the use of papers from carbamoylethylated and related pulps. An unusual chemical modification has been proposed by Neely (78), who treated cellulosic fibers with sucrose in the presence of a dextran sucrase. The enzyme introduces glucosyl side chains into the cellulose. Paper made from the modified fiber has greatly increased strength.

Substitution of a few of the hydroxyl groups in cellulose with any of a great variety of substituents will improve sheet strength. By varying the substituent and the process of substitution it is possible to obtain different combinations of properties. Hydrophilic substituents give particularly good strength properties over a relatively wide range of DS (23, 36, 59, 79).

(b) IMPROVEMENT OF HEAT RESISTANCE AND ELECTRICAL PROPERTIES

For many uses it would be advantageous to have paper with higher resistance to thermal deterioration. A simultaneous improvement in heat resistance and electrical properties is particularly desirable for electrical insulation. It has been known for some time that partial acetylation of cellulosic textile fibers can be used for this purpose (80). The extension of this procedure to paper and the development of other treatments is of more recent origin.

(1) Acetylation

The use of acetylation for paper started about two decades ago for insulating papers. The first processes probably involved acetylation of the paper sheet (81, 82), but the patents involved (83) cover both fiber and sheet acetylations. Acetylation of the preformed sheet is of two types: the sheet may be immersed in the reacting solution or the solution may be pulled through the paper. The solution is usually a mixture of acid and anhydride in the presence of a catalyst, such as sulfoacetic or perchloric acid. Richter and Herdle (84) used the immersion procedure, which has also been used in small-scale acetylations, such as the acetylation of filter paper for chromatographic use; this is discussed in subsection 1(c), "Decreased Affinity for Water." Many of the patents for photographic papers (85, 86) or other types (87) use the pull-through process. The properties of acetylated paper depend on the method of acetylation (88). In the manufacture of paper from acetylated pulp the wood pulp to be acetylated is activated with water or fairly concentrated acetic acid and then acetylated with acetic anhydride and a catalyst (89–91). Rayonier (92) has patented vapor-phase acetylation of pulp with uncatalyzed anhydride (see also Section B in Chapter XVII).

One of the advantages of acetylated paper, by whatever procedure it is made, is the lowered affinity for water at the acetyl content generally used (DS 0.7–1.8%) (22, 81, 82, 93). This aspect is further discussed in subsection 1(c), but it is mentioned here because the slowness of the acetylated paper in picking up moisture is responsible for much of its utility in electical insulation. Not only does insulation resistance increase with increased acetylation at a given moisture content but also the acetylated paper has a lower moisture content. In alternating-current circuits acetylated paper has a lower power factor and a lower dielectric constant, which, in contrast to the behavior of untreated paper, does not change rapidly with relative humidity or with temperature (81, 82). Katsuura and Suzuki (94) have concluded that thorough delignification is more effective than acetylation for improving electrical papers, but Kelk and Wilson (95) have considered the possibility that small amounts of lignin may have some beneficial effects.

In addition to its effect on the electrical properties (96), acetylation also improves heat resistance. Dieterle (81) showed that the drop in tearing strength in air or oil is slower for acetylated paper than for ordinary paper at 150 and 200°C. Back and Klinga (16) also concluded that acetylation stabilizes board against thermal degradation. Korchemkin et al. (97) have corroborated both the thermal stability and the electrical properties. For photographic uses partially acetylated paper may be treated wth antistatic material (98).

There has also been interest in the acetylation of groundwood. Electrical resistance is not a factor in the use of groundwood, but heat resistance may be. Of great importance is the fact that the brightness and the color stability may be improved in this way (99, 100). Loras (99) proposes three possible mechanisms: (a), that carbohydrate acetylation stabilizes against heat, and lignin acetylation against irradiation; (b), that the stabilization in bleaching is due to peroxides formed from the carbohydrate acetates; and (c), that the acetylation of lignin makes photochemical bleaching applicable at lower wavelengths. In this connection it should be noted, however, that Herdle and Griggs (22) also noted improved color stability on acetylation of pulps that did not contain the large amounts of lignin present in mechanical pulp. Manchester and co-workers (100) concluded that acetylation brightens the lignin, but that stabilization by the treatment depends on the hemicellulose portion. They also found that methylation with diazomethane gives substantial brightness improvement on groundwood, but not on holocellulose. Methylation does not improve color stability, however. On the other hand, methylation of acetylated pulps gives improved brightness as well as good stability. Manchester's acetylating bleach reduced pulp strength, and the brightening might arise from increased light scattering due to reduced bonding. Singh (101) benzoylated groundwood for bleaching and for improving brightness stability. He, like Loras (99), holds that resistance to irradiation is a function of the lignin. Methylation with dimethyl sulfate improves the color stability of groundwood, according to Andrews and Des Rosiers (102), in contrast to the work of Manchester (100). The former introduced much less methoxyl than did Manchester.

In summary, the thermal resistance and the behavior of the paper as electrical insulation are both improved by acetylation. This holds true whether the untreated paper is acetylated or prepared from acetylated fibers, providing that proper conditions for treatment are maintained.

(2) Cyanoethylation

Miller and Flowers (103) in 1950 patented dielectric sheet material prepared from cyanoethylated pulps. The pulp was cyanoethylated by adding acrylonitrile to a slurry of pulp in aqueous alkali. The emphasis in this patent is placed on the increased dielectric constant.

In 1959 the manufacture of insulation paper from cyanoethylated pulp by this process was announced (104, 105). It was stated that a 20% increase in peak-load capability is possible without sacrifice of life expectancy. This is attributed largely to increased thermal stability. In the Miller and Flowers process (103) 100 parts of pulp is used to 2000 parts of 10–12% aqueous sodium hydroxide and 30–70 parts of acrylonitrile for 2 hr at 25°C or less. Reference was made in the announcements to a "dry" process—that is, the process of McKay and Bikales (26), who cite the use of 50–400 parts of dilute alkali and 1–100 parts of acrylonitrile per 100 parts of pulp. Later patents (106–108) use about 200 parts of alkali and 15–30 parts of acrylonitrile.

The cyanoethylation reaction is carried out with acrylonitrile and dilute alkali (109). For economy it is advisable to use very low excesses of reagent. Heat and strong alkali should be avoided, for the nitrile group may then be converted to a carboxyl group. This will not impair the strength as such—it was seen above that carboxyethylated fibers have increased strength (73)—but it will impair the electrical properties. Prati (110, 111) has used a ligroin dispersion of alkali and fiber for preparing alkali cellulose in connection with cyanoethylation. Mizutani et al. (27) used for cyanoethylation a two-stage process in which excess alkali was removed, without opening the equipment, before the acrylonitrile was added. The degree of cyanoethylation was greater at short times of exposure. The authors used concentrated potassium hydroxide at 30°C with high concentrations of acrylonitrile.

Because of its increased resistance to rot and acid (104, 109, 112), cyanoethylated paper has been suggested for such things as sandbags, mulch, and filters, but its main use has been for electrical insulation. Dieterle (82) found the dielectric properties to be better for acetylated than for cyanoethylated samples. Mizutani's (27) cyanoethylated paper (DS 0.4) showed better thermal stability at 130 and 170°C than did the untreated kraft. A comparison of heating in air and in transformer oil showed that oxygen exerts a deleterious effect. Experiments indicated that cyanoethylated paper will last four times as long as the usual kraft insulating paper. Earlier patents of the General Electric Company (106, 107) used an acid wash at the end of the cyanoethylation. Later patents (108, 113) eliminated the acid. The water-washed product has an improved power factor and is particularly recommended for use in oil-filled transformers.

Wink and co-workers (31) found that the thermal stability in transformer oil of kraft pulp was greatly improved by cyanoethylation as shown by higher folding endurance, stretch, and tensile energy absorption as compared with these properties in untreated kraft. Tensile strength, however, was hurt by the cyanoethylation and was in most cases lower than that of untreated kraft. Some results are given in Table 2 (see also references 58 and 109).

The manner in which thermal stability and retention of electrical properties are brought about probably involves more than one factor. Certainly, as with acetylated paper, the decreased affinity for water is important, as is the reduced tendency for cellulose to split off water; both effects are due to the blocked hydroxyl groups. Another factor is the occurrence of heat bonding between fibers in the sheets, as observed by Jayne (30) and Wink et al. (31), which may even result in strength improvement rather than strength deterioration under certain test conditions but would probably not account for color stability (see also Section D of Chapter XVII).

TABLE 2
Thermal Stability of Cyanoethylated Kraft Pulps (31)

Sample designation	Extent of cyanoethylation (% N)	Aging time in oil (hr)	Basis weight (g/m²)	Thickness (μ)	Formation[a] (units)	MIT folding endurance (double folds)	Tensile strength (kg/cm)	Stretch (%)	Tensile energy absorption (kg-cm/cm²)
Control	0.0	0	67.4	97	44	601	5.29	2.50	12.9
		24	67.5	97	43	6	4.73	1.31	5.6
123	1.72	0	66.6	127	32	574	3.23	5.00	18.0
		24				295	3.64	3.62	13.7
123[b]	1.72	0	67.3	114	29	1625	4.21	6.02	27.7
		24				692	4.21	3.62	15.7
124	0.99	0	66.3	152	40	27	2.04	2.60	5.8
		24				116	2.80	3.24	9.4
125	1.86	0	64.4	94	34	2525	5.75	4.16	24.6
		24				729	5.36	2.26	11.4

[a] Formation is the uniformity with which the fibers and other solid components of the sheet are distributed.
[b] The pulp used for this treatment was beaten 45 min; for the control and the other treatments it was beaten 30 min.

(3) Hydroxyethylation

Harpham et al. (36) observed that paper from hydroxyethylated linters retains the excellent aging characteristics of the original unetherified linters. The hydroxyethylated sheet after 19 months at 23°C and 50% relative humidity (RH) plus 3 days at 105°C had suffered no appreciable loss in bursting strength or folding endurance and only 5% loss in tearing strength. This has been borne out independently by other workers and extended to much more rigorous conditions. Brummet (114) found that hydroxyethylated kraft paper retained 90% of its tensile strength after 66 hr at 170°C in oil in the presence of copper, whereas ordinary kraft retained only 50%. Tearing strengths showed a similar advantage for the hydroxyethylated material.

Thomas (49) compared hydroxyethylated kraft sheets from pine with corresponding normal sheets beaten to comparable bursting strength. The brightness was better for the hydroxyethylated kraft, and the brightness reversion was less either after 24 hr in an oven at 105°C or after a 5-hr exposure to ultraviolet radiation. The tear factors of the hydroxyethylated sheets showed a greater percentage of decrease but remained higher since they were originally greater. Bursting, folding, and tensile strengths were higher for the hydroxyethylated sheets, and the percentage of strength retained was also greater. In fact both folding endurance and tensile strength showed improvements due to heat curing in the oven tests. Ward et al. (58) tested a number of samples of hydroxyethylated kraft sheets in oil at 170°C. They showed that the improved thermal stability is partly due to heat bonding and partly due to a slower rate of strength deterioration under the conditions used. There is a tendency for strength retained as well as for original strength to increase with MS.

It is not quite easy to see why the hydroxyethylation should produce thermal stability similar to that obtained by acetylation and cyanoethylation. The equilibrium moisture content, for fibers mercerized before etherification at least, is much higher, and although the anhydroglucose hydroxyl groups are blocked, the blocking agent still contains a hydroxyl group. One may speculate that substitution of the hydroxyl groups hinders the formation of the new ring in levoglucosan and so retards the first steps of thermal breakdown (see Section C of Chapter XVIII), but there seems to be no experimental evidence for this. Mitchell (115) proposes that hydroxyethylation attacks sensitive, accessible carbonyl groups and hence hinders brightness reversion.

Hydroxyethylation does not seem to help the dielectric properties. Still, the addition of an epoxy compound to electrical papers has been recommended. It is said to react with the cellulose, as well as to polymerize (116).

(4) Miscellaneous Modifications

There is some evidence that other types of chemical modification may be used either for heat resistance, electrical resistance, or both. Patents assigned to the Union Carbide Corporation (117) indicate that methylcellulose and similar ethers may be used for all or part of the furnish to produce a paper sheet that may be used as the barrier film in dry cells. Russbach and Teiser (118) made insulating paper from partially nitrated pulp. It has already been mentioned that carboxyethylated pulps resist drying at 105°C better than untreated pulps (73).

(c) DECREASED AFFINITY FOR WATER

There seem to be only two important treatments of paper to decrease affinity for water: acetylation and the diketene-based sizes. There are, of course, many cases of crosslinking and of grafting treatments that reduce the affinity for moisture, but these are discussed separately below (see also Section K of this Chapter XIX).

(1) Acetylation

Many of the advantages and proposed uses for acetylated paper arise from the decreased affinity for water above a certain low degree of acetylation. The subject has been already touched on in connection with strength and thermal degradation.

Dimensional stabilization by acetylation of the paper sheet has been studied by Stamm and Beasley (119) and Dieterle (81, 82). The swelling from 50% RH to water soaking can be rapidly and efficiently reduced by 48–89%. The strength properties are adversely affected, less with unbleached than with bleached kraft. Folding endurance is the most greatly reduced of the properties studied. Wet breaking lengths (free-hanging lengths that will cause tensile failure) are considerably increased along with dimensional stability.

In the early 1960s, information began to appear about improvements to be obtained in dimensional stability, wet strength, lack of curl, and resistance to moisture sorption or water penetration by preparing paper from partially acetylated pulp (22). For acetylation to a DS of about 1.0, there are losses in bursting strength and folding endurance, but the wet tensile strength, wet stiffness, and opacity (dry and wet) are greater (Table 3). The sizing is improved, as shown by Valley penetration tests and Cobb size tests. Dimensional stability is also improved, as shown by large decreases in wet expansion, swelling, and bulking. The percentage of improvement in wet expansion due to acetylation is about the same as the percentage of change in expansion between 20 and 85% RH (the latter shown in Table 4). The type of wood pulp used in obtaining the data in Tables 3 and 4 is not stated.

The decreased affinity for water (water retention) is accompanied by an increased affinity for organic liquids. This is evidenced, for instance, by the penetration time for saturation with oils. Partially acetylated paper is uniformly wet after 1.5-sec flooding with transformer oil under conditions that produce only partial penetration of untreated paper after 5 sec. Acetylation of paper in advance will prevent the loss of water repellency caused by certain resin treatments (120).

TABLE 3
Physical Properties of Paper Made from Partially Acetylated Fibers (22)

Property	Wood cellulose	Partially acetylated cellulose	Wood cellulose	Partially acetylated cellulose
Basis weight (lb/100 ft^2)	20	22	11.5	13.25
Thickness (mils)	4.3	4.3	2.7	3.4
Folding endurance (MIT):				
Length	200	69	400	142
Cross	200	56	400	53
Tear (Elmendorf) (g):				
Length	64	81	30	40
Cross	66	86	35	44
Bursting strength (Perkins) (psi)	40	27	30	24
Wet tensile strength (TAPPI) (kg/15 mm)	1.8	2.3	1.2	2.2
Stiffness (Taber):				
Wet length	1.2	2.7	0.6	1.4
Wet cross	0.6	1.3	0.4	0.6
Dry length	3.2	5.4	4.5	2.0
Dry cross	3.8	1.8	1.1	2.0
Opacity (Bausch and Lomb) (%):				
Wet	69	83		
Dry	82	89	74	78
Cross expansion (%; 2 min in H_2O)	2.7	1.55	3.0	1.8
Valley penetration (sec)	150	514	40	265
Swell (%)[a]	50.0	22	46	14
Bulk (%)[b]	17.5	4	15	8
Cobb size (g/100 cm^2; 2 min in H_2O)	0.150	0.09	0.132	0.108

[a]Swell $= \dfrac{\text{wet thickness} - \text{dry thickness}}{\text{dry thickness}} \times 100.$

[b]Bulk $= \dfrac{\text{redried thickness} - \text{dry thickness}}{\text{dry thickness}} \times 100.$

TABLE 4
Humidity Amplitude (22)[a]

Sample	Expansion	Contraction
Nonacetylated	0.24	0.23
Partially acetylated (20% acetyl)	0.14	0.12

[a]Percentage of cross-directional size change per 10% change in relative humidity over 20–85% RH.

The partially acetylated fibers, however, show only slight weight losses on solvent extraction with solvents for cellulose triacetate. Acetic acid dissolves the most, 14%; analysis shows that the 7% material extracted by methylene chloride-methanol (94 : 6) consists largely of highly acetylated hemicelluloses. The resistance of the acetylated fibers to hydrolytic chain scission is about the same as that of the parent fibers. Under alkaline conditions there may be saponification, but there is very little in boiling water for 24 hr or in water at room temperature for 6 weeks. They are stable toward deacetylation at pH 3.

Acetylated paper has been used successfully in paper chromatography (121-125). Most of the papers chemically modified for this purpose have had ionic groups introduced as well, to serve as ion-exchange papers. But the function of acetylation is different; it reduces the absorbency of the paper by blocking the hydroxyl groups. The paper sheet is acetylated, usually with acetic anhydride in the presence of an inert solvent (125-130). Other organic acids besides acetic acid have also been used (131, 132). When half-esters of dibasic acids are produced (132-134), the paper has both reduced absorbancy and ion-exchange properties (see also Section B in Chapter XVII).

(2) Cyanoethylation

Morton and Bikales (25) noticed particularly the increase in dimensional stability when cyanoethylated pulps are used. The average expansion-contraction on going from 20 to 75% RH and back was 0.55% on the untreated pulp; this value dropped linearly with increasing degree of cyanoethylation and was only 0.24% at a DS of 0.60. Cohen and Heaps (135) have used cyanoethylated pulps with rosin size to produce hard-sized paper. Either sulfite or kraft pulp is suitable. Stamm and Cohen (136) also found antishrink efficiency to result from the cyanoethylation process, but their treatments were carried out on preformed paper and the strength properties were decreased by the treatment, at least partly on account of excessive swelling of the sheet in alkali. A process for deswelling chemically modified pulps, especially cyanoethylcellulose, with concentrated salt solutions has been patented (137). (See also Section D of Chapter XVII).

(3) Ketene-Based Sizing

Although "sizing" of paper originally meant reducing ink penetration, it is usually used today to mean improving the resistance to penetration by liquids in general, but especially aqueous liquids. The conventional sizing agents were generally inert to the cellulose of the fibers. Many of them, like starch or rosin, might in theory react chemically with the hydroxyl groups of cellulose, and it might be impossible to prove that an occasional hydroxyl group did not react, but no one has presented any evidence for appreciable reaction. But in 1955 (138, 139) alkylketene dimers (140) were put on the market, and these

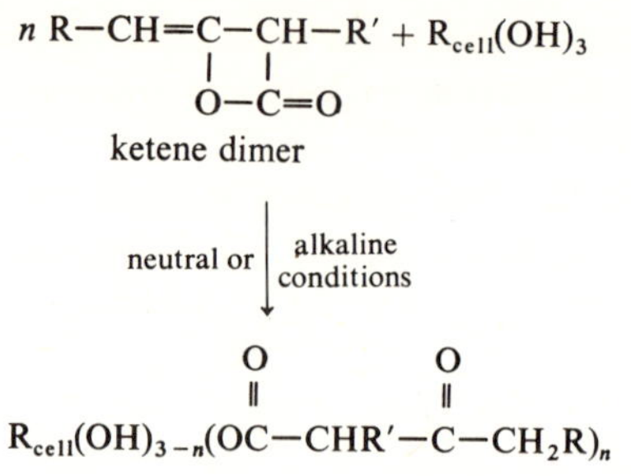

Fig. 9. Reaction of alkylketene dimers with cellulose (133).

are believed to react chemically with the surface of the cellulose, as shown in Figure 9 (141), to change the fiber itself to a more water-repellent material. The evidence for chemical reaction is that organic liquids that dissolve the original dimers extract nothing from the treated fiber, nor does such extraction hurt the sizing of the fiber; moreover, cuprammonium solution dissolves the inner core of the fiber but leaves the reacted outer shell (142). The alkyl groups, R and R′, in the most used dimers are hexadecyl (derived from stearic acid) and tetradecyl (from palmitic acid), but dimers from oleic and myristic acids are also available. The reaction products are esters of dialkylacetoacetic acid; they are inert and resist penetration by both acidic and alkaline liquids, although they are not effective against hot liquids. A mere surface coating of the unreacted dimer does not develop sizing.

The dimer is usually applied as an emulsion (143), although it may be deposited on silica to form a powder (144). The reaction is not immediate and usually requires 1–3 days for completion. The rate increases with pH, time and temperature, but natural aging develops a greater degree of sizing, perhaps because of more uniform distribution. The dimer reacts with water, which results in a loss of efficiency if the rate of drying is too slow.

Some other additives to the furnish may have a deleterious effect on this type of sizing. Alum interferes by reducing the rate of the reaction (145). Amines and ammonia have an adverse effect, and so do some clays and diatomaceous fillers (141, 146). Subsequent application of starch can reduce the effect of the ketene dimer significantly, but this rarely occurs on the machine when the dimer is still mostly unreacted. For internal sizing the dimer is usually applied in an emulsion (147) stabilized with cationic starch (148), polyamide, or other resins (149). The use of various other additives to the emulsion has also been patented (150).

Some of the advantages claimed for this type of sizing are economy (not only direct but also indirect, through the use of less expensive fillers, reduced power requirements in refining, and increased machine speeds) and improved strength, brightness, and brightness reversion. The ketene dimers are used under neutral or alkaline conditions, which permits the use of alkaline fillers like calcium carbonate (146, 151) and increases the permanence, which is of particular importance in book papers (152). The dimer can be applied to the filler before the latter is mixed with the furnish for forming the paper web (153). It also reduces corrosion in the mill. Some additives, such as mannogalactan gums and calcium carbonate, improve the sizing of the ketene dimers. For surface sizing an unmodified emulsion (141) or a solution (154) may be used. Worthen and Oransky (155) have found that printing masters can be made by applying the dimer in a prime coating, the paper web being sized by dimer migrating from this coating. The ketene dimer can also be used as a preliminary sizing treatment (156). In coatings it also can serve as a release agent in drum-finishing the coated paper (157). An interesting application of the ketene dimer that depends on its easy release or peelability is in the manufacture of sausage casings; paper tubing or netting is impregnated with viscose containing the dimer and the viscose is then regenerated (158).

Although several investigations indicate improved permanence with alkaline sizing of this type (152, 159, 160), Dixson and Nelson (161) have concluded that brightness and its retention are improved and that the iron content of the paper is reduced, but that improvement in permanence is secondary.

The Food and Drug Administration permits the use of Aquapel (registered trademark of the alkylketene dimer manufactured by Hercules Incorporated) as a size for milk-carton, drinking-cup, and certain kinds of food-wrapper stock. Water does not extract a measurable quantity of hydrolyzate from waxed milk cartons treated at the 0.2% level with hydrolyzed Aquapel (162).

There are interesting variations of the alkylketene-dimer treatment. Copolymers of polyketenes (such as could be produced from dilinoleic acid) with monoketenes have been suggested (163). The reaction of ketene and its volatile homologs in the gaseous state has also been patented (164). It has also been stated that free acids, anhydrides, amides, chlorides, or aldehydes may be added as extenders (165) or emulsifiers (166).

(4) Miscellaneous Sizing Agents

As already stated, many sizing agents can conceivably react with the hydroxyl groups, but none has attained the commercial importance of the ketene dimers.

Some early patents relating to the alkylketene-dimer sizes contain claims relating to sizing with a number of other reactive derivatives containing long hydrocarbon chains, including acid halides, acid anhydrides, acid azides, and isocyanates (167, 168).

As long ago as the early 1930s, it was proposed to size paper by reacting it with acid anhydrides or chlorides (169–172). In practice, however, many of the organic solvents used were undesirable. Several years ago stabilized emulsions were proposed (170). Sinclair and co-workers (173, 174) have used free fatty acids (behenic acid seems to be the best) for sizing hardboard. The powdered acid is added and migrates as a vapor when heated to size the sheet. Sinclair et al. attribute the sizing effect to chemical reaction. Previous investigators (175) have considered the phenomenon to be related to self-sizing (176) and to be a matter of adsorption. The introduction of perfluoro groups confers both oil and water repellency on cellulose. Most of the work has been done with textiles, but reference is occasionally made to its applicability to paper (177, 178).

Another type of reactive size consists of long-chain esters of propiolic acid. These react with a cellulose to form esters of the corresponding vinyl ethers of cellulose (179).

(d) ENHANCED ION-EXCHANGE PROPERTIES

At least five types of ion-exchange paper are obtainable commercially in small lots (180). Two of these are *cation exchangers*, cellulose phosphate (181) and carboxymethylcellulose. Phosphorylated sheets have been used in the analysis of metallic ions (182–184) and of amino acids (185). Carboxymethylated fibers have been already discussed in this section. Hartel et al. (186) have described the separation of amino acids on papers containing sulfonic acid groups or carboxyls (or quaternary groups), but they do not mention the preparation of the paper.

Three of the commercially available ion-exchange papers are *anion exchangers*. Aminoethylcellulose is weakly basic; it is prepared by the reaction of alkali cellulose with, for example, aminoethyl sulfate or another reactive aminoethyl compound. Diethylaminoethylcellulose, a stronger base, is prepared similarly. It has been useful in biochemistry when enzymes or other proteins, nucleic acids, or related compounds require isolation or purification (185, 187, 188). ECTEOLA-cellulose is prepared by the reaction of epichlorohydrin and triethanolamine with cellulose; it is weakly basic and likewise has found use in biochemistry.

Knight (189) has claimed high selectivity with ion-exchange paper from mixed derivatives of cellulose; one substituent is an ion-exchange group, whereas the other (e.g., acetate) modifies the affinity for water, Micheel and Albers (133) have described similar papers. Also manufactured are ion-exchange papers in which the sheet has been loaded with ion-exchange resins; these are outside the scope of this book, except for those papers that contain chemically modified fibers as well (190). Another form of ion-exchange membrane, which is not strictly paper, is produced by treating cellophane, such as grafting with poly(acrylic acid) or treatment with reactive dyes.

Taylor (191, 192) has disclosed the introduction of ion-exchange sites into parchment paper and then reparchmentizing. For instance, a web is treated with phosphoric acid of at least 80% concentration, and the wet paper is diluted to an acid concentration of about 70%, treated with urea, and heated to phosphorylate the web. Other groups can be introduced by using suitable reagents, such as the anhydrides of polybasic acids, chloroacetic acids, taurine (aminoethanesulfonic acid) and similar amino compounds, and chlorosulfonic acid. Süszei (193) investigated the introduction of various functional groups into cellophane, kraft, and parchment papers for use as membranes. With reactive dyes and related compounds containing basic nitrogen in the ring, the best results were obtained with cellophane; parchment paper was also good, but kraft paper had lower permselectivity. Products from compounds containing acid groups (sulfanilyldichlorotriazine and sulfanilyltrichloropyrimidine) also had lower permselectivity. Basic esters were produced with isatoic anhydride, and acidic ethers with the inner anhydride of hydroxypropanesulfonic acid. Crosslinking with the disodium inner salt of tris(β-sulfatoethyl)sulfonium (Sulfix A, Imperial Chemical Industries) yielded a better product from kraft paper than from the other substrates but the product was brittle, owing to the strong crosslinking. In this connection it might be mentioned that many patents for introducing sulfatoalkylsulfonium and similar compounds into textiles frequently mention their use with paper also (194–196) (see also Section A of this Chapter XIX).

Historically, the first chemically modified papers of this type were simply oxidized to introduce more carboxyl groups in addition to those already present in essentially all papers. Nitrogen oxide oxidation was the first process used (197, 198), and modifications are still being suggested (199–201). Successive oxidations by periodic acid and nitrogen dioxide (202) or chlorous acid (203) have also been considered. The aldehyde groups in oxidized papers can be condensed with various amino compounds to introduce basic or acidic functional groups (204).

Guthrie's production (205) of ion-exchange cottons showed how readily cellulose could be modified to introduce ion-exchange groups, and the preparation of paper containing such groups was subsequently proposed by various investigators (123, 124, 126, 131–134, 189, 204, 206–209).

Diethylaminoethyl pulp fibers have been prepared and used by McKenzie (76) in an interesting study of a series of starch derivatives as beater adhesives. The chemically modified pulp was cationic enough for retention aids to be unnecessary for his starches, thus eliminating a complicating variable from the system. Some patents propose a somewhat similar mechanism for improving the retention of anionic or nonionic resins—for example, by a pretreatment of the fibers with an epoxy resin containing cationic functions (210).

There are some other uses for papers that involve chemical modification to introduce acid or basic groups. One of these provides a solid support for enzymes. Paper is treated with a chlorotriazine; the resulting product will then bind enzymes or other bases (211). Another use is the preparation of soil conditioners (212). Ethylenimine and many of its derivatives are very reactive, and the products with cellulose should have ion-exchange properties. In practice one uses polyethylenimine (213), which is usually considered simply to change the charge on the fiber by sorption, but not by reaction. However, there is at least one patent in which reaction is specifically stated to occur (178). McClure and Goldstein (214, 215) have reviewed the use of ethylenimine in the paper industry.

For a further discussion of ion-exchange celluloses see Section H of this Chapter XIX.

(e) IMPROVED FLAME RESISTANCE

Much attention has been given to the flame resistance of paper, but chemical modification of the fiber to that end has been more a matter of theory than of practice. There are several patents, but most of them are primarily aimed at textiles and mention paper only incidentally. Ordinary phosphorylation of the type we have just discussed for producing ion-exchange papers is one way, but it is less common for paper and board (216, 217) than for textiles.

Condensation with tetrakis(hydroxymethyl)phosphonium salts has been applied to paper in the examples of several patents (218–220). The reagent probably reacts with other components of the furnish besides the fiber, but it is known to react with cellulose also. Other patents cover the production of phosphonomethyl ethers on the fiber (221), the use of tris(1-aziridinyl)-phosphine oxide (APO) (222, 223), the use of boric acid at temperatures sufficient to cause it to react with cellulose (224), and the reaction of mesylated cellulose with metallic halides or phosphorus compounds (396). (See also Section I of this Chapter XIX).

2. Graft Copolymerization

Paper may be modified by the process of grafting copolymers onto the cellulose backbone. In recent years this method of modifying cellulose has received considerable attention, and literally hundreds of papers have been published on the subject. The main body of this work is covered in Section G of Chapter XVII, and we confine our consideration to what is directly pertinent to the modification of paper properties.

The interest that has been generated in the paper field by graft copolymerization has been stimulated by the prospects of attaining desirable new properties. Grafting affords an unusual method of combining the properties of thermoplastic polymers with those of cellulose (225). The principal interest stems from the fact that reaction is largely confined to the accessible regions of the fiber and that levels of addition of several hundred percent are possible without disrupting the basic crystalline structure of the cellulose or the morphology of the fibers. In this way the polymer properties are superimposed on the properties of the paper fiber. Although research continues in the paper industry, little or none of this work seems to have resulted in commercial applications (226).

It is worthy of note here that most of the work in the literature under the heading of graft copolymerization involves only small amounts of true graft (227). Much of the polymer is homopolymer with molecular weights in the range of 100,000–300,000 (228–230). With polymer add-ons as high as 125%, fewer than 1 in 10 of the cellulose chains carry a copolymer graft (229). At these low levels of substitution the properties of the product are not likely to be determined by the number of grafted sites on the cellulose chain. We treat the literature under the title of "graft copolymer," recognizing that the products at best are mixtures of graft copolymer, homopolymer, and cellulose.

Most of the grafting processes on cellulose are of the free-radical type (231, 235), except for a few cases of ionic polymerization (232–234). Imoto and coworkers (236, 237) grafted onto cellulose and silk fibers in the absence of any initiators. The paper used in this work actually contained free radicals, as shown by electron spin resonance studies, but cotton fibers, free of radicals, were also found to form grafts. See also Section I of Chapter XVII.

(a) IMPROVEMENT OF STRENGTH PROPERTIES

(1) Dry Strength

Graft copolymerization may have significant effects on the strength properties of paper. However, many of the claims for dramatic improvements reported in the literature have resulted from the treatment of weak base sheets, such as filter and saturation paper (238–242). It is important to establish the point of reference in interpreting such results.

Schwab, Stannett, and Hermans (238) reported improvements of over 200% in dry tensile and bursting strengths as a result of grafting 100% acrylamide or acrylonitrile on cotton-based filter paper. These authors used the ceric-ion method of Mino and Kaizerman (243) to catalyze the reaction with vinyl monomers. This method utilizes the cellulose as the reducing member of the oxidation–reduction pair and results in the generation of free radicals on the cellulose surface. These radical sites can initiate vinyl polymerization and grafting to the cellulose chains (244, 245). The method is readily carried out in an aqueous medium at room temperature, preferably in an inert atmosphere at a pH of less than 2.

Schwab et al. (239) studied the applicability of the ceric-ion method to many vinyl monomers. They found reactivity to increase with aqueous solubility of the monomer, the acrylic esters and acrylonitrile being the most reactive. Acrylic acid, however, could not be reacted directly (246), whereas vinyl acetate, divinylbenzene, styrene, and diallyl phthalate reacted poorly or not at all.

Guzman (247) found similar reactivitities of various monomers on paper with the ceric ion. Acrylonitrile, acrylamide, acrylic anhydride, and methyl acrylate reacted rapidly, but styrene, vinyl acetate, and methyl methacrylate did not react under the conditions used. The process results in considerable homopolymerization, with up to 40% of the polyacrylonitrile (248) and 66% of the poly(vinyl chloride) (228) reacted on the fiber being solvent extractable.

Probably any monomer that can be grafted to other forms of cellulose (see Chapter XVII, Section G) can be grafted to paper. Among those not mentioned elsewhere in this section are chloroethyl acrylate and methacrylate (249). In addition, other oxidation-reduction systems based on oxidative cations analogous to the ceric ion may be used as initiators. These include salts of iron, manganese, chromium (250, 251), and vanadium (252), and uranyl salts (253).

Using the ceric-ion method, Bergem et al. (248) found that with moderate-strength base sheets from paper- and acetate-grade sulfite pulps breaking length was unaffected by grafts of up to 100% polyacrylonitrile, based on the weight of the cellulose. This contrasts with results previously cited for grafts to filter-paper-base sheets. Lynch (241) confirmed these observations with acrylonitrile treatments of kraft and groundwood sheets. He used benzoyl peroxide to initiate free-radical polymerization of acrylic monomers in paper sheets. Grafts may be possible with this catalyst system (254, 255) through free-radical transfer between growing polymer chains and the cellulose (238). Improved tensile strengths of 70%, however, were observed when 30% methyl methacrylate was polymerized by this method within a kraft sheet. Apparently the nature of the monomer as well as the base sheet can affect the results.

The effects of grafting on the characteristics of paper have been studied in detail (239, 240, 256–258). In general hard polymers, such as those from acrylonitrile and methyl methacrylate, increase the stiffness and the modulus of elasticity but sharply reduce folding endurance. Abrasion resistance, at least in lightly bonded sheets, is improved by both hard and soft polymers. The soft, rubbery polymers, such as those formed from ethyl and butyl acrylates, reduce the elastic modulus and thus increase stretch, toughness, and tear resistance. The softer polymers also have a much lesser adverse effect on folding endurance (248) and, in lightly bonded sheets, actually increase it. If a polyfunctional vinyl compound is present as a component with the vinyl monomer in the grafting system, the resulting grafted product will be crosslinked. A further improvement in strength is obtained by including a fraction of a percent of ethylene dimethacrylate in the butyl acrylate monomer (259, 260).

Trout (261) studied the effect of grafting acrylonitrile on the stiffness of paper and board. Using high-energy electron radiation, he grafted relatively low levels of polymer, in the range of 10–18%, onto a variety of substrates, including bleached and unbleached kraft board, cylinder board, corrugating medium, and filter paper. He found that the modulus of elasticity is slightly reduced at these low levels of grafting if the reaction is carried out in the swollen state in the presence of moisture. Under nonswelling conditions stiffness and modulus of elasticity are increased. Horn and Simmonds (262) also found an increase in stiffness in linerboard and card stock on grafting with acrylonitrile, obtaining a 125% increase in compressive strength in both cases. They also observed a 60% improvement in the dimensional stability of the card stock, although not in the linerboard. Methyl methacrylate grafting has in fact been proposed (263) for improving stiffness in paper and board. The resulting products are recommended for use in making playing cards, punched cards, and cardboard of various sorts.

The effect of radiation grafting of acrylonitrile on the properties of filter paper has been studied by Burke (242), who compared the radiation and chemical methods of free-radical polymerization, using both simultaneous irradiation of monomer and cellulose as well as preirradiation of the sheet followed by monomer contact. Increases in strength properties in agreement with those previously described were observed, with no marked differences in properties resulting from the various radical-initiating methods.

Loos (264) likewise found that the strength of paper, both wet and dry, from a 50% acrylonitrile graft was substantially better than that of the untreated web. Putnam (265) subjected to γ-radiation filter paper immersed in 50% solutions of acrylonitrile in *nonswelling solvents*. The resulting graft polymer was thereby located primarily on the surface. At 50% polyacrylonitrile content most of the strength properties of the web are significantly enhanced. However, solvent extraction by boiling for 3 hr in dimethylformamide removes most of the "graft," indicating a preponderance of homopolymer under these conditions. As would be expected, strength reinforcement disappears with the dissolved polymer. On the other hand, grafting under *swelling conditions* has been found to improve the reaction in paper fibers (266, 267).

In addition to promoting sites for possible graft copolymerization, exposure to ionizing radiation will also cause degradation of the substrate. Degering (268) has found that degradation can be reduced without interfering with grafting if the same total energy is applied in intermittent bursts of small doses, rather than as continuous exposure for the same irradiation time or for a shorter time at a higher dose rate. He used kraft paper with butyl acrylate and styrene (see also Section H in Chapter XVII).

(2) Wet Strength

The wet strength of paper is generally markedly improved by grafting, as long as the grafted polymer is not too water soluble or dependent on hydrogen bonding (239). Several authors have reported wet strengths of 50–100% of the dry strength at grafting levels of acrylonitrile above 35% of sheet weight for both lightly bonded (239, 240, 242) and moderate-strength sheets (248). Lynch (241) confirmed these results with methyl methacrylate on kraft- and groundwood-base sheets, but found acrylonitrile to be effective only in enhancing wet strength on a relatively weak filter-paper-base sheet.

Schwab et al. (239) also compared the effects of copolymer grafts of ethyl acrylate and acrylonitrile with latex saturation of similarly composed polymer on the strength properties of paper. They found that dry tensile strength was somewhat higher at the same polymer level for the latex-treated web compared with the graft. However, grafting was much more effective for improving wet strength. These results are interpreted as indicating that the saturation technique is more efficient in reinforcing *interfiber bonds*. Grafting results in much polymer being buried within the fibers, where it is a less effective reinforcement of fiber-to-fiber bonds. However, in this situation it has a greater influence on reducing fiber swelling and the disruption of interfiber bonds when wetted with water.

It is worthy of note that latex saturation also changes other properties of paper rather dramatically compared with grafting. Daniel et al. (240) reported that at more than a few percent pickup polyacrylonitrile-saturated papers are drastically changed in appearance and hand, are brittle, and are more like resin-coated paper owing to poor penetration into the web. Grafted paper, on the other hand, with uniform intrafiber polymer deposition, resembles untreated paper with regard to fibrous structure and appearance.

(3) Grafting to Pulp Fibers

Bergem et al. (248) compared the effect of grafting pulp fibers in slurry form before sheet fabrication with reaction in the paper web. The ceric-ion method was used in this work. The authors found that moderately refined pulp reacted about twice as fast as paper made from the pulp. It was further found that paper-grade pulps were more reactive than high-alpha pulps and that unbleached kraft pulps could not be grafted at all by the ceric-ion process. It is apparent from these studies that unbonded pulp fibers afford more accessible reaction sites than paper sheets. The presence of hemicelluloses in paper-grade pulps seems to improve the accessibility of reaction sites. Oxidation of the reducing groups in the residual kraft lignin permits subsequent grafting by the ceric-ion process (239). Schurz and co-workers (269–271) studied the effects of reaction variables on the free-radical grafting of vinyl monomers onto wood pulp and presented excellent evidence for true grafts of styrene to cellulose (270). It was also found (235) that the grafting of pulp with styrene, which is recognized to be difficult (272), is made easier by preliminary oxidation. Ogiwara et al. (273) also studied the ease of grafting onto cellulose fibers by the ceric-ion process. They found the ease of reaction to increase going from cotton to kraft, sulfite, and semichemical pulps. The consumption of the ceric ion by the ungrafted pulps was proportional to the copper number.

Grafting onto woodpulp fibers prior to sheet formation was generally detrimental to the papermaking properties of the fiber (248, 262). All the measured dry and wet properties of papers made from acrylonitrile-grafted pulp were significantly lower than those of untreated paper or paper grafted in sheet form. Cornell (228) confirmed these results in the case of ceric-ion grafting of vinyl chloride onto pulp fibers. He found that the adherence of a hydrophobic polymer to the fiber seriously interferes with the response of the pulp to subsequent refining. The blocking of natural bonding sites on the fiber as well as inhibiting the swelling and wet flexibility of the fibers by the hydrophobic polymer interferes with strength development during sheet formation.

If the grafted polymer is hydrophilic, on the other hand, papermaking properties may be enhanced. Adams and Hoftiezer (274) used the iron–peroxide oxidation–reduction system to prepare grafts on paper. This oxidation–reduction system anchors the ferrous ion to the cellulose by ion exchange to permit subsequent polymerization at the cellulose interface (275–277) after introduction of peroxide and monomer. Grafting may occur through radical transfer from growing polymer chains to the cellulose (238, 244). Grafts of acrylic esters, acrylonitrile, vinyl acetate, and the like to wood pulp were prepared and subjected to subsequent hydrolysis (274, 278). The resulting hydrophilic products after converting the grafts to poly(sodium acrylate) or poly(vinyl alcohol) were found to have good bonding capability. These hydrophilic grafted fibers were found to reinforce the dry and wet strengths of paper when admixed with untreated pulp and fabricated into a paper web. The highly swollen fibrous products of this process have the effect of beater adhesives when added to pulp furnishes and can be used to reduce or supplant the beating requirements of paper pulps.

Extensive comparisons of the effects of grafts of hydrophilic and hydrophobic monomers on the papermaking properties of the grafted pulps were made by Lee and Fujii (279–285). Their work (280) showed that hydrophilic grafts to pulp with acrylamide give paper with higher tensile strength and folding endurance, and lower tearing strength. A hydrophobic graft with methyl methacrylate, on the other hand, resulted in reduced tensile strength and tearing strength that goes through a maximum at 10% graft. The tensile strength of single fibers of the grafted pulps increases due to increasing diameter, but strength per unit area decreases. The equilibrium moisture content of the pulp increases with the hydrophilic grafts but decreases with the hydrophobic polymer. Hydrolysis and x-ray diffraction studies indicate that grafting occurs in regions of low order (279). Beating studies on grafted pulps in water, alcohol, and acetone showed that the acrylamide-treated fiber refined in a manner similar to untreated pulp. Corresponding treatments of the fibers grafted with acrylic esters or vinyl acetate indicated that strength development occurs only when the fiber is refined in a swelling medium for the polymer, such as acetone (281, 282). Paper sheets prepared from aqueous slurries of the hydrophobic fibers were weak, porous, and lightly bonded, whereas the hydrophilic grafts gave dense, strong webs (285). The results of these studies are similar to those obtained earlier with partially acetylated pulp fibers (286).

Lee and Fujii (284) found an important difference between paper made from pulps grafted with various methacrylic esters and paper saturated with equivalent polymers. They studied the effect of temperature on tensile stress in the two kinds of paper. Maximum strength in saturated papers occurs at the glass-transition temperature of the polymer but is in the 40–60°C range for the grafted sheets. Folding endurance decreases with increasing temperature, probably because of lower moisture content.

Neimo and Sihtola (287) studied the grafting of acrylamide to wood pulps by the ceric-ion process. They determined that the graft resulted in an increase in the specific volume but a decrease in the surface area of the fibers. They also observed that the hydroxyl groups of the fiber were largely unavailable for further reaction despite the fact that they were mostly in an unreacted state in the grafted pulp. They interpreted their results as indicating that polymerization is initiated at the reducing end of the fiber and polymer growth proceeds helically down the fiber. The resulting polymer is believed to be strongly hydrogen bonded with the cellulose. When acrylamide is grafted to pulp with ozone being used as the free-radical initiator, the ozone causes a severe degradation of the cellulose, but the harmful effect of this degradation on paper strength is compensated for by the graft copolymer (288). The wet strength of paper from the modified pulp is actually significantly improved over that of untreated pulp despite the effect of the ozone.

(b) DECREASED AFFINITY FOR WATER

Schwab et al. (238, 239) showed that the affinity of pulp for water, as measured by moisture content at various relative humidities, could be markedly affected by grafting. At high add-on levels of hydrophobic polymers, such as poly(ethyl acrylate) (250% based on the weight of the cellulose), the regain was reduced 75%. When based on the cellulose content of the fiber, however, the level of reduction for the most effective polymer, polyacrylonitrile, was only 25%. For more hydrophilic polymers, such as from acrylamide, the equilibrium moisture content of the fibers was sharply increased, being as much as double that of the original sheet at 55% polymer content based on dry cellulose.

The hydrophobic polymer grafts were also found to reduce the rate of moisture pickup, particularly at high relative humidities. In addition, both the rate of absorbency and equilibrium pickup of liquid water were sharply reduced by these polymers. These results indicate the reduction in accessibility and rate of swelling of pulp fibers when thoroughly penetrated by nonwetting polymers.

These properties can be utilized for improving the dimensional stability of paper. Acrylic grafts of 100%, based on the weight of the cellulose, were found to reduce the dimensional change in paper by 50–60% over a 60% RH change (239). Daniel et al. (240) reported the dimensional change to be inversely proportional to the logarithm of the polyacrylonitrile content of grafted paper. These properties are closely related to the lower moisture regain and to the slower rate of moisture pickup of the treated fibers. For a given weight of fiber there is less cellulose to absorb moisture and hence less material to swell. In addition, the polymer prevents normal fiber collapse during initial sheet drying, resulting in less reexpansion of the fibers on humidification. Stamm and Cohen (289) found such bulking effects on dimensional stability with preformed polymer inclusions.

Kozmal et al. (290) improved the dimensional stability of paper for map and punched-card applications by grafting styrene onto the paper. Reaction was initiated with benzoyl peroxide in a methanolic solution of the monomer. Grafts of up to 30% polystyrene resulted in a reduction in moisture pickup of 95% without significantly altering the strength of the paper. The lowered moisture regain also enhanced the electrical insulating properties of the paper. Paper of improved thermal and dielectric properties has also been prepared (291) from fiber that had been beaten and then grafted with methacrylic acid.

An unusual hydrophobic graft to cellulose was achieved by grafting polysiloxane onto the fiber (292). A very nonwettable fiber was obtained by this technique. Reaction was achieved by treating alkali cellulose with dimethyldichlorosilane in pyridine and water. Small amounts of methyl methacrylate grafted onto paper can also produce a hydrophobic sheet (291). It is necessary to subject the originally hydrophilic grafted sheet to a subsequent heat treatment (293) or to the action of a solvent for methacrylate that does not swell the cellulose (294).

Rausing and Sunner (234) produced papers and wallboard that were highly impervious to liquid water by graft copolymerization with such olefinic monomers as isobutylene and α-methylstyrene. With 8–10% polymer add-on to the paper, water-penetration time was increased from a few seconds to more than 24 hr. The grafted paper exhibited gas-transmission, printability, and gluing properties similar to those of the untreated paper. Coated paper, by comparison, has greatly altered gas-transmission, printing, and handling qualities.

Resistance to oil is also made possible through the use of fluorochemical grafts (295, 296). Wrapping paper was made oleophobic by γ-ray irradiation of paper impregnated with poly(1,1-dihydroperfluoroisoamyl acrylate). (See also Section K of this Chapter XIX.)

(c) PREVENTION OF DETERIORATION

Paper products grafted with hydrophobic polymers also exhibit enhanced resistance to microbial attack. Daniel et al. (240) made soil-burial tests on paper grafted with a 50% add-on of polyacrylonitrile. The product showed no loss in dry tensile strength after 6 weeks' burial. Untreated paper was completely disintegrated after 2 weeks' exposure. Lynch (241) found similar resistance to attack by microorganisms for paper grafted with methyl methacrylate copolymer. These properties reflect partial blocking by the polymer of the accessible cellulose sites to attack by the hydrolytic enzymes of the microorganisms.

Further evidence of reduced accessibility to deterioration is found in the greater resistance of grafted papers to hydrolytic attack by strong acids and alkalies. Papers grafted with various acrylic polymers and copolymers retained appreciable tensile strength after immersion for 24 hr in 20% mineral acid solutions (239). Untreated paper disintegrated in 1–5 hr. Similar results were obtained with 62% sulfuric acid (240). Resistance to decomposition in 20% sodium hydroxide (239) and 1 *M* cupriethylenediamine (240) was also reported. Pulps grafted with vinyl chloride show similar resistance to chemical attack (253). Also, paper resistant to acidic hydrolysis resulted from a copolymer graft of nonene or butyl acrylate with acrylonitrile (297). (See also Sections A, B, and E in Chapter XVIII).

(d) MISCELLANEOUS PROPERTIES

A process that essentially results in the *encapsulation* of wood-pulp and other fibers was that developed by Herman and co-workers (298–300). It involves the polymerization of an olefinic monomer on the surface of the fiber, resulting in dramatic changes in fiber properties but apparently in no true grafting. Reaction is carried out by a Ziegler-type polymerization using a cellulose–metal alkyl intermediate as the active center for initiating polymerization. The intermediate is formed by first treating the fiber with a titanium or zirconium compound, such as titanium tetrachloride, followed by an organometallic alkylating agent, such as a Grignard reagent. An olefin, such as ethylene, is then polymerized at the reactive sites on the cellulose. Reaction is stopped by inactivating the catalyst with an alcohol. The product of this process consists of polymer-encapsulated fibers that can be formed into a paper web and bonded into an integral structure by heat-sealing the thermoplastic interfiber bonds. These properties are achieved at a polymer level as low as 20% on the fiber. The product has excellent wet strength, dimensional stability, and chemical resistance. It has been recommended for use as separator plates in lead-acid storage batteries and in industrial filters as well as in numerous packaging and decorative applications (298).

Another process that has been used to produce a polymer film around a fibrous paper web is *interfacial polycondensation* (301). Bolker (302) condensed diacyl halides with alkylenediamines to produce polyamide films intimately attached to the surface of paper webs. Although no evidence for grafting was obtained, interaction of acyl halides with cellulose hydroxyls is quite possible under the conditions of reaction (303). The sheet developed good barrier properties for oil, air, and water with films no more than 2.5 μ thick. Films as thick as 10 μ were produced.

Other special products, such as *ion-exchange membranes*, have also been produced by graft copolymerization onto paper. Richards and White (246) prepared permselective ion-exchange membranes by grafting a copolymer of acrylic acid and acrylonitrile to acid-parchmentized paper. Since direct polymerization of acrylic acid by the ceric-ion method was not successful, ethyl acrylate was copolymerized with acrylonitrile and later saponified to the acid.

In addition to ion-exchange materials, grafts of poly(acrylic acid) to cellulose have been recommended for use in surgical dressings and as styptics because of their *hemostatic properties* (304, 305). Grafted fibers are better able to withstand steam sterilization than other carboxylated cellulose derivatives (see also Section J of this Chapter XIX). One process for grafting to pulp involves preparing an intermediate mercapto or thiocarbonate derivative of cellulose, which is then combined with iron and reacted with peroxide in the presence of monomer. The process is believed to improve grafting efficiency and is applicable to a broad spectrum of monomer systems (305–307). It has been claimed (305) that papers with water repellency, thermoplasticity, dimensional stability, improved electrical properties, and resistance to bacterial degradation have been produced by the process.

Chromatographic substrates have been made from paper grafted with polystyrene (308). Artificial leather is another product that has been obtained from paper grafted with synthetic polymers (309). Filter paper and saturating paper grafted with various acrylates by the ceric-ion process have been suggested for this application. Grafted papers are also suitable as a base for photochromic papers according to a recent patent (310). The Oxford Paper Company (311) has developed a process for grafting thin films of polymers onto paper. Their process is based on the generation of free radicals in the paper by dielectric breakdown, in which the paper is used as the dielectric between two electrodes across which is applied a high potential. Paper thus treated is either contacted simultaneously or subsequently with monomer to form the desired graft. The process is particularly well suited to the coating of offset printing plates and for the binding of metal to paper in metallized paper. It is also claimed to be useful in producing glossy printing papers, papers with curl control, vapor-barrier coated papers, and heat-shrinkable packaging materials.

3. Crosslinking

Attempts to reduce the sensitivity of cellulose to moisture by treatment with crosslinking agents have been carried out for over 60 years (312–319). The dimensional stabilization of paper against moisture and the development of wet strength involve treatments with many polyfunctional reactants that are also believed to crosslink cellulose. Although the mechanisms of these reactions are not well established, they are considered here. The reader is referred to Section E in Chapter XVII for a more complete treatment of crosslinking reactions on cellulose and to Section A of this Chapter XIX for textile applications of crosslinking.

(a) IMPROVEMENT OF DIMENSIONAL STABILITY

Stamm and Cohen (289) first reported on the dimensional stabilization of paper by chemical crosslinking. Their process consisted of drying paper sheets from a 1% zinc chloride solution followed by vapor-phase reaction with formaldehyde at 85–100°C for up to 70 hr. The swelling of the paper in the thickness direction on subsequent immersion in water was reduced as much as 80%, with a weight increase from formaldehyde reaction of less than 0.5%. The treated sheets also exhibited up to a 20% reduction in moisture regain and a high degree of water repellency. Although such mechanical properties as bursting and tensile strengths were increased slightly, the sheets became stiffer and more brittle, had lower tearing strength and sharply decreased folding endurance. Wet strengths were increased sharply, approaching dry strength for the longer reaction periods. Reaction times of 20 hr or more, however, were required to achieve the maximum effects.

In later work Stamm (320) found that no crosslinking occurred in the presence of free water, and the reaction rate was slow above 88% RH at 100–120°C. When the swelling capacity of the sheet was reduced more than 60% by crosslinking, there was a general deterioration in strength properties due to embrittlement. Cohen, Stamm, and Fahey (321) found that crosslinking rates could be improved considerably by using small concentrations of hydrochloric acid in the vapor phase to catalyze reaction with formaldehyde. Sharp reductions in swelling capacity were achieved with reaction times of only 15 min. However, sheet strength deteriorated more readily as a function of dimensional stability achieved under these more drastic reaction conditions. It was further established that the cause of embrittlement during formaldehyde crosslinking is related to the nature of the acid used as catalyst. Cohen and Stamm (322–324) were also able to reduce swelling, dry strength, and viscosity in cupriethylenediamine and to improve dimensional stability by catalytic heat treatments with various acidic or basic catalysts *in the absence of formaldehyde.*

Stamm (320) studied the kinetics of the heat-stabilization reaction of paper and concluded that catalysts may be used to accelerate the reaction without changing the activation energy. The formaldehyde crosslinking reaction has a lower activation energy and can be used to impart a higher degree of stabilization with less sheet embrittlement. Back (325) ascribed the reduced water sorption and swelling and the improved wet strength and dimensional stability of heat-treated paper products to the formation of interfiber and intrafiber crosslinks. His results are in agreement with Stamm's (320) and suggest a chemical reaction within the cellulose structure.

The catalytic effect of acidic salts can be interpreted as resulting from the oxidative formation of carbonyl groups in the cellulose, followed by acidic condensations to produce hemiacetal crosslinks between adjacent chains (326, 327).

Heat treatment at about 200°C results in a permanent shrinkage. The sheet partially re-expands when put through subsequent humidification conditioning cycles, thus relieving part of the crosslinking stresses. These crosslinking stresses are of opposite direction to the paper-machine drying stresses, which tend to shrink the sheet when relieved by conditioning (328, 329). A thermal hardening of the sheet within 2–30 sec was observed in the 240–300°C range; this correlates very well with observed reductions in swelling tendency and with wet-strength development (330).

Marek (331) reported that the dimensional stability of paper can be improved by a process that does not result in the marked embrittlement and loss in absorbency associated with the dry-curing processes. He treated paper sheets with formaldehyde and an acid catalyst in low-swelling solvents in the presence of a controlled amount of moisture. Typical reaction conditions were immersion of a paper sheet in an acetic acid bath containing 2–4% formaldehyde, 0.5–2% hydrochloric acid catalyst, and 10% moisture for 20–40 min at 20–30°C. Up to 1% formaldehyde was bound in the web under these conditions. Paper sheets so treated show a 60–80% reduction in the dimensional changes normally incurred in taking paper sheets through a wetting and drying cycle. Such sheets also retain a high proportion of their dry strength and develop substantial wet strength as a result of the proposed crosslinking reaction. Best results were obtained with sheets from moderately beaten stocks; paper webs from bleached kraft pulps so treated retained 75% of their dry tensile strength and developed wet strengths approaching 60% of the dry strengths. Some high-rag-content sheets actually showed a 30% improvement in dry strength on treatment by this process. Improved properties through formaldehyde treatments have also been reported by other investigators (332–334).

The improvements in paper properties result from the reinforcement of hydrogen bonds in the formed paper web by the crosslinking reaction. If the pulp is reacted prior to web formation, practically no improvement in dimensional stability results and serious interference with normal interfiber bonding is observed. It may be assumed that the effectiveness of the dimensional stabilization of paper against moisture changes by crosslinking is highly dependent on the state of swelling of the paper web at the time of reaction (313).

Kress (335) reduced the hygroexpansivity of paper as much as 67% by catalyzed heat treatment with various acetals, prepared by the acid-catalyzed condensation of short-chain aliphatic aldehydes (e.g., paraformaldehyde) with polyfunctional alcohols (e.g., diethylene glycol). In addition to improved dimensional stability, the paper sheets have sharply higher wet strength and remain softer and more absorbent than with conventional wet-strength treatments. Similar crosslinking processes have been described in references 336, 337, and 338.

Paperboard can be stabilized against the effects of moisture by crosslinking with bifunctional and trifunctional isocyanates (261). Such treatments were found to enhance the rigidity of the board under humid and wet conditions. This reaction also improves the dielectric properties of paper (339). The use of divinyl sulfone or its precursors as well as other related sulfonium compounds has been reported to improve the dimensional stability and wet strength of paper (196, 340–343). Improvement in the wet strength and wet stiffness of paper by treating with tris(l-aziridinyl)phosphine oxide has been claimed (344). The process is unusual in that the reagent is applied to the pulp and cured after sheeting of the paper web.

(b) IMPROVEMENT OF WET STRENGTH

It is apparent from the preceding discussion that the methods used for improving dimensional stability in paper and board also increase wet strength. The question naturally arises as to whether or not the conventional means of developing wet strength in paper are also crosslinking reactions. This is a moot point that is still debated. The materials used are the same or very similar to those used in the creaseproofing of cotton textiles. In these cases the evidence strongly suggests the crosslinking mechanism (345–351) (see also Section E in Chapter XVII and Section A in this Chapter XIX).

There is also no doubt that other mechanisms may be involved. In the creaseproofing of cotton, for instance, penetration of a polyfunctional, monomeric reactant, such as dimethylolurea, within the fiber is preferred (352). To improve the wet strength of paper surface deposition of a polymeric resin is most effective for reinforcing the fiber-to-fiber bonds (353–359). It is theorized that this reinforcement may occur through the formation of a crosslinked insoluble polymer around the bonds, which reduces contact with water and inhibits swelling of the fiber-to-fiber bonds. There is even some indirect evidence against crosslinking, based on studies with model compounds that do not form chemical linkages with urea–formaldehyde resins (360). On the other hand, chemical linkages of model compounds with melamine–formaldehyde resins have been reported (348). It is likely that wet-strength development in paper involves both mechanisms to some degree.

Even alternative types of crosslinks between the cellulose and reactive resins have been proposed. There is good evidence for the formation of covalent crosslinks (346, 347, 361) between resin and anhydroglucose units in alternate cellulose chains involving all three available hydroxyl groups (347). Trout (362) presented evidence for ionic crosslinks between polycationic resins and residual carboxyl groups in the cellulose fiber. Carboxyl groups have also been shown to affect retention and bonding of wet-strength resins to cellulose through an ion-exchange mechanism (362, 363). The use of *N*-hydroxymethyl and *N*-alkoxymethyl derivatives of some urea compounds is specifically claimed to involve crosslinks (364, 365).

The use of *glyoxal* to improve the wet strength of paper without impairing absorbency was claimed by Woods and Schenck (366). Later the use of glyoxal for the development of transient wet strength was reported (367–371). Eldred and Spicer (367) have shown that glyoxal in the absence of a catalyst develops a temporary wet strength. This is explained by the observations of Head (372) and Farrow et al. (373) that glyoxal forms only readily hydrolyzed hemiacetal bonds with cellulose in the absence of a catalyst. Heating in the presence of an acid catalyst leads to the formation of permanent acetal crosslinks that are not cleaved by cuprammonium hydroxide. Glyoxal may also be used with starch or galactomannan gums in paper coatings. Simple crosslinking of the coating has been suggested in the case of starch (371), but bonding of the galactomannan to the cellulose through glyoxal is the mechanism proposed in the case of galactomannan (369).

Dialdehyde starch is another polyfunctional aldehyde material that behaves similarly to glyoxal in imparting temporary wet strength to paper (374, 375). Hemiacetal linkages have also been proposed for the reaction of dialdehyde starch with cellulose to develop readily reversible crosslinks (374, 376). Opie and Keen (377) also reported the development of temporary wet strength using dialdehyde galactomannan gums. They similarly propose hemiacetal links with the cellulose, which would afford crosslinks between adjacent chains. The incorporation of dialdehyde groups directly into the cellulose by periodate oxidation of the pulp has also been shown to produce a temporary wet strength (378).

Papers containing both synthetic and pulp fibers have been prepared by using a crosslinking agent that will react with both types of fiber. The papers have improved wet strength (379, 380).

(c) MISCELLANEOUS MODIFICATIONS

Harpham and Turner (381) discovered that when crosslinked under swollen or partially swollen conditions, cellulosic papermaking fibers can be combined to yield paper with improved porosity and absorbency. In addition, they found that crosslinking permitted them to achieve this high degree of porosity without the customary sacrifice in formation usually incurred. This is particularly important in paper webs prepared from cotton linters, with which it is difficult to achieve good formation without partially closing up the web through excess refining. The process consists of reacting the pulp with a polyfunctional crosslinking agent under conditions that do not tender or embrittle the fibers. Typical reactants are epichlorohydrin or dichloroacetone reacted with the fibers in aqueous alkali, or formaldehyde reacted in an acidic acetone solution. The crosslinked pulp is then refined for 1.5–2 hr and made into paper sheets. The use of crosslinked pulp in mixtures with untreated fiber gives improved properties in proportion to the amount of modified fiber used.

Paper with greater bulk, porosity, and absorbency can also be prepared by incorporating into the furnish pulp crosslinked in the collapsed state. Bernardin (382) produced a stiff, resilient fiber by crosslinking pulp in dry fiber form while the fibers are in individual or loosely associated state. This is accomplished by impregnating pulp in sheet form with a crosslinking agent (such as formaldehyde or dimethylolurea) and an acidic catalyst, partially drying the web without crosslinking, mechanically disentangling the web, and heat-curing the separated fibers.

The product from this process consists of stiffened fibers that have reduced moisture regain, more than double the wet compression resistance of untreated pulp, and sharply decreased filtration resistance. The fibers are highly wettable but no longer have any interfiber bonding capacity for themselves. However, they can be bonded into a paper sheet by mixing with untreated pulp fibers and, when used in this manner, produce a soft, porous highly absorbent paper sheet.

Crosslinking may also be involved in certain flame-retardant treatments of paper (383, 384). In these cases the cellulose is impregnated with precondensates of tetrakis (hydroxymethylphosphonium) chloride or its homologs with urea and formaldehyde. Crosslinking with the cellulose may occur on curing. Similarly polymeric phosphonitrilic chlorides containing reactive groups capable of crosslinking with the cellulose are claimed to be effective flame retardants for paper and textiles (385) (see also Section I in this Chapter XIX).

Many other types of paper products that involve saturation of a paper web with monomeric or polymeric materials containing polyfunctional isocyanates have been prepared (386–395). On curing the saturated webs one obtains complex crosslinked polymer–paper combinations that have useful properties. These are specialized products, such as tape backings and reinforced papers, in which crosslinking may contribute to the final web properties.

References

1. *Pulp and Paper,* **44,** 7 (June 25, 1970).
2. J. d'A. Clark, in Chapter VIII, p. 649, of **Part II.**
3. C. J. Malm and G. D. Hiatt, in Chapter IX-C, p. 785, of **Part II.**
4. G. Jayme and D. Froundjian, *Cellulosechem.,* **18,** 9 (1940).

5. J. C. Bletzinger, *Ind. Eng. Chem.*, **35**, 474 (1943).
6. W. H. Aiken, *Ind. Eng. Chem.*, **35**, 1206 (1943).
7. J. A. Walecka, *Tappi*, **39**, 458 (1956).
8. L. O. Klinga and E. L. Back, *Svensk Papperstidn.*, **69**, 64 (1966).
9. L. O. Klinga and H. Tarkow, *Tappi*, **49**, 23 (1966).
10. A. Meller, W. E. Cohen, H. G. Higgins, and G. M. Stewart, *Appita*, **13**, No. 6, xix (1960).
11. H. G. Higgins, K. Harrington, and A. W. McKenzie, *Proc. Australian Pulp and Paper Ind. Co-operative Research Conf.*, **16**, 92 (1957).
12. H. G. Higgins, A. W. McKenzie, and K. J. Harrington, *Tappi*, **41**, 193 (1958).
13. A. W. McKenzie, K. Harrington, and H. G. Higgins, *Proc. Australian Pulp and Paper Ind. Co-operative Research Conf.*, **16**, 123, 124 (1957).
14. A. W. McKenzie and H. G. Higgins, *Australian J. Appl. Sci.*, **6**, 208 (1955).
15. L. O. Klinga, H. Tarkow, and E. L. Back, *Svensk Papperstidn.*, **68**, 583 (1965).
16. E. L. Back and L. O. Klinga, *Svensk Papperstidn.*, **66**, 745 (1963).
17. O. J. Schulwitz, U.S. Patent 2,992,156 (1961) (to Eastman Kodak Co.).
18. C. J. Malm and G. C. Gandy, U.S. Patent 3,057,755 (1962) (to Eastman Kodak).
19. C. J. Malm and G. C. Gandy, U.S. Patent 3,068,116 (1962) (to Eastman Kodak).
20. C. J. Malm, L. E. Herdle, and G. C. Gandy, U.S. Patent 3,047,391 (1962) (to Eastman Kodak).
21. L. E. Herdle and G. C. Milligan, U.S. Patent 3,062,679 (1962); W. H. Griggs and R. D. Zaffrann, U.S. Patent 3,096,231 (1963); W. H. Griggs, A. H. Gray, and R. D. Zaffrann, U.S. Patents 3,103,462 (1963) and 3,132,944 (1964) (all assigned to Eastman Kodak).
22. L. E. Herdle and W. H. Griggs, *Tappi*, **48**, No. 7, 103A (1965).
23. J. A. Harpham, A. R. Reid, and H. W. Turner, *Tappi*, **41**, 629 (1958).
24. J. J. Spadaro, H. J. Janssen, H. L. E. Vix, and F. L. Simons, *Tappi*, **41**, 674 (1958).
25. J. L. Morton, and N. M. Bikales, *Tappi*, **42**, 855 (1959); *Paper Trade J.*, **143**, 43 (Feb. 16, 1959).
26. E. D. McKay and N. M. Bikales, Canadian Patent 622,868 (1961) (to Southern Chemical Cotton Co. and American Cyanamid Co.).
27. H. Mizutani, N. Nakamura, and Y. Kubo, *Toshiba Rev.*, **16**, No. 9, 1161 (1961); through *Abstr. Bull. Inst. Paper Chem.*, **33**, 578 (1962).
28. F. B. Cramer and E. G. Parrish, U.S. Patent 3,266,972 (1966) (to E. I. du Pont de Nemours & Co.).
29. H. Brody, E. G. Parrish, E. H. Man, and I. K. Miller, British Patent 999,587 (1965) (to Du Pont).
30. J. E. Jayne, U.S. Patent 2,994,634 (1961) (to Kimberly-Clark Corp).
31. W. A. Wink, K. Ward, Jr., and H. A. Swenson, *Tappi*, **51**, 445 (1968).
32. G. A. Richter, U.S. Patent 2,038,679 (1936) (to Brown Co.).
33. A. W. Schorger, U.S. Patent 2,533,145 (1950) (to Burgess Cellulose Co.).
34. *Paper Trade J.*, **141**, 23 (Nov. 25, 1957).
35. *World Paper Trade Review*, **149**, 306, 308, 310 (1958).
36. J. A. Harpman, A. R. Reid, and H. W. Turner, *Tappi*, **41**, 758 (1958).
37. K. Ward, Jr., M. L. Murray, A. J. Morak, and M. H. Voelker, *Tappi*, **50**, 462 (1967).

38. J. A. Harpham, ASTM Special Tech. Publ. No. 241, March 1959, p. 43.
39. J. A. Harpham, U.S. Patent 2,916,413 (1959) (to Hercules Powder Co.).
40. H. W. Turner, U.S. Patent 2,856,402 (1958) (to Hercules).
41. R. H. Gillespie, M. Mueller, H. Swenson, and K. Ward, Jr., *Tappi*, **44**, 662 (1961).
42. R. L. Mitchell, R. F. Bampton, and W. H. Wadman, U.S. Patent 2,847,411 (1958) (to Rayonier, Inc.).
43. K. Ward, Jr., *Tappi*, **43**, 54 (1960).
44. K. Ward, Jr., and M. H. Voelker, *Tappi*, 49, 155 (1966).
45. H. P. Didwania, *Tappi*, **52**, 1511 (1969).
46. N. I. Klenkova, *J. Appl. Chem. USSR*, **29**, 429 (1956); English translation in *Soviet Research on the Swelling of Cellulose Materials*, H. Tarkow, Ed., Consultants Bureau Enterprises, New York, p. 160.
47. K. Ward, Jr., A. J. Morak, R. H. Gillespie and M. H. Voelker, *Tappi*, **51**, 218 (1968).
48. M. Plunguian, ATIP Bulletin 1961, No. 4, p. 277.
49. B. B. Thomas, *Tappi*, **49**, 515 (1966).
50. K. Ward, Jr., A. J. Morak, M. L. Murray, and W. B. Thomas, *Tappi*, **48**, 207 (1965).
51. K. A. Jurbergs, *Tappi*, **43**, 554 (1960).
52. J. H. Beyer, J. H. Foreman, and L. R. Parks, U.S. Patent 3,031,371 (1962) (to Buckeye Cellulose Corp.).
53. K. Ward, Jr., A. J. Morak, and M. H. Voelker, *Tappi*, **52**, 479 (1969)
54. D. F. Maskey and T. A. Feazel, *Paper Trade J.*, **142**, 56 (Sept. 22, 1958).
55. E. L. Akim and N. I. Nikitin, *Zh. Prikl. Khim.*, **36**, 1075 (1963); *Tr. Leningrad. Tekhnol. Inst. Tsellyuloz-Burmazh. Prom.*, No. 12, 193, (1964); *Abstr. Bull. Inst. Paper Chem.*, **34**, 1613 (1964); **35**, 1065 (1965); **37**, 343,748 (1967).
56. O. A. Battista, M. M. Cruz, Jr., and F. H. Reichel, Jr., Canadian Patent 705,292 (1965) (to FMC Corp.).
57. B. B. Thomas and G. B. Creamer, U.S. Patent 3,340,139 (1967) (to Rayonier).
58. K. Ward, Jr., H. A. Swenson, W. A. Wink, and R. J. Greisinger, *Tappi*, **52**, 2164, 2336 (1969).
59. J. A. Harpham, Canadian Patent 627,921 (1961) (to Hercules).
60. Rayonier, Inc., British Patent 1,024,096 (1966).
61. K. Ward, Jr., M. L. Murray, and W. B. Thomas, *Tappi*, **49**, 20 (1966).
62. M. L. Murray and K. Ward, Jr., U.S. Patent 2,987,434 (1961) (to Institute of Paper Chemistry).
63. K. K. Talwar, *Tappi*, **41**, 207 (1958).
64. Papeteries de la Chapelle, French Patent 1,299,052 (1962).
65. P. F. Nelson and C. G. Kalkipsakis, *Tappi*, **47**, 107 (1964).
66. P. F. Nelson and C. G. Kalkipsakis, *Tappi*, **47**, 170 (1964).
67. G. Ruffini, *Svensk Papperstidn.*, **69**, 72 (1966).
68. R. T. Cady and J. H. Steinberg, Canadian Patent 632,111 (1961) (to Du Pont).
69. Mishima Paper Mfg. Co., Ltd., British Patent 1,071,706 (1967).
70. *Chem. Week*, Oct. 21, 1967, p. 92.
71. Y. Mizutani, S. Sakamoto, and Y. Ishibashi, British Patent 1,111,273 (1968) (to Mishima Paper Manufacturing Co., Ltd.).
72. M. M. Cruz, Jr., and R. L. McDowell, U.S. Patent 3,116,199 (1963) (to FMC Corp.).

73. D. L. Baker and O. Sepall, *Pulp Paper Mag. Can.*, **66**, T449 (1965).
74. I. K. Miller, U.S. Patent 3,338,884 (1967) (to Du Pont).
75. O. Andersson, *Svensk Papperstidn.*, **64**, 417 (1961).
76. A. W. McKenzie, *Appita*, **19**, 79 (1965).
77. N. M. Bikales and J. R. Stephens, U.S. Patent 3,029,232 (1962) (to American Cyanamid).
78. W. B. Neely, U.S. Patent 3,133,856 (1964) (to Dow Chemical Co.).
79. Hercules Powder Co., British Patent 795,587 (1958); K. Ward, Jr., and J. A. Walecka, U.S. Patent 2,987,433 (1961) (to Institute of Paper Chemistry).
80. A. G. Thaysen, British Patent 399,952 (1933); A. Rheiner, U.S. Patent 2,017,607 (1935) (to Sandoz).
81. W. Dieterle, *Bull. Swiss Electrotechn. Inst.*, **46**, No. 22, 1045 (1955).
82. W. Dieterle, *Insulation*, 19 (1962).
83. H. Leemann, A. Rheiner, and W. Hagenbuch, U.S. Patents 2,357,962 (1944), 2,577,715 (1951), 2,814,617 (1957) (to Sandoz).
84. G. A. Richter and L. E. Herdle, Canadian Patent 615,410 (1961) (to Canadian Kodak Co.).
85. G. G. Gray and J. T. Leone, U.S. Patent 3,015,537 (1962) (to Eastman Kodak).
86. Kodak S. A., Belgian Patent 563,842 (1958).
87. G. P. Touey, U.S. Patent 3,033,209 (1962) (to Eastman Kodak).
88. V. K. Eletskaya and F. I. Korchemkin, *Sb. Tr. Tsentr. Nauch.-Issled. Inst. Bumagi*, No. 2, 87, 288 (1967); through *Abstr. Bull. Inst. Paper Chem.*, **39**, 394 (1968).
89. L. E. Herdle and G. C. Gandy, Canadian Patent 649,270 (1962) (to Canadian Kodak Co.).
90. C. J. Malm and L. W. Blanchard, Jr., Canadian Patent 652,570 (1962) (to Canadian Kodak Co., Ltd.).
91. W. H. Griggs, R. D. Zaffrann, L. E. Herdle, G. C. Gandy, C. J. Malm, and L. W. Blanchard, Jr., British Patent 937,494 (1963) (to Kodak Ltd.); W. H. Griggs and R. D. Zaffrann, Canadian Patent 685,711 (1964) (to Canadian Kodak).
92. D. Edge, Jr., M. D. Fahey, and R. G. Rickey, Canadian Patent 733,013 (1966) (to Rayonier); Rayonier, Inc., British Patents 937,977 (1963) and 964,197 (1964).
93. A. Letenay and E. Oltus, *Sb. Vyskum. Prác Odboru Celulozy Papiera*, No. 10, 135 (1965); through *Abstr. Bull. Inst. Paper Chem.*, **39**, 118 (1968).
94. K. Katsuura and T. Suzuki, *J. Japan Tappi*, **16**, 440 (1962); through *Abstr. Bull. Inst. Paper Chem.*, 34, 831 (1964).
95. E. Kelk and I. O. Wilson, *Proc. Inst. Electr. Engrs. (London)*, **112**, No. 3, 602 (1965).
96. K. Schwaner, *Elektrotechnica (Budapest)*, **59**, No. 11/12, 515 (1966); through *Abstr. Bull. Inst. Paper Chem.*, **38**, 507 (1968); V. Lica, *Elektrotehnica (Bucharest)*, **15**, No. 1, 16 (1967); through *Abstr. Bull. Inst. Paper Chem.*, **38**, 648 (1968).
97. F. I. Korchemkin and A. I. Bobrov, *Bumazh. Prom.*, **39**, No. 1, 15 (1964); through *Abstr. Bull. Inst. Paper Chem.*, 34, 1810 (1964); V. V. Vasil'eva, F. I. Korchemkin, and A. I. Bobrov, *Bumazh. Prom.*, No. 10, 9 (1965); through *Abstr. Bull. Inst. Paper Chem.*, **36**, 1275 (1966); V. T. Renne, L. M. Gulyaeva, A. I. Bobrov, F. I. Korchemkin, V. V. Vasil'eva, and V. K. Eletskaya, *Izv. Vyssikh Ucheb. Zavedenii, Energetika*, **9**, No. 8, 103 (1966); through *Abstr. Bull. Inst. Paper Chem.*, **37**, 696 (1967); F. I. Korchemkin and V. K. Eletskaya, *Sb. Tr. Tsentr. Nauch.-Issled. Inst. Bumagi*, No. 1, 26 (1966); through *Abstr. Bull. Inst. Paper Chem.*, **39**, 33 (1968).

98. F. Y. Chu, U.S. Patent 3,082,123 (1963) (to Eastman Kodak).

99. V. Loras, *Pulp Paper Mag. Can.*, **69**, 57 (Jan. 19, 1968).

100. D. F. Manchester, J. W. McKinney, and A. A. Pataky, *Svensk Papperstidn.*, **63**, 699 (1960).

101. R. P. Singh, *Tappi*, **49**, 281 (1966).

102. D. H. Andrews and P. Des Rosiers, *Pulp Paper Mag. Can.*, **67C**, T119 (1966).

103. H. F. Miller and R. G. Flowers, U.S. Patent 2,535,690 (1950) (to General Electric Co.).

104. *Paper Trade J.*, **143**, 44 (Oct. 19, 1959).

105. *Chem. Week*, **85**, 61 (Aug. 15, 1959).

106. J. C. Leslie, Canadian Patent 678,017 (1964) (to General Electric); French Patent 1,284,771 (1962) (to Compagnie Française Thomson-Houston).

107. General Electric Co., British Patent 975,188 (1964).

108. General Electric Co., British Patent 1,060,983 (1967).

109. N. M. Bikales, in N. M. Bikales, Ed. *Encyclopedia of Polymer Science and Technology*, Vol. 4, Interscience, New York, 1966, p. 533.

110. G. Prati, *Ind. Carta* (N.S.), **1**, No. 11, 457 (1963).

111. G. Prati, *Ind. Carta* (N.S.), **2**, 39 (1964).

112. L. M. Gulyaeva, L. I. Gurevich, and V. T. Renne, *Tr. Leningr. Politekh. Inst.*, No. 276, 18 (1967); through *Abstr. Bull. Inst. Paper Chem.*, **38**, 503 (1968).

113. General Electric Co., British Patent 922,909 (1963).

114. B. D. Brummet, Canadian Patent 692,739 (1964) [to McGraw-Edison (Canada) Ltd.]

115. R. L. Mitchell, U.S. Patent 3,032,549 (1962) (to Rayonier).

116. Le Matériel Électrique, French Patent 1,297,186 (1962).

117. F. L. Granger, Canadian Patent 608,233 (1960) (to Union Carbide and Carbon Corp.); R. Hatfield, U.S. Patent 2,904,615 (1959) (to Union Carbide).

118. W. G. Russbach and S. A. Teiser, U.S. Patent 3,103,549 (1963) (to Du Pont).

119. A. J. Stamm and J. N. Beasley, *Tappi*, **44**, 271 (1961).

120. K. Schwaner and K. Bitay, *Plaste Kautschuk*, **14**, No. 3, 166 (1967).

121. J. V. Kŏstíř and K. Slavík, *Coll. Czech. Chem. Commun.*, **15**, 17 (1950); through *Chem. Abstr.*, **44**, 8817 (1950).

122. A. Rieche and M. Schulz, *Angew. Chem.*, **70**, 694 (1958).

123. F. Micheel and H. Schweppe, *Angew. Chem.*, **66**, 136 (1954).

124. F. Micheel and H. Schweppe, *Mikrochim. Acta*, 53 (1954).

125. *Paper Maker (London)*, 134, International no. 36 (1957).

126. F. Micheel and H. Schweppe, *Naturwissenschaften*, **39**, 380 (1952).

127. E. M. Buras, Jr., and S. R. Hobart, *Anal. Chem.*, **27**, 1507 (1955).

128. J. W. H. Zijp, *Chem. Weekblad*, **51**, 547 (1955).

129. J. Janousek, *Faserforsch. u. Textiltech.*, 8, No. 8, 339 (1957).

130. B. Hocevar, *Vest. Sloven. Kemi Drustva*, **6**, 1 (1959); through *Chem. Abstr.*, **54**, 20414 (1960).

131. F. Micheel and P. Albers, *Mikrochim. Acta*, 489 (1954).

132. F. Micheel, *Acta Chim. Acad. Sci. Hung.*, **12**, 331 (1957); through *Chem. Abstr.*, **52**, 10687 (1968).

133. F. Micheel and P. Albers, *Ber.*, **89**, 140 (1956).

134. F. Micheel and W. Leifels, *Ber.*, **91**, 1212 (1958).
135. O. P. Cohen and J. F. Heaps, U.S. Patent 2,794,736 (1957) (to Monsanto Chemical Co.).
136. A. J. Stamm and W. E. Cohen, *Australian Pulp and Paper Ind. Tech. Assoc. Proc.*, **10**, 366 (1956).
137. E. H. Man, U.S. Patent 3,328,892 (1967) (to Du Pont).
138. Hercules Powder Co., *Chem. Eng. News*, **33**, No. 10, 1018 (March 7, 1955).
139. *Aquapel for the Paper Industry*, Hercules Powder Co., Wilmington, Del., 1955.
140. W. F. Downey, U.S. Patent 2,627,477 (1953) (to Hercules).
141. *Aquapel*, Hercules Powder Co., Wilmington, Del., 1963.
142. J. W. Davis, W. H. Roberson, and C. A. Weisgerber, *Tappi*, **39**, 21 (1956).
143. G. I. Keim and W. D. Thompson, U.S. Patent 2,762,270 (1956) (to Hercules).
144. C. A. Weisgerber, U.S. Patent 2,865,743 (1958) (to Hercules).
145. W. O. Kincannon, Jr., *Paper Mill News*, **86**, No. 51, 28 (1963).
146. *Allg. Papier-Rundschau*, No. 1,19 (1966); *ibid.*, No. 3, 65 (Jan. 20, 1966).
147. W. E. Osberg, Jr., U.S. Patent 2,785,067 (1957) (to Hercules).
148, G. C. Harris and C. A. Weisgerber, U.S. Patent 3,070,452 (1962) (to Hercules); C. H. Chapman, U.S. Patent 3,130,118 (1964) (to Hercules).
149. H. G. Arlt, Jr., U.S. Patent 3,046,186 (1962) (to American Cyanamid); C. A. Weisgerber, U.S. Patents 2,961,366 and 2,961,367 (1960) (to Hercules).
150. C. A. Weisgerber and J. W. Davis, U.S. Patent 2,986,488 (1961) (to Hercules); A. R. Savina, U.S. Patent 3,223,544 (1965) (to American Cyanamid).
151. Hercules Powder Co., British Patent 802,357 (1958); M. O. Schur, U.S. Patent 3,006,806 (1961) (to Olin Mathieson Chemical Corp.).
152. C. A. Weisgerber and C. B. Hanford, *Tappi*, **43**, No. 12, 178A (1960).
153. T. H. Werner, W. H. Marra, and W. B. Gilman, U.S. Patent 2,992,964 (1961) (to S. D. Warren Co.).
154. N. P. Perekal'skii, L. N. Antonovich, Z. M. Kryukov, and Yu. V. Kurylev, *Tr. Leningrad. Tekhnol. Inst. Tsellyuloz-Bumazh. Prom.*, No. 10, 15 (1962); *ibid*, No. 12, 224 (1964); through *Abstr. Bull. Inst. Paper Chem.*, **34**, 1672 (1964); **36**, 209 (1965).
155. S. V. Worthen and R. L. Oransky, U.S. Patent 3,276,359 (1966) (to S. D. Warren Co.).
156. W. H. Roberson, Canadian Patent 567,352 (1958) (to Hercules); C. A. Weisgerber, U.S. Patent 3,212,961 (1965) (to Hercules).
157. R. T. Hart, U.S. Patent 2,776,226 (1957) (to S. D. Warren Co.).
158. E. A. Shiner, British Patent 1,042,182 (1966) (to Union Carbide).
159. W. J. Barrow, Publication No. 13, Library Board of Virginia, 1960, pp. 13, 26.
160. R. W. Church, *Tappi*, **46**, No. 6, 202A (1963).
161. H. P. Dixson, Jr., and J. C. Nelson, *Tappi*, **45**, 753 (1962); *ibid.*, **46**, No. 6, 202A (1963).
162. D. G. Lundgren, R. E. Kelly, and A. K. Wiebe, *Modern Packaging*, **31**, No. 2, 174 (1957).
163. J. P. Bain, U.S. Patent 3,219,517 (1965) (to Glidden Co.).
164. A. Skogseid, U.S. Patent 3,051,543 (1962) (to Jens Villadsen's Fabrikker A/S).
165. R. J. Kulick, E. Strazdins, and A. R. Savina, U.S. Patent 3,311,532 (1967) (to American Cyanamid).

166. H. G. Arlt, Jr., U.S. Patent 2,901,371 (1959) (to American Cyanamid).

167. C. A. Weisgerber, Canadian Patent 611,283 (1960) (to Hercules); Hercules Powder Co., British Patent 804,504 (1956).

168. N. J. Hayes and W. C. Ross, U.S. Patent 2,897,094 (1959) (to W. R. Grace and Co.); R. L. Berger and C. H. Gelbert, *Tappi,* **44**, 467 (1961); H. G. Arlt, Jr., U.S. Patents 3,050,437 (1962) and 3,084,092 (1963) (to American Cyanamid).

169. A. Nathansohn, U.S. Patent 1,996,707 (1935).

170. E. Strazdins and R. J. Kulick, Canadian Patents 710,234 and 710,365 (1965); A. R. Savina, U.S. Patent 3,223,543 (1965); (all assigned to American Cyanamid).

171. O. B. Wurzburg and E. D. Mazzarella, U.S. Patent 3,102,064 (1963) (to National Starch and Chemical Corp.); R. W. Liggett, U.S. Patent 3,130,373 (1964) (to Allied Chemical Corp.); J. E. Berger, Canadian Patent 770,079 (1967) (to Shell Internationale Research Maatschappij N.V.).

172. D. H. Page and J. W. Sargent, *Nature,* **207**, 217 (1965).

173. G. D. Sinclair, R. S. Evans, and H. R. Sallans, *Pulp Paper Mag. Can.,* **62**, No. 1, T23 (1961). G. D. Sinclair, *Tappi,* **47**, 579 (1964).

174. G. D. Sinclair and H. R. Sallans, U.S. Patent 3,306,812 (1967) (to National Research Council, Canada).

175. M. A. Buchanan, S. L. Burson, Jr., and C. H. Springer, *Tappi*, **44**, 576 (1961).

176. J. W. Swanson and S. Cordingly, *Tappi,* **42**, 812 (1959).

177. H. A. Brown and R. J. Koshar, U.S. Patent 3,147,064 (1964) (to Minnesota Mining and Manufacturing Co.).

178. D. K. Pattilloch, C. Polowczyk, U.S. Patent 2,957,796 (1960) (to Michigan Research Labs., Inc., and Electro-Chem Fiber Seal Corp.).

179. L. A. Miller, U.S. Patents 3,100,794 (1963) and 3,218,314 (1965) (to Monsanto).

180. *Whatman Cellulose Ion Exchangers,* Technical Bulletin 1031, H. Reeve Angel & Co., Clifton, N.J.

181. A. J. Head, N. F. Kember, R. P. Miller, and R. A. Wells, *J. Chem. Soc.,* 3418, (1958).

182. C. S. Knight, *Nature,* **183**, 165 (1959).

183. N. F. Kember, *Analyst,* **85**, 449 (1960).

184. O. Huber, H. Kolb, and J. Weigl, *Papier*, **20**, 173 (1966).

185. C. S. Knight, *Nature,* **184**, 1486 (1959); *ibid.,* **194**, 90 (1962).

186. J. Hartel, J. A. Copper, and C. Van Bochove, *Rec. Trav. Chim.,* **82**, 264 (1963).

187. J. R. Sherman, *Anal. Biochem.,* **5**, 548 (1963).

188. J. S. Gulbinsky and W. W. Cleland, *Biochemistry,* **7**, 566 (1968).

189. C. S. Knight, U.S. Patent 3,226,380 (1965) (to W. and R. Balston, Ltd.).

190. C. S. Knight, British Patent 980,174 (1965) (to W. and R. Balston, Ltd.).

191. E. L. Taylor, U.S. Patent 3,238,192 (1966) (to KVP Sutherland Paper Co.).

192. E. L. Taylor, *Textile Res. J.,* **38**, 104 (1968).

193. A. Süszer, *Israel J. Chem.,* **4**, 123 (1966).

194. F. H. Burkitt and R. Tattersall, British Patent 1,036,571 (1966) (to Tootal Broadhurst Lee Co., Ltd.).

195. F. H. Burkitt, J. H. Butterworth, R. E. Hunt, W. H. Watson, and F. Farrington, British Patent 1,047,323 (1966) (to Tootal Broadhurst Lee).

196. L. M. Valentine and H. Fletcher, British Patent 988,511 (1965), (to Tootal Broadhurst Lee).
197. T. Wieland and A. Berg, *Angew. Chem.,* **64,** 418 (1952).
198. U. Ströle, *Z. Anal. Chem.,* **144,** 256 (1955).
199. I. N. Ermolenko and M. L. Longin, *Zh. Anal. Khim.,* **19,** No. 4,425 (1964); through *Abstr. Bull. Inst. Paper Chem.,* **35,** 1267 (1965).
200. I. N. Ermolenko and V. I. Sokolova, U.S.S.R. Patent 189,684 (1966) through *Abstr. Bull. Inst. Paper Chem.,* **37,** 802 (1967).
201. I. N. Ermolenko, M. L. Longin, and G. L. Elina, U.S.S.R. Patent 196,298 (1967); through *Abstr. Bull. Inst. Paper Chem.,* **38,** 463 (1967).
202. V. I. Ivanov, N. Ya. Lenshina, and V. S. Ivanova, *Stärke,* **12,** No. 10, 311 (1960).
203. D. D. Gilboe and R. M. Bock, *J. Chromatog.,* **17,** 149 (1965).
204. M. A. Jermyn and R. Thomas, *Nature,* **172,** 728 (1953).
205. J. D. Guthrie, *Textile Res. J.,* **17,** 625 (1947).
206. W. Lautsch, G. Manecke, and W. Broser, *Z Naturforsch.,* **8b,** 232 (1953).
207. N. F. Kember and R. A. Wells, *Nature,* **175,** 512 (1955).
208. J. D. Guthrie and A. L. Bullock, *Ind. Eng. Chem.,* **52,** 935 (1960).
209. M. A. Mitz and L. Metcalfe, Canadian Patent 610,901 (1960) (to Armour and Co.).
210. J. C. Williams, Canadian Patent 745,528 (1966) (to Hawley Products Co.).
211. G. Kay and E. M. Crook, *Nature,* **216,** 514 (1967); G. Kay, M. D. Lilly, A. K. Sharp, and R. J. H. Wilson, *Nature,* **217,** 641 (1968).
212. G. W. Meadows, Canadian Patent 584,959 (1959) (to Du Pont).
213. K. Randerath, *Angew. Chem., Intern. Ed.,* **1,** 553 (1962).
214. H. H. McClure, *Can. Pulp Paper Ind.,* **17,** 48, 52, 58 (June, 1964).
215. A. Goldstein, in N. M. Bikales, Ed., *Encyclopedia of Polymer Science and Technology,* Vol. 1, Interscience, New York, 1964, p. 734.
216. Statens Skogsindustrier, AB, British Patent 922,674 (1963).
217. N. Ya. Solechnik and A. A. Leonovich, U.S.S.R. Patent 195,626 (1966); (to Leningradskaya Lesotekhnicheskaya Akademiya im. S. M. Kirova) through *Abstr. Bull. Inst. Paper Chem.,* **38,** 465 (1967).
218. H. Coates, U.S. Patent 2,983,623 (1961), (to Albright and Wilson, Ltd.).
219. H. Coates and B. Chalkley, U.S. Patents 3,096,201 (1963) and 3,236,676 (1966) (to Albright and Wilson, Ltd.).
220. R. J. Dearborn, Canadian Patents 631,411 and 631,954 (1961) (to Hooker Chemirad Corp.).
221. Albright and Wilson, Ltd., British Patent 893,283 (1962).
222. Dow Chemical Co., British Patent 1,063,273 (1967).
223. H. Osborg, J. W. Brook, and A. Goldstein, U.S. Patent 3,298,902 (1967) (to Chemirad Corp.).
224. L. R. Dunn and L. W. Eckert, U.S. Patent 2,875,044 (1959) (to Armstrong Cork Co.).
225. E. H. Immergut, in N. M. Bikales, Ed., *Encyclopedia of Polymer Science and Technology,* Vol. 3, Interscience, New York, 1965, pp. 242–284.
226. B. Steenberg, *Pulp Paper Mag. Can.,* **63,** 84 (Nov., 1962).
227. V. Stannett, *Chem. Eng. News,* **40,** 57 (June 18, 1962).

228. R. H. Cornell, *Tappi,* **45,** No. 7, 145A (1962).

229. R. Y. M. Huang, B. Immergut, E. H. Immergut, and W. H. Rapson, *J. Polymer Sci.,* [A], **1,** 1257 (1963).

230. A. Y. Kulkarni and P. C. Mehta, *J. Appl. Polymer Sci.,* **9,** 2633 (1965).

231. V. Stannett, *Tappi,* **47,** No. 3., 58A, 64A, and 72A (1964).

232. A Zilkha, B.-A. Feit, and B.-N. Akiva, U.S. Patent 3,341,483 (1967) (to Yissum Research Development Co., Hebrew University).

233. R. F. Schwenker, Jr., and E. Pacsu, U.S. Patent 3,312,642 (1967) (to Kimberly-Clark); *Tappi,* **46,** 665 (1963).

234. G. Rausing and S. Sunner, *Tappi,* **45,** No. 1, 203A (1962).

235. J. Schurz, *Papier,* **18,** 437 (1964).

236. M. Imoto, K. Takemoto, and M. Kondo, *Makromol. Chem.,* **98,** 74 (1966).

237. M. Imoto, K. Takemoto, H. I. Azuma, N. Kita, and M. Kondo, *Makromol. Chem.,* **107,** 188 (1967).

238. E. Schwab, V. Stannett, and J. J. Hermans, *Tappi,* **44,** 251 (1961).

239. E. Schwab, V. Stannett, D. H. Rakowitz, J. N. Magrane, *Tappi,* **45,** 390 (1962).

240. J. H. Daniel, Jr., S. T. Moore, and N. R. Segro, *Tappi,* **45,** 53 (1962).

241. L. W. Lynch, *Tappi,* **46,** 480 (1963).

242. E. S. Burke, Rock Island Arsenal Laboratory Report No. 59-77, Rock Island, Ill., (1959).

243. G. Mino and S. Kaizerman, *J. Polymer Sci.,* **31,** 242 (1958).

244. G. N. Richards, *J. Appl. Polymer Sci.,* **5,** 539 (1961).

245. F. K. Guthrie, *Tappi,* **46,** 656 (1963).

246. G. N. Richards and E. F. T. White, *J. Polymer Sci.,* [C], **4,** 1251 (1963).

247. G. M. Guzman, *Anales Real Soc. Espan. Fis. Quim. (Madrid),* **58B,** No. 2, 193 (1962), through *Chem. Abstr.,* **57,** 16925 (1962).

248. N. Bergem, O. Ellefsen, K. Kringstad, and P. Kroken, *Norsk Skogind.,* **17,** 347 (1963).

249. T. G. Kulagina, M. A. Askarov, and S. D. Savranskaya, *Uzbek. Khim. Zh.,* **10,** 34 (1966); through *Chem. Abstr.,* **66,** 47137 (1967).

250. Tee-Pak, Inc., British Patents 818,412 (1959) and 1,007,941 (1965).

251. I. Sobolev, Canadian Patent 729, 711 (1966) (to Rayonier).

252. H. Singh, R. T. Thampy, and V. B. Chipalkatti, *J. Polymer Sci.,* A, **3,** 1247 (1965).

253. C. Simionescu, D. Feldman, and S. Oprea, *Rev. Roumaine Chem.,* **9,** 65 (1964); through *Chem. Abstr.,* **61,** 9627 (1964).

254. G. Smets and M. Claesen, *J. Polymer Sci.,* **8,** 289 (1952).

255. H. Mark and E. H. Immergut, *Proc. First Cellulose Conference, Syracuse, N.Y.,* 1958, pp. 2–14.

256. A. H. Gruber, C. E. Light, and S. T. Moore, U.S. Patent 3,130,076 (1964) (to American Cyanamid).

257. J. M. Salsbury, S. Kaizerman, and G. Mino, U.S. Patent 3,046,078 (1962) (to American Cyanamid).

258. C. E. Boesen, *Dansk. Kemi,* **46,** 147 (1965); through *Chem. Abstr.,* **64,** 9941 (1966).

259. T. J. Suen, U.S. Patent 3,065,041 (1962) (to American Cyanamid).

260. American Cyanamid Co., British Patents 884,805 (1961) and 902,788 (1962).

261. P. E. Trout, *Tappi,* **50**, No. 7, 89A (1967).

262. R. A. Horn and F. A. Simmonds, U.S. Forest Service Res. Note FPL-083, March 1965.

263. N. R. Segro, S. T. Moore, and J. H. Daniel, Jr., U.S. Patent 3,081,143 (1963) (to American Cyanamid).

264. W. E. Loos, Ph.D. Thesis, North Carolina State University, 1966; through *Dissertation Abstr.,* **27**, No. 12 (I), 4213B (1967).

265. E. S. Putnam, Rock Island Arsenal Laboratory Report No. 62-554, Rock Island, Ill., Feb. 12, 1962.

266. S. DasGupta, *Can. Spectroscopy,* **10**, 9 (1965).

267. R. W. Hendricks, U. S. Patent 3,101,276 (1963) (to Du Pont).

268. E. F. Degering, U.S. Patent 3,236,584 (1966) (to U.S. Secretary of the Army).

269. M. Rebek, J. Schurz, and W. Jacob, *Monatsh. Chem.,* **97**, 1142 (1966).

270. J. Schurz, M. Rebek, and H. Spork, *Angew. Makromol. Chem.,* **1**, 42 (1967).

271. M. Rebek, J. Schurz, and W. Popp, *Holzforschung,* **21**, 58 (1967).

272. G. M. Guzman, J. M. Gomez Fatou, and E. G. de la Rosa, *Anales Real Soc. Espana Fis. Quim. (Madrid),* **59B**, 519 (1963); through *Chem. Abstr.,* **60**, 13430 (1964).

273. Y. Ogiwara, Y. Ogiwara, and H. Kubota, *Kogyo Kagaku Zasshi,* **70**, 103 (1967); through *Chem. Abstr.,* **67**, 55272 (1967).

274. J. W. Adams and H. W. Hoftiezer, U.S. Patents 3,194,727 (1965) and 3,395,070 (1968) (to Tee Pak, Inc.).

275. G. N. Richards, *J. Soc. Dyers Colourists,* **80**, 640 (1964).

276. G. Landells and C. S. Whewell, *J. Soc. Dyers Colourists,* **67**, 338 (1951).

277. D. J. Bridgeford, *Ind. Eng. Chem., Prod. Res. Develop.,* **1**, 45 (1962); U.S. Patent 3,083,118 (1963) (to Tee Pak, Inc.).

278. J. W. Adams and H. W. Hoftiezer, U.S. Patent 3,256,372 (1966) (to American Can Co.).

279. S. B. Lee and M. Fujii, *J. Japan. Tappi (Kami-pa Gikyoshi),* **18**, 413 (1964); through *Chem. Abstr.,* **62**, 14922 (1965).

280. S. B. Lee and M. Fujii, *J. Japan. Tappi,* **18**, 485 (1964); through *Chem. Abstr.,* **63**, 1988 (1965).

281. S. B. Lee and M. Fujii, *J. Japan. Tappi,* **19**, 18 (1965); through *Chem. Abstr.,* **63**, 1988 (1965).

282. S. B. Lee and M. Fujii, *J. Japan. Tappi,* **19**, 139 (1965).

283. S. B. Lee and M. Fujii, *J. Japan Tappi,* **19**, 431 (1965); through *Chem. Abstr.,* **63**, 16598 (1965).

284. S. B. Lee and M. Fujii, *J. Japan. Tappi,* **20**, 383 (1966); through *Chem. Abstr.,* **65**, 13945 (1966).

285. S. B. Lee and M. Fujii, *J. Japan. Tappi,* **20**, 561 (1966); through *Chem. Abstr.,* **65**, 18829 (1966).

286. J. C. Bletzinger, *Ind. Eng. Chem.,* **35**, 474 (1943).

287. L. Neimo and H. Sihtola, *Paperi Puu,* **47**, 369, 374 (1965).

288. L. Neimo, H. Sihtola, O. Harva, and A. Sivola, *Paperi Puu,* **49**, 509 (1967).

289. A. J. Stamm and W. E. Cohen, *Appita Proc.,* **10**, 346 (1956).

290. F. Kozmal, J. Svitek, A. Pikler, J. Repka, and D. Lodesova, *Papir Celuloza,* **21**, 193 (1966); through *Chem. Abstr.,* **65**, 18829 (1966).

291. Z. L. Zosim, N. I. Levi, S. S. Gusev, and I. N. Ermolenko, *Bumazh. Prom.*, No. 12, 11 (1966); through *Chem. Abstr.*, **66**, 5383 (1967).

292. M. Rebek, J. Schurz, and H. Spörk, *Monatsh. Chem.*, **98**, 1161 (1967).

293. I. Sobolev, Canadian Patent 727,181 (1966) (to Rayonier).

294. I. Sobolev, Canadian Patent 729,069 (1966) (to Rayonier).

295. Heberlein and Co., French Patent 1,269,126 (1961).

296. F. Münzel, German Patent 1,151,485 (1964) (to Heberlein and Co.).

297. R. H. Pfeiffer, U.S. Patent 3,183,056 (1965) (to W. R. Grace and Co.).

298. J. A. Orsino, D. F. Herman, and J. J. Brancato, U.S. Patent 3,121,698 (1964) (to National Lead Co.).

299. D. F. Herman, U. Kruse, and J. J. Brancato, *J. Polymer Sci.*, [C], No 11, 75 (1965).

300. D. F. Herman, in N. M. Bikales, Ed., *Encyclopedia of Polymer Science and Technology*, Vol. 8, Interscience, New York, 1968, pp. 736–752.

301. R. E. Whitfield, L. A. Miller, and W. L. Wasley, *Textile Res. J.*, **31**, 704 (1961).

302. H. I. Bolker, *J. Polymer Sci.*, [C], No. 11, 49 (1965); Technical Report No. 412, Pulp and Paper Research Institute of Canada, June 1965.

303. R. E. Whitfield, L. A. Miller, and W. L. Wasley, *J. Appl. Polymer Sci.*, **8**, 1607 (1964).

304. Z. A. Rogovin, *Svensk Papperstidn.*, **70**, 799 (1967).

305. R. W. Faessinger and J. S. Conte, U.S. Patent 3,330,787 (1967) (to Scott Paper Co.).

306. R. W. Faessinger and J. S. Conte, U.S. Patent 3,359,224 (1967) (to Scott Paper Co.).

307. R. W. Faessinger and J. S. Conte, U.S. Patent 3,357,933 (1967) (to Scott Paper Co.).

308. J. R. A. Anderson, S. Dilli, J. L. Garnett, and E. C. Martin, *Nature*, **201**, 772 (1964).

309. N. R. Segro and J. H. Daniel, Jr., U.S. Patent 2,955,015 (1960) (to American Cyanamid).

310. G. H. Dorion, K. O. Loeffler, and D. H. Rakowitz, U.S. Patent 3,304,180 (1967) (to American Cyanamid).

311. Oxford Paper Co., British Patent 1,069,953 (1967).

312. X. Eschalier, British Patent 25,647 (1906) as cited in W. J. Roff, *J. Textile Inst., Trans.*, 49, T646 (1958).

313. W. A. Reeves, R. M. Perkins, and L. H. Chance, *Textile Res. J.*, **30**, 179 (1960).

314. J. B. Rivlin, *Structural Investigation of a Chemically Crosslinked Cellulose*, doctoral dissertation, Princeton University, 1965.

315. M. Araki, *Chemical Reaction Between Cellulose and Formaldehyde*, Wood Research, Bulletin of the Wood Research Institute, Kyoto University, No. 33, 1 (1964).

316. C. F. Goldthwait, E. M. Buras, Jr., and A. S. Cooper, *Textile Res. J.*, **21**, 831 (1951).

317. W. J. Roff, *J. Textile Inst., Trans.*, **49**, T646 (1958).

318. R. E. Wagner and E. Pacsu, *Textile Res. J.*, **22**, 12 (1952).

319. D. M. Jones, *Textile Res. J.*, **34**, 257 (1964).

320. A. J. Stamm, *Tappi*, **42**, 44 (1959).

321. W. E. Cohen, A. J. Stamm, and D. J. Fahey, *Tappi*, **42**, 934 (1959).

322. W. E. Cohen, *Appita*, **20**, 95 (1967).

323. W. E. Cohen, A. J. Stamm, and D. J. Fahey, *Tappi*, **42**, 904 (1959).

324. W. E. Cohen and A. J. Stamm, *Indian Pulp Paper*, **13**, 555 (1959).

325. E. L. Back, *Transactions of Symposium on Consolidation of the Paper Web, Cambridge, September, 1965,* British Paper and Board Makers' Association, London, 1966, p. 545.
326. P. Luner, *Proc. 20th Anniv. Meeting ESPRI,* (1965), p. 114.
327. I. Yoshino, S. Senda, and Y. Sumikawa, *J. Soc. Textile Cellulose Ind. Japan,* **10**, 8 (1954); through *Chem. Abstr.*, **48**, 6121*f* (1954).
328. E. L. Back and L. O. Klinga, *Svensk Papperstidn.*, **66**, 745 (1963).
329. E. L. Back and L. O. Klinga, *Tappi*, **46**, 284 (1963).
330. E. L. Back, M. J. Htun, M. Jackson, and F. Johanson, *Tappi,* **50**, 542 (1967).
331. B. Marek, *Papier,* **16**, 457 (1962); U.S. Patent 3,230,108 (1966) (to Société de la Viscose Suisse).
332. D. J. Fahey, *Tappi,* **45**, No. 9, 192A (1962).
333. R. M. Reinhardt, R. A. Fujimoto, J. D. Reid, and R. L. Arceneaux, U.S. Patent 3,264,054 (1966) (to U.S. Secretary of Agriculture).
334. J. Russell, W. E. Carlson, and C. H. Mayhood, Jr., U.S. Patent 3,310,363 (1967) (to St. Regis).
335. B. H. Kress, U.S. Patent 2,785,995 (1957) (to Quaker Chemical Products Corp.).
336. B. H. Kress, U.S. Patent 2,903,328 (1959) (to Quaker Chemical).
337. Bradford Dyers Association, Ltd., French Patent 1,318,636 (1963).
338. D. E. Hudgin, Canadian Patent 626,101 (1961) (to Celanese Corp. of America).
339. P. Robinson, U.S. Patent 2,806,190 (1957) (to Sprague Electric Co.).
340. Tootal Broadhurst Lee Co., Ltd., French Patent 1,291,579 (1962).
341. L. M. Valentine and B. F. Sagar, British Patent 915, 572 (1963) (to Tootal Broadhurst Lee).
342. L. M. Valentine, British Patent 915,574 (1963) (to Tootal Broadhurst Lee).
343. L. M. Valentine and H. Fletcher, British Patent 1,000,031 (1965) (to Tootal Broadhurst Lee).
344. Chemirad Corp., British Patent 1,082,554 (1967).
345. R. Steele, *Am. Dyestuff Reptr.*, **54**, 19 (P6) (Jan. 4, 1965).
346. J. J. Willard, R. Turner, and R. F. Schwenker, Jr., *Textile Res. J.* **35**, 564 (1965).
347. J. J. Willard, R. Turner, and R. F. Schwenker, Jr., *Textile Res. J.,* **36**, 1051 (1966).
348. N. A. Bates, *Tappi,* **49**, 184 (1966).
349. B. Chabert, *Bull. Inst. Textile France,* **20**, 205 (1966).
350. P. C. Mehta and R. D. Mehta, *Textile Res. J.,* **30**, 524 (1960).
351. H. Tovey, *Textile Res. J.,* **31**, 185 (1961).
352. I. J. Gruntfest and D. D. Gagliardi, *Ind. Eng. Chem.*, **41**, 760 (1949).
353. K. W. Britt, U.S. Patent 2,325,302 (1943) (to Scott Paper).
354. I. J. Gruntfest and E. M. Young, Jr., *Pulp Paper Mag. Can.,* **51**, Convention Issue, 190, 199 (1950).
355. J. J. Becher, G. R. Hoffman, and J. W. Swanson, *Tappi,* **44**, 296 (1961).
356. B. Steenberg, *Svensk Papperstidn.*, **49**, 311 (1946).
357. A. Jurecic, T. Lindh, S. E. Church, and V. Stannett, *Tappi,* **41**, 465 (1958).
358. V. Stannett, in Tappi Monograph 29, *Wet Strength in Paper and Paperboard,* TAPPI, 1965, Chapter VII, p. 85.

359. J. Russell, O. J. Kallmes, and C. H. Mayhood, *Tappi,* **47**, 22 (1964).

360. A. Jurecic, C. M. Hou, K. Sarkanen, C. P. Donofrio, and V. Stannett, *Tappi,* **43**, 861 (1960).

361. P. B. Taft, *Techn. Assn. Papers,* **24**, 288 (1941); *ibid.*, **26**, 531 (1943); *Paper Trade J.,* **112**, T. S. 122 (1941); *ibid.,* **115**, 206 (1942).

362. P. E. Trout, *Tappi,* **34**, 539 (1951).

363. R. J. Kennedy, *Tappi,* **45**, 738 (1962).

364. M. de Domenico and G.-B. R. de Domenico, French Patent 1,134,393 (1957) (to Toulmin and Toulmin).

365. M. de Domenico and G.-B. R. de Domenico, U.S. Patent 2,965,436 (1960) (to Toulmin and Toulmin).

366. W. E. Woods, and W. A. Schenck, U.S. Patent 2,622,960 (1952) (to A. P. W. Products, Inc.).

367. N. R. Eldred and J. C. Spicer, *Tappi,* **46,** 608 (1963).

368. W. U. Day, H. F. Donnelly, and H. L. Rohs, U.S. Patent 3,096,228 (1963) (to Kimberly-Clark).

369. Nobel-Bozel, French Patent 1,344,235 (1963).

370. R. K. Barnes and J. T. Adams, Canadian Patent 734,513 (1966) (to Union Carbide).

371. Union Carbide, Ltd., *World's Paper Trade Review,* **166,** No. 3,178, 180, 182, 185 (1966).

372. F. S. H. Head, *J. Textile Inst., Trans.,* **49,** T345 (1958).

373. B. Farrow, W. J. Roff and S. C. Simmens, *J. Textile Inst., Trans.,* **49**, T516 (1958).

374. C. L. Mehltretter, in R. L. Whistler and E. F. Paschall, Eds., *Starch Chemistry and Technology,* Vol. II, Academic Press, New York, 1967, pp. 438–442.

375. A. Meller, *Tappi,* **41,** 684 (1958).

376. P. J. Borchert and J. Mirza, *Tappi,* **47**, 525 (1964).

377. J. W. Opie and J. L. Keen, *Tappi,* **47,** 504 (1964).

378. A. Meller, *Tappi,* **41**, 679 (1958).

379. Société de la Viscose Suisse, French Patent 1,297,558 (1962).

380. B. S. V. Marek and L. Gneisz, U.S. Patent 3,141,812 (1964) (to Société de la Viscose Suisse).

381. J. A. Harpham and H. W. Turner, U.S. Patent 3,069,311 (1962) (to Hercules).

382. L. J. Bernardin, U.S. Patent 3,224,926 (1965) (to Kimberly-Clark).

383. Hooker Chemical Corp., British Patents 1,049,341 (1966) and 1,057,829 (1967).

384. G. M. Wagner, Canadian Patent 728,608 (1966) (to Hooker Chemical).

385. C. A. Redfarn and H. Coates, U.S. Patent 2,909,446 (1959) (to Albright and Wilson, Ltd.).

386. C. L. Aldridge, British Patent 885,523 (1961) (to Esso Research and Engineering Co.).

387. J. W. Crone, Jr., U.S. Patent 3,251,713 (1966) (to Shuford Mills).

388. P. T. Gay, U.S. Patent 3,346,445 (1967) (to Hollingsworth and Vose Co.).

389. E. E. Habib and M. Nimoy, U.S. Patent 3,238,010 (1966) (to W. R. Grace and Co.).

390. S. M. Richardson and R. C. Hoch, U.S. Patent 3,269,860 (1966) (to W. R. Grace and Co.).

391. E. M. La Combe and W. P. Miller, Canadian Patent 752,485 (1967) (to Union Carbide).

392. Minnesota Mining and Manufacturing Co., British Patent 1,028,911 (1966).
393. R. E. Zenk and W. E. Lundquist, U.S. Patent 3,248,254 (1966) (to Minnesota Mining).
394. G. R. Nelson, U.S. Patent 3,233,962 (1966) (to Dennison Mfg. Co.).
395. Société Nouvelle des Ateliers de Venissieux S.A., French Patent 1,375,836 (1964).
396. E. Pacsu and R. F. Schwenker, Jr., U.S. Patents 2,990,232 and 2,990,233 (1961) (to Textile Research Institute).

E. FILMS

G. C. Daul and R. L. Mitchell

The basic processes for making the cellulosic derivatives and solutions from which cellulose films are produced have varied little in the last 20 years. Significant advances have been made, however, in improving the economy of production methods and in increasing the utility of the base sheet by coating with improved moisture-proofing polymers, by the development of metallizing, laminating, and other techniques that have led to an enormous variety of new end products and uses.

Several comprehensive articles detailing the progress made in production, evaluation of properties, and uses of cellophane, acetate, and other cellulosic films can be found in various excellent sources. Tables of film properties and an up-to-date listing of test methods may be found in the *Kirk–Othmer Encyclopedia of Chemical Technology* (1) and in the *Modern Plastics Encyclopedia* for 1968 (2). Other descriptive articles are to be found in the *Encyclopedia of Polymer Science and Technology* (3), and numerous end uses for films are described in the *Modern Packaging Encyclopedia* (4).

1. Cellophane

Progress has been made in recent years in the production of improved wood pulps for cellophane. These pulps have been designed to be highly reactive, require a minimum of carbon disulfide for xanthation, and can be converted to especially economical viscose formulations containing relatively high amounts of cellulose (9–10%) and low amounts of caustic soda (5–5.7%). Batch operations in most cellophane plants have been replaced by automated, continuous units for shredding and steeping. In some plants, aging cans have been replaced by continuous-aging units, such as the Seaman and King (5) towers or other continuous equipment (6). Accelerated aging of the alkali cellulose has been possible through the use of oxidation catalysts, such as salts of manganese or cobalt (7). These are commonly applied to the wood pulp as a strip treat by the pulp manufacturers or added to the steeping caustic soda by the film producer. Baratte xanthation is being replaced by wet churns that allow xanthation and mixing in a single vessel. Continuous-xanthation processes are also being used (8, 9). High-shear, high-speed mixing devices that accelerate the dissolution of the cellulose xanthate to form good filtering viscoses, continuous thin-film deaeration, and other expedients are being used to improve the economics of cellophane production.

Resins have been developed to improve the strength of bonding, or "anchoring," of various of the newer coating materials to the base film. The most common of these are the amino-formaldehyde condensates. Polymer coatings are used to give the base cellophane film new or improved properties, such as reduced moisture permeability or selective impermeability, stronger heat sealing, and improved receptivity to printing inks. Coatings used on cellophane include cellulose nitrate-wax-resin compositions, copolymers of vinylidene chloride (10-12), and cellulose nitrate. These are usually applied from solution, but some polymers can be applied from aqueous emulsions. Still others, such as polyolefins, and some more recently developed coatings that are based on poly(vinyl chloride) (13) can be melt-extruded, solution-coated, or laminated. Cellophane is being laminated with polyethylene, aluminum foil, rubber hydrochloride film, paper, wax, etc., to give a combination of properties not available in a single film. The use of coatings containing ultraviolet-radiation absorbers, such as hydroxybenzophenones and substituted triazoles, gives added protection to the base film.

As a result of these improvements, cellophane still commands the largest share of the cigarette-packaging-film market and remains the preferred film for wrapping cereals and hard candies, for potato chip bags, and the like. A further discussion of cellophane may be found in reference 14.

2. Cuprammonium Film

Although of substantially lesser commercial importance than cellophane, cuprammonium film is still being manufactured in Germany for specialty uses (14). Films 0.0004 in. or less in thickness are finding current specialized uses as membranes in "artificial kidney" machines and desalination equipment. The properties of cuprammonium film that particularly distinguish it from cellophane are the absence of sulfur residues and a different internal fine structure, recently described by Jayme and Balser (15). From their electron-microscope studies of cellophane and cuprammonium films they showed that cellophane has a thin cuticle, or skin, of a few fibrillary layers, followed by a spongelike, porous region in which a large number of *vacuoles* exist.

This layered structure is undoubtedly due to the skinning effect of the acid-salt bath during coagulation. The vacuoles are formed by the evolution of gases (hydrogen sulfide and carbon disulfide) during regeneration. This phenomenon is absent in cuprammonium films, which are converted directly to cellulose from the cuam complex without gas evolution.

3. Cellulose Ester Films

Since the 1950s there has been a paucity of published literature on new developments in cellulose-ester-film technology. Major emphasis by the large cellulose ester and film producers has apparently been centered on making improvements in production equipment, increased efficiency of reaction cycles and recovery systems, increased utilization of improved wood pulps as a source of more uniform and reactive cellulose to replace the cotton linters used in the past, optimizing formulations for film casting, and designing better melt-extrusion and film-handling machinery.

A notable new development has been in ultrafiltration, or reverse osmosis, which has experienced intensive development during the past decade (16). Primarily used for desalting water, separation is achieved by forcing a solution under pressure through a membrane that passes solvent more readily than solute. Most of the work in this field has been done with cellulose acetate membranes and capillary tubes. A detailed discussion of this new development may be found in Section F of this Chapter XIX.

The properties of cellulose ester products are fully described with instructions for their use as films and sheets in various well-written technical bulletins issued by the manufacturers (17). Improvements in properties have been obtained through uses of new plasticizer–solvent combinations for casting (18, 19); use of plasticizers with different effects on properties (20–22); modified coatings (23); and additives to prevent light degradation, to impart flame-retardant properties, and the like. A detailed discussion of cellulose ester plastics, including films and sheeting, may be found in reference 24 (see also Section B in Chapter XVII).

4. Cellulose Ether Films

Cellulose ethers permit film manufacturers to utilize materials with a wide range of solubilities, from completely water-soluble materials to water-insoluble as well as solvent-soluble ones, and with other properties not usually found in a single class of polymeric materials. As a result cellulose ethers can be found in such diverse products as water-soluble thickeners for paints, emulsions, or foods, as well as coatings and lacquers. Their film-forming capabilities have likewise been utilized for a variety of specialized products, such as water-soluble packages for foods, detergents, household bleaches, and the like (see also Section C of Chapter XVII).

(a) ETHYLCELLULOSE

Ethylcellulose is made from alkali cellulose and ethylene chloride reacted under pressure. Substituted to a degree of substitution (DS) of 2.3, ethylcellulose is used for plastics, sheetings, or other materials where good solvent resistance is required. At a DS of 2.44–2.50 (ethoxyl content of 48.0–49.5%), it is used mostly for lacquers. Ethylcellulose transparent sheeting is cast from solvents or melt-extruded. The properties of ethylcellulose sheeting are described in the *Encyclopedia of Polymer Science and Technology* (25).

(b) HYDROXYETHYLCELLULOSE

The only really new cellulosic film to be produced commercially in recent years is that made from hydroxyethylcellulose. Production of film from a low-DS hydroxyethylcellulose began on a commercial basis in the 1960s. This process takes advantage of the fact that cellulose substituted with about 3.5–4% hydroxyethyl groups (DS~0.15), is soluble in 6–8% sodium hydroxide to give filterable solutions, a discovery make by Schorger and Shoemaker (26, 27) in the 1930s. The solution of hydroxyethylcellulose in sodium hydroxide can be cast into films by extrusion into acid–salt baths with ordinary cellophane-casting machinery. In their properties such films have several advantages over cellophane and are potentially more economical to produce.

A patent issued to Levesley and Craik (28) in 1936 described the preparation of a solution of hydroxyethylcellulose, having less than one substitutent group per glucose unit, in caustic soda by cooling below 0°C and submitting the mixture to the action of a colloid mill. Such solutions were described as being suitable for the preparation of "clear, strong, flexible films, fibers, threads, etc. by coagulation with acids or salts solutions." Broderick (29) used a tertiary amine catalyst and a mixed solvent system in which sodium hydroxide is slightly soluble to react alkali cellulose with ethylene oxide. The hydroxyethylcellulose so produced remains in the solid state after reaction and is soluble in 7% aqueous sodium hydroxide (30). Klug and Tennent (31) likewise treated alkali cellulose made from cotton linters with ethylene oxide in a solvent system containing *tert*-butyl alcohol. Low-DS alkali-soluble hydroxyethylcellulose is used for casting films of exceptional dimensional stability and clarity (32–34). Solutions are coagulated from acid–salt baths of the type used for rayon or cellophane. Milne (35, 36) claimed the production of hydroxyethylcellulose films with superior wet (gel) strength by casting a 6% solution of alkali-soluble, water-insoluble hydroxyethylcellulose in 8% sodium hydroxide into a coagulation bath containing 35% phosphoric acid and 17% sodium hydroxide at 25°C. A combination of sodium phosphate salts and phosphoric acid could also be used. In 1962 Mitchell and co-workers (37) patented a method for producing hydroxyethylcellulose that consists of reacting shredded alkali cellulose on a spiral vibratory conveyor housed in a gas-type chamber while ethylene oxide is introduced at a controlled rate and recirculated through the reactor at the required temperature and pressure. The vibration of the reactor subjects the alkali cellulose bed, moving down the spiral conveyor, to a pulsating movement that promotes a uniform reaction with the ethylene oxide. The product is aged to the proper degree of polymerization (DP) after reaction (38). The hydroxyethylcellulose thus produced is capable of being neutralized and washed with water without dissolving, dried, and pelletized for later use. Alternatively it is dissolved directly after aging, in 5.6–7% sodium hydroxide solution.

Clear, filterable solutions that are suitable for film casting can also be prepared by a continuous process (39). A fractional turbidimetric method of analyzing the molar substitution of hydroxyethylcellulose, in the ethylene oxide content range of 2–8%, has been reported (40). Hydroxyethylcellulose solution can be cast on standard cellophane-casting equipment with the provision for increased drying capacity due to the higher gel swelling of the hydroxyethylcellulose film (750–850%) relative to cellophane (400–475%). On the other hand, no desulfuring or other chemical purification of the gel film is required as for cellophane. Softeners, such as glycerol or other glycols, may be used.

Because of its high clarity and sparkle, hydroxyethylcellulose film is especially desirable for high-quality printed or colored packaging. It will accept metallic pigments for silver- or gold-colored printing without discoloring from sulfur-derived by-products as does cellophane. Another important advantage of hydroxyethylcellulose film over cellophane is its unusually good dimensional stability on exposure to changes in humidity. It is estimated that 75% of the cigars and cigar packages manufactured in the United States are wrapped in hydroxyethylcellulose film because of this property. The manufacturers of hydroxyethylcellulose film are now looking toward the lucrative cigarette-package-outerwrap market for expanded sales. The need for dimensionally stable film is accentuated by the longer lengths of the newer cigarettes on the market. Some properties typical of uncoated and coated hydroxyethylcellulose film are shown in Table 1 in comparison with the corresponding properties of cellophane.

Electron-microscope studies by Heyn (41) show that hydroxyethylcellulose film is similar in morphology to the cuprammonium films studied by Jayme and Balser (15). There is no distinct outer skin, or cuticle, as in cellophane but instead somewhat denser microfibrils at the surface. The remainder of the film is relatively uniform. There are practically none of the large vacuoles found in cellophane since gas is not evolved during manufacture.

(c) OTHER CELLULOSE ETHER FILMS

Clear, strong, water-soluble films may be cast from aqueous solutions of sodium carboxymethylcellulose solutions and plasticized with glycols, glycerine, ethanolamine, etc. (42, 43). Similarly methylcellulose solutions are cast into strong films that are water soluble through a range of -17 to 11°C. Such films are finding wide acceptance for unitized packages of dry bleach, detergents,

TABLE 1
Properties of Hydroxyethylcellulose and Cellophane Films

Property	Hydroxyethylcellulose film: Uncoated	Hydroxyethylcellulose film: Coated with cellulose nitrate	Cellophane film: Uncoated	Cellophane film: Coated with cellulose nitrate
Thickness (in.)	0.001	0.0011	0.001	0.0011
Tenacity (kg)[a]	12–15	14–17	14–17	16–19
Elongation to break (%)	10–20	10–20	15–25	15–25
Impact strength (kg/cm^2)	6–8	7–9	6–8	7–10
Tearing strength (g)	2–4	3–5	4–7	6–9
Water-vapor-transmission rate ($g/100\ m^2$ 24 hr.)	–	0.5–1.5	–	0.5–1.5
Resistance to oil, grease	Impermeable	Impermeable	Impermeable	Impermeable
Dimensional change (%) at high relative humidity:[b]				
Machine direction	1–3	1–3	2–5	2–3
Transverse direction	2–4	2–3	4–7	3–5
Cuen intrinsic viscosity (dl/g)	2.0	2.0	2.0	2.0
Sulfur content (ppm):				
Total	100–200	100–200	500–900	500–900
Reducible	5–10	5–10	60–80	60–80
Softener, calculated as glycerol	17–19	17–19	15–17	15–17
Substitution (% EtO)[c]	3.8–4.2	3.6–4.0	0	0

[a]Tenacity = kilograms per inch width of 0.001-in. (1-mil) thick film.
[b]Conditioned at 10% RH, measured, cycled to 90%, measured again to find percentage of change.
[c]Percentage by weight of ethylene oxide on total cellulosic component.

soap, dyes, textile adjuvants, etc. Methylcellulose is the only film material currently accepted by the Food and Drug Administration for use as food packaging that may be cooked as is in water (44). Cyanoethylcellulose films have also acquired limited commercial use (45); their preparation and properties are described in Section D of Chapter XVII.

References

1. D. T. Suprenant, in A. Standen, Ed., *Kirk–Othmer Encyclopedia of Chemical Technology,* 2nd ed., Vol. 9, Interscience, New York, 1968, pp. 220–224.
2. *Modern Plastic Encyclopedia 1968,* McGraw-Hill, New York, 1967.
3. L. E. Wolinski in N. M. Bikales, Ed., *Encyclopedia of Polymer Science and Technology,* Vol. 6, Interscience, New York, 1967, pp. 764–794.

4. *Modern Packaging Encyclopedia,* Vol. 40, No. 13A McGraw-Hill, New York, 1967.
5. S. E. Seaman and R. P. King, U.S. Patent 2,490,097 (1949).
6. J. O. Smith and W. R. Weighan, U.S. Patent 2,805,924 (1957) (to Courtaulds Ltd.).
7. R. L. Mitchell, U.S. Patent 2,542,285 (1951) (to Rayonier, Inc.).
8. T. Dokkum, U.S. Patent 2,122,519 (1938) (to American Enka Corp.).
9. F. G. C. Klein, French Patent 696, 411 (1929).
10. E. R. Covington and J. W. Meier, U.S. Patent 2,950,218 (1960) (to E.I. du Pont de Nemours & Co.).
11. W. G. Grantham, U.S. Patent 2,990,391 (1961) (to American Viscose Corp.).
12. C. A. Blood, Jr., and J. R. Fisher, U.S. Patent 3,039,986 (1962) (to Du Pont).
13. H. H. Sineath, *Mod. Packaging,* **36** (12), 127 (1963).
14. V. C. Haskell, in N. M. Bikales, Ed., *Encyclopedia of Polymer Science and Technology,* Vol. 3, Interscience, New York, 1965, pp. 60–79.
15. G. Jayme and K. Balser, *Papier,* **21** 678 (1967).
16. J. J. Perona, F. H. Butt, S. M. Fleming, S. T. Mayr, R. A. Spitz, M. K. Brown, H. D. Cochran, K. A. Kraus, and J. S. Johnson, Jr., *Environmental Science and Technology,* **1**, No. 12, 991 (1967).
17. Various technical bulletins supplied by Eastman Chemical Products, Inc., a subsidiary of Eastman Kodak Co., and by Celanese Plastics Co., a division of the Celanese Corporation of America.
18. Gevaert Photo-Production N. V., Belgian Patent 530,489 (1954).
19. Gevaert Photo-Production N. V., British Patent 792,430 (1958).
20. L. V. Rozental, G. I. Biudygine, E. D. Korneva and F. G. Zhurina, *Khim. Prom.* 367 (1960).
21. A. N. Pudovik, T. M. Moshkina, G. P. Krupnov, A. E. Buken, and L. A. Semenova, *Tr. Vses. Nauch.-Issled. Kinofotoinst,* No. 52, 5 (1965).
22. K. T. Hasselstrom, U.S. Patent 2,678,283 (1954).
23. E.I. du Pont de Nemours & Co., Belgian Patent 666,385 (1966).
24. R. O. Hill, Jr., in N. M. Bikales, Ed., *Encyclopedia of Polymer Science and Technology,* Vol. 3, Interscience, New York, 1965, pp. 354–419.
25. A. B. Savage, in N. M. Bikales, Ed., *Encyclopedia of Polymer Science and Technology,* Vol. 3, Interscience, New York, 1965, pp. 486–487; see also the discussion of other cellulose ethers in this work, pp. 459–549.
26. A. W. Schorger, U.S. Patent 1,863,208 (1932) (to Burgess Laboratories).
27. A. W. Schorger and M. J. Shoemaker, *Ind. Eng. Chem.,* **29**, 114 (1937).
28. A. S. Levesley and J. Craik, U.S. Patent 2,046,174 (1936) (to Imperial Chemical Industries, Ltd.)
29. A. E. Broderick, U.S. Patents 2,173,470 and 2,173,471 (1939) (to Carbide & Carbon Chemicals Corp.); through *Chem. Abstr.,* **34**, 621 (1940).
30. A. E. Broderick, U.S. Patent 2, 682, 535 (1954). (to Carbide & Carbon); through *Chem. Abstr.,* **48**, 11100 (1954).
31. E. D. Klug and H. G. Tennent, U.S. Patent 2, 572, 039 (1951) (to Hercules Powder Co.) through *Chem. Abstr.,* **46,** 1256 (1951).
32. Reference 25 pp. 518–519.
33. D. R. Erickson, U.S. Patent 2,763,030 (1956).

34. R. L. Mitchell and R. F. Bampton, U.S. Patent 2,847,411 (1958) (to Rayonier).

35. D. T. Milne, U.S. Patent 2,902,334 (1959) (to American Viscose).

36. D. T. Milne, U.S. Patent 2,994,924 (1961) (to American Viscose).

37. R. L. Mitchell, A. R. Albohn, and S. G. Palmgren, U.S. Patent 3,045,007 (1962) (to Rayonier).

38. R. L. Mitchell, U.S. Patent 3,296,248 (1967) (to Rayonier).

39. R. L. Mitchell, A. E. Couninis, and C. F. Murphy, U.S. Patent 3,312,685 (1967) (to Rayonier).

40. T. E. Muller and W. J. Alexander, *J. Appl. Polymer Sci.* **10,** 1285 (1966).

41. A. N. J. Heyn, unpublished work.

42. *Sodium Carboxymethylcellulose: Properties and Uses,* Hercules Powder Co., Wilmington, Del., 1963.

43. E. D. Klug, in N. M. Bikales, Ed., *Encyclopedia of Polymer Science and Technology,* Vol. 3, Interscience, New York, 1965, pp. 520–539.

44. Reference 1, p. 232.

45. N. M. Bikales, in N. M. Bikales, Ed., *Encyclopedia of Polymer Science and Technology,* Vol. 4, Interscience, New York, 1966, pp. 533–562.

F. MEMBRANES

F.1. Flat Membranes

Robert E. Kesting

Cellulosic membranes have played a prominent role in the development of membrane technology. In 1855, only 10 years after Schönbein's (1) momentous discovery of the first man-made polymer, cellulose nitrate, Fick (2) performed his classical studies, "*Über Diffusion*." In the same year Lhermite (3) first clearly stated the *solution* theory (i.e., permeation as a result of the interaction of the permeant species with the membrane) of membrane permeability and recognized that the solution and *capillary* theories were not mutually exclusive but merged, without any abrupt change, one into another. In 1860 Schumacher (4) developed collodion membranes in a tubular form; these were to remain popular into the present century. Bechhold (5) first characterized cellulosic membranes by well-defined permeability measurements in which the pressure required to blow air through a water-filled capillary was related to surface tension and the radius of the capillary. The concept of pore-size distribution was developed by Erbe (6), who combined the blow and the Hagen–Poiseuille capillary methods.

Characterization of membrane morphology by direct microscopic observations awaited the advent of the electron microscope. Ardenne's early studies (7) were followed by the investigations of Maier (8) and Helmcke (9–11), which for the first time revealed the true nature of membrane microstructure. Their work was climaxed in 1960 by Maier and Scheuermann's (12) important study, 'Die Bildungweise teildurchlässiger Membranen." With the recent development of the scanning electron microscope further progress in the understanding of gel morphology is to be expected (13).

Early attempts to control and vary membrane porosity were largely empirical in nature. Bechhold (14) noted that permeability varied inversely with the concentration of "membrane substance" in the casting solution. Bigelow and Gemberling (15) noted the effects of varying the drying time before immersion into a nonsolvent, and Malfitano (16) varied the ratio of alcohol to ether in the solvent system and was among the first to employ annealing as a technique for increasing permselectivity and decreasing permeability. Schoep (17) was responsible for two important innovations: the use of glycerol, a nonvolatile swelling agent, to control gel structure; and the use of castor oil, a plasticizer, to improve physical properties. Brown (18) produced a graded series of collodion membranes by swelling initially dense films in varying concentrations of alcohol-water. He (19) was also the first to employ cellulose acetate as a membrane substance and to note the asymmetry in depth that was later to be of such significance to the development of the ultrafiltration process of desalination (reverse osmosis).

In their attempts to produce uniform collodion membranes Bjerrum and Manegold (20) considered thickness. water content, and permeability to be the three most important membrane parameters. Bartell and Van Loo (21), Elford (22), and Grabar (23) pioneered attempts to describe events that occur in the polymer solutions during the gelation process and account for membrane structure. Elford (22) also developed one of the first practical series of graded membranes from solutions of collodion in acetic acid.

Certain separations have necessitated the development of modified cellulosic membranes. For instance, chemical inertness has been imparted to cellulosic membranes by the deesterification of ester derivatives (24). Ion-exchange properties have been introduced into cellulosic membranes both by the oxidation of collodion and by the formation of polymeric blends of cellulosic and noncellulosic materials (25).

Two recent technological developments have added further support to the already strong position of cellulosic membranes. The first was the discovery by Reid and Breton (26) that dense films of secondary cellulose acetate exhibited greater *permselectivity* (i.e., selective solute rejection) than any of the other polymers they surveyed. The second was the development by Sourirajan and Loeb (27) of a swollen membrane of this material, characterized by an inhomogeneity in depth. These two events are of importance because they have brought the ultrafiltration process, now termed *reverse osmosis,* into the realm of economic feasibility for desalination. Ultimately it is the intense interest in reverse osmosis, developed during the last decade under the auspices of the Office of Saline Water (28) to secure additional sources of fresh water, that may be credited for the current general revival of interest in membrane separations.

Since virtually all cellulosic membranes are formed by the phase-inversion process in which a polymer solution, or *sol* (liquid phase), is desolvated (inverted) to form a *gel* (solid phase), it will be expedient to consider the transition from sol to gel as a continuum. According to this concept membranes may be characterized by their *degrees of swelling and order,* and can exist in both solid and liquid states. In other words, membranes may be said to occupy the entire spectrum of variously swollen structures bound by dense films at one end and by polymer solutions at the other. There are several advantages to this point of view. First, control of the phase-inversion process becomes to some extent a matter of adjusting the interactions between the polymer and various swelling agents and solvents so as to obtain the desired degree of swelling and order at the moment of inversion. Second, the close functional relationship between polymer–solvent interactions in the liquid phase and the transport of various permeant species by diffusion through a solid gel becomes more apparent. Consider, for example, the interaction of a dense polymer film with various liquids: a nonsolvent does not interact and will therefore permeate only with great difficulty; swelling agents and solvents do interact and will consequently expand and permeate the film to varying extents. Thus membrane structure becomes a dynamic, rather than a static, concept. Finally, the consideration of sols as the more highly swollen portions of the *sol–gel continuum* rather than as specifically distinct entities makes it possible to consider transport across the recently discovered *liquid membranes* (29) on a common basis with that across the more conventional solid-gel types.

This section concentrates on the structural, rather than the functional, aspects of cellulosic membranes because the latter have been intensively treated in review articles (30–32). The sol–gel continuum is therefore considered in terms of its controlling factors, *swelling* and *order,* by interpreting the phase-inversion process as it proceeds from the sol to the primary and secondary gel phases. A more extensive treatment of the structural aspects of synthetic polymeric membranes is to be found elsewhere (72).

1. Sol Precursors to Primary Gel Structure

(a) THE PRIMACY OF SOL OVER GEL STRUCTURES

Because the initial step in the phase-inversion process involves the preparation of polymer solutions, it is appropriate to begin a discussion of cellulosic membranes with a consideration of sol properties, the two most important of which are swelling and order. Traditionally these properties have been indirectly determined by deductions from morphological investigations of membrane gels (7–13). Thus the existence of swollen and ordered sol structures has been deduced from the presence of swollen* and ordered† gels. Although it has been long realized that swelling and order originate and are much more tractable in the

* *Gel swelling* is an expression of membrane water content, commonly expressed as the ratio of the weight of the wet to that of the dry membrane.

† *Gel order* in the present sense refers to structural uniformity in a plane parallel to the surface and is not to be confused with the problem of asymmetry in depth, which is treated elsewhere in this section. In nonporous (diffusion-type) membranes order is equivalent to uniformity in interchain displacement, whereas in porous membranes it refers to the narrowness of the pore-size distribution. The chief manifestation of gel order is, of course, membrane permselectivity, which increases with increasing order.

sol phase, light scattering (33), viscometry (34), osmometry (35), and electron microscopy (36) have only recently been utilized in the direct characterization of the phase-inversion solutions themselves (see also Chapter XIV). The overall picture of sol structures that has emerged from both direct and indirect studies is one of a system in which the individual constituents of the polymer solution interact with one another to establish given molecular and supermolecular structures whose disposition at the moment of incipient gelation is one of the factors determining the nature of the primary gel structure. From this standpoint sols should be conceded substantial as well as temporal primacy over their gel counterparts, and membrane fabrication can be viewed as the tailoring of structured sols that are subsequently immobilized to produce structured gels. The manner in which the various polymer–solvent and polymer–polymer interaction forces affect sol swelling and order is therefore of considerable importance to those interested in the control of membrane gel structure.

(b) SOL SWELLING

Sol swelling refers to the imbibition of solvent by dissolved polymer. It is a complex function not only of the nature of the polymer itself (polymer–polymer interaction) but also of the interaction between the polymer and the various other constituents of the phase-inversion solution (polymer–solvent interaction). The relative strengths of these various interactions is of importance to membrane technology because they determine the frequency of intermolecular contact between polymer chains both prior to and during phase inversion and are therefore ultimately responsible for swelling and order in the primary gel structure.

Although any attempt at distinguishing between the factors that affect polymer–polymer and polymer–solvent interactions will perforce be somewhat arbitrary, nevertheless it is helpful to consider changes in the polymer itself under the former, and changes in the solvent under the latter, categories. Some of the specific factors that influence the polymer–polymer interaction within cellulosic sols are the inflexibility of the polymer chains, the degree of substitution (DS), the randomness of substitution, the nature of the functional groups, the nature and extent of polymer–solvent interactions, and polymer concentration.

The characteristic rodlike configuration of cellulosic chains is primarily due to the presence of the large rings, which impart stiffness, although the nature of their functional groups, some of which may participate in intramolecular hydrogen bonding, is also believed to be a factor. Above a minimum value, the effects of variations in the degree of polymerization (DP) are relatively minor, but the effect of DS is of considerable importance. In general sol swelling will be more pronounced for derivatives of intermediate DS owing to the sterically decreased possibility for making both intramolecular and intermolecular contacts (polar crosslinks). One important consequence of decreased polymer–polymer interaction for derivatives of intermediate DS is an increase, because of increased accessibility, in the polymer–solvent interaction. For similar reasons solubility and swelling are more pronounced, at a given DS, for mixed than for pure esters.

In the case of a nonrandomly substituted derivative, such as secondary cellulose acetate, which can differ appreciably, depending on the exact manner in which the hydroxyl groups are distributed between primary and secondary positions (37), knowledge of the DS alone is not sufficient to predict sol-swelling characteristics. Cramer and Purves (38) have suggested that the presence of free primary hydroxyl groups is necessary for the acetone solubility of secondary cellulose acetate. It follows that solubility and swelling for samples of a given DS will decrease as the ratio of free primary to secondary hydroxyl groups decreases. This seems to be borne out by the fact that direct acetylation to the diacetate in phosphoric acid solution, a procedure in which the more reactive primary positions would be preferentially esterified, yields a product of much lower acetone solubility than the standard procedure of complete esterification followed by partial hydrolysis (39). In the latter case free primary hydroxyl groups occur because of the greater susceptibility of the primary acetate to hydrolysis. An additional factor of unknown significance is the less than random number of free glycol groups at the $C_{(2)}$ and $C_{(3)}$ positions (40). This irregularity presumably occurs as a result of steric hindrance to the hydrolysis of the glycol diacetate structure. Perhaps because of these difficulties the sol-swelling properties of secondary cellulose acetates varying in their fine structures of substitution have not yet been reported.

The swelling behavior of the other cellulosic sols is not quite so complex as that of the acetates. This may be due to the more nearly random substitution that is achieved in the case of those cellulosics, such as cellulose nitrate or ethylcellulose, that are prepared by quasi-homogeneous reactions. Moore and Russell (35) have made an interesting comparison between the intrinsic viscosities, $[\eta]$ of cellulose nitrate, ethylcellulose, and primary and secondary cellulose acetates in a variety of solvents. All samples were of the same DP, and the DS (for all but the primary acetate) was about 2.5. The values of $[\eta]$, in deciliters per gram, were strongly influenced by the derivative type, with average values ranging from 3.50 for cellulose nitrate, through 2.17 for ethylcellulose, 1.59 for cellulose diacetate, and 1.10 for cellulose triacetate. The values of $[\eta]$ and the volume of precipitant required for initial phase separation likewise indicate that the nitrate is more highly swollen than the acetates or the ether. The greater swelling of the nitrate as compared with the acetates and ether is apparently related to the former's higher polarity, which presumably leads to increased polymer–solvent interaction.

The preceding paragraphs have described the means by which these interactions can be influenced in a given solvent by varying the nature of the dissolved polymer. However, for a given polymer they can also be influenced by varying the nature of the *solvent.* The nature and extent of polymer–solvent interactions is perhaps the single most important factor influencing sol swelling.

Howlett and Urquhart (41), Moore et al. (34, 35), and Spurlin (42) have all reviewed the relationship between the solubility and structure of the various cellulosics. Owing to chain rigidity and the polarity of substituent groups, the forces between individual molecules are specific and strong. Because of this a large amount of energy must be liberated in the formation of the polymer–solvent bonds, which in effect means that solvation will be a prerequisite to solution. In cellulose derivatives, therefore, solution becomes largely a question of the equilibrium between the solvation of polymer molecules and their aggregation. For this reason most attention has been focused on qualitative theories of solubility involving specific chemical groups. Spurlin has pointed out that if the concept of the solubility parameter (δ),—which has been useful as a basis for classifying solvents, swelling agents, and nonsolvents in the case of nonpolar polymers—is applied to cellulosic materials, the δ of the solvent must equal the δ not of the polymer but the *solvated polymer*. Therefore it is purely coincidental if the solvating agent is the best solvent for the solvated polymer, and multicomponent systems may be the most suitable solvents since variation in δ is possible without drastic alteration of solvation. Spurlin (42) has considered solvation in terms of Lewis acid and base interactions and hypothesized that solvents for polar polymers must not only be acidic or basic but also have an appropriate δ.

Semiquantitative estimates of the strength of polymer–solvent interaction and hence of the degree of sol swelling have been derived from: (a) the Huggins polymer–solvent interaction parameter χ, or μ, which decreases as the polymer–solvent interaction becomes stronger (43); (b) the intrinsic viscosity $[\eta]$ (44), which increases with the strength of the polymer–solvent interaction; and (c) the amount of precipitant needed to effect initial separation of polymer from a dilute solution (45): the stronger the polymer–solvent interaction, the greater the amount of precipitant. For instance, in the case of cellulose acetate acidic solvents and aniline give higher values than basic solvents, such as pyridine and the picolines, with the more neutral solvents, such as acetone and methyl acetate, giving intermediate values (35). With increased strength of the polymer–solvent interaction, solvation increases and the polymer molecules become more fully extended (hence the increase in $[\eta]$). Increased polymer–solvent interaction results in decreased polymer–polymer interaction so that compaction during desolvation is minimized. Therefore *the degree of swelling of a phase-inversion gel can be expected to increase with the strength of the polymer–solvent interaction in its sol precursors*. For this reason the utilization of dimethylformamide, a "good" solvent for secondary cellulose acetate, produces more highly swollen membrane gels of this material than does acetone, a "poor" solvent.

As polymer concentration is increased in a given solvent, the polymer–polymer interaction increases in relation to the polymer–solvent interaction. The coagulation of individual molecules into supermolecular aggregates, particularly at high concentrations and in poor solvents, may result. The net effect of such aggregation is increased polymer density and hence decreased swelling, both in the sols and in the resultant gels.

Of great importance to membrane technology has been the formulation of *multicomponent solutions* containing, in addition to polymer and solvent, one or more nonvolatile substances with an affinity for the polymer midway between that of nonsolvent and solvent. These substances function in a concentration-dependent manner in several capacities: at the comparatively low concentration encountered prior to solvent loss they may act to solvate certain functional groups and hence to improve the solvent power of the solvent; at higher concentrations, such as those that occur during desolvation, they often function as weak precipitants to effect a gradual transition from sol to gel, thereby transferring such order as may exist in sol structures into the gel state. Because the net effect of the inclusion of these substances within a phase-inversion solution is to produce more highly swollen gels with decreased resistance to material transport, they are properly designated swelling agents (46).
Because the net effect of the inclusion of these substances within a phase-inversion solution is to produce more highly swollen gels with decreased resistance to material transport, they are properly designated swelling agents (46).

Spurlin (42) has described in general terms the process by which solvent power for polar polymers, such as the cellulosics, can be increased:

> By solvation or other forms of interaction, the polymer is transformed into a substance of lower melting point, or else is provided with ionizable groups which promote solubility. This is accomplished, first, by greatly increasing the number of particles in solution, and, therefore, the entropy of solution; and, second, by providing repulsive forces between the polymer molecules.

An examination of a number of specific swelling agents will help to demonstrate these points. The inorganic salt magnesium perchlorate, used together with water in the now famous *Loeb–Sourirajan process* (28) for producing cellulose acetate desalination membranes, decreased the polymer–polymer interaction within the phase-inversion solution. This was accomplished because the high charge density of the cation coupled with the symmetry and low charge density of the anion permitted the formation of strong ion–dipole interaction complexes between the cation and the nucleophilic hydroxyl and acetate groups of secondary cellulose acetate (46). The charge repulsion between positively charged polymer chains is of such magnitude that cellulose acetate and many other polar polymers will actually dissolve in concentrated aqueous solutions of these salts (47). Zinc chloride, long used as a swelling agent for cellulosic materials, behaves similarly to magnesium perchlorate. Water and organic swelling agents, such as formamide, also increase the solvent power of acetone for cellulose acetate. The swelling capability of both water and the organic swelling agents may be related to their capacity for solvating hydroxyl groups on the polymer backbone by hydrogen bonding therewith.

Because the number of possible interactions within a polymer solution increases geometrically with the number of solution components, swelling in multicomponent systems is more complex than it is in two-component systems. As a result the efficiency of a given swelling agent will depend not only on its interaction with the polymer but also on polymer–solvent and swelling agent–solvent interactions (see also Section H in Chapter XIII).

(c) ORDER IN SOL STRUCTURES

Although the concept of order in sol structures is still in its infancy, there is every indication that the current intense interest in the direct study of the structure of polymer solutions will rapidly bring it to maturity. The concept that varying degrees of order can exist within polymer solutions arises from the fact that a polymer can be dispersed to various extents in different solvent media. Therefore the nature and the extent of polymer dispersion depend on the natures of all the constituents of the phase-inversion solution. In "good" solvents, in which polymer–solvent interaction predominates, complete dispersion of polymer molecules is favored. In poorer solvents, on the other hand, a certain tendency toward polymer–polymer interaction and hence intermolecular association can be anticipated. It appears reasonable to assume that any predisposition to structural order—that is, the existence of what may be termed crystallite bundles in solution—will greatly increase the probability for the appearance of ordered gel structures subsequent to gelation. Therefore, if gelation is controlled by the diffusion of polymer molecules to one another during desolvation, ordered gels will more probably result from solutions of polymer in poor solvents than in good ones.

Although order in sol structures has not yet been unequivocally defined, several a priori assumptions concerning its nature appear to be justified. The most general of these is the above-mentioned consideration that order is likely to be favored in a situation where the polymer–polymer interaction can occur to some extent, that is, in poor or intermediate solvents. As in the case of sol swelling, order in sol structures can be influenced both by variations in the solvent system for a given polymer and by variations in the polymer for a given solvent system. It is apparent that ordered sol structures will be favored at the extremes in DS rather than at intermediate DS values. This is due to the higher crystallinity and insolubility of gels of the trisubstituted and unsubstituted, as compared with those of partially substituted, cellulose derivatives. Furthermore, the fact that solubility reaches a maximum at a DS of about 1.5 is strongly indicative of the presence of a minimum in the polymer–polymer interaction at this point. For example, Reid and Breton (26) observed that membrane permselectivity in reverse osmosis increases with increasing degree of acetylation of the cellulose diacetates and postulated that the more highly substituted acetates produce more highly ordered gels.

The presence of polymer microstructures in solution prior to visible phase separation has been demonstrated by electron microscopy of rapidly quenched solutions (36). *Supermolecular aggregates in solution were found to exist even in the case of amorphous polymers. It is highly significant that the structuring of the sol was morphologically very similar to that subsequently found in the gel phase.* It was also determined that the appearance of sol structures was an equilibrium phenomenon occurring at specific temperatures, polymer concentrations, and precipitant concentrations. Maier and Scheuermann (12) have ascribed viscosity changes resulting from the inclusion of water within cellulose nitrate solutions to preorientation effects.

Implicit in any discussion of the relevance of ordered sol structures to the structural regularity and permselectivity of cellulosic membrane gels is the maintenance of sol order in the transition from the sol state to the gel state. The means whereby this is accomplished will now be considered.

2. Primary Gel Structure

(a) THE GELATION PROCESS

Gelation refers to the transition from the sol state to the primary gel state. An understanding of gelation is vital to the control of membrane-performance characteristics because it is the critical period with respect to the transfer of swelling and order from the sol into the primary membrane-gel structure. Largely on the basis of electron-microscope investigations of membrane-gel morphology, two closely related models have been deduced to explain the sequence of events occurring during the sol-to-gel transition. The first of these, developed by Helmcke (9), envisions the formation of the open-celled, foamlike membrane structure from the point of view of the solvent. In Helmcke's view the dissolved polymer is an almost neutral matrix that is shaped by the manner in which solvent molecules form gaseous pockets during desolvation. The phase-inversion solution is pictured as a yeasty dough that ultimately gels into a porous microstructure of the type found in leavened bread. This model can be used to explain both the formation of micelles within the sols and the origin of gross structural irregularities.

The model proposed by Maier and Scheuermann (12) places greater emphasis on the disposition of polymer molecules within the solution. After noticing the similarities between the electron photomicrographs of cotton or rayon fibers and membrane structure, Maier and Scheuermann hypothesized that fibers and membranes are formed by a common process. The prerequisite for the formation of either fiber or membrane gels is the presence of a soluble fibrous (straight-chain) polymer, of middle or high molecular weight, whose solutions could be desolvated gradually so as to result in a gel. These authors recognized the necessity of "pseudosolvents" in the phase-inversion solution. By this term is meant swelling agents that can swell the resultant gel, although they are themselves unable to dissolve the polymer. The entire membrane-formation process was considered in terms of five discrete steps: loss of volatile solvents, gelation, contraction of the gel network by syneresis, capillary depletion, and drying.

Since the formation of two liquid phases during the course of desolvation is a prerequisite for the gelation model proposed by Maier and Scheuermann, some speculation will be offered as to the origin of the second interdispersed phase. By analogy to the micelle-forming properties of surfactants, two factors appear to be of significance. First, the polymers that can be fabricated into phase-inversion membranes are all polar; that is, they contain at least two separate groups that possess varying affinities for the solvent. In other words, they have some groups that are relatively lyophilic and others that are lyophobic, a condition that would, in the presence of two separate phases, be conducive to surface activity. Second, the predisposition to the formation of two separate phases, so that the inherent surface activity of the polymers can be utilized to orient themselves at interfaces, exists in the form of the *swelling agent* (or weak precipitant) within the solution. The swelling agents apparently solvate certain of the polymer groups by hydrogen bonding or ion–dipole interactions. As the concentration of swelling agent increases during solvent loss, microdroplets of excess swelling agent may separate from the solution. The interface presented by droplets of the second phase causes polymer molecules, because of their surface activity, to concentrate about these droplets. Insofar as possible, the groups solvated by the swelling agent would be expected to orient themselves toward the interior of the micelle, whereas those solvated by the volatile solvent would situate themselves at the exterior surface.

The effect of increasing swelling-agent concentrations in the sol on gel structure is intelligible in light of the preceding argument. If swelling-agent concentration is low—for example, cellulose acetate in 10 mole % formamide in acetone–phase separation will be postponed to a late stage in the desolvation process, by which time the droplets of swelling agent will have increased in size. The presence of large, widely separated droplets will result in the formation of a small number of large, thick-walled droplets (Fig. 1). After gelation the gel structure will be composed of rather large and widely separated units, many of

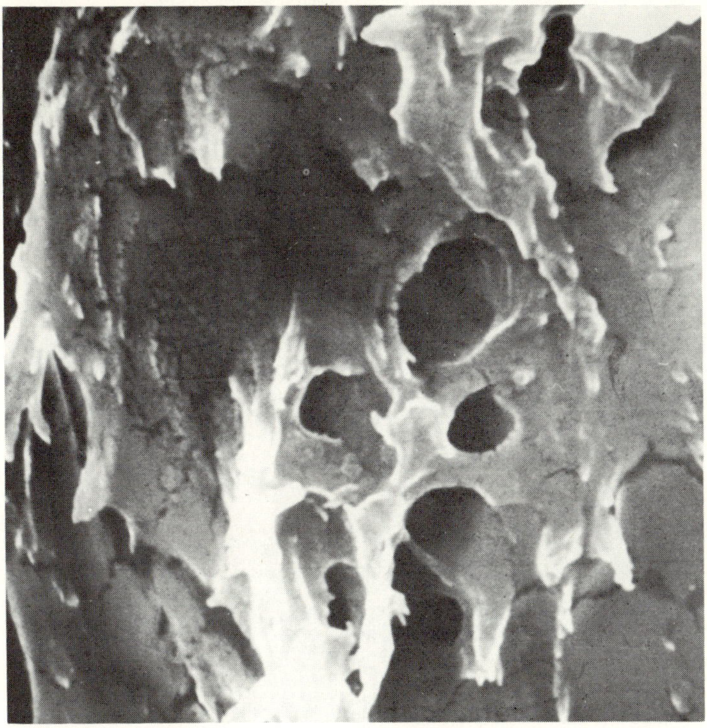

Fig. 1. Cross section of a slightly swollen cellulose acetate membrane (10,000 ×).

which will be closed cells. If, on the other hand, swelling-agent concentration is high—for example, 50 mole % formamide in acetone—phase separation will occur at an earlier stage in desolvation, at which time droplets of the second phase will be small and numerous. Gelation will therefore occur at an early stage, so that the resultant gel will be more swollen and characterized by small, interconnecting, open cells with thin walls (Fig. 2).

Maier and Scheuermann (12) hypothesized that the droplets either exist in the original solution or are formed at an early stage in desolvation. (Fig. 3*a*). They are initially spherical, with the mass of the polymer molecules concentrated at their exterior surface, so that the liquid within the droplets contains only a small amount of polymer. During the course of desolvation individual micelles decrease in diameter, and the thickness of their shells increases (Fig. 3*b*). Eventually the micelles approach (Fig. 3*c*), make contact with one another, and undergo deformation into polyhedra (Fig. 3*d*). As contact becomes more intimate, some interdiffusion of polymer molecules into the walls of neighboring polyhedra occurs (Fig. 3*e*). This interdiffusion apparently occurs at a relatively early stage in desolvation, at a time when fluidity has not been drastically

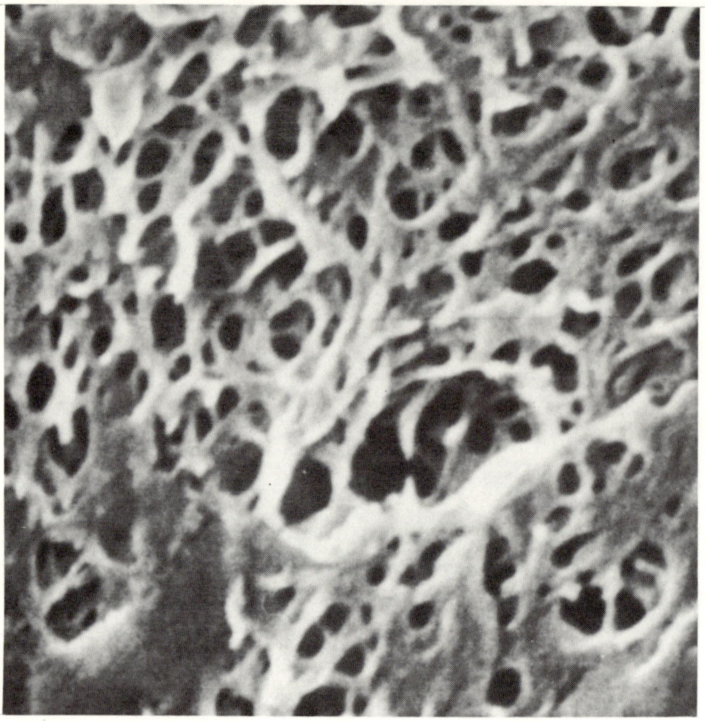

Fig. 2. Cross section of a highly swollen cellulose acetate membrane (30,000 ×).

decreased. If at this stage the solution is immersed into a gelation medium, such as water, desolvation will be accelerated and micellar walls will decrease in depth owing to the loss of solvent being more rapid than micellar contraction. Since the thin walls cannot resist the pressure of contraction, they will tear (Fig. 3*f*) and form a porous, open-cell structure. The contraction with solvent loss is so great that remaining solvent and swelling agent are expressed in the process known as syneresis. After the solvent has been completely removed from the nascent membrane, the formation of the primary gel structure is complete.

3. Secondary Gel Structure

The membrane that appears on completion of the sol-to-gel transition is referred to as a primary gel structure. Such a species can be physically modified by additional treatments to produce secondary gel structures exhibiting altered performance characteristics. These alterations are attributable to changes in both swelling and order within the primary gel.

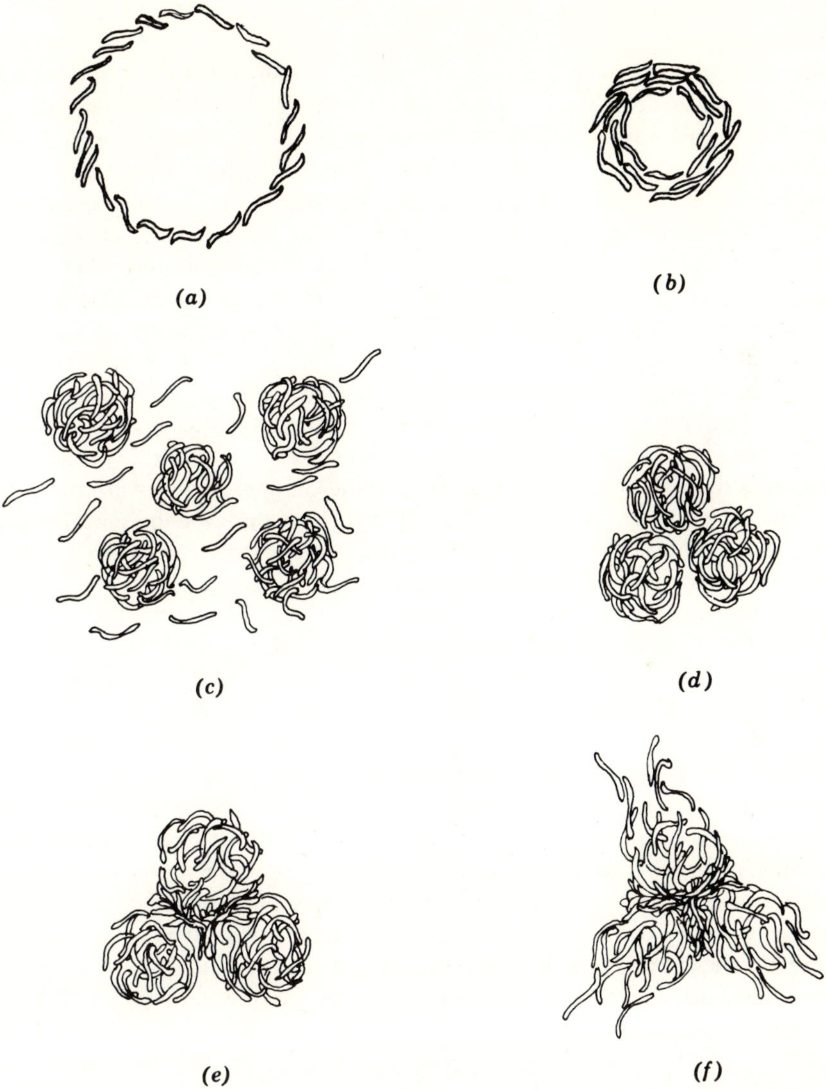

Fig. 3. The gelation process.

(a) MODIFICATIONS OF GEL SWELLING

The degree to which the primary gel is swollen may be either increased or decreased. Increased swelling is of interest in those cases in which it is expedient to produce a swollen membrane gel from an available dense film. The most obvious solution to this problem is also the most effective. The films are immersed in a suitable swelling medium in which the polymer–solvent interaction is sufficient to overcome some, but not all, of the secondary valency forces that constitute the polar intermolecular crosslinks opposing solution. The swelling medium will vary with the nature of the cellulosic film involved: concentrated aqueous solutions of zinc chloride (48) have been utilized for swelling cellulose films, and although they would also serve in this capacity for the cellulosics, mixtures of solvent and nonsolvent [such as ether–ethanol (18) for collodion] are sometimes preferred. The desired degree of swelling is best obtained by varying the concentration of the solvating agent within the swelling medium, but other parameters, such as temperature and immersion time, can frequently be varied to produce similar effects if this is desired.

In the case of swollen, asymmetrical cellulose acetate membranes of the Loeb–Sourirajan type it is usually desirable to deswell the already comparatively unswollen surface layer of the primary gel still further in order to increase permselectivity. Deswelling is desirable for the purpose of removing from the surface layer any capillary water whose presence would favor the solvation and hence permeation of salt ions. The two treatments that have been most widely used to effect deswelling are heating and pressurization. Heating a swollen cellulosic membrane gel results in shrinkage and loss of water for two reasons: first, kinetic energy is added to the system, permitting neighboring polymer chains to make contact with one another so as to facilitate the formation of polar crosslinks; second, the strength of hydrogen bonding decreases with increasing temperature so that water is less strongly bonded and therefore more at liberty to depart. Because water does not strongly interact with the newly formed polar crosslinks (presumably hydrophobic acetate–acetate linkages are involved), the membrane gel does not reswell after cooling. Trudelle and Nicholas (49) have recently shown that the thermal treatment of cellulose acetate reverse-osmosis membranes decreases the water content of the surface layer (from 46 to 38.6%) to a much greater extent than it does that of the substructure (from 63 to 61.8%). Pressurization acts in a similar manner to effect increased density of polar crosslinkages and loss of water.

(b) MODIFICATIONS OF GEL ORDER

Not all differences between the permselectivity of primary and secondary gel structures are related to differences in swelling. For example, although both pressurization and heating can be utilized separately to produce equally swollen membranes of cellulose acetate, differences in the permselectivity of these membranes will be found. These differences are explicable only on the basis of differences in order. One reason for the varying effects of heat and pressurization on order in secondary gel structures is that the latter effects shrinkage primarily in one dimension, that is, in depth. A second reason is that heating in excess of the glass-transition temperature causes segmental motion of anhydroglucose units, sufficient to permit a reorientation and alignment of the various functional groups on neighboring chains. Wasilewski (50) studied, by microwave spectroscopy, the reorientation of acetyl groups during thermal treatment of cellulose acetate membranes. He found that reorientation was 95% complete after the first 5 min of heating above the glass-transition temperature. Therefore the formation of highly ordered amorphous or even crystalline regions is much more likely to result from thermal treatments than from pressurization.

Membrane permselectivity can sometimes be increased by deforming circular pores into ellipses by the application of unidimensional stretching forces (51). This modification is most practical in the case of porous cellulosic membranes that have been swollen by plasticizers (e.g., cellulose membranes swollen by water).

4. Membrane Asymmetry

It has long been recognized that the gel structure at the surface of the membrane that is first exposed to the gelation medium differs considerably from that in the interior of the membrane. This phenomenon is not unique, and in fact closely analogous situations are encountered for the cases of paint skins and skin-and-core structures of viscose-rayon filaments (52). Asymmetry in membrane structure results from both the more rapid desolvation that occurs at the interface between the polymer solution and the gelation medium, and the differences between interfacial and bulk forces. Membranes of the Loeb-Sourirajan type (27) are so asymmetrical that they can be considered as being composed of two distinct layers: the surface, or "active," layer, which is dense and ultrathin and offers effective resistance to solute transport; and the porous substructure, which acts merely as a support. The great advantage of such a bilayered structure is the maintenance of the high degree of permselectivity possessed by dense films while simultaneously attaining the high rates of material transport characteristic of more swollen membranes.

The origin of the *surface layer* in cellulosic membranes can be understood within the framework of the already described gelation model. It will be recalled that the formation of micelles was ascribed to the appearance of a two-phase system at some point during the desolvation process. It is apparent that the surface-active properties of the dissolved polymer molecules situated in the vicinity of the interface between the gelation medium and the polymer solution would effect an alignment with respect to this interface whether or not micelles have any prior existence (Fig. 4). Thus a predisposition toward differentiation between surface and substructure exists in the sol precursors to membrane-gel structure. The swelling and order within the nascent active layer may vary from

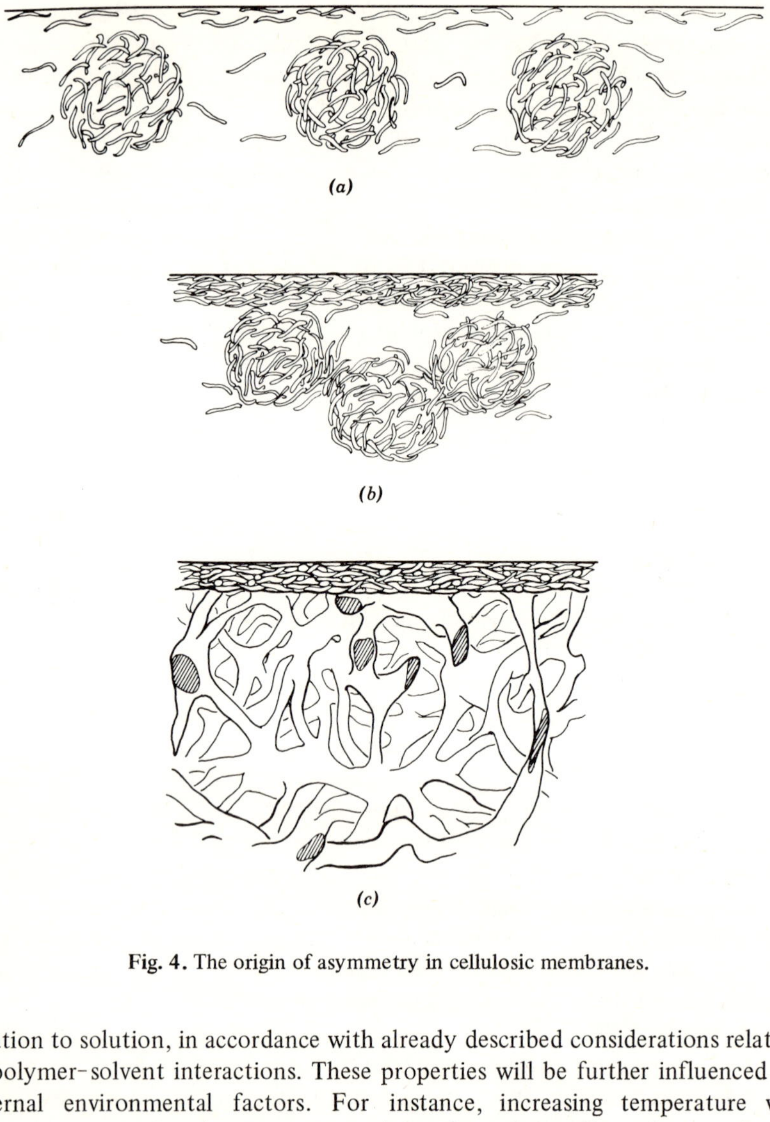

Fig. 4. The origin of asymmetry in cellulosic membranes.

solution to solution, in accordance with already described considerations relating to polymer–solvent interactions. These properties will be further influenced by external environmental factors. For instance, increasing temperature will decrease structural order; increasing relative humidity will accelerate gelation and thus retard compaction and increase solute permeability.

In the traditional method for the fabrication of cellulosic membranes dilute (about 4 wt %) polymer solutions are desolvated to the point of incipient gelation prior to immersion in water or alcohols (22). In such an instance the surface layer will form prior to immersion. In the procedure developed by Loeb and Sourirajan (27) concentrated (about 20 wt %) polymer solutions are allowed to desolvate slightly prior to immersion in a gelation medium. In this process the skin may be formed during the immersion of the still-fluid solution into the gelation medium. The gelation medium can be gaseous, such as air, or liquid, such as water. It must be miscible with all the ingredients of the phase-inversion solution with the exception of the polymer, for which it must be a nonsolvent.

Because of differences in the rates of diffusion, interfacial forces, and polymer–nonsolvent interactions, the nature of the gelation medium will significantly affect a membranes structural and performance characteristics. Only a few generalizations concerning gelation media are warranted here. First, gel swelling increases as the nonsolvent nature of the medium becomes less pronounced. Thus, because ethanol has a greater affinity for cellulose nitrate than does ether, gelation in the former will produce a more swollen gel than in the latter (18). A factor of considerable importance is the temperature of the gelation medium. Higher temperatures will tend to destroy such order as may have been obtained prior to immersion and hence adversely affect permselectivity. By the same token, high gelation-medium temperatures minimize the amount of contraction of the incipient cells that can occur during desolvation and therefore result in more highly swollen gels.

5. Structural Irregularities

One of the principal difficulties encountered by those engaged in membrane technology is the appearance of irregularities in an otherwise homogeneous representative membrane structure. Because permselectivity can be adversely affected by the presence of only a few larger-than-average permeation sites, great care must be taken to minimize their numbers. Close control of such environmental factors as temperature and relative humidity and the elimination of dust and unwanted air currents are fairly common precautions to this end (25). However, even when clean-room techniques are employed, reproducible permselectivity, particularly with small samples, is often an elusive goal.

Helmcke (10) has investigated the nature and origin of such structural inhomogeneities by electron microscopy. He found that irregularities arise after the onset of gelation and occur in conjunction with large voids within the gel. These voids result from cavities in which relatively large amounts of desolvating liquids collect until such a time as they suddenly emerge at the surface of the membrane, thereby causing a portion of this surface to dissolve. Therefore, although the surface layer of the membrane may be homogeneous, as initially formed, it can be subsequently redissolved and re-formed, with resulting changes in the gel structure where this occurs. Helmcke has suggested that the appearance of voids in an irregular fashion might be due to the presence of submicroscopic imperfections in the surface of the substrate on which the polymer solution is cast.

6. Dense and Ultrathin Membranes

No treatment of cellulosic membranes would be complete without a discussion of dense membranes for two reasons. First, they can be regarded as one extreme of the *sol-gel continuum* rather than as specifically distinct entities. Second, they have been considered by some as being identical with the dense "active" layer of reverse-osmosis membranes (53).

The arbitrariness of considering films and membranes as being specifically distinct is particularly apparent in the case of specimens composed of cellulose itself, which might be properly placed in either category, depending solely on the water content of the particular environment in which they are located. In such a case any differences may be primarily semantic in nature (see also Section E of Chapter XIX).

The assumed identity between dense films and the dense "active" layer of reverse-osmosis membranes is a matter of practical concern, both because measurements with dense films have been utilized to calculate "theoretical" salt-rejection capacities of cellulose acetate membranes (53) and because ultrathin, dense membranes are of intrinsic interest (54). From the preceding discussion of order in sol structures it is apparent that the various interactions within a phase-inversion solution will influence the order within the resultant gel structure. Therefore both expanded and dense membranes will possess varying amounts of order owing to their peculiar fabrication histories. Thus identically ordered dense membranes would be unlikely to result from solutions of a given polymer in two different solvents. For the same reason, unless the chemical and physical properties of two different polymers are closely matched, their solutions in a given solvent should produce dense membranes of varying structure.

The development of cellulosic membranes of submicron thickness is currently of interest in attempts to combine the advantages of the thin "active" layer of cellulose acetate desalination membranes with some other porous substrate that is less prone to compaction. If the difficulties in the production, handling, and support of such membranes on a large scale can be overcome, their development may represent a considerable advance in membrane technology.

7. Chemical Modifications of Membrane Structure

Chemical modification of membrane structure presupposes the existence of an intact membrane both before and after a given chemical reaction. It therefore does not include modifications of the polymer made before membrane formation. The common chemical treatments for modifying cellulosic membranes fall into three major categories: hydrolysis, crosslinking, and grafting.

(a) HYDROLYSIS

Because of difficulties in handling and controlling solutions of cellulose itself, the practice of hydrolyzing cellulosic membranes, particularly of the nitrate and the acetate, to membranes of the parent polymer is fairly common. This approach implies that the two membrane structures will be isomorphous—that is, will possess essentially identical morphologies. This is obviously only an approximation, since volume losses and crystallization are apt to accompany the emergence of the cellulose membrane.

The accessibility of dense membranes of cellulose acetate has been increased so as to permit homogeneous deacetylation by swelling in acetone-methanol before immersion in the hydrolysis medium (55). Heterogeneous deacetylation—as for instance, in the stepwise hydrolysis of the dense portion of reverse-osmosis membranes—has been accomplished by the topographical application of an aqueous caustic solution (56).

Water permeability increases with the extent of deesterification both because hydroxyl groups are more hydrophilic than the ester groups they replace and because (in the case of partial hydrolysis) the ester groups occupy a large volume, which is subsequently replaced by the hydrolytic medium. In general permselectivity to solutes that permeate by a solution-diffusion mechanism decreases with increasing extent of deesterification.

(b) CROSSLINKING

Both crystalline modification and the introduction of covalent chemical bonds with polyfunctional reagents have been employed to improve physical properties and permselectivity. Even dense membranes tend to undergo elastic and inelastic deformation under pressure. At a given pressure the extent of compaction depends on the derivative type, DS, degree of swelling, extent of plasticization by the swelling medium, and crystalline structure. Since these parameters can act either to reinforce or oppose one another, the development of a quantitative theoretical approach to correlate the membrane-performance characteristics of cellulosic membranes with crosslinking has not yet been possible. Nevertheless progress has been made in the measurement of physical properties relative to compaction under stress, so that dimensional stability can be correlated, at least qualitatively, with crosslink density (57). The acetates are less susceptible to compaction than are the nitrates and ethers. Resistance to compaction increases with increasing DS, whereas crystallinity decreases with increasing swelling and plasticization.

The crosslinking of cellophane with dialdehyde starch greatly increases its resistance to the permeation of potassium chloride (58). This has been explained on the basis that crosslinking bonds adjacent chains together, thereby creating a more rigid network with increased resistance to the transport of ions by activated diffusion. The crosslinking of methylcellulose with oxalyl chloride has been accomplished (59), but this procedure is inapplicable to cellulose acetate because the hydrochloric acid produced causes excessive hydrolysis. Cellulose acetates have been crosslinked with diisocyanates so extensively that the resultant structures no longer swell in acetone.

Because space does not permit a detailed discussion of the many bifunctional or polyfunctional reagents that are capable of reacting with the hydroxyl groups of the cellulosics, a few generalities must suffice. First, covalent crosslinking of any but the most hydrophilic cellulosics often results in an excessive diminution in water permeability owing both to steric effects and to the replacement of the hydrophilic hydroxyl by some more hydrophobic group. The utilization of very hydrophilic crosslinking agents, on the other hand, may be self-defeating owing to complications arising from internal plasticization. It appears, therefore, that the question of balance in properties between hydrophilic and hydrophobic may be as important in the choice of a crosslinking agent as it is in the initial choice of a membrane polymer. Second, the crosslinking agent should not be chosen without careful consideration of the effect of any by-product on the polymer substrate. The already mentioned case of crosslinking cellulose acetate with oxalyl chloride serves to illustrate this point. Finally, the fact that chemical crosslinking is often a complicated procedure requiring solvent-exchange and other tedious procedures suggests that the formation of virtual crosslinks by controlled crystallization should be investigated where practicable (see also Section E of Chapter XVII).

(c) GRAFTING

No longer a novelty, the modification of cellulosic membranes by graft copolymerization holds considerable promise. True grafting generally involves the *in situ* polymerization of vinyl monomers onto free-radical sites at carbon or oxygen atoms on the anhydroglucose units. Free radicals are produced at these sites by various means: abstraction of hydrogen atoms by free radicals resulting from the decomposition of organic peroxides, high-energy (60) and ultraviolet-light irradiation (61), and a host of chain-transfer reactions. Ceric-ion initiation has also been utilized (62).

The principal purpose of grafting onto a cellulosic membrane is to increase permselectivity. There are several means by which grafting can hope to accomplish this objective: introduction of ionic groups into otherwise neutral substrates can lead to Donnan exclusion of salt ions (63); surface grafting can lead to what is essentially a composite structure consisting of two membranes in series; selective grafting within larger permeation sites might have an obstructive effect, thereby narrowing the equivalent pore-size distribution; asymmetry can be introduced by permitting a gradient of graft copolymer from one membrane surface to the other (64).

The effect of grafting on the hydrophilic-hydrophobic balance of cellulosic membranes is closely akin to that of crosslinking, with one exception: grafting of a hydrophobic material onto a cellulosic may actually increase the water sorption based on the weight of the cellulosic polymer, whereas crosslinking with such a substance invariably decreases water uptake. The origin of this difference lies in the capacity of grafted chains to disrupt crystallites and hence increase accessibility (see also Section G of Chapter XVII).

8. Permeability Considerations

(a) PERMSELECTIVITY

Permselectivity is a measure of the permeation of solute relative to that of solvent. As used here, the terms "solute" and "solvent" refer, respectively, to all suspended and suspending media whether or not they are in solution. All membrane separations are effected on the basis of both *steric* and *solubility* factors.

Solubility factors predominate in "solution-diffusion" membranes in which permeation occurs in four steps as follows;

1. Adsorption of the solute at the surface of the membrane.

2. Solution of the solute in the membrane;.

3. Activated diffusion in the direction of minimized free energy.

4. Desorption of the solute on the other side.

Much confusion about the existence or nonexistence of pores in these membranes has arisen. Rickles and Friedlander (65) have pointed out that although there are voids (on a molecular scale) in any material, the position and size of these voids are constantly changing due to the thermal vibrations of the polymer segments. In dense membranes the diameter of the water-filled pores is so small (20 Å) that there is no longer capillary water but only bound water (66). The latter is water whose dielectric constant has been lowered by hydrogen bonding with certain polar groups (usually hydroxyl) found on the cellulosic molecules. Because of the lower mobility of bound water, it is less able to orient itself about and solvate salt ions, which are consequently less able to dissolve in and diffuse through the membrane. Ions of small size and/or high charge possess a high charge density and require a large number of water molecules to satisfy their hydration requirements. Such ions tend to be strongly attracted to ions of opposite charge, so that their salts will be insoluble unless a large number of water molecules can shield the cation from the anion. Ions of high charge density permeate much less readily than those of low charge density because the probability will be lower that sufficient bound water will be available in any given void to solvate the former.

"Solution-diffusion" membranes can also separate nonionic solutes even when the latter are sterically matched. Separations in such cases appear to be related to the capacity of the polar functional group for hydrogen bonding with the bound water within the polymer matrix. Reid and Breton (26) developed a qualitative bound-water model for water and ion flow across cellulosic membranes. Vincent et al. (67) subsequently enlarged this model and focused attention on the balance between hydrophilic and hydrophobic properties in cellulose acetate, emphasizing hydroxyls as the hydrophilic, and acetates as the hydrophobic groups of secondary cellulose acetates. The hydrophobic acetate groups form polar crosslinks with one another, thereby restricting the amount of water that can associate with intervening hydroxyl groups.

As pore diameter increases, the ratio of bound to capillary water decreases and separations will be effected less on the basis of solubility and more on the basis of steric factors. Separations on the basis of size are readily understood since they simply refer to the ability of a solute particle to be encompassed by a pore. Correlations between molecular weight and permeability are often utilized to demonstrate a separation on this basis. However, molecular weight is not the only steric consideration of importance, since particle shape may be crucial with respect to the separation of substances of a given molecular weight. Permeation is controlled almost exclusively by steric factors for those cases in which solute size is large and/or the extent of interaction with the membrane is small.

(b) INTERFACIAL PHENOMENA

Interfacial phenomena of two types are encountered in membrane-separation processes: surface activity and concentration polarization (boundary-layer effects).

(1) Surface Activity

Surface-active agents concentrate at interfaces because certain portions of their molecules are lyophilic and others are lyophobic. Surface activity will vary with the nature of the solvent system, and hence a cellulosic polymer with considerable surface activity in organic solvents may be insoluble in an aqueous medium. Nevertheless even water-soluble surfactants are highly analogous to cellulosic membranes in that both possess a certain balance between hydrophilic and hydrophobic properties. This balance controls the surface activity of the surfactants and the degree of swelling and permeability of cellulosic membranes.

Because interfacial concentrations are far in excess of bulk surfactant concentration, solutes possessing surface activity can have a very pronounced effect on material transport. This is so because the membrane will be in equilibrium with the concentration of solution with which it is in direct contact, rather than with the bulk solution. This fact offers a possible explanation for hitherto anomalous declines in product flux that occur at low concentrations of certain organic solutes.

Surfactants affect the overall permeability of cellulosic mambranes in two distinct ways, depending on pore diameter. In solution-diffusion membranes surfactants tend to decrease permeability, and in porous membranes they tend to increase it owing to the surfactant's ability to wet pore surfaces.

Another significant surface-active phenomenon is the "feed additive" effect. The addition of a few parts per million of poly(methyl vinyl ether) to the saline feed solution in reverse osmosis effects a large decrease in salt permeability, with only a minor diminution in product flux. This effect has been ascribed to the formation of a *liquid membrane* at the interface between cellulose acetate and the saline solution (29). *Liquid membranes* are believed to function in the same manner as the underlying cellulosic membranes; that is, by binding water and thus limiting its ability to solvate salt ions.

(2) Concentration Polarization (Boundary-Layer) Effects

Sherwood et al. (68) have considered the effect on the permeation of solute and solvent caused by the concentration buildup of solute at the membrane–solution interface owing to the more rapid transport of solvent than that of solute. This effect is of great importance to the economic feasibility of membrane-separation processes owing to the already mentioned fact that the membrane will be in equilibrium with the surface, rather than the bulk, concentration. Increasing the flow rate of solution past the membrane so as to permit operation under conditions of turbulent flow can be utilized to dissipate this boundary layer. An alternative solution is to decrease the height of the fluid channel or the diameter of the tube so as to permit operation in the laminar-flow region. Detailed analytical and experimental solutions to this problem are of critical importance to the economics of those membrane-separation processes in which the feed solution has a high concentration of solute particles.

(c) EXTERNAL FACTORS AFFECTING PERMEABILITY

Irrespective of any effects on membrane structure, temperature, pressure, and solute concentration affect both the overall permeability and the permselectivity of cellulosic membranes for kinetic and thermodynamic reasons. Permeability increases with increasing temperature because of an increase in the kinetic energy available for diffusion. On the other hand, permselectivity decreases with increasing temperature both because of a decrease in the hydrogen-bonding strength of bound water and because such species as ionic solutes will be less strongly hydrated at elevated temperatures. Although permselectivity is not concentration dependent in membrane-separation processes, such as reverse osmosis, nevertheless higher concentrations of solute in the feed solution will obviously result in higher concentration of solute in the product solutions. At the same time, because higher solute concentrations produce an increase in the flow of solvent from the low-concentration to the high-concentration side of the membrane, the effective pressure or driving force available for permeation will decrease with increasing solute concentration.

An important consideration affecting both *capital costs* and the *fluid dynamics* of membrane-separation processes is membrane configuration. Membranes are currently produced in sheet and tubular forms. The former can be utilized in either flat-plate (filter press) or spiral (rolled-up) modules. Tubular membranes have been fabricated in diameters ranging from 5 cm to 2.5×10^{-4} cm. The latter are usually designated capillary membranes. A discussion of the advantages and disadvantages of the various configurations for any given membrane-separation process is beyond the scope of this section. For a discussion of hollow-fiber membranes, see Section F.2 of this Chapter XIX.

9. Practical Applications of Membrane Separations

(a) DESALINATION BY REVERSE OSMOSIS

As mentioned at the beginning of this section, the need for additional sources of fresh water has resulted in the development of an intense interest in reverse osmosis. This process has lower energy requirements than other separation processes, such as distillation, because no change in state is required. In reverse osmosis pressure in excess of the osmotic pressure is applied to a concentrated solution separated from a dilute solution by a semipermeable membrane. Because water, but not salt, will dissolve in and permeate the membrane, a net flow of water from the concentrated to the dilute solution occurs. Within less than a decade, owing to improvements in both membrane technology and module design, this process has been converted from the status of a mere laboratory curiosity to the threshold of economic realization (Table 1). Reverse osmosis has also been investigated for utilization in the concentration of sugar solutions (69), fruit juices, milk, and other food products. The removal of yeast cells from beer to obviate the necessity for pasteurization is now a commercial practice, and the clarification of wine should likewise be practicable by this means. Water purification to produce conductivity water and even pyrogen-free water also appears to be feasible.

TABLE 1
Evolution of Cellulosic Reverse-Osmosis Membranes

Investigators	Cellulosic type	Transport characteristics[a]	
		Product flux (cm/day)	Salt retention (%)
Reid and Breton (26), 1959	Dense di- and triacetates	>1	96
Loeb and Sourirajan (27), 1961	Swollen diacetate	20–40	98.6
Aerojet-General Corp. (70), 1967	Swollen diacetate–triacetate blend	40–60	99

[a] Normalized to 0.6 *M* NaCl feed; 25°C; 102 atm; 2-cm/sec flow rate.

(b) BIOLOGICAL SEPARATIONS

The application of cellulosic membranes to biological separations involving proteins, sera, and bacteria as well as other microorganisms has been covered by Ferry (30). Today what were formerly called filterable viruses can be quantitatively retained on cellulose acetate membranes of the type employed for desalination.

(c) BIOMEDICAL APPLICATIONS

The first and to date the best membranes for artificial kidneys are those of cellulose. Although commercially available regenerated-cellulose films are still the membranes of choice, there has been considerable interest in hydrolyzed cellulosic ester membranes, particularly in those exhibiting asymmetry of the Loeb–Sourirajan type. Attempts are also being made to develop nonthrombogenic surfaces by attaching heparin (a mucopolysaccharide) onto cellulosic membranes (see also Sections F.2 and J of this Chapter XIX).

(d) THE FUTURE OF CELLULOSIC MEMBRANES

Cellulosic membranes have several shortcomings: biodegradability, hydrolytic instability, and compaction under pressure. It should be possible to correct the first of these by cyanoethylation (71), a technique that has been applied with success to cotton textiles (see Chapter XVII-D). Because the other shortcomings are less amenable to correction, fully synthetic polymers are expected to make serious inroads into what has long been the almost exclusive realm of the cellulosics. However, by analogy to cellulose films, the expected decline should be relative only, that is the utilization of fully synthetic membranes will increase more rapidly than that of cellulosic types. Cellulose–synthetic polymer composites are also expected to be of increasing importance; for example, styrene-grafted cellulose acetate of the Loeb-type has been found to have improved resistance to compressive compaction (73, 74). A comprehensive discussion of membranes can be found in reference 72.

References

1. C. Schönbein, British Patent 11,407 (1946).
2. A. Fick, *Ann. Physik Chem.*, **94**, 59 (1855).
3. Lhermite, *Ann. Chim. Phys.* [3], **43**, 420 (1855).
4. W. Schumacher, *Ann. Physik Chem.*, **110**, 337 (1860).
5. H. Bechhold, *Z. Physik. Chem.*, **64**, 328 (1908).
6. F. Erbe, *Kolloid-Z.*, **63**, 277 (1933).
7. M. Ardenne, *Elektronenübermikroskopie,* Springer-Verlag, Berlin, 1940, p.350.
8. K. Maier and H. Beutelspacher, *Naturwissenschaften,* **40**, 605 (1953).
9. J.-G. Helmcke, *Kolloid-Z.*, **135**, 29 (1954).
10. J.-G. Helmcke, *Kolloid-Z.*, **135** 101 (1954).
11. J.-G. Helmcke, *Kolloid-Z.*, **135**, 106 (1954).
12. K. Maier and E. Scheuermann, *Kolloid-Z.*, **171**, 122 (1960).
13. R. Kesting, M. Engdahl, and W. Stone, Jr., *J. Macromol. Sci.-Chem.*, **A3** (1), 157 (1969).
14. H. Bechhold, *Z. Physik. Chem.*, **60**, 257 (1907).
15. S. Bigelow and A. Gemberling, *J. Am. Chem. Soc.*, **29**, 1576 (1907).
16. G. Malfitano, *Z. Physik. Chem.*, **48**, 243 (1910).
17. A. Schoep, *Kolloid-Z.*, **8**, 80 (1911).
18. W. Brown, *Biochem. J.*, **9**, 591 (1915).
19. W. Brown, *Biochem. J.*, **11**, 40 (1917).
20. N. Bjerrum and E. Manegold, *Kolloid-Z.*, **42**, 97 (1927).
21. F. Bartell and M. Van Loos, *J. Phys. Chem.*, **28**, 161 (1924).
22. W. Elford, *Trans. Faraday Soc.*, **33**, 1094 (1937).
23. P. Grabar et al., *Ann. Inst. Pasteur,* **65**, 159 (1939).
24. R. Montanna and L. Jilk, *J. Phys. Chem.*, **45**, 1374 (1941).
25. H. Gregor and K. Sollner, *J. Phys. Chem.*, **50**, 53 (1946).
26. C. Reid and E. Breton, *J. Appl. Polymer Sci.*, **1**, 133 (1959).
27. S. Loeb and S. Sourirajan, UCLA, Department of Engineering Report 60-60, 1961.
28. See the research-and-development progress reports of the Office of Saline Water (OSW) of the U.S. Department of the Interior; available from the Superintendent of Documents, U.S. Government Printing Office, Washington, D.C.

29. R. Kesting and W. Subcasky, *J. Colloid Interface Sci.,* **28,** 156 (1968).
30. D. Ferry, *Chem. Rev.,* **18,** 373 (1936).
31. N. Lakshminarayanaiah, *Chem. Rev.,* **65,** 491 (1965).
32. U. Merten, Ed., *Reverse Osmosis,* M.I.T. Press, Boston, 1967.
33. P. Katti, *Proc. Indian Acad. Sci.,* **38A,** 387 (1953).
34. W. Moore et al., *J. Polymer Sci.,* **23,** 23 (1957).
35. W. Moore and J. Russell, *J. Colloid Sci.,* **9,** 338 (1954).
36. V. Kargin, P. Kozlov, N. Bakeev, and A. Zezin, *Faserforsch. u. Textiltech.,* **18,** 134 (1967).
37. C. Malm, *Svensk Kem. Tidskr.,* **73,** 523 (1961).
38. F. Cramer and C. Purves, *J. Am. Chem. Soc.,* **61,** 3458 (1939).
39. O. Heuser et al., *Ind. Eng. Chem.,* **40,** 1500 (1948).
40. J. Mahoney and C. Purves, *J. Am. Chem. Soc.,* **64,** 9 (1942).
41. F. Howlett and A. Urquhart, *Chem. Ind.,* (London), 82 (1951).
42. H. Spurlin, *J. Polymer Sci.,* **3,** 714 (1948).
43. P. Flory, *Principles of Polymer Chemistry,* Cornell University Press, Ithaca, N.Y., 1953, Chapter 12.
44. F. Billmeyer, Jr., *Textbook of Polymer Chemistry,* Interscience, New York, 1957.
45. A. Doolittle, *Ind. Eng. Chem.,* **36,** 237 (1944).
46. R. Kesting, *J. Appl. Polymer Sci.,* **9,** 663 (1965).
47. A. Dobry, *Bull. Soc. chim.* (5), **3,** 312 (1936).
48. P. Bartl and B. Sedlacek, *Chem. Listy,* **49,** 826 (1955).
49. Y. Trudelle and L. Nicholas, paper presented at Membrane Symposium, Paris, Feb. 1–3, 1967.
50. S. Wasilewski, UCLA, Dept. of Engineering Report 65-10, Feb. 1965.
51. L. Craig and W. Konigsberg, *J. Am. Chem. Soc.,* **65,** 166 (1961).
52. P. H. Hermans, *Textile Res. J.,* **20,** 553 (1950).
53. H. Lonsdale, U. Merten, and R. Riley, *J. Appl. Polymer Sci.,* **9,** 1341 (1965).
54. P. Francis, OSW R & D Report 177, Feb. 1966.
55. M. Odian, Ph.D. dissertation, Columbia University, 1964.
56. R. Kesting, U.S. Patent 3,290,286 (1966).
57. W. Higley, in Office of Saline Water R & D Progress Report No. 154, Nov. 1965.
58. C. Peterson and E. Livingston, *J. Appl. Polymer Sci.,* **3,** 1429 (1964).
59. R. Signer and P. von Tavel, *Helv. Chim. Acta,* **26,** 1972 (1948).
60. R. Huang, *J. Appl. Polymer Sci.,* **10,** 325 (1966).
61. N. Geacintov, V. Stannett, E. Abrahamson, and J. J. Hermans, *J. Appl. Polymer Sci.,* **3,** 54 (1960).
62. G. Mino and S. Kaizerman, *J. Polymer Sci.,* **31,** 242 (1958).
63. W. Baldwin, D. Holcomb, and J. Johnson, *J. Polymer Sci.,* **3,** 833 (1965).
64. C. E. Rogers, *J. Polymer Sci.,* **3,** 93 (1965).
65. R. Rickles and H. Friedlander, *Chem. Eng.,* April 25, 1966, p.163.
66. R. Schultz and S. Asunmaa, Recent Prog. Surface Sci., **3,** 291 (1970).
67. A. Vincent, M. Barsh, and R. Kesting, *J. Appl. Polymer Sci.,* **9,** 2363 (1965).

68. T. Sherwood, P. Brian, R. Fisher, and L. Dressner, *Ind. Eng. Chem., Fundamentals,* **4**, 113 (1965).

69. J. Andelman and M. Suess, *Anal. Chem.,* **38,** 351 (1966).

70. Aerojet-General Corp., paper presented at the 3rd International OSW Conference, Miami, Apr. 21–25, 1969.

71. N. M. Bikales, in N. M. Bikales, Ed., *Encyclopedia of Polymer Science and Technology,* Vol. 4, Interscience, New York, 1966, pp. 533–535.

72. R. Kesting, *Synthetic Polymeric Membranes*, McGraw-Hill, New York, 1971.

73. H. B. Hopfenberg, F. Kimura, P. T. Rigney, and V. Stannett, *J. Polymer Sci.,* [C], No. 28, 243 (1969).

74. H. B. Hopfenberg, V. Stannett, F. Kimura, and P. T. Rigney, *Appl. Polym. Symposia*, No. 13, 139 (1970).

F.2. Hollow-Fiber Membranes

H. I. Mahon and B. J. Lipps

The advent of highly selective cellulosic membranes (see Section F.1 of this Chapter XIX) has generated interest in the development of more efficient membrane devices. One interesting new concept, which provides a thin, self-supporting membrane with a high surface-to-volume ratio, is the fine, hollow fiber. Hollow fibers of cellulose, cellulose acetate, cellulose triacetate, and nylon have been reported (1–4). The fibers range from 45 to 300 μ in outside diameter and have wall thicknesses of 8–25 μ.

The hollow fibers have been assembled in devices for use in reverse osmosis, dialysis, and hemodialysis. In these devices the hollow-fiber wall is the selective membrane. The fluid to be treated is pumped either through the bore of the hollow fiber or over the outer surface; a portion of the fluid selectively permeates the fiber wall, and the permeate is recovered downstream. The driving force for permeation may be either pressure, as in reverse osmosis, or concentration, as in dialysis and hemodialysis.

The hollow fibers are spun by synthetic-fiber-spinning techniques. The polymer melt or solution is pumped through an annular die or spinneret, and an inert gas is injected into the core of the fiber to maintain the tubular shape until the polymer solidifies. Substantial size reduction can be achieved by rapidly drawing the fiber away from the face of the spinneret. There has been reported a process for melt-spinning hollow fibers of cellulose acetate and cellulose triacetate in which an extractable plasticizer is combined with the polymer to reduce the spinning temperature and thereby prevent thermal degradation (5). In addition, extraction of the plasticizer results in a more swollen membrane structure and enhanced water permeability. In one instance cellulose triacetate was plasticized with sulfolane (43 wt %) and spun at 230°C to form hollow fibers with an outside diameter of 63 μ and a wall thickness of 15.5 μ. The sulfolane was then extracted with water.

Hollow fibers appear to be particularly attractive for use in reverse osmosis because of the substantial transmembrane-pressure differences required. The thin, self-supporting structure of the hollow fiber will withstand pressure differences as high as 40 kg/cm^2 (5), whereas the more conventional flat membrane configurations require elaborate backing supports. Cellulose acetate hollow-fiber cells have been tested for the desalination of brackish water by reverse osmosis (2). The fibers were 45 μ in outside diameter and 27 μ in inside diameter. The hollow-fiber wall appears to have the characteristics of a homogeneous (non-Loeb type) membrane, giving water fluxes of 9– 16 l/m^2/day. Since 1 kg of these hollow fibers provides approximately 100 m^2 of membrane area, these fluxes correspond to water-production rates of 900–1600 l/day-kg of fiber. The rejection of hardness ions reported was 92.6% and of total ions 80%.

For use as a dialyzer the water-leached cellulose acetate hollow fiber must be rendered more permeable to water and small solutes. The increased hydrophilicity can be achieved by hydrolyzing the acetate fibers to cellulose with dilute base. Generally the resultant cellulose fibers are replasticized with glycerol and dried to assist dialyzer fabrication and prevent permeability loss. The membrane solute diffusivities and ultrafiltration coefficients (water flux) of cellulose hollow fibers and flat Bemberg PT-150 cellulose membranes are compared in Table 1.

The first commercial hollow-fiber dialyzer was the Dow artificial kidney, which contains about 12,000 cellulose hollow fibers whose ends are potted in silicone rubber-tube sheets. A schematic of the artificial kidney can be seen in Figure 1. The hollow fibers are nominally 215 μ in internal diameter, providing 1.0 m^2 of active dialysis surface area. The blood priming volume, including the blood distribution manifolds, is about 100 ml. The artificial kidney is 6.25 cm in diameter and 20 cm in length. When primed with blood and dialysate, it weighs about 0.7 kg. Comparable flat-membrane kidneys are significantly larger in size (6).

During hemodialysis the patient's blood flows through the inside of the fibers while the dialysate or wash solution is circulated outside the fibers. No blood

TABLE 1

Comparative Transport Properties

Material	Membrane diffusivity 10^{-6} cm^2/sec			Ultrafiltration coefficient 10^{-10} g–cm/(sec-cm^2-mmHg)
	Sodium chloride	Creatinine	Raffinose	
Bemberg PT–150 membrane (flat)	2.6	1.7	0.5	1.2
Cellulose hollow fiber	2.1	1.2	0.6	0.7

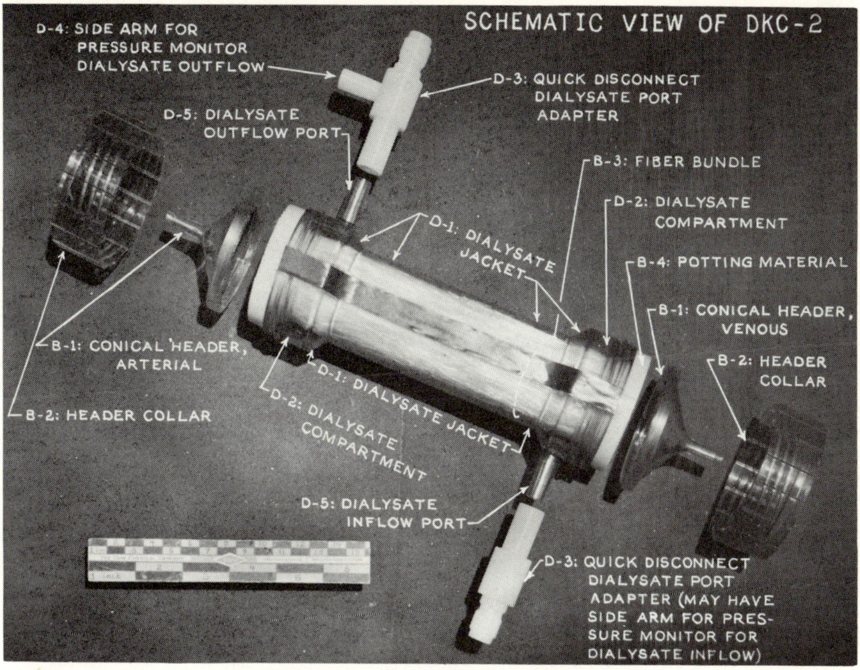

Fig. 1. The Dow artificial kidney.

pumps are required to propel the blood through the dialyzer. Cardiac pressure is sufficient to maintain the requisite 200-ml/min blood-flow rate. The blood-side solute-transfer resistance is greatly reduced with this hollow-fiber hemodialyzer because of the small, well-defined blood channels. The overall transfer capacity of this hemodialyzer is impressive compared with that of flat-membrane dialyzers which have the primary transfer resistance on the blood side.

References

1. B. J. Lipps, R. D. Stewart, H. A. Perkins, G. W. Holmes, E. A. McLain, M. R. Rolfs, and P. D. Oja, *Trans. Am. Soc. Artificial Internal Organs,* **13**, 200 (1967).
2. H. I. Mahon, E. A. McLain, W. E. Skeins, B. J. Green, and T. E. Davis, *Chem. Eng. Progress Symposium Series 65*, 1969, p. 91.
3. H. I. Mahon, *Desalination Research Conference Proceedings*, National Academy of Science–National Research Council, Washington, D.C., 1963, p. 942.
4. R. J. Mattson and V. J. Tomsic, *Chem. Eng. Progress,* **65**, No. 1 (1969).
5. E. A. McLain and H. I. Mahon, U.S. Patent 3,423,491 (1969).
6. F. A. Gotch, B. J. Lipps, J. Weaver, J. Brandes, J. Rosin, J. A. Sargent, and P. D. Oja, *Trans. Am. Soc. Artificial Internal Organs,* **15**, 87 (1969).

G. MICROCRYSTALLINE CELLULOSE

O. A. Battista

Microcrystalline cellulose represents a novel physical state of commercial cellulose (1, 2). The physical and functional properties of microcrystalline cellulose ideally illustrate the useful phenomena that occur when crystalline polymers of high enough molecular weight are transformed into discrete, highly crystalline aggregates of hundreds or thousands of long-chain molecules. The aggregates in general have a high length-to-diameter ratio and are approximately 1 μ in their largest dimension and at least 25 Å, more generally 50 Å or more, in their smallest dimension. In effect the microcrystalline cellulose particles consist of aggregates of molecules joined by strong lateral bonding forces in such a manner as to form truly discrete colloidal particles with dimensions of the same order of polymer micelles and "composite molecular weights" in the many millions.

This author believes that microcrystalline colloidal cellulose is only the first commercial product of many new colloidal macromolecular products to come: new products fashioned from microcrystals of both synthetic and natural high polymers, from organic as well as inorganic high polymers, or combinations of both of them. Analogous colloidal polymer products from other polymer precursors—amylose, asbestos, collagen, and nylon—have already been announced (3, 4). Microcrystals that are to be unhinged and subsequently dispersed must exist within the matrix of a polymer precursor, linked together by primary molecular linkages—"hinges." The polymer precursor may be either natural or synthetic, organic or inorganic, but it must possess a high enough molecular weight to form fibers or films.

It is necessary in the case of each colloidal microcrystalline material to unhinge the individual crystallites or regions of high lateral order from within the matrix of the fiber-forming polymer precursor by a very careful and controlled degradation of the amorphous (disordered) interconnecting regions. This must be done under conditions that preclude excessive swelling of the constituent regions of high lateral order.

During or after the unhinging operation, mechanical action involving high mechanical shear liberates the unhinged submicron polymer microcrystals into a surrounding liquid medium. A minimum concentration of the individual colloidal microcrystals is essential to the formation of these new, stable microcrystal-gel systems.

1. History

The first references to the existence of definite crystalline zones interposed in the amorphous structure of cellulose materials were made by Nageli (5), who in 1877 had already confirmed the optical anisotropy of vegetable products, both in cell walls and in fibers. Herzog (6) and Meyer (7) postulated the crystalline morphology of pure cellulose. The structure of the elementary crystallite in native cellulose was formulated by Meyer and Misch (8), who considered it to be made up of five units of cellulose chain or cellobiose radicals arranged in parallel.

TABLE 1
Chronology of Major Hydrolysis Studies of Cellulose and Cellulose Microcrystals

Brief summary	References
Nageli confirms optical anisotropism in plant cell walls of fibers	5
Girard coins term "hydrocellulose" for acid-degraded vegetable products	15
Herzog, Meyer, and Hess postulate the crystalline nature of cellulose in considerable detail	6–9, 13, 14
Staudinger and Sorkin describe the dual reaction rates of acid attack	16
Beginning in 1941, with major contributions by Nickerson et al., intense activity relating to the use of acid hydrolysis for studying cellulose structure	Chapter III, Section C.2; Chapter XVIII, Section A
Many investigators studied reaction-rate phenomena and the ratio of crystalline to amorphous regions; major investigators and their publications were the following:	
Nickerson and Habrle	17–19
Davidson	20
Battista et al.	21–24
Conrad and Scroggie	25
Philipp, Nelson, and co-workers	26, 27
Pacsu	28
Brenner et al.	29
Roseveare et al.	30, 31
Howsmon	32
Rånby	33, 34
Jørgensen	35–37
Marchessault, Morehead, and co-workers	38–40
Mukherjee and Woods	41
Millett et al.	42

The length of the crystallite along the *b*-axis is 10.3 Å, precisely the length corresponding to the cellobiose unit (9) made up of two anhydroglucose residues, whereas that of the *a*-axis is 8.35 Å, a sufficiently small distance for the space between the proximal hydroxyl groups of different chains to offer the possibility of the formation of strong hydrogen bonds in accordance with Pauling's calculations (10) for this type of bond. Later structural studies, following Meyer's proposals, by Hermans and co-workers (11) and Van der Wyk and Meyer (12) suggested very similar dimensions for the unit cell of native cellulose–dimensions generally accepted today. (See Sections C and G of Chapter XIII.)

"Crystallite zones" also appear, of course, in regenerated, mercerized, and alkali celluloses. Hess (13) and Meyer and Badenhuizen (14) proved that the crystallites of the above forms of cellulose differ in size from those of native celluloses as well as in the angles of the planes of the atoms within the molecules of their unit cells.

Before the advent of commercial microcrystalline colloidal cellulose, investigations involving the treatment of celluloses with various inorganic acids–other than the nitration of cellulose to produce cellulose nitrates–led to few, if any, significant practical or commercial applications. Most of the studies involving the action of acids on cellulose used hydrolytic attack in combination with moisture regain, x-ray diffraction, and electron microscopy to gain insights into the fine structure, the original "molecular architecture" of native- or regenerated-cellulose fibers.

A chronological review of the major studies involving the hydrolysis of cellulose, listing the primary objectives of these studies and the related literature references is presented in Table 1. Because of space limitations, it is impractical to list every publication during this period.

2. Preparation

Microcrystalline cellulose, in its commercial context, got its start from an idea in 1955. The idea was to unhinge and isolate colloidal microcrystals of cellulose and to use these truly colloidal particles to nucleate a viscose spinning solution just prior to regeneration. The goal was a novel route to produce an improved rayon tire cord.

The first step was to convert fibrous cellulose to what was well known at the time as LODP (leveling-off degree of polymerization) (21, 22)* cellulose by a process of severe acid hydrolysis whereby the "hinge," or amorphous, regions in

* The term "leveling-off degree of polymerization" refers to the weight-average degree of polymerization of the macromolecules comprising the discrete, colloidal, unhinged microcrystals of polymers.

the cellulose fibers are attacked, leaving a very highly crystalline residue. A Waring Blendor was used in an attempt to chop the LODP cellulose (comprising millions of aggregates of microcrystals) into a preponderance of isolated or free microcrystals.

When fibrous cellulose fibers, such as wood pulp, are beaten in water (the key step of all papermaking processes with cellulose fibers), very long strands, or fibrils, of cellulose are formed (see Chapter VIII-B). Figure 1 is an electron micrograph of such a sample of beaten wood pulp. Within each long strand, or fibril, of cellulose lie numerous microcrystals interconnected with each other by cellulose molecules. The microcrystals in the cellulose fibrils are therefore hinged together by very strong forces, because the molecules that weave in and out of them are connected by $(1 \rightarrow 4)$-β-glycosidic linkages, true primary linkages. Mechanical beating of cellulose fibers in water serves to expose more and more of the long fibrils shown in Figure 1, but such energy does little or nothing to cause an unhinging of the constituent microcrystals or their liberation from within their original particle matrix. It is almost impossible to introduce enough energy mechanically to break the enormous number of strong primary chemical linkages in the cellulose molecules within the hinge areas, linkages that tie the microcrystals together in a three-dimensional network; stable colloidal dispersions are not formed from fibrous cellulose.

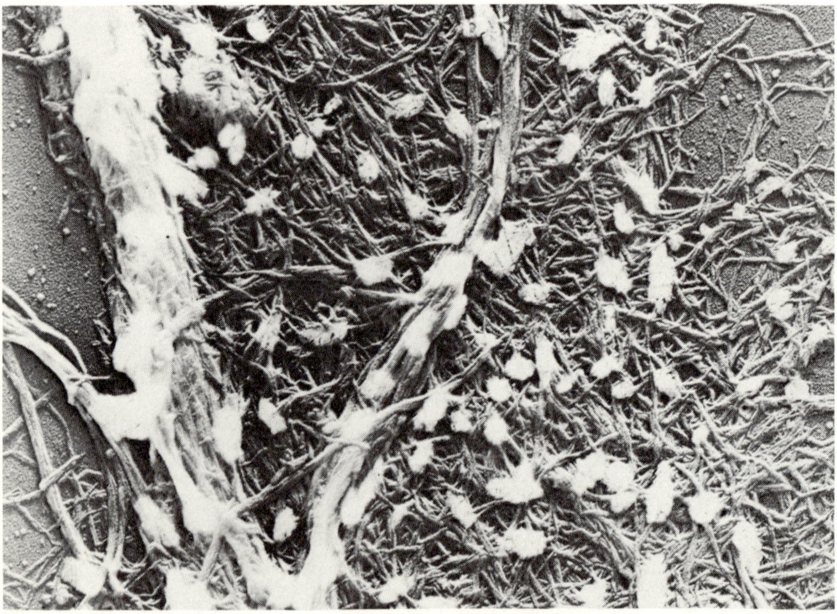

Fig. 1. Beaten α-cellulose pulp fibers (8000 x).

On the other hand, when fibrous cellulose is reduced to its leveling-off DP chemically by treatment with 2.50 *N* hydrochloric acid at 105 ± 1°C for 15 min (1) and then beaten at 5% solids in a Waring Blendor, individual unhinged microcrystals of cellulose peel away from their original architectural positions (see Fig. 2), and when a sufficient number of these microcrystals are freed, a stable colloidal macromolecular gel results. The two-phase water–fiber system is gradually converted to a homogenized-milk consistency, then to a heavy-cream consistency, and finally to a lardlike, opaque, spreadable gel consisting of pure cellulose and water only. This 100% aqueous dispersion, which contains a small percentage of unhinged microcrystals of cellulose and much larger clusters of microcrystals having varying degrees of aggregation, comprises the novel form of microcrystalline cellulose. The length and diameter of the microcrystals are determined largely by their nature in the precursor fibrous cellulose (see Figs. 3 and 4).

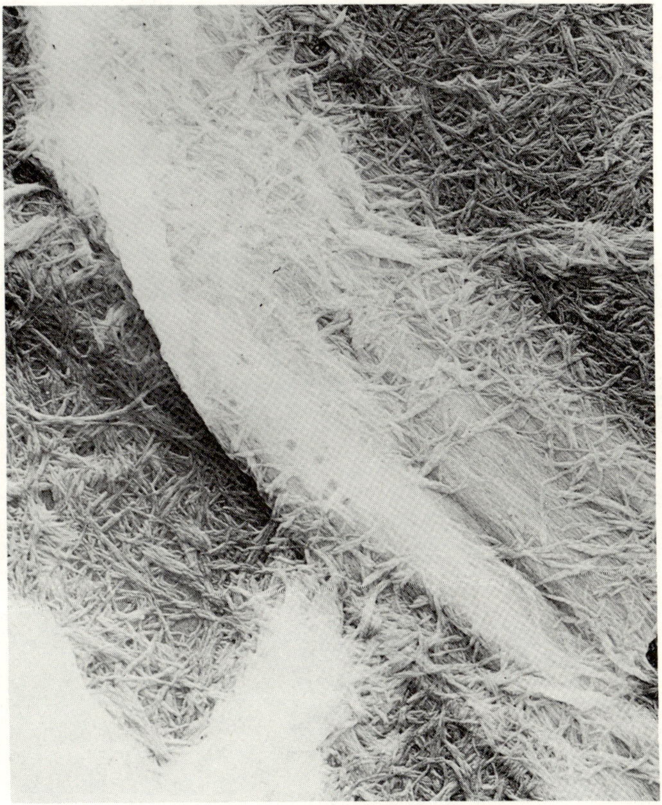

Fig. 2. Microcrystals peeling away from a fiber fragment of LODP cellulose (25,000 ×).

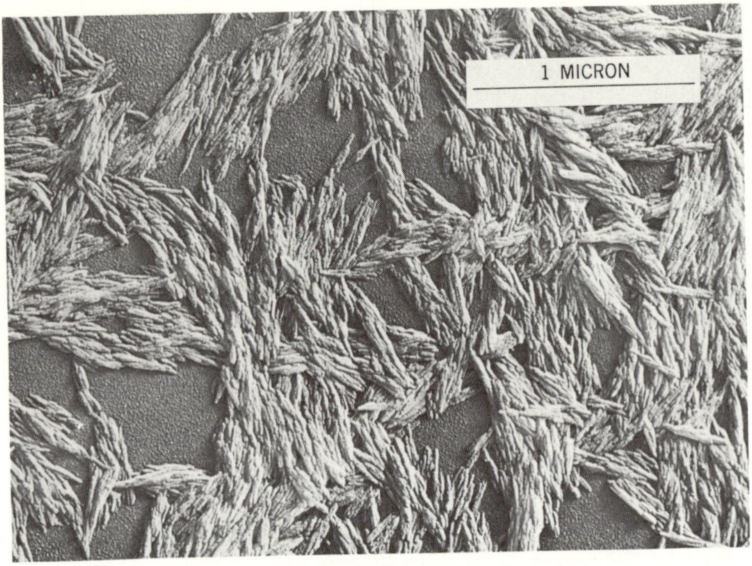

Fig. 3. Microcrystals recovered from cotton linters (25,000 ×).

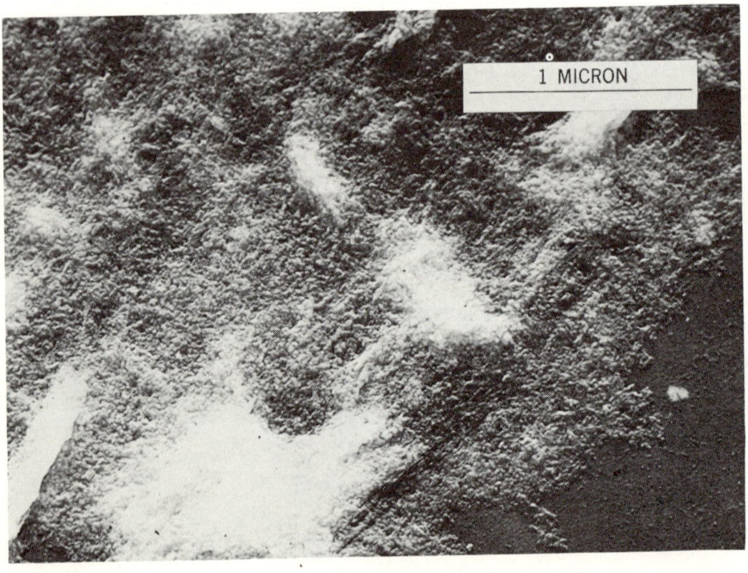

Fig. 4. Microcrystals of rayon tire-cord yarn (25,000 ×).

The *concentration* of LODP cellulose and a *sufficient* input of mechanical energy are crucial variables for producing pure, underivatized colloidal dispersions [without sulfuric acid ester peptization (1,2)] of microcrystalline cellulose. This fact is clearly pinpointed by the original experimental record: in 1955, if 1% LODP cellulose had been dispersed in water (instead of 5%) and the Waring Blendor had been run for only 15 min (instead of over 30 min), the lardlike, pure white gel would not have been produced.

Although many researchers, including ourselves, had studied the isolation of microcrystals (LODP cellulose) from cellulose fibers (2-4, 33, 34, 38-41), their studies were directed primarily at achieving a better understanding of cellulose fine structure. Mukherjee and Woods (41) isolated microcrystals using strong sulfuric acid under conditions that peptized the microcrystal surfaces, no doubt owing to limited formation of sulfate-ester groups on the surface of the microcrystals. Such highly peptized and *surface-modified* (esterified) microcrystals are extremely difficult to purify and, from a practical viewpoint, do not give the same visual and functional properties as underivatized cellulose microcrystals produced with hydrochloric acid. The hydrodynamic characteristics of colloidal dispersions of microcrystals produced by various severe acid-digestion treatments were studied by Marchessault and co-workers (39, 40) and by Hermans and Edelson (43, 44).

3. Utilization

Microcrystalline cellulose, as a snow-white free-flowing powder, is used widely as a pharmaceutical tableting excipient. As a combination binder-disintegrant, it eliminates the granulation step normally required in the production of many active drug ingredients. This permits a more economical direct-compression method of tableting. It also contributes stability to many pharmaceutical formulations. When properly dispersed in water to form stable colloidal gels and dispersions (45), microcrystalline cellulose can be used in cosmetics and pharmaceuticals for both creams and solid suspensions (2, 46-54).

Several new grades of colloidal microcrystalline cellulose have been evolved for different applications. For example, one grade of microcrystalline cellulose that has been developed disperses readily in ice cream and frozen desserts. Tiny, noncaloric, discrete cellulose microcrystal particles are effective at relatively low concentrations for controlling the heat-shock phenomena associated with storing frozen desserts, such as ice milk and ice cream, for long periods of time. The ice crystals that do grow remain smaller.

Microcrystalline cellulose gels permit the preparation of heat-stable oil-in-water emulsions that enable food processors to package ready-to-serve foods, such as canned tuna-fish salads containing mayonnaiselike dressings. Such microcrystalline-cellulose-based dressings permit the cooking of canned fish or other salads under sterilization canning conditions (2 hr or more at 120°C) with the dressing already premixed in the cans. Hollandaise sauces made with microcrystalline cellulose are also on the market. These sauces resist oil separation during subsequent heating—even to a boil—because of the rheological stability imparted by the microcrystals (55–58).

Another new use under active development is in the production of catalysts. For example, microcrystals in suspension are dispersed with fine alumina and the mixture is dried. Subsequent calcining burns out the cellulose and leaves pores of controlled size in the alumina residue. Colloidal cellulose with small microcrystals is finding increasing use in the ceramic and refractory industry. Added in low concentrations to "green" clay mixes, it increases the green strength of wet clays. During calcining the tiny cellulose microcrystals likewise are burned out. In this case, however, the resulting open matrix produces a white, bonelike, pseudotranslucent hue that is characteristic of the more expensive grades of china.

Industrial grades of microcrystalline cellulose that have a high proportion of submicron-size colloidal microcrystals are excellent stabilizing agents for water-based latex paints, as well as for industrial coatings and suspensions.

The formation of a remarkably dense and durable structural form of cellulose by drying an aqueous microcrystalline-cellulose gel (2) is striking evidence of the magnitude of the new hydrogen-bonding forces that become available. For example, when a 15% aqueous gel of microcrystalline cellulose made from the

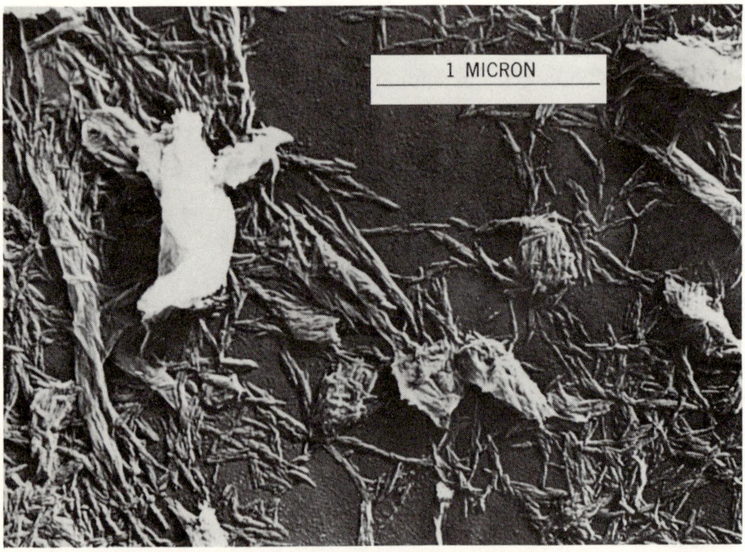

Fig. 5. Reaggregated Avicel microcrystals (25,000 ×) (spray-dried gel).

LODP fraction of wood pulp was allowed to air-dry at atmospheric pressure, two significant observations were made. First, the rate of water loss was extremely slow in comparison with the rate of water loss from a highly overbeaten mat of fibrous cellulose. Second, the gel became transformed into an ivorylike, extremely dense structural form of cellulose with a density of 1.52, versus a theoretical density of 1.56 for a single crystal of cellulose). In essence this product is a novel, dense form of paper, the result of engaging previously unavailable hydroxyl-rich surfaces on the unhinged and exposed microcrystal surfaces (2). Pyrolysis of such structural forms can lead to durable forms of structural carbon or graphite.

When a dilute (5–10%) suspension of individual microcrystals and large aggregates of unhinged microcrystals is spray-dried under proper conditions, the free microcrystals reagglomerate, as shown in Figure 5. The agglomerates possess "holes," varying from 10 to 100 Å in diameter between the microcrystals. This results in a new, porous form of highly crystalline colloidal cellulose capable of absorbing oils, greases, catalysts, etc., in the dried state.

4. Recent developments

The following summary of recent publications reflects the extent to which research and utilization relating to microcrystalline cellulose have advanced.

The original composition-of-matter patent for microcrystalline celluloses (1) has been followed by various process-improvement patents for making microcrystalline cellulose and its derivatives (59–62).

Among the earliest uses of microcrystalline-cellulose gels were coatings (2, 63) on glass (1, 64, 65), modification of polymer surfaces (66), and chromatography (2, 67–81).

Reactions of microcrystalline cellulose proceed with particular ease and speed. It is possible to form topochemical derivatives, including nitrates, carboxymethyl ethers, and oxidized derivatives, that are also colloidal (61). These are entirely new materials, with very different properties and potential application. For example, microcrystalline cellulose nitrate of controlled density and with particles in the 1 to 5 μ range have been produced (61). It is possible to produce dispersions of these novel derivatives, even at 20% solids in water—dispersions that possess different rheological and visible properties from those of untreated microcrystalline-cellulose gels.

Other reported uses of microcrystalline celluloses have been in the preparation of bearings (82), stable suspensions for use in radiology (83), dry-mix paints (66, 84, 85), for nucleating and/or filling films (85), for encapsulating or stabilizing unstable compounds (48, 86), for various kinds of structural products (60, 87–90), and in aerosols (91, 92).

The technology developed for the successful production of microcrystalline celluloses has paved the way for a family of products on which is being built a new field of microcrystalline polymer science (3, 4, 93, 94). Many new polymer products, based on the original concept applied to cellulose, are now emerging (95), and many more are to come.

References

1. O. A. Battista and P. A. Smith, U.S. Patent 2,978,446 (1961).
2. O. A. Battista and P. A. Smith, *Ind. Eng. Chem.* **54**, 20 (1962).
3. O. A. Battista, *The American Scientist*, 151, Summer Issue (1965).
4. O. A. Battista, N. Z. Erdi, C. F. Ferraro, and F. J. Karasinski, *J. Appl. Polymer Sci.*, **11**, 481 (1967).
5. C. Nageli and S. Schwendener, *Das Mikroskop*, 2nd ed., Engelmann, Leipzig, 1877.
6. R. O. Herzog, *Ber.*, **58B**, 1254 (1925).
7. K. H. Meyer, *Angew. Chem.*, **41**, 935 (1928).
8. K. H. Meyer and L. Misch, *Helv. Chim. Acta*, **20**, 232 (1937).
9. K. H. Meyer and L. Misch, *Ber.*, **70B**, 266 (1937).
10. L. Pauling, *The Nature of the Chemical Bond and the Structure of Molecules and Crystals*, 3rd ed., Cornell University Press, Ithaca, N.Y., 1960.
11. P. H. Hermans, J. de Booys, and C. J. Maan, *Kolloid-Z.*, **102**, 169 (1943).
12. A. J. A. Van der Wyk and K. H. Meyer, *J. Polymer Sci.*, **2**, 583 (1947).
13. K. Hess, *Naturwissenschaften*, **22**, 469 (1934).
14. K. H. Meyer and N. P. Badenhuizen, Jr., *Nature*, **140**, 281 (1937).
15. A. Girard, *Compt. Rend.*, **81**, 1105 (1875); *ibid.*, **88**, 1322 (1879).
16. H. Staudinger and M. Sorkin, *Ber.*, **70B**, 1565 (1937).
17. R. F. Nickerson, *Ind. Eng. Chem.*, **33**, 1022 (1941).
18. R. F. Nickerson, *Ind. Eng. Chem.*, **34**, 1480 (1942).
19. R. F. Nickerson and J. A. Habrle, *Ind. Eng. Chem.*, **39**, 1507 (1947).
20. G. F. Davidson, *J. Textile Inst.*, **34**, T87 (1943).
21. O. A. Battista, *Ind. Eng. Chem.*, **42**,, 502 (1950).
22. O. A. Battista, S. Coppick, J. A. Howsmon, F. F. Morehead, and W. A. Sisson, *Ind. Eng. Chem.*, **48**, 333 (1956).
23. O. A. Battista, *Ind. Eng. Chem., Anal. Ed.*, **16**, 351 (1944).
24. O. A. Battista and S. Coppick, *Textile Res. J.*, **17**, 419 (1947).
25. C. C. Conrad and A. G. Scroggie, *Ind. Eng. Chem.*, **37**, 592 (1945).
26. H. J. Philipp, M. L. Nelson, and H. M. Ziifle, *Textile Res. J.*, **17**, 585 (1947).
27. M. L. Nelson and V. W. Tripp, *J. Polymer Sci.*, **10**, 577 (1953).
28. P. C. Mehta and E. Pacsu, *Textile Res. J.*, **18**, 387 (1948).
29. F. C. Brenner, V. Frillette, and H. Mark, *J. Am. Chem. Soc.*, **70**, 877 (1948).
30. W. E. Roseveare, R. C. Waller, and J. N. Wilson, *Textile Res. J.*, **18**, 114 (1948).
31. W. E. Roseveare, *Ind. Eng. Chem.*, **44**, 168 (1952).
32. J. A. Howsmon, *Textile Res. J.*, **19**, 152 (1949).
33. B. G. Rånby, *Acta Chem. Scand.*, **3**, 649 (1949).

34. B. G. Rånby, *Discussions Faraday Soc.*, No. **11**, **158 (1951).**
35. L. Jørgensen, *Acta Chem. Scand.*, **3**, 780 (1949).
36. L. Jørgensen, *Acta Chem. Scand.*, **4**, 185 (1950).
37. L. Jørgensen, *Acta Chem. Scand.*, **4**, 658 (1950).
38. F. F. Morehead, *Textile Res. J.*, **20**, 549 (1950).
39. R. H. Marchessault, F. F. Morehead, and M. J. Koch, *J. Colloid Sci.*, **16**, 327 (1961).
40. R. H. Marchessault, M. J. Koch, and J. T. Yang, *J. Colliod Sci.*, **16**, 345 (1961).
41. S. M. Mukherjee and H. J. Woods, *Biochim. Biophys. Acta*, **10**, 499 (1953).
42. M. A. Millett, W. E. Moore, and J. F. Saeman, *Ind. Eng. Chem.*, **46**, 1493 (1954).
43. J. Hermans, Jr., *J. Polymer Sci.*, C, No. 2, 129 (1963).
44. M. R. Edelson and J. Hermans, Jr., *J. Polymer Sci.*, C, No. 2, 145 (1963).
45. O. A. Battista, U.S. Patent 3,146,170 (1964).
46. L. L. Augsburger and R. F. Shangraw, *J. Pharm. Sci.*, **55**, (4), 418 (1966).
47. O. A. Battista, U.S. Patent 3,146,168 (1964).
48. O. A. Battista and E. G. Fleck, Jr., U.S. Patent 3,297,806 (1967).
49. R. N. Duvall, K. T. Koshy, and R. E. Daskill, *J. Pharm. Sci.*, **54**, 1196 (1965).
50. C. D. Fox et al., *Drug Cosmetic Ind.*, **92**, 161 258 (1963).
51. L. Magid, U.S. Patent 3,332,848 (1967) (to Hoffmann-La Roche, Inc.).
52. G. E. Reier and R. F. Shangraw, *J. Pharm. Sci.*, **55**, 510 (1966).
53. M. D. Richman, C. D. Fox, and R. F. Shangraw, *J. Pharm. Sci.*, in press.
54. R. B. Wortz, *J. Am. Pharm. Assoc.*, in press.
55. *Food Processing*, **28** (6), 100 (1967).
56. O. A. Battista, U.S. Patent 3,023,104 (1962).
57. C. T. Herald and G. E. Raynor, Jr., *Food Processing Marketing*, **27** (11), 54 (1966).
58. T. W. Tusing, O. E. Paynter, and O. A. Battista, *J. Agr. Food Chem.*, **12**, 284 (1964).
59. O. A. Battista, U.S. Patent 3,251,824 (1966).
60. O. A. Battista and P. A. Smith, U.S. Patent 3,259,676 (1966).
61. A. M. Belfort and R. B. Wortz, *Ind. Eng. Chem., Prod. Res. Develop.*, **5**, 115 (1966).
62. E. G. Fleck, Jr., and O. A. Battista, U.S. Patent 3,109,743 (1963).
63. O. A. Battista, U.S. Patent 3,157,518 (1964).
64. H. S. Bondi, U.S. Patent 3,211,575 (1965).
65. O. A. Battista, U.S. Patent 3,218,223 (1965).
66. O. A. Battista, U.S. Patent 3,259,537 (1966).
67. O. A. Battista and J. J. Byrne, U.S. Patent 3,179,587 (1965).
68. O. A. Battista et al., U.S. Patent 3,179,587 (1965).
69. O. A. Battista and C. J. Boone, Belgian Patent 616,233 (1962).
70. A. R. Andres, *Quim. Ind. (Bilbao)*, **14**, (1), 10 (1967).
71. V. Deulofeu, *J. Chromatog.*, in press.
72. FMC Corporation, British Patent 980,022 (1965).
73. D. Giacopello, *J. Chromatog.*, **19**, 172 (1965).
74. P. Grippo, M. Iaccarino, M. Rossi, and E. Scarano, *Biochim. Biophys. Acta*, **95**, 1 (1965).
75. D. Horton, A. Tanimura, and M. L. Wolfrom, *J. Chromatog.*, **23**, 309 (1966).

76. F. D. Houghton, *J. Chromatog.*, **24**, 474 (1966).
77. D. D. Christianson, H. B. Sinclair, and J. W. Paulis, *Biochim. Biophys. Acta,* **121**, 412 (1966).
78. D. L. Simmons, C. K. Koorengevel, R. Kubelka, and P. Seers, *J. Pharm. Sci.* **55**, 219 (1966).
79. M. L. Wolfrom, D. L. Patin, and R. M. de Lederkremer, *Chem. Ind. (London)*, 1065 (1964).
80. M. L. Wolfrom, D. L. Patin, and R. M. de Lederkremer, *J. Chromatog.*, **17**, 488 (1965).
81. M. L. Wolfrom et al., *J. Chromatog.*, **18**, 42 (1965).
82. O. A. Battista, U.S. Patent 3,224,967 (1965).
83. O. A. Battista and G. T. Smith, U.S. Patent 3,192,118 (1965).
84. O. A. Battista, U.S. Patent 3,157,518 (1964).
85. O. A. Battista, U.S. Patent 3,258,354 (1966).
86. H. M. Beal, S. Shah and J. Versano, *J. Am. Pharm. Assoc.,* in press.
87. O. A. Battista and P. A. Smith, U.S. Patent 3,259,676 (1966).
88. O. A. Battista, U.S. Patent 3,357,845 (1967).
89. O. A. Battista, U.S. Patent 3,275,580 (1966).
90. O. A. Battista and P. A. Smith, U.S. Patent 3,278,519 (1966).
91. J. H. Fassnacht and F. A. Bower, *Food Technol.,* **19**, 44 (1965).
92. C. T. Herald, E. J. McGinley, and O. A. Battista, U.S. Patent 3,067,037 (1962).
93. O. A. Battista, *J. Polymer Sci.* C, No. 9, 135 (1965).
94. O. A. Battista, N. Z. Erdi, and M. M. Cruz, *Colloid Interface Sci.,* **28**, 36.
95. O. A. Battista, *Industrial Res.,* 66 (March 1969).

H. ION-EXCHANGE CELLULOSES

John D. Guthrie

Ion-exchange celluloses are made by attaching substituent groups with basic or acidic properties to the cellulose molecule, usually by etherification or esterification reactions. The degree of substitution (DS) is low, since the various uses of ion-exchange celluloses require that they do not dissolve or swell excessively in dilute aqueous acids or bases. Crosslinking of the cellulose prior to chemical modification may be used to control swelling and permit a higher DS (1, 2). Although similar to ion-exchange resins in ion-exchange properties, they are much finer and present a larger surface; furthermore, their porous structure permits the rapid entrance or attachment of large molecules that are not readily adsorbed by the resins. This makes ion-exchange celluloses particularly useful in the chromatographic separation of polyelectrolytes, especially in biochemical research, where they are used in preference to the resins. They can be made in the form of fabric, yarn, fiber, pulp, or paper. The last three forms are made commercially by several companies and are used routinely in biochemical research. Improved products, especially designed for the fractionation of polyelectrolytes, have been described (2).

Many of the ion-exchange celluloses were originally made for purposes other than ion exchange. Cellulose phosphate, in the form of cotton, was made by the urea-phosphate method for the purpose of obtaining a flame-resistant fabric (3). The possible use of cellulose phosphate for the softening of water was patented in 1941 by Urbain and Stemen (4). The use of aminoethylated cellulose (5) and cellulose phosphate (6) to produce seed proteins with low ash content was described in 1949 (7), and the use of various other chemically modified celluloses as ion-exchange materials was reported in 1952 (8). However, the extensive use of ion-exchange celluloses in biochemical research followed the fractionation of proteins on columns of carboxymethylcellulose and of diethylaminoethylcellulose by Sober and Peterson in 1954 (9).

1. Cation-Exchange Celluloses

(a) CELLULOSE PHOSPHATE

Cellulose phosphate can be made in the form of fabric by treating cotton cloth with a mixture of phosphoric acid and urea (3, 10). Cotton sheeting that has been kier-boiled in 2% sodium hydroxide solution is phosphorylated by wetting it with a solution containing 49.6% urea, 18.4% phosphoric acid, and 32% water, squeezing out the excess solution, heating in an oven for 30 min at 150°C, and washing thoroughly. This method has also been applied to cotton in the form of loose fiber, roving, and yarn. The urea-phosphoric acid method has been applied to wood pulp by Head and co-workers (11), who used a liquor-to-cellulose ratio of 3 : 1. The wetted cellulose was heated in a forced-draft oven for 2.5 hr at 130°C. It was found that about 40% of the phosphoric acid groups in the product are diesterified, thereby crosslinking the cellulose. The cellulose phosphate could be converted to essentially the monoesterified form by putting it into the acid form and soaking in water for 7 days. The crosslinked form, because of its restricted swelling, is preferred for use in packing columns for the adsorption of thorium from strongly acidic solutions. Cellulose phosphate has been made by the action of phosphorus oxychloride on cotton in the presence of pyridine (12). About 20% of the phosphoric acid groups are diesterified. Cellulose phosphate can also be made by the action of an ether solution of phosphorus oxychloride on purified wood pulp in the presence of sodium hydroxide (13) (see also Section I of this Chapter XIX).

(b) SULFOETHYLCELLULOSE

Sulfoethylcellulose can be made by reacting 2-chloroethanesulfonic acid ($ClCH_2CH_2SO_3Na$) with cellulose in the presence of a strong sodium hydroxide solution (10, 14):

$$R_{cell}OH + ClCH_2CH_2SO_3Na + NaOH \longrightarrow R_{cell}OCH_2CH_2SO_3Na + NaCl + H_2O \qquad (1)$$

For example, a roll of cotton fabric is passed through a solution composed of 5185 g of water, 2465 g of sodium hydroxide, and 850 g of the sodium salt of 2-chloroethanesulfonic acid (added last to the cooled solution) and then put through squeeze rolls. The fabric is passed over rotating steam-heated cans with a surface temperature of about 85°C at such speed that the fabric is in contact with the cans for about 4 min. The fabric is washed and dried. It is a stiff, organdie-like fabric with a cation-exchange capacity of about 0.5 meq/g.

Sulfoethylcellulose has been made by Timell (15), but his product was water soluble due to its high DS and consequently not useful as a cation-exchange cellulose. In the presence of strong sodium hydroxide, cellulose can also be sulfoethylated by the action of ethylenesulfonic acid ($CH_2=CHSO_3H$) or with ethionic acid ($HO_3SOCH_2CH_2SO_3H$) to produce products similar to those obtained with 2-chloroethanesulfonic acid. Sulfoethylcellulose has been made from shredded cellulose linters by Porath (16). Sodium 2-bromoethanesulfonate is reacted for 3 hr at 120°C in a nitrogen atmosphere with soda cellulose made from the linters. The product is washed successively with alcohol, water, alcohol, ethyl ether, and dried.

(c) SULFOMETHYLCELLULOSE

This cation-exchange cellulose has been made by Porath (16). The required reagent, sodium monochloromethanesulfonate, is made by reacting dichloromethane with sodium sulfite. The reagent is stirred with fibrous cellulose in a strong aqueous solution of sodium hydroxide. The wetted cellulose is heated in an oven at 90–95°C until 80% of the added water has evaporated. The mixture is then washed with acidified ethanol and finally with water. Without having been dried it is mixed with water or buffer and packed into the chromatographic column. Since preparations with an exchange capacity of 0.7 meq/g are water soluble, the exchange capacity is kept at 0.35–0.5 meq/g. The properties of the product are similar to those of sulfoethylcellulose.

(d) CARBOXYMETHYLCELLULOSE

The carboxymethylcellulose used as an ion-exchange material is not the water-soluble product that is used widely for many purposes (see Chapter IX in Part II, pp. 937–945). It is a material of low DS, with a cation-exchange capacity of 0.4–0.7 meq/g. Products of higher capacity swell too much to be used in chromatography, unless they also are crosslinked, as will be discussed later. A product suitable for use as a cation-exchange material can be made by a procedure described by Peterson and Sober (17). Cellulose powder, 60 g, is mixed with 290 g of 30% aqueous sodium hydroxide solution to give a friable mass, which is cooled in an ice bath. A freshly made solution of 30 g of chloroacetic acid in 40 ml of water is added in small portions with mixing. The reaction vessel then is placed in a bath at 70°C for 20 min, mixed well, and returned to the ice bath. The cooled material is partially neutralized with 500 ml of 10% acetic acid, diluted with water, and put through a washing process, which includes a final wash with ethanol. The carboxyl content of the dried product thus obtained is about 0.7 meq/g.

(e) PHOSPHONOMETHYLCELLULOSE

Phosphonomethylcellulose, a bifunctional cation-exchange material, can be made by reacting cellulose with disodium chloromethylphosphonate in the presence of sodium hydroxide (18). The starting reagent may be either chloromethylphosphonic dichloride or chloromethylphosphonic acid, and it is dissolved in enough sodium hydroxide solution to form the disodium salt of the acid, keeping the temperature below 30°C.

$$R_{cell}OH + ClCH_2P(O)(ONa)_2 + NaOH \longrightarrow R_{cell}OCH_2P(O)(ONa)_2 + NaCl + H_2O \qquad (2)$$

The DS can be varied widely by varying the conditions of processing. Since the cellulose can dissolve in water when about 2% phosphorus is introduced, a product having about 1% phosphorus would probably be selected for column chromatography. Crosslinking the cellulose prior to phosphonomethylation can be used to prevent water solubility at higher phosphorus contents, as discussed later in this section.

(f) OXIDIZED CELLULOSE

It is now generally accepted that native cellulose contains no significant number of carboxyl groups. This fact was demonstrated by Van der Wyk and Studer (19), who purified cellulose from cotton and from wood by regeneration from cuprammonium hydroxide solution with exclusion of oxygen and carbon dioxide, and made titrations in a platinum vessel to avoid cation exchange with glass or quartz. Other workers have arrived at similar conclusions. Nevertheless numerous published papers report work in which carboxyl groups of cellulose have been investigated or quantitatively determined.

There are two explanations for this. The first is that native cellulose is accompanied by other substances that contain carboxyl groups or other acidic groups. For example, raw native-cotton fiber contains pectic acid and organic acids, which, unless they are removed by purification processes, may be mistaken for carboxyl groups that are part of the cellulose molecule. The other explanation is that the cellulose may be oxidized during purification processes to produce some carboxyl groups along the cellulose chain molecule. A common method of purifying cellulose is to boil it with a dilute sodium hydroxide solution, but if oxygen is not excluded, some oxidation occurs and the resulting cellulose contains measurable amounts of carboxyl groups. If the cellulose is then bleached with hypochlorites or peroxides, more carboxyl groups are formed. These are especially abundant in absorbent cotton, which is a heavily bleached product frequently used in laboratory experimentation.

Exposure to light and other types of radiation in the presence of oxygen also results in the formation of carboxyl groups in the cellulose. These carboxyl groups undergo cation exchange, and many of the experiments dealing with the cation-exchange properties of cellulose are explained by these groups, which are not present in cellulose as it occurs in plants. It has been pointed out by Pickering (20) that the carboxyl groups present in filter papers play a role in paper chromatography, especially when hydrogen ions and divalent metal ions are involved.

Fibrous products that contain a high proportion of carboxyl groups can be made by oxidizing cotton cellulose with various reagents. Such cottons are of interest as ion-exchange materials. Cotton may be oxidized with nitrogen dioxide to make a fibrous product that contains 12–13 wt % of carboxyl groups (21, 22). This oxidizing agent acts selectively on the primary hydroxyl groups of the cellulose. Products containing less than 13% carboxyl are not soluble in dilute sodium hydroxide, whereas products with more carboxyl groups dissolve quickly in dilute sodium hydroxide, ammonia, or sodium carbonate solutions. A product containing as much as 25% carboxyl has been made.

When cellulose is oxidized with periodate, most of the oxidation takes place at the $C_{(2)}$ and $C_{(3)}$ positions of the glucose residues, and the bond at this point is ruptured with the formation of aldehyde groups. This aldehyde cellulose is unstable, and treatment with dilute alkalies breaks the molecular chain at the point where oxidation has occurred. The product can be stabilized by oxidizing the aldehyde groups to carboxyl groups by means of an acidified solution of sodium chlorite. Carboxyl contents as high as 2.7 meq/g have been obtained in this way (23) (see also Section F of Chapter XVII).

(g) OTHER CATION-EXCHANGE CELLULOSES

The succinic acid half-ester of cotton cellulose can be made by heating fabric with a 10% solution of succinic anhydride in dry pyridine at 60°C for 26 hr and washing thoroughly. The cation-exchange properties of this material have been reported by McIntire and Schenck (24), who have also made the half-ester cellulose derivatives of maleic, glutaric, and phthalic acids and determined their ion-exchange capacities. Rigamonti and Riccio (25) have investigated the cellulose esters of phthalic, maleic, succinic, and adipic acids and have shown that they have cation-exchange capacity. That of the succinate is especially high when the ester is prepared by using boiling succinic anhydride. Cellulose citrate has cation-exchange properties (26).

Sulfatoethylated cotton can be made by the action of 2-chloroethylsulfuric acid ($ClCH_2CH_2OSO_3H$) on fabric by a procedure similar to that described for sulfoethylated cotton. Phosphatoethylated cotton can be made by the use of 2-chloroethylphosphoric acid [$ClCH_2CH_2OPO(OH)_2$]. These two cottons are cation exchangers of the strong-acid type (8).

2. Anion-Exchange Celluloses

(a) AMINOETHYLCELLULOSE

Aminoethylcellulose can be made by reacting cellulose with 2-aminoethylsulfuric acid in the presence of sodium hydroxide (5).

$$R_{cell}OH + H_2NCH_2CH_2OSO_3H + NaOH \longrightarrow R_{cell}OCH_2CH_2NH_2 + NaHSO_4 + H_2O \qquad (3)$$

It has been made in the form of cotton fabric (27). Kier-boiled cotton sheeting is passed through a solution composed of 30 kg of sodium hydroxide, 50 kg of water, and 20 kg of 2-aminoethylsulfuric acid and then put through squeeze rolls (pickup, about 100%). The fabric is then passed for 15 min over rotating cans heated with steam at a pressure of 1.5 atm, washed, and dried. The nitrogen content is 0.8% on a dry basis, and the anion-exchange capacity is 0.6 meq/g. Cotton can also be aminized in the form of loose fiber, sliver, or yarn. The method described by Semenza (28) consists of autoclaving with the same reagents. The product has a capacity of 0.2 meq/g; if crosslinked cellulose is used, it can be processed three times to give a product with a capacity of 0.7 meq/g.

Although one can formulate a reaction by which ethylenimine should react with cellulose to form the partial 2-aminoethyl ether, probably none of this is formed under neutral or alkaline conditions. Under neutral conditions the treatment of cotton cellulose with ethylenimine from the vapor phase probably results in the deposition of polyethylenimine in and on the cellulose accompanied by little, if any, reaction with the cellulose (29). However, the aminoethylation of cellulose with ethylenimine by the use of acidic catalysts has been described (30). Under certain conditions of acidic catalysis in a nonpolar solvent graft copolymerization may occur, forming $R_{cell}–O(CH_2CH_2NH)_n–CH_2CH_2NH_2$, where n is between 4 and 8 (31).

Aminoethylcellulose can be reacted with ethylenimine from the vapor phase to form a fibrous anion-exchange material with an exchange capacity of about 2 meq/g (32).

(b) DIETHYLAMINOETHYLCELLULOSE

Diethylaminoethylcellulose (DEAE-cellulose) was first reported by Hartmann (33), who used cotton as the source of cellulose. Its use as an anion-exchange material was suggested in 1950 (10). Sober and co-workers (34), who prepared it from purified wood pulp, are largely responsible for its extensive use in the ion-exchange chromatography of polyelectrolytes of biochemical origin. It can be prepared from purified cotton linters (35–37) by reacting them with sodium hydroxide solution and 2-chlorotriethylamine (*N*-2-chloroethyl-*N*, *N*-diethylamine) hydrochloride. The product has a nitrogen content of 1.05%.

Diethylaminoethylcellulose containing about 4% nitrogen can be made from purified cotton linters, previously crosslinked with formaldehyde or with 1,3-dichloropropanol (35, 37), by reacting them three times with 2-chlorotriethylamine. Chromatographic columns packed with the products showed satisfactory flow rates, but flow could not be maintained in columns packed with materials made by similarly reacting linters that had not been crosslinked.

The distribution of the diethylaminoethyl groups on the hydroxyl groups of diethylaminoethylcellulose has been determined by Roberts and Rowland (38) and found to be in the proportion of 1.27 : 0.35 : 1.00 for the 2-, 3-, and 6-hydroxyl groups, respectively, when the substitution occurred in aqueous media, but 2.50 : 0.50 : 1.00 in *p*-dioxane.

(c) TRIETHYLAMINOETHYLCELLULOSE

Although triethylaminoethylcellulose (TEAE-cellulose) and related quaternary anion-exchange celluloses have been reported (10, 16) to result from the reaction of DEAE-cellulose with alkyl halides—and a product so named has been available commercially for chromatography—there is considerable evidence that the commercial product is essentially unchanged DEAE-cellulose. However, it has been shown recently that quaternized celluloses with the expected properties can be made by reacting methyl iodide or ethyl bromide under anhydrous conditions with DEAE-cellulose that has been dried in vacuum over phosphorus pentoxide (39). Potentiometric titration curves on the products are similar to those of the known strongly basic anion-exchange resins and different from the curves obtained with DEAE-cellulose, showing them to be strongly basic anion-exchange materials.

(d) ECTEOLA-CELLULOSE

Cellulose can be reacted at low temperatures with a freshly made mixture of epichlorohydrin and triethanolamine in the presence of sodium hydroxide to produce an anion-exchange material, usually designated by the abbreviation ECTEOLA-cellulose (40, 41). It has an exchange capacity of 0.3–0.4 meq/g.

It has been shown that cellulose pretreated with aqueous base and then reacted with a 3 : 1 mole ratio of epichlorohydrin to triethanolamine forms a strongly basic anion-exchange cellulose (42). The properties of this product differ from those of the conventional ECTEOLA-cellulose, and its titration curve is similar to that of cellulose reacted with trimethylglycidylammonium chloride.

(e) OTHER ANION-EXCHANGE CELLULOSES

A variety of chloroalkylamines and aminoalkyl hydrogen sulfates have been reacted with soda cellulose by Jakubovic and Brook (43) to form partial ethers whose anion-exchange properties are similar to those of 2-aminoethylcellulose. A number of amines, especially alkanolamines, have been linked to cellulose by means of epichlorohydrin (44) under conditions similar to those used in making ECTEOLA-cellulose. Guanidinoethylcellulose has been prepared by Semenza (28) by reacting 2-aminoethylcellulose with *O*-methylisourea. The pH at half capacity for the product is about 9.3, showing it to be a strongly basic exchanger. Its low stability must be taken into consideration when it is regenerated. The *p*-aminoacetophenone ether of cellulose, $R_{cell}-OCH_2COC_6H_4NH_2$, has been made by McLaughlin and Mutton (45). A modified cellulose containing iminodiacetate groups has been made by reacting iminodiacetic acid with cellulose tosylate (46). It forms complexes with Cu^{2+} and Ni^{2+} ions.

A cellulose containing one 5-hydroxymethyluracil residue per nine glucose units, called HMU-cellulose, has been made by Barber and Jones (47). It has been used in the fractionation of ribonucleic acids. *p*-Aminobenzylcellulose has been made by reacting cellulose with *p*-nitrobenzyl chloride in the presence of sodium hydroxide, followed by reduction of the product with sodium hydrosulfite (48). It has been used to make immunological adsorbents which are described below.

3. Immunological Adsorbents

Antibodies can be isolated from blood serum by using specific immunological adsorbents made by attaching the corresponding antigens to ion-exchange celluloses and also to unmodified celluloses. Various ways of preparing these specific adsorbents have been described. For example, the antigen, such as crystalline bovine albumin, can be coupled to diazotized *p*-aminobenzylcellulose (48, 49) to yield an adsorbent for the antibody for bovine albumin contained in the blood serum of animals injected with this albumin. The nonspecific proteins present in the serum can be readily washed from the column with 1% sodium chloride solution. The antibody is eluted with glycine-hydrochloric acid buffer at pH 3 and isolated with a purity of about 90%.

An immune adsorbent for the isolation of purine-specific antibodies has been made by reacting aminoethylcellulose with 6-trichloromethylpurine (50). The antiserum was from rabbits immunized with hemocyanin that had been coupled with 6-trichloromethylpurine. When this antiserum was passed through a column of the immune adsorbent, at least 95% of the antibody was retained. Over 82% of the bound antibody was released on elution with 1% sodium chloride solution at pH 2.3. The purity was close to theoretical if nonspecific adsorption was taken into account. This technique is said to be applicable to the isolation of the antibody of any purine or pyrimidine capable of being converted to a derivative that can be coupled to an amine group.

Immunological adsorbents may be made by coupling antigens to cellulose and to ion-exchange celluloses with dicyclohexylcarbodiimide by means of Khorana's reaction (51). Benzidine may be coupled to carboxymethylcellulose by the addition of dicyclohexylcarbodiimide in tetrahydrofuran; the resulting arylamino cellulose is diazotized and coupled to bovine serum albumin. The resulting adsorbent is useful in isolating antibodies from immune serum (52). Dicyclohexylcarbodiimide has also been used to couple polydeoxyribonucleotides to cellulose or to cellulose phosphate to make adsorbents useful in the chromatography of ribonucleotides (53, 54).

The preferential adsorption of antigen or antibody on ion-exchange celluloses under carefully chosen experimental conditions has been used without resort to a chemical reaction leading to covalent attachment (55). In one case the immune gamma globulin was strongly adsorbed on sulfomethylcellulose and then reacted with the antibody. In another example the antigen (R-phycoerythrin) was rather firmly adsorbed on DEAE-cellulose and then reacted with the antiserum. ECTEOLA-cellulose has been used in a similar manner (56). Enzymes, and presumably other proteins, may be coupled to cellulose by using chloro-*s*-triazines (57). A review of early work on immunological adsorbents is available (58) (see also Section J of this Chapter XIX).

4. Crosslinked Ion-Exchange Celluloses

One limitation of most ion-exchange celluloses is their low capacity in comparison with the ion-exchange resins. Low capacity appears to be no disadvantage in chromatographic work with proteins and other high-molecular-weight polyelectrolytes, but it is a disadvantage for some other possible uses. Attempts to increase the ion-exchange capacity by means of more effective or repeated chemical treatments result in excess swelling, which interferes with column operations, or loss of fiber structure, so that the product becomes a soluble polyelectrolyte that is not useful as an ion-exchange material. However, if the cellulose is crosslinked by a bifunctional reagent before introduction of the ionic groups, fiber structure is retained and a much higher exchange capacity can be reached while maintaining properties suitable for chromatography.

Crosslinking with di(2-sulfatoethyl)amine, 1,4-disulfatobutane, 1,3-dichloro-2-propanol, divinyl sulfone, or formaldehyde makes possible an increase in the ion-exchange capacity of aminoethyl-, diethylaminoethyl-, carboxymethyl-, or phosphonomethylcellulose while maintaining fiber properties suitable for chromatography (1). In some instances the crosslinking and attachment of ionic groups can be done in a single treatment by adding the crosslinking agent to the solution used for the chemical modification. With certain compounds, such as dichloroacetic acid, it is possible to attach the ionic group and crosslink at the same time. For a further discussion of crosslinking, see Section E in Chapter XVII.

5. "Microgranular" Ion-Exchange Celluloses

Carboxymethylcellulose and DEAE-cellulose especially designed for the chromatography of polyelectrolytes have been described by Knight (2), who emphasizes careful control of all stages of preparation and handling prior to use. The special ion-exchange celluloses are made from a product that has been called "microgranular" cellulose. This is made by the partial hydrolysis of cotton linters, arresting the hydrolysis before the hydrocellulose stage, or levelling-off degree of polymerization (LODP), is reached. This "microgranular" cellulose is mercerized and crosslinked before conversion to the ion-exchange cellulose to restrict its increased tendency to swell, owing to a reduced molecular weight, and also to avoid loss of accessibility during deswelling by inhibiting the close approach of cellulose chains. When ion-exchange groups are introduced into the undried, crosslinked cellulose, a product with increased uniformity of ion-exchange site distribution and of particle size is obtained. It is suitable for chromatography without further mechanical disintegration and should by used in the never-dried form. If dried, ion-exchange celluloses should be reconstituted prior to use by subjecting them to acid and alkali in the order appropriate for the particular substituent group and then equilibrating the swollen material with the selected buffer. Precycling prior to use is recommended. Due to their high accessibility, the "microgranular" ion-exchange celluloses have a higher capacity for polyelectrolytes, such as insulin, than the conventional ion-exchange resins or cellulose exchangers.

6. Evaluation of Exchange Capacity and Ionic Strength

Prior to being measured for ion-exchange capacity and strength of the acidic or basic groups, ion-exchange celluloses are put into the free-acid or free-base form. The cation-exchange celluloses are usually soaked in 5% hydrochloric acid, followed by thorough washing with distilled water. The anion-exchange celluloses are soaked in 5% sodium hydroxide solution, followed by thorough washing with distilled water. Aminoethylcellulose can be put into the free-base form by soaking in 5% aqueous ammonium hydroxide solution, followed by removal of the excess ammonia by drying in air. The total exchange capacity for small ions can be measured titrimetrically, and the pH at half capacity can be determined in the presence of sodium chloride. The latter value is an approximation of the pK_a of the functional groups (10). The total exchange capacity for small ions may be significantly different from, and correlate poorly with, the capacity to take up polyelectrolytes. This is because the accessibility or the pore size of the exchanger under the particular conditions of preparation and use is the controlling factor with ions of large molecular weight. The ion-exchange capacity for small ions and the ionic strength of some typical ion-exchange celluloses are shown in Table 1.

TABLE 1
Total Exchange Capacity for Small Ions and Ionic Strength of Some Typical Ion-Exchange Celluloses

Material	Total capacity (meq/g)	pH at half capacity
Carboxymethylcellulose	0.7	3.7
Oxidized cellulose	2.7	
Succinic half-ester of cellulose	1.0	4.8
Cellulose citrate	2.0	
Sulfoethylcellulose	0.5	0.5
Cellulose phosphate:		
First hydrogen	1.3	1.5
Second hydrogen	2.6	7.5
Phosphonomethylated cellulose:		
First hydrogen	0.6	2.3
Second hydrogen	1.2	7.0
Diethylaminoethylcellulose	0.8	9.5
Triethylaminoethylcellulose	0.8	
ECTEOLA-cellulose	0.4	7.5
Aminoethylcellulose	0.6	8.0

7. Uses of Ion-Exchange Celluloses

Since well over 500 publications on the uses of ion-exchange celluloses have been published, it is only possible to discuss a few of these to illustrate typical applications. Ion-exchange celluloses have been used extensively in the purification and fractionation of proteins, including enzymes, hemoglobins, hormones, serum proteins, viruses, and seed proteins. They have also been used in the chromatography of peptides, nucleic acids, nucleotides, amino acids, alkaloids, and metallic ions. In one instance they have been used to adsorb and purify rickettsiae (59).

Diethylaminoethylcellulose is the exchanger most frequently used in the chromatography of proteins, including those of special biochemical interest, such as enzymes, hemoglobins, and hormones. Gradient or stepwise elution with buffers of changing pH and ionic strength is used with proteins and other polyelectrolytes. Detailed procedures have been described by Peterson and Sober (60).

The whey proteins of soybean meal have been fractionated into 13 electrophoretically distinct components and 8 ultracentrifugal components on DEAE-cellulose columns (61). Two trypsin inhibitors of high purity were isolated also. Curtain (62) has described a method for chromatographing proteins on DEAE-cellulose in the form of paper; it employs a photoelectrically controlled gradient generator. Mixtures of drugs have been separated into distinct spots on DEAE-cellulose paper by employing chromatography in one direction and then ionophoresis in the other direction on the same paper (63). The separation of amino acids on DEAE-cellulose paper, and also on cellulose phosphate, has been described by Knight (2). Neutral and acidic polysaccharides, such as pectic acid, can be fractionated on DEAE-cellulose columns (64). ECTEOLA-cellulose is frequently selected for the purification of nucleic acids (65, 66), nucleotides (67, 68), and viruses, such as bacteriophages (69).

Carboxymethylcellulose has been frequently utilized for the chromatography of certain proteins (60), including some enzymes and hormones. It has been used also for the separation of the basic amino acids.

Oxidized cellulose has been found very effective in the separation of the active components in crude corticotrophin extracts from pig anterior pituitary glands and has provided a simple method for the preparation of highly active material in almost quantitative yield (70). In the form of filter paper it has been shown to be effective in the chromatographic separation of histidine, lysine, and arginine (71). This ion-exchange paper is also effective in the separation of lead, bismuth, and silver.

Cellulose phosphate in the form of filter paper has been used for the separation of iron, copper, and nickel (72). It has been used for the inactivation of oxidase solutions by the removal of copper (73). The possible use of cellulose phosphate in the form of an endless belt of cotton fabric has been studied for continuous ion exchange in a countercurrent system (74). This process is effective in the removal of copper in copper-sodium exchange. Cellulose phosphate was selected for the purification by chromatography of polysaccharide hydrolases from fungi because of its resistance to hydrolysis by cellulases (75). The recovery of thorium from monazite by means of cellulose phosphate has been studied extensively (76). The monazite ore is broken down with sulfuric acid, and the leach liquor is passed through columns of cellulose phosphate to which the thorium is preferentially adsorbed and from which it is subsequently eluted with 10% ammonium carbonate solution. A typical concentrate contains over 90% thoria and less than 4% phosphorus pentoxide. Aminoethylcellulose forms stable complexes with copper or nickel (77). The complexed copper can be reduced with sodium hydrosulfite to the cuprous form, or to metallic copper, which may be oxidized by the oxygen in tap water. The copper may be repeatedly reduced and oxidized without substantial loss of metal. Lipid extracts from wheat leaves and from algae have been separated by chromatography on aminoethylcellulose (78). It has been used in the separation of dyes (79) and also for the chromatography of a number of proteins.

The commercial suppliers of the various ion-exchange celluloses have issued extensive bibliographies on their use in biochemical investigations and for other applications; these may be consulted for further details.

References

1. J. D. Guthrie and A. L. Bullock, *Ind. Eng. Chem.*, **52**, 935 (1960).
2. C. S. Knight, *Advances in Chromatography*, **4**, 61 (1967).
3. F. M. Ford and W. P. Hall, U.S. Patent 2,482,755 (1949).
4. O. M. Urbain and W. R. Stemen, U.S. Patent 2,265,585 (1941).
5. J. D. Guthrie, *Textile Res. J.*, **17**, 625 (1947).
6. J. F. Jurgens, J. D. Reid, and J. D. Guthrie, *Textile Res. J.*, **18**, 42 (1948).
7. C. L. Hoffpauir and J. D. Guthrie, *J. Biol. Chem.*, **178**, 207 (1949).
8. J. D. Guthrie, *Ind. Eng. Chem.*, **44**, 2187 (1952).
9. H. A. Sober and E. A. Peterson, *J. Am. Chem. Soc.*, **76**, 1711 (1954).
10. C. L. Hoffpauir and J. D. Guthrie, *Textile Res. J.*, **20**, 617 (1950).
11. A. J. Head, N. F. Kember, R. P. Miller, and R. A. Wells, *J. Chem. Soc.*, 3418 (1958).
12. J. D. Reid, L. W. Mazzeno, and E. M. Buras, *Ind. Eng. Chem.*, **41**, 2831 (1949).
13. E. A. Peterson and H. A. Sober, *Biochem. Preparations*, **8**, 47 (1961).
14. J. D. Guthrie, L. H. Chance, and C. L. Hoffpauir, U.S. Patent 2,681,846 (1954).
15. T. Timell, *Svensk Papperstidn.*, **51**, 254 (1948).
16. J. Porath, *Arkiv Kemi*, **11**, 97 (1957).
17. E. A. Peterson and H. A. Sober, *Biochem. Preparations*, **8**, 45 (1961).
18. G. L. Drake, Jr., W. A. Reeves, and J. D. Guthrie, *Textile Res. J.*, **29**, 270 (1959).
19. A. J. A. van der Wyk and M. Studer, *Helv. Chim. Acta*, **32**, 1698 (1949).
20. W. F. Pickering, *J. Chromatog.*, **4**, 477 (1960).
21. C. C. Unruh and W. O. Kenyon, *J. Am. Chem. Soc.*, **64**, 127 (1942).
22. E. C. Yackel and W. O. Kenyon, *J. Am. Chem. Soc.*, **64**, 121 (1942).
23. G. F. Davidson and T. P. Nevell, *J. Textile Inst.*, **46**, T407 (1955).
24. F. C. McIntire and J. R. Schenck, *J. Am. Chem. Soc.*, **70**, 1193 (1948).
25. R. Rigamonti and V. Riccio, *Ann. Chim. (Rome)*, **42**, 283, 580 (1952).
26. G. P. Touey and J. E. Kiefer, U.S. Patent 2,759,787 (1956).
27. W. A. Reeves and J. D. Guthrie, *Textile Res. J.*, **23**, 522 (1953).
28. G. Semenza, *Helv. Chim. Acta*, **43**, 1057 (1960).
29. G. L. Drake, Jr., W. A. Reeves, and J. D. Guthrie, *Textile Res. J.*, **23**, 639 (1953).
30. L. Segal and F. V. Eggerton, *Textile Res. J.*, **33**, 739 (1963).
31. R. J. Hartman and E. J. Fugiwara, U.S. Patent 2,972,606 (1961).

32. G. L. Drake and J. D. Guthrie, U.S. Patent 2,656,241 (1953).
33. M. Hartmann, U.S. Patent 1,777,970 (1930).
34. H. A. Sober, F. J. Gutter, M. M. Wyckoff, and E. A. Peterson, *J. Am. Chem. Soc.,* **78**, 756 (1956).
35. J. D. Guthrie and A. L. Bullock, *Ind. Eng. Chem.,* **52**, 935 (1960).
36. E. A. Peterson and H. A. Sober, *Biochem. Preparations*, **8**, 39 (1961).
37. A. L. Bullock and J. D. Guthrie, in R. L. Whistler, Ed., *Methods in Carbohydrate Chemistry*, Vol. V, Academic Press, New York, 1965, p. 409.
38. E. J. Roberts and S. P. Rowland, *Carbohydrate Res.,* **5**, 1 (1967).
39. R. R. Benerito, B. B. Woodward, and J. D. Guthrie, *Anal. Chem.,* **37**, 1693 (1965).
40. E. A. Peterson and H. A. Sober, *J. Am. Chem. Soc.,* **78**, 751 (1956).
41. E. A. Peterson and H. A. Sober, *Biochem. Preparations,* **8**, 43 (1961).
42. J. B. McKelvey and R. R. Benerito, *J. Appl. Polymer Sci.,* **11**, 1693 (1967).
43. A. O. Jakubovic and B. N. Brook, *Polymer,* **2**, 18 (1961).
44. A. O. Jakubovic, *Polymer*, **1**, 117 (1960).
45. R. R. McLaughlin and D. B. Mutton, U.S. Patent 2,727,034 (1955).
46. Y. A. Chaikina, L. S. Gal'braikh, and Z. A. Rogovin, *Polymer Science* USSR (English translation), 7, (12), 2212 (1965).
47. R. Barber and A. S. Jones, *Nature*, **203**, 45 (1964).
48. D. H. Campbell, E. Lenscher, and L. S. Lerman, *Proc. Natl. Acad. Sci. U.S.,* **37**, 575 (1951).
49. A. Malley and D. H. Campbell, *J. Am. Chem. Soc.,* **85**, 487 (1963).
50. H. H. Weetall and N. Weliky, *Science,* **148**, 1235 (1965).
51. H. G. Khorana, *Chem. Rev.,* **53**, 145 (1953).
52. H. H. Weetall and N. Weliky, *Nature,* **204**, 896 (1964).
53. P. T. Gilham, *J. Am. Chem. Soc,* **84**, 1311 (1962).
54. A. J. Adler and A. Rich, *J. Am. Chem. Soc.,* **84**, 3977 (1962).
55. A. Saba, *J. Chromatog.,* **7**, 155 (1962).
56. N. K. Brown, *Science*, **133**, 331 (1961).
57. G. Kay and E. M. Croak, *Nature,* **216**, 514 (1967).
58. H. C. Isliker, *Advances in Protein Chemistry,* **12**, 387 (1957).
59. B. H. Hoyer, E. T. Bolton, R. A. Ormsbee, G. Le Bouvier, D. B. Ritter, and C. L. Larson, *Science,* **127**, 728 (1953).
60. E. A. Peterson and H. A. Sober, *Methods in Enzymology,* **5**, 3 (1961).
61. J. J. Rackis, H. A. Sasame, R. L. Anderson, and A. K. Smith, *J. Am. Chem. Soc.,* **81**, 6265 (1959).
62. C. C. Curtain, *Nature,* **191**, 1269 (1961).
63. H. V. Street and S. K. Niyogi, *Nature,* **190**, 1199 (1961).
64. H. Neukom and W. Kuendig, in R. L. Whistler, Ed., *Methods in Carbohydrate Chemistry*, Vol. V, Academic Press, New York, 1965, p. 14.
65. A. Bendich, H. B. Pahl, G. C. Korngold, H. S. Rosenkranz, and S. R. Fresco, *J. Am. Chem. Soc.,* **80**, 3949 (1958).
66. K. Randerath, *Angew. Chem.,* **73**, 674 (1961).
67. E. D. Korn, *Biochem. Biophys. Acta,* **32**, 554 (1959).

68. R. Nilsson and M. Sjunneson, *Acta Chem. Scand.*, **15**, 1017 (1961).
69. A. Tausig and E. H. Creaser, *Biochem. Biophys. Acta,* **24**, 448 (1957).
70. E. B. Astwood, M. S. Raben, R. W. Payne, and A. B. Grady, *J. Am. Chem. Soc.*, **73**, 2969 (1951).
71. T. Wieland and A. Berg, *Angew. Chem.*, **64**, 418 (1952).
72. N. F. Kember and R. A. Wells, *Nature,* **175**, 512 (1955).
73. J. C. Arthur, Jr., and T. A. McLemore, *J. Am. Chem. Soc.*, **78**, 4153 (1956).
74. C. H. Muendel and W. A. Selke, *Ind. Eng. Chem.*, **47**, 374 (1955).
75. J. H. McClendon and J. H. Kreisher, *Anal. Biochem.*, **5**, 295 (1963).
76. A. J. Head, N. F. Kember, R. P. Miller, and R. A. Wells, *J. Appl. Chem. (London),* **9**, 599 (1959).
77. W. A. Reeves, O. J. McMillan, Jr., and J. D. Guthrie, *Textile Res. J.*, **23**, 527 (1953).
78. R. O. Mumma and A. A. Benson, *Biochem. Biophys. Res. Commun.*, **5**, 422 (1961).
79. J. C. Brown, *J. Soc. Dyers Colourists,* **76**, 536 (1960).

I. FLAME-RESISTANT TEXTILES

George L. Drake, Jr., and Wilson A. Reeves

The need for reducing the flammability of cellulose has been recognized for several centuries (1–3). At the present time the United States (4) and various other countries have regulations restricting the sale or use of dangerously flammable textiles. Nevertheless, about 150,000 persons are burned in the United States each year in accidents that set clothing afire. (5).

The burning of cellulose is an oxidation process. This process may be accompanied by a flame or glow; most organic fibers undergo a glowing action after the flame has been extinguished. The glow may cause as much damage as the flaming itself, since it can completely consume a textile. Flaming and glowing are distinctly different processes, taking place at different temperatures (see also Section C in Chapter XVIII).

1. Definition of Terms

The term "fireproof" refers to textiles that are essentially unaffected by fire. This means that they do not support fire (flame or glow) and that there is little or no chemical or physical change when the textile is exposed to flames. Few textiles fall in this category; asbestos and glass are the ones in common use today. These fireproof textiles are inorganic materials. A *glow-resistant* textile will not continue to burn by the glow mechanism once the source of ignition has been removed. Glow-resistant textiles may char or melt. Fabric that will not support a flame after the source of ignition has been removed is said to be *flame resistant*. A flame-resistant fabric may continue to burn by the glow mechanism although a flame does not exist. Flame-resistant fabrics can also char or melt. A *fire-resistant* textile is not only flame resistant but also glow resistant. Thus flaming and glowing cease once the source of ignition has been removed. There will be a change in the physical and chemical characteristics; however, in particular, the textile may char or melt.

2. Classes of Fire or Flame Retardants

For practical purposes fire or flame retardants can be divided into four classes: (a) nondurable, (b) semidurable, (c) durable, (d) weather resistant. The nondurable retardants are generally water-soluble compounds that are easily removed by laundering or leaching. The semidurable retardants are generally materials that slowly dissolve and are resistant to leaching and a few mild launderings. Durable retardants are materials that are extremely insoluble and are impervious to leaching, laundering, and dry cleaning for essentially the life of the fabric. Weather-resistant retardants, which are generally of the semidurable or durable type, also resist degradation by sunlight or can readily be protected from actinic degradation by the addition of some material, such as an inorganic pigment.

The interest in and need for the durable type of flame retardant are greater than for the other classes of retardants. They are needed for certain military and industrial uses and for apparel, household, and related uses, including bed linens for hospitals and hotels. Weather-resistant flame retardants are used fairly extensively for tentage materials. One of the primary needs for improvement in this class is effectiveness at lower add-ons.

3. Mechanisms of Fire or Flame Retardancy

In the development of better flame retardants it is helpful to know what chemical reactions occur when cellulose burns and how the reactions are affected when fabrics are treated with a flame retardant. The burning characteristics of cellulose depend to a considerable extent on the chemical and thermal properties of the anhydroglucose units, on the availability of oxygen, and on the nature of noncellulose materials that may also be present. Fiber properties—such as the ignition temperature, the rate and heat of combustion products, and moisture content—also affect the burning characteristics of cellulose. During burning, volatile gases, tars, char, and water are formed. When untreated cellulose burns, the quantity of tar and char is very small; the main final decomposition products are carbon dioxide and water. The presence of a fire retardant can greatly alter the amount and nature of the products.

(a) BURNING OF CELLULOSE

The burning of cellulose takes place in two ways: flaming and glowing. Both processes originate from either an open flame or a hot surface (above 300°C). In the flaming process the thermal decomposition of cellulose results in gaseous, liquid, tarry, and solid products. As the flammable gases burn, the liquids and tars volatilize, and some of these volatile fractions burn, whereas other portions give a carbonized residue that does not readily burn. This process continues until only carbonaceous matter remains. Then the second stage, or glowing process, begins. The residual carbonaceous matter glows, oxidizes, and continues to glow until essentially all organic matter is consumed and only a fluffy ash is left (6). The temperature of flaming celluloses is 400–450°C, whereas afterglow temperature is about 600°C. The low-temperature degradation of cellulose proceeds first by the formation of levoglucosan by scission of the (1→4)-glycosidic linkages of cellulose and subsequently by intermolecular rearrangement of the fragments (7). The levoglucosan is thought to undergo dehydration and polymerization to form char. Volatile organic compounds, gases and water vapor, are also formed by the destructive distillation of some of the levoglucosan. Numerous carbonyl compounds have been identified from the tar and volatile fractions on pyrolysis of cellulose at 420°C (8).

The rate-controlling reaction in the thermal decomposition of cellulose involves the formation of levoglucosan, chemically 1,6-anhydroglucopyranose (9).

The thermal degradation of cellulose is discussed in detail in Section C of Chapter XVIII.

(b) BURNING OF FLAME-RESISTANT CELLULOSE

Theories proposed to explain the function of a flame retardant fall into five categories: coating, gas, thermal, hydrogen bonding, and catalytic dehydration of cellulose to carbon and water.

(1) Coating Theory

According to the coating theory, which originated with Gay-Lussac in 1821 (3), an effective compound forms a glassy skin and stable foam on the fiber surface. Such a flame-retardant coating protects the fabric from the air by serving as a barrier to the flame and by entrapping the volatile tars evolved during combustion. The three requisites for the applied compound are as follows: (a) it must decompose at a relatively low temperature, (b) there must be an evolution of large amounts of gaseous products, and (c) the gas phase must be nonflammable. The flame-retardant coating must melt at a temperature near that at which gas evolution begins, and the foam produced must be minerallike in structure and stable at 500°C. This theory may fit some of the early flame retardants but does not explain the action of most of those available today.

(2) Gas Theory

The gas theory states that the flame retardant decomposes at burning temperatures to give gases which do not burn but which will dilute the flammable gases produced by the decomposition of the cellulose to a concentration below which they can ignite and burn. Effective gases are carbon dioxide, ammonia, hydrogen chloride, sulfur dioxide, and water vapor; these result from the thermal decomposition of such compounds as sodium carbonate, sodium bicarbonate, and ammonium chloride. This theory is not of general applicability but has some merit. Compounds containing halogens are good flame retardants because volatile halides reduce the flammable limits of combustible gases so that flaming of gaseous by-products is greatly hindered. The order of increasing activity is chlorine, bromine, iodine. Fluorine is an exception, possibly due to its high heat of formation.

(3) Thermal Theory

Two thermal mechanisms have been proposed for the retardation of the burning of cellulose. In one it is proposed that the caloric input from a source is dissipated by an endothermic change of retardant, such as fusion and sublimation of the flameproofer. Such endotherms prevent propagation of the flame. In the second mechanism it is proposed that the heat supplied from the source is conducted away from the fibers so rapidly that the fabric never reaches combustion temperatures. Neither mechanism is applicable to the majority of known flame retardants.

(4) Hydrogen-Bonding Theory

It has been suggested that a majority of the flame retardants are strong hydrogen-bonding agents. Therefore, when water (the bridging medium between cellulose hydroxyl groups) is lost at high temperatures, the bridges may be maintained by the strong hydrogen-bonding activity of the flame retardant, thus stabilizing cellulosic fragments by reducing their volatility and hence, their combustibility. Although this mechanism may be a factor, it seems weak. The strength of hydrogen bonds depends on the electronegativity of the atoms involved; even the strongest bonds, such as the F–H–F bond, have bond energies of only 9–10 kcal/mole. Weaker hydrogen bonds (4–6 kcal/mole) would be formed with most flame retardants. It is improbable that such bonds could continue to exist at 400–500°C.

(5) Catalytic Dehydration of Cellulose to Carbon and Water

Ideally a flame retardant would direct the decomposition of cellulose into carbon and water. Theoretically this can be accomplished through dehydration reactions (10), and several dehydrating compounds are known to be good flame retardants for cellulosic fibers. Basically, the path of decomposition is altered, so that the amount of flammable gases and tars is so reduced that combustion will not continue once the source of a flame is removed. This theory is considered by the present authors to be the most generally applicable one. The most recent and widely accepted modification of the theory states that flame retardancy is brought about by the catalytic dehydration of the cellulose through the reaction of a Lewis acid with the cellulose through a carbonium-ion mechanism (6). The flame retardant (Lewis acid) may be either present or produced from its precursor at a temperature close to that of burning cellulose.

In considering the dehydration theory two types of burning will be considered—flaming and glowing. Examination of the decomposition products of burning textiles reveals that the tar from flame-retardant cellulose is similar to that from untreated cellulose (11). Thus reduced flammability is due to a reduction in the amount of combustibles formed as well as to the formation of different chemical decomposition products. The tar from flame-resistant cellulosic fibers contains little or no levoglucosan, whereas that from untreated cellulose contains levoglucosan as a major component. A correlation also exists between the flame resistance of the fabric and the ratio of char to tar or the ratio of carbon monoxide to carbon dioxide degradation. The higher the ratio of carbon monoxide to carbon dioxide or the ratio of char to tar, the better the flame resistance (12–14).

Esteve (15) suggested that prevention of the formation of levoglucosan, the major component of tar from untreated cellulose, or altering the path of decomposition to yield more char would result in flame resistance. Pacsu, Schwenker, Klein, et al. (16–18) showed that mesylation or tosylation of cellulose followed by iodination or bromination to produce a halodeoxycellulose resulted in a flame-retardant cellulosic product. Even though this product did not support a flame, it had an undesirable afterglow, which could be eliminated by treatment with a phosphorus compound, such as diethyl chlorophosphate. It was suggested that the halodeoxycellulose, especially with the halogen on the $C_{(6)}$ position of the anhydroglucose unit, is unable to form levoglucosan on pyrolysis.

Chlorodeoxycellulose, made by reacting thionyl chloride with cotton in the presence of an acid acceptor, is only slightly flame resistant with 6% chlorine (19). Perhaps the lack of flame resistance is due to substitution of the halogen at the $C_{(2)}$ and $C_{(3)}$ positions as well as at the $C_{(6)}$ position. Substitution at the $C_{(2)}$ and $C_{(3)}$ positions would not interfere much with levoglucosan formation. The simple mixing of an organic chlorine-containing compound does not impart significant flame resistance (see also Section F of Chapter XVII).

Holmes and co-workers (20) have shown that the amount of tar and levoglucosan produced by pyrolysis of treated cotton at 420° decreases as the add-on of borax–boric acid increases. When the add-on of retardant is about 8% or greater, little or no levoglucosan is produced. These add-ons correspond closely with the amount needed to make the cotton cloth completely flame resistant.

TABLE 1
Some Decomposition Products of Cotton Treated with Various Amounts of Flame Retardant[a] at 420°C (20)

Add-on of Flame retardant (%)	Time heated (min)	Decomposition products (%)			Flame resistance
		Char	Tar	Levoglucosan	
0	6	8	68	36	increasing ↓
2	6	31.5	33.4	0	
7.2	2	44.6	21.0	0	
11.5	2	50.9	12.5	0	
16.5	2	50.9	9.2	0	

[a] Tris(l-aziridinyl)phosphine oxide–tetrakis(hydroxymethyl)phosphonium chloride.

The prevention of the formation of levoglucosan alone appears to be insufficient to impart flame resistance. As shown in Table 1, levoglucosan is absent when as little as 2% of an effective flame retardant is present in the pyrolyzed fabric. At this low add-on the fabric exhibits excellent glow resistance but very little flame resistance. Complete flame resistance is not normally obtained until about 10% or more of this flame retardant is present. At this add-on of retardant the percentage of char resulting from pyrolysis has about reached a maximum. It seems that formation of large percentages of carbon is much more important for fire resistance than lack of levoglucosan formation. Lack of levoglucosan in the decomposition products is important in that it can influence the amount of carbon resulting from the dehydration reactions.

Comparison of the heats of formation of carbon monoxide and carbon dioxide shows that the oxidation of a given quantity of carbon to carbon dioxide produces almost four times as much heat as oxidation to carbon monoxide:

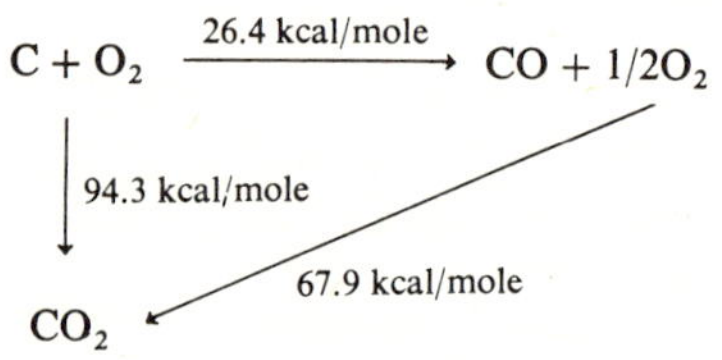

(1)

When the reaction is stopped at carbon monoxide, the combustion is not self-propagating. Consequently one method of imparting fire resistance to cotton is to supply a reagent that will react with carbon monoxide as soon as it is formed and will, in effect, be a carbon monoxide "sink".

4. Testing of Fire Resistance

Methods for testing fire resistance are designed to determine either (a) flame and glow resistance or (b) acceptability of fabrics for various end uses. None of the existing test methods is completely satisfactory, and work in under way by several groups to design new and more appropriate tests. It is generally agreed that (a) ease of ignition, (b) glow time, (c) rate of flame spread, and (d) amount of heat liberated are the essential factors that determine the fire hazard of a fabric. A cellulosic fabric that is allowed to hang without restriction in a vertical position and ignited at the base by flame impinging on both sides is considered to be in the most hazardous position for testing fire resistance. This appears to be logical since the heat lost from the surface is less and the flame and volatile gases from the burning fabric pass up and surround the unburned fabric to promote an atmosphere to cause further ignition. In the case of fabrics that tend to melt and drip when exposed to flames however, portions would drop away from the suspended fabric, preventing further flaming.

The most important tests for measuring fire and flame resistance are the following:

1. *The standard vertical flame test*, in which the length of the char produced by the combined flaming and afterglow is measured after exposure to a standard flame for 12 sec under precisely controlled conditions. Variations of this method are described in references 21-25. It is the most widely used test.

2. *The flammability-test method*, in which a sample is held at an angle of 45°, a flame is applied for 1 sec, and the time required for the flame to travel a distance of 5 in. is recorded (26-29).

3. *The match test*, useful for simple screening, in which an ignited strip of fabric is rotated and the angular position is determined at which the strip must be held for the flame to go out (30). Most flame-resistant fabrics that pass the 135° test will generally pass the standard vertical flame test (31).

4. *The "metal cylinder" carpet-flammability test*, which consists of placing a steel cylinder heated to 800°C on a carpet for 12 sec and measuring the time of flaming and the charred area (32).

Details of the various tests may be found in the references cited.

5. Nondurable Flame Retardants

Nondurable flame retardants are suitable for use on cellulosic fibers that are to be kept under cover. The chemicals wash out of textiles when the goods are exposed to the weather, leached, or laundered. Articles should be treated again after each laundering. Even goods that are not laundered should be retreated about once each 6 months because many of the nondurable-type retardants crystallize on the surface of a fabric after having diffused from the inside of the fiber. Like most other retardants, they are more effective on cotton than on rayon, but they must be uniformly distributed throughout the fabric.

A few good nondurable flame retardants are described below along with some of their effects on fabric properties:

1. Borax-boric acid. This retardant formulation is made by preparing a paste of seven parts of boric acid in a minimum amount of water and then adding this to a water solution of three parts of borax (33). The solution should be essentially clear and be warmed if it becomes cloudy or jellylike on standing. About 10% weighed add-on is adequate for a medium-weight cotton fabric. Fabrics treated with this retardant are flame resistant but have afterglow, which can persist from about 30 sec to several minutes.

2. Borax–diammonium phosphate. This flame-retardant formulation is made by dissolving equal parts of borax and diammonium phosphate in water. It is somewhat less effective on a weight basis than borax–boric acid. However, the presence of the phosphate provides glow resistance.

3. Diammonium phosphate. The fire retardancy imparted by diammonium phosphate is about equivalent to that imparted by the borax-containing phosphate retardant. Diammonium phosphate provides both flame and glow resistance. It is particularly good for fabrics already subjected to a nitrogenous wash-and-wear or durable-press treatment.

4. Ammonium sulfate. In preparing this treating solution the ammonium sulfate is dissolved in water to form a clear solution and then ammonium hydroxide is added to give a faint odor of ammonia. This formulation also provides flame and glow resistance. It may cause a slight reduction in the strength of treated fabric.

Many other nondurable flame retardants are available; some are based on alkyl and aryl phosphates of various types. Still others are based on the products formed by reacting urea or other amides with phosphoric acid and various phosphoric acid derivatives to produce complex water-soluble materials. These are applied to cotton and rayon fabrics from a water solution.

6. Semidurable Flame Retardants

Semidurable flame retardants can be defined as those that withstand one or more laundry cycles, with the upper limit being about 15 mild laundry cycles. The main advantage they have over the durable type, which will be discussed later, is lower cost.

(a) CELLULOSE PHOSPHATES

Apparently, the Joseph Bancroft and Sons Company was the first to be successful in making phosphate derivatives of cotton by a practical method (34). Their process became known as the Ban-flame process. In the original process a solution of urea and phosphoric acid was padded onto fabric; the fabric was then heated at elevated temperatures for about 5 min. A later and more practical process was based on the use of a solution of diammonium phosphate and urea. The fabric was padded, dried, and heated at 175°C for about 7 min. By this technique 10–15% esterification is achieved easily, but cellulosic fibers are severely tendered during the process. It was later found that the tensile-strength loss of treated fabrics could be kept to about 15–20% (35). The better tensile properties are obtained through very careful curing conditions and using formulations that would fix about 3% phosphorus in the fabric. In these instances the cure is for 1–2 hr at 130°C or 5 min at 175°C. Phosphorus attached to cellulose in this fashion is very resistant to removal by boiling 1% sodium hydroxide solution; however, the fabric loses its flame resistance. The loss in flame resistance that is caused by alkaline treatments is due to the formation of sodium salts of the phosphate; these salts are thermally stable at flaming temperature and thus are not flame retardants. Flame retardancy can be regained by rinsing the alkali-treated fabrics in a dilute solution of hydrochloric acid or other strong acids to remove the sodium ions.

One of the earliest modifications of the Ban-flame process was to use cyanamide instead of urea with the phosphoric acid or phosphoric acid salts (36). This modification is said to cause less tendering of fabric. Subsequent improvements have used other amines, such as guanidine, dicyandiamide, and methylolmelamines. Other modifications use in place of urea a mixture of melamine or diguanidine thiosulfate with methylolmelamine (37).

(b) AMMONIUM PHOSPHATE–DICYANDIAMIDE–FORMALDEHYDE

A mixture of methyloldicyandiamide and ammonium phosphate will polymerize to form an insoluble resin in cellulosic fibers when applied to the fabric and baked. Unfortunately at baking temperatures of about 150°C ammonium phosphate decomposes and the liberated phosphoric acid causes excessive tendering of cellulosic fabrics. An ingenious modification of this method was developed by Bowen and co-workers (38). It consists of applying the ammonium phosphate–methyloldicyandiamide mixture to the textile and then precipitating the resin by treatment with alkali. Subsequent drying and heat curing causes polymerization of the resin without excessive tendering of the cellulosic fabric. The finish is resistant to water and mild washing. Fabrics treated in accordance with this method have been used for upholstery. In a somewhat related process a precondensate of urea and ammonium phosphate can be insolubilized with formaldehyde to produce semidurable flame resistance.

(c) CYANAMIDE–PHOSPHORIC ACID PROCESS

An improved procedure makes use of relatively stable commercial solutions of cyanamide (39). It involves padding solutions of cyanamide and phosphoric acid (about 45% solids in a 3 : 1 molar ratio) onto fabrics and curing in ovens at 105–140°C for about 2 min. A strong exotherm (about 120° to 140°C) is observed after about 1 min. The finish so produced imparts to cotton and rayon flame retardancy that withstands some 7–10 laundering cycles.

About one-half the phosphoric acid groups are reacted with nitrogen, and this prevents ion exchange. The other half is cationic and base exchangeable with such cations as sodium and calcium. Tensile strength retained by treated fabric is only about 60% for cotton goods. Good rot resistance and dimensional stability are imparted.

(d) PHOSPHORYL AMIDES

Semidurable flame retardants closely resembling those made from urea and phosphoric acid can be made by reacting ammonia with phosphorus oxychloride (40). Aqueous solutions of the phosphoryl amides are applied to fabric by a pad–dry–cure (5 min at 150°C) process. Treated fabrics withstand laundering in deionized water or deionized water containing a detergent, but not in solutions containing sodium, calcium, or magnesium salts.

7. Durable Flame Retardants

Some of the more important requirements for an ideal durable retardant are the following:

1. Capability to impart fire resistance, that is, both flame and glow resistance.
2. Applicability to any weight fabric and preferably from water solutions.
3. Easy application, preferably on existing finishing equipment without causing hazards.
4. Adequate ability to withstand laundering and dry cleaning.
5. Lack of physiological action.
6. Production of a tough char when exposed to a fire or high-energy radiation.
7. Exertion of little or no adverse effect on the physical properties of the fabric.

8. Resistance to bleaches, particularly hypochlorite types.
9. Reasonable cost.

Durable flame retardants can be applied to the surface of fibers in a fabric and/or put inside the fibers. To be durable on the surface, the retardant will probably have to be a polymer or bonded to the surface with a polymer. Best results from surface deposits are obtained by using thermoplastic polymers. Since the weight add-on required to impart flame resistance by surface deposition ranges upward from about 8% for medium-weight cotton fabrics, the retardants generally impart stiffness or adversely affect some other physical property. The necessary amount of thermoplastic flame retardant can be reduced by using another flame retardant inside the fiber. This approach appears to be the most suitable way of using flame retardants that coat the surface of cellulosic fibers.

Flame retardants that are applied inside cellulosic fibers can be made impervious to laundering by either of three methods: (a) introduction of chemical substituents on the cellulose; (b) crosslinking of cellulose; and (c) *in situ* polymer formation. When chemical substitution is carried out under mild conditions, the tensile properties, stiffness, or hand of the treated fabrics may remain essentially unchanged, provided excessive swelling of the fiber is avoided. Use of flame retardants that crosslink cellulose molecules through primary valence bonds generally decreases tensile properties and increases resilience.

Perhaps the most practical way of fixing flame retardants in cellulosic fibers without adversely affecting strength is through *in situ* polymerization. This method can be carried out in such a way that it actually increases the breaking strength of fabrics without imparting a noticeable amount of stiffness or resilience. Techniques are also available for depositing polymer in fibers while simultaneously producing a few primary-valence-bond crosslinks. When this is done, the treated fabric exhibits resilience and a loss in breaking strength. However, this combination of polymer and a few crosslinks of cellulose causes less loss in breaking strength than crosslinks alone for a given degree of resilience.

Phosphorus-containing materials are generally the most effective fire retardants, they impart both flame and glow resistance. The efficiency of phosphorus-containing retardants is generally affected by the presence of certain other elements and groups. By including materials containing bromine or nitrogen, the amount of phosphorus needed to impart a given degree of fire resistance can be reduced. In one case the phosphorus was reduced from 5 to 1.5% of the weight of fabric by including about 5% triazine nitrogen (31). Generally, however, the addition of about 5% nitrogen or bromine permits a reduction in phosphorus by about 1%. The minimum phosphorus content needed in a medium-weight fabric for fire resistance is about 1% when the optimum amount of nitrogen is present.

The presence of large proportions of organic groups with the phosphorus compound or polymer reduces the effectiveness of the flame retardant. For example, triethyl phosphate is a better fire retardant than tributyl phosphate.

Many flame retardants are durable in the sense that the retardant is not removed by laundering. However, during the laundering operations sodium, calcium, and magnesium ions in the wash solutions exchange for cations in the cellulose or flame retardant. When sufficient metal ion is exchanged in the flame retardant or in the fabric, the resistance to flaming is reduced or essentially eliminated. Thus the flame retardant must not be anionic. The salts formed through the ion-exchange mechanism are thermally stable at flaming temperatures and therefore cannot function as flame retardants since they do not decompose at or below the flaming temperature of cellulose.

(a) FIRE RETARDANTS BASED ON THPC AND RELATED COMPOUNDS

Tetrakis(hydroxymethyl)phosphonium chloride (THPC) is a water-soluble crystalline compound. It is made in very high yields by reacting formaldehyde with phosphine in the presence of hydrochloric acid (41–43). The preparation has been carried out at temperatures ranging from about 25 to 80°C in water systems. The lower temperatures are preferred for the reaction because phosphine is less soluble at high temperatures.

In water solution THPC exhibits acid characteristics. When titrated with base, it has a neutralization equivalence at about pH 8. Apparently THPC dissociates as follows (44):

$$(HOCH_2)_4P^+ \rightleftharpoons (HOCH_2)_3P + CH_2O + H^+ \qquad (2)$$

The equilibrium constant is $pK_a = 7.06 \pm 0.02$ at 25°C (43).

The methylol groups of THPC are very reactive, for example, toward amines and amides:

$$(HOCH_2)_4PCl + RNH_2 \longrightarrow (HOCH_2)_2PCH_2NHR + HCl + HCHO \qquad (3)$$

With amines, the reaction is exothermic and takes place readily at room temperature.

Highly insoluble polymers that contain 15% phosphorus have been made with urea; these closely resemble urea–formaldehyde reaction products, except that they are flame and glow resistant.

Melamine reacts with THPC in mildly acidic solutions, producing initially water-soluble methylol phosphorus derivatives; at least three of the hydrogens on a molecule of melamine can react. The water-soluble derivatives are not easy to stabilize because of their great tendency to continue reaction to form highly insoluble polymers.

THPC undergoes many other reactions. The methylol groups can be esterified by reaction with acid anhydrides. A reaction of particular interest is etherification of cellulose.

$$R_{cell}OH + (HOCH_2)_4PCl \longrightarrow R_{cell}OCH_2P(O)(CH_2OH)_2 + CH_2O + HCl \quad (4)$$

Up to now conditions have not been found for making this reaction profitable commercially. In the reactions given above the phosphonium structure existing in THPC is lost during the course of the reactions and the final products exist as phosphine oxides. The exact mechanism through which THPC reacts is not known but there is strong evidence to indicate that tris(hydroxymethyl)phosphine is the active species when formaldehyde is present.

In the production of fire-resistant cellulosic textiles THPC has been used in water solution at various pH values, ranging from about 4 to 8. The pH has been adjusted with amines, sodium hydroxide, sodium carbonate, and various other compounds, including alkali sulfites. Addition of these compounds has significant effects on the stability of the treating formulation, efficiency of subsequent polymerization with amines and amides, and properties of treated fabrics, as will be seen below. The causes for these effects must be, in part, due to the reactions that occur when various amounts of an alkali are added to THPC. The products of this reaction are not adequately characterized. However, tris(hydroxymethyl)phosphine (THP) and tris(hydroxymethyl)phosphine oxide (THPO) have been isolated. Tetrakis(hydroxymethyl)phosphonium hydroxide (THPOH) has not been isolated, but it would be expected to exist only in solution.

It is known that the *P*-methylol compounds formed in solution by reacting THPC with up to 1 equivalent of a base are very reactive with amines and amides. Some researchers think the product is an equilibrium mixture of THP and THPOH (45). Others have evidence that THPOH is not an intermediate in the formation of THP (46). Furthermore, the formaldehyde produced in the reaction reacts with a *P*-methylol group to form a hemiacetal:

$$(HOCH_2)_3P + CH_2O \rightleftharpoons (HOCH_2)_2PCH_2OCH_2OH \quad (5)$$

Tris(hydroxymethyl)phosphine reacts very slowly with amides and amines unless formaldehyde is present.

When more than 1 equivalent of base is reacted with THPC, the product is THPO. If the intermediate is THPOH, the reaction could be as follows:

$$(HOCH_2)_4P^+ + OH^- \longrightarrow (HOCH_2)_3PO + CH_2O + H_2 \quad (6)$$

If the intermediate is THP, the reaction could be as follows:

$$(HOCH_2)_3P + OH^- + H_2O \longrightarrow (HOCH_2)_3PO + H_2 + OH^- \quad (7)$$

Thus in neutralizing THPC considerable care should be exercised to prevent formation of the less reactive THPO. Excess hydroxide ion can be depleted by reaction with formaldehyde in an oxidation-reduction manner (Cannizzaro) to methanol and formic acid. This reaction, however, is probably much slower than the action of the hydroxide ion on THP to oxidize it to THPO.

The value of THPC as a fire retardant was first recognized in a study of the chemical and physical properties of aminized cotton (30). Many techniques were found for using THPC in the treatment of cellulosic textiles for imparting flame resistance. Some of these processes are described below.

(1) Application by Heat-Cure Processes

THPC-Amide Flame Retardant (47, 48). This flame retardant is applied to the fabric in a pad-dry-cure technique. A typical formulation applied to 8-oz cotton fabric so that it passes the standard vertical flame test is shown in Table 2. After fabric is padded through the formulation to about 70% wet pickup, it is dried at about 90°C and then cured at about 150°C for about 3 min. It is recommended that the fabrics be afterwashed to remove unused chemicals. The fire-retardant finish produced by this treatment is exceedingly impervious to laundering and dry cleaning. The primary limitations of the process are (a) stiffness imparted to some fabrics and (b) reduction in the tensile and tearing strengths of some fabrics. The treatment generally reduces breaking strength by about 30% and increases conditioned wrinkle recovery [warp + fill (W + F)] to about 250°. It also imparts considerable wet wrinkle recovery.

TABLE 2
Typical THPC-Amide Flame-Retardant Formulation[a]

Agent	Concentration in solution (%)
THPC	16
Urea	7
Methylolmelamine	9
NaOH	1.5
Auxiliaries	1.5

[a] Processing steps are padding, drying, and heat curing

The reagents and the concentration of the reagents listed in Table 2 can be varied substantially. Figure 1 illustrates the effects of urea concentration on the ability of the finish to withstand laundering and alkaline boiling; the optimum is about 5% urea. The durability generally decreases with increasing concentration of urea in the treating formulation. However, one cannot easily eliminate the

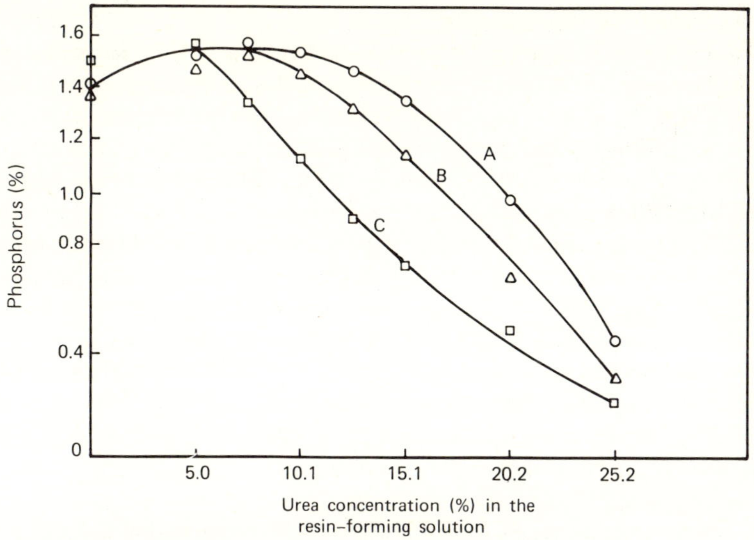

Fig. 1. Effects of urea concentration on the durability of phosphorus in THPC-resin-treated fabric. The resin-forming solution contained 16% THPC, 9.7% trimethylolmelamine and 3% triethanolamine. (*A*) Phosphorus introduced by the resin treatment; (*B*) phosphorus remaining after 10 laundering cycles; (*C*) phosphorus remaining after 3-hr boil in soap and sodium carbonate solution.

urea because without it the tensile and tearing strengths of treated fabrics are lowered greatly. Thiourea can be substituted for urea in about equal weights with a resulting increase in the ability to withstand laundering and alkaline boiling.

Trimethylolmelamine is the preferred triazine for use in this flame-retardant process. When methylated methylolmelamines are substituted for the trimethylolmelamine, treated fabrics have a softer hand, but the amount of phosphorus fixed in the fabric is significantly lower. Generally this same situation results when other *N*-methylolamides are substituted for trimethylolmelamine, including such substances as dimethylolethyleneurea, dimethylolethyltriazone, dimethylol-4,5-dihydroxyethyleneurea, and dimethylol carbamates.

Sodium hydroxide is used in the formulation to neutralize the free hydrochloric acid that is usually present in the THPC. Any excess sodium hydroxide neutralizes or reacts with THPC and converts it to other *P*-methylol compounds, as already discussed. Other bases, including alkanolamines, can be used instead of sodium hydroxide.

Wetting and softening agents are the main auxiliaries used in the formulation. Generally about 0.1% of a wetting agent is used. The amount of softener can vary considerably, depending on the properties of the treated fabric. The softener can have a significant effect on tearing strength; 0.5% cetylamine or stearamide causes an increase of about 25% which is retained through as many as 15 laundry cycles. Usually laundering reduces the tearing strength.

The formulation given in Table 2 has been improved by adding aldehyde-combining materials, such as sodium sulfite (49). By reacting with free formaldehyde they reduce premature polymerization in the treating formulation and produce fabrics with a better hand, less stiffness, and higher tearing strength. The amount of sulfite can vary from about 0.9 up to 2.0 moles per mole of THPC (50). The addition of a polymerization catalyst, such as magnesium chloride or an amine hydrochloride, is also necessary in this formulation.

Modified THPC–Amide Flame Retardants. The THPC–amide flame retardant has been modified in several ways. The objective of these modifications has been at least twofold: (a) to reduce cost and (b) to improve the properties of treated fabric. Some of the more important modifications are the following:

1. Bromoform adduct of triallyl phosphate. The preparation and use of this adduct alone for flameproofing cotton is described in more detail below. The bromoform adduct of triallyl phosphate (BAP) is produced in an aqueous emulsion that can be added directly to THPC–amide flame-retardant formulations (51). The amount of BAP used in this combination finish can vary quite substantially. On a solids basis the amount is generally equal to, or slightly less than, the weight of THPC in the formulation. The THPC–amide retardant penetrates cotton and other cellulose fibers, whereas the adduct is deposited on the surface of these fibers. Thus the combination finish imparts a hand to the fabric that is substantially different from that produced by THPC–amide alone. The adduct is a good flame retardant but is not quite as efficient with respect to weight add-on as the THPC–amide retardant. Because of this fact, slightly greater add-ons are required of the combined finish than of the THPC–amide alone to impart a given degree of flame resistance. This combined flame retardant is applied to fabric by a pad–dry–cure process. The cure is about 3 min at 150°C.

One of the main benefits of the combined finish is greater flame resistance as measured by the match test; narrow strips of fabric exhibit almost no afterflame as well as no glow. A higher total add-on of the combined finish can be applied to some fabric structures, such as a cotton sheeting, without imparting excessive stiffness. Two faults of the surface polymer in this combination finish are that it causes fading of dyed goods and it reduces the resilience normally imparted by the THPC–amide retardant.

2. The bromoform adduct of allyl esters of phosphonitrilic chlorides, known as the PNE–bromoform adduct (52). The preparation and use of this flame retardant alone are described below. The aqueous emulsion of the adduct can be

added directly to the aqueous THPC-amide formulation. The amount of PNE-bromoform used with the THPC-amide is about one-half the weight of THPC. Fabric is padded with the combined formulation to about a 70% wet pickup; then it is dried and subsequently cured for about 3-4 min at 150°C. This combination is especially valuable for treating medium- and light-weight fabrics, which are most difficult to make highly flame resistant. Treated fabrics pass the match test at the 180° angle. The combination flame retardant is more efficient than the THPC-amide retardant; that is, fabric can be made flame resistant with less weight add-on. It does not impart resilience; this is probably due to the physical properties of the PNE-bromoform polymer deposited on the surface of the fibers. This flame retardant, although very effective, has not become important commercially, probably because of its high cost.

3. Antimony trioxide and organic chlorine compounds. Antimony trioxide or chlorinated compounds alone impart little flame resistance to cellulosic fibers. However, if an adequate quantity of suitable halogen-containing compound is present with antimony trioxide, the combination is a good flame retardant. In general halogen is supplied by chlorinated paraffins, poly(vinyl chlorides), and poly(vinylidene chlorides). The purpose of chlorine is to react with the oxide to produce antimony oxychloride. Theoretically only a very small amount of halogen would be needed if all of it reacted to produce antimony oxychloride. Practically this is not the case. The preferred poly(vinyl chlorides) have softening points between about 170 and 200°C (53). When poly(vinyl chloride) is used alone with the THPC-amide flame retardant, the primary benefit seems to be an improvement in the tensile strength of treated fabrics. Since poly(vinyl chloride) in combination with antimony trioxide is a low-cost flame retardant, this combination can substitute for some of the THPC-amide retardant (54), but greater total add-ons must be applied to the fabric to impart a specified degree of flame resistance, because of lower efficiency.

4. Tris(2,3-dibromopropyl) phosphate (55). This compound must be emulsified because it is not soluble in water. A typical formulation for treating polyester-cotton sheets contain 20% THPC, 7% trimethylolmelamine, 5% urea, and 13% tris(2,3-dibromopropyl) phosphate. The brominated phosphate penetrates into both the polyester and the cotton fibers along with the THPC-amide retardant. The retardant is very durable and prevents loss in lightfastness on some dyes (56). Tris(2,3-dibromopropyl) phosphate used in the absence of a binding agent is easily removed from cotton by laundering but is durable in polyester. Incorporation of about 2% poly(vinyl chloride) with the phosphate produces durability in cotton fabrics (57).

"THPOH"-amide heat-cured retardants. As used herein, "THPOH" refers to the solution produced by neutralizing THPC with about 1 equivalent of sodium hydroxide per mole of THPC. It is apparently a mixture of *P*-methylol compounds. When the "THPOH" is produced in methanolic solution, sodium chloride can be filtered from the system. When it is made in aqueous solution, sodium chloride and any other by-products remain in the solution. The reaction of "THPOH" with amides and amines is very similar to that of THPC. A typical "THPOH"-amide flame-retardant formulation for use on an 8-oz cotton fabric contains 15% "THPOH", 7% urea, 9% trimethylolmelamine, and 1.5% auxiliaries, which include wetting and softening agents (58). Fabric is padded with this formulation, dried, and subsequently cured for about 3 min at about 150°C. As with the THPC-amide flame-retardant formulations, the concentrations of urea and melamine can be varied substantially. However, the concentration of these reagents seems to exert less effect upon resistance to laundering than it does in the case of the THPC-amide flame retardant. This can be seen in Table 3. There are several outstanding features of fabrics treated with the "THPOH" amide flame retardant. Breaking-strength retention is generally about 80-90%, but in some cases it is essentially 100%. This is particularly unusual for cotton fabrics exhibiting high degrees of conditioned wrinkle recovery. Wrinkle-recovery values (W + F) in the range of 300° are not unusual. Fabrics exhibit therefore a high degree of smooth-drying and durable-press characteristics. Wash-and-wear ratings for sateen fabrics having these high wrinkle-recovery values are generally about 3 or 4, where a rating of 5 represents the maximum. Another important property that is imparted is minimum yellowing of undyed fabrics on exposure to hypochlorite bleach. The add-on is essentially the same as that required of the THPC-amide retardant, or about 16-18% for an 8-oz cotton-sateen fabric.

(2) Application by Ammonia-Cure Processes

Since all of the flame-retardant processes based on THPC and related *P*-methylol compounds reduce the strength of cellulosic fabrics, techniques were sought to produce flame resistance without causing losses in strength. Most of the above methods make the cellulose insoluble in cuen and impart at least a modest degree of conditioned wrinkle recovery; therefore it was theorized that the processes caused crosslinking of cellulose molecules and hence a reduction in strength. By using methods that would cause the *P*-methylol compounds to be polymerized inside the fiber without reacting with the cellulose, it should be possible to produce flame-resistant fabrics without causing degradation of strength. The following techniques illustrate that this goal has been attained.

THPC-Trimethylolmelamine-Urea Retardant (59, 60). In the process utilizing ammonia to insolubilize *P*-methylol polymers inside cotton the treating solution is made by dissolving 16.2 parts of THPC in water, adjusting the pH to about 7 by adding triethanolamine or alkali hydroxides, and then adding 11.6 parts of trimethylolmelamine and 8.4 parts of urea. After the fabric has been padded in this solution and dried for 4 min at 90°C, it is treated with either ammonia or ammonium hydroxide to form an insoluble polymer inside the fibers. This

TABLE 3
Properties of Cotton Sateen Treated with "THPOH"–Methylolmelamine–Urea at Various Component Molar Ratios[a] (31)

Molar ratio of "THPOH"–methylolmelamine–urea	Resin add-on (%)	Breaking strength[b] (lb)	Elmendorf tearing strength[b] (g)	Wrinkle recovery[c] (degrees)	Char length (in.)			Phosphorus content (%)	
					Orig-nal	After soap–soda boil	Laundered 15 times	Origi-nal	After 15 launderings
2 : 1 : 4	18.4	143.0	2130	283	2.1	2.3	2.2	3.46	58
2 : 2 : 2	21.7	109.4	1240	298	2.4	2.2	2.7	3 31	70
2 : 2 : 4	20.2	131.7	1660	306	2.5	2.2	2.7	3.07	63
2 : 4 : 2	21.5	106.8	1190	306	2.4	2.6	2.7	2.27	68
4 : 2 : 1	18.3	129.8	1800	298	3.1	2.6	2.8	4.84	64
2 : 1 : 2	19.1	132.4	1860	287	2.5	2.1	2.4	4.31	65
Untreated control	–	142.4	2920	166	–	–	–	0.02	–

[a] Fabrics treated with 34% solids solutions.
[b] Warp
[c] Monsanto, warp plus fill.

process produces fabric that passes the standard vertical flame test even after 3 hr of boiling in 1% sodium hydroxide solution. The strength of the treated fabric is essentially unchanged. Other nitrogenous compounds can be used in place of the trimethylolmelamine and urea. A typical formulation of this modification contains 0.25–0.75 equivalent of alkali hydroxide per mole of THPC (60). The alkali stabilizes the treating formulation so as to prevent polymerization prior to the treatment of cellulosic textiles. This formulation can be applied in commercial laundry apparatus.

THPC–Urea Precondensate Process. Water-soluble precondensates made of THPC and urea have been used in Europe for about 10 years (61). They are prepared by refluxing an aqueous solution of THPC and urea for 30 min and then cooling rapidly. Fabric is impregnated to about 70% wet pickup with the solution, then dried at about 90°C, and subsequently exposed first to ammonia vapor for a few seconds to immobilize the reagent and finally to an ammonium hydroxide solution to ensure complete penetration by ammonia and the production of a highly insoluble polymer. The finish is very resistant to boiling soap and sodium carbonate solutions and withstands repeated laundering and dry cleaning. The tearing strength of fabric is slightly reduced, but tensile properties are essentially unchanged.

The need for the gaseous-ammonia treatment can be eliminated by incorporating ammonium sulfite or a mixture of ammonium salt of another weak acid together with a sulfite in to the treating formulation (62).

"THPOH"–NH_3 process. As already described, the reaction of 1 equivalent of sodium hydroxide with THPC produces a mixture referred to as "THPOH." If the amount of sodium hydroxide reacted exceeds 1 equivalent per mole of THPC, the reaction product is THPO, which is not sufficiently reactive with amines and amides even at 150°C to be of economic interest. In the earliest studies 0.5 mole of sodium hydroxide was used per mole of THPC, thus converting approximately one-half the THPC to "THPOH" (63), and the impregnated fabric was dried thoroughly at temperatures ranging up to 180°C before exposure to ammonia.

Subsequently it was found that when about 1 mole of sodium hydroxide is reacted with 1 mole of THPC, the impregnated fabric cannot be dried below 10–20% moisture content before exposure to ammonia because of loss of formaldchyde without severely reducing the efficiency of polymer formation (64). With the larger proportion of sodium hydroxide, polymer formation with ammonia proceeds more rapidly, but the moisture content of the fabric must be carefully controlled.

An 8.5-oz twill or sateen thus treated will pass the standard vertical flame test with about 13% resin; 2-oz fabrics need about 20%. Stiffness is essentially unaffected. Breaking strength is often increased by as much as 25%. When a softener is not employed, tearing strength is reduced by about 30%; with softener the reduction is about 10%.

"THPOH"-Cu-NH_3 process. When certain copper salts, such as copper nitrate, are added to solutions of "THPOH," the *P*-methylol compounds are stabilized, so that they do not react with ammonium hydroxide in solution (65). Stabilization is apparently accomplished through complex formation between copper and phosphorus atoms. Thus in this modified technique "THPOH" is made by reacting about 1 mole of sodium hydroxide per mole of THPC and then adding copper nitrate and subsequently ammonium hydroxide or other ammonium salts. Fabric is then padded through the solution and cured at about 150°C for 3 min. Treated fabrics have a bluish-green color owing to the presence of the copper salts. Trimethylolmelamine and certain other *N*-methylol compounds can be added before or after adding the copper-salt solution. When this is done, the amount of "THPOH" used in the treating solution can be reduced.

"THPOH"-Trimethylolmelamine-NH_3 Process. In the "THPOH"-NH_3 process described above the *P*-methylol conpounds are the most expensive agents used in the process. Generally *N*-methylol amides are substantially cheaper than "THPOH," and trimethylolmelamine can be used together with "THPOH" to reduce the overall cost without great changes in properties (31). In the application of this retardant fabric is padded (70% wet pickup) with the reagents, partially dried (to about 10-20% moisture), exposed to ammonia for a few seconds, and finally heated at 150°C for 3 min. In a typical formulation 17% THPC is reacted with about 1 equivalent of sodium hydroxide and then 8% trimethylolmelamine is added.

(b) PHOSPHATE ESTERS WITH BROMINE AS DURABLE FLAME RETARDANTS

Bromine compounds, in the presence of phosphorus compounds, produce more efficient flame retardancy than phosphorus compounds used alone. The efficiency of compounds containing bromine and phosphorus is about equivalent to that of materials containing phosphorus and nitrogen. Although water-soluble phosphate esters are reasonably easy to hydrolyze by alkaline solutions, insoluble esters are adequately resistant to alkaline laundering to be useful as durable flame retardants.

(1) Bromoform-Triallyl Phosphate Ester Polymers

In the early 1950s it was known that tris(2,3-dibromopropyl) phosphate would impart fire resistance to cotton, but ways were not known at that time for using it in the production of durable flame-resistant finishes for cellulosic fabrics. The first success in this area was through a partial bromination of triallyl phosphate followed by polymerization through the remaining allyl groups (66). These polymers were efficient fire retardants, and the properties of treated fabrics were reasonably good. However, the method of preparing the treating solutions was laborious, and the polymers were applied to the fabric from an organic solvent.

A more practical method for using triallyl phosphate is to make the bromoform adduct simultaneously with allyl polymerization in aqueous emulsions (67). This reaction proceeds by a free-radical-polymerization mechanism producing a stable emulsion of a low-molecular-weight polymer, or telomer, containing about 8% phosphorus and 40% bromine. The polymer retains some unsaturated groups that can be further polymerized on cellulosic fibers by heat to produce a durable finish. For convenience the bromoform–triallyl phosphate telomer is referred to as BAP. The materials used in a typical emulsion are shown in Table 4.

TABLE 4
Materials Required for Preparing Typical Bromoform–Triallyl Phosphate Ester Polymers (67)

	Amount required (parts for 100 parts emulsion)	
Material	33% Emulsion	40% Emulsion
Triallyl phosphate	189	236
Bromoform	132	164
Nonionic emulsifying agent	–	2
Poly(vinyl alcohol) (medium viscosity)	6	8
Sodium bicarbonate	24	32
Potassium peroxydisulfate	6	10
Water	643	548

The emulsion is stable to coagulation and has a density of about 1.14–1.17; the particles have an average diameter of 0.2 μ. The chemical structure of the telomer is thought to be

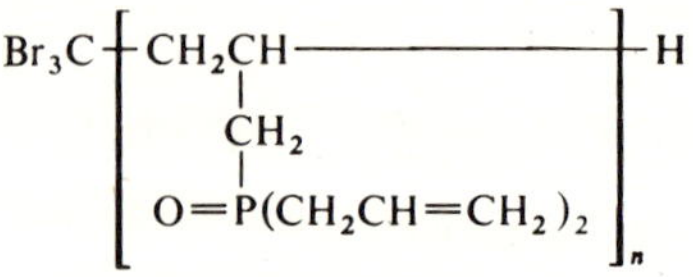

$n = 1.0–2.5$

Treatment of fabric with the emulsion is done on conventional textile-finishing equipment, using a pad–dry–cure procedure with adequate ventilation to remove irritating fumes (probably allyl alcohol).

In the treatment of 8- to 9-oz cotton fabric with a 33% emulsion fabric is padded to about a 75% wet pickup and then dried, usually at about 110°C. Curing is for 4–10 min at about 140°C. After being washed and dried the fabric will have about 18–20% dry add-on. Treated fabrics pass the standard vertical flame test, even after repeated laundering. The treatment causes very little loss in breaking strength and only about 10–30% loss in tearing strength. The hand of the fabric is slightly rubbery. Treated fabrics are truly fire resistant, that is, both flame and glow resistant.

(2) Tris(2,3-dibromopropyl) Phosphate

When tris(2,3-dibromopropyl) phosphate and an appropriate vinyl polymer in which it is soluble are applied to cotton fabric from an organic solvent, they became durably fixed to the surface of the fibers (68). Among the polymers used are poly(vinyl chloride), poly(vinyl acetate), and acrylic polymers.

The finish can be applied to yard goods by utilizing existing plant equipment designed for solvent application, and to garments in conventional dry-cleaning equipment. The solvents used are either xylene, methyl ethyl ketone, or chlorinated solvents, such as perchloroethylene. No cure is required.

The outstanding property of fabric treated with tris(2,3-dibromopropyl) phosphate in combination with poly(vinyl chloride) is high tearing strength. The add-on of the tris(2,3-dibromopropyl) phosphate required to withstand 15 accelerated launderings ranges from 26 to 35% for fabrics weighing about 5–8 oz/yd^2. A unique property of flame retardants that contain large quantities of bromine is that the char length does not decrease with add-on and is usually long. The decomposition products tender or degrade the uncharred portion of the sample, with the result that the tear in the standard vertical flame test is not confined to the charred area.

A typical formulation is 57 parts of xylene, 11 parts of methyl ethyl ketone, 1 part of vinyl chloride–vinyl acetate copolymer, and 20 parts tris(2,3-dibromopropyl) phosphate. The wet pickup must exceed 100%.

Rayon fibers have been made flame resistant with tris(2,3-dibromopropyl) phosphate by dispersing the ester in the viscose before the fibers are spun (69). The amount of ester added should not exceed about 30–40% of the weight of the cellulose. Substantially less than this will impart adequate flame resistance for the fibers to pass the standard vertical flame test. Rayon fibers produced in this way exhibit only about 15% less tensile strength than similar fibers that do not contain the ester.

(3) Bromine-Containing Phosphonitrilates

These materials have already been described as additives to the THPC–amide formulation. The esters used in preparing flame retardants are formed by substituting $CH_2{=}CHCH_2O{-}$ and $CH_2BrCHBrCH_2O{-}$ groups for most of the chlorine atoms in the phosphonitrilic chlorides (phosphazenes) (70). They are made by reacting the corresponding alcohol with phosphonitrilic chloride in the presence of sodium. The bromoform-adduct polymer of allylphosphonitrilate, PNE-bromoform, is perhaps the most satisfactory flame retardant of this group.

A 30% PNE-bromoform emulsion is used for the treatment of 8-oz cotton fabrics. Fabrics are padded to about 70% wet pickup, dried at about 100°C, cured for about 6 min at 140°C, washed, and dried. Twill fabrics treated in this manner contain about 11% add-on of the polymer and pass the standard vertical flame test after repeated laundering. The trapezoid tearing strength is essentially unchanged by the treatment, whereas the Elmendorf tearing strength is about 75% of that of the untreated fabric. The properties of an 8-oz cotton twill treated with the PNE-bromoform adduct are shown in Table 5 along with the properties of a similar fabric treated with partially brominated PNE and the 2,3-dibromopropyl phosphonitrilate (2,3-DBPPN). None of the flame retardants appears to have completely satisfactory resistance to an alkaline soap boil. However, when the PNE-bromoform adduct is used together with the THPC-amide flame retardant, the combined flame retardant has excellent resistance to alkaline soap boil, laundering, and dry cleaning.

TABLE 5
Comparison of the Properties of Flame-Resistant Fabrics (71)

Flame retardant	Add-on (%)	Nitrogen (%)	Phosphorus (%)	Other elements	Char length (warp) (in)[a]
APO[b]	7.4	2.1	1.6	–	2.7
THPC-amide	16.9	6.0	1.6	–	2.4
PNE-bromoform	21.3	1.0	2.1	6.8% Br	3.1
2,3-DBPPN[c]	12.9	0.8	1.6	7.7% Br	4.3
THPC-amide-PNE-bromoform	19.7	4.7	1.7	2.7% Br	1.9
APO-THPC	14.3	1.6	2.9	–	2.2

[a] Maximum permissible value for this fabric weight (8.5 oz/yd^2) is 4.5 in.
[b] Tris(1-aziridinyl)phosphine oxide.
[c] 2,3-Dibromopropyl phosphonitrilate.

(c) AZIRIDINYL COMPOUNDS

The most useful compound of the aziridinyl group is tris(l-aziridinyl)-phosphine oxide (APO). It is prepared by the reaction of phosphorus oxychloride with ethylenimine (72):

$$POCl_3 + 3HNCH_2CH_2 \longrightarrow OP(NCH_2CH_2)_3 \qquad (8)$$

A white, crystalline, hygroscopic compound, APO is soluble in water and in most organic solvents and usually sold commercially as an 80–90% alcoholic solution. This compound and its solutions can be stabilized by the addition of anhydrous methanol, ethanol, peroxides, or dehydrating compounds (73, 74).

The aziridinyl groups in APO are very reactive. They readily undergo ring-opening reactions. In an acid solution APO will homopolymerize or copolymerize with other aziridinyl compounds. This is shown below for any one of the three aziridinyl (A) groups.

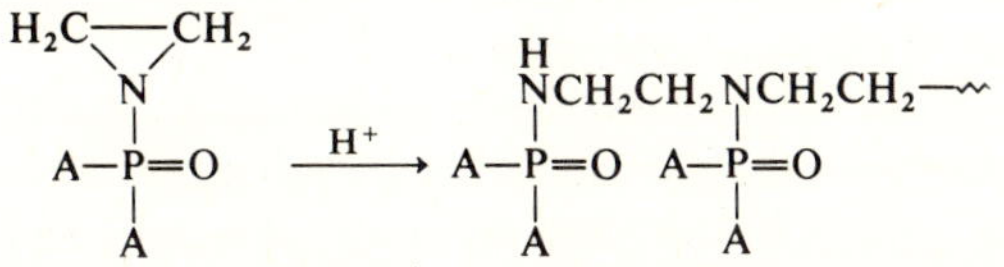

(9)

In actual practice polymerization involves all of the aziridinyl rings of the molecule, and a highly crosslinked polymer results.

In basic media the reaction is thought to involve attack of the nucleophile on the ring carbon, giving rise to an amide ion:

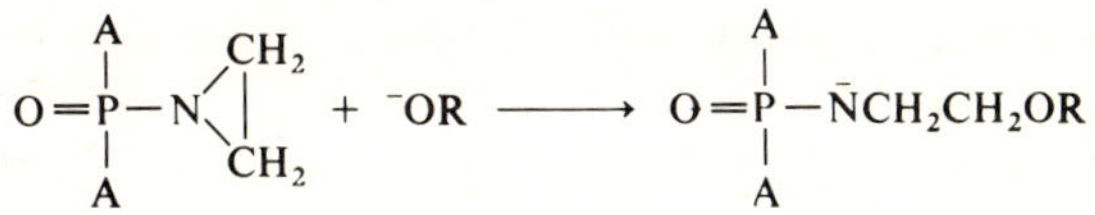

(10)

In the above equation R can represent cellulose; therefore, when more than one aziridinyl ring is involved, crosslinking can occur (see also Section E in Chapter XVII).

When APO is applied to cellulosic materials, it can react with the cellulose, condense with itself to form a polymer, or react with other active compounds to form copolymers. It can react with sodium thiosulfate (75, 76) (this is a good assay method), amide and amine compounds (77, 78), phenols (79), carboxylic acids (80), and polyhydric alcohols (81, 82). APO can be used alone or with a coreactant to produce flame-retardant finishes. It has been used alone to produce a wash-and-wear cotton with excellent smooth-drying and wrinkle-recovery properties, to produce a durable embossed pattern, to attach dyes, and to bind pigments in printing.

Although APO is toxic and must be handled with care, once the monomer is polymerized, the end products are reported to be harmless (83–85).

(1) APO–THPC

An outstanding flame-retardant finish is based on APO and THPC (86, 87). These compounds react with each other and polymerize in cotton to impart a very durable fire-resistant finish with a relatively low resin add-on; they are satisfactory for fabrics of many structures and weights. The first stage of the reaction between APO and THPC can be illustrated as follows:

$$OP(\underbrace{NCH_2CH_2}_{})_3 + (HOCH_2)_4\overset{+}{P}Cl^-$$

$$\downarrow$$

$$(\underbrace{CH_2CH_2N}_{})_2P(O)NHCH_2CH_2OCH_2P(CH_2OH)_2 + HCl + HCHO \qquad (11)$$

During the reaction the phosphonium structure of THPC is converted to the phosphine and eventually to a phosphine oxide.

Application of the APO–THPC flame retardant is accomplished in conventional textile equipment, using a pad–dry–cure procedure. A typical formulation for treating 8-oz cotton fabrics contains about 13% APO and 14% THPC, plus about 1.5% sodium hydroxide, about 0.1% of a wetting agent, and a softening agent. Fabric is padded to approximately a 70% wet pickup, then dried at about 85°C, cured for 4 min at about 145°C, washed, and dried.

The APO–THPC finish can withstand repeated laundering and dry cleaning. Fabrics that are treated with it become both glow and flame resistant, and are physiologically inert. When exposed to flames or intense heat, they form tough pliable and protective chars. In addition, the treated fabrics exhibit little or no adverse change in hand or appearance and are resistant to shrink, rot, and mildew. The treatment reduces breaking strength by about 20% and tearing strength by about 40%. Finished fabrics have about 250° (W + F) wrinkle recovery. One drawback of this finish and of most durable flame retardants is that it yellows when treated fabric is subjected to sodium hypochlorite bleach. Temporary methods for removing the yellowing utilize a reducing bleach like sodium borohydride (88). Less yellowing is observed on dyed fabrics. Fabrics of various weights and constructions treated with APO–THPC retain their fire resistance after numerous launderings and dry cleanings. It should be emphasized that the amount of resin add-on needed in a fabric in order for it to pass the standard vertical flame test after repeated laundering is less than that of most other durable flame retardants. For example, an 8-oz cotton-twill fabric containing about 12% add-on of the resin can be laundered repeatedly by either military, commercial, or household procedures and retain its flame resistance. In one test consisting of 60 home washes the fabric was still flame resistant and its char length was unchanged from that of the unwashed but treated control.

Examination of stained cross sections of fibers treated with the APO–THPC finish and then stained with a dye, such as Kiton blue (which stains the resin but not the cellulosic fiber), shows that most of the resin is deposited inside the fiber.

(2) APS-THPC

The sulfur analog of APO, APS, $SP(\underbrace{NCH_2CH_2})_3$, has been made in good yields by the reaction of ethylenimine with thiophosphoryl chloride. Like APO, it reacts very readily with active hydrogen-containing compounds to form highly crosslinked polymers. It can be applied with THPC to fabric in the same manner as is used for the APO–THPC flame retardant (86). The properties of treated fabrics are also comparable to those imparted by the APO–THPC finish. The APS–THPC solutions are more stable than the APO–THPC solutions because of the greater stability of the aziridinyl ring in APS. Since APS requires the use of thiophosphoryl chloride for its preparation, it is more expensive than APO.

(3) APO

Used alone with a catalyst, APO imparts excellent durable-press and good flame-resistant properties to cotton fabrics (88–90). Fabrics have been produced with over 300° wrinkle-recovery (W + F) values.

The process of applying the APO flame retardant to cellulosic fabrics is very simple and can be carried out on standard textile-resin-finishing equipment, using the pad–dry–cure process. Drying is done at about 85°C, and curing is done at 140°C for 4 min. A typical formulation for treating 8-oz cotton fabric contains 25% APO and 2% zinc fluoroborate plus wetting and softening agents as needed. Heat, with the help of the catalyst, produces an insoluble, highly crosslinked polymer inside the cotton fiber. The resulting fabric is flame and wrinkle resistant. Addition of softeners to the treating bath can keep tearing-strength losses to a minimum, usually about 40%. A detailed discussion of the properties of a cotton fabric treated with this flame retardant can be found in reference 90.

The infrared-absorption band at 12.3 μ, which represents the aziridinyl rings, disappears completely after APO has been thoroughly cured (91). The degree of cure is important because laundering is best withstood by samples that are thoroughly cured; infrared spectroscopy provides a tool for studying the cure.

(4) APO–Diammonium Phosphate

Benzenesulfonic acid, zinc perfluoroborate, and diammonium phosphate have been shown to be effective catalysts for APO cure (91). Due to its low cost, ease of handling, and effectiveness, diammonium phosphate was selected for use with APO in the development of a flame retardant. One feature of this "catalyst" is that it is also a coreactant in that it becomes part of the final polymer; thus it is an inexpensive source of phosphorus. The method for applying this flame retardant is based on a pad–dry–cure process. A typical aqueous formulation for treating 8.5-oz cotton-sateen or twill fabrics to impart sufficient flame resistance to pass the standard vertical flame test consists of 5% diammonium phosphate

(DAP) and 18% APO plus wetting and softening agents as needed. The wet pickup of fabric is about 75%. The fabric is dried for 2 min at about 125°C and cured for 5 min at about 150°C. It shows good retention of flame retardancy through 15 cycles of laundering. The main advantages of this flame retardant over the APO-THPC retardant is lower cost and little or no release of fumes during the drying and curing steps. However, the physical properties (such as tearing strength) of treated fabrics are not as good.

(5) APO-Thiourea

It was noted above that APO copolymerizes with many compounds containing two or more active hydrogens. Clear polymers are formed with urea and thiourea, the latter being more resistant to degradation acids and bases. Thiourea reacts with APO according to the following reaction:

$$OP(\underline{NCH_2CH_2})_3 + H_2NC(S)NH_2$$

$$\downarrow$$

$$(\underline{CH_2CH_2N})_2P(O)NHCH_2CH_2NHC(S)NH_2 \qquad (12)$$

The other aziridinyl groups are capable of reacting with more thiourea or even with cellulose. Treatment of cotton fabrics with a ratio of 2 moles of APO to 1 mole of thiourea produces good flame resistance (92). The method of application consists of a pad-dry-cure (2 min at 160°C) procedure. Treated fabrics exhibit wrinkle and rot resistance, and good flame resistance with an add-on of about 20%. Tensile- and tearing-strength losses are about 20 and 40%, respectively. The finish withstands laundering and dry cleaning. Little or no fumes are given off during application of the flame retardant.

(6) Solvent Application of APO Flame Retardants

The use of solvents other than water in textile wet processing is expected to increase. APO is readily adaptable to application from most organic solvents; the coreactants that have been already described, such as THPC, diammonium phosphate, and thiourea, are not suitable. Several other compounds, however, are suitable coreactants in organic solvent systems: 2,2-bis(bromomethyl)propanediol, 3-bromo-2,2-bis(bromomethyl)propanol, tetrabromobisphenol A, and 2,4,6-tribromophenol (93). One or more of these agents can be combined with APO in an organic solvent and applied to cotton fabrics by a pad-dry-cure process. Drying requires about 2 min, and curing is done at 160°C for about 2 min, followed by washing. The flame retardancy imparted to fabrics is good. Durability is satisfactory through at least 20 laundering cycles. However, the finish applied in this manner usually imparts stiffness to fabrics; of the coreactants studied, 2,2-bis(bromomethyl)propanediol and tribromophenol appeared to impart the least amount of stiffness.

(d) PHOSPHONATES AND PHOSPHINE OXIDES AS FLAME RETARDANTS

These compounds have been used in three ways: (a) in the preparation of polymers that are applied to the surface of the fibers, (b) for making derivatives of the cellulose of fibers, and (c) as monomers that react with cellulose and polymerize in the fiber.

(1) Allyl Phosphonates

Polymers for use on the surface of fibers have been based on chloromethyl diallyl phosphonate (**1**) and cyanoethyl diallyl phosphonate (**2**) (94).

$$(CH_2{=}CHCH_2O)_2\overset{\overset{\displaystyle O}{\|}}{P}CH_2Cl \qquad (CH_2{=}CHCH_2O)_2\overset{\overset{\displaystyle O}{\|}}{P}CH_2CH_2CN$$

1 **2**

These monomers may be polymerized with benzoyl peroxide as catalyst at about 85–100°C. Better flame retardants are prepared, however, when the monomers are first partially brominated and then polymerized. About 25% add-on of the partially brominated polymer products is required for an 8-oz cotton fabric to provide sufficient flame resistance to pass the standard vertical flame test. The relatively high add-ons are probably accounted for by the high percentage of fuel (carbon and hydrogen) that is added per unit of phosphorus and bromine. In general these polymers are about equivalent to the bromoform adduct of triallyl phosphate in flame retardancy. The phosphonate polymers are somewhat more impervious to alkaline laundering than polymers made from phosphates. The phosphonates are applied to fabric from an organic solvent or a mixture of solvents, such as methanol and ethylene dichloride, using a pad–dry–cure procedure. Curing at an elevated temperature is not required. The flame-retardant finishes withstand several accelerated laundering cycles.

(2) Phosphonomethylcellulose

Phosphonomethylcellulose was prepared by reacting the disodium salt of chloromethyl phosphonate with cotton in the presence of a strong base (95). The preferred method for carrying out this reaction is to pad the fabric with a solution containing 25% sodium hydroxide and 15% of the sodium salt of chloromethyl phosphonate, dry the fabric, then heat for 5 min at 140°C. The wet pickup for most cotton fabrics is about 125–140%. Under these conditions nearly 2% phosphorus can be attached to the cotton fiber in a single processing.

The phosphonomethylcellulose produced in this manner exists as the sodium salt and is not very flame resistant. It is necessary to exchange the sodium ions for hydrogen or ammonium ions to obtain high degrees of flame resistance. In the ammonium-salt form only 0.7% phosphorus is required in a cotton sheeting to pass the standard vertical flame test. About 1% phosphorus is required to pass the 180° strip match test. This treatment generally does not reduce breaking strength and reduces tearing strength only slightly (96), but stiffness is increased severalfold. Phosphonomethylcellulose is very resistant to both acid and base hydrolysis, but flame resistance is lost when the material forms a salt with Group I or Group II metal ions (see also Section H of this Chapter XIX).

(3) Tris(chloromethyl)phosphine and Phosphine Oxides

Tris(chloromethyl)phosphine oxide and closely related compounds can be reacted with cotton as an alkaline solution (10% NaOH) that is applied to fabric (97). The wet fabric is dried and heated at about 150°C for 5 min. The amount of phosphorus introduced into a cotton fabric by this process is only about 0.7% because of limited solubility in the alkali solution, and the fabrics show only a low degree of flame resistance.

(4) N-Methylolamides of Phosphines and Phosphine Oxides

Tris(2-carbamoylethyl)phosphine and tris(2-carbamoylethyl)phosphine oxide reacted with formaldehyde and then with cotton fail to produce an adequate level of flame resistance even at high levels of add-on (98).

(5) Phosphonopropionamides

The *N*-methylol derivatives (**3**) of several dimethyl, diethyl, and di(chloroethyl) phosphonopropionamides have been studied as flame retardants for cellulosic fabrics (99).

$$(RO)_2P(O)CH_2CH_2C(O)NHCH_2OH$$

3

These compounds may be prepared by reacting the dialkyl phosphite with acrylamide and methylolating the resulting product.

A typical treating formulation contains 5% of trimethylolmelamine, 1% of urea, 0.1% of a suitable stabilizing agent, 0.5% of polyethylene softener, 0.4% of amine hydrochloride catalyst, and 30% of the *N*-methyloldialkyl phosphonopropionamide. The add-on for an 8-oz cotton fabric to pass the standard vertical flame test is about 25%. Fabrics are padded in the treating solution, dried, and then cured for either 1–2 min at 175°C or 5 min at 160°C. Treated fabrics lose about 30% of their breaking and tearing strengths.

(6) Phosphonoacetamides

Cellulose derivatives very closely related to those based on *N*-methyloldialkyl phosphonopropionamide have been made by using a two-step process (100), as follows:

$$R_{cell}OH + HOCH_2NHCOCH_2I \longrightarrow R_{cell}OCH_2NHCOCH_2I \qquad (13)$$

$$R_{cell}OCH_2NHCOCH_2I + (ClCH_2CH_2O)_3P$$

$$\downarrow$$

$$R_{cell}OCH_2NHCOCH_2P(O)(OCH_2CH_2Cl)_2 + ClCH_2CH_2I \qquad (14)$$

In the first step of this process the *N*-methyloliodoacetamide is applied to cotton in dimethylformamide, using a conventional pad-dry-cure technique with zinc nitrate catalyst. The second step of the process, the Arbuzov reaction, is also carried out in an organic solvent at about 140-150°C. The amount of phosphorus needed for flame resistance can be reduced by applying trimethylol-melamine in another step. Cotton fabrics treated in this manner have excellent retention of flame resistance after laundering with detergent only and limited retention after laundering with chlorine bleach. Alkaline scouring, even for 8 hr, causes only a minimum increase in char length and no significant change in the phosphorus content. The breaking and tearing strengths of treated fabrics are reduced by about one-third, but there is no change in stiffness.

(e) INORGANIC DURABLE RETARDANTS

Many inorganic materials are flame retardants for cellulosic materials. However, few finishes have been found to withstand laundering. The main problem in obtaining durability has been in finding ways of insolubilizing the inorganic materials in or on cellulosic fibers. The compounds that come closest to meeting the requirements for durable flame retardants are based on antimony and titanium.

(1) Antimony Trioxide-Chlorinated Paraffins

Antimony trioxide, Sb_4O_6, in combination with chlorinated paraffins is a good flame retardant (101), used mostly in the production of weather-resistant goods. However, one formulation has been suggested as a durable flame retardant for apparel goods, such as life preservers and work uniforms (102). It consists of 9.6% of *N*-methylolurea, 0.4% of catalyst, 14.4% of chlorinated paraffin (70% chlorine content), 12% of chlorinated paraffin (42.5% chlorine content), 25% of Stoddard solvent, 3.9% of an oil-modified alkyd resin, 12.3% of antimony trioxide, and about 22.3% of water. The emulsion is applied to fabric by a pad-dry-cure procedure. A 30-60% add-on is required.

(2) Antimony–Titanium Complex

Although titanium and antimony oxides impart very little flame resistance in the absence of a halogen, certain compounds have been found to be very efficient flame retardants. A mixture of titanium oxychloride and antimony chloride imparts durable flame resistance to cellulosic fabrics (103). The formulation can be prepared by reacting titanium tetrachloride and antimony trioxide; it must be kept strongly acidic to prevent hydrolysis.

$$6TiCl_4 + Sb_4O_6 \longrightarrow 6TiOCl_2 + 4SbCl_3 \qquad (15)$$

The formulation is applied by padding fabric through the treating solution and then neutralizing it with a strong solution of sodium carbonate. Titanium and antimony are not merely precipitated within the fiber as oxides. They are thought to form a compound (**4**) with the cellulose as indicated here:

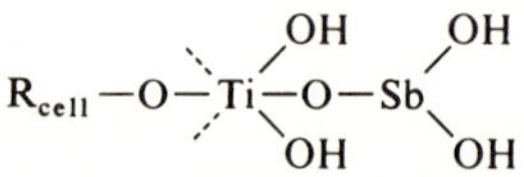

4

The fixation of the titanium and antimony is very efficient when about 15% sodium carbonate solution is used to neutralize the fabric. Cotton fabrics having add-ons of 10–15% pass the standard vertical flame test. Breaking strength is usually about 90% or more of the original untreated fabric, and hand is very good.

8. Weather-Resistant Flame Retardants

The cellulosic fiber substrate and the flame retardant must be resistant to environmental degradation in order for a flame-resistant textile to be considered weather resistant. The principal agents of cellulose destruction are sunlight and microorganisms. In addition, cellulosic fibers can be tendered by the degradation products of the flame retardant; for example, a flame retardant may degrade in sunlight to produce an acid, such as hydrochloric acid, that can readily degrade the cellulose. Sunlight and water are usually the main elements of the weather that destroy the flame retardants.

The only reliable way to evaluate flame-resistant textiles for weather resistance is to expose samples of the fabric in the location where the fabric is to be used. In this test the fabric is held at a 45° angle facing south (in the Northern Hemisphere), and then the extent of degradation is determined by one or more techniques. The degree of flame resistance after exposure is determined by the standard vertical flame test.

Water-soluble flame retardants are not weather resistant because they are readily removed by rain and humid atmospheres.

The weather-resistant flame retardants available today may be divided into two broad groups: (a) inorganic oxides and (b) organophosphorus compounds. The water-insoluble inorganic oxides generally have good resistance to degradation by sunlight; however, they are poor flame retardants and must be used in combination with some other material, such as chlorinated paraffins. Organophosphorus flame retardants are generally degraded by ultraviolet light and therefore must be protected by some additional shielding material.

(a) INSOLUBLE INORGANIC OXIDES

Research leading to the use of insoluble oxides as weather-resistant flame retardants began with the work of Perkin (104), who precipitated hydrous stannic oxide in cellulosic fibers. Ferric hydroxide was later found to be substantially more efficient, requiring only about 25% add-on. The next advance was made by precipitating two oxides that acted synergistically. This combination consisted of tungstic oxide and stannic oxide, obtained as follows:

$$Na_2WO_4 + SnCl_4 + H_2O \longrightarrow 2NaCl + SnO(OH_2)_2 + 2WO_2(OH)_2 \qquad (16)$$

Deposition of the insoluble oxides was brought about by hydrolysis after the soluble precursors were deposited in the fibers. Fabrics treated by any of these processes exhibit considerable afterglow.

(1) Antimony Oxides with Organohalogens

Practically all of the weather-resistant flame retardants being applied to fabrics today are based on a combination of antimony trioxide and an organic halogen compound (101, 105–107). The first real advance in this direction was made by Clayton and Heffner (101), who combined chlorinated paraffins and antimony trioxide. In addition, a plasticizer, colorants, fillers, stabilizers, and fungicides were used. Treatments of this type that impart fire, water, weather, and mildew resistance have become known as FWWMR finishes (107). A typical formulation is shown in Table 6. The finish is applied by a pad-dry technique to add-ons of 40–50% (108). The calcium carbonate is used in this finish to neutralize any hydrochloric acid that might result from actinic degradation of the chlorinated paraffins. Aluminum stearate used in combination with the chlorinated paraffins provides water repellency, and the pentachlorophenol provides mildew resistance.

Outdoor-exposure studies of an FWWMR-finished 10-oz cotton army duck indicates that very little change occurs in the fabric during a 12-month period (102). It is well known from other work that FWWMR-treated fabrics retain their flame resistance and high breaking strength through at least 3 years of severe exposure. Perhaps the greatest disadvantages of FWWMR finishes are

TABLE 6
Typical FWWMR Finish (107)

Component	Amount (%)
Chlorinated paraffin (40% Cl)	13
Chlorinated paraffin (70% Cl)	7
Antimony trioxide	7
Binder	7
Pigment	9
Calcium carbonate	4
Aluminum stearate	1
Pentachlorophenol	1
Organic solvent	51

their high add-on requirements and the fact that they stiffen fabrics. The use of poly(vinyl chloride) instead of part of the chlorinated paraffins reduces the stiffness.

(2) Antimony–Titanium Complex

A weather-resistant flame retardant that requires add-ons of only about 10–15% is based on an antimony–titanium complex (103), described above. Flame resistance and breaking strength of fabrics treated with this flame retardant appear to be adequate after 1 year of outdoor exposure.

(b) ORGANOPHOSPHORUS COMPOUNDS

Organophosphorus finishes are generally much more efficient than the antimony trioxide–chlorinated paraffin type and thus require less add-on. Table 7 provides information on the durability of flame resistance in fabrics treated with various organophosphorus retardants, of the type already discussed, on exposure to weather (71, 109).

All of the organophosphorus flame retardants cited in Table 7 inhibit mildew growth and rotting, thus preventing loss of tensile strength by biological degradation during outdoor exposure.

The standard APO–THPC flame-retardant formulation (mole ratio 1 : 1) is more vulnerable on outdoor exposure than formulations containing a mole ratio of 2 : 1 or higher. By increasing the mole ratio, flame resistance has been extended by about 50%, as measured by the standard vertical flame test (71).

Fabrics treated with various inorganic pigments for protection against actinic degradation (110), finished with APO–THPC, wax–aluminum stearate water repellent, and an antimildew finish, and then coated with a titanium dioxide-pigmented vinyl polymer give different responses on outdoor exposure (Table 8). The retention of flame resistance is best for fabrics with the pigmented vinyl

TABLE 7
Durability of Flame Resistance to Outdoor Exposure (71)

Flame retardant	Maximum allowable exposure (months)[a]
APO	8
THPC–amide	6
PNE–bromoform	12
2,3–DBPPN	18
THPC–amide–PNE–bromoform	18
APO–THPC	9

[a]After which the fabric would still pass the standard vertical flame test.

TABLE 8
Char Length of Flame-Retardant Duck Fabrics (110)[a]

	Duration of exposure (months)									
Fabric	0	2	4	6	8	10	12	18	24	36
Without coating:										
Gray	3.4	2.7	2.9	4.2	(b)	(b)	(b)	(b)	(b)	(b)
Khaki	3.2	3.5	2.8	4.2	5.8	(b)	(b)	(b)	(b)	
Pearl gray	3.4	3.5	2.7	7.8	6.5	8.7	8.3	(b)	(b)	(b)
With vinyl coating:										
Gray	4.0	3.4	2.4	3.2	2.0	3.4	3.1	3.0	3.2	2.8
Khaki	3.5	2.8	3.9	3.9	2.2	3.5	3.1	2.8	3.3	2.7
Pearl gray	2.4	2.0	2.7	2.4	1.3	2.2	2.1	2.4	4.4	1.8

[a]Inches of char in standard vertical flame test after weathering for number of months indicated; 8-oz cotton duck treated with APO–THPC in 1 : 1 ratio.

[b]Burned entire length.

coating. The flame resistance of uncoated fabrics was found to be lost first in the gray fabrics, a little later by the khaki fabrics, whereas the pearl-gray treated fabrics retained flame resistance through 12 months. All of the coated fabrics retained flame resistance and good strength through 36 months (110).

References

1. F. Ward, *J. Soc. Dyers Colourists,* 71, 569 (1955).
2. O. Wyld, British Patent 551 (1735).
3. J. L. Gay-Lussac, *Ann. Chim. Phys.,* **18** (2), 211 (1821).

4. Flammable Fabrics Act Amendments of 1967, 90th Congress, 1st Session, U.S. Senate, Report No. 407, Calendar No. 394.
5. Staff Report, *Textile Ind.,* **132** (1), 103, 105 (1968).
6. H. Schuyten, J. W. Weaver, and J. D. Reid, in *Advances in Chemistry*, Vol. 9, Academic Press, New York, 1954, pp. 7–20.
7. S. L. Madorsky, V. E. Hart, and S. Straus, *J. Res. Natl. Bur. Std.,* **56**, 343 (1956); *ibid.,* **60**, 343 (1958).
8. F. H. Holmes and C. J. G. Shaw, *J. Appl. Chem. (London),* **11**, 210 (1961).
9. J. B. Benkowitz-Mattuck and T. Noguchi, *J. Appl. Polymer Sci.,* **7**, 709 (1963).
10. M. Leatherman, U.S. Department of Agriculture Circular No. 466 (March 1938).
11. W. G. Parks, M. Antoni, A. E. Petrarca, and A. R. Pitochelli, *Textile Res. J.,* **25**, 789 (1955).
12. W. G. Parks, R. Esteve, M. Gollis, R. Guercia, and A. Petrarca, paper presented at the 127th Meeting, American Chemical Society, 1955.
13. R. W. Little, *Textile Res. J.,* **21**, 901 (1951).
14. W. G. Parks, J. G. Erhardt, Jr., and D. R. Roberts, *Am. Dyestuff Reptr.,* **39**, 294 (1950).
15. R. Esteve, paper presented at the 127th Meeting, American Chemical Society, 1955.
16. E. Pacsu, R. F. Schwenker, and P. Way, U.S. Department of the Army Contract No. DA-19-129-QM-122 Progress Report, March–June 1956.
17. R. F. Schwenker and E. Pacsu, *Ind. Eng. Chem.,* **50**, 91 (1958).
18. E. Klein and J. E. Snowden, *Ind. Eng. Chem.,* **50**, 80 (1958).
19. T. L. Vigo and C. M. Welch, *Proc. Ninth Cotton Utilization Research Conference,* Southern Utilization Research and Development Division, Agricultural Research Service, U.S. Department of Agriculture, New Orleans, 1969.
20. G. A. Byrne, D. Gardiner, and F. H. Holmes, *J. Appl. Chem.,* **16**, 81 (1966).
21. "Fire Resistance of Textile Fabrics," Test Method 34-1966, *American Association of Textile Chemists and Colorists Technical Manual*, Howes Publishing, New York, 1968, pp. B-138, 139.
22. *Flame Retardant Properties of Treated Fabrics,* ASTM D626-55 T, American Society for Testing and Materials (1965).
23. *Standard for Fireproof Textiles,* NFPA 701, PR, National Fire Protection Association, 1965.
24. Federal Specification CCC-T-19lb, Method 5903-T, Federal Supply Service Administration, May 15, 1951.
25. *Outline of Proposed Investigation of Factory-Treated, Flame-Resistant Clothing, Bedding, and Drapery Fabrics,* Underwriters' Laboratory, Subject 214, January 1964.
26. "Flammability of Clothing Textiles," Test Method 33-1962, *American Association of Textile Chemists and Colorists Technical Manual,* Howes Publishing, New York, 1968, pp. B-139–142.
27. Commercial Standards 191-53 (revised), U.S. Department of Commerce, available from the Superintendent of Documents, U.S. Government Printing Office, Washington, D.C.
28. Flammable Fabrics Act, Public Law 88, 83rd Congress, Chapter 164, HR 5069, July 1, 1954, amended July 1, 1958.
29. Federal Specification CCC-T-19 lb, Methods 5908 and 5910, Federal Supply Service Administration, May 15, 1951.

30. W. A. Reeves, O. J. McMillan, Jr., and J. D. Guthrie, *Textile Res. J.*, **23**, 527 (1953).
31. J. V. Beninate, R. M. Perkins, G. L. Drake, Jr., and W. A. Reeves, *Textile Res. J.*, **39**, 368 (1969).
32. L. W. Sayers, *Textile Inst. Ind.*, 168 (July 1965).
33. J. E. Ramsbottom and A. W. Snoad, Second Report of the Fabrics Co-ordinating Research Committee, Department of Scientific and Industrial Research, Great Britain. 1930.
34. Joseph Bancroft and Sons Co., British Patent 604,197 (1948).
35. F. V. Davis, J. Findlay, and E. Rogers, *J. Textile Inst.*, **40**, T839 (1949).
36. T. H. Marton, British Patent 634,690 (1950) (to Courtaulds Ltd.).
37. American Cyanamid Co., British Patent 638,434 (1950).
38. H. H. Bowen, V. H. Majerus, W. S. Meals, and S. Kellett, British Patent 586,095 (1947).
39. S. J. O'Brien, *Textile Res. J.*, **38**, 256 (1968).
40. M. L. Nielsen, *Textile Res. J.*, **27**, 603 (1957).
41. A. Hoffman, *J. Am. Chem. Soc.*, **43**, 1684 (1921).
42. W. A. Reeves, F. F. Flynn, and J. D. Guthrie, *J. Am. Chem. Soc.*, **77**, 3923 (1955).
43. L. M. Fodor, Ph.D. Thesis, Cornell University, Ithaca, N.Y., 1963.
44. M. Grayson, *J. Am. Chem. Soc.*, **85**, 79 (1963).
45. N. Filipescu, L. M. Kindley, H. E. Podall, and F. A. Serafin, *Can. J. Chem.*, **41**, 821 (1963).
46. W. J. Vullo, *J. Org. Chem.*, **33**, 3665 (1968).
47. W. A. Reeves and J. D. Guthrie, *U.S. Dept. of Agriculture Ind. Chem. Bull.* AIC-364 1953.
48. J. D. Guthrie, G. L. Drake, Jr., and W. A. Reeves, *Am. Dyestuff Reptr.*, **44**, 328 (1955).
49. G. M. Wagner and P. E. Hoch, U.S. Patent 3,101,279 (1963) (to Hooker Chemical Corp.).
50. R. F. Zimmerman and G. M. Wagner, U.S. Patent 3,247,015 (1966) (to Hooker).
51. J. D. Reid, J. G. Frick, Jr., and R. L. Arceneaux, *Textile Res. J.*, **26**, 137 (1956).
52. C. Hamalainen, W. A. Reeves, and J. D. Guthrie, *Textile Res. J.*, **26**, 145 (1956).
53. G. M. Wagner, U.S. Patent 3,054,698 (1962) (to Hooker).
54. G. M. Wagner, U.S. Patent 3,219,478 (1965) (to Hooker).
55. Piedmont Section, American Association of Textile Chemists and Colorists, *Am. Dyestuff Reptr.*, **57**, 373 (1968).
56. Piedmont Section, American Association of Textile Chemists and Colorists, *Textile Chem. Colorist*, **1**, 124 (1969).
57. T. D. Miles and A. C. Delasanta, *Textile Res. J.*, **38**, 273 (1968).
58. J. V. Beninate, E. K. Boylston, G. L. Drake, Jr., and W. A. Reeves, *Textile Res. J.*, **38**, 267 (1968).
59. W. A. Reeves and J. D. Guthrie, U.S. Patent 2,772,188 (1956) (to U.S. Secretary of Agriculture).
60. G. M. Wagner, U.S. Patent 3,310,419 (1967) (to Hooker).
61. H. Coates, U.S. Patent 2,983,623 (1961) (to Albright and Wilson, Ltd.).

62. H. Coates and B. Chalbley, U.S. Patent 3,096,201 (1963) (to Albright and Wilson, Ltd.).
63. H. Coates, U.S. Patent 3,236,676 (1966) (to Albright and Wilson, Ltd.).
64. J. V. Beninate, E. K. Boylston, G. L. Drake, Jr., and W. A. Reeves, *Am. Dyestuff Reptr.*, **57**, 981 (1968).
65. D. J. Donaldson and D. J. Daigle, *Textile Res. J.*, **39**, 363 (1969).
66. G. E. Walter, I. Hornstein, and C. H. Sheld, U.S. Patent 2,660,543 (1953) (to Glenn L. Martin Co.).
67. J. G. Frick, Jr., J. W. Weaver, and J. D. Reid, *Textile Res. J.*, **25**, 100 (1955).
68. T. D. Miles and A. C. Delasanta, *Textile Res. J.*, **37**, 273 (1967).
69. J. W. Schappel and A. I. Bates, U.S. Patent 3,266,918 (1966) (to FMC Corp.).
70. C. Hamalainen and J. D. Guthrie, *Textile Res. J.*, **26**, 141 (1956).
71. J. B. Bullock, and C. M. Welch, *Textile Res. J.*, **36**, 441 (1966).
72. H. von Bestian, *Ann. Chem.*, **566**, 210 (1950); through *Chem. Abstr.*, **44**, 5805 (1950).
73. H. Dethloff, A. Rückert, and W. Ziegler, German Patent 953,078 (1959) (to Farbwerke Hoechst AG).
74. Dow Chemical Corp., Bulletin No. 164-122, 1963.
75. F. C. Schaefer, J. T. Geoghegan, and D. W. Kaiser, *J. Am. Chem. Soc.*, **77**, 5918 (1955).
76. E. Allen and W. Seaman, *Anal. Chem.*, **27**, 540 (1955).
77. W. A. Reeves, L. H. Chance, and G. L. Drake, Jr., U.S. Patent 2,859,134 (1958) (to U.S. Secretary of Agriculture).
78. W. A. Reeves, L. H. Chance, and G. L. Drake, Jr., U.S. Patent 2,889,289 (1959) (to U.S. Secretary of Agriculture).
79. W. A. Reeves, J. D. Guthrie, and L. H. Chance, U.S. Patent 2,912,412 (1959) (to U.S. Secretary of Agriculture).
80. W. A. Reeves, G. L. Drake, Jr., and J. D. Guthrie, U.S. Patent 2,915,480 (1959) (to U.S. Secretary of Agriculture).
81. L. H. Chance, G. L. Drake, Jr., and W. A. Reeves, U.S. Patent 2,870,042 (1959) (to U.S. Secretary of Agriculture).
82. L. H. Chance, G. L. Drake, Jr., and W. A. Reeves, U.S. Patent 2,886,538 (1959) (to U.S. Secretary of Agriculture).
83. R. B. LeBlanc, *Textile Res. J.*, **35**, 341 (1965).
84. G. L. Drake, Jr., B. M. Kopacz, and F. S. Perkerson, U.S. Department of Agriculture, Agricultural Research Service, ARS 72-14, 1958.
85. R. H. Wilson, *Textile Res. J.*, **32**, 424 (1962).
86. W. A. Reeves, G. L. Drake, Jr., L. H. Chance, and J. D. Guthrie, *Textile Res. J.*, **27**, 260 (1957).
87. G. L. Drake, Jr., J. V. Beninate, and J. D. Guthrie, *Am. Dyestuff Reptr.*, **50**, 129 (1961).
88. R. B. LeBlanc and A. P. Ingram, Jr., *Textile Res. J.*, **32**, 284 (1962).
89. G. L. Drake, Jr., and J. D. Guthrie, *Textile Res. J.*, **29**, 155 (1959).
90. G. L. Drake, Jr., L. H. Chance, J. V. Beninate, and J. D. Guthrie, *Am. Dyestuff Reptr.*, **51**, 40 (1962).
91. T. D. Miles, F. A. Hoffman, and A. Merola, *Am. Dyestuff Reptr.*, **49**, 596 (1960).

92. R. B. LeBlanc, *Textile Res. J.*, **35**, 341 (1965).

93. R. H. Symm, *Text. Chem. Color.*, **1**, (6), 42 (1969).

94. A. D. F. Toy and K. H. Rattenbury, U.S. Patent 2,735,789 (1956); J. R. Costello, Jr., and T. P. Traise, U.S. Patent 2,841,507 (1958); A. D. F. Toy and R. S. Cooper, U.S. Patent 2,867,547 (1959) (all assigned to Victor Chemical Works).

95. G. L. Drake, Jr., W. A. Reeves, and J. D. Guthrie, *Textile Res. J.*, **29**, 270 (1959).

96. S. R. Hobart, G. L. Drake, Jr., and J. D. Guthrie, *Textile Res. J.*, **29**, 884 (1959).

97. L. H. Chance, D. J. Daigle, and G. L. Drake, Jr., *Am. Dyestuff Reptr.*, **56**, 614 (1967).

98. L. H. Chance, W. A. Reeves, and G. L. Drake, Jr., *Textile Res. J.*, **35**, 291 (1965).

99. Ciba Ltd., French Patent 1,395, 178 (1965); R. Aenishanlin, C. Guth, P. Hoffman, A. Maeder, and H. Nachbur, *Textile Res. J.*, **39**, 375 (1969)

100. G. C. Tesoro, S. B. Sello, and J. J. Willard, *Textile Res. J.*, **38**, 245 (1968).

101. E. C. Clayton and L. L. Heffner, U.S. Patent 2,299,612 (1942) (to William E. Hooper & Sons, Inc.).

102. K. S. Campbell and J. E. Sands, *Textile World*, **96**, 118 (1946).

103. H. C. Gulledge and G. R. Seidel, *Ind. Eng. Chem.*, **42**, 440 (1950).

104. W. H. Perkin, Original Commission 8th Intern. Congr. Appl. Chem.; *Met. Chem. Eng.*, **10**, 636 (1912).

105. C. Snyder, U.S. Patent 1,885,870 (1932).

106. M. Leatherman, U.S. Patent 2,012,686 (1935).

107. L. L. Heffner, *Textile Res. J.*, **24**, 272 (1954).

108. R. B. LeBlanc, *Textile Ind.*, **132**, 274 (1968).

109. J. B. Bullock, C. M. Welch, and J. D. Guthrie, *Textile Res. J.*, **34**, 691 (1964).

110. G. L. Drake, Jr., E. K. Leonard, and W. A. Reeves, *Textile Ind.*, **130**, 146 (1962).

J. BIOLOGICALLY ACTIVE CELLULOSIC MATERIALS

Z. A. Rogovin and A. D. Virnik

The preparation of biologically active materials constitutes one of the most interesting trends in the chemical modification of cellulose. A number of recently published investigations are concerned with the production of cellulosic-type materials possessing antibacterial properties by chemically binding bactericides or fungicides to the cellulose macromolecule. To obtain these materials cellulose is subjected to a pretreatment to introduce functional groups capable of further reacting with bactericidal or fungicidal preparations. Cellulose can also be treated directly with preparations provided that they contain functional groups capable of entering into reaction with the hydroxyl groups of cellulose. These methods make possible the preparation of modified cellulose materials whose antibacterial activity persists even after repeated wet treatments.

The biological activity of antibacterial fibers of this kind results from the gradual splitting off of a small amount of the bactericide or fungicide from the cellulose (1, 2). If there is a strong bond between the functional group of the polymer molecule and the bactericide, the latter has no antibacterial efficacy (2). Cellulose derivatives containing chemically bound bactericidal cations (metal ions, quaternary ammonium bases), antibiotics, iodine, and organomercury or organotin residues are known at the present time.

Gagliardi (3, 4) has described the treatment of cellulosic materials with compounds of the general formula $A_2B{-}C{-}M$, where A represents groups capable of reacting with the hydroxyl groups of cellulose, C is an ionogen, M is a cation with antibacterial properties, and B is an aliphatic, aromatic, or heterocyclic residue. Materials containing bactericidal cations or antibiotics can be produced from cellulose by first grafting onto it poly(acrylic acid) (5–7) or treating it with 2,4-dichloro-6-(3′,6′,8′-trisulfo-1′-naphthylamino)-*sym*-triazine (5, 8); cellulose xanthate (9) and carboxymethylcellulose (10) can also be used. The chemical structure of copper xanthate fiber can be represented by the formula $R_{cell}(OH)_{3-x}[OCS_2Cu_{1/2}]_x$. An important disadvantage of the copper xanthate fiber is that it is gradually destroyed on standing, probably because of the occurrence of oxidation–reduction reactions. To produce antibacterial materials containing iodine, use has been made of cellulosic fabrics modified by grafting poly(methylvinylpyridine) or poly(vinylpyrrolidone), which are known to complex with iodine (11). Antibacterial materials containing the residues of organotin compounds have been obtained by treating the graft copolymer of cellulose and poly(acrylic acid) with tributyltin hydroxide (12). Mercury-containing derivatives of cellulose can be prepared by the reaction of cellulose ethers and esters containing aromatic, heterocyclic, or unsaturated aliphatic residues with mercuric acetate (13–17).

Antibacterial cellulose materials containing chemically bound bactericides or fungicides can be used to manufacture clothing, as well as wrappings that allow the sterility of surgical instruments to be preserved for a prolonged period of time (18); they also serve as a filter material for the sterilization of air (19) and disinfection of river water (20).

The water-insoluble graft copolymers of cellulose and biologically active proteins (enzymes, antigens) are of great scientific and practical significance. The graft copolymers of cellulose and enzymes can be employed as specific catalysts that can be removed from the reaction zone at any time. The application of these copolymers eliminates a number of problems that cannot be solved by using ordinary water-soluble enzymes; these include the isolation of pure fermentation products; the isolation and subsequent investigation of intermediates in fermentation processes; the activation of an enzyme with subsequent complete removal of the activating substance; the sorption, subsequent isolation, and investigation of enzyme inhibitors, and so on (21). The water-insoluble graft copolymers of cellulose and antigens, which have been given the name "immunosorbents," are employed for sorption of antibodies for their quantitative determination and their isolation in pure form for subsequent study and application (22–24). To synthesize water-insoluble graft copolymers of biologically active proteins preference should be given to cellulose rather than synthetic polymers since nonspecific adsorption of proteins on cellulosic materials is considerably lower than it is on synthetic polymers.

The addition of enzymes and antigens to the cellulose macromolecule is mainly effected through diazonium-coupling and acylation reactions. Here the starting cellulose materials are the ethers and esters of cellulose that contain aromatic amino groups (22–32) and carboxymethylcellulose (31–35). Gurwich and co-workers (29) have shown that the content of grafted protein can be markedly increased in the copolymer by using, for synthesis, highly dispersed suspensions of cellulose derivatives produced by reprecipitation from a cuprammonium solution.

Among polymer preparations possessing hemostatic properties, monocarboxyl cellulose and its salts are well known (see also Section F in Chapter XIX). A number of patents suggest the use of cellulose sulfates and the ethers from cellulose and aliphatic hydroxycarboxylic acids as hemostatics (36–38). A new hemostatic has been recently developed in the Soviet Union (39); it is the calcium salt of the graft copolymer of cellulose and poly(acrylic acid). This material does not resolve in the body and is used in the form of a gauze for surgical operations to stop parenchymatous and capillary bleeding as well as bleeding from small blood vessels in the wound. In contrast to monocarboxyl cellulose, this new hemostatic is resistant to the action of water at high temperatures, and can therefore be sterilized in an autoclave.

The cellulose-based biologically active materials that have been recently produced also include radioactive β-particle applicators intended for the treatment of skin diseases (40). These materials are prepared by the addition of ions of radioactive isotopes to the ionogens of fabrics made from viscose filaments modified by grafting onto them poly(acrylic acid) or poly(methylvinylpyridine). The addition of the radioactive ions to the functional groups of the graft copolymers provides for uniform irradiation of the entire surface of the applicator, thereby increasing its effectiveness.

References

1. T. A. Mal'tseva, A. D. Virnik, S. A. Dmitriev, G. V. Shcheglova, G. D. Pestereva, L. V. Timofeyev, Z. A. Rogovin, and V. I. Vashkov, *Zh. Mikrobiol. Epidemiol. Immunol.* No. 3, 101 (1967).
2. A. D. Virnik, G. V. Shcheglova, Z. A. Rogovin, and V. I. Vashkov, *Zh. Mikrobiol. Epidemiol. Immunol.*, No. 5, 110 (1967).
3. D. D. Gagliardi, *Am. Dyestuff Rept.*, **51**, 31 (1962).
4. D. D. Gagliardi and W. J. Jutras, Jr., U.S. Patent 2,985,541 (1961).
5. T. A. Mal'tseva, A. D. Virnik, Z. A. Rogovin, G. V. Shcheglova, and V. I. Vashkov, *Tekstil'naya Prom.*, No 4, 15 (1965).
6. N. S. Plotkina and A. D. Virnik, *Antibiotiki*, No. 3, 245 (1967).
7. Z. A. Rogovin, V. I. Vashkov, A. D. Virnik, Yu. S. Kozlova, T. A. Mal'tseva, A. A. Pogadaeva, G. V. Shcheglova, and E. N. Nikiforova, Belgian Patent 681,379 (1966); French Patent 1,499,358 (1967).
8. Z. A. Rogovin, V. I. Vashkov, N. A. Shluger, A. D. Virnik, G. V. Shcheglova, T. A. Mal'tseva, and A. A. Menzheritsky, USSR Author Certificate 176,363; *Byul. Izobr.*, (1965) No. 22.
9. V. I. Nepochatykh, G. G. Finger, E. M. Mogilevsky, and Z. A. Rogovin, *Khim. Volokna*, No. 2, 49 (1966).
10. D. L. Snezhko, A. D. Virnik, G. D. Pestereva, and Z. A. Rogovin, *Izv. Vyssh. Ucheb. Zaved., Tekhnol. Tekst. Prom.*, No. 3, 162 (1968).
11. T. A. Mal'tseva, A. D. Virnik, G. D. Pestereva, and Z. A. Rogovin, *Izv. Vyssh. Ucheb. Zaved., Tekhnol. Tekst. Prom.*, No. 4, 92 (1966).
12. Yu. V. Artemova, A. D. Virnik, N. N. Zemlyansky, and Z. A. Rogovin, *Vysokomol. Soedin.*, **9B**, 797 (1967).
13. E. M. Akovbyan, L. S. Gal'braikh, and Z. A. Rogovin, *Vysokomol. Soedin., Tsellyulozy i ee Proizvodmye, Sb. Statei*, 107 (1963).

14. S. T. Shaposhnikova, O. K. Ioanidis, B. I. Aikhodzhayev, and Yu. L. Pogosov, *Vysokomol. Soedin.*, 7, 1129 (1965).

15. Yu. L. Pogosov, B. I. Aikhodzhayev, S. T. Shaposhnikova, R. P. Svistunova, and I. V. Rakhimova, USSR Author Certificates 167,853 [ref. *Zh. Khim.*, **9C**, 935] (1966); 167,854 [ref. *Zh. Khim.*, **7C**, 357]; (1966) 173,740 [ref. *Zh. Khim.*, **15C**, 345] (1966).

16. S. T. Shaposhnikova, B. I. Aikhodzhayev, and Yu. L. Pogosov, *Vysokomol. Soedin.*, **7**, 1314 (1965).

17. E. F. Sharkova, A. D. Virnik, and G. D. Pestereva, *Zh. Prikl. Khim.*, **40**, 1864 (1967).

18. N. E. Gefen, V. M. Shustikov, N. A. Shluger, A. D. Virnik, A. I. Menzheritsky, T. A. Mal'tseva, and D. L. Snezhko, *Voyenno-Meditsinsky Zh.*, No. 8, 80 (1965).

19. K. M. Smirnova, G. D. Pestereva, E. S. Bylinkina, V. I. Nepochatykh, Z. A. Rogovin, A. G. Nesterenko, V. N. Panchinsky, and T. M. Khorbut, *Khimiko-Farm. Zh.*, No. 12, 44 (1967).

20. N. F. Sokolova, E. I. Molozhavaya, N. F. Leontyeva, A. P. Volkova, N. R. Grin', A. D. Virnik, T. A. Mal'tseva, and Z. A. Rogovin, *Tr. Tsent. Issled. Dezinfekt. Inst.*, **18**, Part 1, 14 (1967).

21. J. N. Silmann and E. Katchalski, *Ann. Rev. Biochem.*, **35**, Part 2, 873 (1966).

22. A. E. Gurvich, *Biokhimiya*, **22**, 1026 (1957).

23. A. E. Gurvich, R. B. Kapner, and R. S. Nezlin, *Biokhimiya*, **24**, 144 (1959).

24. A. E. Gurvich, O. B. Kuzovleva, and A. E. Tumanova, *Biokhimiya*, **27**, 246 (1962).

25. D. H. Campbell, E. Luescher, and L. S. Lerman, *Proc. Natl. Acad. Sci. U.S.*, **37**, 357 (1951).

26. R. S. Nezlin, *Biokhimiya*, **24**, 486 (1959).

27. T. Webb, and C. Lapresle, *Biochem. J.*, **91**, 24 (1964).

28. A. Malley and D. H. Campbell, *J. Am. Chem. Soc.*, **85**, 487 (1963).

29. A. E. Gurvich, O. B. Kuzovleva, and A. E. Tumanova, *Biokhimiya*, **26**, 934 (1961).

30. O. B. Kuzovleva, A. E. Gurvich, Z. A. Rogovin, and A. D. Virnik, *Vopr. Meditsin. Khim.*, **12**, 106 (1966).

31. M. A. Mitz, and L. Y. Summaria, *Nature*, **189**, 576 (1961).

32. M. Gilly, C. Maney, W. Hornby, and E. M. Crook, *Biochem. J.*, **95**, 45 (1965).

33. C. J. Epstein and C. B. Anfinsen, *J. Biol. Chem.*, **237**, 2175 (1962).

34. H. H. Weetall and N. Wiliky, *Nature*, **204**, 896 (1964).

35. N. Wiliky, H. H. Weetall, R. V. Gilden, and D. H. Campbell, *Immunochemistry*, **1**, 299 (1964).

36. I. N. Masci and W. L. George, U.S. Patent 2,772,999 (1956).

37. I. N. Masci, W. H. Aston, and J. Wiland, U.S. Patent 2,773,000 (1956).

38. D. F. Smith, U.S. Patent 2,914,444 (1959).

39. Z. A. Rogovin, A. D. Virnik, T. T. Daurova, and M. A. Penenzhik, USSR Author Certificates 226,793 (1968); 218,871 (1968); 209,439 (1967); French Patent 1,489,205 (1967).

40. A. V. Braitsev, A. D. Virnik, E. T. Sal'nikov, and Z. A. Rogovin, *Meditsin. Radiol.*, No. 8, 25 (1967).

K. DIMENSIONAL STABILIZATION

Harold Tarkow

Traditional cellulosic materials show varying degrees of sensitivity to water. A cotton shirt may suffer from poor launderability, a hemp rope may shorten on wetting, a paper bag loses its strength when wetted, a window may swell and stick in humid weather, the handle of a striking tool may loosen, and the finish on wood siding may deteriorate. Motivation for correcting these deficiencies results from the increasing use of competitive materials, such as synthetic fabrics, nylon rope, aluminum siding, vinyl-type flooring, and others. From a natural resource viewpoint, added motivation is the availability and the abundance of ever-renewable, cellulose-containing plants. The cultivated land area in the world devoted to growing fibers, such as cotton, jute, and hemp, is about 100 million acres; 10 billion acres are classified as forest land (1).

The elementary botanical units, or cells, in seed hairs are independent of each other. In bast and woody tissues they are bonded to each other with lignin, which constitutes 20–30% of the weight of wood; it is considerably less in bast tissue. A portion of the lignin permeates the layered cell walls (2). In this discussion the layered and permeating components of the cell walls are grouped together and termed "cell-wall substance."

Cell-wall substance develops in an aqueous environment; the solid content may be as low as 30–40%. The first step in processing cellulosic materials is to remove most of the water. With cotton, this occurs naturally in the standing plant when the boll opens. Loss of moisture is rapid because of the high external surface area of the separated fibers. During the evaporation, there are appreciable tensile forces resulting from surface tension in the water held in the coarse capillaries. Initially the rigidity of the cell-wall substance is relatively low. As a consequence, on drying, the entire cell collapses or is drawn into the lumen. The fibers distort and twist around their axes. These dimensional instabilities, however, are not serious, because the fibers are subjected to considerable processing before they are converted to textile materials.

A serious dimensional instability resulting from the manner of processing is found with textiles in use. During the spinning and manufacture of cloth, tension is continuously applied. The resulting strains are maintained mechanically by friction or, at the molecular level, by hydrogen bonding. During laundering, especially at elevated temperatures, some of these strains are recovered as a "shrinkage," to the great annoyance of the user. The strains can be partly restored by careful steam-ironing, which reestablishes a portion of the original network of strains. Thus the term "shrinkage," commonly used to describe the undesirable consequences of laundering textiles, has little to do with the normal reversible swelling or shrinking of cell-wall substance in the hygroscopic range below 100% relative humidity (RH). Because of these processing-induced strains, a cotton flannel may shrink more than 12% and a cotton–rayon blanket more than 20% on laundering. These instabilities, of course, are serious.

Learning how to finish textiles with resins (3) is perhaps the most important advance in textile processing. The improved launderability, however, is actually the result of a complex combination of chemical and physical effects, including those of resin treatment and the construction of the fabric (twist, yarn number, weave pattern, and others).

The loss of moisture from a standing tree is prevented or minimized by the thick sheath of bark and by the compact condition of the fibers. Even after felling, loss of moisture is extremely slow, and unless other steps are taken, decay may set in because of the high moisture content. Other steps to hasten reduction in moisture include debarking, converting to boards, and drying naturally (air-drying) or in a forced manner (kiln-drying). Forces resulting from surface tension are again operative, and unless special precautions are taken, especially during kiln-drying (when initial cell-wall rigidities are low), serious distortions may occur even at high moisture contents (Figure 1).

Fig. 1. Distortion from improper drying of green wood.

The distortions that drying causes in wood are serious, for unlike the initial instability of the cotton fiber, which is corrected in subsequent processing, distortions in wood are removed or corrected only with great difficulty and expense (4). The special precautions involve a programmed control of relative humidity and dry-bulb temperature during drying to maximize cell-wall rigidity and minimize tensile forces within the water entrained in the capillaries or lumina. The rheological properties of wood are very important here. Programmed drying schedules are established empirically for different species and even for different board geometries or sizes. Kiln-drying has become a highly developed art.

Dimensional changes are encountered during consumer use, in the hygroscopic range; that is, in the absence of free water. These changes are generally reversible. In Table 1 the equilibrium moisture contents of four cellulosic materials are listed. At relative humidities of between 30 and 90%, cotton adsorbs 8.5% moisture; wood adsorbs 14%.

Because the adsorbed moisture is held within the cell walls and not in the lumen, the cell-wall substance of cotton swells about 11% in volume, and that of wood swells about 20% (or shrinks about 20% between 90 and 30% RH).

The extent to which this change in internal volume is transmitted to the external bulk volume varies among materials, depending on the degree to which the swelling is accommodated by internal swelling into the lumen and, in the case of textiles, into the interfiber regions. With cotton textiles, the external reversible swelling and shrinking are not serious and seldom are recognized, in contrast to the case of the shrinkage that results from laundering.

TABLE 1
Equilibrium Moisture Contents of Cellulose-Containing Materials at 25°C

	Moisture content (%) at relative humidity of			
Material	30%	65%	80%	90%
Cotton	3.5	6	9	12
Rayon	7.5	12	16	24
Wood	6	11	13	20
Paper	5	10	12	17

1. Dimensional Changes in Wood

The transmission of changes in cell-wall dimensions to the bulk dimensions of wood is highly efficient. A simple demonstration is to equilibrate sections of wood to increasing relative humidities and to measure the new bulk volume and weight at each relative humidity. The ratio of volumetric swelling to the volume of adsorbed moisture is computed (the adsorbed moisture is assumed to have a density of 1.0 g/ml, which, above 30% RH, is reasonable). If the adsorbed moisture merely adds its volume to that of the wood, with no change in lumen diameter, the ratio should be unity. If considerable internal swelling occurs (decrease in lumen diameter), the ratio should be less than unity. If the ratio is greater than unity, lumen diameters increase as the wood swells. Ratios for this are listed in Table 2. With the exception of balsa, where considerable changes in lumen diameter occur, the ratio for most species is roughly unity. The transmission of volumetric swelling of the cell-wall substance to the bulk volume of wood is highly efficient.

TABLE 2
Ratio of Absolute Change in Volume of Wood to Volume of Moisture Gained or Lost

	Change in volume (cm^3/cm^3)	
Species	On swelling (30–90% RH)	On shrinking (90–30% RH)
Sitka spruce	1.20	1.25
Hickory	1.0	1.0
Maple	0.9	0.9
Butternut	0.8	0.7
Balsa	2.9	2.9

Even with a swelling ratio of unity, the swelling of wood is not fully described, because the ratio does not convey the magnitude of the volumetric swelling (percentage of swelling). This magnitude is determined by the wood density of the species. This is exemplified by Sitka spruce and sugar maple, with densities of 0.4 and 0.7 g/ml, respectively. Each gains about 14% moisture between 30 and 90% RH. Yet because of the different densities, the volumetric swelling of the maple is about 10%, and that of the spruce about 5.5%. For similar intervals of relative humidity the swelling or the shrinking of wood increases with increasing density of the wood. Thus some judgment in the selection of a species can be made where dimensional change must be kept minimal.

Wood is a highly *anisotropic* material. As a first approximation, swelling in the grain, or longitudinal, direction is negligible. There are, however, conditions under which this can be serious. Volumetric swelling, then, is the consequence of a component in the radial direction (at right angles to the grain and to the annual growth rings), and of a component in the tangential direction (at right angles to the grain and parallel to the growth rings). Here, too, a marked anisotropy exists. The tangential swelling is always greater. The ratio of tangential swelling to radial swelling for Sitka spruce is shown in Table 3. Because it is anisotropic, wood changes its shape as it swells or shrinks. The changes in shape when wood is air-dried from a condition of high moisture content are shown in Figure 2.

TABLE 3
Swelling of Sitka Spruce Showing Anisotropy

Relative humidity interval (%)	Tangential swelling (%)	Radial swelling (%)	Ratio: tangential to radial
30–65	1.62	0.77	2.1
30–80	2.75	1.31	2.1
30–90,	4.15	1.91	2.1
65–90	2.64	1.18	2.2
80–90	1.43	0.59	2.4
80–97	3.55	1.14	3.1
90–97	2.16	0.55	3.9
97–100	0.10	–	74.0

Even though reversible, changes in dimension and shape can be serious, especially where close control of dimensions is necessary. Hence there has always been a strong interest in improving the dimensional stability of wood in its response to changes in relative humidity.

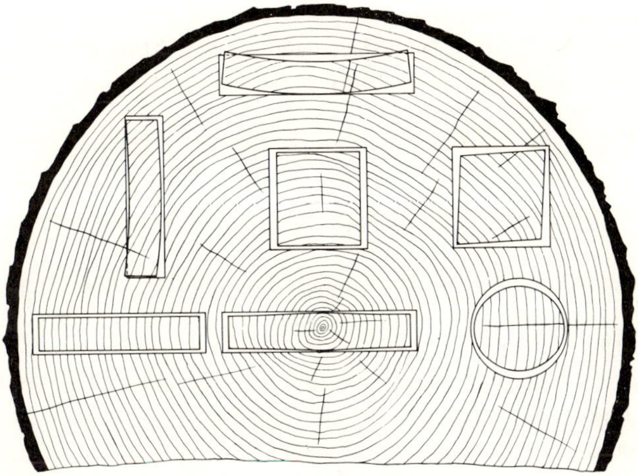

Fig. 2. Wood anisotropy is shown by the change in shape as green wood dries.

2. Improving the Dimensional Stability of Wood

Four methods are suggested for improving the dimensional stability of wood. These are discussed below.

(a) USING WOOD PROPERLY

Ideally wood for interior use at a given geographical location should be dried, before installation, to moisture contents between the expected seasonal extremes. Although the moisture content of such correctly installed material will fluctuate somewhat around some characteristic average throughout the seasons, the pronounced initial change, which otherwise would occur as the moisture content drifted toward the characteristic average, is eliminated. Unfortunately the benefits of this careful seasoning are not widely appreciated.

(b) BREAKING DOWN AND RECONSTITUTING WOOD

There are a number of procedures whereby, although the hygroscopicity of the wood is unaltered, the extent of internally generated swelling transmitted to the external dimensions is reduced. Perhaps the oldest and most successful procedure is that of converting wood to plywood (5). Swelling does occur, but because of the lamellar construction, it is forced to occur primarily internally into the abundant internal void volume. Thus for 30 species the average swelling across the grain of the bulk wood at relative humidities between 30 and 90% was 4.1%. When the wood was converted to plywood (three plies, 1/16-in. veneer), the average swelling across the grain of the face plies was only 0.18%.

Although the external swelling of plywood is small, under certain conditions it may still be accompanied by surface deterioration. This occurs if the plywood is subjected to cyclic wetting and drying. Because of the sharp moisture-content gradients, stresses are very high at and near the surface.

(c) RESTRICTING THE RATE OF MOISTURE ACCESS BY USING FINISHES

Under certain conditions the simplest method for controlling the rate of moisture exchange in wood is by reducing the diffusion rate of moisture by applying finishes to the wood (Figure 3). Dimensional changes do occur, but so slowly that the buildup of serious stresses is prevented. The beneficial effect remains as long as the finish retains its integrity. It should be appreciated that, although surface deterioration is minimized, the normal equilibrium reversible swelling is not altered. Siding and plywood are protected in this manner.

Films prepared from poly(vinyl fluoride) have excellent weathering

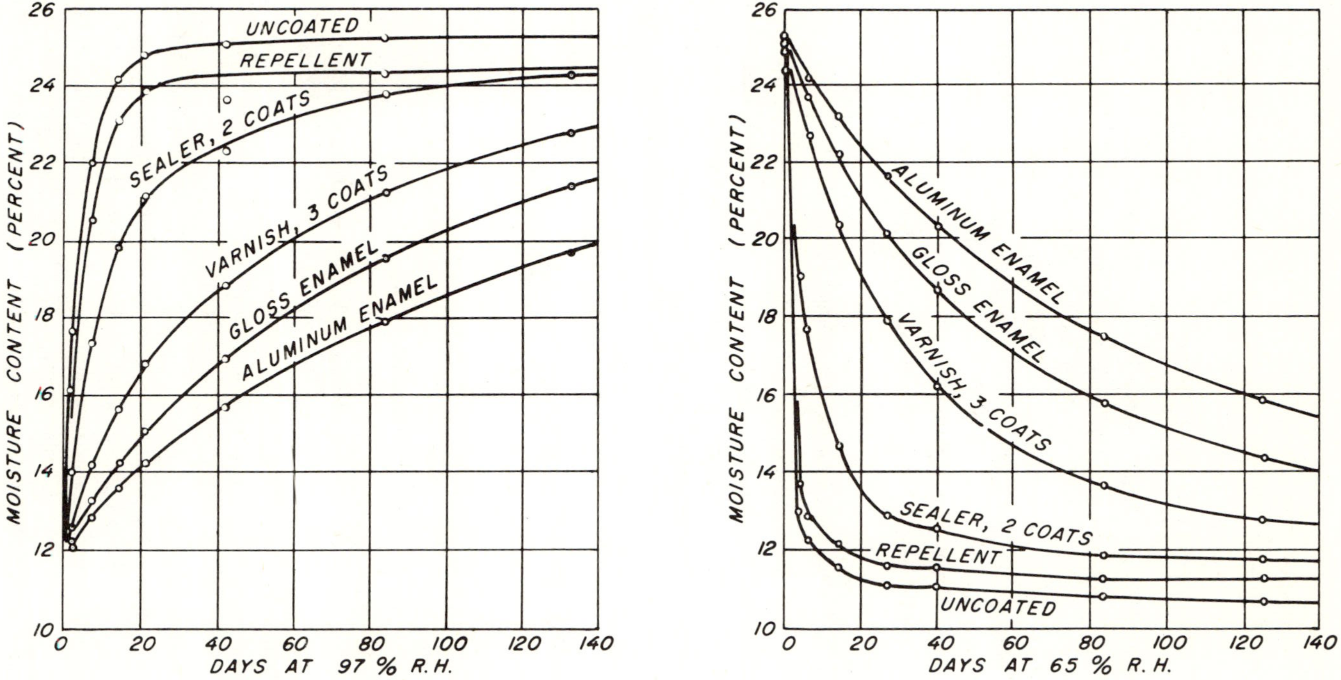

Fig. 3. Effect of various finishes on the rate of gain or loss of moisture from wood: *left*, wood initially at 30% RH; *right*, wood initially at 97% RH.

resistance. Procedures have been developed for bonding such films, suitably pigmented with opaque fillers, to siding and plywood. Several commercial products are being marketed.

(d) IMPREGNATING WOOD WITH CERTAIN CHEMICALS

Sixty to seventy percent of the volume of air-dried wood of most commercial species is that of the optically visible cell lumina approximately 1–50 μ in diameter. The remaining 30–40% is that of the cell-wall substance (6). Access of moisture to the substance is by continuous diffusion through the substance and by diffusion through the complex array of air-filled capillaries. According to Stamm (6), more than 90% of the diffusion constant of moisture in wood is accounted for by diffusion in the lumina. This factoring suggests two general procedures for controlling the response of the wood substance to variations in relative humidity.

(1) Blocking the Air-Filled Lumina

By removing this important pathway for moisture movement, the rate of access of moisture to the cell-wall substance is markedly reduced and swelling rates are very low. Typical procedures consist of impregnating wood under pressure with certain vinyl-type monomers and polymerizing with radiation (7) or heat and catalysts (8) (see Sections G and H in Chapter XVII). A serious limitation is that by filling the void volume of wood of average density with resin the final density may be doubled. This may be tolerable where the main objective is to increase the mechanical properties of wood (density-dependent) rather than to increase the apparent dimensional stability. These procedures are still in the developmental stage.

(2) Modifying Wood Substance

Acetylation of hardwoods to 20% and softwoods to 25% will reduce the equilibrium swelling tendency by 70% (9). However, unlike the acetylation of cotton, in which each fiber is immediately accessible to the reactants, the acetylation of wood is difficult because the compact nature of the fibers makes for poor accessibility. Theoretically, wood substance can be modified with other chemical reactants; however, the same limitation of poor accessibility would exist.

The hygroscopicity of wood tissues is caused by a specific interaction with moisture as well as by a partial solution or dispersion of polymer chains or swelling units within the cell-wall substance. Because of the natural crosslink content (composed of regions of strong and regular hydrogen bonding and even primary valence bonding), the maximum swelling that the plant tissue can undergo in water is roughly fixed by the volume in the original green condition.

In the terminology of the colloid chemist, plant substance is an example of a limited-swelling gel (10). In the terminology of modern polymer chemistry, the configuration assumed by the network system with increasing relative humidity approaches that developed in the original green condition.

Treatment with certain chemicals, such as water-soluble phenolic resins, leaves the wood in a *permanently swollen* condition and preempts the swelling that normally occurs in water (6). Thus when a specimen of wood is treated with a water-soluble phenolic resin, dried, and cured, the air-dry dimensions become permanently increased. When exposed to liquid water, the treated sample swells to about the same final dimension as an untreated control sample. From the air-dry condition, this may amount to about 7% for the control but only to about 2.5% for the specimen properly treated with 30% phenolic resin. The reduction in swelling and shrinking is 65%.

Treatments must be made with aqueous solutions of low-molecular-weight materials to permit easy penetration into the water-swollen cell-wall substance. If the penetrant is a thermosetting resin, heat treatment converts it to a water-insensitive polymer that produces the permanent swelling or bulking action in the cell wall. Although dimensional stability can be imparted to wood by resin treatments, large-scale acceptance has not been achieved because wood, though porous, is relatively impermeable. Limited application is being made in the manufacture of automobile die models and pattern stock—where high dimensional stability is necessary and costs are secondary (11).

These wood-modifying treatments involve simply permeating the cell-wall substance with low-molecular-weight materials. Ideally none of the material remains in the lumen; yet, for good stability, about 30% chemical is still necessary. With phenolic resins, the solute is converted by heat to a water-insensitive material after treatment.

Other materials, such as urea and poly(oxyethylene) glycols, have been proposed. Since these are water soluble, outdoor use is precluded. Furthermore, because they are humectants, the treated wood feels moist to the touch at high relative humidities. Under certain use conditions, such as indoors, prolonged high relative humidities are not encountered, and the humectant properties of these water-sensitive materials within the cell walls are not serious.

The purpose, then, of stabilizing the dimensions of wood if high relative humidities will not be encountered might be questioned. The benefit derived from this type of treatment is that it prevents the shrinkage that otherwise would occur when green wood is dried to equilibrium with some typical relative humidity, for example, 50%. For these purposes a nonvolatile, water-soluble, relatively-low-molecular-weight material is necessary. Poly(oxyethylene) glycols have these properties (6); however, if the molecular weight is too high, penetration into the cell walls cannot occur (12). A good compromise is poly(oxyethylene) glycol-1000 (PEG-1000). Green wood properly treated with

Fig. 4. Effect of poly(oxyethylene) glycol-1000 treatment on reducing surface checking on a yellow-pine board that had been submerged for 250 years in Lake George, New York. *Top*, section in original waterlogged condition; *center*, same section after treating with 50% poly(oxyethylene) glycol-1000 and air-drying; *bottom*, adjacent section (untreated) air-dried from waterlogged condition.

aqueous solutions of PEG-1000 (13) does not shrink as it dries. Two of the uses that can be made of this treating procedure are discussed below.

Carving Green Wood. Green wood is seldom carved because of shrinkage during subsequent air-drying. Advantage can be taken of the ease of carving green wood by treating the green carving with aqueous (30–50%) PEG-1000 (14). Treating times are long (weeks to months), but the treated carving will dry with no shrinkage or distortion.

Preserving Old Waterlogged Wood. The preservation of waterlogged wood from archeological sites has long been a problem. The wood generally develops cracks and flakes while drying and undergoes severe changes in shape and dimension. These can be avoided by immersing the waterlogged archeological wood object in aqueous PEG-1000. After having been dried, the specimen retains the dimensions, shape, and appearance it had when excavated (15) (Fig. 4). This procedure is being used to preserve whole boats brought up from the bottom of the sea after centuries of submersion (16). Treating periods are long because of the compact nature of the wood fibers; hence the procedure is used only for specific cases. The treatment with PEG-1000 is not practical for large-scale applications.

This discussion has considered the problem of dimensional instability encountered primarily with wood. Wood products are often used in sheet or flat form in which dimensional stability is important. Even small percentage changes in dimension may impose a relatively large absolute dimensional change. No one comprehensive procedure is available for correcting this problem. However, the procedures that have been presented can be used successfully in specific situations.

See also Section D in Chapter XVII and Section D in Chapter XIX for discussions of other aspects of dimensional stabilization.

References

1. H. Brown, *The Challenge of Man's Future,* Viking, New York, 1956.
2. I. B. Sachs, I. T. Clark, and J. C. Pew, *J. Polymer Sci.,* [C], No. 2, 203 (1963).
3. J. E. Lynn and J. J. Press, *Advances in Textile Processing,* Interscience, New York, 1961.
4. *Dry Kiln Operator's Manual,* Agricultural Handbook No. 188, U.S. Government Printing Office, Washington, DC., 1961.
5. *Wood Handbook No. 72,* U.S. Government Printing Office, Washington, D.C., 1955.
6. A. J. Stamm, *Wood and Cellulose Science,* Ronald Press, New York, 1964.
7. W. E. Loos, R. E. Walters, and J. A. Kent, *Forest Prod. J.,* 17 (5), 40 (1967).
8. J. F. Siau, J. A. Meyer, and C. Skaar, *Forest Prod. J.,* **15** (4), 162 (1965).
9. H. Tarkow, A. J. Stamm, and E. C. O. Erickson, U.S. of Agriculture, Forest Service, FPL Rep. No. 1593, Forest Products Laboratory, Madison, Wis., 1960.

10. H. Tarkow and W. C. Feist, *Tappi,* **5,** 80 (1968).
11. R. M. Seborg and A. E. Vallier, *Forest Prod. J.,* 4 (5), 305 (1954).
12. H. Tarkow, W. C. Feist, and C. F. Southerland, *Forest Prod. J.,* **16** (10), 61 (1966).
13. A. J. Stamm, *Forest Prod. J.,* **6** (5), 201 (1956).
14. H. L. Mitchell and E. S. Iversen, *Forest Prod. J.,* **11** (1), 6 (1961).
15. R. M. Seborg and R. B. Inverarity, *Science,* **136,** 649 (1962).
16. R. M. Seborg and R. B. Inverarity, *Studies in Conservation,* 7 (4), 111 (1962).

L. PULPING PROCESSES

Norman S. Thompson

In the last decade, since the publication of Holzer's review on pulping technology in Part II of this work (1), important technological advances have occurred in the pulp and paper industry. The reader interested in a detailed study of these changes is referred to the large, sweeping survey by Rydholm (2). This section does not cover the basic aspects of pulping, such as definitions, terminology, and process descriptions, already dealt with by Holzer (1), but attempts to describe some of the recent technological advances and the underlying concepts that are of importance to pulping technology today.

The development of superior bleaching processes alluded to in the review by Hatch (3) and their acceptance by the pulp and paper industry have resulted in the use of kraft pulps for purposes not considered possible a decade and a half ago. Until these mild bleaching sequences were available, use of the generally superior kraft fiber was limited to products in which color was not a major consideration. The demand for kraft pulp brought about by these new products has resulted in the construction of many new kraft mills, and the excess production has encouraged the expansion of kraft pulps into areas previously dominated by sulfite pulps.

As will be seen, technological developments have made it possible for a greater variety of pulps to be prepared by the sulfite process, but the technologically simpler kraft process (4), has been more adaptable to the desirable continuous-pulping techniques (5).

1. Classification of Pulping Processes

Pulping may be considered to involve the steps of fiber separation, removal of unnecessary and/or undesirable components, and the tailoring of the pulp to achieve desired chemical and physical properties. In actual practice commercial pulping processes do not separate these stages from one another and, in many instances, do not even complete them. Other manipulations during the bleaching and refining stages complete the pulping process as it is defined here. Attempts to control the physical properties of pulp fibers are limited to ranges of properties dictated by the pulping process employed.

In general, pulping processes may be considered to be chiefly mechanical or chiefly chemical processes but, as shown in Table 1, no clear-cut distinction occurs between them. It is thought, for example, that chemical reactions developed by frictional forces in the groundwood and refiner wood processes aid fiberization, whereas mechanical action in the form of a blow or of refiner action is very helpful in producing pulp by chemical means. Great advances have been made in understanding and in improving the various mechanical and chemimechanical pulping processes since Holzer's review, and have been described briefly by Tomlinson (6) as well as Rydholm (2).

TABLE 1
Classification of Pulping Processes

Type	Yield (%)	Process
Multistage mechanical	93–98	Grinding of chips
Mechanical	93–98	Grinding of logs
Chemimechanical	80–90	Grinding of chemically pretreated logs
Semichemical	60–90	Chemical pretreatments, followed by attrition
Chemical	40–60	Digestion of chips
Multistage chemical	30–60	Digestion of chemically pretreated chips

Chemical pulping processes may be subclassified according to the type of degradation that is employed to separate the fibers. The retting of flax or other plants to obtain bast fibers is one of the oldest technologies related to pulping and might be considered a biological process. Most chemical pulping procedures today may be considered to be variations of *hydrolytic* processes, whereas most bleaching procedures (which may also be used for pulping under certain conditions) are variations of *oxidative* and *reductive* processes. Some attempts have been made to develop hydrotropic or solvolytic pulping processes, but these have not achieved economic feasibility. Commercial chemical pulping processes may be considered to be mixtures of these general types of pulping classifications; some, such as the acid sulfite process, are predominantly hydrolytic, whereas others, such as the nitric acid process, are mixtures of both hydrolytic and oxidative actions. Most commercial hydrolytic pulping processes may be classified according to the pH of their cooking liquors. Peckham and Van Drunen (7) observed that the pulps prepared by acidic processes are more easily beaten or fibrillated than pulps produced by alkaline processes.

2. Chemistry of Wood

Since the early 1950s the knowledge of the chemistry of wood has expanded greatly. According to Timell (8), the gross structures of all major hemicellulose components are known and from this knowledge many generalizations concerning the amount and type of carbohydrate components in our economically important wood species may be safely made. Electron microscopy, infrared studies, biological and botanical studies coupled with advanced techniques in carbohydrate chemistry have increased our knowledge of the fine structure of cellulose and of the location and synthesis of hemicellulose and lignin themselves within the tracheid.

(a) HEMICELLULOSE

The advent of various forms of chromatography and the development of new processes for the separation of polysaccharides has been the key to our knowledge of hemicelluloses. These researches, described in review articles by Timell and others (8–11) indicate that conifers and deciduous woods contain characteristic amounts of unique polysaccharides. Unlike cellulose, the hemicelluloses are copolymers composed of varying amounts of sugar units. Although it has been established that cellulose may occur in a crystalline state in wood, less is known of the order of hemicelluloses. By using x-ray crystallography techniques, Marchessault, Settineri, and Winter (12) have demonstrated that the "xylan" of certain hardwoods is not crystalline in the native state but that it will crystallize in a holocellulose (or lignin-free residue of wood) when subjected to the action of very dilute alkali. It is not possible to observe the crystallization of the mannose-containing polymers in holocellulose because their x-ray diffraction pattern cannot be separated from that of cellulose (13). Unlike in the case of cellulose, it is not possible to claim with certainty that any particular feature of a fiber is composed only of hemicellulose, although it is known that certain hemicelluloses are concentrated within a particular cell-wall layer.

Every hardwood investigated so far has been found to contain a "*xylan*," or *O*-acetyl-4-*O*-methylglucuronoxylan (8), consisting of a series of (1→4)-*β*-linked D-xylopyranose residues, some of which carry a 4-*O*-methylglucuronopyranose residue attached by an α-glycosidic bond to the second hydroxyl group of some of the xylose units. The amount of uronic acid in the polymer varies from wood to wood, and there is nothing to indicate whether the acid is attached in a random or a complex pattern arrangement along the "xylan" chain. The "xylan" present in the sapwood of deciduous trees has been found to contain about 10–13% acetyl groups, linked chiefly to the third hydroxyl group (although some linkage to the second hydroxyl group occurs) of certain of the xylose units of the polysaccharides (14). There is evidence that, in addition to these short

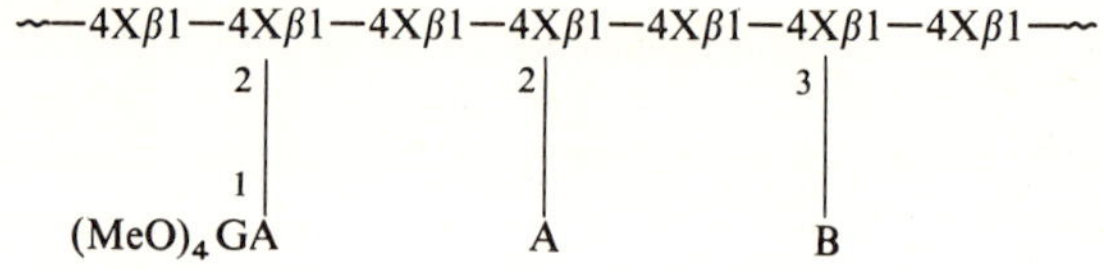

Hardwoods	Softwoods
A = *O*-acetyl group	A = α-L-arabinofuranosyl residue
B = *O*-acetyl group	B = hydroxyl group

Xβ = β-D-xylopyranosyl residue

Fig. 1. A formulation for arboreal xylans.

terminal branches of single anhydrosugar units or acetyl groups, at least two xylose-xylose branches are present in each "xylan" molecule (15). A formulation for arboreal xylans is shown in Figure 1.

The mannose-containing polysaccharides of many hardwoods have been found to consist of chains of D-glucopyranose and D-mannopyranose residues linked by (1→4)-β-glycosidic bonds to one another in an unknown pattern. The molar ratio of glucose to mannose varies from 1 : 1 to 1 : 2 for different woods, and because of experimental difficulties, it is not known whether they are acetylated. A formulation for arboreal glucomannans is shown in Figure 2.

Hardwoods contain numerous other polysaccharides, composed of uronic acid, galactose, arabinose, and rhamnose units. Recently, for example, Banerji and Thompson (16) isolated from aspenwood a polygalacturonan that is composed only of anhydrogalactopyranosyluronic acid residues and resembles degraded polymers from sunflower heads (17), fruit pectin (18), and Engelmann spruce bark (19).

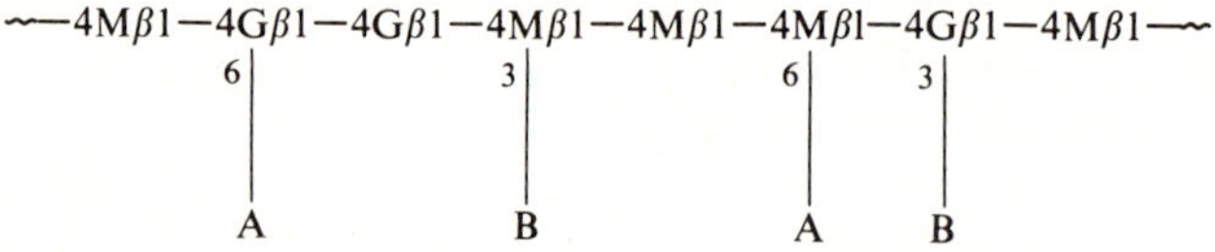

Hardwoods	Softwoods
A = hydroxyl group	A = α-D-galactopyranosyl residue
B = hydroxyl group	B = *O*-acetyl group

Gβ = β-D-glucopyranosyl
Mβ = β-D-mannopyranosyl residue

Fig. 2. A formulation for arboreal glucomannans.

The polysaccharide components of softwoods are more numerous and more complex than those of hardwoods, and evidence for the existence of many as yet unisolated polymers is available (8). In contrast to earlier observations, recent research indicates that significant quantities of xylose-containing polysaccharides may be isolated from some conifers (20, 21) by extracting the lignified wood with alkali. These polysaccharides resemble the less easily isolated "xylans" of conifers, and both differ in many important respects from the "xylan" of hardwoods. Softwood "xylans" are composed of a backbone of (1→4)-*β*-linked D-xylopyranose residues to which are attached 4-*O*-methyl-D-glucuronic acid residues in a fashion similar to that observed in hardwood "xylan." In addition, softwood "xylan" contains L-arabinofuranose residues linked by (1→3)-glycosidic bonds to certain of the D-xylopyranose residues, The polymer does not contain acetyl groups, and there is evidence that it too contains a few branch chains of anhydroxylose units (20).

The most plentiful hemicellulose component of most conifers is the family of *galactoglucomannans* (frequently referred to as "glucomannan" in the literature), which consists of a chain of D-glucopyranose and D-mannopyranose residues linked to one another by (1→4)-*β*-glycosidic bonds in an unknown pattern. Terminal D-galactopyranose residues are linked by (1→6)-*α*-glycosidic bonds to some of the D-mannopyranose residues and the D-glucopyranose residues of the main chain. Fractionation experiments have demonstrated that there can be a wide range of galactose contents (1–20%) within a given sample, but the fraction containing the least amount of galactose is the most plentiful. The galactoglucomannan component of the sapwood of conifers contains acetyl groups attached to either the second or third hydroxyl groups of either mannose of glucose residues (22–24). Since acetyl groups are frequently absent from coniferous and deciduous heartwood (9, 25), and since ester groups can migrate (26, 27), it is possible the amount and the location of the acetyl groups on naturally occurring hemicelluloses of wood may be a function of the location within the tree, the history of the sample, and the treatment used to detect the acetyl groups. The condition of the acetyl groups within wood has been related to profound changes in the physical characteristics of pulps produced by the sulfite process will be discussed later. Several investigators have discovered that it is impossible to remove ligninlike substances or lignin debris from "glucomannan" preparatives (28–31) unless degradative techniques are employed, and it is therefore postulated that there is a chemical bond between "glucomannan" and lignin.

Numerous less plentiful hemicelluloses are present in conifers, including arabinogalactans (32), glucuronoarabinogalactans (33), arabinans (33), and pectic substances (34). These polysaccharides are completely removed during the preparation of low- and moderate-yield pulps by commercial processes (35). A comparison of the polysaccharide components in wood with those of pulps was first begun in Finland (36) and has been since accomplished by a number of researchers (35, 37–40). The results indicate that, apart from cellulose, the xylose- and mannose-containing components are the only carbohydrate components of wood pulp cooked to low lignin contents.

(b) LIGNIN

The knowledge of lignin has also developed rapidly during the last decade, and much of this knowledge has been due to the biological experiments of Freudenberg (41) and the analytical studies of Adler (42). From his experiments, Freudenberg has composed a structural scheme for lignin that takes into consideration both the analytical features and the structural inferences derived from dimer and trimer fragments isolated during lignin synthesis from likely precursors (Figure 3). The recent reviews by Kratzl (43) and by Schubert (44, 45) describe these developments in lignin chemistry.

Lignification in wood takes place after cell division and differentiation of the cells into tracheids has occurred (46). Usually secondary wall formation is well developed before lignification begins first at the corners of the middle-lamella region between tracheids. Lignin then spreads throughout the whole middle-lamella region, penetrates the secondary wall and terminates in the tertiary wall layer. Numerous researchers have shown that the concentration is greatest in the

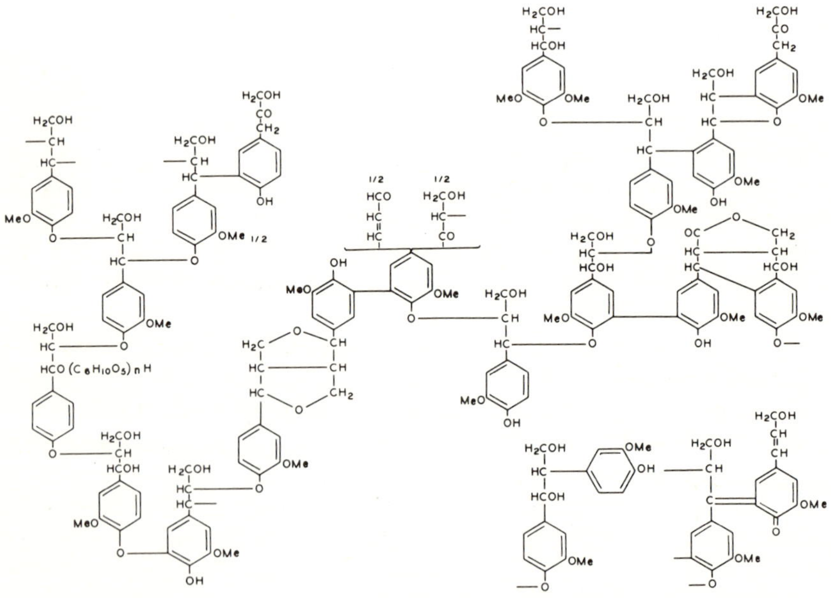

Fig. 3. Formulation for spruce lignin based on that proposed by Freudenberg (41).

middle lamella and diminishes in the secondary wall (47, 48). The amount of lignin in the secondary wall is equal to or greater than the amount in the middle lamella (49, 50). Lignin may be pictured as a complex amorphous substance permeating the mass of the wood.

3. Bonding in Wood

The chemical and physical nature of the forces that hold the fibers of wood together is of obvious importance to the pulp and paper industry and is probably different from the nature of the forces that hold paper together. Research by botanists [e.g., Wardrop (51)] has demonstrated that the bonds of the middle-lamella region between fibers or tracheids of whole wood are stronger than the bonds between the various wall layers of the tracheid. Consequently, when wood or chemically modified wood is converted into a pulp by mechanical means, breakage frequently occurs across cell-wall layers and along the boundaries between the transition lamella and the secondary wall of the fibers (52–54). As wood is delignified, breakage occurs more easily and more frequently in the middle-lamella region (54). Ultimately, if delignification is carried far enough, fiber separation occurs in most pulping processes with very little mechanical assistance at about 55–60% yield. For many types of pulp, such as high-yield semichemical pulps, it is desirable that the yield at this "fiber-liberation point" be as high as possible in order to minimize the amount of mechanical energy necessary to complete the pulping process.

The parallel between the loss of lignin and the ease of fiber separation has led to the generalization that lignin acts as the cementing substance that holds the fibers of wood together (55, 56). Evidence gathered by Dadswell and Ellis (57) indicates that both lignin and nonlignin components hold fibers together and that both of these bonding forces must be destroyed before fiber separation occurs. It has not been possible to isolate this conjectured nonlignin bonding component (58), and other researchers speculate that physical, rather than chemical, factors are involved (59, 60).

A comparison of the effect of the progressive removal of lignin from wood by neutral sulfite and acid chlorite pulping processes has been carried out by Stone and Kallmes (61). It shows that the chlorite-delignified wood is considerably stronger and stiffer than the pulp prepared by the neutral sulfite cooking process. The residual strength of partially delignified wood is destroyed by heating or by prolonged delignification. On the basis of these tests, it was speculated that carbohydrate removal during pulping is just as important as lignin removal for weakening the intermolecular and intramolecular bonds of fibers.

Other researchers have found that when the Klason lignin content of blocks or chips of wood is reduced to about 1% by mild, nonswelling oxidative processes in such a way that most of the carbohydrate components are retained, the residues possess considerable mechanical strength and the tracheids can be separated from one another only by mechanical action, heat treatments (61), or by dilute solutions of certain swelling agents (62, 63). Further investigation has shown that these bonds are destroyed by prolonged chlorite or peracetic oxidation (63–65) but not by relatively nonswelling solvents, such as divalent hydroxides and dimethyl sulfoxide (66). The bonding forces of wood are very complex, and it has been shown that the presence of multivalent ions increases the bonding strength between the thin-walled cellular components of wood, such as ray cells and epithelia (67), and may even help these structures survive the kraft cooking process (68). Additional evidence that nonlignin bonding forces are present in wood is provided by the observation that incompletely developed fibers and tracheids of the growing layer of aspen and pine (66, 69) and aspen and pine holocellulose (16, 66) may not be separated from one another by the techniques used to separate mature tracheids from one another.

4. Technological Advances

Advances in pulping technology begin in the nursery, where seeds and seedlings of economically desirable varieties of plants are selected on the basis of potential physical characteristics (70). The use of genetic selection and the development of seedlings from cultured tissues of plants (71) makes possible plant breeding for special end uses. Forest-management practices have been developed to control, maintain, and safeguard forest crops until they reach maturity (72), and new concepts in harvesting are being developed to transform logging operations into completely mechanical procedures (73). The transportation of the raw material to the pulp mill is also receiving critical attention (e.g. experimental lifting of logs from inaccessible regions with balloons and pipeline transportation of chips, with or without concomitant chemical treatments, to the mill) (74).

The range of raw materials available to the pulp industry has increased, and such wood species as jack pine, aspens, birches, southern pines, and oaks are being pulped by processes previously considered impossible. The range of new raw materials now includes not only waste slabs from sawmills but also waste sawdust (75). The latter development is possible because the design of certain continuous kraft digesters makes good liquor circulation possible and because the pulp may be used as a beneficial additive to other paper pulps and does not decrease the strength of the sheet significantly (76).

The major parameters associated with the penetration of wood by cooking liquor have been reviewed by Holzer (1) and are associated with the preparation of chips of satisfactory size. Critical examination of the action of chippers, the mechanical damage they inflict, and the effect of this damage on pulp properties have been the subjects of many investigations (77, 78). Although very thin chips have been used experimentally for many processes (79), larger scale use of these concepts has not been realized. The Guillet cutter (80) uses the least amount of energy since the plane of cutting is parallel to the fiber direction and the direction of cutting is perpendicular to the fiber direction (81). The newly developed drum chipper (82) is an improvement since it makes more uniform chips with less mechanical damage than conventional chippers.

(a) PULPING EQUIPMENT

The design and construction of reaction vessels used in pulping has undergone continuous change, and the development of entirely new engineering concepts pioneered by Richter (83) has led to the use of continuous digesters. Batch digesters have also been improved, and greater pressures are attainable with less degradation of the digester by using resistant steels composed of alloys of chronium, nickel, and molybdenum (84).

Corrosion problems limited the early use of continuous digesters to the kraft pulping processes. This equipment is available in a variety of designs. The digesters offering the greatest variations of process—as regards cooking conditions, retention time, and introduction of chemicals—are the vertical types, such as the up-flowing Impco digester (85, 86) and the down-flowing Asplund defibrater and Storabrite digesters (87), the Escher-Wyss (88), and the Kamyr (89) digesters. The latter digester is one of the most popular and consists of a presteaming stage into which chips are introduced in a uniform, volumetrically controlled stream. After steaming, the chips are transported under pressure to the top of the vertical digester, where they move from the top, through various cooking and washing zones, to the bottom, where they are discharged after cooling. The digester is very adaptable and may be modified for use by multistage processes as well as by conventional pulping processes (2, 5, 6).

Another class of digesters are the nonvertical types, which require mechanical transport of the chips through them. The Black Clawson digester is essentially a refined development of a semichemical pulper; many features of its operation have been described by Sutherland (90). The Bauer Messing-Durkee (M & D) continuous digester is an outstanding example of this class. It consists of a cylindrical reaction tube set at an angle of 45° and is divided into two chambers by a hollow midfeather partition along the length of the tube. Chips enter and leave at the upper end and are carried through the digester by flights along the midfeather (91). An advantage of this digester is that several may be connected in series to provide an opportunity to separate the stages of the multistage cooking processes. Another of its characteristics is that it permits the circulation of liquor through sawdust and has thus opened up new pulping potentials.

More recent technological advances have resulted in the use of continuous digesters for the sodium bisulfite (5, 6) and magnesium bisulfite (92–94) pulping processes. By adding a preliminary chamber or Mumin tube at the top of the Kamyr digester after the presteaming stage, it became possible to conduct continuous prehydrolyzed kraft, acid sulfite, bisulfite, and two-stage sulfite pulping digestions (95) and to apply full cooking temperatures to the head of the digester, thus avoiding the possibility of eddy currents (2). Techniques for monitoring the progress of the cooking cycle may eventually be coupled with a computer in order to obtain precise quality control, by the use of automatic techniques (96–98).

(b) RECOVERY PROCESSES

After digestion the cooking liquors from the digester and from certain washing stages are evaporated to about 48–54% solids. Additional chemical is added in the case of kraft liquors to make up losses, and the liquor is burnt to yield a smelt, which is added to water to form "green liquor" (91, 99). The green liquor from the kraft recovery process is composed chiefly of sodium carbonate and hydrogen sulfide, and is converted to "white liquor" (a mixture of sodium hydroxide and sodium sulfide) by treatment with calcium oxide.

Additional complications occur in the case of sodium sulfite recovery systems since combustion yields a mixture of sodium carbonate, sodium sulfite, and sodium sulfide. Most recovery processes convert the sulfide to sulfur dioxide and these processes differ from one another chiefly in the manner in which the oxidation is accomplished and in the method of sodium recovery (2, 99, 100). The rising cost of sulfur and the increasing concern over air and water pollution have intensified the search for a truly effective process for the recovery of chemicals from the sodium-base sulfite liquors. One possibility—the cross-recovery system—utilizes the spent liquors from two adjacent mills, one of which is a kraft mill and the other an acid sulfite or neutral sulfite mill. The objection to the system is that the sulfite-mill capacity will be limited to a production where its liquors will make up the sodium losses of the kraft mill (2). Another type of recovery system, the Silfate process, uses spent sulfite liquor to dissolve the smelt from a kraft recovery furnace (101). These liquors are used in the kraft or neutral sulfite semichemical pulping processes. Numerous alternatives to these sodium sulfite recovery systems have been investigated, including ion-exchange techniques (102) and electrodialysis processes (103), but none has become economically feasible.

The use of magnesium base for sulfite pulping has many advantages with respect to pulp quality and recovery of chemicals (6, 104, 105). When the residual liquor is evaporated and burnt in the recovery furnace, the products of combustion are sulfur dioxide and magnesium oxide. These components are recovered by reacting the liberated sulfur dioxide with an aqueous magnesia suspension prepared from the magnesium oxide collected in the dust collectors of the furnace. This process is free from serious technological problems and results in only small heat and chemical losses. Over 20% of the United States sulfite pulp is produced by this process (6).

5. Acidic Pulping Processes

(a) ACID SULFITE

The various commercial sulfite processes are based on the fact that lignin will react with sulfur dioxide (106) or solutions of its salts at different pH levels (107) to form a soluble derivative that can be removed from the wood. At low yields all fiber bonds are destroyed, but at higher yields mechanical action is necessary to complete the pulping operation. Originally calcium acid sulfite liquors were employed because of their cheapness, but the need to reduce pollution and to compete with kraft pulps resulted in the use of more versatile bases, such as sodium, ammonium, and magnesium. The sodium system is the most versatile and adaptable to the complex multistage pulping processes, although mixed sodium-calcium sequences and magnesium multistage processes are now being considered.

The study of the mechanism of reaction between sulfite liquor and lignin has closely paralleled the development of our knowledge of lignin. Early investigators in the United States (108), Canada (109, 110), and Scandinavia (111-113) attacked the problem by kinetic studies. These studies were limited in theoretical significance because of the heterogeneous nature of the reaction, the uncertainties introduced by penetration and diffusion, and by a lack of knowledge about the reactants themselves. Nevertheless, they demonstrated that the rate of delignification is proportional to the excess sulfur dioxide (110, 114) and that the rate approximately doubles for each 10°C rise in temperature. Maass and co-workers (109) speculated that the concentrations of hydronium ions and bisulfite ions were the most important chemical variables in the sulfite process. On the basis of other reactions Hägglund and Waller (115) postulated that the reaction process occurs in two stages, the first of which involves sulfonation of the lignin and the second involves hydrolysis of the sulfonated intermediate to smaller fragments. Hydrolysis was felt to be the rate-controlling step by the Swedish researchers, whereas Maass and co-workers (109) considered that both sulfonation and hydrolysis could be rate controlling. The controversy was finally settled by Rydholm and Lagergren (114) when they demonstrated that the dissolving of lignin involves the concurrent sulfonation and hydrolysis of lignin and that the rate of dissolving is proportional to the concentration of the limiting component (hydronium ion or bisulfite ion) in solution.

From these and other sulfonation studies it was concluded that lignin contains at least two types of functional group capable of reacting with sulfite cooking liquor (116–118). The more reactive configuration, called group A, can be sulfonated by neutral sulfite liquors at 135°C and reacts more quickly with phenols than with bisulfite ions at pH 1.5–2. On the other hand, this group reacts more quickly with bisulfite ions than with phenols when the pH of the liquor is greater than 3.5. Further research showed that group A is a mixture of configurations (118, 119), which could be recognized by the differences in their rate of reaction with sulfite cooking liquor. The more reactive component of group A was called group X, and it sulfonated very rapidly at all pH levels. The comparison of its behavior to sulfite liquor with those of numerous model substances (reviewed in reference 111) suggests it to be similar to benzyl alcohols and to benzyl ethers containing a free *p*-hydroxyl group. The less reactive component of group A, called group Z, sulfonates slowly in neutral sulfite liquor and is thought to contain benzyl alcohols with *p*-hydroxyl ether linkages. The least reactive lignin configuration, called group B, does not react with phenols but does react with sulfite under acidic conditions. Leopold (120) found that group B becomes reactive to sulfonation at neutral pH levels if it is first degraded by a preliminary acid treatment, and it appears that this reaction produces configurations similar to those of groups A and B.

The lignosulfonic acids produced during sulfite pulping are stronger acids than sulfurous acid. The pH of the liquor tends to drop as anions are captured by the strong sulfonic acid and sulfurous acid is liberated. The removal of excess sulfur dioxide by venting the digester increases the pH to satisfactory levels. Because the migration of the calcium ion through wood is slower than that of the bisulfite ion, sulfonic acids formed during the early stages of the cook may not become neutralized (121) until some time after their formation. If the temperature has been increased too rapidly, the newly formed sulfonic acids may catalyze the condensation of lignin into a derivative that will not dissolve on further cooking. The wood chips acquire dark cores and cannot be pulped or bleached satisfactorily.

Kleinert and Marraccini (122) have speculated that lignin becomes plasticized when heated and will coalesce in such a manner as to diminish in surface area. This reduction in surface area [which corresponds to the glass transition of isolated lignin described by Goring (123)] makes subsequent sulfonation difficult. Read and Purves (124) and Richtzenhain (125), on the other hand, have shown that lignin undergoes chemical reactions with itself to form complex aromatic nuclei when subjected to extremes of pH and temperature. The sensitivity of the calcium-base sulfite process to rapid increases in temperature is probably due to both chemical and physical transformations of lignin. It is not known whether these reactions are due to the presence of free radicals in sulfonated lignin (126) and other lignin derivatives (127).

If the temperature of the cooking cycle is not increased too rapidly, the calcium ion can diffuse into the fibers and the freshly formed lignosulfonic acid will become neutralized. Lignin condensation and coalescing reactions are thus avoided, and the lignin preserves its reactivity to pulping chemicals. The problem does not exist in the case of ammonium and sodium sulfite processes since these anions migrate very quickly into the wood. Pulping reactions are therefore very complex (128) and consist of sulfonating reactions occurring simultaneously with the hydrolysis of lignin and its derivatives into smaller, more soluble fragments (129). Lignin and lignin reaction products react with themselves to produce other less soluble aggregates, which can spoil a cooking cycle if left uncontrolled. The observation that the latter reactions are minimized by the presence of the sodium sulfonate group suggests that the reactive group A of the lignin is responsible. Commercial pulping procedures derived by empirical techniques have achieved a maximum lignin removal consistent with acceptable changes in the character of the cellulosic residue. In general the procedure involves heating chips in digesters to about 135°C in the presence of the bisulfite ion, sulfur dioxide, and water with a suitable base to achieve the desired cooking pH for a required period of time of up to several hours (130).

(b) BISULFITE

The reaction of lignin with phenolic substances in preference to a reaction with acid sulfite cooking liquor has made the pulping of extractive-rich woods, such as oaks, pines, and Douglas fir very difficult (131). Erdtman (132) found that a two-stage pulping process involving a neutral sulfite pretreatment stage followed by a terminal acid sulfite cook would produce acceptable pulps from these otherwise intractable woods. Other researchers discovered that the bisulfite cooking process carried out at pH 3–4 was also able to pulp these woods without the formation of hard, difficult-to-bleach products. Since calcium-base cooking could not be carried out at these high pH levels, the use of more soluble chemicals, such as sodium (133), ammonium (134), and magnesium (135) sulfites was mandatory. Although magnesium-base sulfite liquors may be used to cook pines, care is necessary to ensure that the divalent ion will not react with the extractives in the wood to produce pitch problems at other stages of the pulping process. In these instances sodium-base processes, such as the Arbiso (133), are preferable.

At low yields there are only minor differences between pulps produced by the acid sulfite and the bisulfite cooking processes, but at higher total yields the amount of carbohydrates is greater in the case of bisulfite pulps (4, 107). The bisulfite pulps cooked at about pH 4 are more easily converted into fibers, easier to refine for papermaking, and have properties between those of kraft and acid sulfite pulps. Husband (136) has pointed out desirable features of bisulfite semichemical pulps compared with neutral sulfite semichemical pulps.

Bisulfite pulping at pH 5.5–6.0 is difficult to control because of the formation of sodium thiosulfate as a result of side reactions of the cooking liquor with the reducing groups of carbohydrates. In this reaction the thiosulfate formed by the side reaction will break down into elemental sulfur when the bisulfate ion is depleted and the cook will be ruined. Additional complications arise because the formation of acetic acid and other wood acids will change the pH of the liquor during the cook. It is apparently impossible to produce a satisfactory pulp when the pH is held at about 6 (7, 121). Satisfactory pulps may be prepared if this process is used as the terminal stage of a two-stage process having an initial acid-sulfite-cooking stage. In general this type of two-stage pulp will have a lower hemicellulose content and greater tearing strength than acid sulfite and bisulfite pulp. Pulps that have been produced by a single-stage bisulfite process at pH 6 consume greater quantities of chemical than pH 4 cooks; they contain lignosulfonic acids that are not rendered insoluble with 72% sulfuric acid (121) and exhibit severe carbohydrate degradation (7).

(c) CARBOHYDRATE REACTIONS

The carbohydrate components of wood react with the active cooking chemical and with the aqueous medium of the cooking liquor. The technique and knowledge necessary to investigate chemical equilibrium in heated solution have only recently become available (137). Components are dissolved from the wood with little or no chemical degradation in a manner analogous to the extraction of wood with water. As the lignin incrustant is removed, more and more carbohydrate material becomes accessible to the action of the cooking liquor, and, according to the analytical data of Annergren and Rydholm (138), the components of conifer dissolved at this stage probably include pectins. arabinogalactans, arabinans, 4-*O*-methylglucuronoarabinoxylans, and galactose-rich galactoglucomannans. Pectins, arabinogalactans of uncertain structure, as well as 4-*O*-methylglucuronoxylans are probably removed from hardwoods under these conditions.

It is possible, as the delignification proceeds, that cellulosic surfaces will become available for sorption, and many of the hemicelluloses (such as "xylans" and "glucomannans") that have dissolved will tend to be sorbed back onto the fibers (139). Other polymers, such as arabinogalactans, are not readily sorbed onto cellulose, but chemical degradations of polysaccharides during the pulping process may alter them so that they can be sorbed.

The sulfite ion itself does not exert a profound influence on the polysaccharide components of wood. It is possible that bisulfite addition to carbonyl groups to form hydroxysulfonates may occur (140), and analysis of the components of sulfite liquor indicates that oxidation of aldoses to aldonic acids (141, 142), together with the reduction of sulfite to thiosulfite, does occur (143). Other rearrangement products include sugar sulfonates and, in the case of xylose but not hexoses, the formation of α,β-dihydroxy-γ-sulfovaleric acid (144). Some evidence for the decomposition of sugars and the destruction of carbon–carbon bonds exists since methylglyoxal, furfural, formaldehyde, and unidentified trioses have been detected in sulfite liquor (145). Such products as methanol (from the deesterification of pectin) and acetic acid are also readily detected (146, 147).

The most important technological reactions that occur during sulfite pulping are those associated with the cleavage of glycosidic bonds. These reactions, brought about by the hydronium ion in the system, can result in the formation of low-molecular-weight products and in the specific removal of labile components from certain polysaccharides (35, 37, 148). They are dependent on time, temperature, and acidity of the reaction (149, 150). The mechanisms for the hydrolysis of different types of glycosidic bonds by the hydronium ion under idealized circumstances have been studied by many investigators (151), and the degradative mechanisms proposed by Bunton et al. (152) are the basis of most theories. It is presumed that a similar mechanism will occur in wood chips at the high temperatures (~135°C) ordinarily employed for sulfite pulping. Recent advances in measuring acidity at very high temperatures by Ingruber (137, 153) will facilitate the understanding of chemical reactions under these more drastic conditions.

Glycosidic bonds differ greatly in their resistance to acid hydrolysis. The cleavage of the (1→4)-β-linked glycosidic bonds of cellulose has been studied under heterogeneous and homogeneous conditions (see Chapter XVIII-A). It has been established that the (1→4)-β-linked glycosidic bonds of wood "xylans" are less resistant to hydrolysis than the bonds in cellulose or glucomannan (154). Certain glycosidic bonds are very susceptible to degradation by the hydronium ion. Thus 4-*O*-methylglucuronoarabinoxylans may be converted into 4-*O*-methylglucuronoxylans because of the great instability of the glycosidic bond linking the L-arabinofuranosyl residues to the polysaccharide (35). Arabinans, which are composed almost entirely of L-arabinofuranosyl residues linked to various hydroxyl groups of one another, are completely hydrolyzed to L-arabinose by the action of very dilute acids (33). The (1→6)-α-linked D-galactopyranosyl residues are selectively removed from galactoglucomannans by mild acidic treatments (155), and as a result of this instability only a residual glucomannan component can be isolated from acid sulfite pulps of conifers (156, 157). Some glycosidic bonds, such as those linking uronic acids to polysaccharides, are more resistant to hydrolysis than the bonds in cellulose (158). This difference in behavior is responsible for the difficulty encountered in hydrolyzing these units without the concomitant destruction of some of the hydrolysis products (159). It is also responsible for the resistance of the ubiquitous 2-*O*-α-D-(4-*O*-methylglucopyranuronic acid)-D-xylose fragment of wood "xylans" to hydrolysis. The reasons for the resistance of this type of bond have been the subject of numerous controversies (149, 150, 160–162).

The analysis of the 4-*O*-methylglucuronoarabinoxylans isolated from western hemlock, pine, and spruce shows a ratio of one uronic acid for each four to six xylose residues (155, 163). The ratio of uronic acid to xylose residue in the polymer isolated from the sulfite pulps of these woods is different and averages about 1 : 8 (155, 163). This difference in composition suggests that in conifers there is a mixture of "xylan" polysaccharides with widely differing uronic acid contents and/or that the random hydrolysis of glycosidic bonds within the anhydroxylose chain of the polymer during cooking produces low-molecular-weight polymers of varying uronic acid content. Since experiments have shown that xylans of lower uronic acid content sorb more readily onto cellulose (164) in alkaline conditions, it is possible that the mixture present in the wood will become fractionated by preferential sorption under acidic conditions, under which the polymer is even less soluble, and the differences in the relative uronic acid content of the various "xylan" components will become apparent. This rationalization could explain the differences in uronic acid content of the "xylan" of bisulfite pulps described in the literature (165–167) and may even partly explain the preferential retention of glucomannans rather than galactoglucomannans in acid sulfite pulps.

The extent of acid hydrolysis is controlled by the time and temperature of the cooking cycle and by the acidity of the cooking liquor. Hydrolytic degradation must therefore be less for pulps prepared at pH 4 than for those prepared at pH 2 with similar times and temperatures of reaction. In actual practice the viscosity of bisulfite pulps (produced by sulfite cooking at pH 4) is less than that of the sulfite pulps (cooked at pH 2 to the same yield or lignin content), and Hartler and Sundberg (168) have speculated that this difference is due to the longer times and higher temperatures necessary to effect delignification at high pH levels.

Changes in the amount of cooking liquor can have a significant effect on the amount of carbohydrates present in sulfite pulps. Stockman (5) reports that an increase in the chemical charge (chemical-to-wood ratio) increases the yield of carbohydrate retention in bisulfite pulps in relation to lignin dissolution. Chromatographic examination of the pulps showed that the yield increase is due to the retention of "glucomannan" and, to a lesser extent, of "xylan." Although no explanation was given, it may be speculated that the increasing chemical charge will "salt out" hemicelluloses, especially when their solubilizing D-galactopyranosyl and L-arabinofuranosyl branches are removed by hydrolysis.

Softwoods are not as satisfactorily pulped by sulfite liquors of pH 6–9 (measured at room temperature) as the hardwoods because of the difficulty in removing lignin from the cellulosic secondary walls (169). Few, if any, low-yield single-stage processes are conducted at pH 5.5–7, and small-scale cooks described in the literature show that the products from softwoods are unsatisfactory for most purposes (7). Because of the high temperatures and salt concentrations involved during the cooks, it is difficult to estimate the true pH of the process under cooking condition. However, when softwood pulps are cooked by neutral sulfite processes beyond their fiberizing point, losses in viscosity and acid-labile-sugar components are observed. This degradation is not as great when higher yield semichemical pulps (which require mechanical action for fiberization) are prepared. On the other hand, when conifers are cooked in sulfite liquors whose pH at room temperature varies between 8 and 10, no loss of acid-labile sugars occurs even when they have been cooked well past their fiberizing point. In addition, these pulps contain polysaccharide components that are labile to the hydroxide ion, and increasing pH results in an increasing viscosity of the pulps (163). The physical properties are correspondingly better (7) and in some instances are superior to those of kraft pulps. Figure 4 presents a summary of these findings.

(d) MULTISTAGE SULFITE PROCESSES

According to Rydholm (95), when pulp researchers in Sweden were attempting to prepare a spruce sulfite pulp by a continuous pulping process in a Kamyr digester, they found the α-cellulose content of their experimental pulp to be 2–3% lower than anticipated. Further research showed that part of the problem was due to damage caused by blowing hot pulp from the digester. Most of the trouble could not be eliminated by cold blowing and was due to the retention of hemicellulose brought about by the use of a neutral sulfite pretreatment in a high-pressure feeder to the digester. Subsequent research showed that the application of a neutral sulfite pretreatment to the wood before acid sulfite cooking by either the continuous or the batch processes led to excellent papermaking pulps rich in hemicellulose (170).

Previous pulping experience showed that when digesters are charged with chips and ordinary acid sulfite cooking liquor, left at a low temperature (70–80°C) overnight, and then brought to the maximum cooking temperature for a conventional acid sulfite cook, the α-cellulose content of the pulp is unusually low. Analysis of this pulp, as well as of the alkaline pretreated pulp and a Mitscherlich pulp, showed that the decrease in α-cellulose content was in each case due to the retention of "glucomannan" (170). Other experiments showed that there was little difference between the "glucomannan" components of normal and pretreated sulfite pulps (171). Preswelling of wood with alkaline reagents, including calcium oxide (172), but not with dimethyl sulfoxide resulted in "glucomannan" retention after sulfite pulping and suggested that ester hydrolysis is the primary requisite for hemicellulose retention (171). It was shown that the acetyl content of birch "xylan" was not removed by acid sulfite

Coniferous wood containing cellulose, partially acetylated glucomannan, galactoglucomannan, 4-O-methylglucuronoarabinoxylan, glucuronoarabinogalactan, pectic substances, etc.

← Completely delignified* at pH 1–5 by acid sulfite or bisulfite processes
↓
Pulp containing cellulose, glucomannan, galactoglucomannan, 4-O-methylglucuronoarabinoxylan

↓ Completely delignified* at pH 8–11 by neutral sulfite or alkaline sulfite processes
↓
Pulp containing cellulose, partially acetylated glucomannan, 4-O-methylglucuronoxylan

→ Completely delignified* at pH 12–14 by alkaline sulfite, kraft, or soda processes
↓
Pulp containing cellulose, glucomannan, galactoglucomannan

* The chemical transformations of the polysaccharides of partially delignified wood are incomplete and complex mixtures of unreacted, partially transformed, and degraded polysaccharides probably exist in these pulps. Multistage pulping processes introduce additional, as yet incompletely resolved, complications in polysaccharide composition.

Fig. 4. The dependence of the polysaccharide composition of completely delignified coniferous pulps on the pH of the cooking liquor.

pulping (173, 174) but was removed by neutral sulfite and kraft pulping processes (174). After eliminating sulfite and bisulfite ions as contributors to stabilization, Annergren and Rydholm (175) concluded that the cleavage of ester bonds must preclude the stabilization of "glucomannan" and that these ester bonds may be either those of the acetyl groups or of conjectured linkages between the 4-*O*-methylglucuronoarabinoxylan and the galactoglucomannan in the original wood. Croon (176) demonstrated that this retention is proportional to the acetyl content of the wood after the pretreatment stage and not to the alkaline conditions themselves.

A review of the theories explaining "glucomannan" retention has been given by Rydholm (170, 171, 175, 177), who concludes that the pH effect, the weekend cook, and the Mitscherlich cook involve a crystallization on the cellulose microfibrils of linear "glucomannan" fragments formed on cautious hydrolysis (probably by removing branches without greatly diminishing the degree of polymerization) such as occurs at fairly low temperatures in acid sulfite cooking liquor and at higher temperatures in liquor of lower acidity. Croon (176) speculated that the alkaline pretreatment results in the removal of the bulky acetyl groups from the "glucomannan," thus preventing the close contact of the polymer with the cellulosic phase of the fiber. This development of packing order makes the "glucomannan" relatively inaccessible to the hydronium ion during the final sulfite cook, inhibits degradation of the polymer, and leads to a higher yield of pulp. Jansen and Sjöström (178) have demonstrated that hardwood two-stage sulfite pulps have greater yields due to the retention of "xylan," and they show that this retention is not dependent on the loss of acetyl content during the first stage. The compositions of a holocellulose (22–24), an acid sulfite pulp, two Mitscherlich pulps of different yields, and a multistage sulfite pulp (neutral sulfite stage followed by an acid sulfite stage) prepared from Paraña pine are compared in Table 2 (179). The results show, in agreement with the literature, that some acetyl groups survive the acid sulfite process, but no acetyl groups survive the processes (multistage and alkaline sulfites) that include an alkaline treatment. The retention of "glucomannan" and acetyl by the pulps prepared by the Mitscherlich process suggests that "glucomannan" retention does not depend entirely on the removal of acetyl groups but possibly on the removal of any branch, such as terminal D-galactopyranose residues if present. The less highly branched polysaccharide may then come into closer contact with cellulose and be relatively inaccessible to further degradation. This conjecture is supported by the observations that *O*-acetylgalactoglucomannans are present in coniferous woods (22–24), that galactoglucomannans can be isolated from coniferous multistage sulfite pulps (180), and that "*O*-acetylglucomannans" can be detected in Mitscherlich pulps (179), whereas only small amounts of the latter polymer can be isolated from acid sulfite pulps (156, 181). The retention of "xylan" by hardwood

TABLE 2
A Comparison of the Acetyl Content of Paraña Pine Pulps

Parameter	Type of pulp					
	Holocellulose	Acid sulfite	Mitscherlich	Mitscherlich	Multistage sulfite	Alkaline sulfite
Temperature of cook (°C):						
Stage 1	35	120	120		120	180
Stage 2	–	–	–		135	–
pH of liquor:						
Stage 1	4	3	3		8	10
Stage 2	–	–	–		2	–
Cooking chemical:						
Stage 1	Chlorous acid	Bisulfite	Bisulfite		Sulfite	Soda, sulfite
Stage 2	–	–	–		Bisulfite	
Yield (%)	80	47	55	67	55	45
Klason lignin content (%)	7.0	2.7	0.5	8.5	5.9	–
Kappa number	–	14	18	33	34	21
Acetyl content in glucomannan component (%)	8.0	2.0	2.7	2.7	0	0
Relative carbohydrate composition (%):						
Galactose	2.4	0	0	0	Trace	0.9
Glucose	72.4	89.7	85.0	84.6	84.5	89.3
Mannose	16.3	7.5	11.0	11.4	11.2	4.7
Arabinose	2.1	0	0	0	0	0.8
Xylose	6.8	2.8	3.6	3.6	4.1	4.3

TABLE 3
A Comparison of Some Cooking Conditions for Technologically Feasible Sulfite Pulping Processes (184)

Process	First stage	Second stage
Single-stage acid calcium sulfite	None	Total SO_2, 7% Combined SO_2, 1.2% Time to temperature, 5 hr Maximum cooking temperature, 130°C
Single-stage sodium bisulfite	None	Total SO_2, 5% pH 4.5 Time to temperature, 4 hr Maximum cooking temperature, 155°C
Sivola method	Total SO_2, 5% pH 3.5 Time to temperature, 1¾ hr Maximum cooking temperature, 165°C for 1½ hr	Na_2CO_3 added to pH 6–7 Maximum cooking temperature, 165°C
High-opacity magnesium method	Total SO_2, 3.3% pH 4.0 Cooking cycle: 3½ hr to 166°C, then 1½ hr at 160°C	MgO added to pH 6 Maximum cooking temperature, 170°C
Fast-beating magnesium method	Total SO_2, 3.2% pH 6.2 Time to temperature, 4¼ hr Maximum cooking temperature, 150°C for 2 hr	SO_2 added to pH 4 Maximum cooking temperature, 150°C
Stora 6 method	Total SO_2, 5% pH 6.5 Time to temperature, 2 hr Maximum cooking temperature, 110°C for 2 hr	SO_2 added to pH ~2 Time to temperature, 1¾ hr Maximum cooking temperature, 135°C
Stora 8 method	Total SO_2, 5% pH 8 Time to temperature, 2 hr Maximum cooking temperature, 110°C for 2 hr	SO_2 added to pH ~2 Time to temperature, 1¾ hr Maximum cooking temperature, 135°C
Stora-Billerud method	Total SO_2, 5% pH 8 Time to temperature, 2 hr Maximum cooking temperature, 110°C for 2 hr	SO_2 added to pH 4 Time to temperature, 1¾ hr Maximum cooking temperature, 165°C

two-stage sulfite pulps may depend on the removal of both *O*-acetyl groups and the terminal 4-*O*-methylglucopyranuronic acid residues of that polymer.

New multistage sulfite pulping processes have been developed as a result of the observation that pretreatments can stabilize hemicellulose components and yield pulps with different, but useful, papermaking properties. The literature pertaining to these processes has been reviewed by several investigators, their technological merits have been compared (2, 182–184), and liquor compositions and material balances of the processes have been examined (185). Most of these processes involve a pretreatment with sodium sulfite at pH 6–9, but sodium hydroxide (186) and magnesium sulfite (187) solutions have also been employed. The second stage is a conventional acid sulfite or bisulfite cook adjusted for the pulping effects of the first stage. A summary of the more important of these processes [based on the work of Makkonen (184)] is given in Table 3.

In general these pulps beat more quickly than single-stage sulfite pulps and do not usually have as great a tearing strength (188). Greater tearing strengths can be obtained by reversing the pulping sequences so that the first stage is an acid sulfite process and the second stage is conducted at pH 6 with sodium carbonate (189, 190) or magnesium sulfite (187) liquors (two examples are also given in Table 3). Unfortunately these pulps have lower yields at a given lignin content than any other pulps except kraft. An additional pretreatment stage with neutral sulfite in addition to acid sulfite and carbonate stages is sometimes employed to achieve the benefits of sulfonation before cooking (191).

6. Alkaline Pulping Processes

(a) KRAFT AND SODA PULPING

The simplest alkaline pulping process is the *soda process*, for which, because of the high cost of sodium hydroxide, an efficient recovery process was developed at an early time. The pulps produced were light in color and relatively easy to bleach, but they lacked great strength. The superior character of the pulps prepared by other processes led to the eventual displacement of the soda process. The superior character of the pulps prepared by the addition of small amounts of sulfur or sodium sulfide to the liquors, together with the adaptability of the soda-recovery system to the formation and recovery of sodium sulfide, led to the modern *kraft* or *sulfate pulping process*. The already mentioned technologically desirable features have made it feasible to use kraft pulps for the manufacture of many products traditionally made from sulfite pulps. The detrimental characteristics of the process, such as the low yield of useful carbohydrate components and the production of noxious by-products, are serious and research is being conducted to eliminate them.

The kraft process does not suffer from the liquor-penetration problems that are so characteristic of the sulfite processes, and kraft digesters may be brought quickly to reaction temperature when pulps are prepared by conventional techniques. Liquor penetration throughout the chip does become a significant parameter when very rapid pulping conditions are employed or when it is desired to drain the excess liquor after penetration to finish the cook as a vapor-phase operation (192). Unlike the effects of the sulfite process, chip damage is not reflected in inferior kraft pulp properties (193).

The action of alkaline liquors on lignin has not been studied as completely as the action of sulfite liquors on lignin. Studies on milled wood lignin have shown that it will dissolve easily in alkali at room temperature, whereas elevated temperatures ($>150°C$) and prolonged reaction times are necessary to dissolve lignin from hardwoods and softwoods (194, 195). It was argued by Bjorkman that the complex interpenetration of lignin and other components in the wood structure, together with the presence of lignin-carbohydrate bonds, is responsible for this difference in behavior.

The presence of sulfide in alkaline cooking liquor has been shown to accelerate the rate of lignin removal, and earlier researchers (196, 197) have claimed that it does not significantly change the rate of carbohydrate degradation. Due to the transitory nature of many of the sulfur–lignin derivatives produced during the kraft process, it has been impossible to obtain an understanding of the delignification by the analysis of the reaction products, as was done in the case of sulfite delignification (198). Dialkyl sulfides (thioethers) have been suggested as the most probable structural arrangement of the small amount of sulfur that remains in kraft lignin at the end of the pulping process (199). During alkaline pulping processes lignin is known to undergo a decrease in molecular weight (198, 200), cleavage of ether bonds (201), and complex condensation reactions (124, 125). Sulfonation and sulfidation reactions have been found to compete for many of the chemically reactive groups of lignin (201), and at high pH levels the sulfide group can replace sulfite groups from lignosulfonic acids. On the basis of experiments with model substances it has been postulated that only those portions of lignin that contain a *p*-hydroxybenzyl alcohol or an alkyl ether configuration will react with the hydrosulfide ion. The complex reactions that can occur after this conjectured initial reaction are the subject of many investigations (202).

A study of great technological importance to the kraft industry concerns the rate of removal of lignin and carbohydrates under various pulping conditions. A literature review of the factors of importance to the kinetic study of the kraft process has been recently published by Wilder and Daleski (203). This information was utilized as a starting point for their own kinetic evaluation of an idealized kraft pulping process. On the basis of their data (204, 205) it was concluded that delignification during the soda process is proportional to the amount of residual lignin and to the hydroxide-ion cocentration. The kraft system was more difficult to evaluate, but the assumption that the effects of the sulfide and the hydroxide ions were additive, and that the sulfide ion rather than the hydrosulfide ion was the active delignifying agent, led to a mathematical model with generally good correlation of the data. An attempt to apply the equation to conditions of time-varying liquor concentration was not successful. Daleski (206) compared the kraft and soda pulps and found that kraft pulps retain more cellulose at a given cooking temperature and lignin content, but he observed that the strength of kraft pulps decreases with increasing pulping temperatures.

An operations research program was developed by Carroll (207) to simulate the operating interrelationships of the kraft process. This expression and extensive material, energy, and economic balances taken throughout the system allowed net dollar returns (revenues minus costs) to be expressed as a function of the systems controllable operating variables. Carroll also developed an empirical rate expression to relate the major parameters of the kraft process to one another. Hinrichs (208) has described the use of the computer for the regression analysis of the results of cooking under various mill conditions, and this program has led to the mathematical prediction of pulp quality brought about by changing these mill conditions.

(b) CARBOHYDRATE DEGRADATIONS

At the present time the reactions that carbohydrates undergo in alkali appear to be more numerous and complex than those that occur during acid sulfite pulping. The reactions include swelling, dissolution, reprecipitation of components onto wood, saponification reactions, various elimination and related reactions (such as cleavage of glycosidic bonds), and oxidative reactions (particularly in the presence of elements such as oxygen and sulfur). The chemical reactions that occur in the presence of alkali have been reviewed by many researchers (209–212). Although sulfide ions themselves do not cause significant carbohydrate degradation, the change in the rate of delignification (213) and the change in hydroxide-ion concentration (214) does manifest itself in the character of the carbohydrate components (see also Section B in Chapter XVIII).

Swelling of wood occurs the moment contact is made with alkali, and certain irreversible degradations of pectin (210), saponification of esters (35), and neutralization of wood acids and acidlike substances take place (215). The acetyl groups and the lactone and other ester linkages shown to be present in wood "xylans" (216) are also saponified. Mild alkaline pretreatment of wood at room temperature with dilute alkali changes the papermaking properties of the resulting holocellulose from a glassinelike material to a kraftlike material (217);

this suggests that alkali has a profound influence on the physical characteristics of wood.

Caustic solutions are much more effective than water as extractants of wood, and this extraction assists penetration of the chemicals into the wood (218, 219). As delignification proceeds, polysaccharides that had previously been inaccessible to alkali become accessible to it and undergo further reactions. Many polysaccharides that dissolve in the cooking liquor are redeposited onto the fibers during the reaction (220) and, although water soluble, can be removed from the pulp only by concentrated akaline reagents. Thus the changes in overall pulp composition measured during the pulping process measure not only the destruction of polysaccharides but also the equilibrium between the components in the liquor and in the fiber. These analyses show that the polysaccharide composition of a pulp is ultimately reduced to cellulose, a "xylan," and a "glucomannan" of varying galactose content (37, 148).

Cellulose and hydrocellulose have been observed to develop an apparent resistance to alkaline degradation when heated in alkaline solutions at temperatures below 100°C (221, 222). At temperatures considerably in excess of 100°C cellulose and pulps do not exhibit this developed resistance and are progressively degraded in both yield and chain length (223). A mathematical analysis of this type of reaction by Samuelson and Wennerblom (224) led to the conclusion that cleavage of glycosidic bonds at nonterminal positions of polysaccharides must be occurring. Accordingly it was concluded that the glycosidic linkages in all glycosides are cleaved by alkali, but that the rate constants for the degradation of different glycosides vary widely (225).

As a result of their researches on the alkaline degradation of different β-phenyl glycosides of glucose, McCloskey and Coleman (226) suggested a mechanism involving the formation of a 1,2-anhydro-D-glucose intermediate that may react further to form 1,6-anhydro-D-glucose (levoglucosan). This mechanism for phenyl glycoside degradation has received considerable support (227, 228) and has also been used to explain the degradation of glycosidic bonds of polysaccharides at elevated temperatures (229). A comparison of the influence of the various glycosidic components as well as the influence of different aglucones on the lability of the glycosidic bond to alkali has been made by Lindberg and co-workers (230, 231). More recently other researchers have reinvestigated the degradation of methyl β-D-glucoside (232) and methyl β-cellobioside (233) and have shown the dependence of the reaction on pH and temperature. The rate-controlling step was considered to be the reaction of a carbinolate ion to form 1,2-anhydro-D-glucose. It is assumed, but not proved, that the same cleavage mechanisms are as observed for model substances will occur when cellulose and hemicellulose are subjected to similar liquors under the heterogeneous conditions of commercial alkaline pulping processes.

During an investigation of the polysaccharide composition of kraft pulps (37, 148) it was observed that the glycosidic bond linking uronic acids to polysaccharides is more susceptible to alkaline cleavage than the bonds between the nonacidic components of the polysaccharide. After initial difficulties because of imprecise analytical techniques (35, 37), definition of liquor compositions (234, 235), and rates of heating (235), the sensitivity of uronic acid to alkaline degradation became generally recognized (236–239). It was shown that the loss of uronic acid is due not to its exposed position in the polymers but to chemical effects (240). A variety of mechanisms have been postulated to explain this instability (35, 37, 240–242), and the most recent research by Robins and Green (243) shows that the glycosidic bond of the uronic acid is indeed cleaved by alkali and that the reaction occurs by a mechanism that differs from the one that degrades the glycosidic bond between nonacidic sugar units.

According to Tomlinson (6) pulps prepared by, or subjected to, hot alkaline treatments contain less of the original cellulose of wood than do pulps prepared by nonalkaline processes. The lower yield of cellulose for pulps prepared by alkaline processes may be seen in Table 4 as well as in the data of Jorgensen (40) and others (236). This behavior is due to the attack of alkali on the terminal reducing end of cellulose. As Franzon and Samuelson (244) indicate, about 65 glucose units will be removed from a cellulosic chain before the reaction is stopped by the formation of a terminal group more resistant to alkaline degradation. The mechanisms of the cellulosic degradations have been reviewed elsewhere as well as in this book and have been shown to involve the conversion of the terminal glucose unit to a glucoisoaccharinic acid with the simultaneous splitting off of the remaining cellulose and the exposure of a new terminal glucose unit for reaction. Ultimately the rearrangement forms a glucometasaccharinic acid or a saccharinic acid in which the cellulose is not eliminated and an alkali-resistant terminal cellulosic glycoside is formed. The mechanism of formation of these products has been studied by means of model experiments with D-glucose and cellobiose, and the identification of similar degradation products from the alkaline degradation of cotton hydrocellulose (245) supports the concepts proposed by Richtzenhain et al. (246). More recently investigations have been directed to the study of the stepwise removal of sugar units from hemicelluloses (35, 234, 235, 238, 240, 241, 247) and model derivatives of them (248).

The serious loss of carbohydrate yield caused by alkaline pulping has stimulated research to overcome this problem. Two approaches have been attempted. One involves controlling the pH of the cooking liquor so that dissolved hemicellulose (or lignin) is precipitated back onto the fibers. The other approach is the conversion of the terminal aldehydic groups of polysaccharides to an alkali-stable derivative by oxidation to carboxylic acids, reduction to methylol groups, and the formation of derivatives, such as a bisulfite complex, a glycoside, or levoglucosan.

TABLE 4
Polysaccharide Compositions of Various Pulps at Constant Lignin Content[a]

Type of pulp	Yield	Permanganate Number	Cellulose	Glucomannan	Xylan
Acid sulfite	53.3	22.9	46.1	4.5	2.7
Bisulfite	50.3	20.9	42.3	4.9	3.1
Kraft	48.8	21.2	40.8	4.2	3.8

[a]Compositions are expressed as percentage of original oven-dried wood.

(c) YIELD RETENTION IN ALKALINE PULPS

According to a review of the literature (249), the organic substances that are redeposited onto fibers from alkaline cooking liquors are lignin and polysaccharides. The sorbed polysaccharides may be present in a loosely bound form (250–252), or crystallized onto the cellulosic component, or they may even be chemically linked to cellulose by means of a transglycosidation reaction (249, 253, 254). After careful experiments employing tritium-labeled "xylan," Clayton and Stone (255) demonstrated that adsorption of "xylan" takes place during normal pulping operations and that it does so without transglycosidation through the intervention of a lignin–carbohydrate complex. In a later study (256) it was demonstrated that the sorption of spruce "glucomannan" and birch "xylan" onto α-cellulose is due to physical phenomena and that "glucomannan" is sorbed at a greater rate than is "xylan."

Aurell and Hartler (257, 258) found that by decreasing the pH of a birch kraft cook with sulfuric acid toward the end of the cooking cycle, and completing the process at the lower pH, an increase of yield was obtained, amounting to 1–1.5%, when 22.5% effective alkali was employed in the cooking cycle. A greater yield increase could be obtained for birch cooks (but not pine) if cooking liquor was withdrawn during a time of high hemicellulose concentration and then reintroduced toward the end of the cycle (259).

Greater yield increases can be obtained by modifying the reducing end of polysaccharides by oxidation or reduction so that they are resistant to the stepwise depolymerization reaction. Many attempts have been made to stabilize cellulose to alkaline degradation, but only a few of these procedures can be used to stabilize the polysaccharides of wood to alkaline pulping chemicals. Reduction of wood with sodium borohydride has been shown by Hartler (260) to stabilize the glucomannan component of conifers to kraft degradations.

Meller and Ritman (261) have examined the effects of sodium borohydride on the yield and papermaking characteristics of Monterey pine, and Sanyer and co-workers (262) have demonstrated that borohydride reduction stabilized the polysaccharides of hardwoods and softwoods to the degradations that occur during alkaline sulfite pulping. The rapid decomposition of sodium borohydride in hot alkaline solution (263), its potential contamination of recovery processes, and its great expense make it an unsatisfactory pulping reagent at present. Clayton and Marraccini (264) screened a variety of reagents capable of preventing the degradation of polysaccharides during alkaline pulping and found that hydroxylamine is an effective stabilizing reagent, though subject to side reactions with lignin. Their results agree with earlier reactions of hydroxylamine with hydrocellulose (265) but differ from the results on the pretreatment of sprucewood by Chirkin and Tishchenko (266).

Of the many chemical processes available for increasing the yield of alkaline pulps, the various polysulfide processes have the greatest industrial potential. Ever since the earliest observations (267) that the addition of sulfur to kraft cooking liquor has a beneficial effect on yield, researchers have attempted to understand and improve the factors involved (268–271). They discovered that polysulfides formed by reactions between sodium sulfide and sulfur are the effective chemical reagents. Colton, Kilbrick, and Scopp (272) suggested mixing sulfur with white liquor before pulping, or, preferably, mixing it with the dissolved and cooled smelt.

The effect of polysulfide cooking on the pulps prepared from a variety of woods has been studied by many investigators (273–275). It has been found that the increase in yield is due to the retention of "glucomannan" in the case of coniferous pulps and "xylan" in the case of hardwood pulps (276, 277). In general researchers have found up to 8% yield increases and the largest yield per unit weight of added sulfur occur in the region of 1–4% sulfur calculated on the weight of wood (278). The decomposition of polysulfide solutions has been followed during cooking, and the analysis shows more polysaccharides (composed chiefly of xylose-, galactose-, and arabinose-containing substances) to be present in polysulfide liquors than in kraft liquors (279). The claim that the stabilization of hemicellulose and cellulose is due to the oxidation of their end groups to carboxylic acids was verified by Alfredsson, Samuelson, and Sandstig (280). Although most of the yield increase is due to glucomannan stabilization, some investigators have estimated that about one-third of the yield increase is due to cellulose stabilization. Very little additional "xylan" is retained by coniferous polysulfide pulps (214), whereas the yield increase of hardwood polysulfide pulps is due to the retention of hardwood "xylan."

Many technological problems remain to be solved before the polysulfide pulping process is widely accepted by industry. The corrosive nature of polysulfide solutions on killed and rimmed steels has been studied by Rennel and Stockman (281). One of the greatest difficulties facing the process is the development of an effective regeneration of elemental sulfur by a process adaptable to the conventional kraft recovery processes. Venemark (282) has evolved two processes, based on chemical and electrical processes, that can be added to the kraft recovery system. It is possible that the two-stage polysulfide kraft process developed by Toga, Miyao, and Nishi (283) might overcome the need for elemental sulfur recovery. In this process chips are pretreated with polysulfide solutions in a first stage and the liquors are drained off to be used again. The chips are then cooked by a conventional kraft process to give pulps in 5–6% greater yield than the corresponding kraft pulps,

(d) MULTISTAGE ALKALINE PROCESSES

Wenzl (284) has speculated that the alkaline cooking processes consist of three integrated steps: an impregnation stage, a reaction stage, and a final extraction stage in which solubilized lignin and hemicellulose are removed from the pulp. Attempts have been made to separate these stages [most conveniently accomplished in continuous digesters (285)], and optimization of the separated stages in many instances results in better pulp quality and savings in pulping chemicals (286, 287). The multistage process evolved by Ritman (288) employs a first-stage kraft cook, followed by a second-stage alkaline sulfite cook. This process produces pulps that can be beaten easily and develop kraftlike properties.

The need for synthetic fibers during World War II led to the development of the *prehydrolyzed kraft pulping processes.* This type of process, which was based in part on the pioneering studies of Richter (289), removed the hemicelluloses, which were considered to be detrimental to the properties of dissolving pulps. The prehydrolysis stage may involve superheated steam or dilute acids at temperatures up to 170°C for varying lengths of time up to about 1 hr, after which the wood is cooked by a conventional kraft process. The data of Brasch and Free (290) are given in Table 5 and compare the compositions of a prehydrolyzed kraft and a normal kraft, cooked with identical alkaline delignification stages.

TABLE 5
Comparison of Kraft and Prehydrolyzed Kraft Pulps from Monterey Pine (290)

Parameter	Kraft	Prehydrolyzed kraft
Yield (of wood) (%)	46	37
α-Cellulose of pulp (%)	92	96
Pentosan of pulp (%)	7.0	1.8
Klason lignin of pulp (%)	4.6	2.7
Intrinsic viscosity (dl/g)	6.9	5.1

These results show the characteristically lower hemicellulose content, viscosity yield, and the greater α-cellulose content of prehydrolyzed kraft pulps, as compared with conventional kraft pulps. It is claimed (37) that the cleavage of the polysaccharide components of the accessible regions of the wood (chiefly hemicelluloses) by the prehydrolysis stage increases the number of reducing groups that may be attacked by alkali. The shortened hemicellulose polymers are either completely consumed by the stepwise depolymerization reaction or have their molecular size reduced to such a level they would dissolve in the cooking liquor and not be sorbed back onto the fibers. The removal of these extraneous substances has permitted the production of high-tenacity viscose-rayon fibers.

(e) MODIFIED KRAFT PROCESSES

It was early discovered that kraftlike pulping liquors containing a buffering agent, such as sodium carbonate, would yield satisfactory pulps from beechwood (291). At a later date Jayme and co-workers (292) confirmed this observation. The technique was considered to be an uneconomic variation of the neutral sulfite semichemical pulping process. Brauns (293) observed the superior papermaking characteristics of full chemical alkaline sulfite pulps as compared with kraft pulps. At the same time Peckham and Van Drunen (294) investigated the physical properties of pulps cooked in sulfite liquors buffered to various pH levels and found that all full chemical pulps cooked at pH levels from 8 to 14 (measured at room temperature) were similar in physical properties to kraft pulps. It was also observed that pulps cooked in liquors whose pH levels were slightly lower than that of kraft liquor were superior in physical properties to the kraft at all yield levels studied. An investigation of the chemical properties of these pulps (163) showed that the differences in physical behavior could not be attributed to the nature and amount of hemicelluloses present but are approximately related to the viscosity of the pulps. The poorest quality pulps were prepared by cooking at pH levels between 5.5 and 9, and their unusually low viscosity and physical properties were claimed to result from pyrolytic degradation (295). The active pulping chemical of these liquors may consist of various combinations of sodium sulfite and sodium sulfide, but for optimum effectiveness all must be buffered to a pH (at room temperature) between that of kraft liquor and pH 11. Although it produces desirable pulps, the process suffers from the very objectionable odor of its relief gases and from the difficulty in recovering the chemicals from the heavily buffered cooking liquors.

7. Comparison of the Physical Properties of Wood Pulps

A great variety of wood pulps, prepared by different pulping processes, are available today. Although each process at each pulp mill produces a pulp with unique physical characteristics, certain generalizations concerning the papermaking properties of these pulps can be made and have been the subject of many publications (296–302). These show that pulps prepared by acidic pulping processes are fibrillated very easily by mechanical means and have lower tearing strengths than pulps prepared by alkaline processes, which, in general, are more difficult to fibrillate. Critical examination of many of the theories proposed to explain these results can only be made when a deeper understanding of the structure and the chemistry of fibers is available.

The relative papermaking properties of pulps may be qualitatively determined by comparing the change of the tearing strength and tensile strength of paper sheets prepared from pulps after different degrees of fibrillation. This empirical comparison is based on the need in commercial practice to increase the tensile and bursting strengths of raw pulps to greater values. Since the tearing strengths are reduced at the same time, the tear–tensile relationships are necessary to develop the optimum physical characteristics of the pulps. A comparison of the tear–tensile relationships over the useful beating range of pulps prepared by cooking at different pH levels by Peckham and Van Drunen (294) is given in Figure 5. The superior quality of pulps prepared by cooking at pH 11–12 and the inferior quality of pulps cooked at pH 5.5 is at once apparent. Most

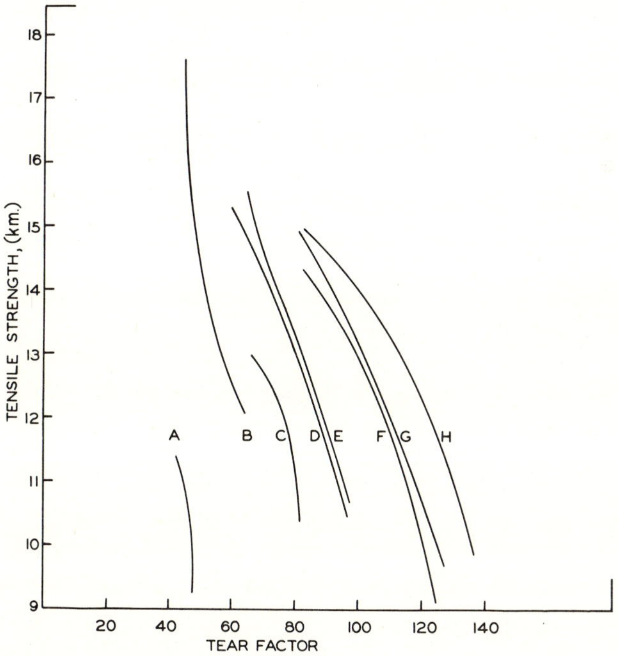

Fig. 5. The tearing and tensile strengths of paper sheets from single-stage pulps cooked at different pH levels to 54–59% yield (294): *A*, pH 5.5; *B*, pH 4; *C*, pH 2; *D*, pH 9.7; *E*, pH 9.2; *F*, pH 12+ (kraft); *G*, pH 12; *H*, pH 11.

frequently bisulfite pulps are superior in papermaking properties to acid sulfite pulps at high yield levels but not at the low yield of the pulp shown in Figure 5.

Diagrams similar to Figure 5 have been published by other investigators. In an an extensive survey Annergren, Rydholm, and Vardheim (276) compared kraft, sulfite, bisulfite, polysulfide, multistage kraft, and multistage sulfite processes at different yield levels. A comparison between the pulps of commercially useful hardwoods and softwoods is also given and illustrates the profound influence of fiber morphology on the physical properties of pulps.

References

1. W. F. Holzer, in E. Ott and H. M. Spurlin, Eds., *Cellulose and Cellulose Derivatives,* Part II, Interscience, New York, 1954, pp, 511–549.
2. S. A. Rydholm, *Pulping Processes,* Interscience, New York, 1965.
3. R. S. Hatch, in E. Ott and H. M. Spurlin, Eds., *Cellulose and Cellulose Derivatives,* Part II, Interscience, New York, 1954, pp. 589–619.
4. H. W. Bialkowski and P. S. Billington, *Tappi,* **44**, 195 (1961).
5. L. Stockman, *Pure Appl. Chem.,* **5**, 165 (1962).
6. G. H. Tomlinson II, *Tappi,* **44**, 133A (1961); *ibid.,* **49**, 195 (1961).
7. J. R. Peckham and E. V. Van Drunen, *Tappi,* **44**, 374 (1961).
8. T. E. Timell, *Advances in Carbohydrate Chemistry,* **19**, 247 (1964); *ibid.,* **20**, 409 (1965).
9. N. S. Thompson, H. H. Heller, J. D. Hankey, and O. Smith, *Pulp Paper Mag. Can.,* **67**, T-341 (1966).
10. G. A. Adams, *Pulp Paper Mag. Can.,* **65**, T-13 (1964).
11. H. Meier, in M. H. Zimmermann, Ed., *The Formation of Wood in Forest Trees,* Academic Press, New York, 1964.
12. R. H. Marchessault, W. Settineri, and W. Winter, *Tappi,* **50**, 55 (1967).
13. R. H. Marchessault, private communication, 1967.
14. E. Hägglund, B. Lindberg, and J. McPherson, *Acta Chem. Scand.,* **10**, 1160 (1960).
15. G. G. S. Dutton and A. M. Unrau, *Can. J. Chem.,* **42**, 348 (1962).
16. N. Banerji and N. S. Thompson, *Cell. Chem. Ind.,* **2**, 655 (1968).
17. C. T. Bishop, *Can. J. Chem.,* **33**, 1521 (1955).
18. V. Zitko and C. T. Bishop, *Can. J. Chem.,* **43**, 3206 (1965).
19. S. S. Bhattasharjee and T. E. Timell, *Can. J. Chem.,* **43**, 758 (1965).
20. M. Zinbo and T. E. Timell, *Svensk Papperstidn.,* **70**, 597 (1967).
21. J. Saarnio, *Suomen Kemistilehti,* **B29**, 35 (1956).
22. T. Koshijima, *J. Japan Wood Res. Soc.,* **6**, 194 (1960); through *Abstr. Bull. Inst. Paper Chem.,* **31**, 825 (1961).
23. H. Meier, *Acta Chem. Scand.,* **15**, 1381 (1961).
24. G. Katz, *Tappi,* **48**, 34 (1965).
25. G. J. Ritter and L. C. Fleck, *Ind. Eng. Chem.,* **14**, 1050 (1922); *ibid.,* **15**, 1055 (1923); *ibid.,* **18**, 508 (1926).

26. E. Pacsu, *Advances in Carbohydrate Chemistry,* **1**, 77 (1945).
27. J. M. Sugihara, *Advances in Carbohydrate Chemistry,* **8**, 1 (1953).
28. W. S. Linnell, N. S. Thompson, and H. A. Swenson, *Tappi,* **49**, 491 (1966).
29. K. Kringstad and O. Ellefsen, *Papier,* **18**, 583 (1964).
30. H. Meier, *Acta Chem. Scand.,* **14**, 749 (1960).
31. I. Croon, B. Lindberg, and H. Meier, *Acta Chem. Scand.,* **13**, 1299 (1959).
32. H. O. Bouveng, *Svensk Kem. Tidskr.,* **73**, 3 (1961).
33. A. Roudier and L. Eberhard, *Bull. Soc. Chem. France,* **1965**, 460.
34. E. Anderson, *J. Biol. Chem.,* **112**, 531 (1936).
35. J. K. Hamilton and N. S. Thompson, *Pulp Paper Mag. Can.* **61**, T-263 (1960).
36. J. Saarnio, K. Wathen, and C. Gustofsson, *Paperi Puu,* **35**, 65 (1953); *ibid.,* **36**, 209 (1954).
37. J. K. Hamilton and N. S. Thompson, *Tappi,* **41**, 803 (1958).
38. J. A. McPherson and O. E. Öhrn, *Svensk Papperstidn.,* **63**, 762 (1960).
39. S. Axelsson, I. Croon, and B. Enstrom, *Svensk Papperstidn.,* **65**, 693 (1962).
40. L. Jorgensen, in F. Bolam, Ed., *Fundamentals of Paper Making Fibers,* Technical Section, British Paper and Board Makers' Association, Inc., Kenley, Surrey, England, 1958.
41. K. Freudenberg, *J. Polymer Sci.,* **48**, 371 (1960).
42. E. Adler, *Papier,* **15**, 604 (1961).
43. K. Kratzl, in W. A. Côté, Ed., *Cellular Ultrastructure of Woody Plants,* Syracuse University Press, Syracuse, N.Y., 1964.
44. W. J. Schubert, *Lignin Biochemistry,* Academic Press, New York, 1965.
45. W. J. Schubert, in N. M. Bikales, Ed., *Encyclopedia of Polymer Science and Technology,* Vol. 8, Interscience, New York, 1968, pp. 233–272.
46. A. B. Wardrop, in W. A. Côté, Ed., *Cellular Ultrastructure of Woody Plants,* Syracuse University Press, Syracuse, N.Y., 1964.
47. P. W. Lange, *Svensk Papperstidn.,* **53**, 749 (1957).
48. N. P. Kutscha, Ph. D. thesis, State University College of Forestry, Syracuse University, 1967.
49. G. P. Berlyn and R. E. Mark, *Forest Prod. J.,* **16**, 140 (1965).
50. W. A. Côté, T. E. Timell, and R. A. Zabel, *Holz Roh-Werkstoff,* **24**, 432 (1966).
51. A. B. Wardrop, *Australian J. Sci., Res.,* **B4**, 391 (1951).
52. J. E. Stone, *Tappi,* **38**, 449, 452 (1955).
53. S. Lagergren, S. Rydholm, and L. Stockman, *Svensk Papperstidn.,* **60**, 632 (1957).
54. C. A. Carlson, and S. Lagergren, *Svensk Papperstidn.,* **60**, 664 (1957).
55. T. E. Timell, *Wood Sci. Technol.,* **1**, 45 (1967).
56. W. H. Aiken, *Tappi,* **50**, 37A (1967).
57. H. E. Dadswell and D. J. Ellis, *J. Council Sci. Ind. Res. (Australia),* **13**, No. 4, 290 (1940).
58 A. W. Mackney, *J. Council Sci. Ind. Res. (Australia),* **13**, No. 4, 299 (1940).
59. W. M. Harlow, *Paper Trade J.,* **10**, 242 (November 1939).
60. A. B. Wardrop and H. E. Dadswell, *Australian Council Sci. Ind. Res. Bull.,* No. 221, 7 (1947).

61. J. E. Stone and O. Kallmes, *Pulp Paper Mag. Can.,* **58**,191 (1957).
62. G. A. Adams and A. E. Castagne, *Can. J. Res.,* **F26**, 151 (1948).
63. N. S. Thompson and O. A. Kaustinen, *Tappi,* **47**, 157 (1964).
64. W. E. Spearin and I. H. Isenberg, *Science,* **105**, 214 (1947).
65. E. L. Lovell, *Ind. Eng. Chem.,* **37**, 1034 (1945).
66. N. S. Thompson, R. E. Kremers, and O. A. Kaustinen, *Tappi,* **51,** 123 (1968).
67. N. S. Thompson and J. D. Hankey, *Tappi,* **51,** 88 (1968).
68. V. Nyren and E. Back, *Norsk Skogind.,* **8**, 267 (1959).
69. R. F. Sultze, *Tappi,* **40**, 985 (1957).
70. D. W. Einspahr, J. P. van Buijtenen, and J. R. Peckham, *Tappi,* **45**, 58 (1962); *Silvae Genet.,* **12**, 51 (1963).
71. M. C. Mathes, *Forest Sci.,* **10**, 35 (1964); *Phyton,* **21**, 137 (1964).
72. R. M. Newnham, *Forest Chrònicle,* **44**, 7 (1968).
73. C. P. Klass, *Paper Trade J.,* **151,** 50 (Aug. 14, 1967).
74. L. R. Thiesmeyer, *Trend,* **4**, 4 (Winter, 1964).
75. K. L. Hackett, *Paper Trade J.,* **151,** 46 (Jan. 9, 1967).
76. A. W. J. Dyck, *Paper Ind.,* **47**, 34 (August 1965).
77. N. Hartler, *Svensk Papperstidn.,* **65**, 313, 351, 397 (1962).
78. J. Stone, *Pulp Paper Mag. Can.,* **57**, 139 (1956).
79. J. d'A. Clark, U.S. Patent 2,735,762 (1956).
80. N. Hartler, *Svensk Papperstidn.,* **64**, 101 (1961).
81. E. Kivimaa, *Holz Roh-Werkstoff,* **10**, 94 (1952).
82. K. C. Logan, O. J. K. Sepall, J. L. Chollet, and T. B. Little, *Pulp Paper Mag. Can.,* **61**, T515 (1960).
83. J. Richter, *Tappi,* **32**, 330 (1949).
84. M. A. Scheil, *Tappi,* **36**, 241 (1953).
85. L. A. Carlsmith and R. H. Rasch, *Tappi,* **43**, 1013 (1960).
86. L. A. Carlsmith and J. P. Rich, U.S. Patent 2,878,116 (1959).
87. A. von Koeppen, *Paper Trade J.,* **145,** (16) 24 (1961).
88. W. Brecht, G. Merckens, and K. Tengler, *Wochbl. Papierfabrik,* **88**, 451 (1960); through *Abstr. Bull. Inst. Paper Chem.,* **31**, 229 (1961).
89. J. Richter, U.S. Patent 2,474,862 (1949); *Svensk Papperstidn.,* **61,** 171 (1948).
90. D. G. Sutherland, *Tappi,* **46**, 178A (1963).
91. H. F. J. Wenzl, *Kraft Theory and Practice,* Lockwood, New York, 1967, Chapter 5.
92. G. Rowlandson, *Pulp Paper Mag. Can.,* **66**, T-71 (1965).
93. H. K. Van Eyken, *Pulp Paper Mag. Can.,* **65**, T-103 (1964).
94. G. G. Arnold, *Paper Trade J.,* **140**, 41 (Oct. 12, 1965).
95. S. A. Rydholm, *Pulp Paper Mag. Can.,* **68**, T-2 (1967).
96. A. J. Walters, *Appita,* **18**, 37 (1964).
97. D. S. Hoag, *Tappi,* **47**, 734 (1964).
98. W. B. Mackay, *Pulp Paper Mag. Can.,* **65**, T-107 (1964).
99. R. O. Donoghue, *Tappi,* **38**, 162A (1955); *Can Chem. Processing,* **43**, 47 (December 1959).

100. H. F. J. Wenzl, *Sulfite Pulping Technology,* Lockwood, New York, 1965.

101. N. Mannbro, *Svensk Papperstidn.,* **54**, 19, 61 (1951); *ibid.,* **55**, 665 (1952); *ibid.,* **58**, 525, 571 (1955).

102. D. F. Manchester and J. P. Termini, *Pulp Paper Mag. Can.,* **62**, T-415, (1961).

103. M. S. Mintz, *Ind. Eng. Chem.,* **55**, 18 (1963).

104. G. H. Tomlinson, *Chem. Ind. (London),* No. 24, 371 (1948); *Pulp Paper Mag. Can.,* **49**, 63 (1948).

105. G. H. Tomlinson and L. S. Wilcoxson, *Tech. Assoc. Papers,* **23**, 448 (1940).

106. N. Hartler, P. Rönström, and L. Stockman, *Svensk Papperstidn.,* **64**, 699 (1961).

107. N. Hartler, L. Stockman, and O. Sundberg, *Svensk Papperstidn.,* **64**, 23, 67, (1961).

108. R. N. Miller and W. H. Swanson, *Paper Trade J.,* **78,** (15) 178 (1924); *Ind. Eng. Chem.,* **17,** 843 (1925).

109. J. M. Calhoun, F. H. Yorston, and O. Maass, *Can. J. Res.,* **B15,** 457 (1937).

110. F. H. Yorston, Studies in Sulfite Pulping, *Dominion Forest Services Bulletin,* 97, (1942).

111. E. Hägglund, *Chemistry of Wood,* Academic Press, New York, 1951.

112. E. Hägglund, *Svensk Kem. Tidskr.,* **37**, 120 (1925).

113. E. Hägglund, T. Johnson, and H. Busch, *Paperi Puu,* **16**, 282 (1934).

114. S. Rydholm and S. Lagergren, *Svensk Papperstidn.,* **62**, 103 (1959).

115. E. Hägglund and A. Waller, *Paperi Puu,* **16**, 383 (1934).

116. H. Erdtman, *Svensk Papperstidn.,* **43**, 255 (1940).

117. B. O. Lindgren, *Acta Chem. Scand.,* **5**, 603 (1951).

118. B. O. Lindgren, *Svensk Papperstidn.,* **55**, 78 (1952).

119. H. J. Mikawa, *Chem. Soc. Japan, Ind. Chem. Sect.,* **54**, 651 (1951).

120. B. Leopold, *Acta. Chem. Scand.,* **6**, 75 (1952).

121. O. V. Ingruber and G. A. Allard, *Tappi,* **50**, 597 (1967).

122. T. N. Kleinert and L. M. Marraccini, *Tappi,* **47**, 605 (1964).

123. D. A. I. Goring, *Pulp Paper Mag. Can.,* **64**, T-517 (1963).

124. D. E. Read and C. B. Purves, *J. Am. Chem. Soc.,* **74**, 116, 120 (1952).

125. H. Richtzenhain, *Acta Chem. Scand.,* **4**, 206, 589 (1950).

126. T. N. Kleinert, *Nature,* **207**, 631 (1965).

127. T. N. Kleinert, *Tappi,* **50**, 120 (1967).

128. K. V. Sarkanen, in B. L. Browning, Ed., *The Chemistry of Wood*, Interscience, New York, 1963, Chapter 6.

129. K. Forss and K. E. Fremer, *Tappi,* **47**, 485 (1964); *Paperi Puu,* **47**, 485 (1965).

130. J. P. Casey, *Pulp and Paper, Chemistry and Technology*, Vol. 1, *Pulping and Papermaking*, Interscience, New York, 1952, Chapter IV.

131. J. C. Pew, *Tappi,* **36**, 39 (1939).

132. H. Erdtman, *Svensk Papperstidn.,* **43**, 255 (1940); *Cellulosechem.,* **18**, 83 (1940).

133. R. M. Dorland, R. A. Leask, and J. W. McKinney, *Pulp Paper Mag. Can.,* **59**, 236 (1958); *ibid.,* **60**, T-37 (1959).

134. W. Giertz and S. Wennerås, *Norsk Skogind.,* **14**, 322 (1960).

135. G. H. Tomlinson, G. H. Tomlinson II, J. R. G. Bryce, and N. G. M. Tuck, *Pulp Paper Mag. Can.,* **59**, 247 (1958).

136. R. M. Husband, *Tappi*, **36**, 529 (1953); *ibid.*, **38**, 577 (1955); *ibid.*, **40**, 412 (1957).
137. O. V. Ingruber, *Pulp Paper Mag. Can.*, **60**, T-215 (1965).
138. G. E. Annergren and S. A. Rydholm, *Svensk Papperstidn.*, **62**, 737 (1959; *ibid.*, **63**, 591 (1960).
139. E. Erickson, O. Samuelson, and A. Viala, *Svensk Papperstidn.*, **66**, 403 (1963).
140. T. Höpner, *Cellulosechem.*, **22**, 38 (1944).
141. O. Samuelson and R. Simonson, *Svensk Papperstidn.*, **65**, 363, 685 (1962).
142. O. Samuelson, K. J. Ljungquist, and C. Parck, *Svensk Papperstidn.*, **61**, 1043 (1958).
143. G. Menzinsky, *Ber.*, **68**, 822 (1935).
144. I. Slavik, *Svensk Papperstidn.*, **64**, 427 (1961).
145. E. Adler, *Svensk Papperstidn.*, **50**, 261 (1947); *ibid.*, **50**, 9 (1947).
146. E. Adler and S. Hernestam, *Acta Chem. Scand.*, **9**, 319 (1955).
147. L. Ahlén and O. Samuelson, *Svensk Papperstidn.*, **58**, 421 (1955).
148. J. K. Hamilton, E. V. Partlow, and N. S. Thompson, *Tappi*, **41**, 811 (1958).
149. T. E. Timell, N. Enterman, F. Spencer, and E. J. Soltes, *Can. J. Res.*, **43**, 2296 (1965).
150. L. K. Semke, N. S. Thompson, and D. G. Williams, *J. Org. Chem.*, **29**, 1041 (1964).
151. J. N. BeMiller, *Advances in Carbohydrate Chemistry*, **22**, 25 (1967).
152. C. A. Bunton, T. A. Lewis, D. R. Llewellyn, and C. A. Vernon, *J. Chem. Soc.*, 4419 (1955).
153. O. V. Ingruber, *Svensk Papperstidn.*, **65**, 448 (1962).
154. H. Meier, *Svensk Papperstidn.*, **65**, 589 (1962).
155. J. K. Hamilton, E. V. Partlow, and N. S. Thompson, *J. Am. Chem. Soc.*, **82**, 451 (1960).
156. J. K. Hamilton, H. W. Kircher, and N. S. Thompson, *J. Am. Chem. Soc.*, **78**, 2508 (1956).
157. T. E. Timell, *Tappi*, **44**, 88 (1961).
158. L. Sands and W. L. Gary, *J. Biol. Chem.*, **101**, 573 (1933).
159. A. Meller, *J. Polymer Sci.*, **A-1**, 1443 (1967).
160. J. K. Hamilton and N. S. Thompson, *J. Am. Chem. Soc.*, **79**, 6464 (1957).
161. D. B. Easty, *J. Org. Chem.*, **27**, 2102 (1962).
162. S. C. McKee and E. E. Dickey, *J. Org. Chem.*, **28**, 1561 (1963).
163. H. Meier, *Svensk Papperstidn.*, **65**, 299 (1962).
164. N. Hartler and A. Lund, *Svensk Papperstidn.*, **65**, 951 (1962).
165. H. Toda and T. Kijima, *J. Japan Tappi*, **15**, 40, 255, 534, 593, 622 (1961).
166. H. Toda, T. Kijima, and T. Hamada, *J. Japan Tappi*, **15**, 322, 381 (1961).
167. E. Sjöström and P. Haglund, *Svensk Papperstidn.*, **65**, 370 (1962).
168. N. Hartler and O. Sundberg, *Svensk Papperstidn.*, **63**, 263 (1960).
169. Reference 2, p. 422.
170. G. E. Annergren and S. A. Rydholm, *Svensk Papperstidn.*, **62**, 737 (1959).
171. G. E. Annergren, I. Croon, B. F. Enström, and S. A. Rydholm, *Svensk Papperstidn.*, **64**, 386 (1961).
172. A. Ogait, *Papier*, **17**, 639 (1963).
173. G. E. Annergren and I. Croon, *Svensk Papperstidn.*, **64**, 618 (1961).

174. Ö. E. Ohrn and I. Croon, *Svensk Papperstidn.*, **63**, 601 (1960).
175. G. E. Annergren and S. A. Rydholm, *Svensk Papperstidn.*, **63**, 591 (1960).
176. I. Croon, *Pulp Paper Mag. Can.*, **66**, T-71 (1965).
177. S. A. Rydholm, *Papier*, **14**, 535 (1960).
178. J. Jansen and E. Sjöström, *Svensk Papperstidn.*, **67**, 764 (1964).
179. N. S. Thompson and J. R. Peckham, unpublished research, Institute of Paper Chemistry, 1965.
180. J. M. Vaughan and E. E. Dickey, *Tappi*, **47**, 142 (1964); *J. Org. Chem.*, **29**, 715 (1964).
181. B. G. Rånby and B. Immergut, *Svensk Papperstidn.*, **60**, 573 (1957).
182. H. Makkonen, O. Järvelä, and N. E. Virkola, *Paperi Puu*, **47**, 59 (1965).
183. N. E. Virkola, O. Järvelä, and V. Vartianinen, *Paperi Puu*, **47**, 575 (1965).
184. H. Makkonen, *Paperi Puu*, **49**, 383, 437 (1967).
185. E. Sjöström, P. Haglund, and J. Jansen, *Svensk Papperstidn.*, **65**, 855 (1962).
186. I. Croon, *Svensk Papperstidn.*, **66**, 1 (1963).
187. E. L. Bailey, C. D. Miller, and S. C. McKee, *Tappi*, **46**, 176A (1963).
188. A. Vethe, V. Lorås, and F. Loschbrandt, *Norsk Skogind.*, **14**, 167 (1960).
189. L. H. Räsänen, and L. I. Luotonen, *Zellstoff Papier*, 9, 340 (1958); *ibid.*, **10**, 375 (1959); *Paper Mill News*, **83**, No. 3, 16 (1960).
190. G. Sivola, U.S. Patent 2,710,763 (1948); Canadian Patent 480,404 (1952).
191. G. Sivola, Swedish Patent 146,535 (1952).
192. T. N. Kleinert, *Pulp Paper Mag. Can.*, **65**, T-275 (1964).
193. S. E. Petersson and S. A. Rydholm, *Svensk Papperstidn.*, **64**, 4 (1961).
194. A. Bjorkman, *Svensk Papperstidn.*, **59**, 477 (1956); *ibid.*, **60**, 243, 329 (1960).
195. B. Person and A. Bjorkman, *Svensk Papperstidn.*, **60**, 158, 285 (1957).
196. T. Enkvist and B. Alfredsson, *Svensk Papperstidn.*, **57**, 159 (1954).
197. E. Hägglund, *Tappi*, **32**, 241 (1949); *Holz Roh-Werkstoff*, **11**, 284 (1953).
198. L. Field, P. E. Drummond, R. H. Higgins, and E. A. Jones, *Tappi*, **41**, 721, 727 (1958).
199. T. Enkvist, *Svensk Papperstidn.*, **51**, 225 (1948); *Tappi*, **37**, 350 (1954).
200. S. K. Gross, K. Sarkanen, and C. Schuerch, *Anal. Chem.*, **30**, 518 (1958).
201. T. Enkvist and B. Alfredsson, *Svensk Papperstidn.*, **53**, 185 (1951).
202. T. Enkvist, *Paperi Puu*, **44**, 395 (1962).
203. H. D. Wilder and E. J. Daleski, *Tappi*, **47**, 270 (1964).
204. H. D. Wilder and E. J. Daleski, *Tappi*, **48**, 293 (1965).
205. H. D. Wilder and E. J. Daleski, paper presented to the 20th Alkaline Pulping Conference of the Technical Association of the Pulp and Paper Industry, Richmond, Va., Sept. 13–16, 1966.
206. E. J. Daleski, *Tappi*, **48**, 325 (1965).
207. C. W. Carroll, *Tappi*, **43**, 305, 573 (1960).
208. D. D. Hinrichs, *Tappi*, **50**, 173 (1967).
209. A. Miller, *Holzforschung*, **14**, 78, 129 (1960).
210. R. L. Whistler and J. N. BeMiller, *Advances in Carbohydrate Chemistry*, **13**, 289 (1958).

211. J. C. Sowden, *Advances in Carbohydrate Chemistry,* **12**, 35 (1957).
212. A. Miller, *Tappi,* **47**, 23 (1964).
213. J. E. Stone and D. W. Clayton, *Pulp Paper Mag. Can.,* **61**, T-307 (1960).
214. R. D. Brooks and N. S. Thompson, *Tappi,* **47**, 729 (1964).
215. T. N. Kleinert, L. M. Marraccini, and E. J. Dostal, *Tappi,* **43**, 201 (1960).
216. P. Y. Wang, H. I. Bolker, and C. B. Purves, *Can. J. Chem.,* **42**, 2434 (1964).
217. N. S. Thompson and O. A. Kaustinen, Canadian Patent 806,573 (1969).
218. S. Yllner and B. Enstrom, *Svensk Papperstidn.,* **60**, 549 (1957).
219. T. N. Kleinert, L. M. Marraccini, and E. J. Dostal, *Tappi,* **44**, 440 (1961).
220. S. Yllner and B. Enstrom, *Svensk Papperstidn.,* **59**, 229 (1956).
221. G. Machell and G. N. Richards, *J. Chem. Soc.,* 4500 (1957).
222. H. Richtzenhain and B. Abrahansson, *Svensk Papperstidn.,* **57**, 538 (1954).
223. W. M. Corbett and G. N. Richards, *Svensk Papperstidn.,* **60**, 872 (1957).
224. O. Samuelson and A. Wennerblom, *Svensk Papperstidn.,* **57**, 827 (1954).
225. B. Lindberg, *Svensk Papperstidn.,* **59**, 531 (1956).
226. C. M. McCloskey and G. H. Coleman, *J. Org. Chem.,* **10**, 184 (1945).
227. C. E. Ballow, *Advances in Carbohydrate Chemistry,* **9**, 59 (1954).
228. R. C. Gasman and D. C. Johnson, *J. Org. Chem.,* **31**, 1830 (1966).
229. J. Jansen and B. Lindberg, *Acta Chem. Scand.,* **13**, 138 (1959).
230. E. Dryselius, B. Lindberg, and O. Theander, *Acta Chem. Scand.,* **11**, 663 (1957); *ibid.,* **12**, 340 (1958).
231. A Dyfverman and B. Lindberg, *Acta Chem. Scand.,* **4**, 878 (1950).
232. R. D. Brooks and N. S. Thompson, *Tappi,* **49**, 362 (1966).
233. E. V. Best and J. W. Green, *Tappi,* **52**, 1321 (1969).
234. I. Croon and B. Enstrom, *Tappi,* **44**, 871 (1961); *Svensk Papperstidn.,* **65**, 595 (1962).
235. N. S. Thompson, O. A. Kaustinen, and R. J. Ross, *Tappi,* **46**, 490 (1963).
236. J. K. Hamilton, *Pure Appl. Chem.,* **5**, 197 (1962).
237. N. S. Thompson, J. R. Peckham, and E. F. Thode, *Tappi,* **45**, 433 (1962).
238. H. Meier, *Svensk Papperstidn.,* **65**, 299, 589 (1962).
239. S. Axelson, I. Croon, and B. Enstrom, *Svensk Papperstidn.,* **65**, 693 (1962).
240. R. J. Ross and N. S. Thompson, *Tappi,* **48**, 376 (1965).
241. R. L. Casebier and J. K. Hamilton, *Tappi,* **50**, 441 (1967).
242. D. W. Clayton, *Svensk Papperstidn.,* **66**, 115 (1963).
243. J. H. Robins and J. W. Green, *Tappi,* **52**, 1346 (1969).
244. O. Franzon and O. Samuelson, *Svensk Papperstidn.,* **60**, 872 (1957).
245. G. Machell, G. N. Richards, and H H. Sephton, *Chem. Ind. (London),* 467 (1957).
246. H. Richtzenhain, B. O. Lindgren, B. Abrahansson, and K. Holmberg, *Svensk Papperstidn.,* **57**, 363 (1954).
247. R. L. Casebier and J. K. Hamilton, *Tappi,* **48**, 664 (1965).
248. B. Lindberg, O. Theander, and J. E. Uddegard, *Svensk Papperstidn.,* **69**, 360 (1966).
249. A. Meller, *Holzforschung,* **19**, 118 (1965).
250. N. Hartler and A. Lund, *Svensk Papperstidn.,* **65**, 951 (1962).
251. S. Yllner, K. Ostberg, and L. Stockman, *Svensk Papperstidn.,* **60**, 795 (1957).

252. I. Croon and B. Enstrom, *Tappi,* **44**, 870 (1961).

253. J. W. McKinney, *Paper Trade J.,* **122**, 58 (1946).

254. S. Haggroth and B. Lindberg, *Svensk Papperstidn.,* **59**, 870 (1956).

255. D. W. Clayton and J. E. Stone, *Pulp Paper Mag. Can.,* **64**, T-459 (1963).

256. D. W. Clayton and G. R. Phelps, *J. Polymer Sci.,* [C], No. 11, 197 (1965).

257. R. Aurell and N. Hartler, *Tappi,* **46**, 209 (1963).

258. R. Aurell, *Svensk Papperstidn.,* **66**, 457 (1963).

259. R. Aurell, *Tappi,* **48**, 80 (1965).

260. N. Hartler, *Svensk Papperstidn.,* **62**, 467 (1959); *ibid.,* **65**, 573 (1962); *Tappi,* **46**, 209 (1963); U.S. Patent 3,042,575 (1962).

261. A. Meller and E. L. Ritman, *Tappi,* **47**, 55, 634 (1964).

262. N. Sanyer, T. Itoh, and E. L. Keller, *Tappi,* **47**, 323 (1964).

263. N. Hartler, *Svensk Papperstidn.,* **65**, 513 (1963).

264. D. W. Clayton and L. M. Marraccini, *Svensk Papperstidn.,* **69**, 311 (1966).

265. A. Meller, *J. Polymer Sci.,* **51**, 99 (1961).

266. G. Chirkin and D. V. Tishchenko, *Zh. Prikl. Khim.,* **34**, 855 (1963).

267. A. Tingle, *Papier,* **34**, 653 (1924).

268. R. R. Fuller and V. Woodside, Canadian Patent 444,274 (1947).

269. R. Berthier, *Assoc. Tech. Ind. Papetière, Bull.,* No. 4, 93 (1953).

270. T. Kurenniemi and J. O. Murto, *Paperi Puu,* **30**, 437 (1928).

271. J. R. Peckham and M. N. May, *Tappi,* **44**, 45 (1960).

272. R. B. Colton, A. C. Kilbrick, and I. W. Scopp, Canadian Patent 572,578 (1959); U.S. Patent 2,944,928 (1960); *Paper Ind.,* **43**, 233 (July 1961).

273. P. Colombo, D. Corbetta, A. Pirotta, and G. Ruffini, *Ind. Carta,* **1**, No. 12, 501 (1963).

274. N. Sanyer and J. F. Laundrie, *Tappi,* **47**, 640 (1964).

275. P. J. Kleppe and K. Kringstad, *Norsk Skogind.,* **18**, 1 (1964).

276. G. Annergren, S. A. Rydholm, and S. Vardheim, *Svensk Papperstidn.,* **66**, 196 (1963).

277. N. Sanyer and G. H. Chidester in B. L. Browning, Ed., *The Chemistry of Wood,* Interscience, New York, 1963, Chapter 10.

278. P. J. Kleppe and K. Kringstad, *Norsk Skogind.,* **17**, 428 (1963).

279. R. Simonsen, *Svensk Papperstidn.,* **67**, 721 (1964).

280. B. Alfredsson, O. Samuelson, and B. Sandstig, *Svensk Papperstidn.,* **66**, 703 (1963).

281. J. Rennel and L. Stockman, *Svensk Papperstidn.,* **66**, 863 (1963).

282. E. Venemark, *Svensk Papperstidn.,* **67**, 157 (1964); *Paper Trade J.,* **147**, (18) 42 (1963).

283. G. Toga, T. Miyao, and T. Nishi, *J. Japan Tappi,* **18**, No. 1, 27 (1964).

284. H. F. J. Wenzl, U.S. Patent 2,874,044 (1959).

285. J. L. Marvin, paper presented at the 20th Alkaline Pulping Conference of the Technical Association of the Pulp and Paper Industry, Richmond, Va., September 1966.

286. A. Meller, *Tech. Assoc. Paper,* **29**, 642 (1946).

287. G. Jayme and L. Groegaard, *Papierfabrikant,* **38**, 149 (1940).

288. E. L. Ritman, *Appita,* **19**, 14 (1965).

289. G. A. Richter, U.S. Patents 1,787,953; 1,816,343; 1,819,002 (1931); 1,880,043 (1932); 2,036,606 (1936).

290. D. Brasch and K. W. Free, *Tappi,* **47**, 186 (1964); *ibid.,* **48**, 245 (1965).
291. E. Heuser, *Wochbl. Papierfabrick,* **24**, No. 2, 229 (1913).
292. G. Jayme, H. Wagenbach, and W. Deloff, *Papierfabrikant,* **37**, 229 (1939).
293. R. V. Brauns, Canadian Patent 617,329 (1961).
294. J. R. Peckham and E. V. Van Drunen, *Tappi,* **48**, 193 (1965).
295. N. S. Thompson and O. A. Kaustinen, *Tappi,* **49**, 550 (1966).
296. A. von Koeppen, *Tappi,* **47**, 589 (1964).
297. H. W. Bialkowsky and P. S. Billington, *Tappi,* **44**, 195 (1961).
298. P. E. Wrist, *Tappi,* **44**, 194A (1961); *ibid.,* **49**, 287 (1966).
299. H. W. Giertz, *Svensk Papperstidn.,* **56**, 893 (1953).
300. A. W. McKenzie and H. G. Higgins, *Appita,* **14**, 127 (1960).
301. N. Hartler, *Svensk Papperstidn.,* **65**, 535 (1962).
302. J. H. Kalisch and W. Beazley, *Pulp Paper Mag. Can.,* **61**, T-452 (1962).

CUMULATIVE INDEX
VOLUME 5, PARTS IV & V